# Cherries

## Botany, Production and Uses

This book is dedicated to Norm Looney, deceased in 2016, and Tony Webster, now retired, who conceived and edited the 1996 book *Cherries: Crop Physiology, Production and Uses*, which rapidly became a worldwide reference for the cherry community.

We also dedicate it to five cherry scientists who unfortunately passed away during the course of COST Action FA1104, and who were extremely active and passionate colleagues: Joerg Samietz, Bernard Blum, Petya Gercheva, Sergiu Budan and Yan Wang.

# Cherries

## Botany, Production and Uses

*Edited by*

**J. Quero-García**

*INRA Bordeaux, France*

**Amy Iezzoni**

*Department of Horticulture, Michigan State University East Lansing, USA*

**Joanna Puławska**

*Research Institute of Horticulture, Skierniewice, Poland*

*and*

**Gregory Lang**

*Department of Horticulture, Michigan State University East Lansing, USA*

CABI

**CABI is a trading name of CAB International**

CABI
Nosworthy Way
Wallingford
Oxfordshire OX10 8DE
UK

Tel: +44 (0)1491 832111
Fax: +44 (0)1491 833508
E-mail: info@cabi.org
Website: www.cabi.org

CABI
745 Atlantic Avenue
8th Floor
Boston, MA 02111
USA

Tel: +1 (617) 682-9015
E-mail: cabi-nao@cabi.org



A catalogue record for this book is available from the British Library, London, UK.

**Library of Congress Cataloging-in-Publication Data**

Names: Quero-García, J. (José), editor.
Title: Cherries: botany, production and uses / edited by J. Quero-García and Amy Iezzoni and Joanna Pulawska and Gregory Lang.
Description: Boston, MA: CABI, [2017] | Includes bibliographical references and index.
Identifiers: LCCN 2017004394| ISBN 9781780648378 (hbk : alk. paper) | ISBN 9781780648392 (e-pub)
Subjects: LCSH: Cherry.
Classification: LCC SB379.C5 C454 2017 | DDC 634/.23--dc23
LC record available at https://lccn.loc.gov/2017004394

ISBN-13: 978 1 78064 837 8

Commissioning editor: Rachael Russell
Editorial assistant: Emma McCann
Production editor: Shankari Wilford

Typeset by SPi, Pondicherry, India
Printed and bound in the UK by Severn, Gloucester

First published in 2017.
Reprinted 2018, 2021.

# Contents

# Contributors


**Apostol, János**, National Agricultural Research and Innovation Centre (NARIC) Fruitculture Research Institute, 1223, Budapest, Park u. 2, Hungary. E-mail: apostolj@vipmail.hu

**Ayala, Marlene**, Departamento de Fruticultura y Enología, Pontificia Universidad Católica, Santiago, Chile. E-mail: mayalaz@uc.cl

**Barreneche, Teresa**, Equipe 'Adaptation du Cerisier au Changement Climatique', UMR 1332 Biologie du Fruit et Pathologie, INRA et Université de Bordeaux, CS20032, 33882 Villenave d'Ornon Cedex, France. E-mail: teresa.barreneche@inra.fr

**Beliën, Tim**, Department of Zoology, Proefcentrum Fruitteelt VZW, Fruittuinweg 1, Sint-Truiden, Belgium. E-mail: tim.belien@pcfruit.be

**Blanke, Michael M.**, INRES Horticultural Science, University of Bonn, Bonn, Germany. E-mail: ulp304@uni-bonn.de

**Børve, Jorunn**, Norwegian Institute of Bioeconomy Research, Ullensvang, 5781 N-Lofthus, Norway. E-mail: jorunn.borve@nibio.no

**Boscia, Donato**, CNR – Institute for Sustainable Plant Protection, Bari, Italy. E-mail: donato.boscia@ipsp.cnr.it

**Bujdosó, Géza**, National Agricultural Research and Innovation Centre Fruitculture Research Institute, 1223, Budapest, Park u. 2, Hungary. E-mail: resinfru@yahoo.com

**Campoy, José Antonio**, Equipe 'Adaptation du Cerisier au Changement Climatique', UMR 1332 Biologie du Fruit et Pathologie, INRA et Université de Bordeaux, CS20032, 33882 Villenave d'Ornon Cedex, France. E-mail: jose-antonio.campoy-corbalan@inra.fr

**Candresse, Thierry**, Equipe de Virologie, UMR 1332 Biologie du Fruit et Pathologie, INRA et Université de Bordeaux, CS20032, 33882 Villenave d'Ornon Cedex, France. E-mail: thierry.candresse@inra.fr

**Charlot, Gérard**, CTIFL, 751, Chemin de Balandran, 30127 Bellegarde, France. E-mail: Charlot@ctifl.fr

**Cieślińska, Mirosława**, Research Institute of Horticulture, Skierniewice, Poland. E-mail: miroslawa.cieslinska@inhort.pl

**Ercisli, Sezai**, Department of Horticulture, Agricultural Faculty, Ataturk University, 25240, Erzurum, Turkey. E-mail: sercisli@gmail.com

**Flores, Ricardo**, Instituto de Biología Molecular y Celular de Plantas (UPV-CSIC), Universidad Politécnica de Valencia, Valencia 46022, Spain. E-mail: rflores@ibmcp.upv.es

**Forge, Thomas**, Summerland Research and Development Centre, Agriculture and AgriFood Canada, Summerland, BC, Canada V0H 1Z0, Canada. E-mail: tom.forge@agr.gc.ca

**Fotirić Akšić, Milica**, Department of Fruit Science, Faculty of Agriculture, University of Belgrade, Belgrade, Zemun, Serbia. E-mail: fotiric@agrif.bg.ac.rs

**Gétaz, Michael**, Zurich University of Applied Sciences, Institute of Natural Resource Sciences, Wädenswil, Switzerland. E-mail: geta@zhaw.ch

**Giovannini, Daniela**, CREA-FRF Council for Agricultural Research and Economics, Fruit Tree Unit of Forlì, Via la Canapona, 1 bis, 47100 Forlì, Italy. E-mail: daniela.giovannini@crea.gov.it

**Gónzalez-Gómez, David**, Department of Didactics of Experimental Sciences and Mathematics, Teacher Training College, University of Extremadura, Cáceres, Spain. E-mail: dggomez@unex.es

**Herrero, Maria**, Pomology Department, Estación Experimental Aula Dei, CSIC, Av. Montañana, 1005, 50059 Zaragoza, Spain. E-mail: mherrero@eead.csic.es

**Höfer, Monika**, Julius Kühn-Institute, Federal Research Centre for Cultivated Plants, Institute for Breeding Research on Fruit Crops, Pillnitzer Platz 3a, D-01326 Dresden, Germany. E-mail: monika.hoefer@julius-kuehn.de

**Hrotkó, Károly**, Faculty of Horticultural Science, Szent István University, 1118 Budapest, Villányi u. 29-43, Hungary. E-mail: hrotko.karoly@kertk.szie.hu

**Hrustić, Jovana**, Institute of Pesticides and Environmental Protection, Belgrade, Serbia. E-mail: Jovana.Hrustic@pesting.org.rs

**Iezzoni, Amy**, Department of Horticulture, Michigan State University, East Lansing, MI 48824, USA. E-mail: iezzoni@msu.edu

**Ippolito, Antonio**, Dipartimento di Scienze del Suolo, della Pianta e degli Alimenti, Università degli Studi Aldo Moro, Bari, Italy. E-mail: antonio.ippolito@uniba.it

**James, Delano**, Sidney Laboratory – Centre for Plant Health, Canadian Food Inspection Agency, North Saanich, BC, Canada. E-mail: delano.james@inspection.gc.ca

**Jelkmann, Wilhelm**, Julius Kuhn Institute, Institute for Plant Protection in Fruit Crops and Viticulture, D-69221 Dossenheim, Germany. E-mail: wilhelm.jelkmann@julius-kuehn.de

**Jensen, Martin**, Institut for Fødevarer/Department of Food Science, Aarhus University, Kirstinebjergvej 10, DK-5792 Årslev, Denmark. E-mail: Martin.Jensen@food.au.dk

**Kałużna, Monika**, Research Institute of Horticulture, Skierniewice, Poland. E-mail: monika.kaluzna@inhort.pl

**Knoche, Moritz**, Institute for Horticultural Production Systems, Leibniz-University Hannover, Hannover, Germany. E-mail: moritz.knoche@obst.uni-hannover.de

**Köppler, Kirsten**, Center for Agricultural Technology Augustenberg (LTZ), Karlsruhe, Germany. E-mail: kirsten.koeppler@ltz.bwl.de

**Koumanov, Kouman S.**, Department of Fruitgrowing Technologies, Fruitgrowing Institute, Agricultural Academy, 12 Ostromila, Plovdiv 4004, Bulgaria. E-mail: kskoumanov@hotmail.com

**Kuzmanović, Nemanja**, University of Belgrade – Faculty of Agriculture, Belgrade, Serbia. E-mail: kuzmanovic1306@gmail.com

**Lang, Gregory A.**, Department of Horticulture, Michigan State University, East Lansing, MI 48824, USA. E-mail: langg@msu.edu

**Long, Lynn E.**, Oregon State University Extension, 400 E. Scenic Drive, Suite 2.278, The Dalles, OR 97058, USA. E-mail: lynn.long@oregonstate.edu

**López-Ortega, Gregorio**, Murcia Institute of Agri-Food Research and Development (IMIDA), Murcia 30150, Spain. E-mail: goyologa@gmail.com

**Lux, Sławomir A.**, inSilico-IPM, Konstancin-Jeziorna, Poland. E-mail: slawomirlux@yahoo.co.uk

**Manganaris, George A.**, Department of Agricultural Sciences, Biotechnology and Food Science, Cyprus University of Technology, Lemesos, Cyprus. E-mail: george.manganaris@cut.ac.cy

**Mari, Marta**, University of Bologna, CRIOF-DipSA, Bologna, Italy. E-mail: marta.mari@unibo.it

**Meland, Mekjell**, Norwegian Institute of Bioeconomy Research – NIBIO Ullensvang, N-5781 Lofthus, Norway. E-mail: mekjell.meland@nibio.no

**Michalecka, Monika**, Research Institute of Horticulture, Skierniewice, Poland. E-mail: monika.michalecka@inhort.pl

**Milatović, Dragan**, Department of Fruit Science, Faculty of Agriculture, University of Belgrade, Belgrade, Serbia. E-mail: mdragan@agrif.bg.ac.rs

**Neilsen, Denise**, Summerland Research and Development Centre, Agriculture and AgriFood Canada, Summerland, BC, Canada V0H 1Z0, Canada. E-mail: denise.neilsen@agr.gc.ca

**Neilsen, Gerry H.**, Summerland Research and Development Centre, Agriculture and AgriFood Canada, Summerland, BC, Canada V0H 1Z0, Canada. E-mail: gerry.neilsen@agr.gc.ca

**Obradović, Aleksa**, University of Belgrade – Faculty of Agriculture, Belgrade, Serbia. E-mail: aleksao@agrif.bg.ac.rs

**Palkovics, László**, Szent István University, Faculty of Horticultural Science, Department of Plant Pathology, Budapest, Hungary. E-mail: laszlo.palkovics@uni-corvinus.hu

**Pallás, Vicente**, Instituto de Biología Molecular y Celular de Plantas (UPV-CSIC), Universidad Politécnica de Valencia, Valencia 46022, Spain. E-mail: vpallas@ibmcp.upv.es

**Papadopoulos, Nikolaos T.**, Department of Agriculture, Crop Production and Rural Environment, University of Thessaly, Volos, Greece. E-mail: nikopap@uth.gr

**Perry, Ronald L.**, Department of Horticulture, Michigan State University, East Lansing, MI 48824, USA. E-mail: perryr@msu.edu

**Poniatowska, Anna**, Research Institute of Horticulture, Skierniewice, Poland. E-mail: anna.poniatowska@inhort.pl

**Pothier, Joël F.**, Zurich University of Applied Sciences, Institute of Natural Resource Sciences, Wädenswil, Switzerland. E-mail: poth@zhaw.ch

**Puławska, Joanna**, Research Institute of Horticulture, Skierniewice, Poland. E-mail: joanna.pulawska@inhort.pl

**Quero-García, José**, Equipe 'Adaptation du Cerisier au Changement Climatique', UMR 1332 Biologie du Fruit et Pathologie, INRA et Université de Bordeaux, CS20032, 33882 Villenave d'Ornon Cedex, France. E-mail: jose.quero-garcia@inra.fr

**Rodrigo, Javier**, Unidad de Hortofruticultura, Centro de Investigación y Tecnología Agroalimentaria de Aragón. Instituto Agroalimentario de Aragón - IA2 (CITA-Universidad de Zaragoza), Av. Montañana, 930, 50059 Zaragoza, Spain. E-mail: jrodrigo@aragon.es

**Rozpara, Elżbieta**, Department of Pomology and Nursery, Research Institute of Horticulture, 96-100 Skierniewice, Poland. E-mail: elzbieta.rozpara@inhort.pl

**Ruinelli, Michela**, Zurich University of Applied Sciences, Institute of Natural Resource Sciences, Wädenswil, Switzerland. E-mail: ruin@zhaw.ch

**Sanzani, Simona Mariana**, Dipartimento di Scienze del Suolo, della Pianta e degli Alimenti, Università degli Studi di Bari Aldo Moro, Bari, Italy, E-mail: simonamarianna.sanzani@uniba.it

**Saponari, Maria**, CNR – Institute for Sustainable Plant Protection, Bari, Italy. E-mail: maria.saponari@ipsp.cnr.it

**Schuster, Mirko**, Julius Kühn-Institut, Federal Research Centre for Cultivated Plants, Institute for Breeding Research on Fruit Crops, Pillnitzer Platz 3a, D-01326 Dresden, Germany. E-mail: mirko.schuster@julius-kuehn.de

**Serradilla, Manuel Joaquín**, Department of Vegetables, Scientific and Technological Research Centre of Extremadura (CICYTEX), Junta de Extremadura, Badajoz, Spain. E-mail: manuel.serradilla@juntaex.es

**Tanović, Brankica**, Institute of Pesticides and Environmental Protection, Belgrade, Serbia. E-mail: brankica.tanovic@pesting.org.rs

**Toivonen, Peter**, Summerland Research and Development Centre, Agriculture and Agri-Food Canada, Summerland, BC, Canada V0H 1Z0, Canada. E-mail: peter.toivonen@agr.gc.ca

**Valero, Daniel**, Department of Food Technology, EPSO, University Miguel Hernández, Alicante, Spain. E-mail: daniel.valero@umh.es

**Végh, Anita**, Szent István University, Faculty of Horticultural Science, Department of Plant Pathology, Budapest, Hungary. E-mail: vegh.anita@kertk.szie.hu

**Vokurka, Aleš**, Department for Plant Breeding and Genetics, Faculty of Agriculture, University of Zagreb, Svetošimunska 25, HR-10,000 Zagreb, Croatia, E-mail: avokurka@agr.hr

**Wang, Yan**, Mid-Columbia Agricultural Research and Extension Center, Oregon State University, Hood River, OR 97031, USA. E-mail: yan.wang@oregonstate.edu

**Wenden, Bénédicte**, Equipe 'Adaptation du Cerisier au Changement Climatique', UMR 1332 Biologie du Fruit et Pathologie, INRA et Université de Bordeaux, CS20032, 33882 Villenave d'Ornon Cedex, France. E-mail: benedicte.wenden@inra.fr

**Whiting, Matthew D.**, Department of Horticulture, Irrigated Agriculture Research and Extension Center, Washington State University, Prosser, WA 99350, USA. E-mail: mdwhiting@wsu.edu

**Winkler, Andreas**, Institute for Horticultural Production Systems, Leibniz-University Hannover, Hannover, Germany. E-mail: andreas.winkler@obst.uni-hannover.de

**Wünsch, Ana**, Unidad de Hortofruticultura, Centro de Investigación y Tecnología Agroalimentaria de Aragón. Instituto Agroalimentario de Aragón - IA2 (CITA-Universidad de Zaragoza), Av. Montañana, 930, 50059 Zaragoza, Spain. E-mail: awunsch@aragon.es

**Zoffoli, Juan Pablo**, Departamento de Fruticultura y Enología, Pontificia Universidad Católica de Chile, Santiago, Chile. E-mail: zoffolij@uc.cl

# Preface

In 1996, Webster and Looney's *Cherries: Crop Physiology, Production and Uses* became a landmark publication for cherry research scientists, growers, packers, processors and marketers worldwide. Perhaps not coincidentally, the international production of cherries has increased dramatically in the two decades since that book was published, as has the associated breadth of advances in fruit science and technology. This book provides an updated overview of all aspects of cherry science and culture since that important contribution to pomology.

Cherries are a special tree fruit, having an attractive appearance and intensely desirable flavour, a nutrient and phytochemical content that provides physiological benefits for human health, and a consumer-friendly size for eating out of hand or incorporating into baked goods and other culinary accents. Consumers are willing to pay well for these traits, which has helped drive the strong increase in worldwide production. One of the greatest advantages of cherries is that they ripen first among temperate tree fruits, and thus they have always been associated, along with strawberries, with the optimism, vitality, happiness and even romance of each new season's bounty of life.

Cherries are, however, a challenging and high-risk fruit crop to grow. Their relatively small fruit size and naturally large tree canopy gave rise to the Latin name for sweet cherry, *Prunus avium* ('bird cherry'), for indeed, traditionally, they are harvested much more easily by birds than by humans with tall ladders. Cherries are susceptible to many damaging insect pests and serious plant diseases, as well as to damage from low winter temperatures, spring frosts due to early bloom timing and rain-cracking of fruits during ripening. Consequently, these are among the key areas of on-going research.

Both sweet and sour cherry production has increased significantly during the past two decades in the traditional leading cherry-producing countries such as Turkey, the USA, Italy, Poland, the Russian Federation, Ukraine and Iran. Also during that time, Chile has emerged from being a minor producer to the largest producer in the southern hemisphere and the third largest sweet cherry exporter, while production in China is also increasing dramatically with the rise of the middle class in the world's most populous nation. Although both sweet and sour cherries are particularly well adapted to temperate Mediterranean climates, attempts to adapt production to low-chilling and subtropical regions have been initiated to expand marketing windows and capture high market values when supplies are usually low or non-existent.

Both sweet and sour cherries are believed to have originated around the Caspian and Black Seas, with the Caucasus area considered to be a major centre of genetic diversity. The domestication of cherries through observational selection can be related closely to the historic civilization of Europe. On the other hand, modern breeding efforts are relatively recent, coincident with the early 20th century. Significant progress has been achieved in sweet cherry scion breeding, with many dozens of new varieties released over the past two decades including many self-compatible genotypes. All generally have a desirable flavour (sugar–acid balance) and are characterized by significant improvements in fruit size and firmness. Constraints in cherry breeding include the difficulty of producing large progenies in certain hybrid combinations, their relatively long juvenile period, and the many labour-intensive processes involved from pollen collection, manual pollination and fruit protection from birds/insects/diseases to hand-harvest. Nevertheless, rapid advances during the past decade in the field of molecular genetics hold great promise for more efficient and broader improved variety traits for cherries. Indeed, marker-assisted selection has recently been implemented for several key traits, such as self-incompatibility and fruit weight, and will presumably extend further to other traits as genomic research is tied to horticultural traits of interest. This will reduce breeding programme operating costs and facilitate more efficient and successful design of future crosses.

Sweet cherry cultivation has advanced rapidly since the 1990s, with the development of intensive orchard systems following the example of apple. This revolution was initiated by the advent of dwarfing or semi-dwarfing precocious rootstocks, which facilitated reduced tree height, higher tree densities, earlier yields and improved labour efficiencies. This rootstock-based alteration of reproductive and vegetative growth habits necessitated physiological studies to determine how to optimize these interrelationships and maintain the fruit quality traits that high-value consumer and export markets demand. These foundational studies have included plant water relations, carbon partitioning, and mineral nutrient acquisition, storage and remobilization. The integration of such physiological knowledge with horticultural orchard innovations has led to new training techniques and canopy architectures that harness the benefits of particular variety/rootstock combinations, adapt to new environments and climates, or even deploy climate-altering orchard protection technologies such as high tunnels or row covers against rain and hail. In the case of sour cherries, along with new varieties, mechanized harvest technologies have revolutionized production for processing purposes. Finally, postharvest packing and shipping technologies have been greatly improved and standardized, allowing some countries such as Chile to significantly expand their cherry production for export to very distant markets, such as China, Singapore and Japan.

Although these achievements were more or less directly fuelled by horticultural research, various challenges remain for cherry growers in both traditional and expanding production regions. One of the overarching threats for sustainable cultivation of cherries (as well as most other crops) is the impact of climate change on natural resources, such as water and nutrient availability, and key physiological processes in the cherry production cycle. The consequence of increased autumn, winter and spring temperatures may be dramatic, resulting in insufficient fulfilment of chilling requirements necessary for full reproductive and vegetative development in spring, synchronized flowering and earlier bloom dates that are at an increased risk of frost damage. Increased summer temperatures have a direct impact on fruit quality through arrested ripening, softer fruit texture and a higher occurrence of double fruits. Climate change is also associated with a higher frequency of extreme events that vary across regions, including drought, hail and damaging levels of rain. A clear biological and mechanistic understanding of physiological processes such as dormancy, flowering, double fruit induction and fruit cracking is essential for the development of novel mitigation strategies, including accurate characterization and utilization of cherry genetic resources for modifying these traits.

An increasingly important threat for cherry cultivation is the introduction of non-native pests and diseases in various production regions, which is a consequence of international transportation of both commodities and humans to the farthest reaches of the planet. A very impactful example is the recent introduction and very fast naturalization of the Asian fruit fly *Drosophila suzukii* in Europe and North America, which has been particularly harmful in small fruits such as cherries, most berries and grapes. Pests and diseases will become more challenging for growers as well because of higher societal pressures to reduce reliance on broad-spectrum pesticides in agriculture. With the advent of 'softer' or organic pesticides plus new invasive pests and diseases, it may be difficult to reduce pest management treatments and costs and achieve sustainable production without genetic solutions from breeding programmes. However, a considerable initial challenge will be to discover whether sources of resistance or tolerance even exist in the cherry genome or may be introduced from related *Prunus* or other sources, or achieved via new gene-editing techniques to selectively modify key biochemical response pathways. Currently, phenotyping protocols for insect/disease resistance/tolerance have yet to be developed as an initial step to eventual utilization of genetic strategies for creating new genotypes having such important traits for resilience to either biotic (pests) or abiotic (climatic/environmental) stresses.

We foresee a higher degree of specialization among future cherry growers, who will have to adopt more complex, site-specific production strategies in terms of plant materials and cultural practices. Current and future climatic conditions, and privatization of some potential genetic advances (such as proprietary varieties for low-chilling regions), may result in limitations for some growers in implementation of certain modern orchard technologies. The amount of worldwide research funds devoted to cherries is rather small in comparison with agronomic crops, or even many other fruit crops, and therefore the need for some coordination in research efforts among the small fraternity of cherry scientists of different countries or regions is highly desirable. This includes the characterization and preservation of the diversity of cherry genetic resources. Towards that end, this book arose from such a scientific coordination, as it is based upon COST Action FA1104 ('Sustainable production of high-quality cherries for the European market'), supported by COST (European Cooperation in Science and Technology). This Action established a large and multi-disciplinary network of cherry specialists from over 30 countries in Europe and around the world, with various goals including adoption of common experimental protocols, exchange of data between research teams, and the implementation of various coordinated research strategies (http://www.bordeaux.inra.fr/cherry). This book is one of the most important outcomes from the Action's 2012–2016 activities.

Finally, the mention of generic or registered plant growth regulators, agrochemicals and/or other agricultural products in the various chapters of this book do not imply any endorsement by the authors nor presumption of legal status for commercial use in cherry production. The primary goal of this book is to present new research results and commercial practices across a wide range of cherry-producing countries and continents. Every country, and even individual states or provinces within a country, may have differing product registrations for commercial use. The reader is always referred to consult local regulatory authorities and to read the label of any product prior to commercial or experimental use.

**J. Quero-García**
**A. Iezzoni**
**J. Puławska**
**G. Lang**

# Acknowledgements

This publication is based upon work from COST Action FA1104, supported by COST (European Cooperation in Science and Technology).

COST (European Cooperation in Science and Technology) is a funding agency for research and innovation networks. Our Actions help connect research initiatives across Europe and enable scientists to grow their ideas by sharing them with their peers. This boosts their research, career and innovation.

EUROPEAN COOPERATION IN SCIENCE & TECHNOLOGY

Funded by the Horizon 2020 Framework Programme of the European Union

# 1 Cherry Production

**Géza Bujdosó[1]* and Károly Hrotkó[2]**
[1]*National Agricultural Research and Innovation Centre Fruitculture Research Institute, Budapest, Hungary;* [2]*Szent István University, Faculty of Horticultural Science, Budapest, Hungary*

## 1.1 Introduction

Sweet (*Prunus avium* L.) and sour (syn. tart, *Prunus cerasus* L.) cherry ripen first among stone fruits, followed by apricot, peach and plum. Because sweet cherry is first on the fresh market, it is in high demand in the late spring and early summer. Sweet cherry cultivars with a red fruit colour dominate the market, while cultivars of yellow, white or blush colour are in less demand. Sour cherries have smaller fruit size and are less firm than sweet cherries. The vast majority of sour cherries are processed; however, sour cherries with higher sugar content are becoming more common on the fresh fruit market in recent decades.

Sweet cherry cultivars span a longer maturity period than sour cherries. In temperate zones of the northern hemisphere, sweet cherry cultivars mature from the end of April (in southern growing regions) to June–July (main season), while the picking season finishes in late August in Norway. In the southern hemisphere, the majority of sweet cherries are harvested in December and January, as this harvest time coincides with lucrative markets, such as those of North America and western Europe, as well as Southeast and East Asia. Sour cherries, which are grown mainly in the northern hemisphere, are harvested beginning in May in the more southern regions, and the season finishes in July to early August in Poland, Germany and Michigan (USA).

It is difficult to obtain accurate cherry production data. In many cases, production for both cherry species are published together under one 'cherry' category. This form of data presentation is more frequent in countries where sour cherry production is negligible, and therefore the production figures can be assumed to represent sweet cherry. In addition, the hectares of cherry production and yields reported are not uniform between countries. There is variation in whether the orchard surface and yields reported include only commercial or commercial plus backyard garden production, bearing or both bearing and non-bearing orchards, and exported production. Therefore, accurate comparisons are difficult to obtain, especially the average yield $ha^{-1}$ data when calculated by total yield of the country divided by total orchard surface.

* resinfru@yahoo.com

## 1.2 Global Sweet Cherry Production

In the 1960s and 1970s, producing fresh market sweet cherries was considered difficult because it required a large labour force to hand-pick the fruit. The hand-picking was also time consuming due to the central leader or modified central leader trees on vigorous seedling rootstocks, which resulted in very large canopies. To decrease the tree size, numerous cherry rootstock breeding programmes were started worldwide (see Chapter 6, this volume), resulting in rootstock series that imparted different levels of scion vigour. As a result, starting in the 1980s and 1990s, intensive orchard systems for hand-picking were developed and increasingly adopted. In the modern 'pedestrian' orchards, growers prefer to use these new size-controlling rootstocks. The tree height has decreased by as much as 2.5–4.0 m, and tree density varies from 667 to 1250 trees $ha^{-1}$. New orchard systems (e.g. Zahn Spindle, Vogel Spindle, Hungarian Spindle, Spanish Bush, Tatura hedge system, Tall Spindle Axe, Super Spindle Axe, Kym Green Bush, Upright Fruiting Offshoots) were developed. The smaller tree sizes made it feasible to use rain covers, hail protection and netting against birds. Innovations such as new pruning methods, irrigation systems, tree support systems, plant nutrition methods, plant protection methods and machines have been adapted to these new orchard systems. Today in the northern part of the northern hemisphere, canopies with central leader (spindle canopies) are planted, while in the southern part of the northern hemisphere, open canopies (e.g. Spanish Bush) are preferred. In the southern hemisphere, the Tatura hedge system, V- or Y-shaped hedges, and multiple leader bush (e.g. Kym Green Bush) canopies are most common. Other new canopies are currently in the trial phase.

There is a keen interest in new cultivars as well. Self-fertile sweet cherry cultivars were not common in the 1990s, and many breeding programmes focused on self-fertility (see Chapter 4, this volume). Producers looked for cultivars that would extend the cherry picking season (Sansavini and Lugli, 2008). Cultivars with very early and late ripening times and a minimum of 26 mm in diameter, dark or light red/burgundy fruit colour, shiny fruit skin, sweet taste, and medium or long stem are most desired. Although cultivars with yellow and blush fruit colour are less important than red/burgundy cultivars, consumers' interest in blush cherries is increasing, especially in China and the USA. Sweet cherry is used mostly for fresh consumption, with a small percentage of production resulting in processed products such as jam, glass-packed or canned products (see Chapter 20, this volume).

The annual global sweet cherry production is about 2.2 million t, and shows a slightly increasing tendency (Table 1.1). The leading sweet cherry-producing country is Turkey, followed by the USA, Iran, Italy, Spain, Chile and Ukraine. During the period of 1980–2013, sweet cherry production in Turkey, the USA, Chile and China showed dramatic increases, while production in France, Germany and Bulgaria exhibited a strong decrease. In terms of production for export, the top three countries are Chile, the USA and Turkey (USDA-FAS, 2016).

In particular, Turkish and Chilean sweet cherry production has increased rapidly, supplying markets mainly in Europe and Russia, and in China and the USA, respectively. Turkey produces almost 500,000 t of cherries annually. Almost 70–80% of the Turkish production is based on the cultivar '0900 Ziraat' grafted on *Prunus mahaleb* and Mazzard (*P. avium*) seedling rootstocks. The second most important cultivar is 'Stark Gold' (6%) followed by 'Regina' (5%). In the near future, increased 'Regina' production and use of clonal rootstocks are projected. The ripening times of the dominant cultivars are extended by establishing orchards in different regions of the country with different climatic conditions. Some orchards are grafted on precocious rootstocks such as 'GiSelA 5' and 'GiSelA 6'. The success of the Turkish sweet cherry industry is based on cheap labour, a good field advisory service, well-developed and well-organized postharvest technology, and excellent export logistics.

The USA annually produces ~300,000 t of sweet cherries on 36,500 ha of commercial orchards. The largest producing state is

**Table 1.1.** The most important countries for sweet cherry production (1000 t). (From FAO, 2015.)

| Country | Year | | | | |
|---|---|---|---|---|---|
| | 1980 | 1990 | 2000 | 2010 | 2013 |
| Turkey | 96 | 143 | 230 | 417 | 494 |
| USA | 155 | 142 | 185 | 284 | 301 |
| Iran | 53 | 85 | 216 | 251 | 200 |
| Italy | 119 | 100 | 125 | 115 | 131 |
| Spain | 79 | 54 | 112 | 85 | 97 |
| Chile | 5 | 13 | 31 | 60 | 91 |
| Ukraine | – | – | 76 | 73 | 81 |
| Russian Federation[a] | – | – | 85 | 66 | 78 |
| Romania | 40 | 40 | 43 | 42 | 42 |
| Poland | 25 | 9 | 35 | 41 | 48 |
| China | – | – | 8 | 28 | 36 |
| France | 112 | 82 | 66 | 44 | 39 |
| Germany | 71 | 80 | 44 | 30 | 24 |
| Bulgaria | 55 | 71 | 28 | 24 | 19 |
| Australia | 4 | 5 | 6 | 13 | 17 |
| Japan | 15 | 16 | 17 | 20 | 18 |
| Canada | – | – | 3.7 | 10 | 12 |
| Portugal | 11 | 11 | 8 | 10 | 11 |
| Greece | 18 | 47 | 57 | 38 | 58 |
| Serbia | – | – | 23[b] | 22 | 28 |
| Hungary | 23 | 27 | 18 | 6 | 5 |
| Bosnia and Herzegovina | – | – | 4.6 | 9.8 | 10.8 |
| Belgium | 11[c] | 10[c] | 8 | 8 | 7 |
| Slovenia[d] | – | – | 3 | 3 | 5 |
| Czech Republic | – | – | 14 | 2 | 2 |
| Austria | – | – | 0.5 | 2 | 2 |
| Norway | 7 | 1 | 1 | 0.9 | 0.7 |
| Latvia | – | – | 0.5 | 0.05 | 0.07 |

–, Data unavailable.
[a]Numbers include Russian production and the amount imported.
[b]In Serbia and Montenegro.
[c]In Belgium and Luxemburg.
[d]Includes production in commercial orchards and backyards.

Washington, followed by California, Oregon and Michigan. The vast majority of the sweet cherries grown on the west coast of the USA are for fresh consumption, while the majority of the sweet cherries grown in Michigan are for processed products (yogurt and maraschino). In California, the main cultivar is 'Bing', followed by 'Burlat', 'Brooks', 'Coral Champagne', 'Chelan', Early Garnet™, 'Garnet', 'Rainier', 'Royal Rainier' and 'Tulare' grafted on Mazzard, 'Colt', 'MaxMa 14', 'Krymsk 5' and 'Krymsk 6' rootstocks. The last three rootstocks are being used more frequently in newer orchards. There is strong interest from growers in low-chilling cultivars, which are the focus of several private breeding programmes for expanding production in the lower San Joaquin valley.

In Washington and Oregon, 'Bing' is the major cultivar grown, followed by a group of cultivars that spread out the harvest season: 'Chelan', 'Santina', 'Tieton', Early Robin™, 'Benton', 'Rainier', 'Attika' ('Kordia'), 'Lapins', 'Skeena', 'Regina' and 'Sumtare' (Sweetheart™). The orchards in Washington, California and Oregon are irrigated, and the majority of new orchards are planted at moderate to high densities due to the increasing use of vigour-limiting rootstocks. The main rootstock used for sweet cherry in Washington and Oregon is Mazzard, followed by 'GiSelA 6', 'GiSelA 5', 'GiSelA 12', 'Krymsk 5' and 'Krymsk 6'. The trees grafted on Mazzard rootstocks are trained to a steep leader and planted at 4–5 m × 5–6 m (between trees × between rows). On vigour-limiting rootstocks, the tree spacing can range from 1.5 to 4 m, and from 4 to 5 m between rows.

Michigan grows primarily the following processing cultivars: 'Emperor Francis', 'Gold', 'Napoleon', 'Sam' and 'Ulster' grafted on Mazzard and *P. mahaleb* and are usually not irrigated as generally there is sufficient rainfall to reach the size desired for processing. For fresh market (and less common), the following cultivars are grown with irrigation: 'Attika' ('Kordia'), 'Benton', 'Cavalier', 'Ulster', 'Summit', 'Hudson' and 'Regina' grafted on 'GiSelA 5', 'GiSelA 6', 'GiSelA 12' and Mazzard. The canopies are trained to a central leader; however, high-density systems are being tested for fresh market production.

Iran has 35,804 ha of sweet cherry orchards producing ~200,000 t yearly. The Iranian sweet cherry industry is based on the cultivars 'Sorati Lavasan', 'Zarde Daneshkade', 'Shishei', 'Siahe Mashhad', 'Bing', 'Lambert' and 'Van' grafted on Mazzard, 'Colt' and 'GiSelA' rootstocks. The centre of Iranian sweet cherry production is around Isfahan, Alborz, Tehran and Khorasan. The trees are planted at 4 × 5–6 m and trained to an open centre canopy (G. Davarynejad, Mashhad, Iran, 2015, personal communication).

In Italy, there are ~30,000 ha of sweet cherry orchards producing about 110,000–120,000 t. This quantity is sold mostly in Italy, and the production trend is stable. Italian production is based on 'Burlat', 'Early Lory', 'Giorgia', 'Van' and 'Ferrovia' grafted on *P. mahaleb*, followed by 'Colt', *P. cerasus*, 'MaxMa 60', 'GiSelA 6' and 'CAB 6P'. The most important production areas are the Puglia, Campania and Basilicata regions in southern Italy, and the Emilia-Romagna and Veneto regions in northern Italy. The trees are trained to open vase and Spanish Bush canopies. In new orchards, planting distance is 3.5–5 × 5 m. New orchards are irrigated by drip irrigation, especially in the southern regions. Older orchards are planted with 6–7 m between trees and between rows (D. Giovannini, Forli, Italy, 2014, personal communication; M. Palasciano, Bari, Italy, 2014, personal communication).

The Spanish sweet cherry industry shows an increasing trend with about 33,000 ha producing ~90,000 t year$^{-1}$. The dominant cultivars for the early ripening season are 'Earlise®Rivedel', 'Burlat', 'Chelan', 'Prime Giant', 'Nimba', 'Pacific Red', 'Frisco' and 'Crystal Champaign' among the new cultivars; for the mid-season are 'New Star', 'Starking Hardy Giant','Santina', '13S 3-13', '4-84', 'Van', Summit' and 'Sunburst'; and for the late ripening season are 'Ambrunes', 'Lambert', 'SP-106', 'Sommerset', 'Lapins', 'Skeena' and 'Sweetheart', among others (Moreno, 2002; Iglesias *et al.*, 2016). The dominant rootstock is INRA 'SL 64', while in some cases 'MaxMa 14' and 'Colt' are used. More recently, 'Adara' alone or 'Mariana 26-24' or 'GF 8-1'/'Adara' as an interstem, commonly named Marilan, are increasingly used. The most important growing areas are Extremadura (Valle del Jerte), followed by Aragón (e.g. La Almunia, Caspe), Catalonia (e.g. Baix Llobregat, Ribera d'Ebre), Andalucía (Granada) and Comunidad Valenciana (Alicante) (Alonso, 2011; Iglesias *et al.*, 2016). The irrigated orchards are planted at 2.5–3 × 4–5 m and are trained to the Spanish or Catalan bush system, and in recent years to the Ebro system, a modification of the Spanish Bush developed in the Ebro Valley to induce early yields.

The Chilean sweet cherry industry is growing rapidly by thousands of hectares each year. Today there are 21,000–23,000 ha of orchards producing 124,000 t. The main production is located between the Valparaíso and Metropolitana regions (33°S) and 350 km south (35°S) in the Maule Region. Cherries are also produced in small microclimate areas just north of Santiago, characterized by a chilling accumulation between 400 and 750 h, and in the XI region located in Patagonia (46°S), where there is a risk of frosts in springtime and rain during harvest. These extreme regions extend the harvest season from early November to mid-January with differences in harvest time along the east and west sides of the valleys, due to the influence of the Andes Mountain chain and the Pacific Ocean. The main producing area is in the VI and VII regions, with 800–1200 chilling hours and harvests in November and December. Almost 75% of the production is shipped to Asia. The production is based on 'Sweetheart' and 'Bing', followed by 'Lapins', 'Santina', 'Royal Dawn', 'Regina', 'Brooks' and 'Rainier'. The predominant rootstocks are Mazzard and *P. mahaleb* seedlings, 'Colt' and 'F 12/1' in the old orchards; the young orchards are grafted on 'Colt', 'MaxMa 14', 'GiSelA 6', 'CAB 6P' and 'GiSelA 5'. High-density orchards and cultivars with early or late maturity times are preferred. The old orchards are planted at 4.5 × 5.25 m with vase-shaped canopies; young orchards are planted at 2 × 4.5 m with central leader and spindle canopies. Orchards are irrigated using micro-sprinklers and drip irrigation (Stehr, 2003; E. Gratacos, Valparaíso, Chile, 2015, personal communication).

In Ukraine, sweet cherry production is mainly for domestic fresh consumption, with 12,400 ha producing 70,000–80,000 t. The most important cultivar is 'Krupnoplidna' (syn. 'Krupnoplodnaja' or 'Krupnoplodnya'), followed by 'Valerii Chkalov' (syn. 'Valerij Tschkalov', 'Valerij Cskalov' or 'Valery Chkalov'), 'Liubava', 'Melitopolska Chorna', 'Donetskyi Uholok', 'Kytaivska Chorna' and 'Burlat' grafted on seedling *P. mahaleb* and Mazzard (70% of production), 'VSL-2' (syn. 'Krymsk 5') and 'GiSelA 5'. The most important sweet cherry production areas are Zaporizhia, Dnipropetrovsk and Kherson. The trees are planted at 3–4 × 6 m on seedling rootstocks and at 2–3 × 4–5 m trained to 'natural' multi-leader or modified central leader canopies. Some orchards grafted on clonal rootstocks are irrigated (Y. Ivanovych, O. Kishchak, S. Vasyuta and V. Vasylenko, Kiev, Ukraine, 2015, personal communication).

The Russian Federation has 2500 ha producing 49,000 t annually, much of which is backyard production (State Commission of the Russian Federation for Selection Achievements Test and Protection, http://en.gossort.com/, accessed 29 October 2015). The Russian Federation also imports about 40,000 t annually, primarily from Iran, Italy, Germany, the Netherlands, Azerbaijan and Syria. Yields are increasing slightly. The most important cultivars, 'Valery Chkalov', 'Denissena sholtaya' (syn. 'Dönissens Gelb'), 'Gold', 'The Gift of Ryazan', 'Sinyavskaya' and 'Chermoshnaya', are grafted on Russian-bred rootstocks such as 'VC-13' ((*P. cerasifera* × *P. maackii*) × *P. cerasus*), 'LC-52' ((*P. cerasifera* × *P. maackii*) × *P. cerasus*), 'Krymsk 5', 'Krymsk 6' and 'Colt'. The trees are most commonly trained to different spindle canopies. The most important sweet cherry-producing areas are the Central region, Central Chernozem region, North Caucasus and Low Volzhsky regions of Russia. Commonly, trees are planted at 1–1.5 × 3–3.5 m. Orchards are mostly irrigated with overhead or drip irrigation (I.M. Kulikov and A.A. Borisova, Moscow, Russia, 2015, personal communication; Association of Fruits, Berries and Planting Material Producers, http://www.asprus.ru, accessed 29 October 2015).

Romanian sweet cherry production shows an increasing tendency. Almost 7000 ha produce ~42,000 t of fruit annually. The major cultivar is 'Van', followed by 'Stella', 'Boambe de Cotnari', 'Hedelfinger' (syn. 'Hedelfingen'), 'Germersdorf' (syn. 'Germersdorfi'), 'Daria', 'Rubin' and 'Rivan' grafted on seedlings of Mazzard (60% of production) and *P. mahaleb* (30%). The remaining trees are grafted on 'GiSelA 5'. The commercial orchards are located in the southern hills of the Carpathian Mountains and the north-east part of the country. Standard orchards are planted at 4–5 × 5–6 m; however, the most intensive orchards are planted at 2 × 4 m. The trees are trained to open vase, central leader and steep leader canopies. Only newly established orchards are irrigated (S. Budan, Pitesti, Romania, 2014, personal communication).

Sweet cherry production in Poland has been increasing, with about 10,000 ha producing 40,000 t annually. Increasingly, newly planted orchards are high-density systems under plastic cover; however, finding less environmentally risky sites is a key production factor. The most common cultivars are 'Burlat', 'Vanda', 'Techlovan', 'Summit', 'Kordia', 'Regina' and 'Sylvia', as well as 'Hedelfinger' and 'Red Buttners' grafted on Mazzard seedlings and 'F 12/1' in older orchards. In modern new orchards, 'GiSelA 5' is very popular, followed by 'PHL-A' and 'Colt'. *P. mahaleb* seedlings are not used as rootstocks for sweet cherry in Poland due to incompatibility symptoms. Some new orchards use Frutana interstock to decrease tree vigour; others use 'GiSelA 5' or 'PHL-A' as an interstock. New scion cultivars bred in Czech Republic, Hungary and Canada are promising, including 'Sandra', 'Kasandra', 'Jacinta', 'Justyna', 'Tamara', 'Debora', 'Vera®', 'Annus®' and 'Staccato'. The most important growing areas are in west (Wrocław, Poznan and Piła regions), south (Tarnów, Busko regions) and central Poland. The non-irrigated orchards grafted on vigorous rootstocks are planted at 4–5 × 5 m; new irrigated orchards grafted on dwarfing rootstocks are planted at 2–2.5 × 3–4 m. Trees are trained to a spindle canopy (A. Glowacka, Skierniewice, Poland, 2014, personal communication; E. Rozpara, Skierniewice, Poland, 2015, personal communication).

Sweet cherry production is increasing rapidly in China, with 141,000 ha producing 36,000 t annually. Production, including local *Prunus pseudocerasus* cultivars, is mostly for fresh consumption (90%); the remaining 10% is for processed products such as cherry liqueur, spirits, canned fruit and fruit juice. The main cultivar is 'Hongde' (~50%), followed by 'Longguan', 'Van', 'Lapins', 'Summit', 'Sunburst', 'Rainier', 'Tieton', 'Red Lantern' and 'Brooks' grafted on 'Daqingye' (*P. pseudocerasus*, ~50%), 'GiSelA' rootstocks (~30%), *P. mahaleb* rootstocks (~20%) and Dongbeishanying (*Prunus serrulata*). The emphasis is on early-ripening cultivars (~40%), with the remaining production proportioned equally between mid- and late-season cultivars. The most important provinces are Liaoning, Shandong (Yantai, Liaocheng), Hebei and Beijing, followed by Shaanxi (Xian, Tongchuan), Henan, Gansu, Anhui, Sichuan, Qinghai and the Xinjiang Uyghur Autonomous region. The trees in traditional orchards are planted at 2–3 × 4–5 m and trained to multi-leader or central leader canopies. In Shaanxi, on loess soil, about 3000 ha of high-density orchards have been planted in the last 10 years on *P. mahaleb* rootstock, trained to central leader, Hungarian Spindle and V-trellis systems. Eighty per cent of the orchards are flood irrigated, but water-saving irrigation methods are becoming used more widely (Z. Huang and Y. Cai, Yangling, China, 2015, personal communication).

French sweet cherry production remains stable or has decreased slightly in recent years to 34,000–36,000 t. The most planted cultivar is 'Burlat' (~16%), followed by 'Belge' (~14%), 'Summit' (~12%), 'Sweetheart' (~6%), 'Napoleon' (~5%), 'Folfer' (~4%) and 'Regina' (~4%) grafted on 'MaxMa 14' (~39%), Mahaleb (~27%), Mazzard (~19%), 'MaxMa 60' (~10%), 'GiSelA 6' (~3%) and other rootstocks ('P-HL-A', 'GiSelA 5', 'PiKu 1', 'Weiroot 158', 'GF8-1' and 'Adara'). The most important growing areas are in Provence-Alpes-Côte d'Azur (in south-east France, near Marseille) and Rhône-Alpes. Almost all orchards are irrigated. Trees are trained to a vase-shaped canopy and planted at 4–6 × 6 m (G. Charlot, Baladran, France, 2016, personal communication).

Sweet cherry production is decreasing in Germany, with ~5500 ha of orchards producing 30,000 t. Production is based on 'Regina' and 'Kordia', followed by 'Bellise®Bedel', 'Sumste' (Samba™), 'Sumete' (Satin™), 'Grace Star', 'Early Korvik', 'Sumgita' (Canada Giant™), 'Karina', 'Vanda' and 'Schneiders' clones, grafted on 'GiSelA 5', 'GiSelA 3' and 'Piku 1' rootstocks. The growing areas are concentrated around the Rhine valley (Rhineland-Palatinate, Baden), as well as in Altes Land, Thuringia and Saxony-Anhalt. The most common canopy type is the spindle, with trees planted at 2.5 × 4.5 m and irrigated (M. Schuster, Dresden, Germany, 2014, personal communication).

In Bulgaria, 12,000–15,000 ha of sweet cherry orchards produce ~20,000 t. 'Burlat', 'Bing', 'Van', 'Kozerska', 'Stella', 'Rainier', 'Lapins', 'Sunburst' and 'Regina' are the major cultivars in recently planted orchards. Most trees are grafted on *P. mahaleb* seedling rootstocks; however, some orchards are grafted on dwarfing rootstocks such as 'GiSelA 5' or 'Weiroot 158', but these must be irrigated (V. Lichev, Plovdiv, Bulgaria, 2004, personal communication). Trees are planted at 3–4 × 5–7 m and trained to central leader canopies.

In Australia, commercial sweet cherry production began just a couple of decades ago. 'Lapins', 'Sweetheart', 'Kordia', 'Van', 'Simone', 'Stella' and 'Merchant' are the major cultivars, grafted on Mazzard and 'Colt'. Currently, there are ~11,000 ha in New South Wales, Victoria and Tasmania provinces producing ~15,000 t year$^{-1}$. The irrigated orchards are trained to multiple leader bush and Tatura trellis canopies (P. Measham, Hobart, Australia, 2014, personal communication).

Japan has 4460 ha of non-irrigated commercial sweet cherry orchards producing ~19,000 t. The most prevalent cultivar is 'Satonishiki' grafted on 'Aobazakura' (*Prunus lannesiana* Wils.). The dominant production area is in Yamagata province. Trees are planted at 7–8 m in rows and between rows, and trained to an open centre canopy (K. Isuzugawa, Sagae, Japan, 2015, personal communication).

Canada has 3500 ha of commercial sweet cherry orchards and production is increasing. Ninety-five per cent of Canadian sweet cherry production is in British Columbia, where one of the most important cherry breeding programmes in the world exists at the Pacific AgriFood Research Centre in Summerland. The Okanagan, Similkameen and Creston valleys comprise an annual production capacity of 12,000–15,000 t, with the main cultivars including 'Lapins', 'Bing', 'Sweetheart', 'Skeena', '13S2009' (Staccato™), 'Rainier' and 'Santina'. The remaining part of Canadian production is in the Niagara area of Ontario. Orchards tend to be high density, with trees on Mazzard and 'GiSelA 6' and trained to central leader and spindle canopies.

In the main western production regions, the orchards are irrigated because of the long dry summer season (N. Ibuki, Summerland, Canada, 2016, personal communication).

The Portuguese sweet cherry industry has 5600–5700 ha producing 10,000–11,000 t of fruit. Growers are interested in intensification of sweet cherry production and in new cultivars with low-chilling requirements. The main cultivars are 'Burlat', 'Brooks', 'Van', 'Summit', 'De Saco', 'Skeena', 'Sweetheart', 'Regina' and 'Sunburst', grafted on 'SL 64', 'Tabel Edabriz', 'CAB6P', 'CAB11E', 'Colt', 'MaxMa 14' and the 'GiSelA' series. Irrigated orchards are planted mostly in the area of Cova da Beira, Resende, Alfandega da Fé, Portalegre at 2 × 4 m in the old orchards trained to an open centre canopy, and at 1.5 × 3 m in the young orchards trained to a central leader (A. Santos, Évora, Portugal, 2015, personal communication).

Sweet cherry planting is increasing in some European countries that historically have had low production. For example, in Greece there are 10,000 ha producing 44,000 t annually. The Central Macedonia Pella and Imathia regions dominate Greek production. The major cultivars are 'Ferrovia', 'Regina', 'Lapins', 'Grace Star', 'Skeena' and 'Burlat', grafted on 'MaxMa 14', 'GiSelA 5' and 'GiSelA 6'. The irrigated orchards trained to a central leader canopy are planted at 2 × 4 m, and those trained to an open vase canopy are planted at 5 × 5 m (K. Sotiropoulos, Naoussa, Greece, 2013, personal communication).

In Serbia, 'Summit', 'Kordia', 'Lapins' and 'Regina' grafted on Mazzard seedlings are the dominant combinations. There are ~4500 ha of commercial orchards producing 28,000 t, grown mainly in the Belgrade region and west Serbia. Very few orchards are irrigated, and those that are have trees grafted on 'GiSelA 5' (S. Radicevic, Čačak, Serbia, 2014, personal communication).

Hungary has 1300 ha of sweet cherry orchards producing 5000 t, which includes home gardens. 'Germersdorfi' (syn. 'Schneider's Späte Knorpelkische') clones, 'Burlat' and 'Van' are the dominant cultivars, but the Hungarian-bred 'Katalin' and 'Linda' are slowly replacing them. Traditional orchards are planted at 5 × 7–8 m and trained to a central leader canopy. The irrigated intensive orchards are planted 2–3 × 4–5 m and trained to different spindle canopies. Growers are interested in new early-ripening cultivars to capitalize on the early market. As a result, the new Hungarian-bred cultivars 'Rita®', 'Vera®', 'Carmen®', 'Sándor®', 'Annus®' and 'Paulus®' grafted on *P. mahaleb* seedling and clonal rootstocks are being planted more frequently.

Bosnia and Herzegovina have ~2700 ha producing 9000–10,000 t. 'Summit', 'Kordia', 'Lapins', 'Regina', 'Napoleon', 'Sweetheart' and 'Sylvia' grafted on *P. mahaleb*, 'SL 64' and 'GiSelA 5' are the most dominant combinations. Irrigated orchards are located in east Herzegovina and the Neretva valley, north-west Bosnia in the Krajina region and north-east Bosnia in the Tuzla region. Trees are planted at 1.5–2 × 4–5 m and trained to a Spanish Bush and modified spindle system (G. Djuric, Banja Luka, Bosnia and Herzegovina 2014, personal communication).

The Belgian sweet cherry industry is small but increasing, with 808 ha producing 6000–8000 t. 'Kordia', 'Regina', 'Lapins' and 'Sweetheart' grafted on 'GiSelA 5' are dominant; however, old orchards are grafted on Mazzard seedlings. Production is located around Limburg. Most orchards are irrigated and planted at 2–3 × 4–5 m (J. Vercammen, Sint-Truiden, Belgium, 2014, personal communication).

In Slovenia, commercial sweet cherry production is limited, with about 150 ha producing 2700 t. However, there are also many trees in backyard gardens and meadow orchards. 'Burlat', 'Van', 'Giorgia' and 'Regina', grafted on Mazzard seedlings and 'GiSelA 5', are typical. The Primorska region is the centre of production. Trees are planted at 2.5–5 × 5–7 m and trained to a modified spindle canopy (V. Usenik, Ljubljana, Slovenia, 2014, personal communication; N. Fajt, Nova Gorica, Slovenia, 2014, personal communication).

The Czech Republic has ~950 ha producing 2595 t. The main cultivars are 'Kordia', 'Regina', 'Burlat' and 'Sam', grafted on 'Colt', 'GiSelA 5', 'P-HL-A' and 'F 12/1'. The most important sweet cherry-producing regions are Hradec Králové, southern Bohemia, central Bohemia and Olomuc. Trees are either untrained in older orchards or trained to a spindle canopy in new orchards. Orchards are not irrigated; trees grown on dwarfing rootstocks are planted at 2–2.5 × 4–5 m and on vigorous rootstocks at 4–5 × 5–6 m (F. Paprstein and J. Sedlak, Holovousy, Czech Republic, 2014, personal communication).

Austria has 230 ha of irrigated commercial sweet cherry orchards, which produce 2000 t, located primarily in the east (Niederösterreich, Burgenland and Steiermark) and Oberösterreich (A. Spornberger, Vienna, Austria, 2014, personal communication). Harvest spans 4–5 weeks, with 'Burlat', 'Kordia' and 'Regina' as the most dominant cultivars, grafted on 'GiSelA 5'.

The Norwegian sweet cherry industry produces 600–700 t from 200 ha, with increasing production. The fruit is sold inland, mainly for fresh local consumption, with no export. The most important cultivar is 'Lapins' (~44%), followed by 'Van', 'Sweetheart' and other late-ripening cultivars, and 'Ulster' and 'Burlat', grafted on 'GiSelA 5' and 'GiSelA 6'. The trickle-irrigated orchards around Ullensvang and Lærdal (western Norway) as well as in Gvarv (eastern Norway) are trained to a Tall Spindle Axe system, and planted at 1–2 × 3.5–4.5 m. All orchards have rain covers, and high tunnel production is increasing (M. Meland, Ullensvang, Norway, 2016, personal communication).

Latvia has 245 ha of commercial orchards producing ~100 t. 'Bryanskaya Rozovaya', 'Iputj' and 'Aiya' grafted on *P. mahaleb* are the important combinations. Orchards are planted at 4 × 5 m, trained to a central leader and not irrigated (S. Ruisa, Dobele, Lativa, 2014, personal communication).

## 1.3 Global Sour Cherry Production

Sour cherry is often called the fruit species of eastern Europe because the most important producing countries are located in this part of the world. Global production is about 1,100,000 t. In countries where there is a keen interest in sour cherry-based products, such as the eastern European countries, production is usually machine harvested and is increasing slightly (Table 1.2). The world's leading sour cherry-producing country is Turkey, followed by the Russian Federation, Poland, Ukraine, Iran, the USA, Serbia and Hungary. Specialized fruit preservation companies purchase a large majority of the production. This industrial demand is relatively stable, but in some years with high yields, sour cherry prices to growers can be very low, and sometimes lower than the cost to produce.

Sour cherries are produced almost solely for processing. Besides canned, bottled or dried end products, preserved and frozen sour cherries are also prepared for secondary food industrial uses, such as baking, dairy and the confectionary industries. Specific cultivars for processing that dominate global sour cherry production include: Serbian landrace 'Oblačinska', German landraces 'Schattenmorelle' and 'Ostheimer', Hungarian-bred 'Újfehértói Fürtös' (syn. 'Ungarische Traubige', Balaton™), 'Érdi Bőtermő' (syn. Danube™), and the old French cultivar 'Montmorency'.

Since the 1970s, machines for mechanical harvesting have been used worldwide. These shaking machines reduced harvest costs significantly, although new challenges arose to preserve the quality of mechanically harvested fruit. Practices such as the plant growth regulator ethephon sprays for acceleration of fruit maturity, tree pruning and postharvest cooling strategies were developed at the same time. As a result, hand-picking is no longer practised in the majority of producing countries. Orchards specifically designed for trunk shakers have cultivars grafted on vigorous seedling rootstocks and planted in wide rows with densities of 210–285 trees $ha^{-1}$. The trees are sometimes trained to an open canopy (either an open vase or a canopy that includes removal of the central leader in years 6–8) to facilitate trunk shaker mechanical harvesters.

Growers in several eastern European countries (Russia, Ukraine, Belarus, Moldavia, Romania and Hungary) inherited sour cherry

**Table 1.2.** The most important countries for sour cherry production (1000 t). (From FAO, 2015.)

| | Year | | | | |
|---|---|---|---|---|---|
| Country | 1980 | 1990 | 2000 | 2010 | 2013 |
| Turkey | 60 | 90 | 106 | 195 | 180 |
| Russian Federation[a] | 162[b] | 221[b] | 200 | 165 | 200 |
| Poland | 42 | 77 | 140 | 147 | 188 |
| Ukraine | – | – | 155 | 155 | 200 |
| Iran | 9 | 19 | 49 | 103 | 106 |
| USA | 99 | 94 | 128 | 86 | 133 |
| Serbia and Montenegro[c] | 58[c] | 120[c] | 59 | 66 | 98 |
| Hungary | 37 | 61 | 49 | 52 | 53 |
| Romania[d] | 27 | 27 | 29 | 28 | 28 |
| Germany | 142 | 118 | 36 | 18 | 13 |
| Czech Republic | – | – | 10 | 5 | 5 |
| Denmark | – | – | 18 | 13 | 9 |
| Italy | 0 | 0 | 10 | 7 | 7 |
| Bosnia and Herzegovina | – | – | 1 | 3 | 5 |
| Belarus | – | – | 16 | 51 | 15 |
| Canada | 10 | 5 | 8 | 6 | 6 |
| Croatia | – | – | 7 | 7 | 10 |
| Norway | – | – | – | – | 0.3 |
| Latvia | – | – | 0.5 | 0.05 | 0.07 |

–, Data unavailable.
[a]Includes commercial orchards and backyard gardens, as well as imports.
[b]Includes the former Soviet Union for 1980 and 1990 data.
[c]Includes the former Republic of Yugoslavia for 1980 and 1990 data.
[d]Estimated from sweet cherry data.

orchards from the previous state-owned socialist establishments following privatization in the 1990s. These plantations were old, often with poor cultural practices and very low production (2–4 t ha$^{-1}$). In these countries, production competitiveness was only achieved by considerable state subsidies. Subsequently, significant industry development became possible only where low labour costs were combined with higher-than-average yields. This prosperous combination has tended to occur where sour cherry production was traditionally strong and where new private companies, together with government-supported innovation efforts, occurred simultaneously (Szabó *et al.*, 2006).

In recent years, some markets have developed for fresh sour cherries. Sour and sour × sweet cherry (duke cherry) genotypes with well-balanced sweet or sweet–acidic taste are suitable for fresh consumption, and consumer demand is growing. Consumers have an increased awareness that cherries have health benefits, being rich in vitamin C, antioxidants and polyphenols (see Chapter 17, this volume). Sour cherries for fresh consumption and hand-picking are usually grown in high-density orchards (600–1000 trees ha$^{-1}$). Cultivars are grafted on strong, medium or semi-dwarf rootstocks having 2.5–3 m maximum final tree height and a spindle canopy, and the fruit is picked with the stem. In Turkey and Hungary, these so-called table sour cherry orchards comprise ~30% of the total production area, along with 10% in Romania and 5% in Serbia and Poland; in Germany and the USA, this production is negligible. In Belarus, 70% of the sour cherries are used for fresh consumption, while 30% are for processing. At present, the Hungarian-bred cultivars 'Érdi Bőtermő' and 'Újfehértói Fürtös', together with a few other local cultivars, are grown throughout the world and play an important role in the global fresh market sour cherry industry. Some breeding programmes aim to develop novel cultivars suitable for fresh consumption (see Chapter 5, this volume). This component of the sour cherry industry is increasing rapidly, supported by active marketing to increase sour cherry consumption as a fresh fruit. Furthermore, efforts are ongoing to produce novel sour cherry-based products (e.g. liqueurs, wines, juices and dried fruit) to increase per-capita consumption (see Chapter 20, this volume).

The world-leading Turkish sour cherry production is ~180,000 t annually and is based on just one cultivar, the late-ripening 'Kutahya', comprising ~90% of production. Afyon county is the most important production region in Turkey (S. Ercisli, Dresden, Germany, 2014, personal communication).

In the Russian Federation, there are 1500 ha of commercial sour cherry orchards producing 23,000 t. Backyard production is also very important. Huge imports (36,000–74,000 t) from Hungary, Poland, Turkey, the USA, Georgia and other countries meet the annual needs of the Russian market. Production and processing is increasing. The major cultivars are 'Zhukovsky', 'Youth', 'Lyubsky', 'Turgenev', 'Rusinka' and 'Enikeev's Memory', grafted on 'Izmaylovsky', 'P-3', 'P-7' and seedlings of the cherry cultivars 'Vladimir' and 'Rastunyi'. In southern Russia, *P. mahaleb* seedlings are used. Own-rooted orchards can also be found. The important growing regions are the Central, Central Chernozem, Northern Caucasus, and Middle and Low Volzhsky regions. Trees are trained to a central leader canopy and planted at 1–1.5 × 3–3.5 m. Some orchards have drip irrigation (I.M. Kulikov and A.A. Borisova, Moscow, Russia, 2015, personal communication).

In Poland, commercial sour cherry orchards comprise ~35,000 ha and total annual production ranges from ~160,000 to 200,000 t, 43% of which is from backyard gardens. The largest percentage of the total production is for the freezing industry, followed by the juice industry. 'Lutówka' (syn. 'Schattenmorelle') is the dominant cultivar (~70–80% of production), followed by 'Kelleris 16', 'Újfehértói Fürtös' and 'Debreceni Bőtermő'. Fifty-five per cent of orchards are grafted on *P. mahaleb* (on 'Piast' and 'Popiel' cultivars), and the remaining 45% on Mazzard ('Alkavo'). The most important growing areas are in central Poland (Grójec and Radom regions) and south-east Poland (Sandomierz, Ostrowiec Świętokrzyski, Ożarów Vistula and Lublin regions). The non-irrigated orchards are planted at 2–2.5 × 4–4.5 m. In old orchards, growers keep the natural tree habit, while in new orchards the trees are trained to spindle canopies. The over-the-row continuously moving mechanical harvester, which was developed in Poland, requires a small spindle tree and high-density planting. About 30% of orchards are irrigated (A. Glowacka, Skierniewice, Poland, 2014, personal communication; E. Rozpara, Skierniewice, Poland, 2015, personal communication).

In Ukraine, there are ~20,000 ha of commercial sour cherry orchards with an annual production of 155,000–200,000 t, which is slowly increasing. Domestic production is not enough to supply domestic consumption, and therefore sour cherry imports are 2.5 times higher than production to meet demand. Industrial processing is the most important use of sour cherry, but fruit from some duke cherry cultivars is used for fresh consumption. The most common cultivar is 'Újfehértói Fürtös' followed by 'Schattenmorelle', 'Melitopolska Desertna', 'Vstriecha', 'Northstar' and 'Shalunia', grafted on *P. mahaleb* seedlings, 'VSL-2' ('Krymsk 5'), 'GiSelA 5', 'Colt' and 'Alfa' (*P. cerasus*) rootstocks. The majority of orchards are not irrigated and are established in the forest-steppe and steppe zones of Ukraine in the Dnipropetrovsk, Lviv, Rivne, Poltava and Khmelnytsky regions. Trees are planted at 2.5 × 5–6 m, and trained to natural round and central leader canopies. Irrigation is used in southern regions (Y. Ivanovych, O. Kishchak, S. Vasyuta, V. Vasylenko, Kiev, Ukraine, 2015, personal communication).

Iran has 17,911 ha of sour cherry orchards producing 94,837 t, with a trend for increasing production. The season is very short, starting in mid-June and finishing in early July. Local cultivars are mainly planted, followed by 'Montmorency', 'Érdi Bőtermő' and 'Cigány Meggy' clones and 'Érdi Jubileum' grafted on *P. mahaleb* and Mazzard rootstocks. The centre of Iranian sour cherry production is around Isfahan, Alborz, Tehran and Khorasan. Trees are planted at 3 × 4 m or 4 × 5 m, trained to an open vase canopy, and irrigation is necessary (G. Davarynejad, Mashhad, Iran, 2015, personal communication). There are many foods in Iranian cuisine that use sour cherry (e.g. juice, jam, concentrate, dried).

The USA is the largest sour cherry-producing country in North America with 11,000 ha producing ~100,000 t, 99.9% of which is processed, sold as frozen, prepared for the baking industry (pies, pastries), made into jam and specialty products, dried, or used for juice and concentrate. The predominant cultivar is 'Montmorency', which is mostly grafted on *P. mahaleb* seedlings. The most important sour cherry-producing states are Michigan (~70%), followed by Utah (~15%) and Washington (~10%). Trees are trained to a central leader or modified central leader canopy with wide branch angles. Orchards are not irrigated in Michigan, but are irrigated in Washington and Utah. Trees are planted at 4.3–4.8 × 6–7 m (USDA,

2013). Research on development of high-density orchards (training systems, rootstocks and cultivars) for continuous over-the-row harvesting began in 2008 and is slowly expanding to commercial-scale early adopters in Michigan.

Serbian sour cherry production is increasing slightly, with almost 100,000 t produced from 14,000 ha. The largest crop use is freezing for export, with the remaining for other purposes and fresh consumption. Own-rooted ‘Oblačinska’ clones are the dominant cultivars (~85% of production), followed by ‘Cigány Meggy’ clones and ‘Újfehértói Fürtös’. There is a keen interest in novel new cultivars such as ‘Feketička’. Most cultivars, except ‘Oblačinska’ clones, are grafted on Mazzard seedlings. In general, the non-irrigated ‘Oblačinska’ clones are trained to a vase canopy, while the grafted cultivars are trained to a spindle canopy. Typical planting distances are 2–3 × 4–5 m, with the smaller distances typical for ‘Oblačinska’ clones. The most important production areas are south-east Serbia (around Niš) and the Vojvodina province (S. Radicevic, Čačak, Serbia, 2014, personal communication).

Hungary has a long tradition of sour cherry growing and consumption, and commercial production is slowly increasing, currently at 13,000–14,000 ha producing 55,000–65,000 t. One-third of Hungarian orchards are outdated, produce low yields and are ready to be replaced. Production is partitioned at 70% for industrial purposes and 30% for fresh consumption. The main cultivars are the late-ripening ‘Újfehértói Fürtös’, ‘Debreceni Bőtermő’ and ‘Kántorjánosi’ (59% in total), and ‘Érdi Bőtermő’ (~24%) grafted on *P. mahaleb* seedling rootstocks. Some orchards are irrigated, especially intensive orchards for fresh market production. Typical densities are 5 × 7–8 m for shaking and 2.5–3 × 5–6 m for hand-picking. The most important production area is located in the north-east part of the country in Szabolcs-Szatmár-Bereg County followed by Pest County (Anon., 2003).

In Romania, ~4700 ha of commercial sour cherry orchards produce ~28,000 t yearly, of which 90% is processed to produce canned fruit, jam and juice. ‘Crişana’ (syn. ‘Köröser’) is the major cultivar, followed by ‘Mocăneşti’, ‘Schattenmorelle’, ‘Nana’, ‘Tarina’ and ‘Ilva’. Approximately 60% of orchards are grafted on *P. cerasus*, with the remaining 30% on *P. mahaleb*. The largest commercial orchards are on hilly slopes of the Carpathian Mountains and in the north-east part of the country in Iasi County. Typical orchards are non-irrigated, trained to a central leader canopy, and planted at 4 × 4 m or 4 × 3 m (S. Budan, Pitesti, Romania, 2014, personal communication).

German sour cherry production has decreased significantly in recent decades; however, Germany imports the largest quantity in Europe for industrial purposes. Production for fresh consumption has just started, and growers are planting high-density orchards using ‘Újfehértó Fürtös’. There are 2291 ha of commercial orchards producing ~20,000 t. ‘Schattenmorelle’ types dominate, with limited plantings of ‘Morellenfeuer’, ‘Fanal’, ‘Safir’, ‘Újfehértó Fürtös’ and ‘Morina’ grafted on *P. avium* ‘Alkavo’ or ‘F 12/1’, or on *P. mahaleb*. Some new orchards are grafted on ‘GiSelA 5’. The most important production areas are Thuringia, Saxony and Rhineland-Palatinate. Non-irrigated orchards are planted at 3 × 4–4.5 m and trained to central leader, spindle or bush canopies (M. Schuster, Dresden, Germany, 2014, personal communication).

The Czech Republic has a small sour cherry industry with a slightly decreasing production trend, and currently has 1647 ha producing ~4300 t. ‘Fanal’, ‘Újfehértói Fürtös’ and ‘Schattenmorelle’ are the main cultivars for use for frozen fruit, juice and jam. The most important producing regions are Hradec Králové, southern Bohemia, central Bohemia and Olomouc. Non-irrigated orchards are trained to a natural canopy habit, planted at 2–4 × 4–6 m (F. Paprstein and J. Sedlak, Holovousy, Czech Republic, 2014, personal communication).

Sour cherry production in western European countries is rather limited. Denmark has 1329 ha, of which 900 ha are bearing orchards producing 9500 t. Between 2008 and 2012, Denmark's production trend decreased, but from 2012 to 2013 it has increased.

About 80% of the crop is exported to Germany, and 95% is used for making juice and pulp, with a small amount used for jam, cherry sauce, wine, liqueur, syrup and dried fruit. The major cultivar is 'Stevnsbaer' (~60%), followed by 'Kelleris' (~35%) grafted on 'Colt', Mazzard and 'Weiroot 10'. The orchards are in Funen and the southern and northern part of Zealand. Semi-intensive and intensive orchards, planted at 3.5–4 × 5–7 m, are trained to central (new plantations) and multiple (old plantations) leader canopies. Young orchards are irrigated only in the establishment years (M. Jensen, K.F. Nielsen and B.H. Pedersen, Aarslev and Odense, Denmark, 2014, personal communication).

The Italian sour cherry industry is based on 1350 ha of commercial orchards producing 7000 t, which has been stable over the past 10 years. Most of the crop is processed for jam and different types of sweets. Trees are grafted on 'CAB 6P' and mostly trained to a vase system with irrigation. Production is mainly in Puglia and Piemonte (D. Giovannini, Forli, Italy, 2014, personal communication).

In Bosnia and Herzegovina, ~2000 ha of sour cherry orchards produced 3300 t in 2012. The fruit are used mostly for fresh consumption and home processing, with a small quantity for industrial purposes. Much of the production is in backyard gardens countrywide, but the most important orchards are in Herzegovina and north-east Bosnia. In old orchards, which are not irrigated, the traditional cultivar 'Marasca' is the most prevalent in Herzegovina, but 'Oblačinska' is the most grown cultivar in other parts of the country. 'Marasca' and 'Oblačinska' are planted on their own roots and trained to a vase architecture, with Spanish Bush as the dominant training system in new orchards. In new intensive orchards, planted at 3 × 4–5 m and irrigated, Hungarian-bred cultivars such as 'Újfehértói Fürtös', 'Kántorjánosi 3', 'Debreceni Bőtermő', 'Early Meteor' (syn. 'Meteor Korai') and 'Érdi Bőtermő' are the dominant cultivars grafted on *P. mahaleb* (G. Djuric, Banja Luka, Bosnia and Herzegovina, 2014, personal communication).

In Belarus, 120 ha of sour cherry orchards produce 720 t annually, mostly for fresh consumption. 'Zhivitsa', 'Vjanok', 'Novodvorskaja' and 'Griot Belorusskij' are the main cultivars, which are grafted on Mazzard seedlings (80%) and *P. mahaleb* (20%). Trees are trained to a modified central leader canopy and planted at 3 × 5 m. The southern and central zones of the country have the best climatic conditions, and orchards are not irrigated (N. Valasevich, Samokhvalovichy, Belarus, 2014, personal communication).

Canada produces 6 t of sour cherries annually and production is showing a decreasing tendency. Production is concentrated in the Okanagan and Creston regions, but is expanding in the prairies with new cultivars that are more shrub-like and have potential for mechanical harvest. The most widely grown cultivar is 'Montmorency'. Irrigation is provided in the dry growing regions (N. Ibuki, Summerland, Canada, 2016, personal communication).

Croatian sour cherry production is ~10,000 t annually on 2700 ha. The most important growing areas are located on the Adriatic coastline and hinterland between Zadar and Omiš, including several Adriatic islands (Mediterranean climatic zone), and also in the north-eastern continental part of Croatia (continental climatic zone). The most prevalent cultivar on the coast is 'Maraska'. The origin of 'Maraska' is not clear, but it might be a cultivar or a *P. cerasus* var. *marasca* that the ancient Romans took to the Adriatic coastline from Anatolia or the Caspian Sea. This genotype has good tolerance to drought and high lime content of the soil, as well as high dry matter, pectin and antioxidant content, making it a good raw material for the processing industry to make alcoholic or non-alcoholic drinks, as well as functional food. The Adriatic coastline has unique climate conditions, and the 'Maraska' cherry can produce this excellent fruit quality in this growing area. The trees are rooted on *P. mahaleb* rootstock, and the orchards are not irrigated. There is a huge demand for 'Maraska' cherry; therefore, the Croatian cherry industry is showing an increasing tendency, especially considering the fact that before the war (1991–1995) the growing areas of 'Maraska' were significantly larger. Another sour cherry cultivar, 'Oblačinska', is the leading cultivar in the continental part of Croatia. It is grown mainly on its own root, and propagation is mainly from numerous root suckers, which is one of the characteristics of this cultivar. Both 'Maraska' and 'Oblačinska' are diverse and their populations are a source of genetic variability in clonal selection (A. Vokurka, Zagreb, Croatia, 2016, personal communication).

Norway has a small, stable sour cherry production of 350 t, used mainly for local fresh consumption, on 50 ha. 'Fanal' and 'Stevnsbaer' grafted on 'Colt' and Mazzard are the most important combinations trained to a spindle-shaped pyramid. Production is concentrated in eastern Norway, around Gvarv, Øvre Eiker and Svelvik. The trickle-irrigated trees are planted at 3 × 5 m (M. Meland, Ullensvang, Norway, 2016, personal communication).

In Latvia and Lithuania, the sour cherry industries are very small and production is based on a couple of hundred hectares and individual trees in the backyard gardens. Twenty per cent of production in Latvia and 5–10% in Lithuania is for fresh consumption, with the remaining proportion for industrial purposes such as jam, juice, yogurt and wine. In Latvia, 'Latvijas Zemais' is the dominant cultivar (~90%). In Lithuania, 'Turgenevka', 'Vytėnų Žvaigždė', 'Žagarvyšne' and 'Molodiznaya' are the most important cultivars. Trees are grafted on *P. mahaleb* seedlings in both countries. In Latvia, *in vitro*-propagated 'Latvijas Zemais' plants also can be used as rootstocks (S. Ruisa, Dobele, Latvia, 2014, personal communication; V. Stanys, Babtai, Lithuania, 2014, personal communication).

## Acknowledgements

The authors of this chapter wish to thank all contributors who provided details of the cherry production in their countries. This work was supported by Mrs Viktória Nabilek-Kanavál, Agricultural Attaché of the Hungarian Embassy in Moscow, who helped significantly in obtaining some data about the Russian cherry production. The authors are grateful to Mrs Marika Asztalos for the Russian translation.

## References

Alonso, J.S. (2011) Producción, comercialización, mercado y oportunidades de la cereza [Sweet cherry production, marketing, and market opportunities]. *Vida Rural* 12, 46–50.

Anon. (2003) Meggyültetvények főbb jellmezői [The most important characteristics of sour cherry orchards]. In: Gyümölcsültetvények Magyarországon [Orchards in Hungary], 2001, Vol. II. KSH, Budapest, Hungary.

FAO (2015) Crops. Food and Agriculture Organization of the United Nations. Available at: http://faostat.fao.org/site/567/default.aspx#ancor (accessed 11 October 2015).

Iglesias, I., Peris, M., Ruiz, S., Rodrigo, J., Malagón, J., Garcia, F., Lopez, G., Bañuls, P., Manzano, M.A., Lopez-Corrales, M. and Rubio, J.A. (2016) Produzione, consumo e mercati della cerasicoltura spagnola. *Frutticoltura* 4, 2–8.

Moreno, M. (2002) Estado actual del cultivo del cerezo en Espana. Seminario 'Cultivo del cerezo en la zona norte de Chile'. Universidad Católica de Valparaiso, Quillota, Chile, 27–28 November 2002.

Sansavini, S. and Lugli, S. (2008) Sweet cherry breeding programs in Europe and Asia. *Acta Horticulturae* 795, 41–58.

Stehr, R. (2003) Süßkirschenanbau in Chile – ein Reisebericht [Sweet cherry growing in Chile – a study trip report]. *Mitteilungen OVR* 58, 130–134.

Szabó, Z., Szabó, T., Gonda, I., Soltész, M., Thurzó, S. and Nyéki, J. (2006) The current situation of sour cherry production and possibilities for development. *Hungarian Agricultural Research* 1, 11–34.

USDA (2013) Noncitrus Fruits and Nuts 2012: Preliminary Summary. US Department of Agriculture, National Agricultural Statistics Service. Available at: http://usda.mannlib.cornell.edu/usda/nass/NoncFruiNu//2010s/2013/NoncFruiNu-01-25-2013.pdf (accessed 1 January 2017).

USDA-FAS (2016) Fresh Peaches and Cherries: World Markets and Trade. US Department of Agriculture, Foreign Agricultural Service. Available at: https://apps.fas.usda.gov/psdonline/circulars/StoneFruit.pdf (accessed 1 January 2017).

# 2 Flowering, Fruit Set and Development

**Maria Herrero,[1]* Javier Rodrigo[2] and Ana Wünsch[2]**
[1]*Estación Experimental Aula Dei, CSIC, Zaragoza, Spain;* [2]*Unidad de Hortofruticultura, Centro de Investigación y Tecnología Agroalimentaria de Aragón, Instituto Agroalimentario de Aragón – IA2 (CITA-Universidad de Zaragoza), Zaragoza, Spain*

## 2.1 Introduction

Cherry trees are a paradigm of how flower biology influences the final crop. While some 8 weeks elapse from flower to mature fruit, the crop load is established very early after flowering, within approximately 4 weeks, although preharvest fruit abscission ('June drop') can modify the apparent crop load in some years. Differences between growing and non-growing flowers are determined as early as 1 week after pollination, a time that is concomitant with fertilization (Hedhly *et al.*, 2007). What occurs during this short bloom time, along with the prebloom stages of flower development, are critical to understand fruit set.

The expansion of cherry growing to new latitudes often results in erratic cropping. Failures in fruit set are usually attributed to environmental factors. However, often it is not easy to determine the real causes behind irregular crops, since it is difficult to pinpoint specific causal environmental effects. A comprehensive study, relating weather conditions to apple production, showed that weather explained a part of the variability; surprisingly, however, the main source of variable cropping was attributed to an unknown cause that appeared to be within the flower (Jackson and Hamer, 1980).

In this chapter, flower biology and fruit set in cherry are examined, paying attention to the critical steps that result in failures in fruit set. For this purpose, we follow a chronological timeline. First, flower development is examined in relation to dormancy. Then, at flowering, the events from pollination to fertilization are described with a particular emphasis on self-incompatibility. Finally, fruit set and fruit development are considered, evaluating the causes behind failures in fruit set and the main determinants of fruit development.

## 2.2 Before Flowering

The ephemeral life of a cherry flower in bloom contrasts with its long period of floral development. As in other temperate fruit tree species, flower bud development in cherry is a long process that takes place over approximately 8 months from midsummer of the previous year to flowering in the following spring (Fadón *et al.*, 2015a). This long time period is partly due to the fact that flower development is halted during the

* mherrero@eead.csic.es

winter, when the flowers enter a dormant period, and growth only resumes after dormancy. Here, we follow flower bud differentiation and examine what is known about dormancy.

### 2.2.1 Flower bud differentiation

Flower differentiation occurs during two main time periods. Flower initiation and primordial development occur in late summer and autumn, and following dormancy, final flower differentiation occurs when temperatures rise again in the spring.

#### *Early flower differentiation*

Flower buds in sweet cherry are borne mainly on spurs that are formed on long-lived shoots (up to 10–12 years old), although some flowers, especially on sour cherry and on sweet cherry on precocious rootstocks, are also borne in axillary or basal positions, respectively, on 1-year-old shoots (Flore *et al.*, 1996; Thompson, 1996). Each flower bud, which is surrounded by several bud scales, contains up to seven flowers but no leaves (Fadón *et al.*, 2015a). Flower induction, the process by which stimuli originating outside the shoot apex induce the formation of flower primordia (Hempel *et al.*, 2000), occurs from mid-stage II of fruit development (Elfving *et al.*, 2003) through the cessation of shoot growth in midsummer. The time of floral induction depends on the cultivar and the physiological condition of the tree, which is affected by weather, site conditions and cultural practices (Thompson, 1996; Guimond *et al.*, 1998a,b; Engin, 2008; Beppu and Kataoka, 2011), and may even differ with shoot location on the tree (Kozlowski and Pallardy, 1997).

After flower induction, the flower buds contain several floral primordia. Early stages of flower differentiation, from flower induction until dormancy, have been characterized using scanning electron microscopy in sour (Diaz *et al.*, 1981) and sweet (Watanabe, 1983; Guimond *et al.*, 1998a; Engin and Unal, 2007) cherry. Flower initiation first becomes apparent at the shoot apex. The initial morphological change from a vegetative to the reproductive stage shows the meristematic apex acquiring a dome shape, which is followed by the formation of flower primordia. Flower buds show between two and four small lateral protuberances, representing primordial bracts that subtend each flower primordium. The sepal, petal, stamen and pistil primordia then differentiate sequentially in a centripetal way. Recently, a detailed description of flower development has been framed in the BBCH scale, which provides a unified standardized approach for phenological comparative studies (Fadón *et al.*, 2015a). Before leaf fall, the inflorescence buds are apparent in the leaf axis (Fig. 2.1A). Inside the bud, each flower primordium appears as a dome with five sepal primordia in a pentagonal whorl (Fig. 2.1B). At the beginning of leaf fall (Fig. 2.1C), the sepal primordia almost cover each flower (Fig. 2.1D). Inside each flower primordium, all whorls are distinguishable. Stamens appear as protrusions showing the characteristic shape of anthers but without filaments, and pistils appear as semi-circular protuberances showing the incipient suture line. During the winter, flower buds remain closed (Fig 2.1E) without any apparent change (Fig 2.1F).

#### *Late flower bud development*

Reproductive stages in sweet cherry have traditionally been characterized using the external phenological stages of buds and flowers (Baggiolini, 1952; Westwood, 1993), following the classical work of Fleckinger (1948). As in other *Prunus* spp., cherry is deciduous, and flower buds open before leaf buds (Fadón *et al.*, 2015a). Examining flower development within the BBCH scale (Fadón *et al.*, 2015a) shows that after chilling requirements are fulfilled, and as temperatures rise in late winter or spring (see Chapter 8, this volume), the buds swell and burst (Fig 2.1G) and flower growth rapidly resumes. At this phase, flowers are completely green, except the petals, which are slightly translucent. The sepals and petals are very short, with the sepals overpassing the petals. Stamens are conspicuous, and anthers have their characteristic oblong shape, topping very short filaments. The pistil is located in the

**Fig. 2.1.** Stages of flower development framed in principal growth stages 9 (senescence, beginning of dormancy) and 5 (reproductive development of sweet cherry) according to the extended BBCH scale. (A, B)

centre of the flower and its length is equivalent to flower height. Pistil parts are incipiently distinguished as the ovary, style and stigma, with the initiation of the stigmatic surface (Fig. 2.1H). Later, the scales separate and light green bud sections are visible (Fig. 2.1I). The sepals enclose the whole flower. The petals turn a pale white as the pistil significantly elongates, but the most striking change is in the colour of the anthers, which become bright yellow and occupy most of the space inside the flower (Fig. 2.1J). In apricot, it was shown that newly formed pollen was behind this change in colour, since meiosis and pollen formation from the previously undifferentiated tissues occur at the start of this transition period (Julian *et al.*, 2014). Recent work confirms a similar situation in sweet cherry, with anther meiosis occurring at the onset of budburst (Fadón, 2015).

When the bud starts to open (stage 54), the flowers are enclosed within light green scales (Fig. 2.2A). The anther filaments are still short and the stigma surpasses the anthers, located at the same height as the petals and sepals (Fig. 2.2B). Over the next days, the single flower buds are visible on short stalks (Fig. 2.2C), and the green scales open slightly (stage 55). Red spots appear in the sepals that enclose the flower. The hypanthium develops as a cup shape around the ovary in which the calyx, corolla and stamens are inserted. Stamen filaments begin to elongate, and the pistil continues growing and reaches the upper part of the flower and can even surpass it. The stigmatic surface is apparent, and the stigma edges start to curve down (Fig. 2.2D).

When the sepals open (stages 56–59), the flower acquires an elongated shape with a narrowing at the hypanthium. The white petals protrude above the sepals, first showing a white tip and then acquiring a balloon shape (Fig. 2.2E). The sepals open and become perpendicular to the hypanthium. Inside the flower, the anthers are grouped in the upper half of the flower, staggered at different heights. The style continues growing over the anthers, reaching its final length. The swollen ovary is laterally placed and completely surrounded by the hypanthium cavity (Fig. 2.2F). Later, the sepals open and the petal tips, still closed, are fully visible (Fig. 2.2G). The petals completely enclose the flower. The anther filaments are significantly elongated reaching their final length. The style also reaches its final length, and the stigma and the anthers are at the same height (Fig. 2.2H). At the balloon stage, the sepals completely open, the petals are completely extended and rounded but still closed (Fig. 2.2I). Finally, at flowering, the petals open completely (Fig. 2.2J), the stigma is receptive and the anthers dehisce their pollen.

Full bloom usually occurs between 3 and 6 weeks after bud burst, depending on temperature. The flowers are hermaphroditic, with a single pistil and numerous stamens that are enclosed within five white petals and five green sepals (Sterling, 1964). Bloom time depends not only on the cultivar, but also on the weather. In warmer regions or during mild winters when winter chilling may be marginal for the transition of endodormancy to ecodormancy, the timespan between full bloom of early- and late-flowering cultivars is more prolonged than in cooler climates where winter chilling is more than sufficient (Thompson, 1996; Guerra and Rodrigo, 2015). Where winter chilling is adequate, warm spring weather reduces the timespan between bloom for early- and late-flowering cultivars.

**Fig. 2.1.** Continued.

Stage 91: inflorescence buds are apparent in the leaf axis (A) and the flower primordium appears as a dome with five sepal primordia in a pentagonal whorl (B); (C, D) stage 93: the beginning of leaf fall (C), with each flower primordium almost covered by the sepal primordia (D); (E, F) stage 50: the flower buds remain closed in winter (E), with no apparent changes (F); (G, H) stage 51: the buds start to swell and burst (G) and the ovary, style and stigma start to be distinguished (H); (I, J) stage 53: the scales separate and light green bud sections become visible (I) and the anthers become bright yellow and occupy most of the space inside the flower (J). Bars, 0.2 mm. (From Fadón *et al.*, 2015a.)

**Fig. 2.2.** Stages of flower development framed in principal growth stage 5 (flower bud development) of sweet cherry according to the extended BBCH scale. (A, B) Stage 54: the bud starts to open, with the flowers

### 2.2.2 Flower bud dormancy

In late autumn, flower buds cease development and enter a dormant stage, called endodormancy, in which they become adapted to survival at low winter temperatures. While endodormancy is a clear adaptation to temperature, the transition into dormancy is not triggered only by decreasing temperature. Endodormancy is established prior to the exposure to low temperatures and deepens progressively (Rohde and Bhalerao, 2007) until reaching a stage in which bud development becomes unresponsive to growth-promoting temperatures. While exposure to a specific range of low 'chilling' temperatures is needed to alleviate endodormancy, the requirements for regaining the capacity for growth and development can vary depending on environment and genotype. Furthermore, once chilling requirements are fulfilled and endodormancy transitions to ecodormancy (Lang, 1987), budburst is not an immediate response. Ecodormancy requires a period of warm temperatures to reactivate final primordial development, bud break and flowering. Although dormancy has well-established physiological stages in cherry growth, and chilling requirements are known for a number of cherry cultivars (see Chapter 8, this volume), dormancy in cherry is still not fully understood. However, elucidating the physiology and genetics of dormancy is an active area of research, with the current understanding described below.

#### *Stage of flower development at dormancy*

The question of whether there is a particular developmental stage at which floral primordia enter endodormancy, or whether flower buds enter dormancy at whatever stage is present when environmental conditions trigger this response, has been recently addressed (Fadón, 2015; Fadón *et al.*, 2017). Comparative studies were performed over multiple years on different sweet cherry cultivars with different chilling requirements and flowering times. The results consistently showed that early and late flower development in the autumn and spring are asynchronous between cultivars and years. Cultivars differ in the timing for entering endodormancy, and the transition to ecodormancy depends on chilling requirements. However, in all circumstances, flower buds enter endodormancy and survive the winter at the same developmental stage. By leaf fall, all floral parts are apparent (Fadón *et al.*, 2015a), and the flower primordia within a bud are at the same developmental stage, although they may differ in size (Fadón, 2015; Fadón *et al.*, 2017).

The first step for the establishment of endodormancy in flower buds is growth cessation, even under conditions conducive to growth. During dormancy, meristems acclimate to cold and remain unresponsive to growth signals (Cooke *et al.*, 2012). The flower buds remain closed and covered by dark brown scales (Fig. 2.1E), and flower primordia enclosed within the sepals stop growing and each flower primordium is completely covered by the sepals (Fig. 2.1F). The anthers have a translucent green appearance, with the four locules clearly differentiated, but the filaments are still undeveloped. In the pistil, an incipient ovary, style and stigma are apparent.

**Fig. 2.2.** Continued.
enclosed within light green scales (A); the anther filaments are short and the stigma surpasses the anthers, located at the same height as the petals and sepals (B); (C, D) stage 55: the green scales open slightly to reveal single flower buds visible on short stalks (C), while the stigmatic surface becomes visible and the stigma edges start to curve down (D); (E, F) stage 56: the white petals protrude above the sepals (E); the swollen ovary is laterally placed and completely surrounded by the hypanthium cavity (F); (G, H) stage 57: the sepals open and the petal tips become fully visible (G), while style reaches its final length, and the stigma and the anthers are at the same height (H); (I, J) stage 59: at the balloon stage, the sepals open completely, and the petals become completely extended and rounded, although still closed (I); finally, at flowering, the petals open completely, the stigma is receptive and the anthers dehisce their pollen (J). Bars, 1 mm (B, D, F); 2 mm (H, I). (From Fadón *et al.*, 2015a.)

#### *Physiology and genetics of flower bud dormancy*

Although the implications of dormancy on fruit production have been known since the early 20th century, the physiological processes and environmental factors that induce and alleviate dormancy are not completely understood (Campoy *et al.*, 2011; Luedeling, 2012). During midwinter, some physiological processes continue, such as starch accumulation (Felker *et al.*, 1983; Fadón, 2015). Likewise, different biological markers have been related to the level of dormancy in different tissues (Lang, 1994), mainly hormonal changes and cell and organ isolation. Auxins, cytokinins and abscisic acid have been related to the control of dormancy (Crabbé, 1994; Vanstraelen and Benkova, 2012). Low levels of gibberellins have been associated with growth cessation at the start of bud dormancy (Olsen *et al.*, 1995; Eriksson and Moritz, 2002). During bud dormancy, changes in the bud water status have been proposed to have an impact on freezing tolerance, since variations in the content of dehydrins – specific proteins that protect the cell against cellular dehydration – have been detected during dormancy (Faust *et al.*, 1997; Arora *et al.*, 2003; Rinne *et al.*, 2011). Although the mechanisms behind such hormonal regulation are not well understood, these changes could regulate specific components of the cell cycle and thereby regulate induction and breaking of dormancy (Horvath *et al.*, 2003). The impact of hormones on the control of dormancy is further apparent by the effects of external applications of plant growth regulators on different cultivars and situations (see Chapter 12, this volume).

The search for genes involved in the regulation of dormancy in woody perennials has involved three main approaches (Fadón *et al.*, 2015a). One is the search for quantitative trait loci (QTLs) (Bielenberg *et al.*, 2008; Castède *et al.*, 2014) (see Chapter 3, this volume). Another is the study of an evergreen peach in which six *DAM* genes showed interactions with seasonal cessation, bud formation and chilling accumulation (Li *et al.*, 2009; Jiménez *et al.*, 2010a,b); interestingly, these genes have also been identified in other fruit tree species (Esumi *et al.*, 2009; Sasaki *et al.*, 2011; Yamane *et al.*, 2011). Finally, a third approach shows that genes involved in flower development play a part in dormancy regulation (Koornneef *et al.*, 1991; Bradley, 1997; Böhlenius *et al.*, 2006; Mohamed *et al.*, 2010; Hsu *et al.*, 2011; Mimida *et al.*, 2011). Recently, QTLs and candidate genes have been identified in sweet cherry for chilling requirement and flowering date, illustrating the polygenic control of both processes (Castède *et al.*, 2014, 2015).

## 2.3 Flowering and Pollination

### 2.3.1 From pollination to fertilization

The process from pollination to fertilization is very well conserved in fruit trees (Herrero, 1992a) and indeed in flowering plants (Herrero and Hormaza, 1996; Herrero, 2003). Following pollination, the pollen tube, in its journey towards fertilization, traverses the stigma and style before entering the ovary.

#### *Stigma and style*

The stigmas in cherry are receptive following flower opening and are covered by a conspicuous secretion (Hedhly *et al.*, 2003). The pollen grains land on the stigma, hydrate and germinate producing a pollen tube. In cherry, this occurs within the first day following pollination (Hedhly *et al.*, 2007). The living part of the pollen grain generates a pollen tube via the formation of an expanded cell wall. The pollen nuclei move to the forefront of the pollen tube. This strategy allows the pollen gamete to move through the style and then into the ovary, while the pollen tube, except for the living tip, is an empty cell wall-encased structure that is left behind.

The pollen tube then enters the style. The style in cherry is of the solid type, with an inner transmitting tissue where the pollen tubes grow between the cells (Hedhly *et al.*, 2007). The pollen tubes derive the carbohydrates necessary for their growth from the cells of the stylar transmitting tissue; these are full of starch, which is depleted as the pollen tubes grow (Herrero and Dickinson, 1979). Intense pollen tube competition occurs in the style, with the possibility of over 100 pollen grains on a cherry stigma, and usually just one to three pollen tubes reach the ovary (Hormaza and Herrero, 1996a, 1999). The number of pollen tubes that reach the ovary also is influenced by the structure of the style, which, in spite of its cylindrical outer appearance, has a funnel shape inside the transmitting tissue that leaves less space and fewer carbohydrates available for the growing pollen tubes as they progress down the style (Herrero, 1992b). In cherry, approximately 3 days after pollination, the successful pollen tubes reach the base of the style where the ovary is located (Hedhly *et al.*, 2007).

#### *Ovary*

In cherry, as in other *Prunus* spp., the ovary contains two ovules, and fertilization of at least one of them is required for fruit set. The ovule is composed of a number of concentric wrappings that consecutively envelop each other. Thus, the female gamete, the egg cell, is inside the embryo sac, which in turn is inside the nucellus. The nucellus is wrapped within the two integuments, constituting the ovule nested within the ovary. The pollen tube, on its way to encounter the female gamete, has to traverse all of these envelopes. The first is the obturator, a protuberance of the ovary placenta that faces the ovule entrance. In peach, on arrival at the obturator, pollen tube growth stops and only resumes 5 days later as the obturator enters a secretory phase (Arbeloa and Herrero, 1987). The pollen tube then faces the ovule and enters the micropyle (formed by closing of the integuments), traverses the nucellus to reach the embryo sac through a synergid and discharges the two spermatids. One spermatid fuses with the egg cell, forming a zygote that develops into an embryo, while the other fuses with the polar nucleus and develops into the endosperm. In cherry flowers, some pollen tubes are arrested and do not achieve fertilization (Hedhly *et al.*, 2007). In other *Prunus* spp., pollen tube arrests are related to maturation of the pistil (Herrero and Arbeloa, 1989) and the production or failure of secretion at these steps (Herrero, 2000, 2001). All these steps illustrate the importance of male–female synchrony as a prerequisite for successful mating (Herrero, 2003). However, prior to completion of a successful fertilization in cherry production, problems associated with pollen–pistil incompatibility can arise.

### 2.3.2 Pollen–pistil incompatibility

Sweet cherry production intensified in the early 20th century, shifting from small orchards with various cultivars to fewer cultivars in larger plantings. This coincided with reports of fruit set problems attributed to inadequate fertilization (Faust and Suranyi, 1997). Pollen compatibility studies conclusively determined that sweet and some sour cherry cultivars are self-incompatible, and that cross-incompatibility was particularly common in sweet cherry (Crane and Lawrence, 1929; Crane and Brown, 1937). Experimental crossings in sweet cherry established that self-incompatibility (SI) is genetically determined by a single locus, called *S*, with multiple alleles, and that this genetic factor determines pollen tube growth (Crane and Lawrence, 1929; Crane and Brown, 1937). SI studies on cherries, and in other temperate fruit tree species (Crane and Lawrence, 1929), laid the foundation for SI knowledge in members of the Rosaceae family. Many studies followed, which contributed to our current understanding of cherry SI (reviewed by Yamane and Tao, 2009; Tao and Iezzoni, 2010; Wu *et al.*, 2013).

SI operating in *Prunus* spp. such as sweet and sour cherry is termed *S-RNase*-based gametophytic SI (Tao and Iezzoni, 2010). This type of SI is genetically determined by the multi-allelic *S*-locus, which encodes a ribonuclease (*S*-RNase) expressed in the style (Boskovic and Tobutt, 1996; Tao *et al.*, 1999; Yamane *et al.*, 2001) and a pollen-expressed protein, with an F-box domain, called *S*-locus F-box protein (SFB) (Yamane *et al.*, 2003a;

Ikeda *et al.*, 2004; Ushijima *et al.*, 2004). Variants of the *S-RNase* and *SFB* genes are known as *S-RNase* and *SFB* alleles, respectively, while variants of the *S*-locus (these two tightly linked genes together) are termed *S*-haplotypes (Tao and Iezzoni, 2010). During pollen tube growth, the style and pollen *S* products, *S*-RNase and SFB, interact in an allele-specific manner to result in the SI reaction. Pollen tube growth is inhibited when the pollen and the style express the same allele. In sweet cherry, fertilization only takes place when the *SFB* allele expressed by the haploid pollen is different from the two *S-RNase* alleles expressed in the style (diploid tissue). Additionally, modifier genes are needed for full expression of the SI reaction (Tao and Iezzoni, 2010). Several models have been proposed to explain how these factors mediate the SI response (Yamane and Tao, 2009; Tao and Iezzoni, 2010; Matsumoto *et al.*, 2012; Wu *et al.*, 2013).

SI has evolved in many plant families to promote outbreeding; however, it is an undesirable trait in cherry production. SI prevents cherry cultivars from self-fertilization, and pollinator trees need to be included to ensure fruit set. Cross-compatibility of the pollinator and cultivar is needed, and their flowering must overlap to ensure timely transport of pollen to the stigma. The use of pollinators results in more erratic production and more dependence on environmental conditions (Tehrani and Brown, 1992), the costs are increased and often phytochemicals need to be used to induce flowering overlap. Self-compatible cultivars are often preferred to avoid the use of pollinizer cultivars. Breeding for self-compatibility (SC) has, therefore, become a main objective in cherry improvement (see Chapter 3, this volume).

### *S*-genotyping and incompatibility groups in sweet cherry

SI in sweet cherry prevents self-fertilization, but also prevents cross-pollination among cultivars with the same *S*-genotype. Orchard management as well as crossing design requires prior knowledge of the genetic cross-compatibility of different cultivars, as well as their relative bloom times, to ensure fruit set. It is therefore necessary to know the *S*-genotype of each cultivar to establish compatible cultivar combinations that can be interplanted. Cultivars with the same *S*-genotype are cross-incompatible and are grouped in the same incompatibility group (IG). In contrast, cultivars with different *S*-genotypes are cross-compatible and are therefore assigned to different IGs. Cultivars with unique *S*-genotypes, and thus compatible with all other cultivars of known *S*-genotype, are classified in group 0 comprising universal donors. Cultivars previously classified in group 0 have been later assigned to other IGs when cultivars with the same *S*-genotype were identified.

Initially, *S*-genotyping and IG assignment were carried out by controlled pollinations to record fruit set (Crane and Brown, 1937; Matthews and Dow, 1969) and/or the observation of pollen tube growth. The drawback of this approach is that crosses are dependent on environmental factors and are restricted to the flowering period. Alternatively, molecular methods (DNA tests) can be carried out for *S*-genotyping. The first molecular methods for *S*-genotyping were based on the *S*-RNase (Boskovic and Tobutt, 1996; Tao *et al.*, 1999). *S-RNase* alleles in sweet cherry were initially detected from stylar proteins separated by isoelectric focusing and stained for ribonuclease activity (Boskovic and Tobutt, 1996; Boskovic *et al.*, 1997). The cloning and sequence characterization of the *S*-RNases of sweet cherry allowed the development of polymerase chain reaction (PCR)-based and restriction fragment length polymorphism (RFLP)-based methods for *S*-allele typing. Tao *et al.* (1999) designed consensus PCR primers in the conserved regions of the sweet cherry *S-RNase* sequences. The amplified fragments spanned the introns found in the sweet cherry *S-RNase* gene (Tao *et al.*, 1999). As each *S-RNase* allele has two introns varying in length, it was possible to differentiate *S-RNase* alleles by detecting intron size polymorphisms by this method. Subsequently, more sweet cherry *S-RNases* were cloned and other PCR primers based on conserved sequence primers and/or allele-specific primers were developed (Sonneveld *et al.*, 2001, 2003; Wiersma *et al.*,

2001; Szikriszt *et al.*, 2013). RFLP profiles were also used for *S-RNase* typing (Hauck *et al.*, 2001). A high-throughput method based on PCR amplification of the *S-RNase* first intron using consensus primers and analysis of the fragments with an automatic sequencer or genetic analyser (Sonneveld *et al.*, 2006) provided a faster and more reliable approach for *S-RNase* typing and the identification of a large number of *S*-haplotypes.

The identification and characterization of the *S*-locus pollen factor (SFB) in sweet cherry (Yamane *et al.*, 2003b; Ikeda *et al.*, 2004) triggered the development of additional molecular *S*-locus genotyping methods using this gene. As *SFB* also has an intron that varies in size in the different alleles, PCR primers spanning this intron were designed to detect *SFB* intron length variation (Vaughan *et al.*, 2006). Since the length variation of the *SFB* fragments is small, in this method, PCR fragments are also detected using a genetic analyser (Vaughan *et al.*, 2006). The combination of genotyping both *S*-locus genes, *S-RNase* and *SFB*, provided an even more reliable *S*-genotyping method and the possibility of discriminating a larger number of *S*-haplotypes (Vaughan *et al.*, 2006, 2008). Additionally, as PCR fragments of both genes can be detected in a genetic analyser, *S*-allele genotyping of both genes can be done simultaneously. Today, these PCR methods are widely used for *S*-allele genotyping. A dot-blot analysis for *S*-locus genotyping using *SFB* sequence polymorphisms also was developed by Kitashiba *et al.* (2008). To date, 31 *S*-haplotypes have been detected in sweet and sour cherry. These *S*-haplotypes are numbered $S_1$ to $S_{38}$ with some ($S_8$, $S_{11}$, $S_{15}$, $S_{23}$, $S_{24}$ and $S_{25}$) being synonymous to others ($S_3$, $S_5$, $S_7$, $S_{14}$, $S_{22}$ and $S_{21}$, respectively). Due to the overlap of various studies, the identification of synonymous types and the clarification of labelling, the most widely used numbers are $S_1$–$S_7$, $S_9$, $S_{10}$, $S_{12}$–$S_{14}$, $S_{16}$–$S_{22}$ and $S_{26}$–$S_{38}$.

Regarding IGs, Crane and Brown (1937) reported 11 IGs from crossings of 45 sweet cherry cultivars. These were later increased with two additional groups and 13 IGs (named I–XIII) were defined (Matthews and Dow, 1969; Tehrani and Brown, 1992). The use of molecular *S*-locus genotyping methods allowed the confirmation of these *S*-genotypes, the rapid *S*-genotyping of new and old cultivars, and the identification of new *S*-haplotypes and IGs. By 2004, Tobutt *et al.* (2004) compiled results from different sweet cherry studies and summarized the *S*-genotypes of 222 SI and 25 SC sweet cherry cultivars into 26 IGs (I–XXVI), with four cultivars included in group 0. In a more recent compilation of *S*-genotyping results, Schuster (2012) reported the *S*-genotypes of 734 sweet cherry cultivars, with 47 IGs (I–XLVII) and 15 unique *S*-genotypes (group 0). Additionally, 44 cultivars were SC (Schuster, 2012). Table 2.1 shows the *S*-genotypes and IGs of 141 sweet cherry cultivars. This table includes recent sweet cherry releases, traditional cultivars still widely cultivated and some relevant landraces. The information in this table can be used to design crosses and to identify cross-compatible cultivars.

### Self-compatibility in sweet cherry

In sweet cherry cultivation, SC cultivars are often preferred to avoid the need to plant a pollinator cultivar and for the potential for higher fruit set compared with SI cultivars. Dozens of SC sweet cherry cultivars have been bred, and local cultivars that are SC have been described. Two SC mutant sweet cherry selections (JI2420 and JI2434) were obtained from pollination of 'Emperor Francis' with pollen of 'Napoleon' that had been mutated by X-irradiation (Lewis and Crowe, 1954). Genetic studies with these accessions revealed that both were pollen mutants (non-functional pollen during SI). In JI2420, SC was linked to *S*-haplotype $S_4$ (named $S_{4'}$). In JI2434, SC was linked to $S_3$ ($S_{3'}$) (Boskovic *et al.*, 2000). The molecular investigation of the *S*-locus genes of these two accessions concluded that JI2420 has a deletion that results in a frameshift mutation of the $SFB_4$ gene (Ushijima *et al.*, 2004; Sonneveld *et al.*, 2005) and that JI2434 presents a complete deletion of the *SFB3* gene (Sonneveld *et al.*, 2005). From these two SC accessions, many sweet cherry cultivars have been bred. The parents of the first commercially released SC sweet cherry cultivar 'Stella' (Lapins, 1975) were 'Lambert' and JI2420. Subsequently,

**Table 2.1.** Incompatibility groups (IGs) of selected sweet cherry cultivars. (From Schuster, 2012; Cachi *et al.*, 2015.)

| IG | *S* genotype | Cultivars |
|---|---|---|
| I | $S_1S_2$ | 'Black Tartarian', Canada Giant™ ('Sumgita'), 'Early Rivers', 'Ferdouce', 'Starking Hardy Giant', 'Summit' |
| II | $S_1S_3$ | 'Black Star', Cristalina™ ('Sumnue'), 'Early Van Compact', 'Giant Red' (Giant Ruby™), 'Lala Star', 'Prime Giant', 'Regina', Samba™ ('Sumste'), Satin™ ('Sumele'), 'Sonnet', 'Sumbola', 'Van', 'Vera', 'Windsor' |
| III | $S_3S_4$ | 'Belge', 'Bing', 'Lambert', 'Napoleon' ('Royal Ann'), 'Somerset', 'Star', 'Ulster' |
| IV | $S_2S_3$ | 'Nimba', 'Coralise' ('Gardel'), 'Sue', 'Vega' |
| VI | $S_3S_6$ | 'Ambrunés', 'Anita', 'Badacsony', 'Fertille', 'Kordia', 'Pico Negro', 'Satonishiki', 'Stark Gold' ('Dönissens Gelbe', 'Gold'), 'Techlovan', 'Duroni 3, 'Ferdiva', 'Fertard' |
| VII | $S_3S_5$ | 'Hedelfinger' |
| IX | $S_1S_4$ | 'Black Giant', 'Black Republican', 'Garnet', 'Hudson', 'King', 'Rainier', 'Salmo', 'Sylvia' |
| X | $S_6S_9$ | 'Folfer', 'Ramón Oliva' |
| XIII | $S_2S_4$ | 'Corum', 'Deacon', 'Patricia', 'Peggy Rivers', 'Royalton', 'Sam', 'Schmidt', 'Vic' |
| XIV | $S_1S_5$ | 'Alma', 'Annabella', 'Blanca de Provenza', 'Seneca', 'Valera' |
| XV | $S_5S_6$ | 'Colney' |
| XVI | $S_3S_9$ | 'Burlat', 'Chelan', 'Moreau', 'Precoce Bernard', 'Tieton' |
| XVII | $S_4S_6$ | 'Larian', 'Merton Glory' |
| XVIII | $S_1S_9$ | 'Bigisol' ('Early Bigi'), 'Brooks', 'Earlise®Rivedel' ('Rivedel'), 'Early Red' ('Early Garnet'), 'Rocket', 'Marvin' ('Niran', '4-70'), 'Sweet Early' ('Panaro 1')' |
| XIX | $S_3S_{13}$ | 'Reverchon' |
| XX | $S_1S_6$ | 'Vanda' |
| XXI | $S_4S_9$ | 'Cashmere', 'Merchant', 'Merpet' |
| XXII | $S_3S_{12}$ | '0900-Ziraat', 'Ferrovia', 'Germersdorfi', 'Noire de Meched', 'Schneiders' |
| XXIV | $S_6S_{12}$ | 'Aida', 'Flamentier' |
| XXV | $S_2S_6$ | 'Fercer' ('Arcina') |
| XXVI | $S_5S_{13}$ | 'Ferbolus' ('Verdel'), 'Goodnestone Black' |
| XXVII | $S_4S_{12}$ | 'Margit', 'Kavics' |
| XXXIII | $S_1S_{14}$ | 'Fermina' |
| XLIII | $S_2S_9$ | 'Primulat' ('Ferprime') |
| SC[a] | $S_1S_{4'}$ | Celeste™ ('Sumpaca'), 'Frisco', 'Lapins', 'Santina', 'Skeena', 'Symphony', 'Stardust' |
| | $S_3S_{4'}$ | 'Compact Stella', 'Index', 'Newstar', 'Sandra Rose', 'Selah' ('Liberty Bell'), Sonata™ ('Sumleta'), Staccato™ ('Summer Charm', '13S2009'), 'Starkrimson', 'Stella', 'Sumesi', 'Sunburst', Sweetheart™ ('Sumtare'), 'Tehranivee' |
| | $S_3S_{3'}$ | 'Axel' |
| | $S_3S_6$ | 'Cristobalina', 'Temprana de Sot' |
| | $S_4S_6$ | 'Blaze Star', 'Blackgold' |
| | $S_4S_9$ | 'Early Star' ('Panaro 2'), 'Pacific Red', 'Sandor', 'Columbia' ('Benton'), 'Grace Star', 'Glacier' |
| | $S_5S_6$ | 'Kronio' |
| 0 | $S_5S_{22}$ | 'Rita' |

[a]All SC cultivars included are also universal donors.

'Stella' and some of its SC descents, such as 'Sumtare' (Sweetheart™) and 'Sunburst', have been used as progenitors in many sweet cherry breeding programmes (Table 2.1) (Schuster, 2012). Similarly, JI2434 has also been used for breeding, being the ancestor of other SC cultivars such as 'Axel' (Schuster, 2012).

Spontaneous mutations for SC have also been found in local cultivars. These are 'Cristobalina' or 'Talegal Ahim' from Spain (Herrero, 1964; Wünsch and Hormaza, 2004; Cachi and Wünsch, 2014) and 'Kronio' from Italy (Calabrese *et al.*, 1984; Marchese *et al.*, 2007). These cultivars are alternative sources for

SC breeding. In order to understand their SC, these cultivars have been studied using genetic and molecular approaches. In all cases, SC is due to non-functional self-recognition of pollen during the SI reaction (Wünsch and Hormaza, 2004; Marchese *et al.*, 2007; Cachi and Wünsch, 2014). In 'Kronio', SC is due to a deletion in the pollen *SFB5* gene that generates a truncated protein (Marchese *et al.*, 2007). In 'Cristobalina' and 'Talegal Ahim', SC seems to be caused by the same mutation (Cachi and Wünsch, 2014); however, in this case SC is not linked to the *S*-locus (Wünsch and Hormaza, 2004; Wünsch *et al.*, 2010) and it is probably due to mutations in a modifier gene. A major locus controlling the SC in 'Cristobalina' was genetically mapped to the lower region of sweet cherry linkage group 3 (Cachi and Wünsch, 2011). In the same chromosome region, modifiers responsible for SC in apricot were also mapped (Zuriaga *et al.*, 2012, 2013). Additional studies have revealed that the growth of SC pollen in 'Cristobalina' is retarded during self-pollination (Cachi *et al.*, 2014). This SC cultivar flowers very early, when mates are scarce. It is therefore believed that SC in this genotype has been selected as a survival trait. In the absence of compatible matings, this genotype can self-fertilize; however, the slow growth rate of SC pollen tubes indicates that cross-compatibility is favoured when compatible outcross pollen is available (Cachi *et al.*, 2014).

#### *Self-(in)compatibility in sour cherry*

Sour cherry is a tetraploid species, and gametophytic SI operates in this species as well as in sweet cherry (Yamane *et al.*, 2001). However, while in sweet cherry SI is most common and SC genotypes are rarely found, in sour cherry both SC and SI selections are found in nature (Tsukamoto *et al.*, 2010). Even more than in sweet cherry, sour cherry cultivation requires SC cultivars to avoid pollination and fruit set problems. As a consequence, sour cherry breeding programmes target SC as a main objective (see Chapter 5, this volume). The presence of SC in sour cherry is not caused by polyploidy per se, but is due to the accumulation of specific *S*-locus mutations, either in the *S-RNase* (style mutations) or in *SFB* (pollen mutations) gene (Hauck *et al.*, 2006; Tsukamoto *et al.*, 2010). A genetic model to explain SC in tetraploid sour cherry was proposed by Hauck *et al.* (2006). Sour cherry is tetraploid and its pollen is 2x. The 'one-allele-match model' (Hauck *et al.*, 2006) predicts that a pollen grain will be self-incompatible if one or both of the two *S*-haplotypes are fully functional and grows down a style that expresses a matching fully functional *S*-haplotype. As a consequence, sour cherry is SC when it accumulates at least two non-functional *S*-haplotypes (Tsukamoto *et al.*, 2008a, 2010).

Phenotyping combined with genetic and molecular analyses have allowed the identification of the *S*-genotype and the SI/SC phenotype of sour cherry cultivars and selections (Yamane *et al.*, 2001; Hauck *et al.*, 2006; Tsukamoto *et al.*, 2006, 2008a,b, 2010; Sebolt *et al.*, 2017). Twelve different *S*-haplotypes have been detected and four ($S_1$, $S_6$, $S_{13}$ and $S_{36}$) present non-functional variants, with mutations either in the style or in pollen (Sebolt *et al.*, 2017). In fact, for $S_{36}$, only non-functional variants were detected. As predicted by Hauck *et al.* (2006), the presence of at least two of these non-functional *S*-haplotypes in the *S*-genotype results in SC (Yamane *et al.*, 2001; Tsukamoto *et al.*, 2006, 2008b, 2010; Sebolt *et al.*, 2017). Because SI/SC in sour cherry depends on the *S*-haplotypes present in each genotype and the presence of the non-functional variants, the study of cross-compatibility or the early detection of SC by molecular methods is more complex than in sweet cherry (see Chapter 3, this volume). The process requires *S*-genotyping that identifies the *S*-haplotype, including whether it is the functional haplotype or a pollen or stylar non-functional variant (Tsukamoto *et al.*, 2008a,b; Sebolt *et al.*, 2017).

## 2.4 From Flower to Fruit

Once fertilization has been completed, fruit development occurs. However, fruit set does not always occur as expected, and a number of factors can jeopardize fruit set.

### 2.4.1 Factors affecting fruit set

Throughout the flower lifespan, alterations in flower biology may lead to failures in fruit set. These may occur during flower development or at flowering.

*During flower development*

The current interest in expanding the sweet cherry harvest and marketing season has resulted in cultivation beyond typical production areas. Erratic cropping related to altered flower bud development can result in less-than-optimal environments. In warm climates or unusually hot summers, high temperatures during the onset of flower differentiation may cause various floral abnormalities. Double pistil formation resulting in unmarketable double fruit (see Chapter 8, this volume) has been related to hot summers in different cultivars and areas (Tucker, 1934, 1935; Southwick *et al.*, 1991; Beppu and Kataoka, 2000; Engin, 2008; Roversi *et al.*, 2008; Beppu and Kataoka, 2011). The occurrence of double pistils can also be affected by the position in the tree canopy, being more frequent in those parts more exposed to solar radiation. Double pistil frequency is also dependent on genotype; 'Bing' and 'Napoleon' are particularly susceptible (Thompson, 1996). Other flower abnormalities related to hot weather during the previous summer include pistil-like and petal-like appendages on the end of stamen filaments replacing anthers (Philp, 1933; Thompson, 1996), or leaf-like development of different floral organs (Engin and Gokbayrak, 2010).

Cultural practices can also affect early flower differentiation. Early summer pruning of current season growth can advance the initiation of flower buds (Guimond *et al.*, 1998a) and increase the number of floral meristems that develop at the base of current season shoots, presumably by shifting allocation of resources from terminal shoot elongation to reproductive meristem development (Guimond *et al.*, 1998b).

Similar to other temperate *Prunus* spp., cherries need a cultivar-specific amount of chilling during endodormancy for the flower buds to develop and flower normally. Chilling requirements also vary by environment where they are evaluated (Viti *et al.*, 2010; Campoy *et al.*, 2012). This is critical to determine whether a cultivar is adapted to a particular area (see Chapter 8, this volume), and is one of the main drawbacks for the extension of species and cultivars to warmer latitudes (Atkinson *et al.*, 2013). Insufficient chilling in regions with mild winters may cause flower bud abscission, flower malformations and/or low fruit set (Weinberger, 1975; Thompson, 1996; Campoy *et al.*, 2011; Luedeling, 2012; Fadón *et al.*, 2015b; Guerra and Rodrigo, 2015). On the other hand, early chilling fulfilment can result in premature bloom, increasing the risk of spring frost damage and crop loss (Rodrigo, 2000).

Low temperature stress resulting in flower bud death is the most important limiting factor for sweet cherry cultivation in some regions (see Chapter 8, this volume). Freeze damage may occur in autumn before buds are acclimatized, during winter dormancy, during late winter after chilling fulfilment, and especially during deacclimation as buds swell and flower (Thompson, 1996; Rodrigo, 2000). However, high temperatures prior to and during flowering can also affect fruit set. In apricot, warm temperatures prior to flower opening not only resulted in early flowering, but fruit set in these early flowers was greatly reduced and associated with smaller pistils, suggesting that flowering was premature with underdeveloped pistils (Rodrigo and Herrero, 2002).

*At flowering*

Some flowers become a fruit, but many fail to further develop and drop. Looking at failures in fruit set from a flower biology perspective, these malfunctions can occur in all three parts of the pistil that the pollen tube must traverse successfully: the stigma, style and ovary.

A good orchard design, with enough pollinizers and pollinating insects, should result in adequate transfer of compatible pollen to the stigma. Surprisingly, a less-than-optimal situation often occurs when cherries are planted in new areas due to mismatched flowering periods of supposedly good pollinator cultivars, in spite of known bloom time overlap in a different geographical area. This is related to our poor understanding of the endodormancy/ecodormancy requirements for cold and warm temperatures, respectively, of cultivars, as bloom time currently is predicted on an empirical basis. As discussed above, work is under way to study biological indicators to gain a better understanding of dormancy (Julian *et al.*, 2011, 2014; Fadón, 2015; Fadón *et al.*, 2015b) and its genetic control (Jiménez *et al.*, 2010a,b; Yamane *et al.*, 2011; Leida *et al.*, 2012; Castède *et al.*, 2014). Another situation, rooted in the same lack of knowledge on dormancy, involves different responses to dormancy-breaking treatments (Rodrigo *et al.*, 2017), with one cultivar blooming ahead of its pollinizer, resulting in uncoupled flowering times and pollination failure. In these circumstances, the use of SC cultivars would solve the problem.

Pollination also can be jeopardized by a short period of stigmatic receptivity that shortens the effective pollination period (Sanzol and Herrero, 2001). Cherry stigmas are particularly vulnerable to warm temperatures that result in a short stigmatic receptivity (Hedhly *et al.*, 2003). Thus, while stigmatic receptivity lasts for 5 days at 10°C, it is reduced to 2 days at 20°C and to 1 day at 30°C. This is particularly problematic where hot spells may occur during flowering, leaving a very short window of time for successful pollination.

Once pollen has germinated on the stigma, the pollen tubes enter the style where pollen tube competition and discrimination occur due to pollen–pistil incompatibility (de Nettancourt, 2001). However, pollen tube selection for other characters is also likely (Mulcahy, 1979; Hormaza and Herrero, 1992, 1994), and perhaps could be used as a breeding tool (Hormaza and Herrero, 1996b). Pollen tube growth is most responsive and vulnerable to temperature for both sweet (Hedhly *et al.*, 2004, 2005) and sour (Cerovic and Ruzic, 1992a) cherry, and genetic differences could reflect adaptations to certain latitudes (Hedhly *et al.*, 2004). Maternal genotype can also affect the pollen–pistil interaction (Hormaza and Herrero, 1999), and it is conceivable that there might be better pollinators or stylar parents for pollen performance (Hedhly *et al.*, 2005).

The ovary plays a critical role in fruit set; however, it is not as well studied as the pollen–style interaction. The occurrence of degenerated ovules at anthesis appears to be a common feature in sweet and sour cherry (Eaton, 1959; Furukawa and Bukovac, 1989; Thompson, 1996; Cerovic and Micic, 1999; Mert and Soylu, 2007), and short ovule and embryo sac longevity may reduce fruit set (Eaton, 1959), particularly when a delay in pollination occurs (Stösser and Anvari, 1982). Temperature can play a major part in this, and a slight increase in temperature during the pollination-to-fertilization phase resulted in a clear decrease in fruit set, which was related to a higher proportion of flowers for which both ovules degenerated (Hedhly *et al.*, 2007). Indeed, ovules are highly vulnerable structures to increasing temperatures in different species (Williams, 1970; Stösser and Anvari, 1982; Postweiler *et al.*, 1985; Cerovic and Ruzic, 1992b; Hedhly *et al.*, 2009; Hedhly, 2011).

Carbohydrate resources accumulated in the flower also impact flower quality. Starch content in apricot flowers was related to the different fates of the two ovules, in which one becomes a seed, while the other degenerates (Rodrigo and Herrero, 1998). This raises the question of whether all flowers have the same starch content when they open. While this has not been examined in cherry, in apricot a wide variability in starch content can be observed at flower opening, and some flowers have tenfold the starch content of others (Rodrigo *et al.*, 2000). Moreover, in avocado, starch content correlates with the ability of the flower to set fruit (Alcaraz *et al.*, 2013). Work is in progress in apricot and cherry to determine when this starch accumulates (Fadón *et al.*, 2017). All of these findings support the need to further understand what elapses from the onset of flower differentiation up to bloom the next spring, and how multiple variables have a bearing on subsequent fruit set (Julian *et al.*, 2010). Work on these phases will further inform our understanding of cherry fruit set. Provided good fruit set is achieved, the next component of a good crop is fruit size, which relies on fruit development.

### 2.4.2 Fruit development

Once fruit set occurs, the ovary starts developing into the fruit. In cherry, fruit size is a particularly important economic trait, and work has been devoted to understanding both the developmental events required for fruit formation and the genetic control of fruit development.

#### *Developmental events for fruit formation*

Following fertilization, fruit growth proceeds rapidly, and 1 week after flowering, some pistils start to senesce and wilt, while others become greener and begin developing into fruit (Hedhly *et al.*, 2007). As in other *Prunus* spp., cherry fruit growth follows a double sigmoid curve (Coombe, 1976) and is divided into three stages. Stage I is characterized by rapid growth. At stage II, fruit growth enters a lag phase as the endocarp hardens and the embryo develops. Finally, growth becomes exponential again at stage III and ends with fruit maturation and harvest. Final fruit size is the result of cell number and cell size (Yamaguchi *et al.*, 2004). Cell division occurs primarily during prebloom through stage I, and cell enlargement mainly contributes to stage III growth (Tukey and Young, 1939; Olmstead *et al.*, 2007).

Since fruit size has a clear economic impact on crop value (Whiting *et al.*, 2005), a number of studies have investigated the factors affecting fruit size. In an evaluation of three cultivars differing in fruit size, cell number and not cell size was found to be the major contributing component to their fruit size differences (Olmstead *et al.*, 2007). Furthermore, fruit cell number was not influenced significantly by the environment, suggesting that cell number is genetically controlled and that cultural practices that affect fruit size mainly do so through differences in cell size (Olmstead *et al.*, 2007). Rootstock, crop load and environmental factors affect fruit size and quality (Whiting and Lang, 2004; Whiting and Ophardt, 2005; Lenahan *et al.*, 2006). In addition, a number of external treatments have been studied and are being used to improve fruit quality (Zhang and Whiting, 2011). Application of gibberellins during the transition from stage II to stage III has been reported to increase fruit size through cell enlargement (Proebsting *et al.*, 1973; Facteau *et al.*, 1985; Lenahan *et al.*, 2006) and has become standard practice in many sweet cherry production industries.

#### *Genetic control of fruit development*

Sweet cherry studies of fruit size and weight, aiming to understand these processes for breeding purposes, have begun to uncover the genetic control of fruit development. Fruit size in sweet cherry varies not only among cultivars, but also within the tree and is affected by environmental conditions. Recent studies, using a candidate gene approach, identified a gene (*PavCNR12*) that possibly regulates fruit cell number in sweet and sour cherry (de Franceschi *et al.*, 2013). This gene is located in one of the genomic regions where fruit weight QTLs have been described (Zhang *et al.*, 2010; see Chapter 3, this volume), and sequence and genetic analyses show its correlation with fruit size (de Franceschi *et al.*, 2013). The *CNR* gene family was selected for studying fruit size in cherry and other *Prunus* spp. because genes of this family control the fruit cell number in other species such as tomato (Frary *et al.*, 2000). In tomato, the gene *FW2.2* modulates cell proliferation in the carpel ovary, and fruit size is therefore determined by the regulation of cell number (de Franceschi *et al.*, 2013). The mode of action of this tomato gene would agree with the hypothesis that cherry fruit size is determined by cell number (Olmstead *et al.*, 2007).

## 2.5 Perspectives

All these factors illustrate the complexity of the process that begins with floral induction and ends with a mature fruit. Flowering is a crucial period, and adequate pollination during the short flower lifespan is essential. Work on SI in cherry is an excellent example of how basic science has resulted in a clear economic impact, and the technology is now available to quickly determine the *S*-allele genotype of any new cultivar or offspring. In addition, SC and its simple genetic inheritance enable the quick incorporation of this characteristic in breeding programmes. Flowering success is dependent on primordial flower development. Most critical are dormancy and chilling requirements, especially with the expansion of cherry production to warmer latitudes, or even in some conventional growing regions that may be impacted by global warming. The search for the biological underpinning of dormancy will provide the basis for understanding the genetic control and for accurately calculating chilling requirements. The combination of understanding flower biology, together with the genetic variation in cherries, will provide new answers to current and future challenges.

## References

Alcaraz, M.L., Hormaza, J.I. and Rodrigo, J. (2013) Pistil starch reserves at anthesis correlate with final flower fate in avocado (*Persea americana*). *PLoS One* 8, e78467.

Arbeloa, A. and Herrero, M. (1987) The significance of the obturator in the control of pollen tube entry into the ovary in peach (*Prunus persica* L. Batsch). *Annals of Botany* 60, 681–685.

Arora, R., Rowland, L.J. and Tanino, K. (2003) Induction and release of bud dormancy in woody perennials: a science comes of age. *HortScience* 38, 911–921.

Atkinson, C.J., Brennan, R.M. and Jones, H.G. (2013) Declining chilling and its impact on temperate perennial crops. *Environmental and Experimental Botany* 91, 48–62.

Baggiolini, M. (1952) Les stades repérés des arbres fruitiers à noyau. *Revue Romande d'Agriculture de Viticulture et d'Arboriculture* 8, 3–4.

Beppu, K. and Kataoka, I. (2000) Artificial shading reduces the occurrence of double pistils in 'Satohnishiki' sweet cherry. *Scientia Horticulturae* 83, 241–247.

Beppu, K. and Kataoka, I. (2011) Studies on pistil doubling and fruit set of sweet cherry in warm climate. *Journal of the Japanese Society for Horticultural Science* 80, 1–13.

Bielenberg, D.G., Wang, Y.E., Li, Z., Zhebentyayeva, T., Fan, S., Reighard, G.L., Scorza, R. and Abbott, A.G. (2008) Sequencing and annotation of the evergrowing locus in peach [*Prunus persica* (L.) Batsch] reveals a cluster of six MADS-box transcription factors as candidate genes for regulation of terminal bud formation. *Tree Genetics and Genomes* 4, 495–507.

Böhlenius, H., Huang, T., Charbonnel-Campaa, L., Brunner, A.M., Jansson, S., Strauss, S.H. and Nilsson, O. (2006) CO/FT regulatory module controls timing of flowering and seasonal growth cessation in trees. *Science* 312, 1040–1043.

Boskovic, R. and Tobutt, K.R. (1996) Correlation of stylar ribonuclease zymograms with incompatibility alleles in sweet cherry. *Euphytica* 90, 245–250.

Boskovic, R., Russell, K. and Tobutt, K.R. (1997) Inheritance of stylar ribonucleases in cherry progenies, reassignment of incompatibility alleles to two incompatibility groups. *Euphytica* 95, 221–228.

Boskovic, R., Tobutt, K.R., Schmidt, H. and Sonneveld, T. (2000) Re-examination of (in)compatibility genotypes of two John Innes self-compatible sweet cherry selections. *Theoretical and Applied Genetics* 101, 234–240.

Bradley, D. (1997) Inflorescence commitment and architecture in *Arabidopsis*. *Science* 275, 80–83.

Cachi, A.M. and Wünsch, A. (2011) Characterization and mapping of non-*S* gametophytic self-compatibility in sweet cherry (*Prunus avium* L.). *Journal of Experimental Botany* 62, 1847–1856.

Cachi, A.M. and Wünsch, A. (2014) Characterization of self-compatibility in sweet cherry varieties by crossing experiments and molecular genetic analysis. *Tree Genetics and Genomes* 10, 1205–1212.

Cachi, A.M., Hedhly, A., Hormaza, J.I. and Wünsch, A. (2014) Pollen tube growth in the self-compatible sweet cherry genotype, 'Cristobalina', is slowed down after self-pollination. *Annals of Applied Biology* 164, 73–84.

Cachi, A.M., Wünsch, A., Negueroles, J. and Rodrigo, J. (2015) Necesidades de polinización en variedades de cerezo. *Revista de Fruticultura* 39, 2–6.

Calabrese, F., Fenech, L. and Raimondo, A. (1984) Kronio: una cultivar di ciliegio molto precoce e autocompatibile. *Frutticoltura* 46, 27–30.

Campoy, J.A., Ruiz, D. and Egea, J. (2011) Dormancy in temperate fruit trees in a global warming context: a review. *Scientia Horticulturae* 130, 357–372.

Campoy, J.A., Ruiz, D., Allderman, L., Cook, N. and Egea, J. (2012) The fulfilment of chilling requirements and the adaptation of apricot (*Prunus armeniaca* L.) in warm winter climates: an approach in Murcia (Spain) and the Western Cape (South Africa). *European Journal of Agronomy* 37, 43–55.

Castède, S., Campoy, J.A., Quero-García, J., Le Dantec, L., Lafargue, M., Barreneche, T., Wenden, B. and Dirlewanger, E. (2014) Genetic determinism of phenological traits highly affected by climate change in *Prunus avium*: flowering date dissected into chilling and heat requirements. *New Phytologist* 202, 703–715.

Castède, S., Campoy, J.A., Le Dantec, L., Quero-García, J., Barreneche, T., Wenden, B. and Dirlewanger, E. (2015) Mapping of candidate genes involved in bud dormancy and flowering time in sweet cherry (*Prunus avium*). *PLoS One* 10, e0143250.

Cerovic, R. and Micic, N. (1999) Functionality of embryo sacs as related to their viability and fertilization success in sour cherry. *Scientia Horticulturae* 79, 227–235.

Cerovic, R. and Ruzic, D. (1992a) Pollen tube growth in sour cherry (*Prunus cerasus*) at different temperatures. *Journal of Horticultural Science* 67, 333–340.

Cerovic, R. and Ruzic, D. (1992b) Senescence of ovules at different temperatures and their effect on the behaviour of pollen tubes in sour cherry. *Scientia Horticulturae* 51, 321–327.

Cooke, J.E.K., Eriksson, M.E. and Junttila, O. (2012) The dynamic nature of bud dormancy in trees: environmental control and molecular mechanisms. *Plant, Cell and Environment* 35, 1707–1728.

Coombe, B.G. (1976) The development of fleshy fruits. *Annual Review of Plant Physiology* 27, 507–528.

Crabbé, J. (1994) Dormancy. In: Arntzen, C. (ed.) *Encyclopedia of Agricultural Science*. Academic Press, New York, pp. 597–611.

Crane, M.B. and Brown, W.G. (1937) Incompatibility and sterility in the sweet cherry, *Prunus avium* L. *Journal of Pomology and Horticultural Science* 7, 276–301.

Crane, M.B. and Lawrence, J.C. (1929) Genetical and cytological aspects of incompatibility and sterility in cultivated fruits. *Journal of Pomology and Horticultural Science* 7, 276–301.

de Franceschi, P., Stegmeir, T., Cabrera, A., van der Knaap, E., Rosyara, U.R., Sebolt, A.M., Dondini, L., Dirlewanger, E., Quero-García, J., Campoy, J.A. and Iezzoni, A.F. (2013) Cell number regulator genes in *Prunus* provide candidate genes for the control of fruit size in sweet and sour cherry. *Molecular Breeding* 32, 311–326.

de Nettancourt, D. (2001) *Incompatibility and Incongruity in Wild and Cultivated Plants*, 2nd edn. Springer, Berlin, Germany.

Diaz, D.H., Rasmussen, H.P. and Dennis, F.G. (1981) Scanning electron-microscope examination of flower bud differentiation in sour cherry. *Journal of the American Society for Horticultural Science* 106, 513–515.

Eaton, G.W. (1959) A study of the microgametophyte in *Prunus avium* and its relation to fruit setting. *Journal of Plant Science* 39, 466–476.

Elfving, D.C., Lang, G.A. and Visser, D.B. (2003) Prohexadione-Ca and ethephon reduce shoot growth and increase flowering in young, vigorous sweet cherry trees. *HortScience* 38, 293–298.

Engin, H. (2008) Scanning electron microscopy of floral initiation and developmental stages in sweet cherry (*Prunus avium* L.) under water deficits. *Bangladesh Journal of Botany* 37, 15–19.

Engin, H. and Gokbayrak, Z. (2010) Phyllody (flower abnormality) in sweet cherry (*Prunus avium* L.). *Journal of Animal and Plant Sciences* 20, 217–219.

Engin, H. and Unal, A. (2007) Examination of flower bud initiation and differentiation in sweet cherry and peach by scanning electron microscope. *Turkish Journal of Agriculture and Forestry* 31, 373–379.

Eriksson, M.E. and Moritz, T. (2002) Daylength and spatial expression of a gibberellin 20-oxidase isolated from hybrid aspen (*Populus tremula* L. × *P. tremuloides* Michx.). *Planta* 214, 920–930.

Esumi, T., Hagihara, C., Kitamura, Y., Yamane, H. and Tao, R. (2009) Identification of an FT ortholog in Japanese apricot (*Prunus mume* Sieb. et Zucc.). *Journal of Horticultural Science and Biotechnology* 84, 149–154.

Facteau, T.J., Rowe, K.E., Chestnut, N.E. (1985) Firmness of sweet cherry following multiple applications of gibberellic acid. *Journal of the American Society for Horticultural Science* 110, 775–777.

Fadón, E. (2015) Desarrollo floral y reposo en cerezo (*Prunus avium*). PhD thesis, University of Zaragoza, Zaragoza, Spain.

Fadón, E., Herrero, M. and Rodrigo, J. (2015a) Flower development in sweet cherry framed in the BBCH scale. *Scientia Horticulturae* 192, 141–147.

Fadón, E., Herrero, M. and Rodrigo, J. (2015b) Flower bud dormancy in *Prunus* species. In: Anderson, J.V. (ed.) *Advances in Plant Dormancy*. Springer, Heidelberg, Germany, pp. 123–135.

Fadón, E., Herrero, M. and Rodrigo, J. (2017) Flower bud development and chilling requirements in sweet cherry cv 'Bing'. *Acta Horticulturae* (in press).

Faust, M. and Suranyi, D. (1997) Origin and dissemination of cherry. *Horticultural Reviews* 19, 263–317.

Faust, M., Erez, A., Rowland, L.J., Wang, S.Y. and Norman, H.A. (1997) Bud dormancy in perennial fruit trees: physiological basis for dormancy induction, maintenance and release. *HortScience* 32, 623–629.

Felker, F.C., Robitaille, H.A. and Hess, F.D. (1983) Morphological and ultrastructural development and starch accumulation during chilling of sour cherry flower buds. *American Journal of Botany* 70, 376–386.

Fleckinger, J. (1948) Les stades végétatifs des arbres fruitiers, en rapport avec les traitements. *Pomologie Française* Suppl, 81–93.

Flore, J.A., Kesner, C.D. and Webster, A.D. (1996) Tree canopy management and the orchard environment: principles and practices of pruning and training. In: Webster, A.D. and Looney, N.E. (eds) *Cherries: Crop Physiology, Production and Uses*. CAB International, Wallingford, UK, pp. 259–278.

Frary, A., Nesbitt, T.C., Grandillo, S., van der Knaap, E., Cong, B., Liu, J.P., Meller, J., Elbert, R., Alpert, K.B. and Tanksley, S.D. (2000) fw2.2: a quantitative trait locus key to the evolution of tomato fruit size. *Science* 289, 85–88.

Furukawa, Y. and Bukovac, M.J. (1989) Embryo sac development in sour cherry during the pollination period as related to fruit set. *HortScience* 24, 1005–1008.

Guerra, M.E. and Rodrigo, J. (2015) Japanese plum pollination: a review. *Scientia Horticulturae* 197, 674–686.

Guimond, C.M., Andrews, P.K. and Lang, G.A. (1998a) Scanning electron microscopy of floral initiation in sweet cherry. *Journal of the American Society for Horticultural Science* 123, 509–512.

Guimond, C.M., Lang, G.A. and Andrews, P.K. (1998b) Timing and severity of summer pruning affects flower initiation and shoot regrowth in sweet cherry. *HortScience* 33, 647–649.

Hauck, N., Iezzoni, A.F., Yamane, H. and Tao, R. (2001) Revising the *S*-allele nomenclature in sweet cherry (*Prunus avium*) using RFLP probes. *Journal of the American Society for Horticultural Science* 126, 654–660.

Hauck, N.R., Yamane, H., Tao, R. and Iezzoni, A.F. (2006) Accumulation of non-functional *S*-haplotypes results in the breakdown of gametophytic self-incompatibility in tetraploid *Prunus*. *Genetics* 172, 1191–1198.

Hedhly, A. (2011) Flowering sensitivity to temperature fluctuation. *Environmental and Experimental Botany* 74, 9–16.

Hedhly, A., Hormaza, J.I. and Herrero, M. (2003) The effect of temperature on stigmatic receptivity in sweet cherry (*Prunus avium* L.). *Plant, Cell and Environment* 26, 1673–1680.

Hedhly, A., Hormaza, J.I. and Herrero, M. (2004) Effect of temperature on pollen tube kinetics and dynamics in sweet cherry (*Prunus avium*, Rosaceae). *American Journal of Botany* 91, 558–564.

Hedhly, A., Hormaza, J.I. and Herrero, M. (2005) Influence of genotype-temperature interaction on pollen performance. *Journal of Evolutionary Biology* 18, 1494–1502.

Hedhly, A., Hormaza, J.I. and Herrero, M. (2007) Warm temperatures at bloom reduce fruit set in sweet cherry. *Journal of Applied Botany and Food Quality* 81, 158–164.

Hedhly, A., Hormaza, J.I. and Herrero, M. (2009) Global warming and sexual plant reproduction. *Trends in Plant Science* 14, 30–36.

Hempel, F.D., Welch, D.R. and Feldman, L.J. (2000) Floral induction and determination: where is flowering controlled? *Trends in Plant Science* 5, 17–21.

Herrero, J. (coord.) (1964) *Cartografía de las Variedades Frutales de Hueso y Pepita*. CSIC, Aula Dei, Zaragoza, Spain.

Herrero, M. (1992a) From pollination to fertilization in fruit trees. *Plant Growth Regulation* 11, 27–32.

Herrero, M. (1992b) Mechanisms in the pistil that regulate gametophyte population in peach (*Prunus persica*). In: Ottaviano, E., Mulcahy, D.L., Sari Gorla, M. and Mulcahy, G.B. (eds) *Angiosperm Pollen and Ovules*. Springer, Berlin, Germany, pp. 377–381.

Herrero, M. (2000) Changes in the ovary related to pollen tube guidance. *Annals of Botany* 85 (Suppl. A), 79–85.

Herrero, M. (2001) Ovary signals for directional pollen tube growth. *Sexual Plant Reproduction* 14, 3–7.

Herrero, M. (2003) Male–female synchrony and the regulation of mating in flowering plants. *Philosophical Transactions of the Royal Society of London Series B: Biological Sciences* 358, 1019–1024.

Herrero, M. and Arbeloa, A. (1989) Influence of the pistil on pollen tube kinetics in peach (*Prunus persica* (L.) Batsch). *American Journal of Botany* 176, 1441–1447.

Herrero, M. and Dickinson, H.G. (1979) Pollen–pistil incompatibility in *Petunia hybrida*: changes in the pistil following compatible and incompatible intraspecific crosses. *Journal of Cell Science* 36, 1–18.

Herrero, M. and Hormaza, J.I. (1996) Pistil strategies controlling pollen tube growth. *Sexual Plant Reproduction* 9, 343–347.

Hormaza, J.I. and Herrero, M. (1992) Pollen selection. *Theoretical and Applied Genetics* 83, 663–672.

Hormaza, J.I. and Herrero, M. (1994) Gametophytic competition and selection. In: Williams, E.G. Knox, R.B. and Clarke, A.E. (eds) *Genetic Control of Self-incompatibility and Reproductive Development in Flowering Plants*. Kluwer Academic Press, Dordrecht, The Netherlands, pp. 372–400.

Hormaza, J.I. and Herrero, M. (1996a) Dynamics of pollen tube growth under different competition regimes. *Sexual Plant Reproduction* 9, 153–160.

Hormaza, J.I. and Herrero, M. (1996b) Male gametophytic selection as a plant breeding tool. *Scientia Horticulturae* 65, 321–333.

Hormaza, J.I. and Herrero, M. (1999) Pollen performance as affected by the pistilar genotype in sweet cherry (*Prunus avium* L.). *Protoplasma* 208, 129–135.

Horvath, D.P., Anderson, J.V., Chao, W.S. and Foley, M.E. (2003) Knowing when to grow: signals regulating bud dormancy. *Trends in Plant Science* 8, 534–540.

Hsu, C.Y., Adams, J.P., Kim, H., No, K., Ma, C., Strauss, S.H., Drnevich, J., Vandervelde, L., Ellis, J.D., Rice, B.M., Wickett, N., Gunter, L.E., Tuskan, G.A., Brunner, A.M., Page, G.P., Barakat, A., Carlson, J.E., DePamphilis, C.W., Luthe, D.S. and Yuceer, C. (2011) *FLOWERING LOCUS T* duplication coordinates reproductive and vegetative growth in perennial poplar. *Proceedings of the National Academy of Sciences USA* 108, 10756–10761.

Ikeda, K., Igic, B., Ushijima, K., Yamane, H., Hauck, N.R., Nakano, R., Sassa, H., Iezzoni, A.F., Kohn, J.R. and Tao, R. (2004) Primary structural features of the *S*-haplotype-specific F-box protein, SFB, in *Prunus*. *Sexual Plant Reproduction* 16, 235–243.

Jackson, J.E. and Hamer, D.J.C. (1980) The cause of year-to-year variation in the average yield of Cox's orange pippin apple in England. *Journal of Horticultural Science* 55, 149–156.

Jiménez, S., Li, Z.G., Reighard, G.L. and Bielenberg, D.G. (2010a) Identification of genes associated with growth cessation and bud dormancy entrance using a dormancy-incapable tree mutant. *BMC Plant Biology* 10, 25.

Jiménez, S., Reighard, G.L. and Bielenberg, D.G. (2010b) Gene expression of *DAM5* and *DAM6* is suppressed by chilling temperatures and inversely correlated with bud break rate. *Plant Molecular Biology* 73, 157–167.

Julian, C., Herrero, M. and Rodrigo, J. (2010) Flower bud differentiation and development in fruiting and non-fruiting shoots in relation to fruit set in apricot (*Prunus armeniaca* L.). *Trees* 24, 833–841.

Julian, C., Rodrigo, J. and Herrero, M. (2011) Stamen development and winter dormancy in apricot (*Prunus armeniaca*). *Annals of Botany* 108, 617–625.

Julian, C., Herrero, M. and Rodrigo, J. (2014) Anther meiosis time is related to winter cold temperatures in apricot (*Prunus armeniaca*). *Environmental and Experimental Botany* 100, 20–25.

Kitashiba, H., Zhang, S.L., Wu, J., Shirasawa, K. and Nishio, T. (2008) *S* genotyping and *S* screening utilizing *SFB* gene polymorphism in Japanese plum and sweet cherry by dot-blot analysis. *Molecular Breeding* 21, 339–349.

Koornneef, M., Hanhart, C.J. and van der Veen, J.H. (1991) A genetic and physiological analysis of late flowering mutants in *Arabidopsis thaliana*. *Molecular and General Genetics* 229, 57–66.

Kozlowski, T.T. and Pallardy, S.G. (1997) *Physiology of Woody Plants*. Academic Press, San Diego, California.

Lang, G. (1987) Dormancy a new universal terminology. *HortScience* 22, 817–820.

Lang, G.A. (1994) Dormancy – the missing links: molecular studies and integration of regulatory plant and environmental interactions. *HortScience* 29, 1255–1263.

Lapins, K.O. (1975) 'Compact Stella' cherry. *Fruit Varieties Journal* 29, 20.

Leida, C., Conesa, A., Llácer, G., Badenes, M.L. and Ríos, G. (2012) Histone modifications and expression of *DAM6* gene in peach are modulated during bud dormancy release in a cultivar dependent manner. *New Phytologist* 193, 67–80.

Lenahan, O.M., Whiting, M.D., Elfving, D.C. (2006) Gibberellic acid inhibits floral bud induction and improves 'Bing' sweet cherry fruit quality. *HortScience* 41, 654–659.

Lewis, D. and Crowe, L.K. (1954) Structure of the incompatibility gene. IV types of mutation in *Prunus avium* L. *Heredity* 8, 357–363.

Li, Z., Reighard, G.L., Abbott, A.G. and Bielenberg, D.G. (2009) Dormancy-associated MADS genes from the *EVG* locus of peach [*Prunus persica* (L.) Batsch] have distinct seasonal and photoperiodic expression patterns. *Journal of Experimental Botany* 60, 3521–3530.

Luedeling, E. (2012) Climate change impacts on winter chill for temperate fruit and nut production: a review. *Scientia Horticulturae* 144, 218–229.

Marchese, A., Boskovic, R.I., Caruso, T., Raimondo, A., Cutuli, M. and Tobutt, K.R. (2007) A new self-compatibility haplotype in the sweet cherry 'Kronio', S59, attributable to a pollen-part mutation in the *SFB* gene. *Journal of Experimental Botany* 58, 4347–4356.

Matsumoto, D., Yamane, H., Abe, K. and Tao, R. (2012) Identification of a Skp1-like protein interacting with SFB, the pollen *S* determinant of the gametophytic self-incompatibility in *Prunus*. *Plant Physiology* 159, 1252–1262.

Matthews, P. and Dow, K.P. (1969) Incompatibility groups: sweet cherry (*Prunus avium*). In: Knight, R.L. (ed.) *Abstract Bibliography of Fruit Breeding and Genetics to 1965: Prunus*. Commonwealth Agricultural Bureaux, Farnham Royal, pp. 540–544.

Mert, C. and Soylu, A. (2007) Possible cause of low fruit set in the sweet cherry cultivar 0900 Ziraat. *Canadian Journal of Plant Science* 87, 593–594.

Mimida, N., Ureshino, A., Tanaka, N., Shigeta, N., Sato, N., Moriya-Tanaka, Y., Iwanami, H., Honda, C., Suzuki, A., Komori, S. and Wada, M. (2011) Expression patterns of several floral genes during flower initiation in the apical buds of apple (*Malus* × *domestica* Borkh.) revealed by *in situ* hybridization. *Plant Cell Reports* 30, 1485–1492.

Mohamed, R., Wang, C.T., Ma, C., Shevchenko, O., Dye, S.J., Puzey, J.R., Etherington, E., Sheng, X.Y., Meilan, R., Strauss, S.H. and Brunner, A.M. (2010) *Populus CEN/TFL1* regulates first onset of flowering, axillary meristem identity and dormancy release in *Populus*. *Plant Journal* 62, 674–688.

Mulcahy, D.L. (1979) The rise of angiosperms: a genecological factor. *Science* 206, 20–23.

Olmstead, J.W., Iezzoni, A.F. and Whiting, M.D. (2007) Genotypic differences in sweet cherry fruit size are primarily a function of cell number. *Journal of the American Society for Horticultural Science* 132, 697–703.

Olsen, J.E., Junttila, O. and Morizt, T.A. (1995) Localized decrease of GA in shoot tips of *Salix pentandra* seedling precedes cessation of shoot elongation under short photoperiod. *Physiologia Plantarum* 95, 627–632.

Philp, G.L. (1933) Abnormality in sweet cherry blossoms and fruit. *Botanical Gazette* 94, 815–820.

Postweiler, K., Stosser, R. and Anvari, S.F. (1985) The effect of different temperatures on the viability of ovules in cherries. *Scientia Horticulturae* 25, 235–239.

Proebsting, E.L., Carter, G.H. and Mills, H.H. (1973) Quality improvement in canned 'Rainier' cherries (*P. avium* L.) with gibberellic acid. *Journal of the American Society for Horticultural Science* 98, 334–336.

Rinne, P.L.H., Welling, A., Vahala, J., Ripel, L., Ruonala, R., Kangasjarvi, J. and van der Schoot, C. (2011) Chilling of dormant buds hyperinduces *FLOWERING LOCUS T* and recruits GA-inducible 1,3-β-glucanases to reopen signal conduits and release dormancy in *Populus*. *Plant Cell* 23, 130–146.

Rodrigo, J. (2000) Spring frosts in deciduous fruit trees – morphological damage and flower hardiness. *Scientia Horticulturae* 85, 155–173.

Rodrigo, J. and Herrero, M. (1998) Influence of intraovular reserves on ovule fate in apricot (*Prunus armeniaca* L.). *Sexual Plant Reproduction* 11, 86–93.

Rodrigo, J. and Herrero, M. (2002) Effects of pre-blossom temperature on flower development and fruit set in apricot. *Scientia Horticulturae* 92, 125–135.

Rodrigo, J., Hormaza, J.I. and Herrero, M. (2000) Ovary starch reserves and flower development in apricot (*Prunus armeniaca*). *Physiologia Plantarum* 108, 35–41.

Rodrigo, J., Espada, J.L., Bernad, D., Martin, E. and Herrero, M. (2017) Influence of Syncron® and Nitroactive® on flowering and ripening time in two sweet cherry cultivars. *Acta Horticulturae* (in press).

Rohde, A. and Bhalerao, R.P. (2007) Plant dormancy in the perennial context. *Trends in Plant Science* 12, 1360–1385.

Roversi, A., Monteforte, A.D.P. and Folini, L. (2008) Observations on the occurrence of sweet cherry double-fruits in Italy and Slovenia. *Acta Horticulturae* 795, 849–854.

Sanzol, J. and Herrero, M. (2001) The effective pollination period in fruit trees. *Scientia Horticulturae* 90, 1–17.

Sasaki, R., Yamane, H., Ooka, T., Jotatsu, H., Kitamura, Y., Akagi, T. and Tao, R. (2011) Functional and expressional analyses of *PmDAM* genes associated with endodormancy in Japanese apricot. *Plant Physiology* 157, 485–497.

Schuster, M. (2012) Incompatible (*S*-) genotypes of sweet cherry cultivars (*Prunus avium* L.). *Scientia Horticulturae* 148, 59–73.

Sebolt, A.M., Tsukamoto, T. and Iezzoni, A.F. (2017) *S*-Genotyping of cultivars and breeding selections of sour cherry (*Prunus cerasus* L.) in the Michigan State University sour cherry breeding program. *Acta Horticulturae* (in press).

Sonneveld, T., Robbins, T.P., Boskovic, R. and Tobutt, K.R. (2001) Cloning of six cherry self-incompatibility alleles and development of allele-specific PCR detection. *Theoretical and Applied Genetics* 102, 1046–1055.

Sonneveld, T., Tobutt, K.R. and Robbins, T.P. (2003) Allele-specific PCR detection of sweet cherry self-incompatibility (*S*) alleles S1 to S16 using consensus and allele-specific primers. *Theoretical and Applied Genetics* 107, 1059–1070.

Sonneveld, T., Tobutt, K.R., Vaughan, S.P. and Robbins, T.P. (2005) Loss of pollen-*S* function in two self-compatible selections of *Prunus avium* is associated with deletion/mutation of an *S* haplotype-specific F-box gene. *Plant Cell* 17, 37–51.

Sonneveld, T., Robbins, T.P. and Tobutt, K.R. (2006) Improved discrimination of self-incompatibility *S*-RNase alleles in cherry and high throughput genotyping by automated sizing intron PCR products. *Plant Breeding* 125, 305–307.

Southwick, S.M., Shackel, K.A., Yeager, J.T., Asai, W.K. and Katacich, M. (1991) Over-tree sprinkling reduces abnormal shapes in 'Bing' sweet cherries. *California Agriculture* 45, 24–26.

Sterling, C. (1964) Comparative morphology of the carpel in the Rosaceae. I. Prunoideae: *Prunus*. *American Journal of Botany* 51, 36–44.

Stösser, R. and Anvari, S.F. (1982) On the senescence of ovules in cherries. *Scientia Horticulturae* 16, 29–38.

Szikriszt, B., Doğan, A., Ercisli, S., Akcay, M.E., Hegedűs, A. and Hálasz, J. (2013) Molecular typing of the self-incompatibility locus of Turkish sweet cherry genotypes reflects phylogenetic relationships among cherries and other *Prunus*. *Tree Genetics and Genomes* 9, 155–165.

Tao, R. and Iezzoni, A.F. (2010) The *S*-RNase-based gametophytic self-incompatibility system in *Prunus* exhibits distinct genetic and molecular features. *Scientia Horticulturae* 124, 423–433.

Tao, R., Yamane, H., Sugiura, A., Murayama, H., Sassa, H. and Mori, H. (1999) Molecular typing of *S*-alleles through identification, characterization and cDNA cloning for *S*-RNases in sweet cherry. *Journal of the American Society for Horticultural Science* 124, 224–233.

Tehrani, G. and Brown, S. (1992) Pollen-incompatibility and self-fertility in sweet cherry. *Plant Breeding Review* 9, 367–388.

Thompson, M. (1996) Flowering, pollination and fruit set. In: Webster, A.D. and Looney, N.E. (eds) *Cherries: Crop Physiology, Production and Uses*. CAB International, Wallingford, UK.

Tobutt, K.R., Sonneveld, T., Bekefi, Z. and Boskovic, R. (2004) Cherry (in)compatibility genotypes – an updated cultivar table. *Acta Horticulturae* 663, 667–672.

Tsukamoto, T., Hauck, N.R., Tao, R., Jiang, N. and Iezzoni, A.F. (2006) Molecular characterization of three non-functional *S*-haplotypes in sour cherry (*Prunus cerasus*). *Plant Molecular Biology* 62, 371–383.

Tsukamoto, T., Potter, D., Tao, R., Vieira, C.P., Vieira, J. and Iezzoni, A.F. (2008a) Genetic and molecular characterization of three novel S-haplotypes in sour cherry (*Prunus cerasus* L.). *Journal of Experimental Botany* 59, 3169–3185.

Tsukamoto, T., Tao, R. and Iezzoni, A.F. (2008b) PCR markers for mutated *S*-haplotypes enable discrimination between self-incompatible and self-compatible sour cherry selections. *Molecular Breeding* 21, 67–80.

Tsukamoto, T., Hauck, N.R., Tao, R., Jiang, N. and Iezzoni, A.F. (2010) Molecular and genetic analysis of four non-functional S haplotype variants derived from a common ancestral S haplotype identified in sour cherry (*Prunus cerasus* L.). *Genetics* 184, 411–427.

Tucker, L.R. (1934) Notes on sweet cherry doubling. *Proceedings of the American Society for Horticultural Science* 32, 300–302.

Tucker, L.R. (1935) Additional notes on sweet cherry doubling. *Proceedings of the American Society for Horticultural Science* 33, 237–239.

Tukey, H.B. and Young, J.O. (1939) Histological study of the developing fruit of the sour cherry. *Botanical Gazette* 100, 723–749.

Ushijima, K., Yamane, H., Watari, A., Kakehi, E., Ikeda, K., Hauck, N.R., Iezzoni, A.F. and Tao, R. (2004) The *S* haplotype-specific F-box protein gene, *SFB*, is defective in self-compatible haplotypes of *Prunus avium* and *P. mume*. *Plant Journal* 39, 573–586.

Vanstraelen, M. and Benkova, E. (2012) Hormonal interactions in the regulation of plant development. *Annual Review of Cell and Developmental Biology* 28, 463–487.

Vaughan, S.P., Russell, K., Sargent, D.J. and Tobutt, K.R. (2006) Isolation of *S*-locus F-box alleles in *Prunus avium* and their application in a novel method to determine self-incompatibility genotype. *Theoretical and Applied Genetics* 112, 856–866.

Vaughan, S.P., Boskovic, R., Gisbert-Climent, A., Russell, K. and Tobutt, K.R. (2008) Characterization of novel *S*-alleles from cherry (*Prunus avium* L.). *Tree Genetics and Genomes* 4, 531–541.

Viti, R., Andreini, L., Ruiz, D., Egea, J., Bartolini, S., Iacona, C. and Campoy, J.A. (2010) Effect of climatic conditions on the overcoming of dormancy in apricot flower buds in two Mediterranean areas: Murcia (Spain) and Tuscany (Italy). *Scientia Horticulturae* 124, 217–224.

Watanabe, S. (1983) Scanning electron microscope observations of flower bud differentiation in sweet cherry (*Prunus avium*). *Journal of the Yamagata Agriculture and Forestry Society* 77, 15–18.

Weinberger, J.H. (1975) Plums. In: Janick, J. and Moore, J.N. (eds) *Advances in Fruit Breeding*. Purdue University Press, Lafayette, Indiana, pp. 336–347.

Westwood, M.N. (1993) *Temperate-zone Pomology: Physiology and Culture*. Timber Press, Portland, Oregon.

Whiting, M.D. and Lang, G.A. (2004) 'Bing' sweet cherry on the dwarfing rootstock 'Gisela 5': crop load affects fruit quality and vegetative growth but not net $CO_2$ exchange. *Journal of the American Society for Horticultural Science* 129, 407–415.

Whiting, M.D. and Ophardt, D. (2005) Comparing novel sweet cherry crop load management strategies. *HortScience* 40, 1271–1275.

Whiting, M.D., Lang, G. and Orphardt, D. (2005) Rootstock and training system affect cherry growth, yield, and fruit quality. *HortScience* 40, 582–586.

Wiersma, P.A., Wu, Z., Zhou, L., Hampson, C. and Kappel, F. (2001) Identification of new self-incompatibility alleles in sweet cherry (*Prunus avium* L.) and clarification of incompatibility groups by PCR and sequencing analysis. *Theoretical and Applied Genetics* 102, 700–708.

Williams, R.R. (1970) Factors affecting pollination in fruit trees. In: Luckwill, L.C. and Cutting, C.V. (eds) *Physiology of Tree Crops*. Academic Press, London, pp 193–207.

Wu, J., Chao, G., Awais Khan, M., Wu, J., Gao, Y., Wang, C., Schuyler, S.K. and Zhang, S. (2013) Molecular determinants and mechanisms of gametophytic self-incompatibility in fruit trees of Rosaceae. *Critical Reviews in Plant Sciences* 32, 53–68.

Wünsch, A. and Hormaza, J.I. (2004) Genetic and molecular analysis in Cristobalina sweet cherry, a spontaneous self-compatible mutant. *Sex Plant Reproduction* 17, 203–210.

Wünsch, A., Tao, R. and Hormaza, J.I. (2010) Self-compatibility in 'Cristobalina' sweet cherry is not associated with duplications or modified transcription levels of *S*-locus genes. *Plant Cell Reports* 29, 715–721.

Yamaguchi, M., Sato, I., Takase, K., Watanabe, A. and Ishiguro, M. (2004) Differences and yearly variation in number and size of mesocarp cells in sweet cherry (*Prunus avium* L.) cultivars and related species. *Journal of the Japanese Society for Horticultural Science* 73, 12–18.

Yamane, H. and Tao, R. (2009) Molecular basis of self-(in)compatibility and current status of *S*-genotyping in rosaceous fruit trees. *Journal of the Japanese Society for Horticultural Science* 78, 137–157.

Yamane, H., Tao, R., Sugiura, A., Hauck, N.R. and Iezzoni, A.F. (2001) Identification and characterization of *S*-RNases in tetraploid sour cherry (*Prunus cerasus*). *Journal of the American Society for Horticultural Science* 126, 661–667.

Yamane, H., Ikeda, K., Hauck, N.R., Iezzoni, A.F. and Tao, R. (2003a) Self-incompatibility (*S*) locus region of the mutated *S*-6-haplotype of sour cherry (*Prunus cerasus*) contains a functional pollen *S* allele and a non-functional pistil *S* allele. *Journal of Experimental Botany* 54, 2431–2437.

Yamane, H., Ikeda, K., Sassa, H. and Tao, R. (2003b) A pollen expressed gene for a novel protein with an F-box motif that is very tightly linked to a gene for *S*-RNase in two species of cherry, *Prunus cerasus* and *P. avium*. *Plant Cell Physiology* 44, 764–769.

Yamane, H., Ooka, T., Jotatsu, H., Hosaka, Y., Sasaki, R. and Tao, R. (2011) Expressional regulation of *PpDAM5* and *PpDAM6*, peach (*Prunus persica*) dormancy-associated MADS-box genes, by low temperature and dormancy-breaking reagent treatment. *Journal of Experimental Botany* 62, 3481–3488.

Zhang, C. and Whiting, M.D. (2011) Improving 'Bing' sweet cherry fruit quality with plant growth regulators. *Scientia Horticulturae* 127, 341–346.

Zhang, G., Sebolt, A.M., Sooriyapathirana, S.S., Wang, D.C., Bink, M.C.A.M., Olmstead, J.W. and Iezzoni, A.F. (2010) Fruit size QTL analysis of an $F_1$ population derived from a cross between a domesticated sweet cherry cultivar and a wild forest sweet cherry. *Tree Genetics and Genomes* 6, 25–36.

Zuriaga, E., Molina, L., Badenes, M.L. and Romero, C. (2012) Physical mapping of a pollen modifier locus controlling self-incompatibility in apricot and synteny analysis within the Rosaceae. *Plant Molecular Biology* 79, 229–242.

Zuriaga, E., Muñoz-Sanz, J.V., Molina, L., Gisbert, A.D., Badenes, M.L. and Romero, C. (2013) An *S*-locus independent pollen factor confers self-compatibility in 'Katy' apricot. *PLoS One* 8, e53947.

# 3 Biodiversity, Germplasm Resources and Breeding Methods

**Amy Iezzoni,[1]* Ana Wünsch,[2] Monika Höfer,[3] Daniela Giovannini,[4] Martin Jensen,[5] Jose Quero-García,[6] Jose Antonio Campoy,[6] Aleš Vokurka[7] and Teresa Barreneche[6]**

*[1]Michigan State University, Michigan, USA; [2]Unidad de Hortofruticultura, Centro de Investigación y Tecnología Agroalimentaria de Aragón, Instituto Agroalimentario de Aragón – IA2 (CITA-Universidad de Zaragoza), Zaragoza, Spain; [3]Julius Kühn-Institute, Federal Research Centre for Cultivated Plants, Institute for Breeding Research on Fruit Crops, Dresden, Germany; [4]CREA-FRF Council for Agricultural Research and Economics, Fruit Tree Unit of Forlì,Forlì,Italy; [5]Institut for Fødevarer/Department of Food Science, Aarhus University, Årslev,Denmark; [6]UMR 1332 Biologie du Fruit et Pathologie, INRA et Université de Bordeaux, Villenave d'Ornon, France; [7]Department for Plant Breeding and Genetics, Faculty of Agriculture, University of Zagreb, Zagreb, Croatia*

## 3.1 Taxonomy of Cherry Species in the Scion and Rootstock Gene Pool

Cherries are members of the Rosaceae family and Spiraeoideae subfamily (Potter *et al.*, 2007). Within the Spiraeoideae, cherry is in the tribe Amygdaleae and genus *Prunus* with other crop members of the stone fruits: peach/nectarine (*Prunus persica* (L.) Batsch), apricot (*Prunus armeniaca* L.), almond (*Prunus amygdalus* Batsch) and plum (*Prunus domestica* L. and *Prunus salicina* Lindl.). Cherries are further placed within two subgenera (*Cerasus* Pers. and *Padus* (Moench) Koehne) (Table 3.1) (Rehder, 1974). The *Cerasus* Pers. subgenus and *Cerasus* Koehne section contain the diploid sweet cherry (2n = 2x = 16, *Prunus avium*) and the tetraploid sour cherry (2n = 4x = 32, *Prunus cerasus*) and ground cherry (*Prunus fruticosa* Pall.). Based on breeding behaviour and cytogenetic studies, it was proposed that the wild diploid cherry species *Prunus canescens* also belongs in this section (Schmidt, 1973; Schuster, 2005). Also in the subgenus *Cerasus* is *Prunus mahaleb*, which is one of the major cherry rootstock species besides wild sweet cherry. *Prunus maackii*, from the Padus subgenus, is used in rootstock breeding as it can be crossed with both sweet and sour cherry. Other cherry species from both subgenera are grown for their fruit on a very limited basis. These include *Prunus tomentosa* and *Prunus pseudocerasus* in China, and domesticates of *Prunus serotina* (capulin cherry) in South America (Popenoe *et al.*, 1989). Many of the other cherries, notably those in the *Pseudocerasus* section, are valued as blooming ornamentals.

* iezzoni@msu.edu

**Table 3.1.** Systematic classification of *Prunus* cherry species according to Rehder (1974) and species distribution.

| Subgenus | Sections | Species | Distribution |
|---|---|---|---|
| *Cerasus* Pers. | *Microcerasus* Webb[a] | *P. besseyi* Bailey | Canada, USA |
| | | *P. glandulosa* Thunb. | China, Japan |
| | | *P. humilis* Bge | North China |
| | | *P. incana* (Pall.) | South-eastern Europe, West Asia |
| | | *P. jacquemontii* Hook. | North-western Himalayas |
| | | *P. japonica* Thunb. | Central China, East Asia |
| | | *P. microcarpa* C.A. Mey | Asia Minor |
| | | *P. prostrata* Labill. | Mediterranean, West Asia |
| | | *P. pumila* L. | USA |
| | | *P. tomentosa* Thunb. | Northern and western China, Japan, Himalayas |
| | *Pseudocerasus* Koehne | *P. campanulata* Maxim | South Japan, Formosa |
| | | *P. cerasoides* D. Don | Himalayas |
| | | *P. incisa* Thunb. | Japan |
| | | *P. kurilensis* (Miyabe) Wils. | Japan |
| | | *P. nipponica* Matsum | Japan |
| | | *P. sargentii* Rehd. | Japan |
| | | *P. serrulata* Lindl.[b] | Japan, China, Korea |
| | | *P. sieboldii* (Carr.) | Japan |
| | | *P. subhirtella* Miq. | Japan |
| | | *P. yedoensis* Matsum | Japan |
| | *Lobopatalum* Koehne | *P. cantabrigiensis* Stapf. | China |
| | | *P. involucrata* Koehne | Central China |
| | | *P. pseudocerasus* Lindl. | Northern China |
| | | *P. dielsiana* | China |
| | *Cerasus* Koehne | *P. avium* L. | Europe, West Asia, Caucasus |
| | | *P. cerasus* L. | West Asia, south-eastern Europe |
| | | *P. fruticosa* Pall. | Central and eastern Europe, Siberia |
| | | *P. canescens* Bois[c] | Central and western China |
| | *Mahaleb* Focke | *P. emarginata* (Hook.) Walp. | USA |
| | | *P. mahaleb* L. | Europe, West Asia |
| | | *P. pensylvanica* L. | Canada, USA |
| | | *P. prunifolia* (Greene) Shafer | Canada, USA |
| | *Phyllocerasus* Koehne | *P. pilosiuscala* Koehne | Central and western China |
| | *Phyllomahaleb* Koehne | *P. maximowiczii* Rupr. | Manchuria, Korea, and Japan |
| | | *P. pleiocerasus* Koehne | |
| *Padus* (Moench) Koehne | | *P. alabamensis* Mohr. | USA |
| | | *P. buergeriana* Miq. | Japan, Korea |
| | | *P. grayana* Maxim. | Japan |
| | | *P. maackii* Rupr. | Manchuria, Korea |
| | | *P. padus* L. | Europe, North Asia, Korea, Japan |
| | | *P. serotina* Ehrh. | Canada, USA |
| | | *P. ssiori* F. Schmidt | North-east Asia, Japan |
| | | *P. virens* (Woot. & Standl.) | USA |
| | | *P. virginiana* L. | Canada, USA |

[a]*Microcerasus* has been recently included in the subgenus *Prunus* using phylogenetic analysis on 12 chloroplast regions and three nuclear genes (Shi *et al.*, 2013).
[b]*P. lannesiana*, listed as a species used in rootstock breeding, is considered to be a subspecies of *P. serrulata*.
[c]According to its breeding behaviour, it is proposed that *P. canescens* belongs to the Cerasus section (Schmidt, 1973).

## 3.2 Origin and Domestication

Europe is considered the native territory of sweet cherry (Faust and Surányi, 1997), which is found wild in mainland Europe, from Sweden down to Greece, Italy and Spain and into areas of northern Africa (Hedrick *et al.*, 1915; EUFORGEN, 2009). It is believed that sweet cherry may have originated around the Caspian and Black Seas and later spread throughout Europe (Hedrick *et al.*, 1915; Webster, 1996; Dirlewanger *et al.*, 2009). Birds are thought to have dispersed sweet cherry across Europe, predating human migration (Webster, 1996). The original habitat of sour cherry is believed to have been south-eastern Europe near Asia with the centre of origin being the south border of the Black Sea along Anatolia and the south Caucasus to Iran (Hedrick *et al.*, 1915). The tetraploid sour cherry (2n=4x=32) was formed by hybridization between sweet cherry and ground cherry (Olden and Nybom, 1968), with the hybrid origin supported by genetic studies (Beaver and Iezzoni, 1993; Brettin *et al.*, 2000; Schuster and Schreiber, 2000; Tavaud *et al.*, 2004). Sour cherry originated in regions where the distributions of the progenitor species overlap (Zhukovsky, 1965). This resulted in continual crossing between sour cherry and its two progenitor species. The presence of cherries throughout Europe gave rise to ecotypes adapted to different climatic conditions (Iezzoni *et al.*, 1990).

The domestication of cherries followed the history of civilization in Europe step by step (Hedrick *et al.*, 1915). It is known that cherries were consumed by early inhabitants of Europe in 4000–5000 BC (Brown *et al.*, 1996; Webster, 1996), and domestication of cherries in the Danube Valley is believed to have occurred in the Neolithic Period, 4000 years ago (Faust and Surányi, 1997). Several archaeological findings of cherry pits in Europe (Switzerland, France, Italy, Hungary, England and Austria) from the Neolithic Period to the Bronze Age have been reported (Faust and Surányi, 1997). Sour cherry seems to have been introduced by Slavic people from western Asia to eastern and south-east Europe and from there to Middle Europe in the time of the Migration period (6th to 8th centuries). This is supported by the observation that all East and West Asian languages have the same word stem for sour cherry: wischene (vişne) in Turkish, wishnja in Russian, wisnah in Persian, wisnia (wiśnia) in Polish and Weichsel in German (M. Schuster, Dresden, Germany, 2016, personal communication). An exception is the Hungarian word for sour cherry: meggy. It has been suggested that this term for sour cherry comes from an original Finn-Ugric word for 'mol' for blood berry, as this is the language family to which Hungarian belongs (Faust and Surányi, 1997).

Webster (1996) indicated that Albanians knew of sweet cherry before the Greek civilization. It is likely that cherries were cultivated in Greece for wood and fruit (Hedrick *et al.*, 1915; Iezzoni *et al.*, 1990; Webster, 1996). Seeds of cultivated cherries from the Roman period have been found in central Europe (Switzerland, Hungary, Austria and Germany), and a Roman mosaic from the 3rd century found in Cologne (Germany) is one of the earliest representations of cherries (Faust and Surányi, 1997). The first written references to cherry cultivation also come from the Romans, with Theoprastus in 300 BC (Brown *et al.*, 1996). Additional findings document the cultivation of cherries in central Europe (Germany, Czech Republic and Poland) in the Middle Ages up to the 15th century (Faust and Surányi, 1997). Faust and Surányi (1997) also reported fewer findings of cultivated cherries in the northern and most southern areas of Europe, indicating that the first improvement of cherries may have occurred in Middle Europe. Cherries were also commonly grown in England by the 14th century (Faust and Surányi, 1997). Cherry cultivation increased from the 16th century, with most intensity in central Europe (Watkins, 1976). Studies on sweet cherry stones from an archaeological site (Hotel-Dieu, Tours, France) suggested that different cultivars were already cultivated in the 16th century (Burger *et al.*, 2011). Some landraces are still grown today and are also used as parents contributing to modern cultivars (Iezzoni *et al.*, 1990; Brown *et al.*, 1996). Cherries were brought to America by the 19th century, and from the east coast they were taken by early pioneers to the American west (Brown *et al.*, 1996; Faust and Surányi, 1997).

Although cherries have been cultivated for more than 2000 years, cherry breeding is quite recent. Hedrick *et al.* (1915) described sweet cherry breeding starting around the early 1800s, but even today, some modern cultivars are just a few generations away from early ancestors (Iezzoni *et al.*, 1990). It is thought that the short postharvest life and difficulties transporting the cherry fruit limited its use to home and local consumption (Webster, 1996). Today, cherry breeding is carried out in many countries, and sweet cherry cultivars are continuously being released. Some early seedling selections such as 'Bing', selected by Seth Lewelling (a private nursery) in Oregon in 1875, are still cultivated and have been widely used as breeding progenitors (Brown *et al.*, 1996). Sweet cherry modern improvement began in different countries by private nurseries and institutional research stations, mostly during the 20th century. Breeding programmes were initiated in Geneva (New York, USA) by 1911, in Vineland (Ontario, Canada) by 1915 and in Summerland (British Columbia, Canada) by 1924 (Faust and Surányi, 1997). Other European breeding programmes followed, such as the programme at John Innes in England that began by 1925 (Faust and Surányi, 1997). Private breeding by Luther Burbank in California and Ivan Vladimirovich Michurin in Russia released new cultivars as early as the 1910s, and these were followed by the continuous release of new cultivars during the last century with a greater increase during the second half (see Chapter 4, this volume).

The adaptation of cherries across Europe evolved into a wide pool of genetic diversity. The development of ecotypes adapted to the different regions and the initial propagation of selected trees with distinguishing characteristics led to the preservation of this local diversity (Iezzoni *et al.*, 1990). Many of these landraces were used in breeding and are part of the local production or the breeding history of modern cultivars (Iezzoni *et al.*, 1990). The wide genetic diversity of these landraces has been reflected in diversity studies carried out in recent years using both morphological traits and molecular markers. For example, Rodrigues *et al.* (2008) and Ganopoulos *et al.* (2011) characterized Portuguese and Greek sweet cherry landraces and local cultivars, respectively. Pérez-Sánchez *et al.* (2008) morphologically evaluated a small local population in the Spanish region of Castilla y León and detected genotypes less susceptible to fruit cracking. In Chile, Joublan *et al.* (2005) evaluated old trees of the cherry cultivars introduced by settlers during the first half of the 20th century. Diversity assessments of cherry germplasm using microsatellite markers include studies from Turkey (sour cherry: Kaçar *et al.*, 2006; sweet cherry: Demir *et al.*, 2009; wild populations: Ercisli *et al.*, 2011) and Spain (Wünsch and Hormaza, 2004a), and studies of elite wild populations from Greece (Avramidou *et al.*, 2010), wild populations from Italy, Croatia and Slovenia (Guarino *et al.*, 2009), and cherry germplasm in Latvian and Swedish collections (Lacis *et al.*, 2009). More recently, a study of a medium-density single-nucleotide polymorphism (SNP) array was used to assess the diversity of the Institut National de la Recherche Agronomique (INRA) sweet cherry collection including cultivars and landraces from 16 countries (Campoy *et al.*, 2016). Other assessments of molecular diversity have been based on self-incompatibility alleles (*S*-alleles) and have included local or wild populations of cherries (Wünsch and Hormaza, 2004a; Schuster *et al.*, 2007; Stanys *et al.*, 2008; Cachi and Wünsch, 2014a). As is characteristic for plant populations, these studies consistently showed a reduction in genetic diversity from wild to landrace to modern bred sweet cherry cultivars for both microsatellites and *S*-alleles (Mariette *et al.*, 2010).

As cherry has a long juvenile period (the time from germination to first flowers), the introduction of quality traits in improved cultivars is a long process. This factor may have promoted the use of the same progenitors in breeding programmes. One example is the continuous use of the cultivar 'Stella' and its descendants in breeding for self-compatibility. As a result, the genetic diversity of sweet cherry cultivars is limited compared with the available diversity. A study of sweet cherry clones from North American breeding programmes revealed a high level of co-ancestry and inbreeding (Choi and Kappel, 2004). The finding that the paternal parent of 'Bing' is 'Napoleon' (Rosyara *et al.*, 2014) increases the inbreeding in bred cherries in North America from what was previously reported by Choi and Kappel (2004). The introduction of more exotic germplasm in newer breeding programmes, including landrace selections such as the Spanish landraces 'Cristobalina' and 'Ambrunes', can widen this gene pool (Cabrera *et al.*, 2012).

### 3.3 Preservation of Germplasm Resources

Cherries are vegetatively propagated heterozygous fruit crops; therefore, maintaining the genetic diversity as specific genotypes, such as old landrace cultivars, is more demanding than for most inbred seed-producing plants. Since cherry is native in countries from Europe to West Asia, the vast majority of conservation efforts for cherry are in these countries. In Europe, the *ex situ* and *in situ* conservation of cherry genetic resources is facilitated by the European Cooperative Programme for Plant Genetic Resources (ECPGR, http://www.ecpgr.cgiar.org/). This organization was founded in 1980 based on recommendations from the United Nations Development Programme, the Food and Agriculture Organization of the United Nations, and the GenBank Committee of the European Association for Research on Plant Breeding.

The main scope of ECPGR is to facilitate the effective long-term *ex situ* and *in situ* conservation of plant genetic resources and to promote their characterization and evaluation, and encourage their exchange and utilization. ECPGR operates through 18 Crop Working Groups (WGs) and three Thematic WGs (Maggioni and Engels, 2014). The focus of their collaborative actions has been on the documentation of the collections conserved in European countries. The *ex situ* collections are available online through the European Internet Search Catalogue (EURISCO, http://eurisco.ipk-gatersleben.de), which is a compilation of national European inventories. EURISCO currently includes 4667 sweet cherry accessions and 804 sour cherry accessions (status: 19 August 2015). In recent years, ECPGR has been putting forth efforts towards A European Genebank Integrated System (AEGIS), which is an initiative to improve coordination and share the responsibilities of the conservation, management, and access of and to genetic resources in Europe (Engels and Maggioni, 2011). Countries and institutions selected to be part of the European Collection will adhere to the AEGIS programme and assume long-term conservation responsibilities for AEGIS accessions (Maggioni and Engels, 2014). This approach offers great potential to harmonize and improve the quality of germplasm collection management and its use.

The ECPGR *Prunus* WG was established in 1983 as one of the initial six WGs. Currently, 39 countries participate in the WG activities (Benedikova and Giovannini, 2013). The objectives of the *Prunus* WG are: (i) to efficiently conserve the national European collections; (ii) to completely document passport and characterization data in the European *Prunus* Database (EPDB); and (iii) to define and establish the European *Prunus* collection according to the rules of AEGIS.

A recent survey prepared by Benedikova and Giovannini (2013) revealed that ~90% of member countries have developed a national programme coordinating the activities of conservation and management of *Prunus* genetic resources. In some countries, such as France (Balsemin *et al.*, 2005), Germany (Flachowsky and Höfer, 2010), Italy (Giovannini and Engel, 2006) and Switzerland (Kellerhals *et al.*, 1999), the conservation of fruit genetic resources is organized through a decentralized network, with central coordination. With the aim of harmonizing the documentation, characterization and evaluation activities of European Union *Prunus* collections, the *Prunus* WG agreed to develop a list including multi-crop and *Prunus*-specific passport descriptors, as well as morphological, molecular and evaluation descriptors (Zanetto *et al.*, 2002). A set of 16 unlinked, polymorphic microsatellite markers and eight reference accessions were recommended to standardize fingerprinting of ECPGR sweet cherry collections (Clarke and Tobutt, 2009); the original set is currently being revised within the *Prunus* WG in order to improve discrimination effectiveness.

To facilitate access to information and foster the use of the ECPGR collections, the *Prunus* WG has developed the EPDB, managed by INRA-Bordeaux (France) since 1993. The EPDB can host passport, characterization and evaluation data, photographs and molecular data for each accession. Currently, the database for Cherry Genetic Resources (eucherrydb) contains data for 5585 accessions, maintained by 42 institutes from 17 countries (http://www.bordeaux.inra.fr/eucherrydb/; status: 19 August 2015). Of the 5585 accessions, 3688 are sweet cherry, 1553 are sour cherry and 42 are hybrids between sweet and sour cherry (e.g. *Prunus* × *gondouinii*). The Russian Federation is also a member of the *Prunus* WG of the ECPGR; however, the Russian collections are not currently included in the database. As a result of the work of Vavilov who was responsible for the extensive and valuable collections assembled at the former All-Union Institute of Plant Industry at St Petersburg, the N.I. Vavilov Institute of Plant Genetic Resources is the central institution responsible for collecting and conserving global diversity in Russia.

In the USA, the US Department of Agriculture's Agricultural Research Service (USDA-ARS) National Plant Germplasm System has three different cherry collections. The National Clonal Germplasm Repository (NCGR) in Davis, California, was established in 1981 to preserve *Prunus* crop germplasm. It currently maintains 57 sweet cherry accessions along with some wild cherry species. The Plant Genetic Resources Unit in Geneva, New York, maintains 81 sour cherry accessions and other tetraploid cherry accessions. The ornamental cherries are held at the National Arboretum in Washington, DC. Characterization data on cherry accessions maintained in the corresponding collections can be found in the relevant agency's GRIN-Global database (http://www.grin-global.org/).

In Japan, fruit trees are conserved mainly in the Morioka Branch of the Fruit Tree Research Station (Moriguchi *et al.*, 1994). According to the website, 56 accessions of sweet and sour cherry are preserved in Japan; passport and evaluation data can be accessed at the Genebank Project, National Agriculture and Food Research Organization (NARO, http://www.gene.affrc.go.jp/databases_en.php; status: 19/08/2015).

Cherry collections are maintained in the field as active collections where the accessions are available for comprehensive characterization, evaluation and distribution. However, there are several disadvantages that limit the efficiency of active collections and threaten their security. The genetic resources are exposed to pests, diseases and natural abiotic hazards. Field genebanks require considerable input in the form of land, labour, management and materials, and therefore their capacity to ensure the maintenance of the diversity present in a species is limited (Engelmann and Engels, 2002). Additionally, backups for these plant materials are needed to provide security in case of a disease or environmental disaster. A security backup collection comprises accessions of the active collection, but at a different location (Engels and Visser, 2003). Duplication can also be provided using *in vitro* culture or cryopreservation storage (Reed *et al.*, 2004). Currently, no cherry collection is backed up with an *in vitro* collection. However, in Japan for example, the intention is to provide cryopreserved plants as backup for field genebanks for all genetic resources of vegetatively propagated crops (Okuno *et al.*, 2005). Furthermore, in the USA, the NCGR cooperates with the National Center for Genetic Resource Preservation at Fort Collins, Colorado, to develop cryogenic storage protocols for *Prunus*. At present, seven cherry accessions are preserved in the USDA-ARS collection (http://www.ars-grin.gov/npgs/, see collection at Ft Collins; status: 19 August 2015).

*In vitro* storage of *Prunus* spp. is not as well as advanced as in other plants; however, there are examples of successful storage. Cryopreservation in liquid nitrogen is the preferred option currently available for the safe and cost-efficient long-term storage of germplasm of problem species, including vegetatively propagated plants (Engelmann, 2012). This method requires a minimum of space, labour, medium and maintenance. Cryopreservation of cherry has been performed using different donor explants and different techniques. Alginate-coated shoot tips from *in vitro*-grown shoots were successfully cryopreserved following dehydration from sweet cherry (Shatnawi *et al.*, 2007). The so-called 'vitrification technique' (desiccation of the tissue to the point at which no water is available for ice formation resulting in high viscosity of cells) has been most successful and is continually being improved (Lambardi *et al.*, 2009). Successful cryopreservation of *in vitro* shoot tips has been reported by several groups (sweet cherry: Niino *et al.*, 1997; Shatnawi *et al.*, 2007; sour cherry: Barraco *et al.*, 2012). The highest recovery rate reported was 80% (Niino *et al.*, 1997). Cryopreservation could also be used for disease-tested meristem-derived cultures for maintaining pathogen-free stock (de Boucaud *et al.*, 2002). Cryopreservation has also been used successfully for wild sweet cherry embryos using a one-step freezing procedure (de Boucaud *et al.*, 2002; Grenier-de March *et al.*, 2005) and for dormant vegetative buds of sour cherry (Towill and Forsline, 1999).

Access to use of the cherry germplasm in the world's genebanks for breeding depends upon the collection. The cherry germplasm in the US National Germplasm System is freely available for research/breeding purposes as long as the phytosanitary requirements of the recipient can be met. However, access to the use of cherry germplasm from other countries can be more restricted. *Prunus* is not included in the Treaty on Plant Genetic Resources for Food and Agriculture. As a consequence, access to *Prunus* germplasm in other countries may fall under the Nagoya Protocol on Access to Genetic Resources and the Fair and Equitable Sharing of Benefits Arising from their Utilization to the Convention on Biological Diversity (http://www.cbd.int/abs/). The protocol entered into force on 12 October 2014. Details of each country's involvement in the Nagoya Protocol and the established special access legislation can be found at https://absch.cbd.int. For example, international researchers wishing to use materials in the Fruit Genebank in Dresden-Pillnitz have to sign two agreements. The first agreement is the Standard Material Transfer Agreement (SMTA) of the International Treaty (http://www.fao.org/plant-treaty). The second Agreement is to agree that the SMTA shall be applicable to the transfer of Plant Genetic Resources for Food and Agriculture other than those listed in Annex I of the International Treaty on Plant Genetic Resources for Food and Agriculture. The cherry germplasm preserved in the INRA collections is freely available as long as it is not covered by national or international protection and if the phytosanitary requirements of the recipient can be met. Germplasm queries may be made at https://urgi.versailles.inra.fr/siregal/siregal/welcome.do, where the ordering procedure is precisely explained. Recently, a core collection of INRA's sweet cherry germplasm has been selected to perform association mapping studies (Campoy *et al.*, 2016).

## 3.4 Breeding Methods

### 3.4.1 Hybridization

Cherry breeding begins with crossing in the spring to generate segregating populations, with the hybridization strategy depending on whether the maternal parent is self-compatible or self-incompatible. If the maternal parent is self-incompatible, it is not necessary to emasculate the flowers; however, the flowers need to be protected to exclude pollinators. Protective coverings such as pollination bags are slipped over the flowers to be pollinated before any of the flowers are open. As the flowers open, the bags are temporarily slipped off, pollen is applied and the bags are slipped back on. The protective coverings should be removed after petal fall and when all of the pistils have desiccated. If the maternal parent is self-compatible, the flowers need to have their anthers removed when the flowers reach the balloon stage. Emasculation is done by removing the upper part of the perianth tube by making a cut either with fingernails or notched scissors right below the anther whorl and gently pulling the whorl over the pistil. As the petals are also removed in this process, it is generally considered that the flowers are rarely visited by bees; therefore, the flowers are generally not covered because the exposed pistils are exceedingly tender. However, as much as 15% outcrossing is possible; therefore, for genetic studies it is important to confirm parentage with the use of DNA tests. An option to avoid outcrossing can be to cover the flowers after both emasculation and pollination.

Once a cross is chosen, breeders determine the direction of the cross based on bloom time, fruit set potential of the parents and knowledge of the parent's self-(in)compatibility alleles. Generally: (i) the parent that blooms later is chosen as the maternal parent to make it easier to prepare the pollen prior to the cross; (ii) the parent that is later ripening is used as the maternal parent to increase the likelihood of hybrid seed with well-developed embryos; and (iii) individuals with known low fruit set are used as the pollen parent. In addition, selections with pollen-part or stylar-part mutants may require special considerations. For example, since $S_{4'}$ is a pollen-part mutant in sweet cherry, the cross $S_3S_4 \times S_3S_{4'}$ results in seed set, as the $S_{4'}$ pollen can grow down the styles of an $S_3S_4$ plant; however, the reciprocal cross ($S_3S_{4'} \times S_3S_4$) is incompatible, resulting in no seed set.

Pollen is collected from unopened flowers that are generally at the balloon stage, right before the flowers open. The anthers are rubbed out of the flowers, either by hand or by running the flowers over a mesh wire on to a paper tray. The stamen filaments and any other floral debris are then removed with tweezers. The anthers are dried for at least 24 h at 22°C away from natural sunlight. During that time, the anther lobes will dehydrate, split open and expose the pollen. Pollen can be left drying for up to 2–3 days at moderate temperatures if the maternal parent's stigmas are not yet ready for pollination. In addition, pollen can be frozen with desiccant (anhydrous $CaSO_4$) for use in the following year. For both fresh and frozen pollen, the most important criteria is to keep the pollen dry. Checking pollen germination prior to use is a good practice, as pollen viability can vary widely between samples due to possible genotype and environmental effects. Germination can be tested by germinating pollen in a liquid medium containing 10 p.p.m. boron and 15–20% sucrose for approximately 2 h at room temperature.

Pollen should be applied to the stigma of the flowers of the maternal parent as soon as the flowers become receptive and when the temperature is over 12°C. The proper timing is identified by the presence of a sticky exudate on the stigma surface that coincides with anthesis. Pollination timing depends on the weather, but is usually 1 or 2 days after emasculation and no later than 1 day after anthesis to assure that the pollen has sufficient time to reach the ovary. Pollinating on two different days can also be beneficial to account for the slightly different stages of the flowers. Pollen can be applied to the stigmas with a glass rod, finger or brush.

Hybridizations can also be done using bumblebees, where one possibility is to isolate the maternal parent in the field using cages. In this case, the parental parent can be either a potted tree or a set of cut branches that are placed in a bucket with water. The roof of the cage has to be waterproof to avoid rain or hail, whereas the walls of the cage should preferably be built with insect-proof nets that allow air circulation. If plastic is used for the sides of the cage, temperature and humidity might become too high and *Monilinia* blossom blight will quickly develop on the flowers. The second option is to use potted trees in greenhouses. One suggestion is to use two different cultivars, and if both are self-incompatible, the fruit can be harvested from both cultivars, the progeny then being from reciprocal crosses. If a self-compatible cultivar is used, then it has to be the paternal parent. It is crucial to control temperature within the greenhouse as an excess of heat, temperatures over 30°C, will lead to heavy fruit drop.

When working with either cages or greenhouses, it is highly convenient to use large double-walled refrigerated storage where potted trees (or branches) can be stored. This allows breeders to synchronize the bloom dates between the two desired parents. If greenhouses can be compartmentalized, then breeders can conduct several different crosses each year. In addition, when using different cultivars that are self-incompatible together, with one cultivar that can pollinate them, one can simultaneously realize different cross combinations. A last option that has not yet been widely used in cherries is the poly-cross, where several self-compatible cultivars are crossed altogether.

A strategy for obtaining a large number of hybrids from the cross between two cherry cultivars, and which does not involve any artificial pollination activities, is to collect fruit from trees in relatively well-isolated orchards, in which only one cultivar with another pollinator cultivar are planted. This approach requires collaboration with cherry growers or experimental stations, and the number of possible combinations is relatively limited. It has nevertheless already been used by breeders in France and Chile (J. Quero-García, Bordeaux, France, 2016, personal communication). Finally, in cherry breeding, open pollination has also been widely used by selecting only the maternal parents, based on their phenotypic characteristics or on their value as genitors if already known. Many successful cherry cultivars have been created with this method (see Chapters 4 and 5, this volume).

### 3.4.2 Seed anatomy and seed extraction

The outermost surrounding of the sweet and sour cherry seed is the endocarp, called the stone or pit, which consists of two carpels glued together with cellulose and hemicellulose. Inside the endocarp is the testa, or true seedcoat, which is a very thin layer of maternally derived tissue that turns from white to brown during late seed maturation. The testa covers an almost invisible remnant layer of endosperm that is consumed almost entirely by the embryo during seed development to fuel the growth of the cotyledons. Small parts of the endosperm are most easily seen near the radicle tip, sometimes adhering to the testa tissue. The embryo itself consists of a radicle (primary root), embryo axe, cotyledons and the epicotyl bud.

Fruit for seed extraction is collected just prior to optimal harvest maturity; however, normal developing sweet cherry seeds have been shown to germinate successfully even if harvested several weeks before optimal fruit maturation (Jensen and Eriksen, 2001). Cherry fruit easily starts fermenting, which will damage seeds, and it is therefore recommended to extract the stones immediately after harvest. Stones are extracted from the fruit manually or by fruit maceration and seed-washing machines. After removal of all the flesh, the stones can be sterilized by soaking them in a dilute solution of chlorine and bleach for 2–5 min and/or a fungicide solution for 1 h followed by a rinse in distilled water. Empty stones (with no embryo) can generally be separated and therefore discarded by floating freshly harvested moist seeds in water, as empty stones will normally float whereas good seeds will sink.

### 3.4.3 Breaking seed dormancy

All parts of the seed (endocarp, testa, endosperm and embryo) contribute to dormancy. Embryo dormancy is reported to be induced several weeks before full fruit maturity and just before the immature seed acquires the ability to germinate (Jensen and Eriksen, 2001). The cherry seed in an intact stone is deeply dormant when fully mature and will only be able to germinate if the endocarp is removed or split open and the seed and embryo is then exposed to moist chilling. This moist chilling triggers cold-sensitive gene expression and will eventually establish a hormonal balance and hormonal sensitivity that will allow the embryo to germinate. In nature, dispersed moist stones are exposed to 1–3 months of late summer temperatures and to microorganisms present in the environment. These microorganisms produce enzymes that degrade the cellulose between the carpels, and the stone cracks open allowing water and air exchange with the surroundings.

Then, in winter, the moist seeds are exposed to a minimum of 3–4 months of chilling at temperatures between 1 and 6°C. These temperatures break dormancy and allow the embryo to germinate and grow.

### 3.4.4 Germination of large quantities of seeds

Where immediate germination is not needed, seeds may be dried in trays at room temperature for about a week and then stored for a short time at 6–8% moisture content at room temperature. For longer storage, temperatures from –10 to –18°C in sealed airtight containers are preferred. However, immature seeds may not survive desiccation well and therefore they should be stratified and germinated immediately.

An intact endocarp imposes an exogenous dormancy mechanism that needs to be broken before chilling exposure can occur. Manual removal of the endocarp may only be cost-efficient for a small number of seeds. In breeding programmes with many thousands of seeds per year, a non-sterile warm stratification method may be the most cost-efficient way to overcome the endocarp dormancy. Stones previously soaked in cold running tap water for 24 h and then drained are mixed with moist sand, vermiculite or peat moss, and warm stratified for 2–6 weeks at 20°C, and occasionally up to 12 weeks, which will allow all stones to split the endocarp and thus become receptive for chilling. Good air exchange should be ensured during warm stratification by holding the seeds in open boxes and turning/mixing the seeds weekly. Within breeding programmes, however, seeds from certain cherry cultivars and other parents are known to have good germination without the need for a warm stratification period before the chilling treatment. The reason for this is not known.

Chilling of seeds in the same medium or without medium is most efficient at temperatures of 3–5°C and will normally take 12–16 weeks before seeds start to germinate in the cold. Chilling-mediated dormancy breakage of endogenous dormancy can only occur if the stone moisture content is above ~30% of fresh weight. Once germination begins, all the seeds in the seed lot can be sown and preferably placed at a low temperature (not exceeding 15°C) until all seeds have germinated, since there is a risk of inducing secondary dormancy in non-germinated seeds if exposed to high temperatures (20–25°C). These methods generally result in a high germination percentage in normal well-developed seeds and are equal to methods currently used in seedling nurseries (Suszka *et al.*, 1996; Jensen and Eriksen, 2001; Iliev *et al.*, 2012).

### 3.4.5 Controlled germination of smaller quantities of seeds

For smaller quantities of seeds, manual removal of the endocarp is recommended using a vice or knife. This can be done on moist seeds just after harvest, but also on dry seeds that have been stored. The naked seeds with the testa still attached are then sterilized in a 1–2% sodium hypochlorite (NaOCl) solution for 10 min, followed by rinsing three times in pure water (Jensen and Eriksen, 2001; Jensen and Kristiansen, 2009). The bleaching will change the colour of the testa to almost white, which, by experience, has been shown to support a high germination percentage. The seeds may be treated with a mild fungicide if necessary. The seeds are then placed on top of moist paper in germination boxes or stored in bags of moist sterile vermiculite, sand or sphagnum moss, and chilled at 3–5°C. After 3–4 months, the seeds are checked periodically for germination, and seedlings are transplanted to the greenhouse when the radicle is 1 cm in length. This method eliminates the need for warm stratification and often gives slightly better final survival percentages than using warm stratification and nursery methods.

This method may also be used on isolated normally developed embryos where the testa has been removed. However, since naked embryos are more susceptible to fungus attack, the risk of losing embryos is considered higher without the testa. Moist seeds with testa can be soaked in water for 2–4 h to allow the testa to swell up and loosen from the embryo. This makes it easier to remove the testa by making a small scar in the testa at the distal end of the cotyledons and then putting gentle pressure on the seed between two fingers from the radicle end to slide off the testa while avoiding damage to the embryo. Removal of the testa normally shortens the chilling requirement of the embryo. Additionally, as the testa seems to be somewhat semi-impermeable, if externally applied hormones are used to enhance embryo germination, removing the testa often results in better effects.

### 3.4.6 Seed germination via embryo culture

Embryo culture can be used to obtain plants from cherry seeds that may be abnormal or would abort due to insufficient embryo growth during seed development, or seeds from difficult interspecific or different ploidy level crosses that display problems in survival and germination. Early-maturing cultivars (in particular those with a ripening time similar to or earlier than 'Burlat') are more likely to have seeds that exhibit insufficient embryo growth resulting in little or very poor germination (Tukey, 1933a). In this case, the embryos are harvested very immature, from as early as 21 days after full bloom (Fathi *et al.*, 2009) and before they abort (Tukey, 1933b). To obtain seedlings from these undeveloped seeds, the naked embryos are excised, surface sterilized only if necessary due to the risk of damage, and cultured on nutrient agar media. Nutrient agar media for cherries currently used at INRA-Bordeaux include the macroelements: 400 mg $l^{-1}$ $NH_4NO_3$, 1800 mg $l^{-1}$ $KNO_3$, 1200 mg $l^{-1}$ $Ca(NO_3).4H_2O$, 360 mg $l^{-1}$ $MgSO_4.7H_2O$ and 270 mg $l^{-1}$ $KH_2PO_4$; and the microelements: 1 mg $l^{-1}$ $ZnSO_4.7H_2O$, 1 mg $l^{-1}$ $H_3BO_3$, 0.1 mg $l^{-1}$ $MnSO_4.4H_2O$, 0.03 mg $l^{-1}$ $AlCl_3$, 0.03 mg $l^{-1}$ $NiCl_2$, 0.01 mg $l^{-1}$ KI and 1 mg $l^{-1}$ $FeCl_3.6H_2O$ (J. Quero-García, Bordeaux, France, 2016, personal communication). In the very young heterotrophic embryo stage, hormones seem not to be important, but use of 'nursing endosperm' from normal seeds with energy, nutrients and normal ovule osmotic environment is highly important (Fathi and Jahani, 2012). Such young embryos may sometimes be more safely rescued by culturing the entire ovule. In the later autotrophic stage, requirements are less complex, and energy becomes more important for embryo development. Very young embryos (not dormant) do not normally benefit from chilling, whereas more mature embryos are chilled at 5°C for 2–4 months in the dark to overcome dormancy (Bassi *et al.*, 1984, cited in Balla and Brozik, 1996; Bargioni, 1996). Following this, they are transferred to 16 h/8 h light/dark photoperiod and 2000 Lux intensity (Balla, 2012). Embryo culture is most successful if the embryos have a length of at least 3–4 mm (Ivanicka and Pretova, 1986), but even embryos of 1 mm can be cultured with success. In general, these seeds are of very poor quality and thus the success rate is extremely variable (Balla and Brozik, 1996). Dormant embryos sometimes germinate at 20°C, but will show abnormal development with stunted dwarf-like and rosette-like growth, and white leafspots or whole leaves due to lack of chlorophyll production (Jensen and Kristiansen, 2009). These abnormalities are normally overcome by chilling plants for 2–3 months.

### 3.4.7 Embryo rescue without germination: *in vitro* shoot culture and somatic embryogenesis

In some cases, it is not possible to obtain germination of embryos; however, rescuing seedlings that may be very early maturing, for example, would be extremely valuable. In such cases, these seedlings can be rescued using *in vitro* shoot culture where adventitious shoots are induced from cotyledon tissue (Schmidt and Ketzel, 1993; de Rogatis and Fabbri, 1997). These cotyledon-derived shoots may be rooted on a different medium and later transferred to the greenhouse. Sometimes this may involve de-differentiation of tissue into a callus culture before reaching a shoot culture. Using the right hormones in balanced concentrations plays a crucial role in successful cultivation. By using different hormone profiles, it is possible to obtain somatic embryogenesis from cotyledons or embryo axes tissue. Somatic embryos may be germinated *in vitro* and give rise to new plants (Tang *et al.*, 2000; Gutierrez and Rugini, 2004). Embryo culture and rescue techniques are quite expensive procedures and are only relevant in very special cases.

### 3.4.8 Planting in soil

Seedling vigour and survival is maximized if the seedlings are not removed from stratification until the radical reaches 0.5–1 cm in length. As the seedlings grow very weakly, they should be planted with just a light covering layer of soil mix or vermiculite. Commercial trays or plant banks that are ~6 cm in diameter and 25 cm deep are ideal for initial planting. This container maximizes the space for root growth and minimizes the possibility that the seedlings will be overwatered, as excess moisture can quickly lead to seedling death. This initial planting container requires that the seedling be repotted or planted into a seedling nursery prior to field planting. After field planting, sour and sweet cherry seedlings typically begin flowering in years 3 and 4, respectively.

### 3.4.9 Rapid cycling

A rapid cycling procedure was developed and demonstrated in sour cherry that can shorten the time from cross-pollination and seed set to flowering in the progeny (Jensen and Kristiansen, 2009). The first step in this procedure reduces the time needed for dormancy breakage and germination, and two methods were developed. In the first method, the stones are extracted right after normal harvest and the endocarp and testa are removed under aseptic conditions. The distal two-thirds of the cotyledons are then removed with a sterilized razor blade, which removes the germination inhibitors located in the cotyledons and allows the seed to germinate and grow normally. The embryos are placed on moist filter paper in germination boxes where moisture is supplied by filter paper wicks that are in contact with a water reservoir. The seeds are then sprayed with a fungicide solution to prevent fungal growth and the boxes are held at 20°C and at 12 h photoperiod at 30 μmol $m^{-2}$ $s^{-1}$ white fluorescent light. Using this procedure, most of the sour cherry seeds that germinate do so within 2 weeks. Within 4 weeks, a maximum percentage germination of 90% for shallowly dormant seeds and 65% for deeply dormant embryos were recorded (Jensen and Kristiansen, 2009). When the radicles were 1 cm long, the seedlings were transplanted into plugs with peat moss and moved to the greenhouse. All the seedlings obtained grew normally. Similar results were reported by Szymajda and Zurawicz (2014) using the same protocol. In the second method for accelerating germination, isolated moist embryos from deeply dormant seeds are soaked in 10–200 mg $l^{-1}$ of the cytokinin 6-benzylaminopurine for 30 min and then germinated on top of paper in germination boxes held at 20°C under the same conditions as above. After 3 weeks of this procedure, seeds exhibited 70–80% germination and plants grew normally. Applications of gibberellic acid (GA3) and cytokinin have previously been shown to induce fast germination by overcoming dormancy in some *Prunus* spp. (Lin and Boe, 1972; Abou-Zeid *et al.*, 1977). The effects of GA3 and cytokinin are most easily obtained in excised embryos without a testa (Lin and Boe, 1972).

The second step in the rapid cycling protocol is to overcome juvenility and obtain flowering and fruiting as fast as possible. A large tree size can be obtained in a short time by applying the optimal environmental conditions irrespective of the time of year. Germinated seedlings can be grown in a greenhouse or growth room for the first 8–12 months at 20°C with a 20 h day length with optimal irrigation and fertilization, which keeps the plant actively growing year round. This growing regime can be used for seedlings obtained by fast germination methods in late July to the beginning of

August (1 month after fruit harvest) and with seedlings obtained as early as October after normal seed chilling. In May, the fast-tracked trees may reach 1.5–2 m height with some side branches. Seedlings can be planted in the field in the spring when the risk of spring frost is past. If transplant shock is avoided, the seedlings will grow during the summer and produce a small tree during this first year in the field. Only a few trees flower in this first field year; however, in the second year in the field, significant flowering and fruiting occur (Jensen, 2012). Thus, a protocol for rapid cycling could be performed in 27 months from cross-pollination in the spring of year 1 to harvest of a substantial amount of fruit in the summer of year 3. This strategy has so far only been tried in sour cherry and not in sweet cherry.

### 3.4.10 Following field planting

Superior seedlings are selected from segregating populations. Phenotyping protocols for sweet and sour cherry are generally breeding programme specific (see Chapters 4 and 5, this volume); however, published protocols can be used as points of reference (sweet cherry: Chavoshi *et al.*, 2014; sour cherry: Stegmeir *et al.*, 2014b). In sweet cherry, the most important traits for a new cultivar include desired maturity time and fruit size and quality. For sour cherry, yield, fruit quality (including pit characteristics such as shape and freestone/clingstone) and disease susceptibility are most important. Spraying young seedlings with inoculum of *Blumeriella jaapii*, *Monilinia laxa*, or with other fungi or bacteria in tunnels with high humidity or in the orchard can be done to obtain a fast response for early selection of the most resistant and tolerant progeny (Szôdi *et al.*, 2008; Schuster, 2013; Szügyi *et al.*, 2014). Evaluation of seedling resistance or tolerance to diseases over several years is also done under non-sprayed orchard conditions where inplanted susceptible cultivars ensure the presence of inoculum. Laboratory methods using artificial inoculum of *B. jaapii* on detached leaves have also been developed (Wharton and Iezzoni, 2005; Schuster, 2013). Ensuring the presence of inoculum and the optimal humid environment for getting infections is critical to obtain an accurate evaluation. Selection for frost hardiness may be done in the field over several years, but results depend on the weather conditions. Alternatively, laboratory methods for testing frost tolerance of cherry seedlings have been developed and may provide a comparison of relative winter hardiness in autumn, midwinter and spring (Liu *et al.*, 2012; Jensen and Kristiansen, 2014) (see Chapter 8, this volume). These superior selections are then propagated on one or multiple rootstocks to make trees available for field testing at multiple locations.

## 3.5 DNA-informed Breeding

### 3.5.1 Quantitative trait loci

In order to implement marker-assisted selection strategies in sweet or sour cherry, prior knowledge of linkage relations between marker loci and important agronomic traits is required. The first methodology developed in the 1980s that enabled marker-assisted selection was based on the use of genetic maps to identify chromosome regions containing genes controlling both qualitative and quantitative traits. For the latter, the term quantitative trait locus (QTL) is generally used. In sweet cherry, contrary to other *Prunus* species such as peach, most of the traits of agronomic interest are quantitatively inherited.

A recent review summarized the construction of linkage maps and the main QTL studies conducted on *Prunus* spp. (Salazar *et al.*, 2014). Several linkage maps were built in the subgenus *Cerasus* involving inter-specific crosses (Boskovic and Tobutt, 1998; Clarke *et al.*, 2009). However, QTL detection analyses were only initiated by using sweet cherry intraspecific genetic maps. Dirlewanger *et al.* (2004) developed a map from the cross between modern cultivars ‘Regina’ and ‘Lapins’, whereas Olmstead *et al.* (2008) built a map from reciprocal crosses between the great-grandparent of 'Lapins' ('Emperor Francis' (EF)) and a mazzard (wild forest cherry) accession, 'NY 54'. Both maps contained mainly microsatellite or simple sequence repeat (SSR) markers. Subsequently, a consensus sweet cherry map was constructed using Rosaceae Conserved Orthologous Set (RosCOS) SNP and SSR markers, by working with four populations: 'Regina' × 'Lapins', 'NY 54' × EF, 'Namati' × 'Summit', and 'Namati' × 'Krupnoplodnaya' (Cabrera *et al.*, 2012). With the advent of next-generation sequencing technologies, high-throughput SNP genotyping technologies have become available for sweet cherry. Klagges *et al.* (2013) developed the first sweet cherry SNP maps using a cherry 6K SNP array developed within the US RosBREED project (Peace *et al.*, 2012) using two unrelated populations: 'Black Tartarian' × 'Kordia' (BT×K) and 'Regina' × 'Lapins' (R×L). As a result, 723 and 687 markers were mapped into eight linkage groups (LGs) in BT×K and R×L, respectively. The genetic maps were highly saturated, spanning 752.9 and 639.9 cM with an average distance of 1.1 and 0.9 cM between adjacent markers in BT×K and R×L, respectively. More recently, Guajardo *et al.* (2015) reported the construction of another sweet cherry intraspecific linkage map using SSR and SNP markers detected by genotyping-by-sequencing. The cross used was 'Rainier' × 'Rivedel' and the consensus map spanned 731.3 cM with an average distance between markers of 0.7 cM. Finally, Wang *et al.* (2015) reported a genetic map built using a cross between the cultivars 'Wanhongzhu' and 'Lapins'. The authors used the specific-locus amplified fragment sequencing technique to generate 701 informative SNPs. The consensus map spanned 849 cM with a mean intermarker distance of 1.18 cM. In sour cherry, the first linkage maps were constructed from 86 individuals derived from the cross of cultivars 'Rheinische Schattenmorelle' (RS) × 'Érdi Bőtermő' (EB) (Wang *et al.*, 1998). Since sour cherry is a tetraploid, informative restriction fragment length polymorphisms (RFLPs) were scored as single-dose restriction fragments (SDRFs). The RS genetic map consisted of 126 SDRF markers assigned to 19 LGs covering 461 cM, whereas the EB linkage map had 95 SDRF markers assigned to 16 LGs covering 279 cM. Due to the limited number of shared markers between these maps and other reference *Prunus* maps, putative homologous LGs could only be identified for the *Prunus* LG2, -4, -6 and -7. The other LGs were arbitrarily numbered from longest to shortest. The difficulty of identifying SDRFs and eliminating progeny resulting from non-homologous pairing for the LG under study illustrates the complexity of linkage mapping in a segmental allopolyploid such as sour cherry.

QTL detection analyses have been conducted mainly on traits related either to phenology, such as bloom time and maturity date, or to fruit quality, such as fruit weight and firmness, and fruit skin colour. In sour cherry, two bloom time QTLs, *blm 1* and *blm2*, were detected in the RS×EB population; unfortunately the genetic effects of these two QTL alleles from EB were to induce early bloom (Wang *et al.*, 2000). More recently, Dirlewanger *et al.* (2012) performed a QTL analysis for bloom time and maturity date in mapping progenies of sweet cherry (R×L), peach and apricot. For bloom time, major QTLs were detected on LG4 for apricot and sweet cherry and on LG6 for peach, whereas for maturity date, a major QTL was detected on LG4 for all three species. Castède *et al.* (2014) performed a 3-year study on sweet cherry, by dissecting bloom time into chilling and heat requirements. A second mapping population derived from the modern cultivars 'Regina' × 'Garnet' (R×G) was used and this study was complemented with a QTL analysis of bloom time performed over 5 years in R×G and 6 years in the aforementioned mapping population R×L. One stable QTL with a major effect was detected for chilling requirement and bloom date in the same region as LG4. For heat requirement, no stable QTL was detected. Candidate genes underlying the major QTL on LG4 were investigated and key genes were identified for chilling requirement and bloom date. These results provided a foundation for the identification of genes involved in chilling requirement and bloom date in sweet cherry, which could be used to develop ideotypes adapted to future climatic conditions (see Chapter 8, this volume).

Regarding fruit traits, Zhang *et al.* (2010), working with EF × 'NY 54' and 'NY 54' × EF populations, investigated QTLs for fruit weight, length and diameter; mesocarp cell number and length; and pit length and diameter. Fruit size QTLs were found on LG2 on the EF map (EF 2) and LG-2 and -6 on the 'NY 54' map (NY 2 and NY 6). On NY LG6, pit length and diameter QTLs clustered with those for fruit size, suggesting that the underlying morphological basis of this QTL is the difference in pit size. On EF LG2, a cell number QTL clustered with the fruit size QTL, suggesting that increased fruit size is associated with an increase in mesocarp cell number. Based on these results, de Franceschi *et al.* (2013) performed a study of cell number regulator (*CNR*) genes in *Prunus*, providing new evidence for the control of fruit size in sweet and sour cherry. Two of these *CNR* genes were located within the confidence intervals of the major QTL previously discovered on LG2 and -6 in sweet cherry, named *PavCNR12* and *PavCNR20*, respectively. Rosyara *et al.* (2013) detected new regions of the genome controlling fruit weight in sweet cherry using a five-generation pedigree consisting of 23 founders and parents and 424 progeny individuals from the four full-sibling families that had previously been used for the construction of a consensus map (Cabrera *et al.*, 2012). A Bayesian approach implemented in FlexQTL™ software was used (Bink *et al.*, 2002). Six QTLs were identified: three on LG2 with one each on LG1, -3 and -6. Of these QTLs, the second QTL on LG2 and the QTL on LG6 had been discovered previously (Zhang *et al.*, 2010), while the other QTLs were novel. Finally, Campoy *et al.* (2015) conducted QTL analyses on the populations R×L and R×G for fruit weight and fruit firmness. A new major fruit weight QTL was detected at the bottom of LG5 in the 'Regina' parent. An interesting result was the significant and negative correlation between fruit weight and fruit firmness, observed in both R×L and R×G progenies. Hence, numerous QTL co-localizations were found for both traits. This result is important for breeders and shows the difficulty in selecting simultaneously for large and firm fruit when working with certain genetic backgrounds.

Sooriyapathirana *et al.* (2010), working on the EF × 'NY 54' and 'NY 54' × EF populations, found a major QTL on LG3 for fruit skin and flesh coloration in sweet cherry. A candidate gene, *PavMYB10*, homologous to apple *MdMYB10* and *Arabidopsis AtPAP1*, was suggested to be the major determinant of these traits. Other important agronomic traits, such as rain-induced fruit cracking (Quero-García *et al.*, 2014), productivity, sugar content and acidity (J. Quero-García, Bordeaux, France, 2016, personal communication) are under investigation, and phenotyping has been conducted over several years for these traits on the progeny of R×L and R×G crosses. These QTLs could be incorporated relatively soon into marker-assisted selection approaches.

To date, the only QTL reported for disease resistance is for cherry leaf spot, a major disease on sour cherry in all humid growing regions. Cherry leaf spot is caused by the fungal pathogen *B. jaapii* and can result in severe premature defoliation if not controlled by numerous fungicide applications. The wild cherry species *P. canescens* and hybrids selected from *P. canesens* are also resistant, indicating that the alleles that confer resistance have a dominant gene action. One of the major QTLs controlling *P. canescens*-derived cherry leaf spot resistance maps to LG4, *CLSR_G4* (Stegmeir *et al.*, 2014a). In studies in both sweet and sour cherry, all resistant progeny had the *P. canescens* alleles for that locus; however, presence of this allele alone did not confer resistance.

### 3.5.2 Diagnostic DNA tests

Diagnostic DNA tests are being developed and utilized for loci and major QTLs identified in cherry. Use of this information can increase breeding efficiency by: (i) providing information about new germplasm; (ii) predicting the best parents to select and which crosses to make; and (iii) discarding undesirable seedlings before field planting

(Dirlewanger *et al.*, 2009). However, developing a predictive test from a known causal gene or QTL is time consuming, as new DNA markers frequently need to be designed that uniquely distinguish among the available alleles. As a result, the number of DNA tests that are currently available is minimal; however, the number is expected to grow rapidly in the coming years. Below, we describe the currently available diagnostic DNA tests.

### *S-(in)compatibility loci*

Self-incompatibility (SI) is an undesirable trait in sweet and sour cherry production (see Chapter 2, this volume, for further details of aspects described in this section). SI prevents sweet and sour cherry cultivars from self-fertilization, and pollinator trees need to be included to ensure fruit set. To avoid the use of pollinator trees, self-compatible (SC) cultivars are often preferred in cherry production. DNA markers for self-(in)compatibility are based on known base pairs and insertion/deletion differences in the sequences of the two *S*-locus genes, the stylar *S-RNase* and the pollen *S*-locus F-box gene (*SFB*). In cherry breeding, the DNA markers for self-(in)compatibility can be used to determine if a cross is compatible, to test seedling parentage in cases where contamination may have occurred during crossing and to select for self-fertility at the seedling stage.

In sweet cherry, cross-compatible cultivars have been identified and assigned to different incompatibility groups. Incompatibility group assignment is most easily done using DNA markers for the *S*-alleles (*S*-genotyping) and reports of the *S*-genotypes of many sweet cherry cultivars have been published (Schuster, 2012). In addition to the cataloguing of the incompatibility groups, a group of self-fertile cultivars, termed 'universal donors', has also been identified.

Breeding for SC is a priority in sweet cherry breeding programmes, and marker-assisted selection for SC can be carried out using DNA markers. Depending on the genetic nature of SC in each parental cultivar, different molecular markers have been developed for selection. The predominant SC mutation used is the self-fertile pollen-part mutant $S_{4'}$. This mutation was induced by radiation (Lewis, 1949). The $S_{4'}$ is widespread in sweet cherry cultivars due to the common ancestry of the SC cultivar 'Stella', which derives the mutant allele from JI2420 (Lapins, 1975). Specific molecular DNA markers for $S_{4'}$ detection take advantage of the 4 bp deletion in the $S_{4'}$ *SFB* (Sonneveld *et al.*, 2005). Ikeda *et al.* (2004) developed a DNA marker to detect the $S_{4'}$ mutation by polymerase chain reaction (PCR) followed by polyacrylamide gel electrophoresis, or by PCR followed by restriction digestion (known as a derived cleaved amplified polymorphic sequences assay or dCAPS). This assay allows distinction between the wild-type SI $S_4$ and the mutated SC $S_{4'}$ haplotype. Zhu *et al.* (2004) also developed an efficient dominant DNA marker for $S_{4'}$ selection. In this case, two rounds of PCR (nested PCR) allow detection of $S_{4'}$ but not $S_4$. The other SC mutation induced by radiation in sweet cherry is found in accession JI2434 (Sonneveld *et al.*, 2005). This ancestor has been used to a lesser extent in sweet cherry breeding and is found in fewer cultivars (Schuster, 2012). This SC mutation, $S_{3'}$, is due to the complete deletion of the *S*-locus gene $SFB_3$. Progeny derived from JI2434 can be selected by detecting this *SFB* deletion through RFLP or PCR (Sonneveld *et al.*, 2005). Other sources of SC in sweet cherry are the local Italian sweet cherry cultivar 'Kronio' (Calabrese *et al.*, 1984) and 'Cristobalina' or 'Talegal Ahin' from Spain (Wünsch and Hormaza, 2004b; Cachi and Wünsch, 2014a). In 'Kronio', the SC mutation is named $S_{5'}$ due to a premature stop codon in $S_5$ *SFB* (Marchese *et al.*, 2007). A microsatellite found in $S_5$-*RNase* showing length polymorphism between $S_5$- and $S_{5'}$-*RNase* can be used as a marker to distinguish both haplotypes and to select for SC (Marchese *et al.*, 2007). In 'Cristobalina' and 'Talegal Ahin', SC is due to a mutation in gene(s) unlinked to the *S*-locus, and molecular markers (e.g. SSRs) linked to the trait can be used for selection of this trait (Cachi and Wünsch, 2011; Cachi and Wünsch, 2014b). The 142 bp allele of the SSR EMPaS02 (Vaughan and Russell, 2004) can be used to select SC offspring from

'Cristobalina' with high efficiency (Cachi and Wünsch, 2011).

In sour cherry, most cultivars are SC, but SI also exists. Sour cherry cultivation requires SC cultivars, and thus breeding for SC is a priority (Sebolt *et al.*, 2017). As sour cherry is a tetraploid and pollen grains contain two *S*-alleles, the genetics of SI/SC is different from that in sweet cherry. In sour cherry, SC is due to the presence of mutations in the *S*-locus genes, which result in non-functional *S*-haplotypes (Hauck *et al.*, 2006). The 'one-allele-match' model predicts that if one fully functional *S*-allele in the pollen matches one functional *S*-haplotype in the style, the pollen will be SI (Hauck *et al.*, 2006). In sour cherry breeding, it is necessary to identify SI and SC cultivars to select SC progeny (Tsukamoto *et al.*, 2008). For this purpose, *S*-genotyping and determining the presence of pollen or style mutations of the specific *S*-haplotypes is needed.

Molecular *S*-genotyping of sour cherry cultivars is carried out as in sweet cherry by targeting the *S-RNase* and *SFB*. However, *S*-genotyping in sour cherry is more complicated than in sweet cherry because it involves distinguishing among as many as three variants for a single ancestral *S*-allele haplotype. For example, in sour cherry, there is a wild-type functional $S_{13}$ allele, but also a non-functional pollen-part mutant of this haplotype, $S_{13'}$, and a non-functional stylar-part mutant of this haplotype, $S_{13m}$ (Sebolt *et al.*, 2017). Therefore, the discrimination of certain *S*-haplotypes requires using allele-specific markers from sweet or sour cherry (Sebolt *et al.*, 2017). Once the *S*-haplotypes present in each selection are identified, additional tests may be required for detection of the non-functional mutants for that haplotype. This can be done using dCAPS markers designed specifically for each mutation (Tsukamoto *et al.*, 2008; Sebolt *et al.*, 2017).

### Fruit size

Fruit size is one of the most important attributes for a commercial sweet cherry cultivar and is an important consideration for breeders who are introgressing traits such as disease resistance from small-fruited wild relatives. In sweet cherry, a major QTL for fruit size was identified on LG2 (Zhang *et al.*, 2010). The small-fruited allele for this locus is from wild sweet cherry, suggesting that this locus may be associated with domestication. This locus is flanked by two highly polymorphic SSR markers (CPSCT038 and BPPCT034), and the alleles for these two SSRs are used to define the QTL haplotypes (Zhang *et al.*, 2010). The notation for these haplotypes lists the SSR allele size for CPSCT038 followed by the SSR allele size for BPPCT034, and so far seven and four alleles, respectively, have been identified. Subsequent genetic studies suggest that there may be more than one fruit size QTL in this SSR-flanked region (Rosyara *et al.*, 2013), with one of the causal genes being a *CNR* gene (de Franceschi *et al.*, 2013).

Several of the LG2 haplotypes are significantly associated with small fruit size, and therefore individuals with these alleles are unlikely to have fruit size sufficient for a commercial sweet cherry cultivar. For example, if parents are used that have a small-fruited haplotype (such as CPSCT038-192 and BPPCT034-225), marker-assisted seedling selection to identify these seedlings and eliminate them prior to field planting would provide significant cost savings to the breeding programme. In contrast, several haplotypes are significantly associated with large fruit size. These include CPPCT038-190 and BPPCT034-255 (Rosyara *et al.*, 2013). A QTL for fruit firmness also maps to this LG2 region (Campoy *et al.*, 2015). Therefore, knowledge of the haplotype's effect on both fruit size and firmness is important.

The SSR markers that are used to define the LG2 haplotype are approximately 16 cM apart (Klagges *et al.*, 2013). Therefore, when genotyping for this region, approximately one-sixth of the seedlings will have inherited a recombinant haplotype. Recombination in the LG2 region has contributed to the modern cultivars grown today. For example, an LG2 recombinant gamete from 'Lambert' gave rise to 'Stella', and a recombinant gamete from 'Van' gave rise to 'Lapins' (Rosyara *et al.*, 2013). Future fine-mapping studies are needed to locate more precisely the loci controlling fruit size and firmness in this 16 cM region, followed by the development of additional DNA markers that span this region at smaller centiMorgan intervals.

### *Skin and flesh colour*

Cherry skin and flesh colour can define market classes for sweet cherry, and differentiate cherries for processing. A major QTL for cherry skin and flesh colour was identified that co-locates with the anthocyanin transcription factor (*PavMYB10*) on sweet cherry LG3 (Sooriyapathirana *et al.*, 2010). This QTL behaves as a single locus, with mahogany fruit colour (such as 'Bing') dominate to blush colour fruit (such as 'Rainier'). Other minor modifier loci are located on LG6 and -8 (Sooriyapathirana *et al.*, 2010).

In sour cherry, SNP haplotypes built from SNPs genotyped using the 6K Illumina® II SNP array (Peace *et al.*, 2012) have been used to define 13 different haplotypes spanning the *MYB10* region (Stegmeir *et al.*, 2015). An SSR marker was developed that uniquely identifies the *D1* haplotype that is significantly associated with dark purple flesh colour in sour cherry. This marker can be used to select for dark purple colour or for lighter flesh colour similar to the 'Montmorency' bright red skin colour. In sweet cherry, a different SSR marker, Pav-Rf-SSR, was designed for the *PavMYB10* region and can be used to predict whether a selection has mahogany- or blush-coloured fruit (Sandefur *et al.*, 2016).

### *Cherry leaf spot resistance*

A genetic test has been developed for the major QTL controlling *P. canescens*-derived cherry leaf spot resistance (CLSR_G4) (Stegmeir *et al.*, 2014a). Four SSR markers were developed that tag the *P. canescens* allele for the CLSR_G4 (CLS004, CLS005, CLS026 and CLS028). Any one of these markers can be used for marker-assisted seedling selection for cherry leaf spot resistance in progeny derived from *P. canescens*, as progeny not having this allele are predicted to be susceptible and therefore can be discarded prior to field planting.

## References


Abou-Zeid, A., Hedtrich, C.M. and Neumann, K.H. (1977) Untersuchungen zum Nachreifeprozess bei Samen von *Prunus avium*. *Angewandte Botanik* 51, 37–45.

Avramidou, E., Ganopoulos, I.V. and Aravanopoulos, F.A. (2010) DNA fingerprinting of elite Greek wild cherry (*Prunus avium* L.) genotypes using microsatellite markers. *Forestry* 83, 527–533.

Balla, I. (2012) Embryo rescue of early ripening sweet and sour cherry hybrids. In: *Proceedings of the EU COST Action FA1104: Sustainable Production of High-quality Cherries for the European Market.* COST meeting, Palermo, Italy, November 2012. Available at: http://www.bordeaux.inra.fr/cherry/docs/dossiers/Activities/Meetings/21-23_11_2012_WG_and_2nd_MC_Meeting_Palermo/Day_2/presentations/WG1/Balla_Palermo_06.pdf (accessed 29 January 2016).

Balla, I. and Brozik, S. (1996) Embryo culture of sweet cherry hybrids. *Acta Horticulturae* 410, 385–386.

Balsemin, E., Christmann, H. and Barreneche, T. (2005) Un réseau de partenaires pour la sauvegarde du patrimoine de fruits à noyau (*Prunus*). *Fruits Oublies* 36, 7–8.

Bargioni, G. (1996) Sweet cherry scions. Characteristics of the principal commercial cultivars, breeding objectives and methods. In: Webster, A.D. and Looney, N.E. (eds) *Cherries: Crop Physiology, Production and Uses.* CAB International, Wallingford, UK, pp. 73–112.

Barraco, G., Chatelet, Ph., Balsemin, E., Decourcelle, T., Sylvestre, I. and Engelmann, F. (2012) Cryopreservation of *Prunus cerasus* through vitrification and replacement of cold hardening with preculture on medium enriched with sucrose and/or glycerol. *Scientia Horticulturae* 148, 104–108.

Bassi, D., Gaggioloi, G. and Montalti, P. (1984) Chilling effect on development of immature peach and sweet cherry embryos. In: Lange W., Zeven A.C. and Hogenboom N.G. (eds) *Efficiency in Plant Breeding. Proceedings of the 10th Congress of European Research on Plant Breeding, EUCARPIA*, 19–24 June 1983, Pudoc, Wageningen, The Netherlands, p. 293.

Beaver, J.A. and Iezzoni, A.F. (1993) Allozyme inheritance in tetraploid sour cherry (*Prunus cerasus* L.). *Journal of the American Society for Horticultural Science* 118, 873–877.

Benedikova, D. and Giovannini, D. (2013) Review on genetic resources in the ECPGR *Prunus* Working Group. *Acta Horticulturae* 981, 43–51.

Bink, M.C.A.M., Uimari, P., Sillanpaa, M.J., Janss, L.L.G. and Jansen, R.C. (2002) Multiple QTL mapping in related plant populations via a pedigree-analysis approach. *Theoretical and Applied Genetics* 104, 751–762.

Boskovic, R. and Tobutt, K.R. (1998) Inheritance and linkage relationships of isoenzymes in two interspecific cherry progenies. *Euphytica* 103, 273–286.

Brettin, T.S., Karle, R., Crowe, E.L. and Iezzoni, A.F. (2000) Chloroplast inheritance and DNA variation in sweet, sour, and ground cherry. *Journal of Heredity* 91, 75–79.

Brown, S.K., Iezzoni, A.F. and Fogle, H.W. (1996) Cherries. In: Janick, J. and Moore, J.N. (eds) *Fruit Breeding*, Vol. I. *Tree and Tropical Fruits*. Wiley, New York, pp. 213–255.

Burger, P., Terral, J.F., Ruas, M.P., Ivorra, S. and Picq, S. (2011) Assessing past agrobiodiversity of *Prunus avium* L. (*Rosaceae*): a morphometric approach focused on the stones from the archaeological site Hôtel-Dieu (16th century, Tours, France). *Vegetation History and Archaeobotany* 20, 447–458.

Cabrera, A., Rosyara, U.R., de Franceschi, P., Sebolt, A., Sooriyapathirana, S.S., Dirlewanger, E., Quero-García, J., Schuster, M., Iezzoni, A.F. and van der Knaap, E. (2012) Rosaceae conserved orthologous sequences marker polymorphism in sweet cherry germplasm and construction of a SNP-based map. *Tree Genetics and Genomes* 8, 237–247.

Cachi, A.M. and Wünsch, A. (2011) Characterization and mapping of non-*S* gametophytic self-compatibility in sweet cherry (*Prunus avium* L.). *Journal of Experimental Botany* 62, 1847–1856.

Cachi, A.M. and Wünsch, A. (2014a) *S*-Genotyping of sweet cherry varieties from Spain and *S*-locus diversity in Europe. *Euphytica* 197, 229–236.

Cachi, A.M. and Wünsch, A. (2014b) Characterization of self-compatibility in sweet cherry varieties by crossing experiments and molecular genetic analysis. *Tree Genetics and Genomes* 10, 1205–1212.

Calabrese, F., Fenech, L. and Raimondo, A. (1984) Kronio: una cultivar di ciliegio molto precoce e autocompatibile. *Frutticoltura* 46, 27–30.

Campoy, J.A., Le Dantec, L., Barreneche, T., Dirlewanger, E. and Quero-García, J. (2015) New insights into fruit firmness and weight control in sweet cherry. *Plant Molecular Biology Reporter* 22, 783–796.

Campoy, J.A., Lerigoleur-Balsemin, E., Christmann, H., Beauvieux, R., Girollet, N., Quero-García, J., Dirlewanger, E. and Barreneche, T. (2016) Genetic diversity, linkage disequilibrium, population structure and construction of a core collection of *Prunus avium* L. landraces and bred cultivars. *BMC Plant Biology* 16, 49.

Castède, S., Campoy, J.A., Quero-García, J., Le Dantec, L., Lafargue, M., Barreneche, T., Wenden, B. and Dirlewanger, E. (2014) Genetic determinism of phenological traits highly affected by climate change in *Prunus avium*: flowering date dissected into chilling and heat requirements. *New Phytologist* 202, 703–715.

Chavoshi, M., Watkins, C., Oraguzie, B., Zhao, Y., Iezzoni, A. and Oraguzie, N. (2014) Phenotyping protocol for sweet cherry (*Prunus avium* L.) to facilitate an understanding of trait inheritance. *Journal of the American Pomological Society* 68, 125–134.

Choi, C. and Kappel, F. (2004) Inbreeding, coancestry, and founding clones of sweet cherries from North America. *Journal of the American Society for Horticultural Science* 129, 535–543.

Clarke, J.B. and Tobutt, K.R. (2009) A standard set of accessions, microsatellites and genotypes for harmonizing the fingerprinting of cherry collections for the ECPGR. *Acta Horticulturae* 814, 615–618.

Clarke, J.B., Sargent, D.J., Bošković, R.I., Belaj, A. and Tobutt, K.R. (2009) A cherry map from the inter-specific cross *Prunus avium* 'Napoleon' × *P. nipponica* based on microsatellite, gene-specific and isoenzyme markers. *Tree Genetics and Genomes* 5, 41–51.

de Boucaud, M.T., Brison, M., Helliot, B. and Herve-Paulus, V. (2002) Cryopreservation of *Prunus*. In: Towill, L.E. and Bajaj, Y.P.S. (eds) *Biotechnology in Agriculture and Forestry*, Vol. 50. *Cryopreservation of Plant Germplasm II*. Springer, Berlin/Heidelberg, Germany, pp. 287–311.

de Franceschi, P., Stegmeir, T., Cabrera, A., van der Knapp, E., Rosyara, U., Sebolt, A., Dondini, L., Dirlewanger, E., Quero-García, J., Campoy, J. and Iezzoni, A. (2013) Cell number regulator genes in *Prunus* provide candidate genes for the control of fruit size in sweet and sour cherry. *Molecular Breeding* 32, 311–326.

de Rogatis, A. and Fabbri, F. (1997) Shoots *in vitro* regeneration on immature cotyledons of *Prunus avium* L. *Annali dell'Intituto Sperimentale per la Selvicoltura* 28, 3–8.

Demir, T., Demirsoy, L., Demirsoy, H., Kaçar, Y.A., Yilmaz, M. and Macit, I. (2009) Molecular characterization of sweet cherry genetic resources in Giresun, Turkey. *Fruits* 66, 53–62.

Dirlewanger, E., Graziano, E., Joobeur, T., Garriga-Calderé, F., Cosson, P., Howad, W. and Arús, P. (2004) Comparative mapping and marker-assisted selection in Rosaceae fruit crops. *Proceedings of the National Academy of Sciences USA* 101, 9891–9896.

Dirlewanger, E., Claverie, J., Iezzoni, A.F. and Wünsch, A. (2009) Sweet and sour cherries: linkage maps, QTL detection and marker assisted selection. In: Folta, K.M. and Gardiner, S.E. (eds) *Genetics and Genomics of Rosaceae. Plant Genetics and Genomics: Crops and Models*, Vol. 6. Springer, New York, pp. 291–313.

Dirlewanger, E., Quero-García, J., Le Dantec, L., Lambert, P., Ruiz, D., Dondini, L., Illa, E., Quilot-Turion, B., Audergon, J.-M., Tartarini, S., Letourmy, P. and Arùs, P. (2012) Comparison of the genetic determinism of two key phenological traits, flowering and maturity dates, in three *Prunus* species: peach, apricot and sweet cherry. *Heredity* 109, 280–292.

Engelmann, F. (2012) Germplasm collection, storage and preservation. In: Altmann, A. and Hazegawa, P.M. (eds) *Plant Biotechnology and Agriculture – Prospects for the 21st Century*. Academic Press, Oxford, pp. 255–268.

Engelmann, F. and Engels, J.M.M. (2002) Technologies and strategies for *ex situ* conservation. In: Engels, J.M.M., Rao, V.R., Brown, A.H.D. and Jackson, M.T. (eds) *Managing Plant Genetic Diversity*. CAB International, Wallingford, UK, and IPGRI, Rome, pp. 89–104.

Engels, J.M.M. and Maggioni, L. (2011) *AEGIS: A Rationally Based Approach to PGR Conservation*. CAB International, Wallingford, UK.

Engels, J.M.M. and Visser, L. (2003) *A Guide to Effective Management of Germplasm Collections*. IPGRI Handbooks for Genebanks No. 6. International Plant Genetic Resources Institute, Rome.

Ercisli, S., Agar, G., Yildirim, N., Duralija, B., Vokurka, A. and Karlidag, H. (2011) Genetic diversity in wild sweet cherries (*Prunus avium*) in Turkey revealed by SSR markers. *Genetics and Molecular Research* 10, 1211–1219.

EUFORGEN (2009) Distribution map of wild cherry (*Prunus avium*). European forest genetic resources programme, Maccarese, Italy. Available at: http://www.euforgen.org/species/prunus-avium/ (accessed 4 January 2017).

Fathi, H. and Jahani, U. (2012) Review of embryo culture in fruit trees. *Annals of Biological Research* 3, 4276–4281.

Fathi, H., Arzani, K., Ebadi, A and Khalighi, A. (2009) Production of sweet cherry hybrids (Silej-Delamarka and Zard-Daneshkadeh) using embryo culture. *Seed and Plant Improvement Journal* 25, 51–64.

Faust, M. and Surányi, D. (1997) Origin and dissemination of cherry. In: Janick, J. (ed.) *Horticultural Reviews*, Vol. 19. Wiley, New York, pp. 263–317.

Flachowsky, H. and Höfer, M. (2010) Die Deutsche Genbank Obst, ein dezentrales Netzwerk zur nachhaltigen Erhaltung genetischer Ressourcen bei Obst. *Journal für Kulturpflanzen* 62, 9–16.

Ganopoulos, I.V., Kazantzis, K., Chatzicharisis, I., Karayiannis, I. and Tsaftaris, A. (2011) Genetic diversity, structure and fruit trait associations in Greek sweet cherry cultivars using microsatellite based (SSR/ISSR) and morpho-physiological markers. *Euphytica* 181, 237–251.

Giovannini, D. and Engel, P. (2006) Status of *Prunus* collections in Italy. In: Maggioni, L. and Lipman, E. (eds) *Report of a Working Group on Prunus*. Sixth Meeting, 20–21 June 2001, Budapest, Hungary, and Seventh Meeting, 1–3 December 2005. Biodiversity International, Rome, pp. 61–65.

Grenier-de March, G., de Boucaud, M.T. and Chmielarz, P. (2005) Cryopreservation of *Prunus avium* L. embryogenic tissues. *CryoLetters* 26, 341–348.

Guajardo, V., Simon, S., Sagredo, B., Gainza, F., Munoz, C., Gasic, K. and Hinrichsen, P. (2015) Construction of a high density sweet cherry (*Prunus avium* L.) linage maps using microsatellite markers and SNPs detected by genotyping-by-sequencing (GBS). *PLoS One* 10, e0127750.

Guarino, C., Santoro, S., de Simone, L. and Cipriani, G. (2009) *Prunus avium*: nuclear DNA study in wild populations and sweet cherry cultivars. *Genome* 52, 320–337.

Gutierrez, P. and Rugini, E. (2004) Influence of plant growth regulators, carbon source and iron on the cyclic secondary somatic embryogenesis and plant regeneration of transgenic cherry rootstock 'Colt' (*Prunus avium* × *P. pseudocerasus*). *Plant Cell Tissue and Organ Culture* 79, 223–232.

Hauck, N.R., Yamane, H., Tao, R. and Iezzoni, A.F. (2006) Accumulation of non-functional *S*-haplotypes results in the breakdown of gametophytic self-incompatibility in tetraploid *Prunus*. *Genetics* 172, 1191–1198.

Hedrick, U.P., Howe, G.H., Taylor, O.M., Tubergen, C.B. and Wellington, R. (1915) *The Cherries of New York*. Report of the New York Agricultural Experiment Station for the Year 1914 II. J.B. Lyon, Albany, New York.

Iezzoni, A., Schmidt, H. and Albertini, A. (1990) Cherries (*Prunus*). In: Moore, J.N. and Ballington, J.R. Jr (eds) *Genetic Resources of Temperate Fruit and Nut Crops*, Vol. 1. International Society for Horticultural Science, Wageningen, The Netherlands, pp. 111–173.

Ikeda, K., Watari, A., Ushijima, K., Yamane, H., Hauck, N.R., Iezzoni, A.F. and Tao, R. (2004) Molecular markers for the self-compatible $S^{4'}$-haplotype, a pollen-part mutant in sweet cherry (*Prunus avium* L.). *Journal of the American Society for Horticultural Science* 129, 724–728.

Iliev, N., Petrakieva, A. and Milev, M. (2012) Seed dormancy breaking of wild cherry (*Prunus avium* L.). *Forestry Ideas* 18, 28–36.

Ivanicka, J. and Pretova, A. (1986) Cherry (*Prunus avium* L.). In: Bajai, Y.P.S. (ed.) *Biotechnology in Agriculture and Forestry*, Vol. 1, Trees I. Springer, Berlin, pp. 154–169.

Jensen, M. (2012) Sour cherry fast cycling breeding. From pollination to fruit set in offspring. In: *Proceedings of the EU COST Action FA1104 Sustainable production of high-quality cherries for the European market.* COST meeting, Palermo, Italy, November 2012. Available at: https://www.bordeaux.inra.fr/cherry/docs/dossiers/Activities/Meetings/21-23_11_2012_WG_and_2nd_MC_Meeting_Palermo/Day_2/presentations/WG1/Jensen_Palermo_04.pdf (accessed 29th January 2016).

Jensen, M. and Eriksen, E.N. (2001) Development of primary dormancy in seeds of *Prunus avium* during maturation. *Seed Science and Technology* 29, 307–320.

Jensen, M. and Kristiansen, K. (2009) Removal of distal part of cotyledons or soaking in BAP overcomes embryonic dormancy in sour cherry. *Propagation of Ornamental Plants* 9, 135–142.

Jensen, M. and Kristiansen, K. (2014) Aspects of freezing tolerance in sour cherry. *Proceedings of the EU COST Action FA1104: Sustainable Production of High-quality Cherries for the European Market.* COST meeting, Plovdiv, Bulgaria, May 2014. Available at: http://www.bordeaux.inra.fr/cherry/docs/dossiers/Activities/Meetings/2014%2005%2026-27%20WG1-WG3%20Meeting_Plovdiv/Presentations/Jensen_Plovdiv2014.pdf (accessed 29 January 2016).

Joublan, J.P., Serri, H. and Ocompo, J. (2005) Evaluation of sweet cherry germplasm in southern Chile. *Acta Horticulturae* 667, 69–74.

Kaçar, Y.A., Çetiner, M.S., Cantini, C. and Iezzoni, A.F. (2006) Simple sequence repeat (SSR) markers differentiate Turkish sour cherry germplasm. *Journal of the American Society for Horticultural Science* 60, 136–143.

Kellerhals, M., Goerre, M., Rsterholz, P., Gersbach, K. and Bossard, M. (1999) Obstsorten: das reiche genetische Erbe sichern. *Schweizerische Zeitschrift für Obst- und Weinbau* 23, 561–564.

Klagges, C., Campoy, J.A., Quero-García., J, Guzman, A., Mansur, L., Gratacos, E., Silva, H., Rosyara, U.R., Iezzoni, A., Meisel, L.A. and Dirlewanger, E. (2013) Construction and comparative analyses of highly dense linkage maps of two sweet cherry intra-specific progenies of commercial cultivars. *PLoS One* 8, e54743.

Lacis, G., Rashal, I., Ruisa, S., Trajkovski, V. and Iezzoni, A.F. (2009) Assessment of genetic diversity of Latvian and Swedish sweet cherry (*Prunus avium* L.) genetic resources collections by using SSR (microsatellite) markers. *Scientia Horticulturae* 121, 451–457.

Lambardi, M., Benelli, C., de Carlo, A. and Previati, A. (2009) Advances in the cryopreservation of fruit plant germplasm at the CNR-IVALSA. *Acta Horticulturae* 839, 237–243.

Lapins, K.O. (1975) 'Compact Stella' cherry. *Fruit Varieties Journal* 29, 20.

Lewis, D. (1949) Structure of the incompatibility gene. II. Induced mutation rate. *Heredity* 3, 339–355.

Lin, C.F. and Boe, A.A. (1972) Effects of some endogenous and exogenous growth regulators on plum seed dormancy. *Journal of the American Society for Horticultural Science* 97, 41–44.

Liu, L., Pagter, M. and Andersen, L. (2012) Preliminary results on seasonal changes in flower bud cold hardiness of sour cherry. *European Journal of Horticultural Science* 77, 109–114.

Maggioni, L. and Engels, J. (2014) Networking for plant genetic resources. *Pan European Networks: Science and Technology* 10, 285–287.

Marchese, A., Bošković, R.I., Caruso, T., Raimondo, A., Cutuli, M. and Tobutt, K.R. (2007) A new self-compatibility haplotype in the sweet cherry 'Kronio', $S_{5'}$, attributable to a pollen-part mutation in the *SFB* gene. *Journal of Experimental Botany* 58, 4347–4356.

Mariette, S., Tavaud, M., Arunyawat, U., Capdeville, G., Millan, M. and Salin, F. (2010) Population structure and genetic bottleneck in sweet cherry estimated with SSRs and the gametophytic self-incompatibility locus. *BMC Genetics* 11, 77.

Moriguchi, T., Teramoto, S. and Sanada, T. (1994) Conservation system of fruit tree genetic resources and recently released cultivars from fruit tree research station in Japan. *Fruit Varieties Journal* 48, 73–80.

Niino, T., Tashiro, K., Suzuki, M., Ohuchi, S., Magoshi, J. and Akihama, T. (1997) Cryopreservation of *in vitro* grown shoot tips of cherry and sweet cherry by one-step vitrification. *Scientia Horticulturae* 70, 155–163.

Okuno, K., Shirata, K., Nino, T. and Kawase, M. (2005) Plant genetic resources in Japan: platforms and destinations to conserve and utilize plant genetic diversity. *Japan Agricultural Research Quarterly* 39, 231–237.

Olden, E.J. and Nybom, N. (1968) On the origin of *Prunus cerasus* L. *Hereditas* 59, 327–345.

Olmstead, J., Sebolt, A., Cabrera, A., Sooriyapathirana, S., Hammar, S., Iriart, G., Wang, D., Chen, C.Y., van der Knapp, E. and Iezzoni, A.F. (2008) Construction of an intra-specific sweet cherry (*Prunus avium* L.) genetic linkage map and synteny analysis with the *Prunus* reference map. *Tree Genetics and Genomes* 4, 897–910.

Peace, C., Bassil, N., Main, D., Ficklin, S., Rosyara, U.R., Stegmeir, T., Sebolt, A., Gilmore, B., Lawley, C., Mockler, T.C., Bruant, D.W., Wilhelm, L. and Iezzoni, A. (2012) Development and evaluation of a genome-wide 6K SNP array for diploid sweet cherry and tetraploid sour cherry. *PLoS One* 7, e48350.

Pérez-Sánchez, R., Gómez-Sánchez, M.A. and Morales-Corts, R. (2008) Agromorphological characterization of traditional Spanish sweet cherry (*Prunus avium* L.), sour cherry (*Prunus cerasus* L.) and duke cherry (*Prunus* × *gondouinii* Rehd.) cultivars. *Spanish Journal of Agricultural Research* 6, 42–55.

Popenoe, H., King, S.R., Leon, J. and Kalinowksi, L.S. (1989) Capuli cherry. In: *Lost Crops of the Incas: Little Known Plants for the Andes with Promise for Worldwide Cultivation*. National Academy Press, Washington, DC, pp. 223–227.

Potter, D., Eriksson, T., Evans, R.C., Oh, S., Smedmark, J.E.E., Morgan, D.R., Kerr, M., Robertston, K.R., Arsenault, M., Dickinson, T.A. and Campbell, C.S. (2007) Phylogeny and classification of Rosaceae. *Plant Systematics and Evolution* 266, 5–43.

Quero-García, J., Fodor, A., Reignier, A., Capdeville, G., Joly, J., Tauzin, Y., Fouilhaux, L. and Dirlewanger, E. (2014) QTL detection of important agronomic traits for sweet cherry breeding. *Acta Horticulturae* 1020, 57–64.

Reed, B.M., Engelmann, F., Dullo, M.E. and Engels, J.M.M. (2004) *Technical Guidelines for the Management of Field and In Vitro Germplasm Collections*. IPGRI Handbooks for Genebanks No. 7. International Plant Genetic Resources Institute, Rome.

Rehder, A. (1974) *Manual of Cultivated Trees and Shrubs Hardy in North America*. Macmillan, New York.

Rodrigues, L.C., Morales, M.R., Fernandes, A.J.B. and Ortiz, J.M. (2008) Morphological characterization of sweet and sour cherry cultivars in a germplasm bank at Portugal. *Genetic Resources and Crop Evolution* 55, 593–601.

Rosyara, U., Bink, C.A.M., van de Weg, E., Zhang, G., Wang, D., Sebolt, A., Dirlewanger, E., Quero-García, J., Schuster, M. and Iezzoni, A. (2013) Fruit size QTL identification and the prediction of parental QTL genotypes and breeding values in multiple pedigreed populations of sweet cherry. *Molecular Breeding* 32, 875–887.

Rosyara, U., Sebolt, A., Peace, C. and Iezzoni, A. (2014) Identification of the paternal parent of 'Bing' sweet cherry and confirmation of descendants using single nucleotide polymorphism markers. *Journal of the American Society for Horticultural Science* 139, 148–156.

Salazar, J., Ruiz, D., Campoy, J., Sánchez-Pérez, R., Crisosto, C., Martínez-García, P., Blenda, A., Jung, S., Main, D., Martínez-Gómez, P. and Rubio, M. (2014) Quantitative trait loci (QTL) and Mendelian trait loci (MTL) analysis in *Prunus*: a breeding perspective and beyond. *Plant Molecular Biology Reporter* 32, 1–18.

Sandefur, P., Oraguzie, N. and Peace, C. (2016) A DNA test for routine prediction in breeding of sweet cherry fruit color, Pav-$R_f$-SSR. *Molecular Breeding* 36, 33.

Schmidt, H. (1973) Investigations on breeding dwarf rootsocks for sweet cherries. I. Flower differentiation and development in *Prunus* species and forms. *Zeitschrift für Pflanzenzucht* 70, 72–82.

Schmidt, H. and Ketzel, A. (1993) Regeneration of adventitious shoots *in vitro* in cherries. 4. Adventitious shoots regeneration of cotyledons and embryo rescue as tools in sweet cherry breeding. *Gartenbauwissenschaft* 58, 64–67.

Schuster, M. (2005) Meiotic investigations in a *Prunus avium* × *P. canescens* hybrid. *Acta Horticulturae* 667, 101–102.

Schuster, M. (2012) Incompatible (*S*-) genotypes of sweet cherry cultivars (*Prunus avium* L.). *Scientia Horticulturae* 148, 59–73.

Schuster, M. (2013) Resistance breeding in cherries – goals and results. In: *Proceedings of the EU COST Action FA1104: Sustainable Production of High-quality Cherries for the European Market*. COST meeting, Pitesti, Romania, October 2013. Available at: https://www.bordeaux.inra.fr/cherry/docs/dossiers/Activities/Meetings/15-17%2010%202013_3rd%20MC%20and%20all%20WG%20Meeting_Pitesti/Presentations/16_Schuster_Mirko.pdf (accessed 29 January 2016).

Schuster, M. and Schreiber, H. (2000) Genome investigations in sour cherry, *P. cerasus* L. *Acta Horticulturae* 538, 375–379.

Schuster, M., Flachowsky, H. and Köhler, D. (2007) Determination of self-incompatible genotypes in sweet cherry (*Prunus avium* L.) accessions and cultivars of the German Fruit Gene Bank and from private collections. *Plant Breeding* 126, 533–540.

Sebolt, A.M., Iezzoni, A.F. and Tsukamoto, T. (2017) *S*-Genotyping of cultivars and breeding selections of sour cherry (*Prunus cerasus* L.) in the Michigan State University Sour Cherry breeding program. *Acta Horticulturae* (in press).

Shatnawi, M.A., Shibli, R., Qrunfleh, I., Bataeineh, K. and Obeidat, M. (2007) *In vitro* propagation and cryopreservation of *Prunus avium* using vitrification and encapsulation dehydration methods. *Journal of Food, Agriculture and Environment* 5, 204–208.

Shi, S., Li, J., Sun, J., Yu, J. and Zhou, S. (2013) Phylogeny and classification of *Prunus sensu lato* (Rosaceae). *Journal of Integrative Plant Biology* 55, 1069–1079.
Sonneveld, T., Tobutt, K.R., Vaughan, S.P. and Robbins, T.P. (2005) Loss of pollen-*S* function in two self-compatible selections of *Prunus avium* is associated with deletion/mutation of an *S* haplotype-specific F-box gene. *Plant Cell* 17, 37–51.
Sooriyapathirana, S.S., Khan, A., Sebolt, A.M., Wang, D., Bushakra, J.M., Lin-Wang, K., Allan, A.C., Gardiner, S.E., Chagne, D. and Iezzoni, A.F. (2010) QTL analysis and candidate gene mapping for skin and flesh color in sweet cherry fruit (*Prunus avium* L.). *Tree Genetics and Genomes* 6, 821–832.
Stanys, V., Stanytė, R., Stanienė, G. and Vinskienė, J. (2008) *S*-Allele identification by PCR analysis in Lithuanian sweet cherries. *Biologija* 54, 22–26.
Stegmeir, T., Schuster, M., Sebolt, A., Rosyara, U., Sundin, G.S. and Iezzoni, A. (2014a) Cherry leaf spot resistance in cherry (*Prunus*) is associated with a quantitative trait locus on linkage group 4 inherited from *P. canescens*. *Molecular Breeding* 34, 927–935.
Stegmeir, T., Sebolt, A. and Iezzoni, A. (2014b) Phenotyping protocol for sour cherry (*Prunus cerasus* L.) to enable a better understanding of trait inheritance. *Journal of the American Pomological Society* 68, 40–47.
Stegmeir, T., Cai, L., Basundari, R.A., Sebolt, A.M. and Iezzoni, A.F. (2015) A DNA test for fruit flesh color in tetraploid sour cherry (*Prunus cerasus* L.). *Molecular Breeding* 35, 149.
Suszka, B., Muller, C. and Bonnet-Masimbert, M. (1996) *Seeds of Forest Broadleaves, from Harvest to Sowing*. Institut National de la Recherche Agronomique, Paris, pp. 213–233.
Szôdi, S., Rozsnyay, Z., Rózsa, E. and Turóczi, G. (2008) Susceptibility of sour cherry cultivars to isolates of *Monilia laxa* (Ehrenbergh) Saccardo et Voglino. *International Journal of Horticultural Science* 14, 83–87.
Szügyi, S., Rozsnyay, Z., Apostol, J. and Békefi, Z. (2014) Breeding and evaluation of Hungarian bred *Monilia laxa* resistant sour cherry genotypes. In: *Proceedings of the EU COST Action FA1104: Sustainable Production of High-quality Cherries for the European Market*. COST meeting, Plovdiv, Bulgaria, May 2014. Available at: https://www.bordeaux.inra.fr/cherry/page6-1-13.html (accessed 29 January 2016).
Szymajda, M. and Zurawicz, E. (2014) Effect of postharvest treatments on the germination of three sour cherry (*Prunus cerasus* L.) genotypes. In: *Proceedings of the EU COST Action FA1104: Sustainable Production of High-quality Cherries for the European Market*. COST meeting, Novi Sad, Serbia, September 2014. Available at: https://www.bordeaux.inra.fr/cherry/page6-1-15.html (accessed 29 January 2016).
Tang, H.R., Ren, Z.L. and Krczal, G. (2000) Somatic embryogenesis and organogenesis from immature embryo cotyledons of three sour cherry cultivars (*Prunus cerasus* L.). *Scientia Horticulturae* 83, 109–126.
Tavaud, M., Zanetto, A., David, J.L., Laigret, F. and Dirlewanger, E. (2004) Genetic relationships between diploid and allotetraploid cherry species (*Prunus avium, Prunus* × *gondouinii* and *Prunus* cerasus). *Heredity* 93, 631–638.
Towill, L.E. and Forsline, P. (1999) Cryopreservation of sour cherry (*Prunus cerasus* L.) using a dormant vegetative bud method. *CryoLetters* 20, 215–222.
Tsukamoto, T., Tao, R. and Iezzoni, A.F. (2008) PCR markers for mutated *S*-haplotypes enable discrimination between self-incompatible and self-compatible sour cherry selections. *Molecular Breeding* 21, 67–80.
Tukey, H.B. (1933a) Embryo abortion in early-ripening varieties of *Prunus avium*. *Botanical Gazette* 94, 433–468.
Tukey, H.B. (1933b) Artificial culture of sweet cherry embryos. *Journal of Heredity* 24, 7–12.
Vaughan, S.P. and Russell, K. (2004) Characterization of novel microsatellites and development of multiplex PCR for large-scale population studies in wild cherry, *Prunus avium*. *Molecular Ecology Notes* 4, 429–431.
Wang, D., Karle, R., Brettin, S. and Iezzoni, A.F. (1998) Genetic linkage map in sour cherry using RFLP markers. *Theoretical and Applied Genetics* 97, 1217–1224.
Wang, D., Karle, R. and Iezzoni, A.F. (2000) QTL analysis of flower and fruit traits in sour cherry. *Tree Genetics and Genomes* 100, 535–544.
Wang, J., Zhang, K., Zhang, X., Yan, G., Zhou, Y., Feng, L., Ni, Y. and Duan, X. (2015) Construction of commercial sweet cherry linkage maps and QTL analysis for trunk diameter. *PLoS One* 10, e0141261.
Watkins, R. (1976) Cherry, plum, peach, apricot and almond. In: N.W. Simmonds (ed.) *Evolution of Crop Plants*. Longman, New York, pp. 242–247.
Webster, A.D. (1996) The taxonomic classification of sweet and sour cherries and a brief history of their cultivation. In: Webster, A.D. and Looney, N.E. (eds) *Cherries: Crop Physiology, Production and Uses*. CAB International, Wallingford, UK.
Wharton, P. and Iezzoni, A. (2005) Development of a protocol for screening cherry germplasm for resistance to cherry leaf spot. *Acta Horticulturae* 667, 509–514.

Wünsch, A. and Hormaza, J.I. (2004a) Molecular evaluation of genetic diversity and *S*-allele composition of local Spanish sweet cherry (*Prunus avium* L.) cultivars. *Genetic Resources and Crop Evolution* 51, 635–641.

Wünsch, A. and Hormaza, J.I. (2004b) Genetic and molecular analysis in Cristobalina sweet cherry, a spontaneous self-compatible mutant. *Sexual Plant Reproduction* 17, 203–210.

Zanetto, A., Maggioni, L., Tobutt, K.R. and Dosba, F. (2002) *Prunus* genetic resources in Europe: achievement and perspectives of a networking activity. *Genetic Resources and Crop Evolution* 49, 331–337.

Zhang, G., Sebolt, A.M., Sooriyapathirana, S.S., Wang, D., Bink, M.C.A.M., Olmstead, J.W. and Iezzoni, A.F. (2010) Fruit size QTL analysis in an $F_1$ population derived from a cross between a domesticated sweet cherry cultivar and a wild forest sweet cherry. *Tree Genetics and Genomes* 6, 25–36.

Zhu, M., Zhang, X., Zhang, K., Jiang, L. and Zhang, L. (2004) Development of a simple molecular marker specific for detecting the self-compatible $S_{4'}$ haplotype in sweet cherry (*Prunus avium* L.). *Plant Molecular Biology Reporter* 22, 387–398.

Zhukovsky, P.M. (1965) Main gene centers of cultivated plants and their wild relations within territory of the USSR. *Euphytica* 14, 177–188.

# 4 Sweet Cherry Varieties and Improvement

José Quero-García,[1]* Mirko Schuster,[2] Gregorio López-Ortega[3] and Gérard Charlot[4]

[1]*UMR 1332 Biologie du Fruit et Pathologie, INRA et Université de Bordeaux, Villenave d'Ornon, France;* [2]*Julius Kühn-Institut, Federal Research Centre for Cultivated Plants, Institute for Breeding Research on Fruit Crops, Dresden, Germany;* [3]*Murcia Institute of Agri-Food Research and Development (IMIDA), Murcia, Spain;* [4]*CTIFL, Bellegarde, France*

## 4.1 History of Improvement

Today, hundreds of sweet cherry cultivars are available for growers. A large diversity of landraces are preserved and many have been used for production at a local level or more recently in modern breeding programmes. Sweet cherry is in general a diploid species (2n = 2x = 16), and only in a very few cases have triploids or tetraploids been reported (Fogle, 1975, cited by Bargioni, 1996). The observation of one triploid individual resulting from a natural cross between one haploid and one diploid gamete might also be due to an incorrect chromosome count. As for tetraploids, they are only the result of artificial polyploidization (M. Schuster, Dresden, Germany, 2016, personal communication).

There have been attempts to propose classification of sweet cherry cultivars based on important and relatively simple phenotypic traits such as skin colour, colour and firmness of the fruit flesh, juice colour and time of ripening (Bargioni, 1996). For instance, Zwitzscher (1961), according to a previous classification by Truchseß (1819), classified cherries into two groups: the soft-fleshed heart cherries and the firm-fleshed Bigarreau cherries. Both groups were further subdivided in dark red-coloured fruit with staining juice and uncoloured fruit with no staining juice. Nevertheless, classification into clear-cut groups is always difficult as there is a vast continuum of morphological diversity, and many traits, including firmness, may have an important influence by climate, crop load, rootstock or any other cultural practice.

Although the number of available commercial sweet cherry cultivars has increased significantly in recent decades, it is noteworthy that in many countries a large percentage of production still relies on a small number of cultivars. Some of these are very old selections such as 'Bing' or 'Burlat', or even old cultivars of unknown origin, such as '0900 Ziraat' in Turkey.

## 4.2 Sweet Cherry Breeding

Genetic improvement of sweet cherry based on controlled breeding is relatively recent compared with other fruit crop species. As well as a lesser economic profitability in comparison with species such as apple or peach, other biological characteristics must

* jose.quero-garcia@inra.fr

also be taken into account: (i) a gametophytic self-incompatibility system (see Chapter 2, this volume); (ii) a high dependence of fruit set on climatic conditions during bloom time (see Chapter 8, this volume); (iii) the high vigour and lack of precocity of most sweet cherry cultivars; (iv) until recently, the lack of dwarfing rootstocks allowing intensive orchard production (see Chapter 6, this volume); and (v) other specific agronomic problems such as bird damage to fruit and rain-induced fruit cracking (see Chapter 7, this volume).

Despite these special features, numerous breeding programmes were initiated after the 1950s in almost every sweet cherry-producing country. According to Sansavini and Lugli (2008), during the period 1991–2004, the breeding of sweet cherry was, within the stone fruits, second only to peach. During that period, 230 new cultivars were released, with 116 being bred in Europe, 71 in North America and 33 in Asia. During the period 2002–2016, in the USA alone, 88 new sweet cherry cultivars were published (or will soon be published) in the Register of New Fruit and Nut Cultivars (G. Lang, East Lansing, Michigan, 2016, personal communication).

Sansavini and Lugli (2008) identified 27 European leading public and private sweet cherry breeding programmes and conducted a survey for which they obtained 20 responses. Kappel (2008) described seven breeding programmes from the USA, Canada and Australia, while in this chapter, we will present 22 breeding programmes from 15 different countries, including relatively young programmes initiated during the year 2000, in particular in highly dynamic countries with respect to cherry cultivation, such as Chile, China and Spain. As mentioned in Chapter 3 (this volume), the genetic base used in most breeding programmes has been extremely narrow, with each programme focusing on a very limited number of main progenitors due to the specific characteristics of these cultivars. Nevertheless, interest is increasing today towards the use of more diversified sources of genetic diversity, mainly from the cultivated gene pool but without excluding the possibility of searching for interesting alleles in wild relatives. As an example, breeders from France, Germany and Spain recently collected stones from landraces and wild materials of sweet and sour cherries in different regions of Azerbaijan, within the supposed area of origin of cherries (López-Ortega, 2015).

### 4.2.1 Objectives in sweet cherry breeding

The main breeding goals in sweet cherry have been described previously (reviewed by Bargioni, 1996; Sansavini and Lugli, 2008; Kappel *et al.*, 2012). In this section, the most important goals, along with relatively recent breeding objectives, will be reviewed.

#### *Tree and fruiting structure*

Excessive tree vigour has traditionally been considered one of the main problems for sweet cherry intensive growing throughout the world. Attempts to create cultivars with compact habit were made by conventional hybridization and by selection from natural mutations and ionizing radiations (Bargioni, 1996). Some examples of produced dwarf types are 'Compact Stella' or 'Compact Lambert'. With the advent of a new generation of dwarfing and semi-dwarfing rootstocks, along with the adoption of new pruning techniques and training systems, breeding for scion dwarf types is not a major goal nowadays. However, an excess of vigour is avoided, and genotypes with semi-upright (with high spur density) or spreading habit and good branching (with more loose distribution of spurs) are preferred.

Yield precocity, productivity and, more recently, regularity of production, are currently key selection criteria. Nevertheless, it is important to take into account that all of these characteristics may be highly influenced by the rootstock, the planting and the training systems. For precocity, cultivars such as 'Sweetheart' can be used as a standard for a sweet cherry cultivar considered as quite precocious (Kappel *et al.*, 2012). Productivity can be rated with a qualitative scale at the seedling stage, but only on advanced selections planted in replicated trials can yield measurements be made. Regularity of production can be linked to environmental adaptability, in relation to the phenotypic plasticity of a given genotype. Sweet cherry cultivars are extremely dependent on climatic conditions for a regular and proper fruit set. Only through multi-year observations or through the use of multi-site trials with a marked difference in climatic conditions between sites will breeders obtain an accurate evaluation of environmental adaptability.

### Flower characteristics

One of the major achievements in sweet cherry breeding was the development of self-fertility. Today, self-fertility is incorporated as a major breeding goal in almost every programme, even more since the use of molecular markers for selecting self-fertile seedlings became extremely reliable and cost-efficient. Of the 119 cultivars described in this chapter 21 are self-fertile, whereas Bargioni (1996) reported only eight self-fertile cultivars from a total of 83. All of the commercially released self-fertile cultivars can trace their ancestry back to ‘Stella’ with a mutated $S_{4'}$ allele. An exception is the Hungarian cultivar ‘Axel’ with a mutated $S_{3'}$ allele ($S_3S_{3'}$) (Kappel *et al.*, 2012). However, there is growing concern to diversify the sources of self-compatibility, and self-compatible landraces, such as ‘Cristobalina’ or ‘Kronio’, are being used as parents in several breeding programmes.

### Tolerance to abiotic and biotic stresses

The most damaging abiotic stress to sweet cherry profitability is undoubtedly rain-induced fruit cracking. As this phenomenon is highly complex, no reliable laboratory or field phenotyping protocol has yet been designed to evaluate varietal tolerance to cracking. Hence, only multi-year field observations in sites with sufficient rainfall during the harvest period allow an evaluation of cracking tolerance of hybrids. After the first round of hybrid selection, a clear advantage of conducting multi-site evaluation is the multiplication of chances of observing rainfall events at harvest time for a determined hybrid. Section 4.3 provides information, where available, about the cracking susceptibility of the described cultivars. Very few of the currently most commonly grown commercial cultivars exhibit a high tolerance to cracking. Two of these cultivars are ‘Regina’ and ‘Fermina’. In order to combine different types of genetic tolerance to cracking, breeders should diversify the pool of genitors used, including cultivars with other undesired fruit quality traits.

Resistance to winter frost has been sought in countries at the margins of traditional production areas, such as Latvia and Russia, or in very cold continental areas of countries, such as Ukraine, Romania and Germany. Molecular and physiological aspects of cold hardiness are reviewed in Chapter 8 (this volume). Concerning the evaluation of cultivar resistance to frost, a number of studies have been carried out in field conditions or after artificial freezing tests. One of the most comprehensive works conducted in Germany using 131 sweet cherry cultivars identified nine winter-hardy cultivars (Fischer and Hohlfeld, 1998), and Bargioni (1996) cited ‘Windsor’, ‘Black Eagle’, ‘Vic’, ‘Kristin’ and ‘Hudson’ as cold hardy. In some extremely cold areas, such as northern Russia and Latvia, various sweet cherry cultivars of western European origin were introduced in the 1920s, but the orchards were destroyed after very severe winters. Hence, local cultivars producing very small fruit were identified as extremely winter hardy and were used in breeding programmes (D. Feldmane, Riga, Latvia, 2016, personal communication). In breeding programmes in the former Soviet Union, the German cultivar ‘Dönnissens Gelbe’ was used as a donor for winter hardiness. Today, ‘Aiya’, ‘Bryanskaya Rozovaya’ and ‘Ipujt’ are grown in Latvia (see section 4.3), and a new breeding generation of cultivars from Russia (‘Radica’, ‘Ovstuzhenka’ and ‘Tyutchevka’), producing larger fruit, has been introduced recently for testing.

In contrast to other fruit species such as apple or peach, sweet cherry cultivation has not yet been attempted in subtropical areas. Indeed, the large majority of commercial cultivars, but also landraces, have winter chilling requirements not adapted to these latitudes (see Chapter 8, this volume). However, in recent decades, there has been a growing interest to adapt sweet cherry growing to regions characterized by mild winters, such as south-east Spain, California, central parts of Chile and even North African countries such as Tunisia, Algeria and Morocco. Several commercial cultivars, such as 'Lapins', 'Brooks' and 'Rainier', are regularly productive, even during particularly mild winters. Nevertheless, very few cultivars are clearly low chilling, although one exception might be the landrace 'Cristobalina', which, besides being self-fertile, has an extremely early bloom time. In California, Zaiger Genetics has recently released several cultivars that also have a very early blooming time and, presumably, very low-chilling requirements, such as 'Royal Hazel' and 'Royal Marie'. Whether 'Cristobalina' or another related cultivar was used in the breeding of these new cultivars remains unknown. Another interesting landrace called 'Bouargoub' was recently reported from Tunisia; as with 'Cristobalina', it flowers very early, is self-fertile and produces rather small fruit (T. Azizi, Tunis, Tunisia, 2016, personal communication). Ideally, breeders will seek cultivars with low-chilling requirements for flowering but with sufficient heat requirements in order not to flower too early and avoid the risk of frost damage. To the best of our knowledge, this 'ideotype' has not yet been reached for sweet cherry.

The last abiotic stress that will become increasingly serious due to global warming is the formation of double fruit. As described in Chapter 8 (this volume), different studies have allowed the identification of several cultivars with low potential for double fruit formation. No studies have yet been conducted to investigate the genetic determinism of this trait.

One of the most serious diseases in sweet cherry is bacterial canker (caused by *Pseudomonas* spp.), in particular in the cooler and wetter areas of production (see Chapter 15, this volume). Two methods of inoculation are used to test for resistance. The leaf node method involves inoculating young seedlings with a mixture of bacteria (Krzesinska and Azarenko, 1992). The second method is bark inoculation on older trees (Kappel *et al.*, 2012). More recently, rapid screening methods based on immature fruitlet tests have been tested (Kałużna and Sobiczewski, 2014; Ozaktan, 2015). A breeding programme carried out at the John Innes Institute, UK, released a number of cultivars resistant to bacterial canker caused by *Pseudomonas syringae* pv. *morsprunorum*, such as 'Merla', 'Mermat', 'Merpet' and 'Inge' (Matthews and Dow, 1978, 1979, 1983, cited in Kappel *et al.*, 2012). Unfortunately, some of these cultivars later showed some susceptibility to canker, due to infection by new more virulent bacterial strains (Bargioni, 1996). While no total resistance to canker might be achieved, breeders could include tolerant cultivars in their pools of genitors, such as 'Colney', 'Hertford', 'Merton Glory' or 'Vittoria' (Bargioni, 1996).

Brown rot caused by *Monilinia* spp. can cause severe damage on flowers and fruit, and fungicide control is difficult (see Chapter 14, this volume). The first artificial inoculation tests on twigs in the laboratory and in the field were described by Schmidt (1937). Several authors have evaluated a number of cultivars but only found different levels of susceptibility, and in no case could resistance be proved (Brown and Wilcox, 1989; Kappel and Sholberg, 2008; Kappel *et al.*, 2012). Within the cultivars described in section 4.3, only 'Regina', 'Early Korvik', 'Melitopolska Chorna' and 'Valerij Chkalov' are reported as having good levels of tolerance to brown rot.

The most damaging pests in sweet cherry are cherry fruit fly (*Rhagoletis* spp.), black cherry aphid (*Myzus cerasi* Fab.) and, in recent years, the fly *Drosophila suzukii* (see Chapter 13, this volume). While no tolerance or resistance has been identified to the cherry fruit fly or *D. suzukii*, breeding research carried out at East Malling in the UK has focused on resistance to *M. cerasi*. Clones from the species *Prunus canescens*, *Prunus incisa*, *Prunus kurilensis* and *Prunus nipponica* all showed resistance to colonization. Some crosses were conducted between these clones and cultivar 'Napoleon', and some of the hybrids proved to be tolerant but not fully resistant to colonization (Bargioni, 1996).

### *Fruit quality*

The main quality traits evaluated by sweet cherry breeders are fruit size, fruit firmness, skin and flesh colour, sugar content and flavour. Many other morphological and biochemical traits can be evaluated in more advanced stages of the selection process, and can be related to the skin, flesh, juice, pit or stalk. Tremendous progress has been achieved in many breeding programmes in terms of fruit size and firmness, with cultivars that can regularly produce very firm fruit of more than 12 g. Nevertheless, fruit weight and fruit firmness may be negatively correlated in certain genetic backgrounds (Campoy *et al.*, 2015), and more research is needed to disentangle the complex genetic determinism of these traits. Both appear as highly heritable, although very polygenic; hence, the fine mapping of the main quantitative trait locus involved would allow a more accurate strategy of marker-assisted selection. Concerning colour, routine DNA tests are now available to differentiate red skin and blush-type cherries (see Chapter 3, this volume). Fruit taste is usually determined by a combination of sensory panels and objective measurements, such as soluble solids content and titratable acidity (Kappel *et al.*, 2012). Although all measurements related to fruit quality are now standardized in most breeding programmes, climatic and cultural practice differences between sites, as well as different choices in terms of technology (e.g. Durofel® versus Firmtech® for firmness evaluation), render values from different programmes difficult to compare.

### *Extension of harvest period*

Many breeding programmes, in particular in Europe, have set as one of their primary objectives the breeding of early genotypes ripening before the benchmark cultivar 'Burlat'. However, no cultivar with a ripening date close to or earlier than 'Burlat' has proven significantly better than 'Burlat' in terms of fruit size, taste, cracking susceptibility and flesh firmness. Although, as mentioned above, very early cultivars are now available, breeders have not yet managed to combine such flowering earliness with a short fruit ripening phase and premium fruit quality traits. Extending the end-of-season calendar with extra-late cultivars has proved more successful, and further improvements are expected.

### *Suitability for mechanical harvesting*

Mechanical harvesting has been developed mainly for processing cherries, but interest in a market of stemless sweet cherries for fresh produce is growing (Kappel *et al.*, 2012). Stemless cultivars require an abscission site between the pedicel and the fruit that produces a hardened scar tissue to prevent juice loss, oxidation and pathogen attacks (Sansavini and Lugli, 2008). Cultivars suitable for this type of harvest would be 'Ambrunés', 'Cristalina', 'Fermina', 'Linda', 'Sumste' and 'Vittoria' (Bargioni, 1996; G. Charlot, Balandran, France, 2016, personal communication).

## 4.2.2 Methods of sweet cherry breeding

The most common methods used in sweet cherry breeding are clonal selection, or selective breeding, and cross-combination breeding. Clonal selection methods in cherry do not differ from those employed for other species. This technique has been employed in natural populations (e.g. 'Kordia') and within pools of cultivars (e.g. 'Burlat', 'Napoleon' and 'Germersdorfer') (Bargioni, 1996). Strategies and methodologies involved in cross-combination are described fully in Chapter 3 (this volume). Interspecific hybridization has been far less active in sweet cherry breeding compared with sour cherry or cherry rootstock breeding, but some examples were described in section 4.2.1.

During the 1960s, ionizing radiation techniques were used in several breeding programmes, in particular in Summerland, Canada, with the objective of producing dwarf sweet cherry cultivars, such as 'Compact Lambert'. This methodology was described extensively by Bargioni (1996) but has been progressively abandoned since. Other breeding tools such as the exploitation of stable somatic variants from cell and *in vitro* tissue cultures have been explored but without any practical results as yet (Sansavini and Lugli, 2008). Finally, genetic transformation of sweet cherry, as in *Prunus* spp. in general, has proven very difficult to achieve (Kappel *et al.*, 2012). Several regeneration protocols have been reported for sweet and sour cherry cultivars, but stable transgenic plants have only been produced for the sour cherry 'Montmorency', and for cherry rootstocks such as 'GiSelA 6' and 'Colt' (reviewed in Song, 2014).

## 4.3 Sweet Cherry Breeding Programmes

The sweet cherry breeding programmes available in different countries are described below.

### 4.3.1 Bulgaria

*Breeding institute (or company): Fruit Growing Institute (FGI)*

Sweet cherry breeding activities in Bulgaria started in 1951 at 'Vasil Kolarov' Higher Institute of Agriculture, Plovdiv, when S. Popov developed the cultivar 'Hebros' by crossing 'Hedelfinger' with mixed pollen of 'Ranna Cherna' and '11 May'. Later, in 1953, V. Georgiev started his breeding activities at the FGI – Kyustendil (currently named the Institute of Agriculture). The scope of his work was: conventional breeding with genetic studies in $F_1$ progeny and mutation breeding, and the introduction of valuable cultivars. Cultivars 'Pobeda', 'Kyustendilska Hrushtyalka', 'Cherna Konyavska', 'Bulgarska Hrushtyalka', 'Mizia', 'Danelia', 'Stefania', 'Vasilena' and 'Superstar' were created.

The sweet cherry breeding programme of the FGI, Plovdiv, started in 1987. The major breeding objectives were self-fertility, compact habit, early ripening and large fruit size. The following cultivars were used: 'Van', 'Stella', 'Compact Van', 'Compact Stella', 'Rivan', 'Sunburst', 'Bigarreau Burlat', 'Early Cherna', 'Ohio Beauty', 'Early Rivers', 'Compact Lambert', 'Starkrimson', 'Germersdorfer', 'Badacsony', 'Ferrovia Spur', 'Lapins' and 'Sweet September'. Evaluation of the seedlings, selection and propagation of elites started in the period 1996–2000. The creation of second-hybrid generations started within the period 1996–2000 and the process is still continuing. Up until 2012 inclusive, the number of the parent combinations used throughout the years has surpassed 120. Today, the main breeding objectives are: extension of the maturity period, filling in the harvest window between 'Bigarreau Burlat' and 'Bing', self-fertility, intensive red and light fruit colour, firm texture, resistance to cracking, resistance to biotic and abiotic stresses, and poor or moderate growth vigour. Until 2011, four candidate cultivars were officially recognized: 'Kossara', 'Rosita', 'Rosalina' and 'Trakiiska Hrushtyalka'. Seven elite selections are currently under evaluation.

### 4.3.2 Canada

*Breeding institute (or company): Agriculture and Agri-Food Canada*

The sweet cherry breeding programme in Summerland, British Columbia, began in 1936. The first cultivar to be released was 'Van' in 1944. Some early cultivar releases were cherries bred for canning (e.g. 'Star'), an industry that no longer exists in the region. Three particularly important cultivars were 'Stella', 'Lapins' and 'Sumtare' (Sweetheart™). 'Stella', released in 1968 by K. Lapins, has been fundamentally important as a parent for self-fertility in many breeding programmes, including that at Summerland. 'Lapins', released in 1984 by David Lane, was named in honour of K. Lapins. It has found commercial success in British Columbia and in several other production regions around the world. 'Sumtare' (Sweetheart™), named in 1994, was the first of a series of late-harvest cultivars that have been instrumental in improving profitability for the British Columbia cherry industry.

The programme uses hand-pollination hybridization techniques and open pollinations. Sometimes tree enclosures are used for crosses if the female parent is not self-fertile. Molecular markers are used for determination of *S*-alleles and self-fertility. Other markers suitable for screening purposes are under development.

The main goals are, among classical traits: fruit size, fruit firmness, fruit taste, low rain splitting, freedom from postharvest defects such as pitting, stem quality, precocity, productivity, self-fertility, a range of maturity periods, good tree growth habit, strong fruit stem attachment, and tree and flower bud winter hardiness adequate for Canada's production regions. Other goals are: suitability for long-distance shipping, improved powdery mildew resistance, low stem detachment force for mechanical harvest potential, new cultivars that ripen in what are now gaps in the cherry maturity season and filling the maturity season for blush cherry cultivars.

Cultivars created include the following: 'Van', 'Sparkle', 'Star', 'Sam', 'Sue', 'Compact Lambert', 'Stella', 'Salmo', 'Compact Stella', 'Summit', 'Lapins', 'Sunburst', 'Newstar', 'Sylvia', 'Newmoon', 'Sumpaca' (Celeste™), 'Sumtare' (Sweetheart™), 'Sumnue' (Cristalina™), 'Sumste' (Samba™), 'Sandra Rose', 'Santina', 'Skeena', 'Sumleta' (Sonata™), 'Summer Jewel', 'Symphony', 'Sonnet', 'Sumele' (Satin™), '13S2009' (Staccato™), '13N0770' (Stardust™), '13S2101' (Sovereign™), 'SPC043' (Sentennial™), 'SPC207' (Starblush™), 'SPC046' (Sofia™), '13N0739' (Starletta™) and 'SPC046' (Suite Note™).

### 4.3.3 Chile

*Breeding institute (or company): Politécnica Universidad Católica de Valparaíso (PUCV)*

The first breeding programme was established in Chile in 2007, with funding from INNOVA-CORFO (Corporación de Fomento de la Producción). This programme involved researchers from PUCV, Universidad Andrés Bello, as well as members from the private companies Vivero Sur, Vivero Rancagua and Agricola Garcés. The project was implemented in different areas of Chile: Quillota, San Francisco de Mostazal, Rancagua, Curicó, Molina and Angol.

The first step was to establish a germplasm collection of 58 sweet cherry cultivars, which were analysed with 327 simple sequence repeat (SSR) markers to check authenticity and evaluate genetic distances between them. These potential genitors were selected for their fruit size and firmness, productivity and chilling requirements. Controlled crosses were made between genitors of similar harvest periods that were cross-compatible and without any legal restriction. Today, 8000 hybrids have been produced and the first flowers were observed in 2012–2013.

*Breeding institute (or company): INIA–Biofrutales*

In 2010, Biofrutales, a Chilean consortium involving six public and seven private partners, applied to a partial governmental support, in order to develop a cherry breeding programme together with Instituto Nacional de Investigaciones Agropecuarias (INIA), the main Chilean agricultural research institution. The target of this programme is to create Chilean cherry cultivars, evaluated for long postharvest conditions, with a high standard of fruit production and fruit quality. Another challenge is to have low-chilling requirements in order to extend the harvest period, especially with early-harvest cultivars. The programme began working with open-pollination hybrids, and, year by year, controlled crosses are increasing the pool of hybrids to be evaluated. The conventional breeding method is beginning to be complemented by biotechnological tools. There are three areas where hybrids are evaluated: La Serena, Buin and Rengo, corresponding to less than 400 chilling hours up to 700 or more chilling hours, respectively. The programme has more than 19,000 hybrids, distributed in the three fields, and around 6000 have been evaluated so far, at least for two seasons. The first 11 advanced selections were grafted on 'MaxMa 14' rootstock, for a second evaluation. The programme expects to release the first cultivar, at least, within the next 10 years.

*Breeding institute (or company): Consorcio Tecnológico de la Fruta S.A. and Pontificia Universidad Católica de Chile ('PUC')*

In 2010, the 'Programa de mejoramiento genético de cerezos' ('PMGCe') was launched to generate new cultivars for Chile, producing high-quality fruit (large, firm and sweet), adapted to Chilean edaphoclimatic conditions and having a longer postharvest life. In addition, the 'PMGCe' is focused on expanding the commercial window for exports of early and late cultivars, which would be adapted to the warm areas of the Central Valley and the rainy areas of the south of Chile, respectively. The programme has used different cultivars such as: 'Brooks', 'Early Burlat', 'Lapins', 'Bing', 'Regina', 'Ruby' and 'Tulare' and some selected Hungarian germplasm. Each season, the 'PMGCe' establishes between 3000 and 4000 hybrids in the field. Seedlings are obtained by seed stratification or by embryo rescue. So far, more than 20,000 hybrids have been established in the field. Hand pollination, cages with bumblebees and open pollination are all used to generate new hybrids.

### 4.3.4 China

*Breeding institute (or company): Institute of Pomology (IP), Dalian Academy of Agricultural Sciences (DAAS)*

The cherry programme of IP, DAAS, is engaged mainly in sweet cherry breeding and cultivation technology research. A sweet cherry cultivar collection was established in 1958. The main collected cultivars were 'Black Tartarian', 'Napoleon', 'Governor Wood', 'Crystal' and 'Suburban Bing'. Cross-breeding began in 1963. The first-generation breeding nursery was established in the 1960s, the second generation in the 1970s–1980s, the third generation in the 1990s and the fourth generation from 2005 to 2009. The fifth-generation breeding nursery is currently being established. The cultivars released were 'Hongdeng', 'Juhong', 'Jiahong', 'Hongyan', 'Wanhongzhu', 'Zaohongzhu', 'Zaolu' and 'Mingzhu'. Among these, 'Hongdeng' was the first-generation main planted sweet cherry cultivar in China. The main breeding goals are productivity, fruit weight, tasting quality, firmness and disease resistance.

*Breeding institute (or company): Zhengzhou Fruit Research Institute, Chinese Academy of Agricultural Sciences (CAAS)*

The cherry programme of CAAS was initiated in the early 1980s. Thirty-two cherry cultivars were successively imported from the UK, New Zealand, Italy and the USA from 1981 to 1983. Crosses were mainly made between 'Black Tartarian' and 'Napoleon', and several cultivars were released, such as 'Longguan' and 'Longbao', in 1996. More than 220 cherry germplasm resources were collected. New cultivars, such as 'Chunxiao', 'Chunyan' and 'Chunxiu', were released between 2009 and 2012. In recent years, each key technology in sweet cherry breeding has been analysed systematically, according to the soil and climate characteristics of the central–western regions of China. The seed germination rate of precocious sweet cherry is up to 70–100%. The main breeding goals are good quality, high yield, resistance to high temperature and high humidity, and adaptation to soil and climatic conditions from warm temperate regions.

*Breeding institute (or company): Institute of Pomology and Forestry (IPF), Beijing Academy of Agriculture and Forestry Science (BAAFS)*

The Cherry Group of IPF, BAAFS, was founded in 1997. Their research is focused on sweet cherry and rootstock breeding, as well as marker-assisted selection (MAS). Nearly 200 cherry cultivars were collected and evaluated. Self-fertility is the main breeding goal, along with productivity and excellent fruit quality. MAS is used to determine the *S*-genotypes of the main cultivars and selections used in cross-design. Recently, a fruit colour molecular marker has been developed and used. 'Stella' and 'Lapins' were selected as the main parents. More than 10,000 hybrids and nearly 20 selections have been obtained from cross- or open pollinations in the past 18 years. To date, five sweet cherry cultivars have been released: 'Caihong', 'Caixia', 'Zaodan', 'Xiangquan No.1' and 'Xiangquan No.2'. Of these, 'Xiangquan No.1' is the first released self-fertile sweet cherry cultivar in China.

### 4.3.5 Czech Republic

*Breeding institute (or company): Research and Breeding Institute of Pomology Holovousy Ltd (RBIPH)*

Breeding activities in RBIPH started in the 1960s. The parent combinations consisted of cultivars from sweet cherry germplasm and landraces, and in some cases seedlings were also used. In the 1970s, a programme of mutation breeding was also developed. A cultivar with compact growth derived from 'Büttners Späte Knorpelkirsche' was the output of this programme. During this period, an important seedling, which was later registered under the name 'Kordia' ('Attika' in USA), was discovered in Techlovice near Hradec Kralove. Currently, it is a cultivar considered worldwide to be a standard for fruit quality and late-ripening sweet cherries. Breeding targets were extending the harvest period, high-quality fruit, cultivars with resistance to diseases and rain-induced cracking, and resistance of flowers to late spring frosts. Annually, there are ~12,000 flowers pollinated. Concerning hybridization techniques, hand pollination is predominant. At present, there are more than 25 sweet cherry cultivars registered or at the stage of application for registration. These cultivars cover most of the maturity period range. In the 1990s, cultivars 'Techlovan' and 'Vanda' were commercially used in plantations. Today, cultivars 'Early Korvik' and 'Tamara' have been assigned plant protection rights in USA. At present, a new programme is being launched aimed at the implementation of MAS strategies and utilization of old landraces.

The main cultivars released for commercial use are 'Early Korvik', 'Christiana', 'Justyna', 'Kasandra', 'Tamara', 'Techlovan', 'Tim' and 'Vanda'.

### 4.3.6 France

*Breeding institute (or company): Institut National de la Recherche Agronomique (INRA) – CEP Innovation*

The first breeding programme (1968–1980) was based on the use of a reduced number of cultivars (mainly 'Burlat' and 'Hedelfinger'). No new cultivars were released. From 1980 onwards, a large collection of more than 400 cultivars was constituted in order to enlarge the genetic base used in the breeding programme. The first INRA cultivars were created, among which 'Arcina®Fercer' was the first to produce very large and firm fruit. Hence, in a second phase of this programme, 'Arcina®Fercer' was heavily used as genitor, and at the end of the 1990s, INRA released a series of cultivars covering most of the maturity period range in French conditions. More recently, a new programme was launched aimed at the implementation of MAS strategies and at the widening of the genetic base used so far, in particular by the utilization of old landraces and exotic germplasm. MAS has recently been implemented for fruit weight and *S*-allele typing, and it should be further developed for other traits such as bloom and maturity dates and fruit-cracking tolerance. Concerning hybridization techniques, hand pollination and open pollination have been predominant, and from 2010, controlled pollination with the use of bumblebees and potted trees were conducted. CEP Innovation is a consortium of French nurserymen. Until 2011, it had the responsibility of commercializing INRA cultivars, and in 2011 it also became the co-owner of the newly released ones.

The main selection criteria of the current breeding programme are fruit weight and firmness, productivity, yield precocity, tasting quality, maturity period range, tolerance to rain-induced fruit cracking, adaptation to climate change (e.g. low-chilling cultivars) and self-fertility. In the medium to long term, new traits such as tolerance/resistance to pests and diseases will be incorporated. Priority will be given to brown rot disease caused by different *Monilinia* spp. and to bacterial canker.

The first four cultivars released by INRA were 'Ferbolus', 'Fernola', 'Fernier' and 'Arcina®Fercer'; only the last two met commercial success. The second series of INRA cultivars comprised: 'Primulat®Ferprime', 'Folfer', 'Ferpin', 'Fertard', 'Fermina', 'Ferdiva', 'Feria', 'Ferpact', 'Ferlizac', 'Fertille', 'Ferdouce' and 'Ferobri'. Today, the main INRA cultivars that are commercialized are 'Folfer', 'Ferdouce', 'Fertille', 'Fermina', 'Ferdiva' and 'Fertard'. In terms of selection rate, a total of 15,000 hybrids were evaluated to finally select ten high-level cultivars, which represents a success rate of 1 out of 1500.

The most promising hybrids are grafted as two clonal copies and evaluated at three different sites. After a second selection phase, the best hybrids are grafted as five to ten clonal copies and evaluated at five sites representing the main French sweet cherry-producing areas. The best hybrids of this second phase enter precommercial trials. This evaluation network is coordinated by the Centre Technique Inter-professionnel des Fruits et Légumes (CTIFL).

### 4.3.7 Germany

*Breeding institute (or company): Julius Kühn-Institut (JKI), Institute for Breeding Research on Fruit Crops*

The first breeding activities were started by M. Schmidt at the Kaiser-Wilhelm-Institut in Müncheberg in the 1930s. The main breeding goals were: large, yellow and firm sweet cherries with properties of good transportability, early ripening time, self-fertility and tolerance to frost. After World War II, M. Zwitzscher continued the sweet cherry breeding in West Germany at the Max Planck Institut in Vogelsang, Cologne, and selected the cultivars 'Primavera' and 'Sekunda'. In 1953, a new breeding programme was started at the Obstbauversuchsanstalt Jork (OVA) in Jork by E.L. Loewel, E.V. Vahl and F. Zahn. The goal of this programme was to select well-adapted cultivars for the climatic conditions of the northern part of Germany. As a result, the cultivars 'Erika', 'Valeska', 'Annabella', 'Oktavia', 'Alma', 'Viola', 'Bianca' and 'Regina' were selected and released. These breeding activities were finished in the 1980s. From 1982 to 1999, H. Schmidt started a new breeding programme at Ahrensburg.

In East Germany, sweet cherry breeding was started by H. Mihatsch in Naumburg in 1958. The main goals were high and stable fruit set, early ripening time, large and firm fruit, low cracking, tolerance to spring frost and suitability for mechanical harvesting. From 1971 to 1990, these breeding activities were continued at the Institute for Fruit Research in Kauscha, Dresden. At this time, M. Fischer, R. Posselt and R. Kaltschmidt worked together with H. Mihatsch in the sweet cherry breeding programme. Their additional research topics were studies of incompatibility, self-fertility, resistance to bacterial canker (*Pseudomonas* spp.) and low growth habit. This long breeding work resulted in the release of a series of four 'Na' cultivars: 'Namosa', 'Nadina', 'Namare' and 'Namati'. Since 2001, sweet cherry breeding has been continued by M. Schuster at the Julius Kühn-Institut in Pillnitz, Dresden. In recent years, the cultivars 'Narana', 'Areko', 'Swing' and 'Habunt' have been released.

The main breeding goals are: fruit quality, fruit set, ripening time (early and late ripening), resistance to biotic and abiotic stresses (*Blumeriella jaapii*, *Monilinia laxa*, spring frost), self-fertility and fruit-cracking tolerance. An important objective of the breeding research is to increase the genetic diversity of breeding material. Seedlings of collected stones from sweet cherry genotypes of the gene centre of sweet cherry in Asia Minor and progeny of interspecific hybridization with *P. canescens* and *Prunus tomentosa* have been evaluated and used in the breeding programme. For the characterization of breeding materials, molecular markers are utilized for *S*-allele determination, fruit size characterization and cherry leaf spot resistance.

### 4.3.8 Hungary

*Breeding institute (or company): NARIC Fruitculture Research Institute (FRI)*

In Hungary, the first breeding activities (1950–1985) started after World War II by the late S. Brózik and were continued by J. Apostol and more recently also by Z. Békefi. Three main approaches were followed: clonal and landrace selection, artificial cross-breeding and introduction of foreign cultivars. The main objectives are: extending the maturity time, with a focus on precocity and late ripening, excellent fruit quality for both the fresh market and the canning industry, large fruit size (26–28 mm or larger), good firmness with high sugar content and pleasant sugar/acid balance, low sensitivity to rain cracking, good shelf-life, a long green flexible stem, a round or wide-shouldered shape, self-fertility, winter hardiness, and tolerance or resistance to diseases/leaf spot, brown rot, *Cytospora* canker.

From 1950 to 1953, the breeding programme was aimed at selecting country sorts from the three Hungarian native sweet cherry growing areas. Three cultivars, 'Solymári Gömbölyű', 'Pomázi Hosszúszárú' and 'Szomolyai Fekete', resulted from this landrace selection work. The basis of the clonal selection work was using the cultivar 'Germersdorfer' from which three selected clones (numbers 1, 3 and 45) are still in propagation.

In the first crossing programme (1953–1972), the cultivar 'Germersdorfer' was the main female parent. 'Hedelfinger' and 'Pojebrad' and some selected landraces were the male parents. Four cultivars were registered: 'Margit', 'Linda', 'Katalin' and 'Kavics'. The second crossing programme (1972–1985) aimed to reach self-fertility. Cultivars 'Vera' (2002) and 'Axel' ('Alex', 1999) were released. The third crossing period (1986–2000) covered the raising of the $F_2$ offspring generation of previous crossings and further cross-breeding with self-fertile foreign cultivars such as 'Stella' and 'Sunburst'. These crossings were based on the female parents 'Bigarreau Burlat', 'Van', 'Ljana', 'Trusenszkaja 2', 'Trusenszkaja 6', 'Yellow Dragan' and 'Hedelfinger'. As male parents, some $F_1$ hybrids (H-2, H-3, H-203, H-236) and cultivars 'Stella', 'Bigarreau Burlat' and 'Van' were used. The released cultivars were 'Rita', 'Carmen', 'Sándor', 'Petrus', 'Aida', 'Paulus', 'Annus' ('Anita') and 'Tünde'. The fourth crossing period has been going on since 2001. Mainly $F_2$ and $F_3$ progenies and the cultivars 'Kordia' and 'Regina' have been used as female parents, with the male parents 'Sweetheart' and 'Sunburst' and Hungarian fertile cultivars such as 'Axel', 'Sándor', 'Paulus' and 'Petrus'. An application was made recently for six candidate cultivars to be included in the Hungarian National Variety List. Today, there are 8500 hybrids under evaluation.

### 4.3.9 Italy

*Breeding institute: Bologna University, Department of Agricultural Sciences*

The breeding programme began in 1983 and was developed in two phases, 1983–2007 and 2005 onwards. During the first phase, the 'Star' series was developed. The main goals were: self-fertility, early maturity and consistent high yield. Classical hybridization was followed by three selection steps: seedling assessment, propagation of selected seedlings for S2 seedling assessment in small trial plots, and genotype selection followed by comparative field trials of S3 selections grafted on several rootstocks and then commercial trials. This process lasted at least 15 years. Crosses were made mostly between American self-fertile cultivars such as 'Lapins', 'Stella' and 'Sunburst' and European cultivars such as 'Burlat', 'Giorgia' and 'Ferrovia'. About 8000 seedlings were produced, and seven cultivars were patented: 'Early Star®Panaro 2', 'Blaze Star', 'LaLa Star', 'Sweet Early®Panaro 1', 'Grace Star', 'Black Star' and 'Big Star'.

The second phase corresponds to the 'Sweet' series. This new suite of cultivars is the result of crosses begun in the early

2000s at Bologna's former Department of Arboreal Crops (DCA), now called the Department of Agricultural Sciences. The specific intent was to cover a marketing calendar of 40–50 days by developing six or seven new cultivars, all with a common denominator of top-quality, high-performance trees producing fruit of large size and firm flesh, and of excellent appearance (shiny red colour), flavour and taste, that were readily recognizable. The first crosses involved several old local cultivars, such as 'Duroni di Vignola', which are prized for their sweetness, firmness and fragrance, and several American cultivars renowned for their excellent appearance, large size and shiny colour.

The University of Bologna has undertaken this second phase of the programme under a public–private financing partnership including more than five partners. The selection of seedlings began using strict rating parameters of minimum threshold in the screening process (e.g. ≥28 mm as the prevalent size, 4–6 on the CTIFL skin colour scale). This selection screened the number of potentially viable seedlings down to 12 in 2004, from an initial 3000, and these 12 went straight to precommercial Selection III in three trial fields set up in 2008 for extensive testing. Molecular markers such as SSRs were used to determine sterility factors (allelic groups) and to produce a genetic ID card of the first five cultivars. Other qualitative factors of fruit were also investigated using a threefold approach: (i) sensory evaluation via panel tests involving consumers and experts to determine preferences and appeal; (ii) laboratory biochemical analysis via gas chromatography to identify aromatic compounds; (iii) general nutraceutical properties.

The final selection stage led to the release of the six new cultivars under the names of the breeders' children: 'Sweet Aryana®PA1UNIBO', 'Sweet Lorenz®PA2UNIBO', 'Sweet Gabriel®PA3UNIBO', 'Sweet Valina®PA4UNIBO', 'Sweet Saretta®PA5UNIBO' and 'Sweet Stephany®PA7 UNIBO'. Another, non-Sweet cultivar, called 'Marysa®PA6UNIBO', has also been developed, patented and licensed for commercial sale.

### 4.3.10 Japan

*Breeding institute: Horticultural Experiment Station, Yamagata Integrated Agricultural Research Center*

The first breeding programme was started in 1957 at the Okitama branch of the Yamagata Agricultural Experiment Station. In 1978, the breeding programme was transferred to the Horticultural Experiment Station, Yamagata Integrated Agricultural Research Center. The objective of the first breeding programme was to breed early- and late-ripening cultivars of high quality, because 'Satonishiki' matures at mid-season (middle to late June maturity). Today, the cultivars sought are early and late ripening with high-quality fruit (weight and firmness, tasting quality), bright red skin colour, white flesh colour, high productivity and self-fertility. Recently, MAS has been launched to select seedlings with white flesh colour, early maturity and self-fertility. The main released cultivars are: 'Nannyo' (first programme), 'Benisayaka' and 'Benishuhou' (second programme) and 'Benikirari' and 'Beniyutaka' (third programme).

### 4.3.11 Romania

*Breeding institute (or company): Research Institute for Fruit Growing (RIFG)*

The first sweet cherry breeding programme started in 1951 at the Research Station for Fruit Growing, at Bistriţa (RSFG Bistriţa) and expanded in 1967 at the RIFG, Pitesti, and the Research Station for Fruit Growing (RSFG), Iasi. The objective of the first phase, conducted until 1970, was the breeding of new cultivars that were early or late ripening, with fruit of superior quality, and these were registered as the cultivars 'Negre de Bistriţa' and 'Uriase de Bistriţa'. After 1970, the volume of crossings increased considerably. A comprehensive collection and study of valuable biotypes of bitter cherry from spontaneous flora was conducted, enriching the available germplasm (to more than 360 genotypes), which included the most valuable cultivars and selections of Russian, European and North American origin. By 1990, the cultivars 'Cerna', 'Ponoare', 'Colina', 'Izverna' (at RIFG Pitesti), 'Timpurii de Bistriţa', 'Rosii de Bistriţa', 'Jubileu 30' and 'Rubin' (RSFG Bistriţa) were released and the selections of bitter cherry 'Silva' as well as 'Amara' (RIFG Pitesti). After 1990, the objectives of the programme were oriented towards precocity, self-fertility, high productivity and high fruit quality. Currently, breeding aims are to find sources of resistance to leaf spot (*B. jaapii*), blossom blight (*Monilinia* spp.) and fruit cracking. The following cultivars have been released: 'Severin', 'Daria', 'Tentant', 'Clasic', 'Superb', 'Sublim', 'Simbol', 'Spectral', 'Special' (RIFG Pitesti), 'Splendid' (RSFG Cluj), 'Roze', 'Someşan', 'Iva', 'Ana' (RSFG Bistriţa) and 'Bucium', 'Catalina', 'Cetatuia', 'Golia', 'Iasirom', 'Maria', 'Marina', 'Stefan', 'Tereza', 'Oana', 'Radu', 'Lucia', 'George', 'Paul', 'Margo', 'Mihai', 'Cociu', 'Iosif', 'Ludovic', 'Anda', 'Alex', 'Andrei', 'Amar de Maxut' and 'Amar de Galata' (RSFG Iasi).

Breeding methods were and still are the classical ones, consisting of selection and testing of valuable genotypes from local landraces (especially for bitter cherry) and artificial pollination made by hand or by bumblebees. Today, the main cultivars that are commercialized are 'Bucium', 'Maria', 'Ludovic', 'Daria', 'Severin', 'Rubin', 'Special' and 'Amar Galata'.

### 4.3.12 Spain

*Breeding institute (or company): Centro de Investigaciones Científicas y Tecnológicas de Extremadura (CICYTEX-La Orden)*

The 'Picota' cherry cultivars, which are harvested naturally stemless, are a hallmark of the Jerte Valley, Extremadura. Despite this importance, several technical problems restrict the profitability of 'Picota' in the Jerte Valley. Moreover, the replacement of traditional cultivars by foreign ones is provoking a loss of the 'Picota' differentiating factor together with a high risk of losing the traditional cultivars. In order to provide solutions to this problematic situation, in 2006 the Breeding Program of Sweet Cherries and 'Picota' of the Jerte Valley was initiated. The main objectives of the programme are to improve certain physicochemical properties of native cultivars cultivated in the Jerte Valley and to diversify this product. Hence, cultivars from the Jerte Valley Germplasm Bank and from breeding programmes from the USA and Canada, with interesting traits, such as earliness, high productive potential, self-fertility, tolerance to rain-induced cracking, large size, flesh firmness and good organoleptic quality, are used as genitors. Currently, the number of seedlings in the field is 1828 and the selection process based on fruit quality traits began in 2010. According to our previous results, ten potential new cultivars have been selected, four of which correspond to crosses with the cultivar type 'Picota' as the parent, for which, in 2013, an application was made for Community Plant Variety Rights.

*Breeding institute: Murcia Institute of Agri-Food Research and Development (IMIDA)*

The IMIDA sweet cherry breeding programme began in 2006. Around 2000 hybrids older than 5 years, from 42 crosses, are currently under evaluation. The programme follows a classical methodology with techniques such as hand pollination, emasculation and enclosure of female parents if they are not self-fertile, but also uses seed from open-pollinated trees. The main objectives are: extending the maturity time, with a focus on precocity ripening and low-chilling requirements, excellent fruit quality with a large fruit size (up to 28 mm), good firmness with a pleasant sugar/acid balance, self-fertility, and good behaviour in warm climates with special reference to double fruit production and pronounced fruit suture. The main female parents are 'Brooks', 'Burlat', 'Early Lory', 'Early Bigi' and 'Ferprime', and as male parents, the Spanish landrace cultivar 'Cristobalina' and Canadian cultivars such as 'Lapins' and 'Newstar', are used mainly to confer self-fertility. The selection process based on maturity date, fruit size and firmness, double fruit production and fruit suture began in 2013. A total of 43 selected genotypes are in a second phase of study, while 12 of them are proposed as advanced preselections because they cover most of the maturity period range with enough fruit quality and good agronomic performance.

### 4.3.13 Turkey

*Breeding institute (or company): Atatürk Horticultural Central Research Institute*

In Turkey, there were over 60 local sweet cherry cultivars, which were believed to have originated from human selections among large-fruited semi-wild sweet cherries grown in different parts of Anatolia for centuries. However, among these 60 local cultivars, '0900 Ziraat' gained importance and popularity due to its superior fruit characteristics. The first planned breeding programme was started in 2001 with the aim of improving fruit quality and also the self-incompatible characteristics of this cultivar. Cross-breeding and mutation methods are used to obtain new cultivars. Through mutation breeding of '0900 Ziraat', two new cultivars, 'Burak' and 'Aldamla', were released in 2013. Cross-breeding was done between '0900 Ziraat' and 'Stella', and six promising self-fertile cultivar candidates that were obtained are now in the registration stage. MAS is used to determine self-fertile hybrids. The main selection criteria of the current breeding programme are self-fertility, fruit weight and firmness, productivity, yield precocity, tasting quality and maturity period range.

### 4.3.14 UK

*Breeding institute (or company): East Malling Research (EMR)*

Cherry breeding in the UK started in the 1920s and was originally divided between two sites. Scion breeding was based at the John Innes Institute, while rootstock breeding took place at East Malling, Kent. In the 1980s, both programmes merged at EMR. Genetic diversity of cherry germplasm was greatly improved with the introduction of cultivars and species material from around the world. The main breeding objectives were late-ripening cherries with low susceptibility to cracking and rots. Introgression of resistance/tolerance to bacterial canker (*Pseudomonas* spp.) and black fly (*Myzus cerasi*) into cultivated cherries, as well as the development of molecular tools to characterize and elucidate the mechanisms of self-(in)compatibility were also key to the programme. Hand pollination of emasculated flowers, seedling evaluation on their own roots and trials using 'Colt' as a common rootstock were common practice during that period, and the scion cultivars 'Penny' (2001) and 'Zoë' (2008) were released as a result.

Following the withdrawal of public funding for fruit tree breeding in the UK, the East Malling Cherry Group (EMCG) was formed in 2010 to utilize the existing germplasm pipeline at EMR and engage in further breeding. This programme is fully privatized and vertically integrated as a partnership between EMR, Univeg UK Ltd and the Associated International Group of Nurseries. The EMCG aims to develop cherry cultivars of outstanding fruit quality that are adapted to intensive orchards. Season extension (for both early and late ripening), novelty types and improved storage ability are also main breeding targets. A range of pollination techniques are deployed, including the use of natural pollinators in cages, and open-pollinated seed from selected trees is also introduced into the pipeline. Seedlings are currently evaluated on a common rootstock in the first instance, and the most promising materials are evaluated for four to five seasons. Advanced selections are then distributed to the partners for multi-site trials in various countries. The implementation of MAS is under evaluation; parental germplasm and selections are being characterized with existing published markers of interest and a cost–benefit analysis of seedling preselection will be carried out in the near future.

### 4.3.15 Ukraine

*Breeding institute (or company): Institute of Horticulture of National Academy of Agrarian Sciences of Ukraine (IH NAAS)*

IH NAAS was founded in Kiev in 1930. Under the guidance of the institute, a network of regional research stations and bases function in different zones of fruit growing in Ukraine. Significant progress in sweet cherry breeding was achieved at the Melitopol Research Station of Horticulture, where selection started in 1928–1929, using different methods including chemical and physical mutagenesis. Many cultivars were selected at the Artemivsk Nursery Research Station, Opytne, where selection of sweet and sour cherry began in 1952. The cultivar 'Melitopolska Chorna' obtained in 1933 became an important cultivar. Cultivars 'Drogans Gelbe' and 'Valerii Chkalov' were heavily used as genitors, and many modern cultivars are hybrids from these parents. Concerning methods of hybridization, in the early stages of breeding, free pollination and pollen mixtures of different cultivars, and in later phases, controlled hand pollination, were used. Recently, MAS has been launched at IH NAAS for fruit weight. Important sweet cherry breeding centres that are not a part of IH NAAS are the Institute of Pomology NAAS, Mliiv, and the Nikita Botanical Garden, Nikita, Crimea. In these institutions, breeding started in the early 20th century.

The main breeding goals are: fruit weight, productivity, yield precocity, firmness, tasting quality, maturity period range and vigour (tree size). Other desired traits are adaptability to different climatic conditions (high winter and drought hardiness, tolerance to spring frosts and dry winds), tolerance/resistance to *Monilinia* blossom blight and cherry leaf spot diseases, tolerance to rain-induced fruit cracking and fruit–pedicel separation.

The best cultivars created at the Melitopol Research Station of Horticulture are 'Valerij Chkalov', 'Melitopolska Chorna', 'Prysadybna', 'Dachnitsja', 'Chervneva Rannya', 'Krupnoplidna', 'Talisman', 'Anons', 'Kazka', 'Prostir', 'Anshlah', 'Era' and 'Liubymytsia Turovtseva'; at the Artemivsk Nursery Research Station are 'Annushka', 'Vasylysa Prekrasna', 'Etyka', 'Otrada', 'Donetska Krasavytsia' and 'Proshchalna Taranenko'; and at IH NAAS are 'Nizhnist' and 'Liubava'. Among the best sweet cherry cultivars bred at Institute of Pomology, NAAS, are 'Lehenda Mliieva' and 'Dar Mliieva'.

### 4.3.16 USA

*Breeding institute (or company): Washington State University (WSU)*

The Sweet Cherry Breeding Program at WSU has been the first to routinely implement MAS. The programme's releases include cultivars such as 'Rainier', 'Chelan', 'Cashmere', 'Index', 'Benton' and 'Selah'. The breeding programme came to a hiatus in the mid-1980s following the retirement of the former breeder T. Toyama. Evaluation of crosses continued, which led to further releases including 'Kiona' and 'Cowiche' in 2007. However, active breeding resumed in 2004 with the involvement of the Washington Tree Fruit Research Commission and the Oregon Sweet Cherry Commission.

The goal of the breeding programme is to develop new high-quality cultivars with high consumer appeal suitable for production in the Pacific Northwest growing regions of the USA. The ultimate goal is to produce cultivars that fit into distinct target market classes. One major objective is to develop a suite of new early-, mid- and late-ripening self-fertile cultivars with a range of skin colours, to extend the market window of Pacific Northwest cherries. DNA information guides breeding decisions and improves the efficiency and cost-effectiveness of breeding operations. Genetic tests for self-(in)compatibility, fruit size, firmness, ripening date and fruit colour are currently used to select parents with a high probability of producing progeny with the desired attributes and for culling inferior seedlings prior to field planting. Seedlings possessing desirable marker genotypes are planted into the field at the optimal planting time and cultivated according to best horticultural practices for the Pacific Northwest growing area. MAS for other traits including pedicel fruit retention force, powdery mildew resistance and bacterial canker resistance will be deployed as soon as new DNA tests are available.

Currently, the programme is evaluating the merit of 40 advanced selections representing six target market classes: (i) mid-season, self-fertile, mahogany; (ii) late-season, self-fertile, mahogany; (iii) late-season, self-fertile, blush; (iv) early-season, self-fertile, mahogany; (v) early-season, self-fertile, blush; and (vi) early- to mid-late season, self-fertile, mahogany. These should be suitable for mechanical harvest. Each target market class is populated with a current leading cultivar possessing a trait(s) to be improved upon through judicious breeding.

*Breeding institute (or company): Cornell University*

The cherry breeding programme at Cornell University, Geneva, New York, began in the early 1900s with R. Way, who was also an apple breeder. In 1964, Way released 'Ulster', which was adapted to harsh colder climates, as well as the late-ripening 'Hudson'. Way named 'Kristin' cherry in 1982. Subsequently, the programme was led by S. Brown and then by B. Anderson, until he retired in 2004. During this time, four cultivars were released: 'Hartland' (1992), 'Royalton' (1991), 'BlackGold®' (2002) and 'WhiteGold®' (2000). The Cornell Center for Technology Enterprise and Commercialization contracted the commercialization of all Cornell's cherries for North America to International Plant Management, and in 2008 four new fresh-market cherries were released: 'BlackPearl®', 'RadiancePearl™', 'EbonyPearl™' and 'BurgundyPearl™'.

## 4.4 Characteristics of Sweet Cherry Cultivars

In this section, we give the main characteristics of 119 cultivars of commercial importance. Twenty-nine of them were described previously by Bargioni (1996), who documented 83 cultivars; this means that, in the last 20 years, 54 of these cultivars are no longer considered to be commercially interesting, showing the dynamics of varietal improvement over the last two decades. Nevertheless, it is interesting to recall several positive agronomic characteristics of some of these cultivars that could be useful in breeding programmes, such as:

- low susceptibility to rain-induced fruit cracking: 'Adriana', 'Annabella', 'Castor', 'Della Rocca', 'Early Rivers', 'Merton Heart', 'Mora di Cazzano', 'Namosa', 'Royalton', 'Viscount', 'Vittoria' and 'Viva';
- low susceptibility to bacterial canker: 'Colney', 'Hertford', 'Merton Glory' and 'Vittoria';
- low susceptibility to winter frost: 'Büttners Späte Rote Knorpelkirsche' and 'Grosse Schwarze Knorpelkirsche';
- early maturity: 'Ramon Oliva', 'Nalina' and 'Sandra';
- late maturity: 'Hudson';
- winter hardiness: 'Kristin';
- mechanical harvest (stemless): 'Namosa' and 'Vittoria'.

### 4.4.1 Sweet cherry cultivars with global importance

In this category are listed cultivars with global importance that are important contributors to production in several countries. Also included are several cultivars planted in only one country but with a proven growing potential in other regions. Several cultivars included in this list have been identified as very close, genetically speaking, in recent studies conducted with molecular markers. Campoy *et al.* (2016) analysed a collection of genetic resources with the 6K single-nucleotide polymorphism chip developed within the RosBREED project and found that 'Belge' and 'Ferrovia' were identical, and that 'Schneiders Späte Knorpel' was very similar. The cultivars 'Stark Lambert', 'Gege', 'Noire de Meched' and 'Badacsoni' were also identical to 'Belge' and 'Ferrovia'. Another study, conducted with 12 SSR markers (Schüller, 2013), could not differentiate cultivars 'Germersdorfer', 'Noire de Meched' and 'Schneiders Späte Knorpel'.

Although quantitative data are available for many important agronomic traits such as fruit weight (or size) and firmness, we decided to use only qualitative descriptors due to high genotype × environment and/or genotype × cultural practices interactions. Concerning bloom and maturity time, as far as possible, a comparison has been made with the reference cultivar 'Burlat'. Finally, among the fruit characteristics, the colour is given only for the fruit skin.

Two very important cultivars from Iran, 'Shishei' and 'Siahe Mashhad' are not included in this list as little information is available on their characteristics. Both cultivars produce dark red cherries on medium-vigour trees. 'Siahe Mashhad' has the *S*-genotype $S_3S_{12}$, is tolerant to bacterial canker and could be a synonym of 'Noire de Meched' (M. Schuster, Dresden, 2016, personal communication).

### *'Ambrunés'*

Origin: Extremadura, Spain.
Parentage: unknown, old cultivar.
Tree growth: upside-down heart-shaped canopy, medium vigour.
*S*-alleles: $S_3S_6$.
Productivity: moderate.
Blooming time: early.
Ripening time: 30 days after 'Burlat'.
Fruit characteristics: heart-shaped, large and very firm fruit, dark rose.
Resistances/specifics: marketed without stalk ('Picota' type).

### *'Bedel' (Bellise™)*

Origin: selected by Pierre Argot, France.
Parentage: 'Starking Hardy Giant' × 'Burlat'.
Tree growth: semi-upright, spreading, medium to strong vigour.
*S*-alleles: $S_1S_9$.
Productivity: very good.
Blooming time: early.
Ripening time: 4–8 days after 'Burlat'.
Fruit characteristics: reniform, medium to large, firm, light red.
Resistances/specifics: very susceptible to cracking and double fruit; not susceptible to mild winters.

### *'Belge'*

Origin: France.
Parentage: unknown.
Tree growth: upright, medium vigour.
*S*-alleles: $S_3S_{12}$.
Productivity: good.
Blooming time: late.
Ripening time: 23–27 days after 'Burlat'.
Fruit characteristics: heart-shaped, large, firm, light red.
Resistances/specifics: susceptible to cracking and to mild winters.

### *'Bing'*

Origin: Oregon, USA (1875).
Parentage: 'Black Republican' × 'Napoleon'.
Tree growth: erect branches, vigorous.
*S*-alleles: $S_3S_4$.
Productivity: very good.
Blooming time: early/mid-season.
Ripening time: 20–25 days after 'Burlat'.
Fruit characteristics: roundish, large to very large, very firm, red to black.
Resistances/specifics: very susceptible to cracking and double fruit.

### *'Black Star'*

Origin: University of Bologna, Italy.
Parentage: 'Lapins' × 'Burlat'.
Tree growth: semi-upright, medium vigour.
*S*-alleles: $S_1S_3$.
Productivity: very good.
Blooming time: early.
Ripening time: close to 'Summit'.
Fruit characteristics: round, large to very large, very firm, red to black.
Resistances/specifics: low to average susceptibility to cracking.

### *'Boambe de Cotnari'*

Origin: Romania.
Parentage: unknown.
Tree growth: semi-upright, low to medium vigour.
*S*-alleles: unknown.
Productivity: very good.
Blooming time: 1 day before 'Burlat'.
Ripening time: 15–16 days after 'Burlat'.

Fruit characteristics: reniform to cordate, medium to large, medium firm, blushed type.
Resistances/specifics: susceptible to cracking.

### *'Brooks'*

Origin: University of California-Davis, USA.
Parentage: 'Rainier' × 'Burlat'.
Tree growth: upright, vigorous.
*S*-alleles: $S_1S_9$.
Productivity: good.
Blooming time: same as 'Burlat'.
Ripening time: 10–17 days after 'Burlat'.
Fruit characteristics: reniform, large, firm, red to dark red.
Resistances/specifics: extremely susceptible to cracking; low-chilling requirements.

### *'Burlat'*

Origin: selected by Léonard Burlat, France.
Parentage: unknown.
Tree growth: upright, strong to very strong vigour.
*S*-alleles: $S_3S_9$.
Productivity: very good.
Blooming time: early/mid-season.
Ripening time: very early.
Fruit characteristics: oblate, medium size, medium to low firmness, red.
Resistances/specifics: susceptible to cracking, double fruit and mild winters.

### *'Büttners Späte Rote Knorpelkirsche'*

Origin: Halle, Germany.
Parentage: unknown.
Tree growth: upright, very vigorous.
*S*-alleles: $S_3S_4$.
Productivity: very good.
Blooming time: early/mid-season.
Ripening time: 23–29 days after 'Burlat'.
Fruit characteristics: heart-shaped, medium size, very firm, blush type.
Resistances/specifics: very susceptible to cracking.

### *'Carmen'*

Origin: Fruitculture Research Institute, NARIC, Hungary.
Parentage: 'Yellow Dragan' ('Dogans Gelbe') × 'H-303' ('Germersforfer' × open pollination (o.p.)).
Tree growth: semi-upright, moderate vigour.
*S*-alleles: $S_4S_5$.
Productivity: medium to good.
Blooming time: early/mid-season.
Ripening time: 10–12 days after 'Burlat'.
Fruit characteristics: flattened, round, very large, firm, deep red.
Resistances/specifics: susceptible to cracking.

### *'Chelan'*

Origin: Washington State University, USA.
Parentage: 'Stella' × 'Beaulieu'.
Tree growth: moderate branching (compared with 'Bing').
*S*-alleles: $S_3S_9$.
Productivity: very good.
Blooming time: early.
Ripening time: 10–12 days after 'Burlat'.
Fruit characteristics: roundish, medium to large, firm, red.
Resistances/specifics: incompatible with Mahaleb rootstock.

### *'Early Korvik'*

Origin: RBIPH, Holovousy, Czech Republic.
Parentage: mutant of 'Korvik' ('Kordia' × 'Vic').
Tree growth: spreading, medium vigour.
*S*-alleles: $S_2S_6$.
Productivity: very good.
Blooming time: mid- to late season.
Ripening time: 3 weeks after 'Burlat'.
Fruit characteristics: heart-shaped, large, firm, dark to black-red.
Resistances/specifics: tolerant to cracking, good tolerance to leaf spot and *Monilinia* brown rot.

### *'Emperor Francis' (syn. 'Kaiser Franz (Josef)')*

Origin: Germany/Austria.
Parentage: unknown, very old cultivar.
Tree growth: spreading, medium vigour.
*S*-alleles: $S_3S_4$.
Productivity: good.
Blooming time: mid-season.
Ripening time: 18 days after 'Burlat'.
Fruit characteristics: roundish heart-shaped, medium size, firm, blush type.
Resistances/specifics: redder and firmer than 'Napoleon'; mainly used for the processing industry.

### 'Ferrovia'

Origin: Italy.
Parentage: unknown.
Tree growth: upright, low to medium vigour.
*S*-alleles: $S_3S_{12}$.
Productivity: good.
Blooming time: late.
Ripening time: 20–22 days after 'Burlat'.
Fruit characteristics: heart-shaped, large, firm, pink.
Resistances/specifics: susceptible to cracking and mild winters.

### 'Folfer'

Origin: INRA, France.
Parentage: 'Arcina®Fercer' × o.p.
Tree growth: semi-upright, medium to strong vigour.
*S*-alleles: $S_6S_9$.
Productivity: very good.
Blooming time: very early.
Ripening time: 7–13 days after 'Burlat'.
Fruit characteristics: round, large to very large, very firm, red.
Resistances/specifics: susceptible to cracking, double fruit and mild winters.

### 'Germersdorfer'

Origin: Guben, Germany.
Parentage: unknown.
Tree growth: upright, semi-strong to strong vigour.
*S*-alleles: $S_3S_{12}$.
Productivity: very good.
Blooming time: late.
Ripening time: 23–27 days after 'Burlat'.
Fruit characteristics: heart-shaped, large, firm, crisp, light red.
Resistances/specifics: susceptible to cracking.

### 'Giorgia'

Origin: Verona, Italy.
Parentage: 'ISF 123' × 'Caccianese'.
Tree growth: upright, vigorous.
*S*-alleles: $S_1S_{13}$.
Productivity: very good.
Blooming time: early/mid-season.
Ripening time: 7–11 days after 'Burlat'.
Fruit characteristics: heart-shaped, large, firm, red.
Resistances/specifics: moderately susceptible to cracking.

### 'Grace Star'

Origin: University of Bologna, Italy.
Parentage: 'Burlat' × o.p.
Tree growth: semi-upright, strong growth.
*S*-alleles: $S_4S_9$ (self-fertile).
Productivity: very good.
Blooming time: mid-season.
Ripening time: 14–20 days after 'Burlat'.
Fruit characteristics: reniform, very large, medium firm, light red.
Resistances/specifics: moderately susceptible to cracking.

### 'Hedelfinger'

Origin: Hedelfingen, Germany.
Parentage: unknown.
Tree growth: upright to spreading, strong vigour.
*S*-alleles: $S_3S_5$.
Productivity: very good.
Blooming time: medium-late to late.
Ripening time: 25 days after 'Burlat'.
Fruit characteristics: ovoid to heart-shaped, medium to large, firm, brown to red.
Resistances/specifics: medium susceptibility to cracking, quite resistant to frost.

### 'Hongdeng'

Origin: DAAS, Dalian, China.
Parentage: 'Napoleon' × 'Governor Wood'.
Tree growth: upright, strong growth.
*S*-alleles: $S_3S_9$.
Productivity: very good.
Blooming time: 'Burlat'.
Ripening time: 3 days after 'Burlat'.
Fruit characteristics: reniform, large, medium firm, red.
Resistances/specifics: susceptible to cracking.

### 'Kordia'

Origin: Bohemia, Czech Republic.
Parentage: unknown.
Tree growth: upright, pyramidal with spreading branches, medium vigour.
*S*-alleles: $S_3S_6$.

Productivity: very good.
Blooming time: late.
Ripening time: 18–25 days after 'Burlat'.
Fruit characteristics: heart-shaped, large, firm, red to dark violet.
Resistances/specifics: relatively tolerant to cracking, susceptible to high-temperature changes in winter time.

### 'Krupnoplidna' (syn. 'Krupnoplodnaja')

Origin: Melitopol, Ukraine.
Parentage: 'Bigarreau Napoleon Blanc' × pollen mix ('Valerij Chkalov' + 'Elton' + 'Jaboulay').
Tree growth: round crown, vigorous.
*S*-alleles: $S_5S_9$.
Productivity: very good.
Blooming time: early.
Ripening time: 17–23 days after 'Burlat'.
Fruit characteristics: flat-round, large to very large, firm, red.
Resistances/specifics: very susceptible to cracking; high winter and drought hardiness.

### 'Lambert'

Origin: Oregon, USA.
Parentage: 'Napoleon' × 'Black Heart'.
Tree growth: upright, vigorous.
*S*-alleles: $S_3S_4$.
Productivity: very good.
Blooming time: medium-late.
Ripening time: 20 days after 'Burlat'.
Fruit characteristics: heart-shaped, medium to large, firm, purple red.
Resistances/specifics: susceptible to cracking; slow to enter into production.

### 'Lapins'

Origin: Summerland, Canada.
Parentage: 'Van' × 'Stella'.
Tree growth: very upright, vigorous.
*S*-alleles: $S_1S_{4'}$ (self-fertile).
Productivity: very good.
Blooming time: early.
Ripening time: 25–28 days after 'Burlat'.
Fruit characteristics: roundish to heart-shaped, medium to large, firm, purple red.
Resistances/specifics: susceptible to cracking, low-chilling requirements, regularly productive.

### 'Melitopolska Chorna'

Origin: Melitopol, Ukraine.
Parentage: 'Frantsuzka Chorna' × o.p.
Tree growth: semi-upright, vigorous.
*S*-alleles: unknown.
Productivity: very good.
Blooming time: early/mid-season.
Ripening time: 30–35 days after 'Burlat'.
Fruit characteristics: round, medium size, firm, dark red.
Resistances/specifics: low susceptibility to cracking; winter hardy; resistant to *Monilinia* blossom blight.

### 'Merchant'

Origin: John Innes, UK.
Parentage: 'Merton Glory' × o.p.
Tree growth: spreading, medium vigour.
*S*-alleles: $S_4S_9$.
Productivity: very good.
Blooming time: early/mid-season.
Ripening time: 15 days after 'Burlat'.
Fruit characteristics: moderate size, black.
Resistances/specifics: moderate susceptibility to cracking; good resistance to bacterial canker.

### 'Narana'

Origin: JKI, Dresden, Germany.
Parentage: 'Knauffs Schwarze' × 'St. Charmes'.
Tree growth: spreading, medium vigour.
*S*-alleles: $S_2S_9$.
Productivity: very good.
Blooming time: early.
Ripening time: 2 days before 'Burlat'.
Fruit characteristics: flat-round, medium to large, medium firm, dark red.
Resistances/specifics: tolerant to cracking.

### 'Napoleon' (syn. 'Royal Ann')

Origin: Germany.
Parentage: unknown.
Tree growth: medium upright, vigorous.
*S*-alleles: $S_3S_4$.
Productivity: good to very good.
Blooming time: medium to late.
Ripening time: 18–22 days after 'Burlat'.
Fruit characteristics: heart-shaped, medium size and firmness, blush type.

Resistances/specifics: mainly used for processing; susceptible to double fruit.

### *'Rainier'*

Origin: Washington State University, USA.
Parentage: 'Bing' × 'Van'.
Tree growth: upright, poor branching, vigorous.
*S*-alleles: $S_1S_4$.
Productivity: very good.
Blooming time: early.
Ripening time: 18–22 days after 'Burlat'.
Fruit characteristics: slightly obovate, large, firm, blush type.
Resistances/specifics: susceptible to cracking and *Monilinia* blossom blight; low-chilling requirements; regularly productive.

### *'Regina'*

Origin: Jork, Germany.
Parentage: 'Schneiders Späte Knorpel' × 'Rube'.
Tree growth: pyramidal with spreading branches, vigorous.
*S*-alleles: $S_1S_3$.
Productivity: very good.
Blooming time: very late.
Ripening time: 28–35 days after 'Burlat'.
Fruit characteristics: flat-round to round, medium to large, very firm, dark red.
Resistances/specifics: very tolerant to cracking and *Monilinia* blossom blight; susceptible to bacterial canker in certain environments.

### *'Rivedel' (syn. 'Early Lory'; Earlise™)*

Origin: selected by Pierre Argot, France.
Parentage: 'Starking Hardy Giant' × 'Burlat'.
Tree growth: semi-upright, medium to strong vigour.
*S*-alleles: $S_1S_9$.
Productivity: very good.
Blooming time: early.
Ripening time: 0–4 days before 'Burlat'.
Fruit characteristics: reniform round, large, low firmness, red.
Resistances/specifics: very susceptible to cracking, not susceptible to mild winters.

### *'Rubin'*

Origin: Bistriţa, Romania.
Parentage: 'Hedelfinger' × 'Germersdorfer'.
Tree growth: spreading, good branching, low to medium vigour.
*S*-alleles: $S_3S_{12}$.
Productivity: very good.
Blooming time: very late.
Ripening time: 25–30 days after 'Burlat'.
Fruit characteristics: cordate elongated, large, firm, light red.
Resistances/specifics: low susceptibility to cracking; susceptible to mild winters.

### *'Sandra Rose'*

Origin: Summerland, Canada.
Parentage: '2C-61-18' ('Star' × 'Van') × 'Sunburst'.
Tree growth: spreading, vigorous.
*S*-alleles: $S_3S_{4'}$ (self-fertile).
Productivity: very good.
Blooming time: mid-season.
Ripening time: 16–24 days after 'Burlat'.
Fruit characteristics: reniform, large, medium firm, dark red.
Resistances/specifics: susceptible to cracking.

### *'Santina'*

Origin: Summerland, Canada.
Parentage: 'Stella' × 'Summit'.
Tree growth: spreading, vigorous.
*S*-alleles: $S_1S_{4'}$ (self-fertile).
Productivity: very good.
Blooming time: mid-season.
Ripening time: 2–10 days after 'Burlat'.
Fruit characteristics: reniform, medium to large, firm, red.
Resistances/specifics: very susceptible to cracking.

### *'Satonishiki'*

Origin: Yamagata, Japan (1928).
Parentage: 'Royal Ann' × 'Governor Wood'.
Tree growth: vigorous.
*S*-alleles: $S_3S_6$.
Productivity: very good.
Blooming time: 5 days after 'Burlat'.
Ripening time: 15–24 days after 'Burlat'.

Fruit characteristics: heart-shaped, medium size and firmness, blush type.
Resistances/specifics: relatively tolerant to cracking and *Monilinia* blossom blight; susceptible to double fruit.

### 'Schneiders Späte Knorpel' (syn. 'Kozerska')

Origin: Guben, Germany.
Parentage: unknown.
Tree growth: strong growth, high pyramidal crown.
*S*-alleles: $S_3S_{12}$.
Productivity: very good.
Blooming time: medium-late to late.
Ripening time: 17 days after 'Burlat'.
Fruit characteristics: cordate, large, firm, brown red.
Resistances/specifics: long blooming time.

### 'Skeena'

Origin: Summerland, Canada.
Parentage: '2N-60-07' ('Bing' × 'Stella') × '2N-38-22' ('Van' × 'Stella').
Tree growth: spreading, vigorous.
*S*-alleles: $S_1S_{4'}$ (self-fertile).
Productivity: very good.
Blooming time: early.
Ripening time: 25–33 days after 'Burlat'.
Fruit characteristics: elongated, large, very firm, dark red.
Resistances/specifics: very susceptible to cracking and *Monilinia* blossom blight; fruit sensitivity to sunburn.

### '13S2009' (Staccato™)

Origin: Summerland, Canada.
Parentage: 'Sweetheart' × o.p.
Tree growth: spreading, medium to strong vigour.
*S*-alleles: $S_3S_{4'}$ (self-fertile).
Productivity: very good.
Blooming time: mid-season/late.
Ripening time: 37–42 days after 'Burlat'.
Fruit characteristics: elongated, medium to large, firm, dark red.
Resistances/specifics: susceptible to cracking.

### 'Stella'

Origin: Summerland, Canada.
Parentage: 'Lambert' × 'JI 2420' ('Emperor Francis' × 'Napoleon' X-rayed pollen).
Tree growth: upright, vigorous.
*S*-alleles: $S_3S_{4'}$ (self-fertile).
Productivity: very good.
Blooming time: semi-early.
Ripening time: 15–20 days after 'Burlat'.
Fruit characteristics: heart-shaped, medium size, medium firm to firm, dark red.
Resistances/specifics: intermediate susceptibility to cracking; susceptible to double fruit.

### 'Sumele' (Satin™)

Origin: Summerland, Canada.
Parentage: 'Lapins' × '2 N 39-05' ('Van' × 'Stella').
Tree growth: semi-upright, strong to very strong vigour.
*S*-alleles: $S_1S_3$.
Productivity: good.
Blooming time: 'Burlat'.
Ripening time: 14–24 days after 'Burlat'.
Fruit characteristics: cordate elongated, large, very firm, dark red.
Resistances/specifics: very good shelf-life; sensitive to mild winters.

### 'Summit'

Origin: Summerland, Canada.
Parentage: 'Van' × 'Sam'.
Tree growth: upright, poor branching, vigorous.
*S*-alleles: $S_1S_2$.
Productivity: medium to very good.
Blooming time: late to very late.
Ripening time: 15–22 days after 'Burlat'.
Fruit characteristics: heart-shaped, very large, medium firm, pale pink.
Resistances/specifics: variable susceptibility to cracking (high at onset of ripening, low at maturity); high chilling requirements; susceptible to *Monilinia* blossom blight and brown rot.

### 'Sumnue' (Cristalina™)

Origin: Summerland, Canada.
Parentage: 'Star' × 'Van'.

Tree growth: spreading, medium to strong vigour.
*S*-alleles: $S_1S_3$.
Productivity: very good.
Blooming time: mid-season.
Ripening time: 9–17 days after 'Burlat'.
Fruit characteristics: reniform elongated, large, medium firm, black.
Resistances/specifics: intermediate susceptibility to cracking; good for stemless harvest.

### *'Sumste' (Samba™)*

Origin: Summerland, Canada.
Parentage: '2S 84–10' ('Stella 35 A' × o.p.) × 'Stella 16 A7'.
Tree growth: semi-upright, medium-strong vigour.
*S*-alleles: $S_1S_3$.
Productivity: good.
Blooming time: 2–6 days before 'Burlat'.
Ripening time: 15–25 days after 'Burlat'.
Fruit characteristics: elongated, large, firm, harvest when dark red.
Resistances/specifics: medium susceptibility to cracking; possible harvest stemless.

### *'Sumtare' (Sweetheart™)*

Origin: Summerland, Canada.
Parentage: 'Van' × 'New Star' (the paternal parent may be 'Lapins' and not 'New Star', Iezzoni A., East Lansing, Michigan, 2016, personal communication).
Tree growth: spreading, strong vigour.
*S*-alleles: $S_3S_{4'}$ (self-fertile).
Productivity: very good.
Blooming time: mid-season.
Ripening time: 30–35 days after 'Burlat'.
Fruit characteristics: elongated, medium to large, very firm, red.
Resistances/specifics: susceptible to cracking and *Monilinia* blossom blight and brown rot.

### *'Sunburst'*

Origin: Summerland, Canada.
Parentage: 'Van' × 'Stella'.
Tree growth: semi-upright, wide spreading, medium vigour.
*S*-alleles: $S_3S_{4'}$ (self-fertile).
Productivity: medium to good.
Blooming time: medium to late.
Ripening time: 15–24 days after 'Burlat'.
Fruit characteristics: round to reniform, large, medium firm, pale red.
Resistances/specifics: very susceptible to cracking and *Monilinia* blossom blight.

### *'Sylvia' (syn. '4 C-17-31')*

Origin: Summerland, Canada.
Parentage: 'Van'× 'Sam'.
Tree growth: semi-compact, low vigour.
*S*-alleles: $S_1S_4$.
Productivity: good to very good.
Blooming time: late.
Ripening time: 16–20 days after 'Burlat'.
Fruit characteristics: reniform, medium size, firm, bright red.
Resistances/specifics: tolerant to cracking; excellent short-term storage; susceptible to double fruit.

### *'Techlovan'*

Origin: RBIPH, Holovousy, Czech Republic.
Parentage: 'Van' × 'Kordia'.
Tree growth: spreading, medium vigour.
*S*-alleles: $S_3S_6$.
Productivity: medium to good.
Blooming time: early.
Ripening time: 14–25 days after 'Burlat'.
Fruit characteristics: heart-shaped, large to very large, very firm, dark red.
Resistances/specifics: very susceptible to cracking; susceptible to mild winters.

### *'PC 7144.6' (Tieton™)*

Origin: Washington State University, USA.
Parentage: 'Stella' × 'Early Burlat'.
Tree growth: upright, strong to very strong vigour.
*S*-alleles: $S_3S_9$.
Productivity: medium to good.
Blooming time: early.
Ripening time: 7–14 days after 'Burlat'.
Fruit characteristics: elongated, very large, medium firm, dark red.
Resistances/specifics: susceptible to cracking, *Monilinia* blossom blight and double fruit.

### 'Van'

Origin: Summerland, Canada.
Parentage: 'Empress Eugenie' × o.p.
Tree growth: upright, medium to strong vigour.
*S*-alleles: $S_1S_3$.
Productivity: very good.
Blooming time: mid-season.
Ripening time: 16–24 days after 'Burlat'.
Fruit characteristics: reniform, medium size, firm, bright red.
Resistances/specifics: very susceptible to cracking; susceptible to *Monilinia* blossom blight, bacterial canker and double fruit.

### 'Vanda'

Origin: RBIPH, Holovousy, Czech Republic.
Parentage: 'Van' × 'Kordia'.
Tree growth: semi-upright, medium to strong vigour.
*S*-alleles: $S_1S_6$.
Productivity: good.
Blooming time: early.
Ripening time: 12–16 days after 'Burlat'.
Fruit characteristics: reniform, medium to large, firm, brown red.
Resistances/specifics: tolerant to cracking.

### '0900 Ziraat'

Origin: Aegean region, Manisa, Turkey.
Parentage: unknown.
Tree growth: upright, medium vigour.
*S*-alleles: $S_3S_{12}$.
Productivity: good.
Blooming time: late.
Ripening time: 19 days after 'Burlat'.
Fruit characteristics: heart-shaped, large, firm, dark red.
Resistances/specifics: relatively tolerant to cracking.

## 4.4.2 Sweet cherry cultivars with local importance and/or promising cultivars

### 'Aiya'

Origin: Dobele, Latvia.
Parentage: unknown.
Tree growth: semi-upright, strong vigour.
*S*-alleles: $S_1S_7$.
Productivity: very good.
Blooming time: 1 day after 'Burlat'.
Ripening time: 16–24 days after 'Burlat'.
Fruit characteristics: round, small, soft, light red.
Resistances/specifics: tolerant to cracking.

### 'Alex'

Origin: RIFG, Pitesti, Romania.
Parentage: 'Lijana' × o.p.
Tree growth: semi-upright, medium to strong vigour.
*S*-alleles: unknown.
Productivity: good to very good.
Blooming time: early/mid-season.
Ripening time: 5–14 days after 'Burlat'.
Fruit characteristics: cordate, large, firm, bright red.
Resistances/specifics: tolerant to cracking; recommended pollinators: 'Van', 'Stella' and 'Maria'.

### 'Andrei'

Origin: RIFG, Pitesti, Romania.
Parentage: 'HC 27/4' × 'Boambe de Cotnari'.
Tree growth: semi-upright, medium vigour.
*S*-alleles: unknown.
Productivity: very good.
Blooming time: 4 days before 'Burlat'.
Ripening time: 5–14 days after 'Burlat'.
Fruit characteristics: cordate, large, firm, brown red.
Resistances/specifics: susceptible to cracking; recommended pollinators: 'Stella' and 'Maria'.

### 'Areko'

Origin: JKI, Dresden, Germany.
Parentage: 'Kordia'× 'Regina'.
Tree growth: pyramidal with spreading branches, medium vigour.
*S*-alleles: $S_1S_3$.
Productivity: very good.
Blooming time: late.
Ripening time: 16 days after 'Burlat'.
Fruit characteristics: cordate, large, firm, brown red.
Resistances/specifics: low susceptibility to cracking.

### 'Axel'

Origin: Fruitculture Research Institute, NARIC, Hungary.
Parentage: 'Van'× 'JI 2434' ('Emperor Francis' × 'Napoleon' X-rayed pollen).
Tree growth: semi-upright, moderate vigour.
*S*-alleles: $S_3S_{3'}$ (self-fertile).
Productivity: very good.
Blooming time: early.
Ripening time: 40–45 days after 'Burlat'.
Fruit characteristics: roundish to elongated, medium to large, medium firm, deep purple.
Resistances/specifics: not sensitive to cracking.

### 'Benisayaka'

Origin: Yamagata, Japan.
Parentage: 'Satonishiki' × 'Seneka'.
Tree growth: slightly upright, vigorous.
*S*-alleles: $S_1S_6$.
Productivity: very good.
Blooming time: unknown.
Ripening time: early.
Fruit characteristics: blunt to cordate, medium to small size, firm, blush type.
Resistances/specifics: unknown.

### 'Benishuhou'

Origin: Yamagata, Japan.
Parentage: 'Satonishiki' × 'Tenkonishiki'.
Tree growth: slightly spreading, medium vigour.
*S*-alleles: $S_4S_6$.
Productivity: very good.
Blooming time: unknown.
Ripening time: late.
Fruit characteristics: reniform, medium to large, firm, blush type.
Resistances/specifics: good postharvest behaviour.

### 'Benton'

Origin: Washington State University, USA.
Parentage: 'Stella' × 'Beaulieu'.
Tree growth: upright, strong vigour.
*S*-alleles: $S_{4'}S_9$ (self-fertile).
Productivity: medium to good.
Blooming time: 4 days before 'Burlat'.
Ripening time: 19 days after 'Burlat'.
Fruit characteristics: reniform, large to very large, firm, mahogany red.
Resistances/specifics: susceptible to cracking.

### 'Bryanskaya rozovaya'

Origin: Bryansk, Russia.
Parentage: 'Muskatnaya Chornaya' × o.p.
Tree growth: spreading, strong vigour.
*S*-alleles: $S_3S_6$.
Productivity: very good.
Blooming time: 2–3 days after 'Burlat'.
Ripening time: 24–31 days after 'Burlat'.
Fruit characteristics: oblate, small, firm, blush type.
Resistances/specifics: tolerant to cracking, winter hardy.

### 'Cambrina'

Origin: Cornell University, Geneva, New York, USA.
Parentage: 'Prosser I 633' × 'NY 5656'.
Tree growth: semi-upright, strong vigour.
*S*-alleles: $S_1S_{13}$.
Productivity: very good.
Blooming time: 0–4 days after 'Rainier'.
Ripening time: 15–22 days after 'Burlat', 0–5 days before 'Rainier'.
Fruit characteristics: reniform, medium to large, firm, blush type.
Resistances/specifics: very susceptible to cracking; susceptible to mild winters.

### 'Cavalier'

Origin: Dorr, Michigan, USA.
Parentage: chance seedling from an orchard of 'Schmidt' cherry.
Tree growth: upright spreading canopy, moderately vigorous.
*S*-alleles: $S_2S_3$.
Productivity: moderate, not precocious.
Blooming time: early to mid-season.
Ripening time: 13–17 days after 'Burlat'.
Fruit characteristics: globose to slightly oblate, medium to large, firm, dark red.
Resistances/specifics: moderately resistant to cracking.

### *'Chunxiu'*

Origin: CAAS, Zhengzhou, China.
Parentage: 'Bing' × o.p.
Tree growth: semi-upright, medium vigour.
*S*-alleles: $S_4S_6$.
Productivity: very good.
Blooming time: 4 days after 'Burlat'.
Ripening time: 14 days after 'Burlat'.
Fruit characteristics: heart-shaped, large, firm, red purple.
Resistances/specifics: unknown.

### *'Coral Champagne'*

Origin: University of California-Davis, USA.
Parentage: 'Rainier' × 'Burlat'.
Tree growth: semi-upright, medium vigour.
*S*-alleles: $S_1S_3$.
Productivity: very good.
Blooming time: 3 days before 'Burlat'.
Ripening time: 10 days after 'Burlat'.
Fruit characteristics: reniform, large, very firm, bright red.
Resistances/specifics: very susceptible to cracking and bacterial canker; susceptible to double fruit.

### *'Danelia'*

Origin: Kiustendil, Bulgaria.
Parentage: 'Hedelfinger' × 'Germersdorfer'.
Tree growth: semi-upright, medium vigour.
*S*-alleles: unknown.
Productivity: very good.
Blooming time: mid-late.
Ripening time: 10–15 days after 'Burlat'.
Fruit characteristics: cordate, medium size, firm, dark red.
Resistances/specifics: high resistance to winter cold and late spring frost.

### *'Doty' (trademark: Early Robin™)*

Origin: Mattawa, Washington, USA.
Parentage: reported to be a whole-tree mutation of 'Rainier' but unlikely (inconsistent *S*-alleles).
Tree growth: upright to spreading canopy, moderately vigorous.
*S*-alleles: $S_1S_3$.
Productivity: moderate.
Blooming time: medium early.
Ripening time: 12–15 days after 'Burlat'.
Fruit characteristics: rounded heart shape, large and firm, blush type.
Resistances/specifics: rather susceptible to cracking and bacterial canker.

### *'Early Red' (syn. 'Maraly', Early Garnet™)*

Origin: selected by Marvin Nies, USA.
Parentage: 'Garnet' × 'Ruby'.
Tree growth: semi-upright, strong vigour.
*S*-alleles: $S_1S_9$.
Productivity: very good.
Blooming time: early.
Ripening time: 4–8 days after 'Burlat'.
Fruit characteristics: reniform, large, medium firm, light red, very short peduncle.
Resistances/specifics: very susceptible to cracking and *Monilinia* blossom blight; not susceptible to mild winters or double fruit.

### *'Ferdiva'*

Origin: INRA, France.
Parentage: 'Arcina®Fercer' × o.p.
Tree growth: semi-spreading, strong to very strong vigour.
*S*-alleles: $S_3S_6$.
Productivity: good.
Blooming time: late.
Ripening time: 28–35 days after 'Burlat'.
Fruit characteristics: elongated, flattened cordiform, large, very firm, red.
Resistances/specifics: susceptible to cracking; high chilling requirements; susceptible to bacterial canker.

### *'Ferdouce'*

Origin: INRA, France.
Parentage: 'Rainier' × 'Arcina®Fercer'.
Tree growth: spreading, medium vigour.
*S*-alleles: $S_1S_2$.
Productivity: very good.
Blooming time: early (1–6 days before 'Burlat').
Ripening time: 8–15 days after 'Burlat'.
Fruit characteristics: elongated, large to very large, firm, light red.
Resistances/specifics: susceptible to cracking; medium susceptibility to bacterial canker; good pollinator for 'Folfer'.

### *'Fermina'*

Origin: INRA, France.
Parentage: 'Vittoria' × o.p.
Tree growth: semi-upright, medium to strong vigour.
*S*-alleles: $S_1S_{14}$.
Productivity: medium to good.
Blooming time: mid-season to late.
Ripening time: 15–22 days after 'Burlat'.
Fruit characteristics: round, large, very firm, red to dark red.
Resistances/specifics: tolerant to cracking; sensitive to mild winters; excellent for stemless harvest.

### *'Fertard'*

Origin: INRA, France.
Parentage: 'Sunburst' × o.p.
Tree growth: main branches upright, laterals spreading, medium to strong vigour.
*S*-alleles: $S_3S_6$.
Productivity: medium to good.
Blooming time: late to very late.
Ripening time: 31–39 days after 'Burlat'.
Fruit characteristics: elongated to heart-shaped, large to very large, very firm, dark red.
Resistances/specifics: susceptible to cracking, high chilling requirements, sensitive to mild winters.

### *'Fertille'*

Origin: INRA, France.
Parentage: 'Arcina®Fercer' × 'Van'.
Tree growth: semi-upright to spreading.
*S*-alleles: $S_3S_6$.
Productivity: very good.
Blooming time: 1–3 days after 'Burlat'.
Ripening time: 16–20 days after 'Burlat'.
Fruit characteristics: reniform, large to very large, very firm, red to dark red.
Resistances/specifics: moderately susceptible to cracking; sensitive to mild winters.

### *'Frisco'*

Origin: sdR Fruit LLC, California, USA.
Parentage: unknown.
Tree growth: semi-upright to upright, low vigour.
*S*-alleles: $S_1S_4$.
Productivity: good.
Blooming time: early.
Ripening time: 4–7 days after 'Burlat'.
Fruit characteristics: reniform, very large, very firm, red to dark red.
Resistances/specifics: susceptible to cherry leaf spot.

### *'Gold' (syn. 'Dönissens Gelbe', 'Stark Gold')*

Origin: Germany.
Parentage: unknown.
Tree growth: semi-upright, medium vigour.
*S*-alleles: $S_3S_6$.
Productivity: good.
Blooming time: 4–5 days after 'Burlat'.
Ripening time: late (mid-July in the USA).
Fruit characteristics: reniform to cordiform, small, firm, yellow.
Resistances/specifics: particularly winter hardy; used for brining and processing.

### *'Iputj'*

Origin: Bryansk, Russia.
Parentage: 'Zhabule' ('Leningradska Chornaya' × 'Pobeda') × o.p.
Tree growth: spreading, medium vigour.
*S*-alleles: $S_3S_{13}$.
Productivity: very good.
Blooming time: 1 day after 'Burlat'.
Ripening time: 7 days after 'Burlat'.
Fruit characteristics: elliptic, small, firm, dark red.
Resistances/specifics: very susceptible to cracking; winter hardy.

### *'Jiahong'*

Origin: DAAS, Dalian, China.
Parentage: 'Bing' × 'Xiang Jiao'.
Tree growth: semi-upright, strong vigour.
*S*-alleles: $S_4S_6$.
Productivity: medium to good.
Blooming time: 1 day after 'Burlat'.
Ripening time: 14 days after 'Burlat'.
Fruit characteristics: wide cordiform, large to very large, middle firm, yellowish red.
Resistances/specifics: susceptible to cracking.

### 'Katalin'

Origin: Fruitculture Research Institute, NARIC, Hungary.
Parentage: 'Germersdorfer' × 'Podjebrad'.
Tree growth: upright when young, spreading when adult, medium vigour.
*S*-alleles: $S_4S_{12}$.
Productivity: very good.
Blooming time: medium to late.
Ripening time: late (7 weeks after 'Burlat').
Fruit characteristics: cordate, medium to large, firm, light red.
Resistances/specifics: low susceptibility to cracking.

### 'Kossara'

Origin: FGI, Plovdiv, Bulgaria.
Parentage: 'Ranna Cherna' × 'Bigarreau Burlat'.
Tree growth: semi-upright, high vigour.
*S*-alleles: unknown.
Productivity: very good.
Blooming time: 2 days before 'Burlat'.
Ripening time: 10 days before 'Burlat'.
Fruit characteristics: cordate, medium to large, soft, dark red, deep sweet–sour taste.
Resistances/specifics: tolerant to cracking.

### 'Linda'

Origin: Fruitculture Research Institute, NARIC, Hungary.
Parentage: 'Hedelfinger' × 'Germersdorfer'.
Tree growth: spreading, medium vigour.
*S*-alleles: $S_3S_{12}$.
Productivity: very good.
Blooming time: late.
Ripening time: mid-season to late.
Fruit characteristics: elongated, medium size, firm, deep mahogany.
Resistances/specifics: low susceptibility to cracking.

### 'Longguan'

Origin: CAAS, Zhengzhou, China.
Parentage: unknown (superior seedling).
Tree growth: upright, strong vigour.
*S*-alleles: $S_4S_9$.
Productivity: very good.
Blooming time: 2 days after 'Burlat'.
Ripening time: 5 days after 'Burlat'.
Fruit characteristics: wide cordiform, medium size and firmness, purple red.
Resistances/specifics: susceptible to cracking.

### 'Ludovic'

Origin: RFIG, Pitesti, Romania.
Parentage: 'Van' × 'Boambe de Cotnari'.
Tree growth: spreading, medium vigour.
*S*-alleles: unknown.
Productivity: good to very good.
Blooming time: mid-season.
Ripening time: 7–20 days after 'Burlat'.
Fruit characteristics: reniform, very large, firm, dark red.
Resistances/specifics: intermediate susceptibility to cracking.

### 'Maria'

Origin: RFIG, Pitesti, Romania.
Parentage: 'Van' × 'Stella'.
Tree growth: spreading, medium vigour.
*S*-alleles: presumably $S_1S_{4'}$ or $S_3S_{4'}$ (self-fertile).
Productivity: good to very good.
Blooming time: 1–2 days after 'Burlat'.
Ripening time: 10–15 days after 'Burlat'.
Fruit characteristics: cordate, large, firm, dark red.
Resistances/specifics: intermediate susceptibility to cracking.

### 'Namare'

Origin: JKI, Dresden, Germany.
Parentage: 'Grosse Schwarze Knorpel' × o.p.
Tree growth: spreading, medium vigour.
*S*-alleles: $S_3S_4$.
Productivity: very good.
Blooming time: medium to late.
Ripening time: 14 days after 'Burlat'.
Fruit characteristics: flat to round, medium size and firmness, dark red.
Resistances/specifics: medium susceptibility to cracking.

### 'Namati'

Origin: JKI, Dresden, Germany.
Parentage: 'Boppader Kracher' × o.p.

Tree growth: spreading, medium vigour.
S-alleles: $S_1S_4$.
Productivity: very good.
Blooming time: very late (as for 'Regina').
Ripening time: 21 days after 'Burlat'.
Fruit characteristics: flat to round, medium size, firm, dark red.
Resistances/specifics: low susceptibility to cracking.

### *'Naprumi'*

Origin: JKI, Dresden, Germany.
Parentage: 'Hedelfinger' × 'St Charmes'.
Tree growth: spreading, medium to strong vigour.
S-alleles: $S_3S_9$.
Productivity: very good.
Blooming time: early/mid-season.
Ripening time: 1 day after 'Burlat'.
Fruit characteristics: flat to round, medium size and firmness, dark red.
Resistances/specifics: low susceptibility to spring frost.

### *'Oktavia'*

Origin: Jork, Germany.
Parentage: 'Schneiders Späte Knorpel' × 'Rube'.
Tree growth: spreading, medium vigour.
S-alleles: $S_1S_3$.
Productivity: very good.
Blooming time: medium to late.
Ripening time: 16–22 days after 'Burlat'.
Fruit characteristics: cordate, large, very firm, red.
Resistances/specifics: susceptible to cracking.

### *'PA1UNIBO' (Sweet Aryana™)*

Origin: University of Bologna, Italy.
Parentage: unknown.
Tree growth: upright, medium to strong vigour.
S-alleles: $S_3S_4$.
Productivity: very good.
Blooming time: early.
Ripening time: 3–5 days after 'Burlat'.
Fruit characteristics: heart-shaped, large, firm, shiny dark red.
Resistances/specifics: average susceptibility to cracking.

### *'PA2UNIBO' (Sweet Lorenz™)*

Origin: University of Bologna, Italy.
Parentage: unknown.
Tree growth: upright, low vigour.
S-alleles: $S_3S_4$.
Productivity: very good.
Blooming time: early.
Ripening time: 8–10 days after 'Burlat'.
Fruit characteristics: heart-shaped, very large, very firm, shiny dark red.
Resistances/specifics: low to medium susceptibility to cracking.

### *'PA3UNIBO' (Sweet Gabriel™)*

Origin: University of Bologna, Italy.
Parentage: unknown.
Tree growth: semi-upright, low to medium vigour.
S-alleles: $S_1S_4$.
Productivity: very good.
Blooming time: mid-season.
Ripening time: 11–14 days after 'Burlat'.
Fruit characteristics: heart-shaped, very large, firm, shiny dark red.
Resistances/specifics: susceptible to cracking.

### *'Paulus'*

Origin: Fruitculture Research Institute, NARIC, Hungary.
Parentage: 'Burlat' × 'Stella'.
Tree growth: upright, moderate vigour.
S-alleles: $S_4S_9$ (self-fertile).
Productivity: very good.
Blooming time: mid-early.
Ripening time: 10 days after 'Burlat'.
Fruit characteristics: flat to round, large, firm, dark red.
Resistances/specifics: not susceptible to cracking or to *Cytospora* canker.

### *'Penny'*

Origin: East Malling Research, UK.
Parentage: 'Colney' × 'Inge'.
Tree growth: semi-upright, medium to strong vigour.
S-alleles: $S_6S_9$.
Productivity: good to very good.

Blooming time: very late (as 'Regina').
Ripening time: 25–35 days after 'Burlat'.
Fruit characteristics: flat to round, medium to large, firm, red.
Resistances/specifics: susceptible to cracking and to bare wood (base of branches).

### *'Prime Giant' (syn. 'Giant Red', 'Mariant'; Giant Ruby™)'*

Origin: selected by Marvin Nies, USA.
Parentage: Hybrid ('Lodi' × 'Ruby') × o.p.
Tree growth: semi-upright, medium to strong vigour.
*S*-alleles: $S_1S_3$.
Productivity: very good.
Blooming time: 'Burlat'.
Ripening time: 7–18 days after 'Burlat'.
Fruit characteristics: reniform to roundish, very large, firm, dark red.
Resistances/specifics: very susceptible to cracking; average susceptibility to double fruit; susceptible to bacterial canker; not susceptible to mild winters.

### *'Rita'*

Origin: Fruitculture Research Institute, NARIC, Hungary.
Parentage: 'Trusenskaja 2' × 'H-2' ('Germersdorfer' × o.p.).
Tree growth: little pendula, medium vigour.
*S*-alleles: $S_5S_{22}$.
Productivity: very good.
Blooming time: early.
Ripening time: 7–14 days before 'Burlat'.
Fruit characteristics: flat to round, medium size, firm, deep red.
Resistances/specifics: highly susceptible to cracking; susceptible to *Leucostoma* canker.

### *'Rocket'*

Origin: SMS Unlimited, California, USA.
Parentage: unknown.
Tree growth: semi-upright to upright, low vigour.
*S*-alleles: $S_1S_9$.
Productivity: medium to good.
Blooming time: early.
Ripening time: 4–7 days after 'Burlat'.
Fruit characteristics: cordiform, large to very large, very firm, red.
Resistances/specifics: susceptible to cherry leaf spot.

### *'Rosie'*

Origin: Zaiger Genetics, Modesto, California, USA.
Parentage: '181LB359' × o.p.
Tree growth: semi-upright to upright, vigorous.
*S*-alleles: $S_1S_3$.
Productivity: very good.
Blooming time: early (0–3 days before 'Rainier').
Ripening time: 10–18 days before 'Rainier'.
Fruit characteristics: round reniform, large to very large, very firm, blush type.
Resistances/specifics: very low susceptibility to marks; very susceptible to leaf spot; not susceptible to mild winters.

### *'Royal Bailey'*

Origin: Zaiger Genetics, Modesto, California, USA.
Parentage: '22ZB383' × o.p.
Tree growth: upright, low vigour.
*S*-alleles: $S_1S_3$.
Productivity: good.
Blooming time: very early.
Ripening time: 7–13 days after 'Burlat'.
Fruit characteristics: reniform, very large, firm to very firm, red.
Resistances/specifics: very susceptible to cracking and *Monilinia* blossom blight.

### *'Royal Dawn'*

Origin: Zaiger Genetics, Modesto, California, USA.
Parentage: '32G500' × o.p.
Tree growth: upright, vigorous.
*S*-alleles: $S_3S_4$.
Productivity: very good.
Blooming time: early.
Ripening time: 16 days before 'Bing'.
Fruit characteristics: globose, slightly flattened, medium to large, firm, red.
Resistances/specifics: very high susceptibility to cracking; susceptible to bacterial canker.

### 'Royal Edie'

Origin: Zaiger Genetics, Modesto, California, USA.
Parentage: '92LB341' ('Bing' × 'Royal Dawn') × o.p.
Tree growth: semi-upright to upright, vigorous.
*S*-alleles: $S_1S_{4'}$ (self-fertile).
Productivity: good to very good.
Blooming time: 'Burlat'.
Ripening time: 23–27 days after 'Burlat'.
Fruit characteristics: round to reniform, large to very large, very firm, red.
Resistances/specifics: susceptible to cracking.

### 'Royal Helen'

Origin: Zaiger Genetics, Modesto, California, USA.
Parentage: '92LB341' ('Bing' × 'Royal Dawn') × o.p.
Tree growth: semi-upright to upright, vigorous.
*S*-alleles: $S_1S_{4'}$ (self-fertile).
Productivity: very good.
Blooming time: 'Burlat'.
Ripening time: 23–27 days after 'Burlat'.
Fruit characteristics: round to reniform, large to very large, very firm, red.
Resistances/specifics: susceptible or very susceptible to cracking.

### 'Royal Rainier'

Origin: Zaiger Genetics, Modesto, California, USA.
Parentage: 'Stella' × o.p.
Tree growth: upright canopy, vigorous.
*S*-alleles: unknown.
Productivity: high.
Blooming time: mid-season.
Ripening time: 17–20 days after 'Burlat'.
Fruit characteristics: globose to slightly oblate, medium to large and firm, blush type.
Resistances/specifics: moderately low-chilling requirement (reported to be 500 h).

### 'Rozalina'

Origin: FGI, Plovdiv, Bulgaria.
Parentage: 'Van' × o.p.
Tree growth: semi-upright.
*S*-alleles: unknown.
Productivity: very good.
Blooming time: 2 days before 'Burlat', extended blooming period.
Ripening time: 1 week before 'Van'.
Fruit characteristics: reniform, large, very firm, blush type, very dense texture.
Resistances/specifics: susceptible to bacterial canker.

### 'Sam'

Origin: Summerland, Canada.
Parentage: 'V-160140' ('Windsor' × o.p.) × o.p.
Tree growth: upright but later spreading, vigorous.
*S*-alleles: $S_2S_4$.
Productivity: good.
Blooming time: late.
Ripening time: 9 days after 'Burlat'.
Fruit characteristics: cordiform, medium size, medium firm, fully black.
Resistances/specifics: tolerant to cracking.

### 'Sándor'

Origin: Fruitculture Research Institute, NARIC, Hungary.
Parentage: 'Burlat' × 'Stella'.
Tree growth: little upright, vigorous.
*S*-alleles: $S_{4'}S_9$ (self-fertile).
Productivity: very good.
Blooming time: early.
Ripening time: 4–6 days before 'Burlat'.
Fruit characteristics: cordiform, medium to small, medium firm, light red.
Resistances/specifics: susceptible to cracking.

### 'Severin'

Origin: RFIG, Pitesti, Romania.
Parentage: 'Thurn und Taxis' × 'Germersdorfer'.
Tree growth: semi-upright, medium to strong vigour.
*S*-alleles: unknown.
Productivity: good.
Blooming time: 2–4 days after 'Burlat'.
Ripening time: 12 days after 'Burlat'.
Fruit characteristics: flat to round, large, medium firm, bright red skin.
Resistances/specifics: intermediate susceptibility to cracking.

### 'Simone'

Origin: Australia.
Parentage: unknown.
Tree growth: responds well to bush style, branches well.
*S*-alleles: self-fertile.
Productivity: very good.
Blooming time: early.
Ripening time: a little earlier than 'Lapins'.
Fruit characteristics: similar to 'Lapins'.
Resistances/specifics: less susceptible to cracking than 'Lapins'.

### 'SPC103' (Sentennial™)

Origin: Summerland, Canada.
Parentage: 'Sweetheart' × o.p.
Tree growth: spreading, low to medium vigour.
*S*-alleles: $S_3S_{4'}$ (self-fertile).
Productivity: very good.
Blooming time: early (similar to 'Lapins').
Ripening time: 5 days after 'Staccato' (very late) (close to 'Sweetheart' in France; G. Charlot, Balandran, France, 2016, personal communication).
Fruit characteristics: similar to 'Sweetheart' in shape, medium to large, firmer than 'Staccato'.
Resistances/specifics: susceptible to cracking.

### '13S2101' (Sovereign™)

Origin: Summerland, Canada.
Parentage: 'Sweetheart' × o.p.
Tree growth: spreading, low to medium vigour.
*S*-alleles: $S_3S_{4'}$ (self-fertile).
Productivity: very good.
Blooming time: mid-season.
Ripening time: 5 days after 'Staccato' (very late) (close to 'Staccato' in France; G. Charlot, Balandran, France, 2016, personal communication).
Fruit characteristics: elongated, medium to large, very firm, red.
Resistances/specifics: moderately susceptible to cracking.

### 'SPC136' (Suite Note™)

Origin: Summerland, Canada.
Parentage: '2S-36–36' × 'Summit'.
Tree growth: upright, medium vigour.
*S*-alleles: $S_2S_4$.
Productivity: very good.
Blooming time: late.
Ripening time: a few days before 'Van'/'Bing'.
Fruit characteristics: elongated, medium to large, firm, bright red.
Resistances/specifics: very susceptible to cracking.

### 'Starking Hardy Giant'

Origin: Wisconsin, USA.
Parentage: unknown.
Tree growth: spreading, medium to strong vigour.
*S*-alleles: $S_1S_2$.
Productivity: good to very good.
Blooming time: early/mid-season.
Ripening time: 12–15 days after 'Burlat'.
Fruit characteristics: round to heart-shaped, medium to large, firm, red to dark red.
Resistances/specifics: very susceptible to cracking and *Monilinia* blossom blight.

### 'Stefania'

Origin: Kiustendil, Bulgaria.
Parentage: 'Compact Lambert' × 'Stella 35B-11'.
Tree growth: spreading, high vigour.
*S*-alleles: unknown.
Productivity: very good.
Blooming time: 3 days after 'Burlat'.
Ripening time: several days after 'Bing'.
Fruit characteristics: broad cordate, medium to large, firm, dark red.
Resistances/specifics: high resistance to winter cold and late spring frost.

### 'Sumgita' (Canada Giant™)

Origin: Summerland, Canada.
Parentage: unknown.
Tree growth: upright, vigorous.
*S*-alleles: $S_1S_2$.
Productivity: good.
Blooming time: mid-season to late.
Ripening time: 15 days after 'Burlat'.
Fruit characteristics: reniform, large to very large, firm, red.
Resistances/specifics: susceptible to cracking at the pistillar end; susceptible to double fruit.

### 'Summer Sun'

Origin: John Innes, UK.
Parentage: 'Merton Glory' × o.p.
Tree growth: spreading, vigorous.
*S*-alleles: $S_4S_9$.
Productivity: very good.
Blooming time: mid-season.
Ripening time: 29 days after 'Burlat'.
Fruit characteristics: cordiform, large, medium firm, purple black.
Resistances/specifics: very susceptible to cracking.

### 'Sumpaca' (Celeste™)

Origin: Summerland, Canada.
Parentage: 'Van' × 'Newstar'.
Tree growth: highly upright, semi-compact, medium vigour.
*S*-alleles: $S_1S_4$, (self-fertile).
Productivity: very good.
Blooming time: early.
Ripening time: 12–14 days after 'Burlat'.
Fruit characteristics: reniform, large to very large, medium firm, dark red.
Resistances/specifics: susceptible to cracking.

### 'Tulare'

Origin: Bradford, USA.
Parentage: unknown.
Tree growth: semi-upright, medium vigour.
*S*-alleles: $S_1S_2$.
Productivity: very good.
Blooming time: early (8 days before 'Burlat').
Ripening time: 10 days after 'Burlat'.
Fruit characteristics: heart-shaped, medium-large, firm, fire engine red.
Resistances/specifics: moderately susceptible to cracking.

### 'Ulster'

Origin: Cornell University, Geneva, New York, USA.
Parentage: 'Schmidt' × 'Lambert'.
Tree growth: semi-upright, vigorous.
*S*-alleles: $S_3S_4$.
Productivity: good.
Blooming time: mid-season.
Ripening time: 19–21 days after 'Burlat'.
Fruit characteristics: cordiform, medium size and firmness, purple.
Resistances/specifics: tolerant to cracking and to frost.

### 'Valerij Chkalov'

Origin: Melitopol, Ukraine.
Parentage: 'Rozozva' × o.p.
Tree growth: spreading (when adult), vigorous.
*S*-alleles: $S_1S_9$.
Productivity: very good.
Blooming time: early.
Ripening time: close to 'Burlat'.
Fruit characteristics: heart-shaped, medium-large, firm, dark red.
Resistances/specifics: susceptible to cracking; tolerant to *Monilinia* blossom blight; low susceptibility to leaf spot.

### 'Vera'

Origin: Fruitculture Research Institute, NARIC, Hungary.
Parentage: 'Ljana' × 'Van'.
Tree growth: semi-upright, moderate vigour.
*S*-alleles: $S_1S_3$.
Productivity: good.
Blooming time: early.
Ripening time: 10–12 days after 'Burlat'.
Fruit characteristics: flat-round, large, firm, deep red.
Resistances/specifics: medium susceptibility to cracking and *Cytospora* canker.

### 'Wanhongzhu'

Origin: DAAS, Dalian, China.
Parentage: unknown (superior seedling).
Tree growth: spreading, strong vigour.
*S*-alleles: $S_6S_9$.
Productivity: very good.
Blooming time: 'Burlat'.
Ripening time: 27 days after 'Burlat'.
Fruit characteristics: wide cordiform, large, firm, red.
Resistances/specifics: susceptible to cracking.

### 'Xiangquan 1'

Origin: BAAFS, Beijing, China.
Parentage: 'Van' × 'Stella'.

Tree growth: semi-upright, medium vigour.
*S*-alleles: $S_3S_4$, (self-fertile)
Productivity: very good.
Blooming time: 2 days after 'Burlat'.
Ripening time: 16 days after 'Burlat'.
Fruit characteristics: nearly circular, medium to large, firm, blush type.
Resistances/specifics: high tolerance to frost.

## Acknowledgements

The authors thank the following additional contributors: S. Malchev, Fruit Growing Institute, Plovdiv, Bulgaria; C. Hampson, Agriculture and Agri-Food Canada, Summerland, British Columbia, Canada; E. Gratacós, Centro Regional de Innovación Hortofrutícola de Valparaiso, Chile; G. Lemus and J. Donoso, Instituto de Investigaciones Agropecuarias, Rayentué, Chile; M. Ayala, M. Gebauer and J.P. Zoffoli, Facultad de Agronomía e Ingeniería Forestal, Pontificia Universidad Católica de Chile, Chile; K. Zhang and J. Wang, Institute of Forestry and Pomology, Beijing, China; J. Sedlak and F. Paprstein, Research and Breeding Institute of Pomology, Holovousy, Czech Republic; J. Apostol, NARIC Fruitculture Research Institute, Budapest, Hungary; G.H. Davarynejad, Faculty of Agriculture, Mashhad, Iran; S. Lugli, Bologna University, Department of Agricultural Sciences, Bologna, Italy; I. Makoto, Yamagata Integrated Agricultural Research Center, Yamagata, Japan; D. Feldmane, Latvia State Institute of Fruit-Growing, Dobele, Latvia; S. Budan, Research Institute for Fruit Growing Pitesti, Pitesti, Romania; M. López-Corrales, Centro de Investigación Finca la Orden-Valdesquera, Guadajira, Spain; Y. Ammari and T. Azizi, INRGREF Laboratoire Ecologie Forestière, Tunis, Tunisia; S. Ercişli, Ataturk University Agricultural, Erzurum, Turkey; Y. Ivanovych, NAAS Institute of Horticulture, Novosilky, Ukraine; F. Fernandez and M. Lipska, NIAB EMR, East Malling, UK; N. Oraguzie, Washington State University, Prosser, Washington, USA.

## References

Bargioni, G. (1996) Sweet cherry scions. Characteristics of the principal commercial cultivars, breeding objectives and methods. In: Webster, A.D. and Looney, N.E. (eds) *Cherries: Crop Physiology, Production and Uses*. CAB International, Wallingford, UK, pp. 73–112.

Brown, S.K. and Wilcox, W.F. (1989) Evaluation of cherry genotypes for resistance to fruit infection by *Monilinia fructicola* (Wint.) Honey. *HortScience* 24, 1013–1015.

Campoy, J.A., Le Dantec, L., Barreneche, T., Dirlewanger, E. and Quero-García, J. (2015) New insights into fruit firmness and weight control in sweet cherry. *Plant Molecular Biology Reporter* 22, 783–796.

Campoy, J.A., Lerigoleur-Balsemin, E., Christmann, H., Beauvieux, R., Girollet, N., Quero-García, J., Dirlewanger, E. and Barreneche, T. (2016) Genetic diversity, linkage disequilibrium, population structure and construction of a core collection of *Prunus avium* L., landraces and bred cultivars. *BMC Plant Biology* 16, 49.

Fischer, M. and Hohlfeld, B. (1998) Resistenszprüfungen an Süsskirschen (*Prunus avium* L.) Teil 5: evaluierung des Süsskirchensortiments der Genbank Obst Dresden Pilnitz auf Winterfrost Resistens. *Erwerbsobstbau* 40, 42–51.

Fogle, H.W. (1975) Cherries. In: Janick, J. and Moore, J.N. (eds) *Advances in Fruit Breeding*. Purdue University Press, West Lafayette, Indiana, pp. 348–366.

Kałużna, M. and Sobiczewski, P. (2014) Bacterial canker of cherry – methods of susceptibility testing. In: *Proceedings of the EU COST Action FA1104: Sustainable Production of High-quality Cherries for the European Market*. COST meeting, Plovdiv, Bulgaria, May 2014. Available at: https://www.bordeaux.inra.fr/cherry/docs/dossiers/Activities/Meetings/2014%2005%2026-27%20WG1-WG3%20Meeting_Plovdiv/Presentations/Kaluzna_Plovdiv2014.pdf (accessed 7 June 2016).

Kappel, F. (2008) Breeding cherries in the 'New World'. *Acta Horticulturae* 795, 59–70.

Kappel, F. and Sholberg, P.L. (2008) Screening sweet cherry cultivars from the Pacific Agri-Food Research Centre Summerland breeding program for resistance to brown rot (*Monilinia fructicola*). *Canadian Journal of Plant Science* 88, 747–752.

Kappel, F., Granger, A., Hrotkó, K. and Schuster, M. (2012) Cherry. In: Badenes, M.L. and Byrne, D.H. (eds) *Fruit Breeding, Handbook of Plant and Breeding 8*. Springer Science + Business Media, New York, pp. 459–504.

Krzesinska, E.Z. and Azarenko, A.N.M. (1992) Excised twig assay to evaluate cherry rootstocks for tolerance to *Pseudomonas syringae* pv. *syringae*. *HortScience* 27, 153–155.

López-Ortega, G. (2015) Expedition and field explorations of natural populations of sweet and sour cherries in Azerbaijan. In: *Proceedings of the EU COST Action FA1104: Sustainable Production of High-quality Cherries for the European Market*. COST meeting, Dresden, Germany, July 2015. Available at: https://www.bordeaux.inra.fr/cherry/docs/dossiers/Activities/Short%20Term%20Scientific%20Missions/STSM%20Scientific%20Report_Lopez-Ortega%202.pdf (accessed 2 May 2016).

Matthews, P. and Dow, P. (1978) Cherry breeding. In: *John Innes Institute 68th Annual Report for 1977*. John Innes Centre, Norwich, UK, p. 34.

Matthews, P. and Dow, P. (1979) Cherry breeding. In: *John Innes Institute 69th Annual Report for 1978*. John Innes Centre, Norwich, UK, p. 38.

Matthews, P. and Dow, P. (1983) Cherries. In: *John Innes Institute 72nd Report for the two years 1981–1982*. John Innes Centre, Norwich, UK, p. 151.

Ozaktan, H. (2015) Screening the susceptibility of some sweet cherry cultivars to *Pseudomonas syringae* pv. *syringae* isolates by immature fruitlet test. In: *Proceedings of the EU COST Action FA1104: Sustainable Production of High-quality Cherries for the European Market*. COST meeting, Plovdiv, Bulgaria, May 2014. Available at: https://www.bordeaux.inra.fr/cherry/docs/dossiers/Activities/Meetings/2014%2005%2026-27%20WG1-WG3%20Meeting_Plovdiv/Presentations/Ozaktan_Plovdiv2014.pdf (accessed 24 March 2016).

Sansavini, S. and Lugli, S. (2008) Sweet cherry breeding programmes in Europe and Asia. *Acta Horticulturae* 795, 41–58.

Schmidt, M. (1937) Infektionversuche mit *Sclerotina cinerea* an Süß- und Sauerkirschen. *Gartenbauwisswissenschaft* 11, 167–182.

Schüller, E. (2013) Genetic fingerprinting of old Austrian cherry cultivars. In: *Proceedings of the EU COST Action FA1104: Sustainable Production of High-quality Cherries for the European Market*. COST meeting, Pitesti, Romania, October 2013. https://www.bordeaux.inra.fr/cherry/docs/dossiers/Activities/Meetings/15-17%2010%202013_3rd%20MC%20and%20all%20WG%20Meeting_Pitesti/Presentations/15_Schueller_Elisabeth.pdf (accessed 13 January 2017).

Song, G.Q. (2014) Recent advances and opportunities in cherry biotechnology. *Acta Horticulturae* 1020, 89–98.

Truchseß, C. (1819) *Systematische Classification und Beschreibung der Kirschensorten*. Cottaische Buchhandlung, Stuttgart, Germany.

Zwitzscher, M. (1961) Kirschen. In: Kappert, H. and Rudolf, W. (eds) *Handbuch der Pflanzenzüchtung*, Vol. V. Züchtung der Sonderkulturpflanzen, Paul Parey Verlag, Berlin, pp. 573–602.

# 5 Sour Cherry Varieties and Improvement

Mirko Schuster,[1*] Janos Apostol,[2] Amy Iezzoni,[3] Martin Jensen[4] and Dragan Milatović[5]

[1]*Julius Kühn-Institut, Dresden, Germany;* [2]*NARIC Fruitculture Research Institute, Budapest, Hungary;* [3]*Michigan State University, East Lansing, Michigan, USA;* [4]*Aarhus University, Årslev, Denmark;* [5]*University of Belgrade, Belgrade, Serbia*

## 5.1 History of Improvement

The variability in tree morphology and fruit characteristics is very high in sour cherry, especially in germplasm from the native regions in Eastern Europe and Asia Minor (Faust and Surányi, 1997). In these regions, sour cherry is not reproductively isolated from its progenitor species, and this continual gene flow has contributed to this high level of diversity. For example, sour cherry individuals with more sweet cherry- or ground cherry-like traits occur and probably represent individuals that have resulted from a 'backcross' to one of the two progenitor species (Hillig and Iezzoni, 1988). From this rich genetic diversity, human selection has resulted in the proliferation of many local landraces. For example, local landraces were selected in Hungary ('Pándy' (syn. 'Köröser Weichsel', 'Crişana'), 'Cigány' (syn. 'Zigeunerkirsche', 'Gypsy Cherry') and 'Újfehértói Fürtös'), Romania ('Mocăneşti'), Serbia ('Oblačinska' and 'Feketićka'), Croatia ('Marasca'), Germany ('Strauchweichsel' and 'Weinweichsel'), Russia ('Vladimirskaya') and Turkey ('Kutahya'). These landraces have been very important for breeding, as most of the common cultivars grown in these regions today are selections or hybrids of these original landraces.

Truchseß (1819) and Hedrick (1915) divided sour cherry into two groups based on fruit characteristics. These groups vary in both tree habit and fruit characteristics but have a constant difference in only a single, very easily distinguished character: juice colour. Sour cherries with red- to dark red-coloured juice are described as Morellos (Griottes, Weichsel). Morello fruit are very dark red with a spherical or cordate shape. Cherries with colourless juice are described as Amarelles (Kentish). Amarelles have pale red fruit and are more or less flattened at the end. As an additional division, Hedrick (1915) described the Marasca cherry. This cherry is a native of Dalmatia near Zadar in Croatia, where the tree grows wild and now is sparingly cultivated. Similar sour cherry genotypes have been described in northern Serbia in Feketic and near Novi Sad (G. Barać and V. Ognjanov, Novi Sad, Serbia, 2014, personal communication). The tree and fruit characteristics of the Danish local cultivar 'Stevnsbaer' are very similar. It is possible that 'Stevnsbaer' originated from the Marasca cherry (Stainer, 1975).

* mirko.schuster@julius-kuehn.de

Currently, sour cherry production is dominated by a small number of cultivars. In most cases, these cultivars are landrace cultivars or clonal selections of regional cultivars. In central Europe, the main sour cherry cultivar is 'Schattenmorelle' with many local synonyms, such as 'Łutovka' in Poland, 'Griotte du Nord' or 'Griotte Noire Tardive' in France, and 'Benelux' and occasionally 'English Morello' in the UK. The origin for this cultivar is probably the Chateau de Moreille in France. In the USA, the 400-year-old cultivar from France, 'Montmorency', is the dominant cultivar. The landrace cultivar 'Pándy' (syn. 'Crişana', 'Köröser') and related cultivars are very popular in Hungary and Romania.

## 5.2 Sour Cherry Breeding

Sour cherry improvement began in the Middle Ages with the selection and propagation of valuable sour cherry genotypes. With the development of fruit growing in the 19th century, the first artificial crosses were made between selected parents. As a result, a large collection of cultivars were developed. Today, sour cherry breeding is concentrated in middle and south-eastern Europe, with more limited efforts in North America. The new cultivars resulted from the clonal selection of landraces or from crossing programmes.

In areas with extensive diversity of native sour cherry, sour cherry breeding has been ongoing for more than half a century. In Poland, most new cultivars have resulted from artificial crosses with 'Łutówka' (syn. 'Schattenmorelle'). In the German and Hungarian programmes, the new cultivars originated from crosses where the landrace 'Köröser Weichsel' (syn. 'Köröser', 'Pándy') was one parent (Schuster and Wolfram, 2005; Apostol, 2011). In Hungary, Romania, Serbia and Denmark, many new cultivars resulted from regional clonal selection of the landraces 'Pándy', 'Mocăneşti', 'Oblačinska' and 'Stevnsbaer', respectively, or are hybrids between landraces and cultivars (Apostol, 2005; Budan *et al.*, 2005a; Miletić *et al.*, 2008). Cultivars released in Russia and Canada may be interspecific hybrids of sour cherry with *fruticosa* because of the need to incorporate cold hardiness (Zhukov and Kharitonova, 1988; Bors, 2005). As with the Canadian cultivars, US new cultivars are derived from European germplasm, as sour cherry is not native to North America (Iezzoni *et al.*, 2005).

### 5.2.1 Objectives in sour cherry breeding

Traditionally, sour cherries are used as processed products. Therefore, the main breeding goals include superior fruit characteristics, high productivity, tolerance/resistance to biotic and abiotic stresses, suitability for mechanical harvesting and extension of the harvest period. Currently, there is a growing interest in breeding sour cherries for the fresh market with increased fruit size, firmness and a pleasant cherry flavour.

#### *Tree and fruiting structure*

There are wide ranges of tree and fruiting habits in sour cherry. Tree sizes vary from an upright and vigorous growth habit such as sweet cherry to dwarf or bushy tree types more similar to *P. fruticosa*. For high-density planting, especially for mechanical harvesting with over-the-row harvesters, less vigorous cultivars are required. Therefore, one goal in sour cherry breeding is to develop cultivars with low vigour. In Sweden, the low-vigour cultivars 'Kirsa' and 'Nordia' were selected, which grow as shrubs to a height up to 2 m (Trajkovski, 1996). In Canada, a collection of dwarf sour cherries that are very good for over-the-row mechanical harvesting were developed from interspecific hybridization of *P. fruticosa* and *P. cerasus* (Bors, 2005; Montgomery, 2009). Another way of reducing the tree growth habit is the use of dwarfing rootstocks. Initial investigations in Germany with sour cherry have shown a reduction in tree size and an increase in fruit set with the use of dwarfing rootstocks.

Sour cherry yield is a complex trait that depends on many factors, such as the characteristics of the fruiting branches, density of flower buds, number of flowers in the bud, degree of fertility, fruit set, fruit size, environmental factors and growing practices. Sour cherry cultivars produce most of their fruit on 1-year-old shoots, and only a small percentage of fruit is on spurs, which are formed on older branches. Cultivars that mainly produce fruit on 1-year-old shoots are prone to form bare wood; an example is the cultivar 'Schattenmorelle'. The tendency to produce bare wood is low in cultivars where the fruit are borne primary on spurs or a bouquet of spurs on older wood. Therefore, in some breeding programmes, selection of new cultivars is focused on trees with upright growth and fruit produced primarily on spurs. This tree habit is advantageous for most harvesting techniques and requires minimal pruning due to a reduced need to remove the bare wood. 'Achat', 'Meteor Korai' and 'Piramis' are examples of cultivars that have fruit produced primarily on spurs.

### Flower characteristics

The primary limiting factor in developing a new sour cherry cultivar is obtaining selections with high yield. High-yielding cultivars include 'Montmorency' in the USA and 'Schattenmorelle' in Europe. Most sour cherry seedlings produce low yields, mainly due to poor fruit set.

High fertility and self-compatibility are important for good fruit set in sour cherry. Sour cherries are frequently considered to be self-compatible, although self-incompatible and partly self-compatible cultivars exist. Redalen (1984) regarded cultivars with a final fruit set of more than 15% after self-pollination as self-compatible. Self-incompatible cultivars may in some years set few fruit. Cultivars with a final fruit set of between 1 and 14% have been characterized as partially self-compatible. Certain pairs of cultivars are cross-incompatible, reciprocally or unilaterally (Bošković *et al.*, 2006). Similar results have been obtained in progenies of cross-populations in sour cherry breeding. This cross-incompatibility is due to the gametophytic self-incompatibility system that exists in sour cherry (Yamane *et al.*, 2001; Tobutt *et al.*, 2004), in which many of the *S*-alleles are shared with sweet cherry. Self-compatibility in sour cherry requires that the individual has a minimum of two non-functional alleles at the *S*-locus (Hauck *et al.*, 2006).

Even in self-compatible cultivars, low fruit set may often occur. An application of compatible pollen does not necessarily always increase fruit set on these low-fertility selections, and it is likely that the low fruit set is due to problems with the ovule or resulting embryo. Possible causes include susceptibility to early ovule degeneration, or ovule or zygotic abortion due to aneuploid gametes resulting from meiotic instability caused by intra- or interspecific crosses or inbreeding effects. Understanding the yield potential of any new selection is critical, as successful cultivars must have a high fruit set potential.

### Tolerance to abiotic and biotic stress

Winter cold hardiness is one of the most important goals of sour cherry breeding in areas with cold climates, such as Russia and Canada. Sour cherry cultivars vary widely in their winter cold hardiness. In some Russian cultivars, flower buds can tolerate temperatures down to –38°C, while in some European cultivars the critical temperature is –20°C (Iezzoni, 1996). Venjaminov (1954) distinguished sour cherry cultivars for their winter cold hardiness in three groups. The first group are cultivars of ground cherries, *P. fruticosa*, and related hybrids such as the Russian cultivars 'Antonovska Kostychevskaya', 'Plodorodnaya Michurina', 'Zakharovskaya', 'Ideal', 'Polzhir' and 'Polevka'. This group of cherries showed no damage after the winters of 1939–1942 in Russia with temperatures as low as –50°C. The second group are sour cherries that show medium resistance to winter frost. Examples are the Russian cultivars 'Vladimirskaya', 'Lyubskaya' and 'Shubinka' and the western European cultivars 'Ostheimer' and 'Kentish'; however both western European cultivars do not reach the level of the Russian cultivars. The third group are winter frost-susceptible cultivars. This group includes most European cultivars such as 'Schattenmorelle' and 'Podbelskaya', the Russian synonym for 'Koch's Ostheimer Weichsel' (Symyrenko, 1963). Donors of winter cold hardiness used in Russia include the sour cherry cultivars 'Pamyati Vavilova' and 'Plodorodnaya Michurina' and the Manchurian cherry, *P. maackii*, the $F_1$ hybrid 'Padocerus' (*P. maackii* × 'Plodorodnaya Michurina') and $F_2$ backcross populations of the $F_1$ hybrid (Zhukov and Kharitonova, 1988). In Canada, interspecific hybrids of *P. fruticosa* × *P. cerasus* were used as a source to increase the frost tolerance in sour cherry. From $F_2$ backcross populations, six sour cherry cultivars have been released since 1999 (Bors, 2005).

Because of the early blossom time of cherries, spring frost can damage buds, flowers and young fruit. Selection of late-blooming genotypes, with high-chilling requirements and tolerance to spring frost can reduce the risk. Avoidance of late spring frost damage can be increased by using cultivars with a late blooming time. Examples for late-blooming sour cherries are the Russian cultivars 'Plodorodnaya Michurina', 'Lyubskaya' and 'Kisloyakovka' (Venjaminov, 1954). In Michigan (USA), 'Plodorodnaya Michurina' blooms 6–9 days after the cultivar 'Montmorency' (Iezzoni, 1996).

Blossom blight and brown rot, caused by *Monilinia laxa* (Aderh. & Ruhl.) Honey, and cherry leaf spot, *Blumeriella jaapii* (Rehm) Arx., are the main fungal diseases of sour cherry. These diseases can significantly reduce the yield in sour cherry orchards. There are no known wild cherry *Prunus* spp. resistant to blossom blight. The symptoms caused by *M. laxa* and the degree of susceptibility depend on the climatic conditions and the virulence of specific local races (Budan *et al.*, 2005b). The German cultivars 'Jade' and 'Achat' and the Hungarian cultivars 'Csengődi' and 'Piramis' show a high level of tolerance to blossom blight and can be used in resistance breeding programmes.

Cherry leaf spot is common in most cherry-growing areas in North America and Europe. When not controlled by fungicides, cherry leaf spot can cause early leaf defoliation. This can result in decreased fruit quality and a reduction of winter hardiness. According to studies by Schuster (2004) and Budan *et al.* (2005b), only a few sour cherry cultivars are tolerant to cherry leaf spot. *P. maackii* (a tetraploid species) and *P. canescens* (a diploid species) exhibit a high level of resistance to *B. jaapii* (Wharton *et al.*, 2003; Schuster, 2004). Resistance breeding programmes in Russia, Germany and the USA have used *P. maackii* as a donor for cherry leaf spot resistance (Schuster *et al.*, 2013), while breeding programmes in Germany and the USA have used *P. canescens*. In the Hungarian breeding programme, the cultivar 'Csengődi' has been used as donor for blossom blight and cherry leaf spot resistance. The cultivar 'North Star', bred at the University of Minnesota (USA), exhibits tolerance to cherry leaf spot (Sjulin *et al.*, 1989) and is also used as a parent. Artificial inoculation protocols have been used for evaluation of the reaction of cherry genotypes to cherry leaf spot (Wharton *et al.*, 2003; Schuster, 2004).

### *Fruit quality*

The importance of fruit quality has increased in recent years. The major quality parameters in sour cherry are soluble solids content, titratable acidity, fruit and juice colour, firmness and good taste. The characteristics for the different parameters vary according to the utilization of the fruit. Most fruit are used for processing purposes, such as juice, canning, jam, drying and wine. Only a small percentage of sour cherries are produced for the fresh market. The quality of cherry products is determined by their appearance and sensory properties. These characteristics are determined by the colour, acidity and sugar content of the fruit and their concentrations of volatile compounds. The ideal fruit characteristics for processing are a fruit diameter of 21–24 mm, a dark red-coloured juice with a high staining intensity (except in the USA), a high soluble solids content (>15°Brix) and acidity (>20 g $l^{-1}$ malic acid) combined with a good aroma. For juice production and fresh consumption, an increased fruit size is desired. Additionally, for fresh consumption, high sugar and lower acidity are preferred. In the past, many studies investigated the anthocyanin and aroma components in sour cherry during the ripening season (Schmid and Grosch, 1986; Poll *et al.*, 2003; Šimunic *et al.*, 2005). Anthocyanins from sour cherry have been shown to possess strong antioxidant and anti-inflammatory activities (Wang *et al.*, 1999) (see Chapter 17, this volume). Stone characteristics are also important. For processing, the stone has to be small (ideally no more than 7% of fresh weight) and round and should be removed easily from the fruit flesh (see Chapter 20, this volume).

#### *Extension of harvest period*

The ripening period of sour cherry cultivars may range over a 4-week period in any one country; however, sour cherry production is mostly dominated by one or only a few cultivars in specific cherry-growing areas. Therefore, the harvest period is very short in most growing areas (1–2 weeks). Extension of the harvest period could increase the utilization of machinery and decrease the labour costs.

#### *Suitability for mechanical harvesting*

Most sour cherries that are grown for processing are mechanically harvested. This harvest technique requires specific characteristics of the cherry fruit and trees: firm fruit that are highly resistant to bruising, a low fruit retention force from the stem and a dry stem scar to reduce juice loss, uniformity of ripening, an upright and stable trunk, and a maximum tree height of around 3–3.5 m. Sour cherry cultivars differ considerably in their suitability for mechanical harvesting. Brown and Kollar (1996) reviewed cultivars suited to mechanical harvesting. Some of the cultivars with good suitability for mechanical harvesting unfortunately have low fruit set. Examples include 'Ujferhertoi Fürtös', 'Morina' and 'Pándy'. The use of chemicals, many of which contain ethylene, to lower the retention force of the fruit from the stem (which encourages the formation of abscission layers) is restricted in many countries of the world.

### 5.2.2 Methods of sour cherry breeding

Two main breeding methods are used in the sour cherry. The first method is termed selective breeding, or clonal selection, and is carried out among natural variants appearing in nature and within traditional cultivars. The second method, cross-combination breeding, utilizes controlled crosses by selection of genotypes that possess desirable characteristics from different parents.

Selective breeding is the oldest method in fruit breeding generally and uses native local populations as the basis for selection. The domestication of sour cherry was based on this method. In the traditional growing areas in central, eastern and southern Europe, landraces such as 'Schattenmorelle', 'Vladimirskaya', 'Pándy', 'Cigány', 'Újfehértói Fürtös', 'Kántorjánosi' 'Debreceni Bőtermő', 'Petri', 'Csengődi', 'Oblačinska', 'Kutahya' and 'Marasca' arose because of selective breeding. These landraces and related selections show a high adaptation to the local climatic and growing conditions. Due to their special characteristics, landraces have formed the basis for modern sour cherry breeding. Currently, only a few breeding programmes in Hungary, Serbia, and Turkey are focused on selection in local native sour cherry populations.

Today, the most common breeding method in sour cherry is cross-combination breeding by controlled crosses or open pollination. The aim of cross-combination breeding is to bring together desired traits found in different cherry genotypes into seedling populations via cross-pollination. Most breeding programmes are applying this method to generate new populations in order to develop and release new cultivars.

Interspecific hybridization is an important part of cross-combination breeding. Some breeding programmes have utilized interspecific hybridizations with wild *Prunus* spp. as donors for different traits of interest to enlarge the genetic diversity of cherry. One of the first scientists who used wild species in cherry breeding was the Russian breeder I.V. Michurin. He made interspecific crosses with *P. maackii* and *P. fruticosa* to increase the resistance to winter frost and diseases in sour cherry. He created the interspecific hybrids 'Cerapadus no. 1' (*P. fruticosa* × *P. maackii*), 'Cerapadus Krupny' and 'Cerapadus Sladki' ('Ideal', (*P. fruticosa* × *P. pensylvanica*) × *P. maackii*) and 'Plodocerus' (*P. maackii* × *P. cerasus*) within his sour cherry breeding programme (Mitschurin, 1951; Shukov and Charitonova, 1988). Different interspecific hybridizations have been carried out to provide resistance to cherry leaf spot (*B. jaapii*) by Ukrainian and Russian cherry breeders in recent decades. Progeny of crosses between *P. maackii*, *P. fruticosa* and *P. cerasus*, and hybrids of *P. cerasus* × *P. avium* were used as resistance donors (Shukov and Charitonova, 1988). In 1944, L. Kerr began to hybridize *P. fruticosa* and *P. cerasus* to increase the cold hardiness in the Canadian sour cherry breeding programme. This programme was continued by S. Nelson and R.H. Bors (Bors, 2005). In the German cherry rootstock breeding programmes at Pillnitz (Dresden) and Giessen, a number of interspecific hybrids between *P. avium* and *P. cerasus* and the dwarf species *P. canescens*, *P. incisa* and *P. tomentosa* were produced (Webster and Schmidt, 1996). Some of these rootstock clones were used as donors for disease resistance in breeding programmes in the USA and Germany (Wharton *et al.*, 2003; Schuster and Wolfram, 2005). In a cherry breeding programme at Dresden-Pillnitz, different interspecific hybridizations and backcrosses were done between sour cherry cultivars and the tetraploid *Prunus* species *P. maackii*, *P. spinosa*, *P. padus* and *P. serotina* in the period from 2000 to 2009. The goal for the development of these hybrids was to transfer new sources of resistance to biotic and abiotic stresses as well as interesting fruit characteristics such as anthocyanins in the sour cherry genome (Schuster *et al.*, 2013). In Michigan (USA), the sour cherry breeding programme developed interspecific hybrids between the diploid species *P. canescens* and sour cherry cultivars in order to transfer cherry leaf spot resistance to the tetraploid sour cherry genome. The source of resistance was derived from the tetraploid *P. canescens* disease-resistant selection Q39515, and these breeding populations were used to develop molecular markers for resistance to cherry leaf spot (Stegmeir *et al.*, 2014).

## 5.3 Sour Cherry Breeding Programmes

The main sour cherry breeding programmes have been started in Germany, Hungary, Romania, Russia and Ukraine. Breeding activities exist also in Belarus, Canada, Denmark, Poland, Serbia and the USA.

### 5.3.1 Belarus

The breeding work of sour cherries in Belarus was begun in 1927 by E.P. Syubarova. In this first programme, seeds from open-pollination of local and western European cultivars were used, which resulted in development of the cultivars 'Seedling No.1' and 'Novodvorskaya'. In the period from 1965 to 1982, R. Sulimova continued the breeding work and selected the cultivars 'Zhyvitsa', 'Zaranka', 'Vyanok' and 'Glubokskaya'. After a long period of evaluation of local and foreign sour cherry cultivars in 2000 at the Institute for Fruit Growing in Samokhvalovichi, a new breeding programme was started by M.I. Wyshynskaya and continued by A. Taranov. Emphasis was on Belarussian and foreign sour cherry cultivars and interspecific hybrids with *P. fruticosa*, *P. avium* and *P. maackii*. The main breeding goals are high productivity, fruit quality, resistance to diseases and high tolerance to winter frost. Some recently released cultivars are 'Griot Belorusskij' (2004), 'Lasukha' (2008), 'Konfityur' (2013), 'Milavitsa' (2013) and 'Nesvizhskaya' (2013).

### 5.3.2 Canada

Breeding for sour cherries is conducted at the University of Saskatchewan in Saskatoon. Breeding work was begun by L. Kerr in the 1940s, crossing *P. fruticosa* with *P. cerasus*. The main objectives of the crosses were creating dwarf trees suitable for mechanized harvesting, good fruit quality (high sugar content) and resistance to frost. The first cultivar developed from this programme was introduced in 1999 and was named 'SK Carmine Jewel'. In 2004, the Romance series of dwarf sour cherries was released. These include 'Juliet', 'Romeo', 'Cupid', 'Valentine' and 'Crimson Passion' (Bors, 2005).

### 5.3.3 Denmark

A continuous breeding programme has never existed in Denmark. Some breeding and selection activities have, however, been carried out since the 1960s but to date have been based on work in a number of single projects. J.V. Christensen, at the Danish research organization at Blangstrupgaard, which today is part of Research Centre Aarslev, Aarhus University, started a programme to compare the different local landraces of wild sour cherry of the type 'Stevnsbaer' produced under different regional names in the 1960s (Christensen, 1976). Between 1971 and 1981, a cultivar of 'Stevnsbaer', called 'Viki', was selected at Blangstedgaard (Christensen, 1983, 1986). A cultivar of 'Stevnsbaer' called 'Birgitte' was selected as the highest yielding cultivar in Aarslev from 1983 to 1994 (Christensen, 1995a,b). The result of a cross-breeding programme by Christensen resulted in the selection of three additional cultivars, 'Tiki', 'Miki' and 'Siki' in the early 1980s. These cultivars are distinguished by ripening times from early to very late, respectively.

From 2004 to 2008, a project at Aarslev, Aarhus University, by K. Kristiansen, M. Jensen and B.H. Pedersen focused on developing rapid-cycling breeding methods in sour cherry (Kristiansen and Jensen, 2009) and at the same time generating breeding progeny. Currently, new sour cherry clones have been selected for Danish fruit growing by M. Jensen.

### 5.3.4 Germany

The first sour cherry breeding programme was initiated by M. Schmidt at the Kaiser-Wilhelm Institute in Müncheberg in the 1930s. After World War II, M. Zwitzscher started a new breeding programme with breeding material from Müncheberg and seedling progenies of self-pollinated 'Schattenmorelle' at the Max Planck Institute in Vogelgesan, Cologne, and the following cultivars were developed: 'Mailot', 'Cerella', 'Nabella', 'Successa' and 'Bonnie' (Zwitzscher, 1964, 1968, 1969).

A second breeding programme was initiated by B. Wolfram in Müncheberg in 1965 and was continued at the Institute for Fruit Research from 1971 and by M. Schuster from 2000 at the Julius Kühn-Institut in Pillnitz, Dresden. The Hungarian cultivar 'Köröser Weichsel' was used as the primary crossing parent to increase fruit quality and tolerance to diseases in sour cherry. From 1965 to 1991, the cultivars 'Karneol', 'Korund', 'Morina', 'Safir' and 'Topas' (Wolfram, 1990) were released, and from 1991 to 2004, the cultivars 'Achat', 'Jade', 'Coralin', 'Spinell', 'Jachim' and 'Boas' were released (Schuster, 2009, 2011, 2012; Schuster *et al.*, 2014).

In a small temporary breeding programme at the Fruit Research Institute in Geisenheim, W. Jacob selected 'Gerema' and 'Geresa' from open-pollinated sour cherries in the last decade of the 20th century (Jacob, 1994).

### 5.3.5 Hungary

Sour cherry breeding work in Hungary began in 1950 at the Érd and Újfehértó experimental stations of the Horticultural Research Institute (today called the National Agricultural Research and Innovation Center, Fruticulture Research Institute). Three main breeding strategies have been used: clonal selection of 'Pándy' types, the selection of native landrace populations, and a cross-combination breeding program.

Three cultivars were developed and released from the 'Pándy' clonal selection by S. Brózik and Z. Éles: 'Pándy 48', 'Pándy 279' and 'Pándy Bb.119'. Because 'Pándy' is self-incompatible, an additional four cultivars, 'Cigány 7', 'Cigány 59', 'Cigány 3' and

'Cigány C.404', were selected from the landrace 'Cigány' as pollinator.

The selection of native landrace populations was initiated by F. Pethő in the north-eastern region of Hungary in 1950, and was continued by T. Szabó at the Újfehértó Research Station, located in Újfehértó. The aims of the landrace selection were high productivity, self-compatibility, a dry abscission layer between pedicel and fruit, and ripening period. The cultivars 'Újfehértói Fürtös', 'Debreceni Bőtermő', 'Kántorjánosi 3', 'Éva' and 'Petri' were released from this breeding programme (Szabó, 1996; Szabó *et al.*, 2008b). In the central Hungarian region, J. Apostol conducted native selections that resulted in the development of 'Csengődi' and 'Ducat'.

The cross-combination breeding programme was initiated by P. Maliga in 1950, and in 1967 was continued by J. Apostol and from 2009 by S. Szügyi. Seven cultivars, 'Meteor Korai', 'Favorit', 'Érdi Jubileum', 'Korai Pipacs', 'Érdi Nagygyümölcsű', 'Érdi Bőtermő' and 'Maliga Emléke', were selected and released from 1950 to 1967. The cultivars 'Érdi Ipari' and 'Piramis' were released between 1967 and 1979, and 'Érdi Korai', 'Érdi Kedves' and 'Érdi Bíbor' were selected and released after 1980. From a 'Feketićka' landrace population at the border region between Hungary and Serbia, the cultivar 'Prima' was selected by J. Apostol.

### 5.3.6 Poland

The sour cherry breeding programmes were initiated at the Agricultural Universities in Lublin and Poznań by S. Zaliwski and M. Mackowiak and at the Research Institute of Pomology (today called the Research Institute of Horticulture) in Skierniewice in the 1950s. The cultivar 'Nefris' was selected in Lublin. In Poznań, the cultivars 'Agat', 'Ametyst', 'Diament' and 'Dradem' were released in 1997. The breeding activities at the universities in Lublin and Poznań were discontinued at the end of 20th century. The breeding programme at the Institute of Pomology and Floriculture in Skierniewice was initiated by S. Zagaja, A. Buczek-Jackiewicz, A. Wojniakiewicz, and later continued by Z.S. Grzyb and T. Jakubowski. The goal of the sour cherry breeding programmes in Poland are to develop new cultivars with good fruit characteristics for processing and fresh consumption, and that are suitable for mechanical harvesting. The cultivars 'Lucyna', 'Sabina', 'Wanda', 'Koral', 'Winer', 'Wilena', 'Wilga' and 'Mazowia' were selected and released between 1950 and 1992. Since 2006, E. Żurawicz and M. Szymajda have continued the sour cherry breeding programme in Skierniewice. The results of these activities have been the release of the cultivars 'Galena' and 'Granda'.

### 5.3.7 Romania

Sour cherry breeding began in 1956 at the Research Station for Fruit Growing (RSFG) in Cluj-Napoca and was continued at three institutes: the Research Institute for Fruit Growing (RIFG) in Pitesti and the RSFG in Focşani and in Iasi. The largest number of cultivars were developed by the selection of natural clonal populations of 'Crişana', 'Mocăneşti' and other local landraces. The main breeding objectives were to develop new cultivars which have different ripening times, self-compatibility, high productivity, intense coloured fruit and ideal attributes for processing, and a medium tree growth vigour. The clones 'Mocăneşti 16' and 'Crişana 2' and the cultivars 'Vrâncean', 'Scuturător', 'Timpurii de Osoi', 'Timpurii de Pitesti', 'Bucovina', 'De Botoşani', 'Timpurii de Tg. Jiu' and 'Pitic' were released. In 1971, at the Fruit Research Institute, Pitesti, cross-breeding programmes began developing cultivars resistant to *B. jaapii* that were self-compatible, with good fruit quality and suitable for mechanized harvesting (Budan and Stoian, 1996). After 1990, the cultivars 'Amanda', 'Nana', 'Dropia', 'Ilva', 'Timpurii de Cluj', 'Tarina', 'Rival', 'Sătmărean' and 'Stelar' were released.

### 5.3.8 Russia

Russia is the leading country in the world in terms of the number of newly developed sour cherry cultivars. As a result of their breeding work, a number of scientific research institutes and experimental stations located in, for example, Orel, Moscow, Kazan and Michurinsk have released more than 100 new cultivars of sour cherry (e.g. 'Lyubskaya', 'Zhukovskaya', 'Pamyati Vavilova', 'Standart Urala', 'Ural'skaya Ryabinovaya', 'Gurt'evka', 'Turgenevka', 'Komsomolskaya', 'Studencheskaya' and 'Polevka'). The main objectives of the sour cherry breeding programme in Russia are: frost resistance, resistance to diseases (*B. jaapii*, *M. laxa*), self-compatibility, high yield and good fruit quality. Fruit size is not a priority in breeding, and the majority of new cultivars have a smaller fruit size of 3–4 g.

In addition to the breeding of common sour cherry, in Russia the breeding of steppe cherry (*P. fruticosa*) has also been conducted, especially in Siberia (Barnaul, Omsk, Yekaterinburg and Novosibirsk). Cultivars that originate from the steppe cherry are characterized by higher frost resistance, less vigour, better precocity and higher resistance to diseases compared with *P. cerasus* cultivars. However, they have small fruit and the taste is very sour, often with a hint of astringency or bitterness.

In the eastern part of Russia, especially in the Far East, breeding of the Nanking cherry (*P. tomentosa*) is also conducted. The leading institution is the Far East Experimental Station VNIIR in Vladivostok. The most important breeding objectives are: resistance to mid-winter low temperatures and late autumn frost, resistance to diseases, self-compatibility, high yield and good fruit quality (dark red colour, firm flesh, small pit that is easily separated from the flesh and good taste).

### 5.3.9 Serbia

The sour cherry breeding programme at the Fruit Research Institute at Čačak is focused on self-compatible cultivars with high productivity, superior fruit quality suitable for processing and fresh consumption, ease of mechanical harvesting, a range of ripening times and tolerance/resistance to the economically most important pests and diseases (Nikolić and Cerović, 1998; S. Radičević and R. Cerović, Čačak, Serbia, 2014, personal communication). Two cultivars, 'Čačanski Rubin' and 'Šumadinka', were selected by A. Stančević and released in 1973 and 1984, respectively. Three new cultivars 'Nevena', 'Iskra' and 'Sofija', all have very large fruit (7–8 g) and were released in 2014. From the Institute PKB in Belgrade, the cultivar 'Lara' was released in 1993. From the Faculty of Agriculture in Belgrade, the cultivar 'Lenka' was released in 2014. In addition to the cross-breeding programme, a second breeding programme, located at the Universities of Belgrade and Novi Sad, is focused on the selection of sour cherry genotypes from the native landrace populations of 'Oblačinska' and 'Feketićka' (Nikolić *et al.*, 2005; Rakonjac *et al.*, 2010).

### 5.3.10 Ukraine

In Ukraine, sour cherries have been bred at four institutions of the National Academy of Agrarian Sciences of Ukraine (NAAS). Since 1920, these breeding programmes have been selecting sour cherries for the climatic conditions of Ukraine and former Soviet Union. The Institute of Horticulture (IH) NAAS is the leading research institute in fruit breeding in Kiev. Currently, V.I. Vasylenko continues the breeding work of N.V. Moiseichenko. In 2014 the cultivar 'Bohuslavka' was released.

In recent years, the Melitopol Research Station of Horticulture in Melitopol and the Artemivsk Nursery Research Station in Opytne have been integrated at the IH NAAS. At the Melitopol Research Station of Horticulture, the breeding goals are to develop sour cherry cultivars with tolerance to winter frost and humid climate conditions for the Black Earth Region (Chernozem) in Ukraine and the former Soviet Union. Between 1933 and 1965, M.T. Oratovskyy and D.A. Batyuk developed the cultivar 'Melitopolska Desertna'. Between 1966 and 2000, M.I. Turovtsev and V.O. Turovtseva established a cross-combination breeding programme that included interspecific hybridizations to develop new breeding material (Turovtsev and Turovtseva, 2002). Since 2006, A.N. Shkinder-Barmina has continued the breeding work. Between 1990 and 2006, 17 sour and duke cherry cultivars were released, including 'Vstriecha', 'Hriot Melitopolsky', 'Ihrushka', 'Ozhydaniie', 'Prymitna', 'Solidarnist' and 'Shalunia' (Turovtseva *et al.*, 2013).

At the Artemivsk Nursery Research Station, L.I. Taranenko led the cherry breeding in cooperation with the Melitopol Research Station for over 65 years (Taranenko, 2004). Currently, V.V. Yarushnikov continues the sour cherry breeding. The most valuable selected cultivars are 'Kseniia', 'Nochka', 'Chudo Vyshnia', 'Shpanka Donetska', 'Donetskyi Velykan', 'Doch Yaroslavny' and the newest release 'Altruistka' (Yarushnykov, 2013). From the oldest fruit breeding institute, the Institute for Pomology 'L.P. Symyrenko' in Mlijv, the cultivars 'Alfa' and 'Zhadana' were released. Currently, L.S. Yuryk continues the breeding work.

### 5.3.11 USA

The sour cherry breeding programme at Michigan State University (MSU, East Lansing, Michigan) began with pollinations made in the spring of 1983. As sour cherry is not native to the USA, the germplasm used was collected in Europe through collaborations with the existing breeding programmes (Iezzoni, 2005). Cooperation between the MSU programme and the breeding programmes in Érd and Újfehértó, Hungary, resulted in the commercialization of three Hungarian sour cherry cultivars in the USA ('Újfehértói Fürtös', 'Érdi Bőtermő' and 'Érdi Jubileum'). The MSU programme's goals are to develop new sour cherry cultivars that are high yielding and superior to 'Montmorency' due to cherry leaf spot resistance, reduced freeze susceptibility and improved fruit quality, such as improved fruit firmness, while retaining the brilliant red fruit colour characteristic of 'Montmorency' (Iezzoni *et al.*, 2005).

## 5.4 Characteristics of Sour Cherry Cultivars

### 5.4.1 Sour cherry cultivars with global importance

#### *'Érdi Bőtermő' (syn. Danube™)*

Origin: selected by P. Maliga and J. Apostol at the Horticulture Research Institute Budapest, Hungary.
Parentage: 'Pándy' × 'Nagy Angol'.
Tree growth: medium growth with a round crown shape.
Fertility/fruit set: self-compatible; under Hungarian conditions, a high and regular fruit set, but in Germany the fruit set is only medium to low (Szabó *et al.*, 2008a).
Blooming time: early, about 5–6 days before 'Schattenmorelle'.
Ripening time: early, about 3 weeks before 'Schattenmorelle'.
Fruit characteristics: medium to large, dark red skin, medium firm.
Fruit flesh: dark red.
Juice: slightly staining.
Resistance: sensitive to cherry leaf spot and blossom blight.
Utilization: fresh consumption and processing.

#### *'Fanal' (syn. 'Heimanns Konservenkirsche', 'Heimann 23')*

Origin: selected by Heimann, Blankenburg/Harz, Germany.
Parentage: unknown.
Tree growth: medium to strong growth vigour, upright with good branching, crown shape round to wide pyramidal.
Fertility/fruit set: self-compatible, high and regular fruit set.
Blooming time: early, about 5–6 days before 'Schattenmorelle'.
Ripening time: medium, some days before 'Schattenmorelle'.
Fruit characteristics: medium to large, brown-red skin, medium firm.
Fruit flesh: red to brown-red, soft.
Juice: purple red.
Resistance: susceptible to bacterial canker.
Utilization: fresh consumption and processing.

### *'Kelleriis 16' (syn. 'Morellenfeuer')*

Origin: selected by D.T. Poulsen, Kvistgaard, Sealand, Denmark, around 1940.
Parentage: ('Ostheimer Weichsel' × 'Früheste der Mark') × open pollination (o.p.).
Tree growth: upright, vigorous growth with thicker shoots than 'Stevnsbaer' and rounded crown.
Fertility/fruit set: self-compatible, high fruit set.
Blooming time: mid-early, before 'Schattenmorelle'.
Ripening time: mid-early from late July to early August, about 2–3 weeks before 'Stevnsbaer' in Denmark.
Fruit characteristics: medium to large, round, dark red skin, medium firm, sugar, acid and colour content are low to medium.
Fruit flesh: dark red.
Juice: slightly staining.
Resistance: susceptible to blossom blight and *Prunus* necrotic ringspot virus.
Utilization: processing.

### *'Montmorency'*

Origin: old cultivar from France, cultivated in the USA since at least the early 20th century.
Parentage: unknown.
Tree growth: medium growth vigour.
Fertility/fruit set: self-compatible, high fruit set.
Blooming time: mid-season.
Ripening time: medium, about 1 week before 'Schattenmorelle'.
Fruit characteristics: medium size, round, light red skin, medium firm.
Fruit flesh: yellow.
Juice: no staining.
Resistance: susceptible to cherry leaf spot; susceptible to blossom blight only when environmental conditions are very favourable for infection.
Utilization: processing.

### *'Oblačinska'*

Origin: landrace selected from natural sour cherry populations in South Serbia (between Niš and Prokuplje), mixture of different clones, propagated by suckers.
Parentage: unknown.
Tree growth: weak to medium growth vigour with a rounded and dense crown.
Fertility/fruit set: self-compatible, regular and abundant.
Blooming time: mid-season, a few days before 'Schattenmorelle'.
Ripening time: 1 week before 'Schattenmorelle'.
Fruit characteristics: small to medium, round, dark red skin, medium firm, high sugar and acidity content.
Fruit flesh: dark red.
Juice: coloured intensely red with a high content of anthocyanins.
Resistance: good tolerance to frost and drought.
Utilization: processing.

### *'Pándy' (syn. 'Köröser', 'Pándy Meggy', 'Szentesi Meggy', 'Köröser Weichsel', 'Kereska', 'Crişana')*

Origin: landrace clonal population with a wide range of differences in tree vigour, blooming and ripening time, fertility and some fruit characteristics; native in the Carpathian basin (Faust and Surányi, 1997).
Parentage: unknown.
Tree growth: upright, strong growth vigour.
Fertility/fruit set: self-incompatible, low to medium fruit set.
Blooming time: mid-season, a few days before 'Schattenmorelle'.
Ripening time: 1 week before 'Schattenmorelle'.
Fruit characteristics: medium to large, round, dark red skin, medium firm, high sugar and acidity content.
Fruit flesh: dark red.
Juice: coloured intensely red.
Resistance: medium tolerance to cherry leaf spot and sensitive to blossom blight.
Utilization: fresh market, processing.

### *'Schattenmorelle' (syn. 'Große Lange Lotkirsche', 'Łutówka', 'Łutovka', 'Griotte du Nord', 'Moreillska', 'Skyggemorel')*

Origin: has been growing in Germany and in some other countries in central and eastern Europe since the 18th century.

Parentage: unknown.
Tree growth: medium to strong growth vigour with a spherical crown shape and pendulous fruit branches with a tendency to produce blind wood.
Fertility/fruit set: self-compatible, regular high fruit set.
Blooming time: late, end of sour cherry blooming.
Ripening time: late, at the end of July to beginning of August in central Europe.
Fruit characteristics: medium to large, round, brown-red skin, medium firm, high sugar and acidity content.
Fruit flesh: dark red, medium firm.
Juice: dark red.
Resistance: highly susceptible to blossom blight and cherry leaf spot.
Utilization: processing.

#### *'Stevnsbaer' (cultivars 'Viki', 'Birgitte')*

Origin: wild seedling-based landrace type known at least back to the 16th century in Denmark and for more than 200 years in selected regions of Denmark; cultivar 'Viki' was selected from a trial of 20 promising 'Stevnsbaer' clones at Blangstedgaard from 1971 to 1981 (Christensen, 1983, 1986). Cultivar 'Birgitte' was originally found in an orchard near Stevns and selected by J.V. Christensen.
Parentage: unknown.
Tree growth: vigorous pyramidal trees with slender hanging branches.
Fertility/fruit set: self-compatible, regular high fruit set.
Blooming time: mid-early, about 5 days before 'Schattenmorelle'.
Ripening time: at the beginning of August, a few days later than 'Schattenmorelle' in central Europe.
Fruit characteristics: small, flat-round shape, dark red skin, medium firm, very high sugar and acidity content.
Fruit flesh: dark red.
Juice: dark red, staining intensely with a high content of anthocyanins.
Resistance: resistant to rain-induced cracking, susceptible to bacterial canker.
Utilization: processing.

#### *'Újfehértói Fürtös' (syn. Balaton™, 'Ungarische Traubige')*

Origin: selected by F. Pethő and T. Szabó from an unknown seedling in the north-eastern part of Hungary.
Parentage: unknown.
Tree growth: medium growth vigour with slender hanging branches.
Fertility/fruit set: partly self-compatible.
Blooming time: medium, a few days before 'Schattenmorelle'.
Ripening time: medium, in the middle of July about 10 days before 'Schattenmorelle' in central Europe.
Fruit characteristics: medium to large, flat-round shape, dark red skin, medium firm, high sugar and acidity content.
Fruit flesh: brown-red, firm.
Juice: red, moderate staining.
Resistance: susceptible to blossom blight, moderately susceptible to cherry leaf spot.
Utilization: fresh consumption and processing.

### 5.4.2 New sour cherry cultivars and cultivars with local importance

#### *'Achat'*

Origin: selected by B. Wolfram as clone 'F5,5,55' at the Fruit Breeding Institute in Pillnitz, Dresden, Germany.
Parentage: 'Köröser' × 'B7,2,40' ('Fanal' × 'Kelleriis 16').
Tree growth: strong growth vigour with average branching, forming flower buds on bouquet spurs on older wood in the same way as sweet cherry.
Fertility/fruit set: self-compatible, high fruit set similar to 'Schattenmorelle'.
Blooming time: early, about 1 week before 'Schattenmorelle'.
Ripening time: mid-season, about 2 weeks before 'Schattenmorelle'.
Fruit characteristics: large, round, dark red skin, medium firm, high sugar and acidity content.
Fruit flesh: red, medium firm.
Juice: red, moderate staining.
Resistance: high tolerance to blossom blight, susceptible to cherry leaf spot.
Utilization: processing and fresh consumption.

### *'De Botoşani'*

Origin: selected by L. Petre and E. Cârdei from a local sour cherry population of north-eastern Romania.
Parentage: unknown.
Tree growth: medium growth vigour, with fruit borne on 1-year-old wood and spurs of older wood.
Fertility/fruit set: partly self-compatible, medium fruit set.
Blooming time: mid-season.
Ripening time: mid-season, in the middle to end of June in Romania.
Fruit characteristics: large, round, dark red skin, medium firm, high sugar and acidity content.
Fruit flesh: dark red.
Juice: dark red, intensely staining.
Resistance: intermediate susceptibility to blossom blight and cherry leaf spot.
Utilization: processing and fresh consumption.

### *'Bucovina'*

Origin: selected by C. Radulescu at the RSFG, Falticeni, Romania.
Parentage: unknown.
Tree growth: medium growth vigour and a globular canopy.
Fertility/fruit set: self-compatible, high fruit set.
Blooming time: medium.
Ripening time: medium to late, in the middle of July in Romania.
Fruit characteristics: medium size, flat-round shape, dark red skin.
Fruit flesh: dark red, soft.
Juice: dark red.
Resistance: tolerant to blossom blight and sensitive to cherry leaf spot.
Utilization: processing.

### *'Carmine Jewel'*

Origin: originated from a cross between *P. fruticosa* × *P. cerasus* initiated by L. Kerr at the University of Saskatchewan, Canada, in 1966.
Parentage: 'Kerr's Easypick' (*P. cerasus* × *P. fruticosa*) × 'Northstar'.
Tree growth: dwarf sour cherry type, tree grows like a bush.
Fertility/fruit set: self-compatible, medium fruit set.
Blooming time: no information.
Ripening time: late July to early August in Saskatoon, Canada.
Fruit characteristics: small, flat-round shape, dark red skin, medium firm.
Fruit flesh: dark red.
Juice: dark red.
Resistance: tolerance to withstand –40°C winter frost.
Utilization: processing.

### *'Cigány'(syn. 'Cigány Meggy', 'Gipsy Cherry', 'Ziegeunerkirsche'), clonal selections 'Cigány 7', 'Cigány 59', 'Cigány 404'*

Origin: landrace originating in the Carpathian basin and existing as a native clonal population.
Parentage: unknown.
Tree growth: medium growth vigour, tree grows like a bush.
Fertility/fruit set: self-compatible or self-incompatible according to the clone, high fruit set.
Blooming time: different blooming times according to the clone.
Ripening time: different ripening times according to the clone.
Fruit characteristics: small, flat-round shape, dark red skin, medium firm to soft.
Fruit flesh: dark red.
Juice: dark red.
Resistance: moderate sensitive to blossom blight.
Utilization: processing.

### *'Coralin'*

Origin: selected by B. Wolfram and M. Schuster at the Julius Kühn-Institut in Pillnitz, Dresden, Germany.
Parentage: 'Kelleriis 16' × 'P10,17,5' ('Köröser' × 'Schattenmorelle').
Tree growth: upright growth with good branching.
Fertility/fruit set: self-compatible, high fruit set.
Blooming time: late, at the same time as 'Schattenmorelle'.
Ripening time: late, a few days before 'Schattenmorelle'.

Fruit characteristics: medium size, flat-round shape, black red skin, firm.
Fruit flesh: black red, medium firm.
Juice: dark red, very dark colour caused by high anthocyanin content.
Resistance: tolerant to leaf diseases caused by fungi such as cherry leaf spot and shot hole.
Utilization: processing, fresh market.

### *'Crişana 2'*

Origin: selected by V. Cociu at the RIFG, Pitesti, Romania.
Parentage: unknown.
Tree growth: strong growth vigour and upright canopy.
Fertility/fruit set: self-sterile, low fruit set.
Blooming time: medium.
Ripening time: medium, in the middle of June in Romania.
Fruit characteristics: large, flat-round shape, red skin, high sugar and acidity content.
Fruit flesh: red, soft.
Juice: pink.
Resistance: tolerant to blossom blight and cherry leaf spot.
Utilization: fresh consumption and processing.

### *'Csengődi'*

Origin: selected by J. Apostol from local cultivars in southern Hungary.
Parentage: unknown.
Tree growth: upright and vigorous growth with good branching.
Fertility/fruit set: partly self-compatible, medium to high fruit set.
Blooming time: late, at the same time as 'Schattenmorelle'.
Ripening time: mid-season.
Fruit characteristics: medium size, flat-round shape, deep purplish-red skin, firm.
Fruit flesh: dark red, medium firm.
Juice: dark red.
Resistance: resistant to cherry leaf spot and blossom blight.
Utilization: processing.

### *'Debreceni Bőtermő'*

Origin: selected from local cultivars at the Debrecen region, Hungary.
Parentage: unknown.
Tree growth: medium to strong growth vigour, cone-shaped crown habit.
Fertility/fruit set: partly self-compatible, medium to high fruit set.
Blooming time: medium, about 5–6 days before 'Schattenmorelle'.
Ripening time: early, at the beginning of July, 3–5 days before 'Újfehértói Fürtös' in Hungary.
Fruit characteristics: medium size, flat-round shape, dark red skin.
Fruit flesh: dark red, firm.
Juice: red with a low staining intensity.
Resistance: susceptible to blossom blight and cherry leaf spot.
Utilization: fresh consumption and processing.

### *'Kántorjánosi 3'*

Origin: selected by T. Szabó from local cultivars in the Màtèszalka region, Hungary.
Parentage: unknown.
Tree growth: strong growth vigour with a wide spreading crown, high tendency to produce blind wood.
Fertility/fruit set: partly self-compatible, medium to high fruit set.
Blooming time: early, about 1 week before 'Schattenmorelle' in Germany.
Ripening time: medium to late, a few days before 'Schattenmorelle'.
Fruit characteristics: medium to large, round, dark red skin.
Fruit flesh: red, firm.
Juice: red with a low staining intensity.
Resistance: moderately susceptible to cherry leaf spot, higher susceptibility to blossom blight.
Utilization: fresh consumption and processing

### *'Kutahya'*

Origin: landrace, originated in Turkey and exists as a native clonal population.
Parentage: unknown.
Tree growth: medium to strong growth vigour, cone-shaped crown habit.
Fertility/fruit set: self-compatible, medium to high fruit set.
Blooming time: late.
Ripening time: very late.

Fruit characteristics: medium to small, flat-round shape, dark red skin, high content of sugar and acidity.
Fruit flesh: dark purplish, firm.
Juice: dark red.
Resistance: tolerant to blossom blight.
Utilization: fresh consumption and processing.

### 'Gerema'

Origin: selected by W. Jacob at the Fruit Research Institute, Geisenheim, Germany.
Parentage: 'Kelleriis 14' × o.p.
Tree growth: weak growth vigour, upright crown habit, flower buds borne on 1-year-old wood and bouquet spurs.
Fertility/fruit set: self-compatible, high fruit set.
Blooming time: late, similar to 'Schattenmorelle'.
Ripening time: very late, after 'Schattenmorelle'.
Fruit characteristics: medium size, flat-round shape, dark red skin, high content of sugar and acidity.
Fruit flesh: dark red, firm.
Juice: dark red, high staining intensity.
Resistance: tolerant to blossom blight, susceptible to cherry leaf spot.
Utilization: processing.

### 'Ilva'

Origin: selected from a local population by I. Ivan and N. Minoiu, Romania.
Parentage: unknown.
Tree growth: medium growth vigour, upright crown habit, flower buds borne on 1-year-old wood and bouquet spurs.
Fertility/fruit set: self-compatible, high fruit set.
Blooming time: late.
Ripening time: first week of July in Romania.
Fruit characteristics: medium size, round, dark red skin, slight sour taste.
Fruit flesh: red, firm.
Juice: red.
Resistance: tolerant to blossom blight and cherry leaf spot.
Utilization: fresh consumption and processing.

### 'Jachim'

Origin: selected by M. Schuster at the Julius Kühn-Institut in Pillnitz, Dresden, Germany.
Parentage: 'Köröser Gierstädt' × 'Safir'.
Tree growth: pillar type, characterized by distinctive upright growth with narrower branch angles and shorter internodes, flowers borne on spurs and 1-year-old wood.
Fertility/fruit set: self-compatible, high fruit set.
Blooming time: late, like 'Schattenmorelle'.
Ripening time: mid-season, 1 week before 'Schattenmorelle'.
Fruit characteristics: medium to large, flat-round shape, dark red skin.
Fruit flesh: dark red, firm.
Juice: red.
Resistance: tolerant to leaf diseases caused by fungi such as cherry leaf spot and shot hole.
Utilization: fresh consumption and processing.

### 'Jade'

Origin: selected by B. Wolfram as clone 'F5,19,130' at the Fruit Breeding Institute in Pillnitz, Dresden, Germany.
Parentage: 'Köröser' × 'Röhrigs Weichsel'.
Tree growth: medium growth vigour with good branching.
Fertility/fruit set: self-compatible, high fruit set.
Blooming time: medium to late, some days before or at the same time as 'Schattenmorelle'.
Ripening time: mid-season, 1 week before 'Schattenmorelle'.
Fruit characteristics: medium to large, flat-round shape, dark red skin, very good, balanced sweet-acidic and aromatic taste.
Fruit flesh: dark red, firm.
Juice: red.
Resistance: highly tolerant to blossom blight and low susceptibility to cherry leaf spot.
Utilization: processing and fresh consumption.

### 'Latvijas Zemais' (syn. 'Lietuvas Zemais', 'Žagarvyšnė', 'Läti Madalkirss')

Origin: landrace originating in the Baltic states and existing as a native clonal population.

Parentage: unknown.
Tree growth: weak to medium growth, a dense canopy with pendulous fruit branches tending to produce blind wood.
Fertility/fruit set: partly self-compatible, medium fruit set.
Blooming time: medium.
Ripening time: medium, in the middle of July in Latvia.
Fruit characteristics: medium to small, round, dark red skin.
Fruit flesh: dark red, soft.
Juice: dark red.
Resistance: good winter hardiness.
Utilization: processing.

### 'Lyubskaya'

Origin: local cultivar of Kursk region, widespread in Russia.
Parentage: unknown.
Tree growth: weak growth vigour and a spreading tree habit.
Fertility/fruit set: self-compatible, high fruit set.
Blooming time: medium to late.
Ripening time: late, the end of July to early August in central Russia.
Fruit characteristics: medium size, round shape, dark red.
Fruit flesh: dark red, medium firm, very juicy, sour, mediocre flavour.
Juice: light red to red.
Resistance: susceptible to blossom blight and cherry leaf spot.
Utilization: processing.

### 'Marasca'

Origin: unknown, but has been grown in Dalmatia (region of Croatia along the Adriatic Sea) for centuries. Distribution area includes northern and central Dalmatia from Zadar to Makarska. 'Marasca' is considered to be a special type of sour cherry, *P. cerasus* var. *marasca*.
Parentage: unknown.
Tree growth: weak to moderate growth vigour, with two growing types: *P. cerasus* var. *marasca pendula* with a drooping tree habit and *P. cerasus* var. *marasca recta* with an upright tree habit.
Fertility/fruit set: type *pendula* is self-incompatible, type *recta* is partly self-compatible, medium fruit set.
Blooming time: medium, a few days before 'Schattenmorelle'.
Ripening time: late, end of June to beginning of July in Croatia.
Fruit characteristics: small, round, dark red to black skin, high content of soluble solids and anthocyanins.
Fruit flesh: dark red, firm.
Juice: dark red, high staining intensity.
Resistance: no information.
Utilization: processing.

### 'Mocăneşti 16'

Origin: selected by V. Cociu at the RIFG Pitesti, Romania.
Parentage: unknown.
Tree growth: medium to strong growth vigour and a globular canopy.
Fertility/fruit set: self-sterile, medium fruit set.
Blooming time: medium.
Ripening time: medium, towards the end of the middle of June in Romania.
Fruit characteristics: medium size, flat-round shape, dark red skin.
Fruit flesh: red, soft.
Juice: red.
Resistance: sensitive to blossom blight and cherry leaf spot.
Utilization: processing.

### 'Morina'

Origin: selected by B. Wolfram at the Fruit Research Institute, Pillnitz, Dresden, Germany.
Parentage: 'Köröser' × 'Reinhards Ostheimer'.
Tree growth: medium growth vigour with good branching, flower buds borne on 1-year-old wood and bouquet spurs.
Fertility/fruit set: partly self-compatible, medium fruit set.
Blooming time: medium, some days before 'Schattenmorelle'.
Ripening time: mid-season, 1 week before 'Schattenmorelle'.
Fruit characteristics: medium to large, round, dark red skin, sour to sweet-sour taste with a high content of sugar and acidity.

Fruit flesh: dark red, firm.
Juice: dark red, high staining intensity.
Resistance: highly tolerant to blossom blight and leaf diseases.
Utilization: processing and fresh consumption.

### 'Nana'

Origin: selected by P. Popa at the RSFG, Băneasa, Bucharest, Romania.
Parentage: 'Crişana' × o.p.
Tree growth: weak to medium growth vigour and a globular spreading canopy.
Fertility/fruit set: self-compatible, high fruit set.
Blooming time: medium.
Ripening time: medium, in the middle of June in Romania.
Fruit characteristics: medium size, round, red skin.
Fruit flesh: pink-red, soft.
Juice: pink.
Resistance: sensitive to blossom blight and cherry leaf spot.
Utilization: fresh consumption and processing.

### 'Nefris'

Origin: selected at the Agricultural University in Lublin, Poland.
Parentage: probably a seedling of 'Łutówka'.
Tree growth: upright to spreading, medium growth vigour.
Fertility/fruit set: self-compatible.
Blooming time: about 2–3 days before 'Łutówka' in Poland.
Ripening time: mid-season, in the middle of July in Poland.
Fruit characteristics: medium to large, round, dark red skin.
Fruit flesh: dark red, firm.
Juice: dark red.
Resistance: high winter hardiness, high susceptibility to bacterial canker and blossom blight.
Utilization: processing and fresh consumption.

### 'Northstar'

Origin: selected at the University of Minnesota, USA, in 1933.
Parentage: reported as 'English Morello' × 'Serbian Pie No. 1'.
Tree growth: medium vigour with a small stature.
Fertility/fruit set: self-compatible.
Blooming time: medium to late.
Ripening time: mid-season.
Fruit characteristics: medium to small, roundish heart-shaped, dark red.
Fruit flesh: yellowish red to dark red.
Juice: dark red.
Resistance: resistant to cherry leaf spot.
Utilization: processing and fresh consumption.

### 'Petri'

Origin: selected by F. Szöke at Lövöpetri, Hungary.
Parentage: unknown.
Tree growth: medium to strong growth with compact and a spherical crown with good branching.
Fertility/fruit set: partly self-compatible.
Blooming time: late.
Ripening time: mid-season, beginning of July in Hungary.
Fruit characteristics: medium size, round, brown-red skin.
Fruit flesh: red, firm.
Juice: red, low staining intensity.
Resistance: unknown.
Utilization: fresh consumption and processing.

### 'Rival'

Origin: selected by S. Budan at the RIFG, Pitesti, Romania.
Parentage: 'Griot Moscovski' × 'Nana'.
Tree growth: medium growth vigour and a spreading canopy.
Fertility/fruit set: partly self-compatible, good fruit set.
Blooming time: late.
Ripening time: at the end of June in Romania.
Fruit characteristics: medium size, elliptic, red skin.
Fruit flesh: red, soft.
Juice: red, low staining intensity.
Resistance: highly tolerant to cherry leaf spot and susceptible to blossom blight.
Utilization: processing.

### *'Sabina'*

Origin: selected at the Research Institute of Horticulture in Skierniewice, Poland.
Parentage: 'Łutówka' × 'Schirpotreb'.
Tree growth: strong growth vigour with a spreading tree habit.
Fertility/fruit set: self-compatible, high fruit set.
Blooming time: about 4–5 days before 'Łutówka' in Poland.
Ripening time: early to mid-season, beginning of July in Poland.
Fruit characteristics: medium size, flat-round shape, red skin.
Fruit flesh: red.
Juice: red.
Resistance: good winter hardiness.
Utilization: processing.

### *'Safir'*

Origin: selected by B. Wolfram at the Fruit Research Institute, Pillnitz, Dresden, Germany.
Parentage: 'Schattenmorelle' × 'Fanal'.
Tree growth: medium to strong growth with a round crown and high branching.
Fertility/fruit set: self-compatible, high fruit set.
Blooming time: medium to early, some days before 'Schattenmorelle'.
Ripening time: mid-season, 1 week before 'Schattenmorelle'.
Fruit characteristics: medium to large, round, dark red skin, sweet-sour taste.
Fruit flesh: dark red, soft.
Juice: dark red.
Resistance: low susceptibility to blossom blight.
Utilization: processing and fresh consumption.

### *'Sătmărean'*

Origin: selected by T. Gozob and M. Raduc from the RIFG, Pitesti, Romania.
Parentage: 'Anglaise Hâtive' × 'Visin Tufa'.
Tree growth: medium to strong growth vigour and an upright canopy.
Fertility/fruit set: partly self-compatible, medium fruit set.
Blooming time: early.
Ripening time: early, at the beginning of June in Romania.
Fruit characteristics: medium size, flat-round shape, dark red skin.
Fruit flesh: red, soft.
Juice: pink.
Resistance: tolerant to blossom blight and cherry leaf spot.
Utilization: fresh consumption.

### *'Spinell'*

Origin: selected by B. Wolfram at the Fruit Research Institute, Pillnitz, Dresden, Germany.
Parentage: 'Köröser' × 'B7,2,40' ('Fanal' × 'Kelleriis 16').
Tree growth: upright, moderate growth vigour.
Fertility/fruit set: partly self-compatible, good fruit set.
Blooming time: medium-early, a few days before 'Schattenmorelle'.
Ripening time: early, 2 weeks before 'Schattenmorelle'.
Fruit characteristics: very large, kidney shaped, dark red skin, excellent sweet-sour taste.
Fruit flesh: dark red, firm.
Juice: dark red, high anthocyanin content.
Resistance: unknown.
Utilization: fresh consumption.

### *'Stelar'*

Origin: selected by S. Budan at the RIFG, Pitesti, Romania.
Parentage: 'Mocăneşti 16' × 'Anglaise Hâtive'.
Tree growth: medium to strong growth vigour and an upright spreading canopy.
Fertility/fruit set: partly self-compatible, medium fruit set.
Blooming time: early.
Ripening time: early, at the beginning of June in Romania.
Fruit characteristics: medium to large, round-elongated shaped, dark red skin.
Fruit flesh: dark red, soft.
Juice: pink.
Resistance: tolerant to blossom blight and cherry leaf spot.
Utilization: fresh consumption.

### 'Šumadinka'

Origin: selected by A. Stančević and P. Mišić at the Fruit Research Institute, Čačak, Serbia.
Parentage: 'Pándy' × 'Fanal'.
Tree growth: weak to medium-strong growth vigour and a drooping tree habit.
Fertility/fruit set: self-compatible, high fruit set.
Blooming time: medium.
Ripening time: late, at the beginning of July in Serbia.
Fruit characteristics: medium to large, round, dark red.
Fruit flesh: dark red, medium firm, acid.
Juice: ruby red.
Resistance: unknown.
Utilization: processing.

### 'Tarina'

Origin: selected by V. Cociu and T. Gozob at the RIFG, Pitesti, Romania.
Parentage: 'Anglaise Hâtive' × 'Visin Tufa'.
Tree growth: medium growth vigour and an upright canopy.
Fertility/fruit set: partly self-compatible, medium fruit set.
Blooming time: early.
Ripening time: early, at the beginning of June in Romania.
Fruit characteristics: medium size, round, dark red skin.
Fruit flesh: red, soft.
Juice: pink.
Resistance: tolerant to blossom blight and cherry leaf spot.
Utilization: fresh consumption.

### 'Tiki'

Origin: selected by J.V. Christensen, Denmark, in the mid-1990s.
Parentage: 'Stevnsbaer' × 'Fanal'.
Tree growth: vigorous growth, round crown shape.
Fertility/fruit set: self-compatible, higher and more regular fruit set than 'Stevnsbaer'.
Blooming time: medium-early in Denmark.
Ripening time: medium, August in Denmark.
Fruit characteristics: medium to small, round, high firmness.
Fruit flesh: Purple to dark red, soft juicy flesh.
Juice: high in acidity, medium high in sugar, very high in colour and total phenolics, strong aroma, some bitterness.
Resistance: relatively healthy trees.
Utilization: processing for wine and juice.

### 'Timpurii de Osoi'

Origin: selected by I. Bodi, O. Bazgan and G. Dumitrescu at the RSFG, Iasi, Romania.
Parentage: unknown.
Tree growth: medium to strong growth vigour and a globular canopy.
Fertility/fruit set: partly self-compatible, medium fruit set.
Blooming time: early.
Ripening time: early, at the beginning of June in Romania.
Fruit characteristics: medium size, round, dark red skin.
Fruit flesh: red, soft.
Juice: red.
Resistance: tolerant to blossom blight and cherry leaf spot.
Utilization: fresh consumption and processing.

### 'Vladimirskaya'

Origin: unknown, but has been grown in Russia for centuries and exists as a native clonal population.
Parentage: unknown.
Tree growth: weak to medium growth, round canopy with pendulous fruit branches tending to produce blind wood.
Fertility/fruit set: self-compatible, fruit set depends on pollination (a good pollinator is 'Lyubskaya').
Blooming time: medium.
Ripening time: medium, mid-July in central Russia.
Fruit characteristics: small, flat-round shape, dark red skin, sour-sweet taste.
Fruit flesh: dark red, firm, juicy, very good sour-sweet harmonious flavour.
Juice: dark red.
Resistance: susceptible to cherry leaf spot.
Utilization: processing and fresh consumption.

### 'Zhivitsa'

Origin: selected at the Fruit-Growing Institute in Belarus.

Parentage: 'Griotte d'Ostheim' × 'Dönissens Gelbe'.
Tree growth: vigorous, round canopy.
Fertility/fruit set: self-incompatible.
Blooming time: medium.
Ripening time: early, at the beginning of July in Belorussia.
Fruit characteristics: medium to small, round, dark red skin.
Fruit flesh: dark red, medium firm, small stone that separates easily from the flesh.
Juice: dark red, sour-sweet taste.
Resistance: tolerant to cherry leaf spot.
Utilization: processing and fresh consumption.

## Acknowledgements

The authors thank the following additional contributors: A. Taranov, Institute for Fruit Growing, Samohvalovichi, Belarus; T. Szabó, S. Szügyi, NARIC Fruitculture Research Institute, Budapest, Hungary; D. Feldmane, Latvia State Institute of Fruit-Growing, Dobele, Latvia; M. Szymajda, Research Institute of Horticulture, Skierniewice, Poland; S. Budan, RIFG Pitesti, Pitesti, Romania; S. Ercişli, Ataturk University Agricultural, Erzurum, Turkey.

## References

Apostol, J. (2005) New sour cherry varieties and selections in Hungary. *Acta Horticulturae* 667, 123–126.

Apostol, J. (2011) Breeding sweet and sour cherry in Hungary. In: Zbornik Radova III. Savetovanja 'Inovacije u Vocarstvu', Beograd, 10 February, pp. 49–57.

Bors, R.H. (2005) Dwarf sour cherry breeding at the University of Saskatchewan. *Acta Horticulturae* 667, 135–140.

Bošković, R., Wolfram, B., Tobutt, K.R., Cerovic, R. and Sonneveld, T. (2006) Inheritance and interactions of incompatibility alleles in tetraploid sour cherry. *Theoretical and Applied Genetics* 112, 315–326.

Brown, G.K. and Kollar, G. (1996) Harvesting and handling sour and sweet cherries for processing. In: Webster, A.D. and Looney, N.E. (eds) *Cherries*. CAB International, Wallingford, UK.

Budan, S. and Stoian, I. (1996) Genetic recources in the Romanian sour cherry breeding. *Acta Horticulturae* 410, 81–86.

Budan, S., Mutafa, I., Stoian, I. and Popescu, I. (2005a) Screening of 100 sour cherry genotypes for *Monilia laxa* field resistance. *Acta Horticulturae* 667, 145–151.

Budan, S., Mutafa, I., Stoian, I. and Popescu, I. (2005b) Field evaluation of cultivars susceptibility to leaf spot at Romania's sour cherry genebank. *Acta Horticulturae* 667, 153–157.

Christensen, J.V. (1976) Beskrivelse af Surkirsebærsorten 'Stevnsbaer'. Report 1322, *Statens Forsøgsvirksomhed i Plantekultur*.

Christensen, J.V. (1983) Kloner af surkirsebærsorten Stevnsbaer. *Statens Planteavlsforsøg, Meddelelse* no. 1714.

Christensen, J.V. (1986) Clones of the sour cherry Stevnsbaer. *Acta Horticulturae* 180, 69–72.

Christensen, J.V. (1995a) Kloner af Stevnsbaer. *Frugt og Baer* 24, 44–45.

Christensen, J.V. (1995b) Klonselektion bei der Sauerkirschensorte 'Stevnsbaer'. *Erwerbsobstbau* 37, 34–36.

Faust, M. and Surányi, D. (1997) Origin and dissemination of cherry. *Horticultural Reviews* 19, 263–317.

Hauck, N.R., Yamane, H., Tao, R. and Iezzoni, A.F. (2006) Accumulation of non-functional *S*-haplotypes results in the breakdown of gametophytic self-incompatibility in tetraploid *Prunus*. *Genetics* 172, 1191–1198.

Hedrick, U.P. (1915) *The Cherries of New York*. J.B. Lyon Company, Albany, New York.

Hillig, K.W. and Iezzoni, A.F. (1988) Multivariate analysis of a sour cherry germplasm collection. *Journal of the American Society for Horticultural Science* 113, 928–934.

Iezzoni, A.F. (1996) Sour cherry cultivars: objectives and methods of fruit breeding and characteristics of principal commercial cultivars. In: Webster, A.D. and Looney, N.E. (eds) *Cherries: Crop Physiology, Production and Uses*. CAB International, Wallingford, UK.

Iezzoni, A.F. (2005) Acquiring cherry germplasm from Central and Eastern Europe. *HortScience* 40, 304–308.

Iezzoni, A.F., Sebolt, A.M. and Wang, D. (2005) Sour cherry breeding program at Michigan State University. *Acta Horticulturae* 667, 131–134.

Jacob, H. (1994) 'Gerema'. *Obstbau* 7, 352–353.

Kristiansen, K. and Jensen, M. (2009) Towards new cultivars of 'Stevnsbær' sour cherries in Denmark. *Acta Horticulturae* 814, 277–284.

Miletić, R., Žikić, M., Mitić, N. and Nikolić, R. (2008) Identification and *in vitro* propagation of promising 'Oblačinska' sour cherry selections in Eastern Serbia. *Acta Horticulturae* 795, 159–162.

Mitschurin, I.W. (1951) *Ausgewählte Schriften*. Verlag Kultur und Fortschritt, Berlin.

Montgomery, A. (2009) *Sour Cherries in Canada*. Statistics Canada. Available at: http://www.statcan.gc.ca/pub/96-325-x/2007000/article/10775-eng.htm (accessed 9 January 2017).

Nikolić, D., Rakonjac, V., Milutinović, M. and Fotirić, M. (2005) Genetic divergence of Oblačinska sour cherry (*Prunus cerasus* L.) clones. *Genetika* 37, 191–198.

Nikolić, M. and Cerović, R. (1998) A new promising sour cherry hybrid XII/57. *Acta Horticulturae* 468, 199–202.

Poll, L., Peterson, M.B. and Nielson, G.S. (2003) Influence of harvest year and harvest time on soluble solids, titrateable acid, anthocyanins content and aroma components in sour cherry (*Prunus cerasus* L. cv. 'Stevnsbaer'). *European Food Research and Technology* 216, 212–216.

Rakonjac, V., Fotirić-Akšić, M., Nikolić, D., Milatović, D. and Čolić, S. (2010) Morphological characterization of 'Oblačinska' sour cherry by multivariate analysis. *Scientia Horticulturae* 125, 679–684.

Redalen, G. (1984) Fertility in sour cherries. *Gartenbauwissenschaft* 49, 212–217.

Schmid, W. and Grosch, W. (1986) Identifizierung flüchtiger Aromastoffe mit hohen Aromawerten in Sauerkirschen (*Prunus cerasus* L.). *Zeitschrift Lebensmittel Untersuchung Forschung* 182, 407–412.

Schuster, M. (2004) Investigation on resistance to leaf spot disease, *Blumeriella jaapii* in cherries. *Journal of Fruit and Ornamental Plant Research* 12, 275–279.

Schuster, M. (2009) Sour cherries *Prunus cersasus* L. with columnar tree habit. *Acta Horticulturae* 814, 325–328.

Schuster, M. (2011) Sauerkirsche 'Achat'. *JKI Datenblätter – Obstsorten*. Heft 2. doi: 10.5073/jkidos.2012.001.

Schuster, M. (2012) Sauerkirsche 'Jade'. *JKI Datenblätter – Obstsorten*. Heft 3. doi: 10.5073/jkidos.2012.002.

Schuster, M. and Wolfram, B. (2005) Sour cherry breeding at Dresden-Pillnitz. *Acta Horticulturae* 667, 127–130.

Schuster, M., Grafe, C., Hoberg, E and Schütze, W. (2013) Interspecific hybridization in sweet and sour cherry breeding. *Acta Horticulturae* 976, 79–86.

Schuster, M., Grafe, C., Wolfram, B. and Schmidt, H. (2014) Cultivars resulting from cherry breeding in Germany. *Erwerbs-Obstbau* 56, 67–72.

Shukov, O.S. and Charitonova, E.N. (1988) *Sour Cherry Breeding*. Agropromizdat, Moscow.

Šimunic, V., Kovač, S., Gašo-Sokač, D., Pfannhauser, W. and Murkovic, M. (2005) Determination of anthocyanins in four Croatian cultivars of sour cherries (*Prunus cerasus*). *European Food Research and Technology* 220, 575–578.

Sjulin, T.M., Julin, T.M., Jones, A.L. and Andersen, R.L. (1989) Expression of partial resistance to cherry leaf spot in cultivars of sweet, sour, duke, and European ground cherry. *Plant Disease* 73, 56–61.

Stainer, R. (1975) "Stevnsbaer" – eine interessante Sauerkirsche für die Safterzeugung. *Obstbau/Weinbau* 5, 142–145.

Stegmeir, T., Schuster, M., Sebolt, A., Rosyara, U., Sundin, G.W. and Iezzoni, A. (2014) Cherry leaf spot resistance in cherry (*Prunus*) is associated with a quantitative trait locus on linkage group 4 inherited from *P. canescens*. *Molecular Breeding* 34, 927–935.

Symyrenko, L.P. (1963) *Pomology*. Vol. 3 *Stone Fruits, Quince, South Rowan, Dogwood, Medlar, Garden and Forest Hazelnut*. Gov. Pub. of Agricul. Lit. Kiev.

Szabó, T. (1996) Results of sour cherry clonal selection in the north-eastern region of Hungary. *Acta Horticulturae* 410, 97–100.

Szabó, T., Nyéki, J., Soltész, M. and Hilsendegen, P. (2008a) Sortenwahl in Ungarn und Deutschland. In: Nyéki, J., Soltész, M., Szabó, T., Hilsendegen, P. and Hensel, G. (eds) *Sauerkirschanbau*, UD RDI, Germany.

Szabó, T., Inántsy, F. and Csiszár, L. (2008b) Results of sour cherry clone selection carried out at the Research Station of Újfehértó. *Acta Horticulturae* 795, 369–372.

Taranenko, L.I. (2004) Plant breeding and testing of sour cherry cultivars in conditions of Donbass. *Sadovodstvo I Vinogradarstvo* 6, 17–20.

Tobutt, K.R., Bošković, R., Cerović, R., Sonneveld, T. and Ružić, D. (2004) Identification of incompatibility alleles in the tetraploid species sour cherry *Theoretical and Applied Genetics* 108, 775–785.

Trajkovski, V. (1996) A review of the cherry breeding program in Sweden. *Acta Horticulturae* 410, 387–388.

Truchseß, C. (1819) *Systematische Classification und Beschreibung der Kirschensorten*. Cottaische Buchhandlung, Stuttgart, Germany.

Turovtsev, M.I. and Turovtseva, V.O. (2002) *Fruit and Small-fruit Crop Regionalized Varieties of Institute of Irrigated Fruit Growing Selection*. Agrarna Nauka, Kiev.

Turovtseva, V.O., Turovtsev, M.I., Shkinder-Barmina, A.N. and Turovtseva, N.M. (2013) Creation of sour and duke cherry varieties on the south of Ukraine. In: *Scientific Publications of FSBSO NCRRIH&V* 1, pp. 135–142.

Venjaminov, L.N. (1954) *Selekcija Vishni, Slivy i Aprikosa v Uslovijakh Spednej Polosy SSSR.* Gosudarstvennoe Izdatelstvo Seleskokhozjajstvennoj Literatuyri, Moscow.

Wang, H., Nair, M.G., Strasburg, G.M., Chang, Y., Booren, A.M., Gray, J.I. and DeWitt, D.L. (1999) Antioxidant and antiinflammatory activities of anthocyanins and aglycon, cyanidin, from tart cherries. *Journal of Natural Products* 62, 294–296.

Webster, A.D. and Schmidt, H. (1996) Rootstocks for sweet and sour cherries. In: Webster, A.D. and Looney, N.E. (eds) *Cherries: Crop Physiology, Production and Uses.* CAB International, Wallingford, UK, pp. 127–163.

Wharton, P.S., Iezzoni, A. and Jones, A.L. (2003) Screening cherry germ plasm for resistance to leaf spot. *Plant Disease Journal* 87, 471–477.

Wolfram, B. (1990) Kurzinformation über Sauerkirschenneuzüchtungen 'Korund' und 'Karneol'. *Erwerbsobstbau* 32, 204–205.

Yamane, H., Tao, R., Sugiura, A., Hauck, N.R. and Iezzoni, A.F. (2001) Identification and characterization of *S*-RNases in tetraploid sour cherry (*Prunus cerasus*). *Journal of the American Society for Horticultural Science* 126, 661–667.

Yarushnykov, V.V. (2013) The selection of the genepool accessions with valuable traits from the collection of sour cherry germplasm at the Artemivsk experimental station of nursery culture. *Plant Genetic Resources* 12, 73–83.

Zhukov, O.S. and Kharitonova, E.N. (1988) *Selektzja Vishni.* Agroproizdat, Moscow.

Zwitzscher, M. (1964) Die Sauerkirsche 'Mailot'. *Der Erwerbsobstbau* 6, 149–151.

Zwitzscher, M. (1968) Die Sauerkirsche 'Cerella'. *Der Erwerbsobstbau* 10, 108–110.

Zwitzscher, M. (1969) 'Nabella', eine Großfrüchtige Sauerkirsche für die Industrielle Verwertung. *Der Erwerbsobstbau* 11, 145–148.

# 6 Rootstocks and Improvement

**Károly Hrotkó[1]* and Elżbieta Rozpara[2]**
[1]*Faculty of Horticultural Science, Szent István University, Budapest, Hungary;*
[2]*Research Institute of Horticulture, Skierniewice, Poland*

## 6.1 Introduction

Rootstocks for cherries are chosen from among taxa showing appropriate graft compatibility. Considering this criteria, *Prunus avium*, *Prunus cerasus*, *Prunus mahaleb* and *Prunus fruticosa*, as well as their hybrids and some related taxa, can be used as rootstocks.

Further factors that influence rootstock use are the diverse pedoclimatic conditions in the different sites. Rootstocks remain important tools for extending the site adaptability of sweet and sour cherry cultivars, and also allow growers to plant cherries in suboptimal sites. Although modern sweet cherry orchard systems, the so-called 'pedestrian orchards', require dwarfing and precocious rootstocks, the rootstocks used in sweet and sour cherry orchards are still diverse. The training system and rootstock must be considered together, and matched properly with the vigour of the soil fertility and climate of the orchard site. Growers of intensive orchards, producing hand-picked cherries for fresh market, prefer dwarfing rootstocks, which allow planting densities of up to 1000–5000 trees $ha^{-1}$ (Robinson, 2005; Sansavini and Lugli, 2014; Musacchi *et al.*, 2015). However, several training systems with special training and pruning protocols (e.g. Kym green bush, Spanish bush, steep leader, tall spindle axe and upright fruiting offshoots) allow the establishment of pedestrian or semi-pedestrian orchards on semi-vigorous or vigorous rootstocks (Negueroles, 2005; Robinson, 2005; Ercisli *et al.*, 2006; Iglesias and Peris, 2008; Hrotkó, 2010; Lang, 2011; Long *et al.*, 2015).

Although modern hand-picked orchard systems benefit from growth control and precocious rootstocks, orchards producing for the processing industry and harvested by limb and trunk shakers still use vigorous rootstocks that can tolerate the mechanical stress of harvesters.

## 6.2 Sweet and Sour Cherry Rootstock Breeding

Major species in the parentage of rootstocks include *P. avium* L., *P. cerasus* L., *Prunus canescens* Bois, *P. fruticosa* Pall. and *P. mahaleb* L. Other species that have either been used as rootstocks or used in rootstock breeding programmes include: *Prunus* × *dawyckensis* Sealy, *Prunus incisa* Thunb.,

* hrotko.karoly@kertk.szie.hu

 

*Prunus concinna* Koehne, *Prunus serrula* Franch., *Prunus subhirtella* Miq., *Prunus pseudocerasus* Lindl., *Prunus tomentosa* Thunb. and *Prunus serrulata* Lindl. (Cummins, 1979a,b; Webster and Schmidt, 1996). Several interspecific hybrids were created between the above species and were tested for rootstock breeding (Kappel *et al.*, 2012). Another rootstock ('Adara') and interstock (Myrobalan 'R1') for cherries belong to the subgenus *Prunophora* (*Prunus cerasifera* Ehrh.) (Moreno *et al.*, 1996; Lang, 2006).

### 6.2.1 Objectives in rootstock breeding

The breeding objectives listed by Perry (1987) (size reduction, increased scion precocity and cropping, wide range of compatibility, uniformity in performance, cold hardiness, adaptation to a wide range of soils, and disease and pest tolerance) are still of importance, although the order of priorities may vary. The target of the majority of cherry rootstock breeding programmes is a dwarfing rootstock. Currently, there is a range of vigour among cherry rootstocks, but they are not all truly satisfactory. With variable soil fertility in the various cherry-growing regions and the uncertainties surrounding climate change, greater emphasis needs to be placed on soil and climate adaptability without losing sight of the need for vigour control.

#### *Effect of rootstock on scion vigour and growth habit*

Although the mechanism of the rootstock effect on scion vigour and growth habit is still not fully understood, there have been diverse hypotheses and contradicting results. One of the major components is the hormonal interactions between the rootstock and scion, similar to other composite trees (Feucht, 1982; Faust, 1989; Treutter *et al.*, 1993; Webster, 1998). Numerous elements of growth control can be explained by hormonal control. However, the whole complex, involving the rootstock effects on assimilate partitioning, water and nutrient uptake and translocation, as well as the mechanism of rootstock effect on precocity and cropping efficiency, need further intensive research.

Based on the results of breeding programmes (Wolfram, 1971; Trefois, 1980; Gruppe, 1985; Wolfram, 1996), there is no doubt that, within the section *Eucerasus*, *P. fruticosa* and *P. cerasus*, as well as *P. canescens*, are the main sources for vigour control. Further sources may be found in the species of section *Pseudocerasus* (*P. pseudocerasus* and *P. serrulata*), but the hardiness and drought tolerance of these hybrids in continental climates may not be acceptable (Cummins, 1979a,b).

No dwarfing *P. avium* genotypes have been identified as potential rootstocks, although the possible utilization of genetic dwarfs in further breeding (Webster and Schmidt, 1996) and the effect of inbreeding have not been fully investigated. A range of vigour from standard to medium is found in *P. mahaleb* (Hrotkó, 2004; Hrotkó and Magyar, 2004; Lang, 2006; Sotirov, 2012; Barać *et al.*, 2014) and genetic dwarf genotypes of *P. mahaleb* may be sources of scion vigour control.

Branch angle can also be affected by rootstock. Webster and Schmidt (1996) reported that some *P. avium* and *P. pseudocerasus* clones caused the scion to develop wide branch angles. Hrotkó *et al.* (1999) also observed that scions on *P. mahaleb* 'Magyar' showed a wider crotch angle, while on 'MxM 14' and 'MxM 97', the crotch angle was narrower.

#### *Effect of rootstocks on precocity, cropping and fruit quality of scion cultivars*

Precocity, abundance and consistency of yield as well as fruit quality are affected by rootstocks, but there is considerable interaction between the rootstock, training and pruning, tree spacing and nutrition. Perry (1987) reported that scions on Mahaleb seedling rootstock produced fruit 1–2 years earlier than on Mazzard rootstocks. This was confirmed in several trials (Hrotkó, 1990; Hrotkó *et al.*, 2008; Stachowiak *et al.*, 2014). Intensive orchards with close spacing of trees and fruiting wood management can also contribute to precocity (Meland, 1998; Hrotkó *et al.*, 2009a,c; Hrotkó, 2010). Rootstocks in each vigour class can improve precocity, but it is not necessarily linked to dwarfing. Screening for scion productivity can only be determined with field trials. A relationship between yield efficiency, crop load and leaf area can affect fruit size (Edin *et al.*, 1996; Simon *et al.*, 2004; Whiting and Lang, 2004; Cittadini *et al.*, 2007; Gyeviki *et al.*, 2012). Highly efficient dwarfing rootstocks can increase the fruit-to-leaf area ratio and thereby reduce fruit size and quality. Clonal *P. cerasus* rootstocks, both as root and interstem, showed larger fruit size compared with trees on Mahaleb cherry (Magyar and Hrotkó, 2008).

### Graft compatibility

Intraspecific grafts of *P. avium* (i.e. sweet cherry cultivars on Mazzard) are usually compatible. Graft incompatibility occurs only when the composite tree is produced from two or more different species (e.g. sweet cherry/*P. cerasus*, *P. mahaleb* or interspecific hybrids). Sour cherry cultivars usually show good compatibility on *P. mahaleb* and *P. avium*; however, 'Schattenmorelle' shows selective compatibility on *P. mahaleb* and even on sour cherry seedlings. As a result, some nurseries prefer *P. avium* as rootstock for this cultivar.

Within cherries, different species are grafted on to each other; hence, many combinations are heterospecific grafts with diverse metabolic systems. The metabolic system of heterospecific grafts may be more or less stressed (Feucht, 1982; Treutter *et al.*, 1993; Feucht *et al.*, 2001). Incompatibility symptoms may not appear under optimal growth conditions; however, when the tree is subjected to environmental stress, the underlying incompatibility will be revealed.

Incompatibility symptoms can include: poor bud take, the scion snapping off at the bud union, small yellow leaves, stunted growth, early reddening and fall of leaves in the autumn, scion or rootstock overgrowth, excessive rootstock suckering and excessive early fruiting (Perry, 1987; Webster and Schmidt, 1996). Rootstocks may not be compatible with all cultivars within a species (Webster and Schmidt, 1996).

### Propagation opportunities and nursery value of rootstock plants

Selected seed orchards produce more uniform populations when compared with seedlings of unknown origin. Seed propagation is relatively straightforward when using appropriately selected seed sources of *P. avium* or *P. mahaleb*. The germination capacity of *P. avium* scion cultivars is very low and is variable in *P. cerasus*. Vegetative propagation provides uniform rootstock material, and therefore the adventitious root production capacity becomes an essential trait for rootstock candidates. The rooting capacity of *P. avium* is low, and only the clones 'F 12/1' and 'Charger' have been successfully propagated by layering (Howard, 1987; Webster, 1996). 'Colt', a hybrid of *P. avium*, which forms adventitious roots at the base of 1-year-old shoots, propagates readily as hardwood cuttings or by layering. Similarly, 'IP-C4' (*P. avium* × *P. pseudocerasus*) and 'IP-C5' (*P. avium* × *Prunus nipponica kurilensis*) can be propagated by layering. Hybrids of *P. cerasus* and *P. canescens* with *Prunus wadai* (*P. pseudocerasus* × *P. subhirtella*) (Wolfram, 1971; Gruppe, 1985) are also readily propagated as cuttings or by layering. Softwood cuttings of *P. mahaleb* clonal rootstocks form adventitious roots easily (Sarger, 1972; Hrotkó, 1982; Szabó *et al.*, 2014), but hardwood cuttings and layering usually fail. Many of the commercially available rootstocks from interspecific hybrids are micropropagated (Dradi *et al.*, 1996; Druart, 1998; Muna *et al.*, 1999; Dorić *et al.*, 2014), although growth rate in the nursery may be an issue (i.e. length of time before shoots can be budded). Growers who decide to grow sweet cherries on 'GiSelA 5' rootstocks should take into account the poor growth of trees in the nursery and apply appropriate aftercare in the first years after the establishment of an orchard (Baryła *et al.*, 2014). When top grafting scion cultivars or ornamentals, only *P. avium* seedlings and 'F 12/1' are recommended, since they form appropriate trunk quality.

### *Tolerance to environmental conditions (climate, soil) and nutrient and water supply*

Cold hardiness is an important attribute of rootstocks, and rootstocks can also affect the response of the scion to cold temperatures (Howell and Perry, 1990). *P. cerasus* and *P. fruticosa* are considered the hardiest rootstocks, and Mahaleb is hardier than Mazzard. Within the Mahaleb species, the broad-leaved subspecies is hardier than the small-leaved subspecies (Hrotkó, 2004). *P. avium* is the least hardy species (Perry, 1987) within the *Eucerasus* section, although Küppers (1978) reported differences in hardiness of Mazzard selections. In the nursery, 'Colt' can show sensitivity to early frost, but no injury has been observed on sweet cherry trees budded on 'Colt'.

Drought and heat tolerance of rootstocks is essential in many cherry-growing regions, and this attribute may be linked to root depth. Due to climate change, the importance of these rootstock traits is expected to increase. Shallow-rooted dwarfing rootstocks (some dwarfing interspecific hybrids between *P. cerasus* and *P. fruticosa*) are more susceptible to drought and heat injury. The most tolerant rootstocks appear to be the *P. mahaleb* selections and their hybrids (Mazzard × Mahaleb (M×M) series).

Adaptability to different soil conditions is considered an important rootstock trait. *P. mahaleb* and its derivatives are best suited to light sandy or gravelly soils with free drainage, and they can tolerate a high lime content and pH (pH 7.8–8.5). Mahaleb seedlings ('Cema') proved to be tolerant to calcareous and high pH soils in the north-west provinces of China, where in the summer during the rainy season, anaerobic conditions may cause iron chlorosis when using *P. pseudocerasus* as a rootstock (Faust *et al.*, 1998; Cai *et al.*, 2007; Hrotkó and Cai, 2014).

Rootstocks may influence the nutrient supply efficiency (Jiménez *et al.*, 2007). Moreno *et al.* (1996) found that on dry, gravelly calcareous soil, all rootstocks induced low leaf iron concentrations, although visual chlorosis symptoms were not observed. With the cultivar 'Van', 'Adara' rootstock followed by 'CAB 6P' and 'GiSelA 5' showed the most balanced nutritional values. On the other hand, 'SL 64' had leaf mineral element concentrations below the optimum, probably due to bad adaptation to heavy soil conditions. Sitarek *et al.* (1998) showed that leaves of trees on the dwarf rootstocks 'P-HL-A' and 'P-HL-C' contained less calcium than those of control trees. Stachowiak *et al.* (2015) also confirmed the higher iron supply on Mahaleb roots on sandy soil in Poland. Hrotkó *et al.* (2014) confirmed the rootstock effect on nutrient supply, and emphasized that Mahaleb rootstocks provided a balanced nutrient supply.

### *Tolerance or resistance to pests and diseases*

Several nematode species attack the roots of cherry trees. With regard to the sensitivity of rootstocks, rather contradictory results have been reported in the literature. According to Webster and Schmidt (1996), *P. avium* and *P. cerasus* are sensitive to root-lesion nematodes (*Xiphinema* and *Pratylenchus* spp.), while Zepp and Szczygiel (1985) found that *Pratylenchus penetrans* attacks *P. mahaleb* roots more readily than those of Mazzard and *P. cerasus*. Mazzard and *P. cerasus* are reported to be more tolerant than *P. mahaleb* to *Meloidogyne incognita* (Webster and Schmidt, 1996). In contrast, Hartmann *et al.* (2002) found Mahaleb roots to be more resistant to the root-lesion nematode *Pratylenchus vulnus* than Mazzard. Mahaleb rootstocks are also resistant to the root-knot nematode *M. incognita* but susceptible to *Meloidogyne javanica* (Hartmann *et al.*, 2002).

*Phytophthora* spp. may cause serious tree decay on heavy soils with low drainage capacity, and *P. cerasus* and Mazzard are more tolerant than the susceptible *P. mahaleb* (Wicks *et al.*, 1984; Cummins *et al.*, 1986). *P. canescens* ('Camil') and its hybrids are also sensitive to *Phytophthora* (Webster and Schmidt, 1996). All rootstocks are sensitive to *Verticillium* spp., and there is no source of resistance. In the USA where the honey fungus *Armillaria mellea* can cause root damage, *P. mahaleb*, *P. cerasus*, 'Colt' and 'Inmil' were found to be sensitive,

Mazzard was less sensitive and 'MxM 60' showed the least sensitivity (Proffer *et al.*, 1988). Testing of new series of dwarfing rootstocks was started in Michigan, USA, with the aim of selecting *Armillaria*-resistant rootstocks (Olmstead *et al.*, 2011).

Leaf spot caused by *Blumeriella jaapii* can cause severe leaf fall in nursery liners. Only *P. mahaleb* is tolerant, whereas *P. avium*, *P. cerasus* and their derivatives are more or less sensitive. 'VP1' (*P. cerasus* × *P. maacki*) is reported to be tolerant (Yoltuchovski, 1977; Micheyev *et al.*, 1983). According to G. Mladin (Pitesti, Romania, 2011, personal communication), 'IP-C4' and 'IP-C5' are resistant to *Blumeriella* leaf spot. There are a number of wild cherry species with a high level of resistance to leaf spot that could be used in breeding programmes (Wharton *et al.*, 2003; Schuster, 2004; Budan *et al.*, 2005). In north-west Europe, the fungus *Thielaviopsis basicola* can cause severe replant problems, and hybrids of *P. avium* × *P. pseudocerasus* can be used as a source of resistance for this threat.

Bacterial diseases that create problems include crown gall (*Agrobacterium tumefaciens*), which infects trees in the nursery as well as in orchards where it can reduce growth and productivity. 'Colt' and the Mazzard clone 'F 12/1' are both sensitive, whereas Mahaleb rootstocks and *P. fruticosa* hybrids are less sensitive. On 'Colt' rootstocks, galls are formed even on graft unions. *Pseudomonas syringae* pv. *morsprunorum* and *P. syringae* pv. *syringae* cause bacterial canker, which is a particularly damaging disease in the humid zones of temperate areas (see Chapter 15, this volume). Mahaleb rootstocks are known to be tolerant, whereas Mazzard genotypes are considered susceptible. The clonal rootstock 'Charger' and some *P. avium* × *P. pseudocerasus* or *P. avium* × *P. incisa* hybrids can be used as a source of resistance (Webster and Schmidt, 1996).

There is no known resistance to viruses or phytoplasmas (see Chapter 16, this volume), although there are considerable differences in sensitivity. *P. fruticosa* and its derivatives are particularly hypersensitive to viruses. Some clones of *P. cerasus* have shown higher sensitivity to prune dwarf virus (PDV) whereas *P. canescens* is more sensitive to *Prunus* necrotic ringspot virus (PNRSV) (Lang *et al.*, 1998). Lankes (2007) found that 'Colt', 'GiSelA 5' and 'Piku 1', 'Piku 3' and 'Piku 4' are tolerant, while 'GiSelA 3' and 'GiSelA 6' are partially sensitive to PDV and PNRSV. In France, *P. mahaleb* rootstocks are more susceptible to leafhopper-transmitted phytoplasma (Molière's disease) compared with trees on Mazzard. Western X-disease in the USA, which is also transmitted by leafhoppers, can infect trees on Mazzard and 'Colt', whereas *P. mahaleb* rootstocks are hypersensitive.

## 6.3 Programmes of Rootstock Breeding and Major Breeding Achievements

### 6.3.1 Achievements of clonal rootstock selection and creation of their interspecific hybrids

Clonal rootstocks were selected from the three major cherry rootstock species (*P. avium*, *P. mahaleb* and *P. cerasus*) and from among their interspecific hybrids.

The main advantages of clonally selected rootstocks from *P. avium* ('F 12/1', 'Charger') are the propagation possibility in layer beds (Howard, 1987) and the uniform plant material for nurseries and fruit growers.

Interspecific hybrids with *P. avium* (Table 6.1) provide a wide range of vigour, adaptability and tolerance to diseases (James *et al.*, 1987; Wolfram, 1996; Grzyb *et al.*, 2005; Hrotkó *et al.*, 2009a). The first *P. avium* hybrid rootstock was 'Colt' (*P. avium* × *P. pseudocerasus*) (Webster, 1980). Dwarf mutants were also produced using colchicine from 'Colt' rootstock (James *et al.*, 1987). The hexaploid (6n = 48) 'Colt' is now involved in several test series in Europe and proved to be a little bit more growth reducing but not more precocious than the original (Hrotkó, 2008; Lugli and Sansavini, 2008; Meland, 2015).

The P-HL- series is supposed to be *P. avium* × *P. cerasus*, and the dwarf 'P-HL-A'

**Table 6.1.** Vegetatively propagated *P. avium* rootstocks and interspecific hybrids (created with female parent *P. avium*)

| Variety | Brief description | Reference(s) |
|---|---|---|
| ***P. avium*** | | |
| 'F 12/1' | Vigorous, good trunk, resistant to bacterial canker | Webster and Schmidt (1996) |
| 'Charger' | Vigorous, higher resistance to bacterial canker | Webster and Schmidt (1996) |
| 'Cristimar' | Landrace selection, said to be less vigorous | Cireasa and Sardu (1985); Cireasa *et al.* (1993) |
| **Interspecific hybrids** | | |
| 'Colt' | *P. avium* × *P. pseudocerasus*, 2n = 24; easy to propagate, 80% vigour, flat branching, limited adaptability to drought and lime soils | Webster (1980) |
| Hexaploid 'Colt' | *P. avium* × *P. pseudocerasus*, 6n = 48; easy to propagate, 75% vigour | James *et al.* (1987) |
| 'P-HL-A' ('PLH 84') | Supposedly *P. avium* × *P. cerasus*; promising dwarf rootstock in Czech Republic and Poland, limited climate adaptability | Blazek (1983); Anon. (2003); Blažková and Hlušičkova (2004); Grzyb *et al.* (2005); Paprstein *et al.* (2008) |
| 'P-HL-B', 'P-HL-C' | Supposedly *P. avium* × *P. cerasus*; semi-dwarf rootstock in Czech Republic, limited climate adaptability | Blazek (1983); Anon. (2003); Paprstein *et al.* (2008) |
| 'PiKu 1' ('PiKu 4.20') | *P. avium* × (*P. canescens* × *P. tomentosa*); moderate vigour, high productivity, adaptability, tolerant to prune dwarf virus and *Prunus* necrotic ringspot virus | Wolfram (1996); Hilsendegen (2005); Lankes (2007); Hrotkó *et al.* (2009a); Spornberger *et al.* (2015) |
| 'IP-C4' | *P. avium* × *P. pseudocerasus* | G. Mladin (Pitesti, Romania, 2011, personal communication) |
| 'IP-C5' | *P. avium* × *P. pseudocerasus* | G. Mladin (Pitesti, Romania, 2011, personal communication) |
| 'IP-C7' | (*P. avium* × *P. nipponica* var. *kurilensis*) × (*P. avium* 77-33-26 × *P. pseudocerasus*) | G. Mladin (Pitesti, Romania, 2011, personal communication); Tanasescu *et al.* (2013) |
| 'GiSelA 4' ('Gi 473/10') | *P. avium* × *P. fruticosa*; dwarf, suckers badly | Gruppe (1985); Stehr (2005); Kappel *et al.* (2005); Hrotkó *et al.* (2006); Lichev and Papachatzis (2009) |

was reported as very promising in the Czech Republic and Poland but was less successful or disappointing in other countries (Anon., 2003; Grzyb *et al.*, 2005). 'PiKu 1' (*P. avium* × (*P. canescens* × *P. tomentosa*)) (Wolfram, 1996) also proved to be promising as a moderately vigorous or semi-dwarf rootstock in orchard trials (Hrotkó *et al.*, 2009a), although recently low cropping efficiency and fruit quality were reported by Spornberger *et al.* (2015).

Among the cherry rootstock series created in Pitesti (Romania), 'IP-C4' (*P. avium* 77-36-26 × *P. pseudocerasus*), 'IP-C5' and 'IP-C7' (both *P. avium* 77-36-26 × *P. nipponica* var. *kurilensis*) could be classified into this group, with *P. avium* as the female parent. 'IP-C4' and 'IP-C5' are vigorous or moderately vigorous rootstocks with good compatibility, productivity and resistance to *Coccomyces* fungal infection.

One of the most promising rootstock groups is the vegetatively propagated *P. cerasus*, well known as a source of dwarfing rootstocks for sweet cherries, but the variable compatibility and suckering of trees budded on this rootstock is a major disadvantage (Perry, 1987; Granger, 2005) (Table 6.2).

**Table 6.2.** Vegetatively propagated *P. cerasus* rootstocks and derivatives (*P. cerasus* as female parent, except for 'GiSelA 12')

| Variety | Brief description | Reference(s) |
|---|---|---|
| ***P. cerasus*** | | |
| 'Vladimirskaya' (syn. 'Vladimir') | Landrace from Russia, dwarfing, suckers badly, poor anchorage | Perry (1987); Sotirov (2012) |
| 'Stockton Morello' | Landrace from USA, dwarfing, suckers, tolerates clay, shallow roots, poor anchorage | Perry (1987); Granger (2005) |
| 'Oblačinska', 'OV 11', 'OV 18' | Landrace from Serbia, dwarfing, suckers, shallow roots, poor anchorage | Ognjanov *et al.* (2012) |
| 'CAB 6P', 'CAB 11E' | Landrace selection from Italy, moderate dwarfing, few suckers, shallow roots | Faccioli *et al.* (1981); Sansavini and Lugli (1996) |
| 'Weiroot 10', 'Weiroot 13' | Landrace selections from Germany, vigorous, few suckers, adaptability to clay soils, good compatibility and fruit size | Treutter *et al.* (1993); Schimmelpfeng (1996); Hrotkó *et al.* (2006); Lichev and Papachatzis (2009) |
| 'Weiroot 154', 'Weiroot 158' | Hybrids of landraces selected in Germany, semi-dwarf, few suckers, adaptability to clay soils, good compatibility and fruit size | Treutter *et al.* (1993); Schimmelpfeng (1996); Bujdosó *et al.* (2004); Stehr (2005); Lichev and Papachatzis (2009) |
| 'Weiroot 72', 'Weiroot 53', 'Weiroot 720' | Hybrids of landraces selected in Germany, dwarf, few suckers, variable compatibility, low soil adaptability, poor anchorage | Treutter *et al.* (1993); Schimmelpfeng (1996); Bujdosó *et al.* (2004); Lichev and Papachatzis (2009); Neumüller (2009) |
| 'Edabriz' | Selected from Iranian wild genotypes, dwarfing, suited on fertile loam and clay | Edin *et al.* (1996); Hilsendegen (2005); Hrotkó *et al.* (2009c) |
| 'Victor' | Selected in Italy, dwarf to semi-dwarf | Battistini and Berini (2004) |
| **Interspecific hybrids** | | |
| 'GiSelA 3', ('Gi 209/1') | *P. cerasus* × *P. canescens*, very dwarf, partially sensitive to PDV and PNRSV | Gruppe (1985); Franken-Bembenek (2004); Lankes (2007); Sotirov (2015) |
| 'GiSelA 5' ('Gi 148/2') | *P. cerasus* × *P. canescens*, dwarf, tolerates PDV and PNRSV, suited to fertile loam, needs irrigation, good compatibility and productivity, precocious, early senescence | Gruppe (1985); Franken-Bembenek (2005); Hrotkó *et al.* (2009c); Lankes (2007); Lichev and Papachatzis (2009); Sotirov (2015) |
| 'GiSelA 6' ('Gi 148/1') | *P. cerasus* × *P. canescens*, semi-dwarf, partially sensitive to PDV and PNRSV, suited to fertile loam, needs irrigation, precocious | Gruppe (1985); Kappel *et al.* (2005); Stehr (2005); Lankes (2007); Sotirov (2015) |
| 'GiSelA 7', ('Gi 148/8') | *P. cerasus* × *P. canescens*, moderate vigour, high soil adaptability, good precocity and productivity | Gruppe (1985); Kappel *et al.* (2005); Hrotkó *et al.* (2006) |
| 'GiSelA 8' ('Gi 148/9') | *P. cerasus* 'Schattenmorelle' × *P. canescens*, moderate vigour, high soil adaptability, good precocity and productivity | Gruppe (1985); Kappel *et al.* (2005); Hrotkó *et al.* (2006) |
| 'GiSelA 12' ('Gi 195/2') | *P. canescens* × *P. cerasus* 'Leitzkauer', semi-vigorous, high soil adaptability, good precocity and productivity | Gruppe (1985); Kappel *et al.* (2005); Hrotkó *et al.* (2006); Lichev and Papachatzis (2009) |
| 'Gi 195/20' | *P. cerasus* × *P. canescens*, semi-dwarf, good precocity and productivity | Hilsendegen (2005) |
| 'IP-C1' | *P. cerasus* 'Mocăneşti T1' × *P. avium* 77-33-26, sel. Romania, semi-dwarf, fewer suckers, tolerates wet soil, resistant to *Agrobacterium* | Parnia *et al.* (1997); G. Mladin (Pitesti, Romania, 2011, personal communication) |

*Continued*

**Table 6.2.** Continued.

| Variety | Brief description | Reference(s) |
|---|---|---|
| 'IP-C2' | *P. cerasus* 'Mocăneşti T1' × *P. subhirtella*, moderate vigour, easy rooting, heavy producer | G. Mladin (Pitesti, Romania, 2011, personal communication) |
| 'IP-C3' | *P. cerasus* 'Crişana B' × *P. subhirtella*, semi-dwarfing, compatible, very productive trees | G. Mladin (Pitesti, Romania, 2011, personal communication) |
| 'Krymsk 6' ('LC-52') | *P. cerasus* × (*P. cerasus* × *P. maackii*), semi-dwarf, good precocity and productivity | Eremin and Eremin (2002); Maas *et al.* (2014) |

PDV, prune dwarf virus; PNRSV, *Prunus* necrotic ringspot virus.

Breeding programmes in several countries have selected clones from landraces of sour cherry that resulted in non-suckering *P. cerasus* as the rootstock for a wide range of vigour and compatibility (Schimmelpfeng and Liebster, 1979; Faccioli *et al.*, 1981; Schimmelpfeng, 1996). In addition, a considerable advantage is the positive effect on the fruit size of the sweet cherry scion, which was reported for the majority of *P. cerasus* rootstocks. Despite the intense selection work, there have only been a few targeted crosses within *P. cerasus*, except for the Weihenstephan breeding programme, where the second and third generations are crosses between low-vigour landraces (Schimmelpfeng, 1996). Among the 'Oblačinska' sour cherry landraces, moderately vigorous clones ('OV11' and 'OV18') were selected in Serbia (Ognjanov *et al.*, 2012).

The most productive and extensive interspecific hybridization programme was carried out in Giessen, Germany (Gruppe, 1985). Recent results from various national rootstock trials suggest that the most promising hybrids are from *P. cerasus* × *P. canescens* and reciprocal crosses (Walther and Franken-Bembenek, 1998; Franken-Bembenek, 2004; Franken-Bembenek, 2005; Kappel *et al.*, 2005). Agronomic shortcomings are limiting the commercial potential of a number of candidates, but they represent an important genetic resource. Some of the hybrids from the 'GiSelA' series have spread all over the world as promising dwarfing rootstocks (Table 6.2).

In the former USSR, the 'VP 1' hybrid (*P. cerasus* × *Padus maacki*) showed promising results, and was later introduced as rootstock 'Krymsk 6' ('LC 52') and tested in western Europe and North America (Eremin *et al.*, 2000; Eremin and Eremin, 2002; Maas *et al.*, 2014).

Among the cherry rootstock series created in Pitesti (Romania), 'IP-C1', 'IP-C2' and 'IP-C3' are derivatives of *P. cerasus* (see Table 6.2). They are precocious, semi-dwarfing or moderately vigorous rootstocks with good compatibility, productivity and easy rooting, and 'IP-C2' and 'IP-C3' are resistant to *Blumeriella* leaf spot (G. Mladin, Pitesti, Romania, 2011, personal communication).

The first *Prunus mahaleb* L. clonal rootstock, 'Sainte Lucie 64' ('SL 64'), was selected in France for its ease of propagation and compatibility with sweet cherries, as well as its productivity in orchard conditions (Thomas and Sarger, 1965; Sarger, 1972). *P. mahaleb* selections that are semi-vigorous and precocious have also been developed (Table 6.3) (Hrotkó, 1982; Giorgio and Standardi, 1996; Misirli *et al.*, 1996; Hrotkó and Magyar, 2004; Lang, 2006; Sotirov, 2012; Koc and Bilgener, 2013; Barać *et al.*, 2014).

Successful hybridization of *P. mahaleb* with *P. avium* was carried out in Oregon, USA (Westwood, 1978; Perry, 1987) resulting in the M×M series and 'OCR 2' and 'OCR 3' (Table 6.3). Two of these are semi-vigorous ('MaxMa 14' and 'MaxMa 97'), and at certain sites both are considered as possible rootstocks for semi-intensive or intensive orchards (Edin *et al.*, 1996; Hrotkó *et al.*, 1999). Crosses between *P. mahaleb* and *P. fruticosa* were reported by de Palma *et al.* (1996) and Hrotkó (2004), but the evaluation

**Table 6.3.** Vegetatively propagated *P. mahaleb* rootstocks and derivatives (*P. mahaleb* as female parent)

| Variety | Brief description | Reference(s) |
|---|---|---|
| ***P. mahaleb*** | | |
| 'SL 64' | Selected in France from wild genotypes, vigorous, easy to propagate, good compatibility and productivity with sweet and sour cherries | Thomas and Sarger (1965); Claverie (1996); Edin *et al.* (1996); Hrotkó *et al.* (1999) |
| 'Bogdany' | Selected from roots of an old and productive sweet cherry tree, vigorous, wide crotch angles, good compatibility and productivity | Hrotkó and Magyar (2004); Hrotkó *et al.* (2009a,c); Magyar and Hrotkó (2013) |
| 'Egervár', 'Magyar' | Moderate vigorous clones selected in Hungary, good compatibility and productivity, wide crotch angles | Hrotkó (1993); Hrotkó and Magyar (2004); Magyar and Hrotkó (2013); Bujdosó and Hrotkó (2014) |
| 'Bonn 60', 'Bonn 62' | Vigorous clones selected in Germany, did not get into commercial propagation | Baumann (1977) |
| 'UCMH 55', 'UCMH 56' | Vigorous clones selected in USA, University of California, propagated by softwood cuttings | Lang (2006) |
| 'UCMH 59' | Medium vigorous clones selected in USA, University of California, propagated by softwood cuttings | Lang (2006) |
| 'Mahaleb 20-86' | Moderate vigorous, selected in Bulgaria | Sotirov (2012) |
| 'Mahaleb MG 1', 'Mahaleb MG 2', 'Mahaleb MG 3', 'MG 1KB', 'MG 2KB' | Medium vigorous, propagated by softwood cuttings or *in vitro* | Barać *et al.* (2014); Dorić *et al.* (2014) |
| **Interspecific hybrids** | | |
| 'MaxMa 2' ('MxM 2'), 'MaxMa 60' ('MxM 60') | *P. mahaleb* × *P. avium*, USA, Oregon, very vigorous, adaptability as on Mahaleb, good compatibility, resistant to *Phytophthora cambivora* and *Phytophthora megasperma*, narrow crotch angle, more precocious than seedling, good productivity | Westwood (1978); Perry (1987); Hrotkó *et al.* (2006) |
| 'MaxMa 14' ('MxM 14') and MaxMa 97' ('MxM 97') | *P. mahaleb* × *P. avium*, USA, Oregon, moderate vigour, adaptability like on mahaleb, good compatibility, resistant to *Phytophthora cambivora* and *megasperma*, narrow crotch angle, more precocious than seedling, good productivity | Westwood (1978); Perry (1987); Edin *et al.* (1996); Hrotkó *et al.* (1999); Hrotkó *et al.* (1999, 2006, 2009a,c); Santos *et al.* (2014) |

of these hybrids is in the early stages. Lichev *et al.* (2014) evaluated a hybrid rootstock between *P. mahaleb* and 'GiSelA 5' ('Hybrid 2/10') as an interstock.

### Other clonal rootstocks and interspecific hybrids

*P. fruticosa* Pall., as a low-growing cherry shrub (0.3–1 m), has attracted the attention of cherry rootstock researchers as a potential dwarfing rootstock since the early decades of the last century. Spontaneous hybrids of *P. fruticosa*, which form shrubs 1–2 m tall (Wójcicki, 1991; Hrotkó and Facsar, 1996), are found in several central European countries and have been the subject of rootstock selection. The 'Oppenheimer Selection' in Germany, (Plock, 1973; Hein, 1979), 'Prob' in Hungary (Hrotkó, 2004; Hrotkó and Magyar, 2004), 'Frutana' in Poland (Rozpara and Grzyb, 2004) and 'SV 5', 'SV 6', 'SV 7' and 'SV 11' in Serbia (Barać *et al.*, 2014) were reported as selected rootstocks/interstocks from *P. fruticosa*. The main shortcomings, namely variable compatibility, poor anchorage, severe trunk shrinking, suckering and virus sensitivity, have not yet been overcome (Table 6.4). From the 'GiSelA' series, the *P. fruticosa* derivatives 'GiSelA 1' and 'GiSelA 10' have been withdrawn. In the former USSR, 'VSL-2' (*P. fruticosa* × *P. lannesiana*, syn. 'Krymsk 5') showed promising results (Eremin and Eremin, 2002), although Maas *et al.* (2014) reported shortcomings similarly to other *P. fruticosa* derivatives.

In Japan, the rootstock 'Chishimadai 1 Go' was selected from among 114 seedlings of the local species 'Chishimazakura' (*P. nipponica* Matsum var. *kurilensis* Wilson), which is native to Hokkaido. 'Chishimadai 1 Go' forms small trees and shows higher cold resistance than 'Aobazakura' (*P. lannesiana forma mutiplex* Miyos), which is widely used as a cherry rootstock in Japan (Muramatsu *et al.*, 2004).

Some interspecific hybrids with *Prunus* spp. of East Asian origin were created and tested as cherry rootstocks. The Gembloux Series, 'Inmil' (*P. incisa* × *P. serrula*), 'Damil' (*P. canescens* × *P. dielsiana*) and 'Camil' (*P. canescens*), did not get through in a large-scale commercial propagation. From among this series, 'Damil' and 'Camil' were the most tested but are not widely used. Although these rootstocks are very sensitive to drought stress (Vercammen, 2004), Druart (1998) and Magein and Druart (2005) reported further selection among hybrids of the Gembloux Series rootstocks.

### 6.3.2 Achievements in seed tree selections

Seed-sourced/mother trees selected for superior phenotypic traits have been released from several countries (Table 6.5). The advantages of seed orchards include the potential for a virus-free seed source, higher germination capacity, hybrid seed of known

**Table 6.4.** Further vegetatively propagated interspecific hybrids

| Variety | Brief description | References |
|---|---|---|
| 'Frutana' | *P. fruticosa*, recommended as dwarfing interstock, precocious, early senescence | Rozpara and Grzyb (2004); Grzyb *et al.* (2005) |
| 'Krymsk 5', ('VSL-2') | *P. fruticosa* × *P. serrulata* var. *lannesiana.*, dwarf, good productivity, precocious, smaller fruit size, suckering, early senescence | Eremin *et al.* (2000); Eremin and Eremin (2002); Maas *et al.* (2014) |
| 'P 3', 'P 7' | *Cerapadus* × (*P. cerasus* L. × *P. avium* L.), suckering | Lanauskas *et al.* (2004, 2014) |
| *P. pseudocerasus* | *P. pseudocerasus*, widespread, planted as seedling rootstock in China, on lime soil suffers from iron chlorosis | Faust *et al.* (1998); Cai *et al.* (2007) |
| 'Krymsk 7' ('L-2') | *P. serrulata* var. *lannesiana*, vigorous | Eremin and Eremin (2002) |
| 'Aobazakura' | *P. serrulata var. lannesiana forma mutiplex* Miyos, variable compatibility and low hardiness | Muramatsu *et al.* (2004) |
| 'Chishimadai 1 Go' | *P. nipponica* Matsum var. *kurilensis* Wilson, more winter hardiness, medium vigour | Muramatsu *et al.* (2004) |
| 'Inmil' | *Prunus incisa* × *P. serrula*, very dwarf | Trefois (1980); Druart (1998); Vercammen (2004); Magein and Druart (2005) |
| 'Damil' ('GM 61/1') | *P. dawyckensis* × (*P. canescens* × *P. dielsiana*), semi-dwarf | Trefois (1980); Druart (1998); Vercammen (2004); Magein and Druart (2005) |
| 'Camil' ('GM 79') | *P. canescens*, medium vigour | Trefois (1980); Druart (1998); Vercammen (2004); Magein and Druart (2005) |

**Table 6.5.** Selected seed tree clones of Mazzard and Mahaleb cherry

| Country | Selected seed tree clones | Reference(s) |
|---|---|---|
| **Mazzard** | | |
| Bulgaria | 'N 123' ('Dryanovo') | Webster and Schmidt (1996) |
| Czech Republic | 'P-TU 1', 'P-TU 3' | Anon. (2003) |
| France | 'Pontavium' ('Fercahun'), 'Pontaris' ('Fercadeu') | Edin and Claverie (1987) |
| Germany | 'Hz 170', 'Hz 53' (clonal derivatives of 'Limburger'); 'Gi 81', 'Gi 84', 'Gi 90', 'Gi 94'; 'Alkavo' ('K 2/4', 'K 4/2', 'K 4/23', 'K 5/28', 'K 5/38') | Funk (1969b); Küppers (1978) |
| Hungary | 'C 2493', 'Altenweddingen' (clonal derivative of 'Alkavo') | Nyújtó (1987) |
| Ukraine | Mazzard No. 3, No. 4 and No. 5; 'Susleny', 'Napoleon' | Yoltuchovski (1977); Tatarinov and Zuev (1984) |
| USA | Mazzard No. 570 (derivative of 'Harz'), 'Sayler', 'OCR 1' | Perry (1987) |
| Romania | 'F 12/1' and 'Dönissens Gelb' (cross-pollinated) | Webster and Schmidt (1996) |
| **Mahaleb** | | |
| Bulgaria | 'IK-M9' ('Kyustendil'), 'P 1' ('Plovdiv') | Sotirov (2005); Lichev and Papachatzis (2009) |
| France | 'SL 405' (self-fertile) | Claverie (1996) |
| Germany | 'Heimann X' (self-fertile); 'Alpruma' ('AF 5/19', 'AF 3/9', 'AF 6/16' and 'PB 9') | Heimann (1932); Funk (1969a); Küppers (1978) |
| Hungary | 'C 500' ('Cema'), 'C 2753' ('Cemany'), 'Érdi V.' (cross-pollinated); 'Korponay' (self-fertile) | Nyújtó (1987); Hrotkó (1990); Hrotkó (1996) |
| Ukraine | Mahaleb No. 24 | Tatarinov and Zuev (1984); Webster and Schmidt (1996) |
| USA | Nos 902, 904, 908 and 916, cross-pollinated, in commercial use known as Mahaleb 900 | Perry (1987) |
| Poland | 'Piast', 'Popiel', physiological incompatibility with sweet cherry | Grzyb (2004); Rozpara (2005) |
| Moldavia | 'Rozovaya Prodolgovataya', 'Chernaya Kruglaya iz Bykovtsa', 'Nr 1 iz Solonchen' | Yoltuchovski (1977); Tatarinov and Zuev (1984) |

parents and improved uniformity of orchard trees compared with open-pollinated seeds.

The first systematic seed tree selection can be traced back to the first half of the last century (Maurer, 1939; Küppers, 1964, 1978). In the first seed orchards, seedlings of wild populations were planted (e.g. 'Harzer Hellrindige Vogelkirsche', 'Limburger Bosqkriek'). In the second stage (around 1935–1970), selected seed orchards were planted where the seed tree clone was selected by its phenotype, and subsequently the vegetatively propagated clone was planted into the seed orchard. In the third stage of seed tree selection, it was possible to evaluate the progeny of the seed tree clone pollinated with known pollinators. The evaluation of seedlings involved both their nursery and orchard tree rootstock value (Hrotkó and Erdős, 2006). The flower fertility of these genotypes determined the mating options of genotypes within the orchard and accordingly the genetic composition of the seedling progeny. Most seed orchards with cross-pollination consist of three to five clones that pollinate each other well (Funk, 1969a,b; Nyújtó, 1987; Perry, 1987; Hrotkó and Erdős, 2006).

Mazzard seed sources from wild populations include 'Harzer Hellrindige Vogelkirsche' from Germany and 'Merisier Commun' from the forests of Massif Central in France. As derivatives from the Harzer Mazzard ecotype, selected seedlings were planted to produce seed in Belgium, and this is known as 'Limburger Bosqkriek', or, in German, as 'Limburger Vogelkirsche' (Table 6.5).

For *P. mahaleb*, the major seed sources are still wild seedling populations, but the use of seeds from selected seed orchards is increasing. Most hybrid seed orchards are planted with clone-group pollination (Hrotkó and Erdős, 2006), which results in hybrid seed (family of siblings). Self-fertile seed orchards are known from Germany ('Heimann X'), France ('SL 405') and Hungary ('Korponay'). Progeny evaluation of cross-pollinated or self-fertile orchards was carried out in France (Claverie, 1996), Germany (Funk, 1969b) and Hungary (Nyújtó, 1987; Hrotkó, 1990; Hrotkó, 1996; Hrotkó, 2008).

## 6.4 Characteristics of Sweet and Sour Cherry Rootstocks

### 6.4.1 Rootstocks of global importance

#### *'Colt'*

'Colt' was obtained in 1958 at East Malling, UK, and was introduced to nurseries in the 1970s. It is now quite widely used in many European countries. Sweet cherry trees on 'Colt' are vigorous, exhibiting growth vigour at many sites similar to that of sweet cherries grafted on Mazzard seedlings or 'F 12/1' (Rozpara, 2013). A disadvantage of 'Colt' is its low resistance to frost (Grzyb, 2012). In cooler regions, during harsh and snowless winters, the roots of young trees may freeze. Fruit growers should protect themselves against this negative characteristic by mulching the trees in young orchards. The advantages of 'Colt' are that it roots exceptionally easily and has a beneficial influence on the health and vigour of sweet cherry trees, as well as on the quality of the fruit. 'Colt' is most commonly propagated in stool beds. Trees grafted on 'Colt' require good, fertile and humid soils. Shallow, dry and calcareous soils are inadequate for this rootstock.

#### *'F 12/1'*

'F 12/1' is a clonal rootstock selected in the UK in 1944 from a population of *P. avium* seedlings. It is propagated vegetatively by layering, as well as by hardwood and softwood cuttings. It is relatively resistant to frost and exhibits physiological compatibility with major cultivars. The advantage of 'F 12/1' rootstocks in comparison with Mazzard seedlings is the very uniform growth of the trees grafted on it and the possibility of producing virus-free nursery material. Cherry trees on 'F 12/1' grow vigorously, reaching dimensions similar to those of trees grafted on *P. avium* seedlings, and even larger where there is moist, humid and fertile soil. The disadvantage of 'F 12/1' is its high susceptibility to crown gall. Trees require fertile soils with a good water supply. It is not recommended for light soils, or for heavy and overly moist soils.

#### *'GiSelA 5' (syn. 'Gi 148/2')*

This is a dwarfing rootstock recommended for sweet cherry (Table 6.2). The importance of this rootstock in the world is growing steadily due to its dwarf vigour and usefulness for modern, intensive commercial orchards. In the new sweet cherry plantings of numerous countries, especially in north-western Europe (Germany and the Netherlands), 'GiSelA 5' has already become the standard. Compared with *P. avium* seedlings or 'F 12/1', it reduces the growth of cherry trees by 30–50%, depending on the cultivar, soil type, tree age and other factors. Usually, the vigour reduction is stronger in southern countries than in northern countries (Sansavini and Lugli, 2014) but its adaptability is not satisfying in Mediterranean warm areas, making its use here unadvisable (López-Ortega *et al.*, 2016). Trees on this rootstock are precocious and very productive (Walther and Franken-Bembenek, 1998; Franken-Bembenek, 2005; Kappel *et al.*, 2005). The rootstock can be propagated by softwood cuttings but most easily by tissue culture. Other advantages of this rootstock include the high resistance of its root system to frost, compatibility with the majority of cultivars recommended for cultivation and tolerance to viral diseases, such as PDV and PNRSV. Observations conducted in the difficult climatic conditions of central Poland indicated that this rootstock is more resistant to frost than *P. avium* seedlings and also 'P-HL-A' (Rozpara, 2013;

Rozpara and Głowacka, 2014). Sweet cherry trees grafted on to 'GiSelA 5' require a better site than trees grafted on *P. avium* seedlings. Irrigation, fertilization and appropriate humus content in the soil are essential. Dwarf trees tend to overcrop, which may lead to small fruit size. Excessive crop load and early senescence should be overcome by specific fruiting wood management (Lang, 2011).

### *'GiSelA 6' (syn. 'Gi 148/1')*

'GiSelA 6' is a promising semi-dwarf rootstock for sweet cherry that has not been evaluated extensively in Europe (Table 6.2). It results in trees that are more vigorous than 'GiSelA 5'. In field experiments in the USA, the vigour of cherry trees on 'GiSela 6' was variable across different soil and climatic conditions (Kappel *et al.*, 2005). It was also found that 'GiSela 6' reduced the growth of trees of different cultivars to varying degrees (Webster and Schmidt, 1996; Bujdosó and Hrotkó, 2014). Trees on 'GiSela 6' bear fruit 2–3 years earlier than on standard rootstocks, and regularly produce very good crops. The 'GiSela 6' rootstock is also tolerant to virus diseases such as PDV and PNRSV (Lang *et al.*, 1998). 'GiSela 6' is a promising rootstock recommended for semi-intensive sweet cherry orchards.

### *'MaxMa 14' (syn. 'MxM 14', 'Brokforest', 'MaxMa Delbard 14')*

This is one of the rootstocks that was obtained in Brooks Nurseries in Oregon, USA (Table 6.3). Recently, this rootstock has become more common for cultivation in France than in the USA, because of its good behaviour on lime soils. Trees grafted on 'MaxMa 14' grow about 20–30% less vigorously than on 'SL 64' rootstock and 30–40% less than on Mazzard, so it is considered a semi-standard rootstock. The growth vigour of trees grafted on this rootstock depends on soil fertility, but different interactions between this rootstock and cultivars are also observed. Trees grafted on 'MaxMa 14' are usually healthy and resistant to lime-induced chlorosis. They are more precocious than trees on 'SL 64', usually yield well and do not need support. This rootstock is moderately susceptible to *Phytophthora* infection and is more tolerant to bacterial canker than Mazzard. It is, however, sensitive to water shortage in the soil and requires irrigation during periods of drought. Very little suckering was observed in some experiments (Hrotkó *et al.*, 1999, 2009b). It is best to propagate it by tissue culture or softwood cuttings. In the nursery, it has low susceptibility to cherry leaf spot and crown gall.

### P. avium *seedling (syn. Mazzard)*

Mazzard is still a widely used standard-vigour rootstock for sweet and sour cherry throughout the world. It has the advantage of developing a good graft union with almost all sweet and sour cherry cultivars. *P. avium* develops a deep and extensive root system. Trees grow well in warm, even fairly light soils but with enough moisture, with a loamy subsoil, and where groundwater occurs at a depth of at least 1.60–1.80 m below the ground surface. They do not grow successfully in soils that are too moist, with a high groundwater level. They also grow poorly in sandy, permeable and dry soils. In some regions, *P. avium* is not sufficiently frost resistant. This trait often shows up in the nursery or in the first years after planting in the orchard, especially during harsh snowless winters, when the as-yet shallow and poorly developed root system suffers frost damage (Rozpara, 2005). Trees on Mazzard are late to begin bearing, so they are not suitable for modern intensive orchards.

### P. mahaleb *seedling (syn. Mahaleb, Saint Lucie cherry)*

Mahaleb is a widespread rootstock used for cherry trees worldwide, especially for sour cherries in, for example, Poland, Germany, Hungary and the USA. However, sweet cherry trees grafted on Mahaleb are also commonly grown in some countries, such as Hungary, France, Turkey and Ukraine. Mahaleb as a rootstock is useful on light, sandy, calcareous soils, and in arid, continental climates. It is commonly believed that

Mahaleb increases the resistance of sweet cherry trees to frost during harsh winters (Rejman, 1987). Many studies indicate, however, that sweet cherry trees grafted on *P. mahaleb* seedlings grow in the orchard for a considerably shorter time than trees grafted on *P. avium*. Some authors explain this phenomenon by the occurrence of 'late physiological incompatibility', which often does not appear until after a few years of growth in the orchard (Webster and Schmidt, 1996). In some countries, selection activities have been conducted within seedling populations of *P. mahaleb* as rootstock for sweet or sour cherry orchards (Tables 6.3 and 6.5).

#### *'Saint Lucie 64' (syn. 'SL 64')*

This is a vigorous *P. mahaleb* clonal rootstock obtained in 1954 in Bordeaux, France. Its main advantage is its compatibility with the majority of sweet cherry cultivars. It is propagated by hardwood, semi-hardwood and softwood cuttings. Generally, sweet cherry trees grafted on 'SL 64' grow, depending on the cultivar, about 20% less vigorously than on 'F 12/1'. However, their vigour also depends on the type of soil. On sandy, gravelly soils, the trees are less vigorous, but on deep fertile soils they grow more vigorously. Regardless of growth vigour, trees grafted on 'SL 64' usually show good productivity and precocity.

### 6.4.2 Rootstocks of local importance

#### *'Adara'*

'Adara' is an all-purpose rootstock because it can be used for the production of sweet cherry, sour cherry, peach, nectarine and plum trees. 'Adara' is *P. cerasifera*, and propagates well by hardwood cuttings. Experiments conducted on heavy loam soils with a high pH and high organic matter content have demonstrated the high adaptability of 'Adara' to such conditions. Trees grafted on this rootstock develop a very extensive, deep root system, which enables them to take up nutrients easily from the soil. Sweet cherry trees grafted on 'Adara' grow vigorously, more strongly than those on 'Colt' and 'SL 64'. They are characterized by good health and produce very few root suckers. This rootstock will probably not find wider application in fruit-growing practice. It may, however, play some role in those parts of Europe where heavy alkaline soils prevent the cultivation of sweet cherries on other rootstocks.

#### *'Camil' (syn. 'GM 79')*

'Camil' rootstock is resistant to frost and exhibits physiological compatibility with most cultivars grown in the USA, Belgium and other western European countries (Table 6.4). However, there have been reports of its physiological incompatibility with the cultivar 'Summit'. Trees grafted on 'Camil' are semi-dwarf, and they can be planted without support. These trees come into fruit early and are very productive. 'Camil' rootstock is not suitable for soils that are too moist. It is susceptible to fungal diseases caused by *Phytophthora*. Another disadvantage is the difficult propagation by conventional methods, and the production of numerous root suckers in orchard conditions.

#### *'Damil' (syn. 'GM 61/1')*

This rootstock produces sweet cherry trees that are more than 40% smaller than trees grafted on the 'F 12/1' rootstock (Table 6.4). It is propagated by softwood cuttings. It forms a relatively large root system, but in the first years after planting, the trees should be supported with stakes. Trees on this rootstock are resistant to frost and produce very few root suckers. A disadvantage of the 'Damil' rootstock is its high susceptibility to crown gall.

#### *'Edabriz' (syn. 'Tabel® Edabriz')*

'Edabriz' is a rootstock of French/Iranian selection, obtained from a population of *P. cerasus* seedlings (Table 6.2). Compared with *P. avium* seedlings it reduces tree vigour by more than 60%. 'Edabriz' is propagated by tissue culture, as traditional methods of rootstock propagation do not give good results. Because of the very shallow root system, it should be planted in very fertile soils, rich in nutrients and water; otherwise the fruit is too small. Sweet cherry trees on 'Edabriz' require support and produce root suckers. The results obtained in Poland indicate that 'Edabriz' is suitable for very intensive plantings, especially on very fertile soils (Sitarek and Grzyb, 2010).

#### *'GiSelA 3' (syn. 'Gi 209/1')*

'GiSelA 3' rootstock produces trees that are dwarf, healthy, resistant to frost, come early into fruiting and are very productive, while at the same time producing no reduction in fruit size (Table 6.2). This rootstock does not produce root suckers in the orchard. Long-term trials conducted in Germany have shown that trees of many cultivars grafted on 'GiSelA 3' grow less vigorously than those on 'GiSelA 5'. Therefore, on fertile soils, and with the help of irrigation, 'GiSelA 3' may prove to be a more valuable rootstock than 'GiSelA 5' (Franken-Bembenek, 2004).

#### *'GiSelA 7' (syn. 'Gi 148/8')*

This is a semi-dwarfing to moderately vigorous rootstock known for having good site adaptability (Table 6.2). It tolerates heavy clay and moist soils. Trees grafted on 'GiSelA 7' grow a little more vigorously than those on 'GiSelA 5' but less vigorously than those on 'GiSelA 6'. This rootstock has not yet been thoroughly tested.

#### *'GiSelA 8' (syn. 'Gi 148/9')*

'GiSelA 8' is a semi-dwarfing to moderately vigorous rootstock with good site adaptability, precocity and productivity. It has not yet been widely tested.

#### *'GiSelA 12' (syn. 'Gi 195/2')*

This is a rootstock that produces trees that are semi-dwarfing or semi-vigorous, come into fruit early and produce good yields without deterioration in fruit quality (Table 6.2). 'GiSelA 12' is not susceptible to viral diseases (Lang *et al.*, 1998) and is not very demanding in terms of soil requirements. It has been observed that on poorer soils the vigour control is more pronounced. Trees grafted on 'GiSelA 12' do not produce root suckers in the orchard and are healthy, but have a quite poorly developed root system and require support. 'GiSelA 12' is propagated by tissue culture.

#### *'Krymsk 5' (syn. 'VSL-2')*

'Krymsk 5' is considered to be the most interesting dwarfing rootstock bred in Krymsk, Russia (Table 6.4). It is reported to be tolerant to frost and resistant to bacterial canker and diseases of the bark and wood (Eremin *et al.*, 2000). According to breeders, this rootstock exhibits physiological compatibility with all sweet cherry cultivars. It is easily propagated by horizontal layering and by softwood cuttings. In the nursery, trees have a healthy appearance, are not susceptible to leaf spot and retain the activity of the cambium for a long time. Maiden trees have a very well-developed root system and grow to a relatively large size. They come into fruit in the second or third year after planting and bear fruit abundantly and regularly. In the orchard, they retain good viability for 15–18 years. This rootstock is useful for establishing sweet cherry orchards on heavy soils (Eremin *et al.*, 2000). Maas *et al.* (2014), however, reported a smaller fruit size and early senescence of trees on this rootstock.

#### *'Krymsk 6' (syn. 'LC-52')*

This is a dwarf rootstock obtained in Krymsk, Russia (Table 6.2) (Eremin *et al.*, 2000). Sweet cherry trees grafted on this rootstock grow 40% less vigorously than on *P. avium* seedlings, come into fruit 2–3 years earlier and are more productive. 'Krymsk 6' is easily propagated by layering in stool beds and from cuttings. It develops an extensive root system and has a low susceptibility to diseases. Its compatibility with sweet cherry cultivars needs to be tested.

### 'MaxMa 2' (syn. 'MxM 2') and 'MaxMa 60' (syn. 'MxM 60')

These are standard-vigour rootstocks bred in Brooks Nurseries in Oregon, USA, by crossing Mazzard with Mahaleb cherry (Mazzard × Mahaleb (M × M)) (Table 6.3). Except for the state of Oregon in the USA, these rootstocks have not found wider recognition in the world. 'MxM 2' is able to tolerate the dry and rocky soils of Oregon. These rootstocks can be planted on heavier and waterlogged soils because they are resistant to bacterial canker and are not susceptible to crown gall and *Phytophthora* root and crown rot.

### 'P-HL-A' (syn. 'PHL 84')

'P-HL-A', when grafted with sweet cherry scions, produces trees that generally grow 30–50% less vigorously than on *P. avium* and form well-proportioned crowns with wide crotch angles (Table 6.1). Trees on 'P-HL-A' come into bearing very early and produce abundant crops. The 'P-HL-A' rootstock is difficult to propagate by layering but is relatively easy by softwood cuttings and tissue culture. Trees on 'P-HL-A' form an extensive but fairly shallow root system. They grow well on fertile soil, rich in humus and water. In an intensive orchard, it is necessary to support them with stakes and to provide irrigation. This rootstock ensures very good growth and fruiting for the cultivars 'Burlat', 'Karesova', 'Vanda', 'Techlovan', 'Büttner's Red', 'Kordia' and other large-fruited sweet cherries (Rozpara *et al.*, 2004; Rozpara, 2013). It is not recommended for small-fruited cultivars, nor should it be used for the cultivar 'Hedelfingen' due to the occurrence of severe symptoms of physiological incompatibility between the two components. It should be stressed that young sweet cherry trees grafted on 'P-HL-A' are sensitive to magnesium deficiency in the soil (Rozpara, 2005).

### 'PiKu 1' (syn. 'PiKu 4.20')

'PiKu 1' is a German rootstock that was first reported to reduce the growth vigour of sweet cherry trees by as much as half in relation to trees grafted on *P. avium* (Table 6.1). However, further trials showed that it was less dwarfing. So far, there have been no indications of its incompatibility with cultivated varieties. This rootstock is best propagated by softwood cuttings or by tissue culture. It is resistant to frost and tolerates both sandy and compact soils well. To obtain very high yields and good-quality fruit on light soils, it needs irrigation.

### 'Weiroot 10' and 'Weiroot 13'

These are both vigorous rootstocks, similar to 'Colt'. However, sweet cherry trees grafted on to them cropped better than those grafted on 'Colt', Mazzard and Mahaleb in orchards located in different sites (Wertheim, 1998). The trees were healthy and gave good fruit quality. These rootstocks are recommended for extensive sweet cherry orchards set up on unfertile soils (Heyne, 1994; Vogel, 1997). The disadvantage of these rootstocks is their tendency to produce root suckers (Webster and Lucas, 1997).

### 'Weiroot 53'

'Weiroot 53' is one of several rootstocks selected at the Technical University of Munich (TUM), Freising, Weihenstephan, Germany, from among local *P. cerasus* types (Table 6.2). 'Weiroot 53' is a dwarfing rootstock from the second generation of their selection programme. Sweet cherry trees on this rootstock are productive, but not as productive as trees on 'GiSelA 5'. Trees grafted on 'Weiroot 53' should be supported. This is a rootstock for fertile but well-drained soils because it does not tolerate excess water (Baumann, 1997; Vogel, 1997).

### 'Weiroot 72'

This is one of the second-generation rootstocks obtained at TUM (Table 6.2). This rootstock reduces the growth vigour of cherry trees more than 'Weiroot 53'. However, sweet cherry trees grafted on 'Weiroot 72', at least in the first years after planting, are not as productive as those grafted on 'Weiroot 53' or 'GiSelA 5'. 'Weiroot 72' requires well-aerated, rich soils with proper water relationships. Trees should be supported with stakes. This rootstock is recommended in the south of Germany but only on fresh cherry lands and only as virus-free trees.

### 'Weiroot 720'

'Weiroot 720' is a third-generation rootstock of the series from TUM and exhibits improved rootstock traits compared with 'Weiroot 72' (Table 6.2) (Neumüller, 2009).

### 'Weiroot 154'

This is a semi-dwarfing rootstock that comes from the second-generation rootstocks obtained at TUM (Table 6.2). Trees grafted on 'Weiroot 154' do not require support, but they produce numerous root suckers. Outside Germany, knowledge about this rootstock is insufficient.

### 'Weiroot 158'

'Weiroot 158' is a semi-dwarfing rootstock, coming from the second-generation rootstocks obtained at TUM (Table 6.2). It is less vigorous than 'Weiroot 154' but has a very positive influence on fruiting. In one of the German experiments in which 23 different rootstocks were studied, trees of the cultivar 'Regina' on 'Weiroot 158' ranked second, just behind 'GiSelA 5', in terms of crop yields. Sweet cherry trees grafted on 'Weiroot 158' are well rooted in the soil and can grow without support. This rootstock is recommended for southern regions of Germany. Preliminary observations conducted in Poland with cultivars 'Burlat' and 'Vanda' indicate that it may also be useful in Polish conditions (Sitarek and Grzyb, 2010).

## 6.4.3 Dwarfing Interstocks for Sweet Cherry

In the absence of a sufficient number of dwarfing rootstocks, it is possible to obtain low-vigour sweet cherry trees using a dwarfing interstock between a vigorous rootstock and scion. Tukey (1964) suggested that *P. fruticosa* be used as an interstem because, when used as a rootstock, it considerably weakened the growth of sweet cherry trees, but caused a reduction in fruit size and produced numerous root suckers. Stortzer and Grossman (1988) attempted, successfully, to weaken the growth vigour of sweet cherry trees using the new Czech rootstocks 'PHL 84' and 'PHL 4' as interstocks. Several trials confirmed the dwarfing effect of *P. cerasus* and *P. fruticosa* interstems (Hrotkó *et al.*, 2008; Magyar and Hrotkó, 2008).

In Poland, research on interstocks for sweet cherry began in the 1970s. In one 20-year trial, the usefulness of more than a dozen interstocks for two sweet cherry cultivars was assessed. The use of 'Northstar' sour cherry, and a few Polish types of *P. fruticosa* as the interstocks, markedly weakened the growth of trees but improved their productivity (Rozpara, 1994; Rozpara *et al.*, 1998; Rozpara and Grzyb, 1999, 2004, 2006). The trees with the interstocks were healthy, came into fruit early, gave a good yield and produced high-quality fruit. The best of the tested cultivars proved to be *P. fruticosa* 'No. 8'. This interstock was entered with the name 'Frutana®' in the Polish National List of Fruit Plant Varieties and the Register of Plant Breeders' Rights. It is now recommended in Polish conditions, especially for the sweet cherry cultivars 'Burlat', 'Vega', 'Van', 'Kordia', 'Vanda', 'Techlovan' and 'Büttner's Red'. 'Frutana' is a spontaneous hybrid of *P. fruticosa* × *P. cerasus*. Trees with 'Frutana' interstock grafted on *P. avium* seedlings are comparable with trees grafted on semi-dwarf or dwarf rootstocks, come into fruit early and are characterized by very good health and quality fruit. They are also more winter hardy than trees grafted just on *P. avium* seedlings without interstocks. Additionally, they have lower soil requirements in comparison with sweet cherries grafted on low-vigour rootstocks.

Negueroles (2005) reported the results of rootstock trials in Spain with 'Mariana' plum. Recent results (Frutos *et al.*, 2014; López-Ortega, 2016; López-Ortega *et al.*, 2016) showed that in such soil conditions, where Mahaleb cherry does not perform well (in most soils of Murcia) because of root and crown rot incidence, Mahaleb could be substituted by 'Mariana 2624' or peach × almond hybrids (several selections) using *P. cerasifera* 'Adara' as an interstock. In the USA, *P. cerasifera* 'RI-1', used as a dwarfing, precocity-inducing, cross-species-compatible interstock, was registered by Zaiger Genetics (Modesto, California) (Lang, 2006). It induces early fruiting of sweet cherry scions when used as an interstock on 'Citation®' (*Prunus salicina* × *Prunus persica*) compared with Mazzard.

## References


Anon. (2003) *National List of Varieties Inscribed in the State Variety Book of the Czech Republic by August 1, 2003.* Central Institute for Supervising and Testing in Agriculture, Brno, Czech Republic.

Barać, G., Ognjanov, V., Obreht, D., Ljubojević, M., Bošnjaković, D., Pejić, I. and Gasić, K. (2014) Genotypic and phenotypic diversity of cherry species collected in Serbia. *Plant Molecular Biology Reporter* 32, 92–108.

Baryła, P., Kapłan, M. and Krawiec, M. (2014) The effect of different types of rootstock on the quality of maiden trees of sweet cherry (*Prunus avium* L.) cv. 'Regina'. *Acta Agrobotanica* 67, 43–50.

Battistini, A. and Berini, E.S. (2004) Agronomic results of Victor, a semi-dwarf cherry rootstock. *Acta Horticulturae* 658, 111–113.

Baumann, G. (1977) Clonal selection in *Prunus mahaleb* rootstocks. *Acta Horticulturae* 75, 139–148.

Baumann, W. (1997) Süsskirschen auf Gisela und Weiroot – Erfahrungen vom Bodensee. *Obstbau* 22, 188–190.

Blazek, J. (1983) Stand der Forschung beim Anbau von Süßkirsch-Niederstämmen und ihre Einführung in die Praxis der CSSR, in Beiträge zur Industriemäßigen Obstproduktion '83. *Martin Luther Universität Halle-Wittenberg Wissenschaftliche Beiträge* 46, 21–31.

Blažková, J. and Hlušičkova, I. (2004) First results of an orchard trial with new clonal sweet cherry rootstocks at Holovousy. *Horticultural Science* 31, 47–57.

Budan, S., Mutafa, I., Stoian, I. and Popescu, I. (2005) Field evaluation of cultivar susceptibility to leaf spot at Romania's sour cherry genebank. *Acta Horticulturae* 667, 153–157.

Bujdosó, G. and Hrotkó, K. (2014) Preliminary results on growth, yield and fruit size of some new precocious sweet cherry cultivars on Hungarian bred mahaleb rootstocks. *Acta Horticulturae* 1058, 559–564.

Bujdosó, G., Hrotkó, K. and Stehr, R. (2004) Evaluation of sweet and sour cherry cultivars on German dwarfing rootstocks in Hungary. *Journal of Fruit and Ornamental Plant Research* 12, 233–244.

Cai, Y.L., Cao, D.W. and Zhao, G.F. (2007) Studies on genetic variation in cherry germplasm using RAPD analysis. *Scientia Horticulturae* 111, 248–254.

Cireasa, V. and Sardu, V. (1985) Researches regarding the reduction of sweet cherry height. *Acta Horticulturae* 169, 163–168.

Cireasa, V., Cireasa, E. and Gavrilescu, C. (1993) 'Cristimar' – the latest cultivar of dwarfish cherry tree. *Acta Horticulturae* 349, 283–284.

Cittadini, E.D., van Keulen, H., Peri, P.L. and Ridder, N. (2007) Designing a "Target-Tree" for maximizing gross value of product in Patagonian sweet cherry orchards. *International Journal of Fruit Science* 6, 3–22.

Claverie, J. (1996) New selections and approaches for the development of cherry rootstocks in France. *Acta Horticulturae* 410, 373–375.

Cummins, J.N. (1979a) Exotic rootstocks for cherries. *Fruit Varieties Journal* 33, 74–84.

Cummins, J.N. (1979b) Interspecific hybrids as rootstocks for cherries. *Fruit Varieties Journal* 33, 85–89.

Cummins, J.N., Wilcox, W.F. and Forsline, P.L. (1986) Tolerance of some new cherry rootstocks to December freezing and to *Phytophthora* root rots. *Compact Fruit Tree* 19, 90–93.

de Palma, L., Palasciano, M. and Godini, A. (1996) Interspecific hybridization program aimed at obtaining dwarfing and non-suckering rootstocks for sweet cherry. *Acta Horticulturae* 410, 177–181.

Dorić, D., Ognjanov, V., Ljubojević, M., Barać, G., Dulic, J., Ppanjic, A. and Dugalic, K. (2014) Rapid propagation of sweet and sour cherry rootstocks. *Notulae Botanicae Horti Agrobotoanici Cluj-Napoca* 42, 488–494.

Dradi, G., Vito, G. and Standardi, A. (1996) *In vitro* mass propagation of 11 *Prunus mahaleb* ecotypes. *Acta Horticulturae* 410, 477–484.

Druart, P. (1998) Advancement in the selection of new dwarfing cherry rootstocks in the progeny of Damil® (GM 61/1). *Acta Horticulturae* 468, 315–320.

Edin, M. and Claverie, J. (1987) Porte-greffes du cerisier: Fercahun-Pontavium, Fercadeu-Pontaris, deux nouvelles selections de merisier. *CTIFL Infos* 28, 11–16.

Edin, M., Garcin, A., Lichou, J. and Jourdain, J.M. (1996) Influence of dwarfing cherry rootstocks on fruit production. *Acta Horticulturae* 410, 239–243.

Ercisli, S., Esitken, A., Orhan, E. and Ozdemir, O. (2006) Rootstocks used for temperate fruit trees in Turkey: an overview. Scientific Works of the Lithuanian Institute of Horticulture and Lithuanian University of Agriculture. *Sodininkyste Ir Darzininkyste* 25, 27–33.

Eremin, V. and Eremin, G. (2002) The perspective of clonal rootstocks for *Prunus* at Krymsk Breeding Station, Russia. In: *First International Symposium for Deciduous Fruit Tree Species*, Zaragoza, 11–14 June, 2002, Abstract S5-5.

Eremin, G.V., Provorčenko, A.V., Gavriš, V.F., Podorožnij, V.N. and Eremin, V.G. (2000) *Kostočkovie Kuljturi: Viraščivanie na Klonovih Podvojah i Sobstvenih Kornjah*. Feniks, Rostov na Donu, Russia.

Faccioli, F., Interieri, G. and Marangoni, B. (1981) Portinnesti nanizzanti del ciliego: le selezione CAB. Atti. Giorn. *Sulle Scelte Varietali in Frutticoltura* 19, 125–128.

Faust, M. (1989) *Physiology of Temperate Zone Fruit Trees*. Wiley, New York.

Faust, M., Deng, X. and Hrotkó, K. (1998) Development project for cherry growing in Shaanxi province of China P.R. *Acta Horticulturae* 468, 763–769.

Feucht, W. (1982) *Das Obstgehölz*. Eugen Ulmer Verlag, Stuttgart, Germany.

Feucht, W., Vogel, T., Scimmelpfeng, H., Treutter, D. and Zinkernagel, V. (2001) *Kirschen und Zwetschenanbau*. E. Ulmer GmbH & Co., Stuttgart, Germany.

Franken-Bembenek, S. (2004) GiSelA 3 (209/1) – a new cherry rootstock clone of the Giessen series. *Acta Horticulturae* 658, 141–143.

Franken-Bembenek, S. (2005) Gisela 5 Rootstock in Germany. *Acta Horticulturae* 667, 167–172.

Frutos, D., López, G., Carrillo, A., Cos, J., García, F., Guirao, P., López, D. and Ureña, R. (2014) Preliminary works to consolidate sweet cherry crop (*Prunus avium* L.) in certain areas of the Murcia region, Spain. *Acta Horticulturae* 1020, 471–475.

Funk, T. (1969a) Bericht über die Selektion bei *Prunus mahaleb* L. in der Deutschen Demokratischen Republik. *Archive für Gartenbau* 4, 219–237.

Funk, T. (1969b) Virusgetestete 'Kaukasische Vogelkirsche' als neue Unterlage. *Obstbau* 9, 140–142.

Giorgio, V. and Standardi, A. (1996) Growth and production of two sweet cherry cultivars grafted on 60 ecotypes of *Prunus mahaleb*. *Acta Horticulturae* 410, 471–475.

Granger, A.R. (2005) The effect of three rootstocks on yield and fruiting of sweet cherry. *Acta Horticulturae* 667, 233–238.

Gruppe, W. (1985) An overview of the cherry rootstock breeding program at Giessen. *Acta Horticulturae* 169, 189–198.

Grzyb, Z.S. (2004) New rootstocks of stone fruit trees selected in Skierniewice, Poland. *Acta Horticulturae* 658, 487–490.

Grzyb, Z.S. (2012) Na podkładce 'Colt'. *Miesięcznik Praktycznego Sadownictwa Sad* 12, 18–23.

Grzyb, Z.S., Sitarek, M. and Guzowska-Batko, B. (2005) Results of a sweet cherry rootstock trial in northern Poland. *Acta Horticulturae* 667, 207–210.

Gyeviki, M., Hrotkó, K. and Honfi, P. (2012) Comparison of leaf population of sweet cherry (*Prunus avium* L.) trees on different rootstocks. *Scientia Horticulturae* 141, 30–36.

Hartmann, H.T., Kester, D.E., Davies, F.T. and Geneve, R.L. (2002) *Plant Propagation*. Prentice-Hall, New Jersey.

Heimann, O.R. (1932) Zur Frage der Selektion der Steinweichsel *Prunus mahaleb* als Veredlungsunterlage für Kirschen. *Der Obst und Gemüsebau Bulletin* 78, 138.

Hein, K. (1979) Zwischenbericht über eine Prüfung der Steppenkirsche (*P. fruticosa*) und anderen Süsskirschenunterlagen und Unterlagenkombinationen. *Erwerbsobstbau* 21, 219–219.

Heyne, P. (1994) Zehnjähriger Vergleich der Sűsskirschenunterlagen. *Obstbau* 19, 304–306.

Hilsendegen, P. (2005) Preliminary results of a national German sweet cherry rootstock trial. *Acta Horticulturae* 667, 179 188.

Howard, G. (1987) Propagation. In: Rom, R.C. and Carlson, R.F. (eds) *Rootstocks for Fruit Crops*. Wiley, New York, pp. 29–78.

Howell, G.S. and Perry, R.L. (1990) Influence of cherry rootstocks on the cold hardiness of twigs of the cherry scion cultivar. *Scientia Horticulturae* 43, 103–108.

Hrotkó, K. (1982) Sajmeggy alanyklónok szaporítása zölddugványozással [Propagation of mahaleb cherry by leafy cuttings]. *Kertgazdaság* 4, 45–50.

Hrotkó, K. (1990) The effect of rootstocks on the growth and yield of 'Meteor korai' sour cherry variety. In: *XXIII International Horticultural Congress*, Firenze, Italy, 27 August–1 September. Abstracts 2, Poster 4165.

Hrotkó, K. (1993) *Prunus mahaleb* Unterlagenselektion an der Universität für Gartenbau und Lebensmittelindustrie in Budapest. *Erwerbsobstbau* 35, 39–42.
Hrotkó, K. (1996) Variability in *Prunus mahaleb* L. for rootstock breeding. *Acta Horticulturae* 410, 183–188.
Hrotkó, K. (2004) Cherry rootstock breeding at the department of Fruit Science, Budapest. *Acta Horticulturae* 658, 491–495.
Hrotkó, K. (2008) Progress in cherry rootstock research. *Acta Horticulturae* 795, 171–178.
Hrotkó, K. (2010) Intensive cherry orchard systems and rootstocks from Hungary. *Compact Fruit Tree* 43, 5–10.
Hrotkó, K. and Cai, Y.L. (2014) Development in intensive cherry orchard systems in Hungary and China. In: *Proceedings of the EU COST Action FA1104: Sustainable Production of High-quality Cherries for the European Market*. COST meeting, Bordeaux, France, October 2014. Available at: https://www.bordeaux.inra.fr/cherry/docs/dossiers/Activities/Meetings/2014%2010%2013-15_4th%20MC%20and%20all%20WG%20Meeting_Bordeaux/Presentations/Hrotko_Bordeaux2014.pdf (accessed 25 May 2016).
Hrotkó, K. and Erdős, Z. (2006) Floral biology of tree fruit rootstocks. *International Journal of Horticultural Science* 12, 153–161.
Hrotkó, K. and Facsar, G. (1996) Taxonomic classification of Hungarian populations of *Prunus fruticosa* (Pall.) Woronow hybrids. *Acta Horticulturae* 410, 495–498.
Hrotkó, K. and Magyar, L. (2004) Rootstocks for cherries from department of fruit science, Budapest. *International Journal of Horticultural Science* 10, 63–66.
Hrotkó, K., Magyar, L. and Simon, G. (1999) Growth and yield of sweet cherry trees on different rootstocks. *International Journal of Horticultural Science* 5, 98–101.
Hrotkó, K., Gyeviki, M. and Magyar, L. (2006) A 'Lapins' cseresznyefajta növekedése és termőre fordulása 22 alanyon [Growth and yielding of Lapins on 22 rootstocks]. *Kertgazdaság* 38, 14–21.
Hrotkó, K., Magyar, L. and Gyeviki, M. (2008) Evaluation of native hybrids of *Prunus fruticosa* Pall. as cherry interstocks. *Acta Agriculturae Serbica* 13, 41–45.
Hrotkó, K., Magyar, L. and Gyeviki, M. (2009a) Effect of rootstock on vigour and productivity in high density sweet cherry orchard. *Acta Horticulturae* 825, 245–250.
Hrotkó, K., Magyar, L. and Gyeviki, M. (2009b) Effect of rootstocks on growth and yield of 'Carmen'® sweet cherry trees. *Bulletin UASVM Horticulture* 66, 143–148.
Hrotkó, K., Magyar, L., Hoffmann, S. and Gyeviki, M. (2009c) Rootstock evaluation in intensive sweet cherry (*Prunus avium* L.) orchard. *International Journal of Horticultural Science* 15, 7–12.
Hrotkó, K., Magyar, L. Borsos, G. and Gyeviki, M. (2014) Rootstock effect on nutrient concentration of sweet cherry leaves. *Journal of Plant Nutrition* 37, 1395–1409.
Iglesias, I. and Peris, M. (2008) Speciale Ciliego. La produzione spagnola vince grazie a precocitá, qualitá e organizzazione tecnico-commerciale. *Frutticoltura* 3, 20–26.
James, D.J., MacKenzie, K.A.D. and Malhotra, S.B. (1987) The induction of hexaploidity in cherry rootstocks using *in vitro* regeneration techniques. *Theoretical and Applied Genetics* 73, 589–594.
Jiménez, S., Pinochet, J., Gogorcena, Y., Betrán, J.A. and Moreno, M.A. (2007) Influence of different vigour cherry rootstocks on leaves and shoots mineral composition. *Scientia Horticulturae* 112, 73–79.
Kappel, F., Lang, G., Anderson, L., Azarenko, A., Facteau, T., Gaus, A. and Southwick, S. (2005) NC-140 regional sweet cherry rootstock trial (1998) – results from Western North America. *Acta Horticulturae* 667, 223–232.
Kappel, F., Granger, A., Hrotkó, K. and Schuster, M. (2012) Cherry. In: Badenes, M.L. and Byrne, D.H. (eds) *Fruit Breeding, Handbook of Plant & Breeding 8*. Springer Science + Business Media, New York, pp. 459–504.
Koc, A. and Bilgener, S. (2013) Morphological characterization of cherry rootstock candidates selected from Samsun Province in Turkey. *Turkish Journal of Agriculture and Forestry* 37, 1–10.
Küppers, H. (1964) *Prunus avium*, Hüttner 170x53'. *Deutsche Baumschule* 2, 34–36.
Küppers, H. (1978) Problematik der Veredlungsunterlagen für Sauer- und Süßkirschen im Spiegel von 250 Jahren. *Deutsche Baumschule* 11, 350–359.
Lanauskas, J., Kviklys, D. and Uselis, N. (2004) Evaluation of rootstocks for sweet cherry cv. 'Vytėnų Rožinė'. *Acta Horticulturae* 732, 335–339.
Lanauskas, J., Kviklys, D., Uselis, N., Viškelis, P., Kvikliene, N. and Buskiene, L. (2014) Rootstock effect on the performance of sweet cherry (*Prunus avium* L.) cv. 'Vytėnų rožinė'. *Zemdirbyste-Agriculture* 101, 85–90.
Lang, G. (2006) Cherry rootstocks. In: Clark, J.R. and Finn, C.E. (eds) *Register of New Fruit and Nut Cultivars List 43. HortScience* 41, 1109–1110.
Lang, G. (2011) Producing first-class sweet cherries: integrating new technologies, germplasm and physiology into innovative orchard management strategies. In: *Proceedings of the 3rd Conference 'Innovation in Fruit Growing'*, Belgrade, pp. 59–74.

Lang, G., Howell, W. and Ophardt, D. (1998) Sweet cherry rootstock/virus interactions. *Acta Horticulturae* 468, 307–314.
Lankes, C. (2007) Testing of *Prunus* rootstock clones for virus tolerance. *Acta Horticulturae* 732, 351–354.
Lichev, V. and Papachatzis, A. (2009) Results from the 11-year evaluation of 10 rootstocks of the sweet cherry cultivar 'Stella'. *Acta Horticulturae* 825, 513–520.
Lichev, V., Botu, M. and Papachatzis, A. (2014) First results from the examination of three interstocks for the sweet cherry cultivar 'Stella'. *Acta Horticulturae* 1020, 381–384.
Long, L., Lang, G., Musacchi, S. and Whiting, M. (2015) *Cherry Training Systems (PNW 667)*. Oregon State University Extension Service, Oregon.
López-Ortega, G. (2016) Fisiología del cerezo (*Prunus avium* L.) en las condiciones climáticas de la Región de Murcia. PhD thesis, Universidad Politécnica de Cartagena, Murcia Spain.
López-Ortega, G., García-Montiel, F., Bayo-Canha, A., Frutos-Ruíz, C., Frutos-Tomás, D. (2016) Rootstock effects on the growth, yield and fruit quality of sweet cherry cv. 'Newstar' in the growing conditions of the Región of Murcia. *Scientia Horticulturae* 198, 326–335.
Lugli, S. and Sansavini, S. (2008) Preliminary results of a cherry rootstock trial in Vignola, Italy. *Acta Horticulturae* 795, 321–326.
Maas, F.M., Balkhoven-Baart, J. and van der Steeg, P.A.H. (2014) Evaluation of Krymsk® 5 (VSL-2) and Krymsk® 6 (LC-52) as rootstocks for sweet cherry 'Kordia'. *Acta Horticulturae* 1058, 531–536.
Magein, H. and Druart, P. (2005) Growth, flowering, and fruiting potentials of new dwarfing rootstocks in the progeny of Damil. *Acta Horticulturae* 667, 253–260.
Magyar, L. and Hrotkó, K. (2008) *Prunus cerasus* and *Prunus fruticosa* as interstocks for sweet cherry trees. *Acta Horticulturae* 795, 287–292.
Magyar, L. and Hrotkó, K. (2013) The effect of rootstock and spacing on the growth and yield of 'Kántorjánosi' sour cherry variety in intensive orchard. *Acta Horticulturae* 981, 373–378.
Maurer, E. (1939) *Die Unterlagen der Obstgehölze*. Paul Parey, Berlin.
Meland, M. (1998) Yield and fruit quality of 'Van' sweet cherry in four high density production systems over 7 years. *Acta Horticulturae* 636, 425–432.
Meland, M. (2015) Performance of the sweet cherry cultivar 'Lapins' on 27 rootstocks growing in a Northern climate. In: Proceedings of the EU COST Action FA1104: Sustainable production of high-quality cherries for the European market. COST meeting, Trebinje, Bosnia and Herzegovina, February 2015. Available at: https://www.bordeaux.inra.fr/cherry/docs/dossiers/Activities/Meetings/2015%2002%2010-11%20WG2%20Meeting_Trebinje/Presentations/Meland_Trebinje2015.pdf (accessed 18 April 2017).
Micheyev, A.M., Revyakina, N.T. and Drozdova, L.A. (1983) Klonoviye podvoyi vishnyi I osobennosty ich razmnosheniya. *Sadovodstvo* 7, 28–29.
Misirli, A., Gülcan, R. and Tanrisever, A. (1996) The relation between tree vigour of *Prunus mahaleb* L. types and sieve tube size in phloem tissue. *Acta Horticulturae* 410, 227–232.
Moreno, M.A., Montanes, L., Tabuenca, M.C. and Cambra, R. (1996) The performance of Adara as a cherry rootstock. *Scientia Horticulturae* 65, 85–91.
Muna, A.S., Ahmad, A.K., Mahmoud, K. and Abdul-Rahman, K. (1999) *In vitro* propagation of a semi-dwarfing cherry rootstock. *Plant Cell, Tissue and Organ Culture* 59, 203–208.
Muramatsu, H., Shirogane, S., Ogano, R., Sawada, K., Inagawa, Y., Uchida, T. and Inoue, T. (2004) A new cherry [*Prunus avium*] rootstock variety Chishimadai 1 Go. In: Bulletin of the Hokkaido Prefectural Agricultural Experiment Stations, Japan.
Musacchi, S., Gagliardi, F. and Serra, S. (2015) New training systems for high-density planting of sweet cherry. *HortScience* 50, 59–67.
Negueroles, P.J. (2005) Cherry cultivation in Spain. *Acta Horticulturae* 667, 293–301.
Neumüller, M. (2009) Kleine Baume, grosse Früchte. *Obstbau*, 257–259.
Nyújtó, F. (1987) Az alanykutatás hazai eredményei. *Kertgazdaság* 19, 9–34.
Ognjanov, V., Ljubojević, M., Ninić Todorović, J., Bošnjaković, D., Barać, G., Čukanović, J. and Mladenović, E. (2012) Morphometric diversity in dwarf sour cherry germplasm in Serbia. *Journal of Horticultural Science and Biotechnology* 87, 117–122.
Olmstead, J.W., Whiting, M.D., Sebolt, A.M. and Iezzoni, A.F. (2011) Selecting and fingerprinting the next generation of size-controlling rootstocks for sweet cherry. *Acta Horticulturae* 903, 235–240.
Paprstein, F., Kloutvor, J. and Sedlak, J. (2008) P-HL dwarfing rootstocks for sweet cherries. *Acta Horticulturae* 795, 299–302.
Parnia, P., Mladin, G., Dutu, I., Movileanu, M. and Slamnoiu, T. (1997) Ameliorarea portaltoilor pentru cires si visin. In: Braniste, N. and Dutu, I. (eds) *Contributii Romanesti la Ameliorarea Genetica a Soiurilor si*

*Portaltoilor de Pomi, Arbusti Fructiferi si Capsuni*. Institutul de Cercetare – Dezvoltare pentru Pomicultura, Pitesti-Maracineni, Romania, pp. 141–146.

Perry, R.L. (1987) Cherry rootstocks. In: Rom, R.C. and Carlson, R.F. (eds) *Rootstocks for Fruit Crops*. Wiley, New York, pp. 217–264.

Plock, H. (1973) Die Bedeutung der *Prunus fruticosa* Pall. als Zwergunterlage für Süss- und Sauerkirschen. *Mitteilungen Klosterneuburg* 23, 137–140.

Proffer, T.J., Jones, A.L. and Perry, R.L. (1988) Testing of cherry rootstocks for resistance to infection by species of Armillaria. *Plant Disease* 72, 488–490.

Rejman, A. (1987) *Szkółkarstwo Roślin Sadowniczych*. PWRiL, Warsaw.

Robinson, T.L. (2005) Developments in high density sweet cherry pruning and training system around the world. *Acta Horticulturae* 667, 269–272.

Rozpara, E. (1994) Wpływ różnych wstawek na wzrost, owocowanie i wytrzymałość na mróz dwóch odmian czereśni. In: *Praca doktorska wykonana w Instytucie Sadownictwa i Kwiaciarstwa w Skierniewicach*, pp. 1–72.

Rozpara, E. (2005) *Intensywny sad Czereśniowy*. Hortpress, Warsaw.

Rozpara, E. (2013) Intensyfikacja uprawy czereśni (*Prunus avium* L.) w Polsce z zastosowaniem nowych odmian, podkładek i wstawek. In: *Zeszyty Naukowe Instytutu Ogrodnictwa. Monografie i Rozprawy*. Wyd. Instytutu Ogrodnictwa, Skierniewice, Poland, pp. 1–118.

Rozpara, E. and Głowacka, A. (2014) Jeśli czereśnie to nowocześnie. *Sad Nowoczesny* 11, 7–12.

Rozpara, E. and Grzyb, Z.S. (1999) Możliwość wykorzystania wstawek skarlających w intensywnej produkcji czereśni. *Zeszyty Naukowe Akademii Rolniczej w Krakówie* 351, 291–295.

Rozpara, E. and Grzyb, Z.S. (2004) Frutana® – a new interstock for sweet cherry trees. *Acta Horticulturae* 658, 247–250.

Rozpara, E. and Grzyb, Z.S. (2006) The effect of the 'Northstar' interstem on the growth, yielding and fruit quality of five sweet cherry cultivars. *Journal of Fruit and Ornamental Plant Research* 14, 91–96.

Rozpara, E., Grzyb, Z.S. and Zdyb, H. (1998) Growth and fruiting of two sweet cherry cultivars with different interstems. *Acta Horticulturae* 468, 345–352.

Rozpara, E., Grzyb, Z.S., Omiecińska, B. and Czynczyk, A. (2004) Results of 8 years of research on the growth and yield of three sweet cherry cultivars grafted on 'P-HL A' rootstock. *Acta Horticulturae* 663, 965–968.

Sansavini, S. and Lugli, S. (1996) Performance of the sweet cherry cultivar 'Van' on new clonal rootstocks. Acta *Horticulturae* 410, 363–371.

Sansavini, S. and Lugli, S. (2014) New rootstocks for intensive sweet cherry plantations. *Acta Horticulturae* 1020, 411–434.

Santos, A., Cavalheiro, J., Santos-Ribeiro, R. and Lousada, J.L. (2014) Rootstock and plant spacing influence sweet cherry growth and yield, under different soil and water conditions. *Acta Horticulturae* 1020, 503–511.

Sarger, J. (1972) Le bouturage ligneux de l'espèce *Prunus mahaleb*. In: *Convegno del Ciliegio*, Verona, 14–16 June, pp. 419–431.

Schimmelpfeng, H. (1996) Unterlagenzüchtung für Süßkirschen in Deutschland – die Weihenstephaner Arbeiten. *Schweizerische Zeitschrift für Obst- und Weinbau* 132, 331–334.

Schimmelpfeng, H. and Liebster, G. (1979) *Prunus cerasus* als Unterlage. *Gartenbauwissenschaft* 44, 55–59.

Schuster, M. (2004) Investigation on resistance to leaf spot disease, *Blumeriella jaapii* in cherries. *Journal of Fruit and Ornamental Plant Research* 12, 275–279.

Simon, G., Hrotkó, K. and Magyar, L. (2004) Fruit quality of sweet cherry cultivars grafted on four different rootstocks. *International Journal of Horticultural Science* 10, 59–62.

Sitarek, M. and Grzyb, Z.S. (2010) Growth, productivity and fruit quality of 'Kordia' sweet cherry trees on eight clonal rootstocks. *Journal of Fruit and Ornamental Plant Research* 18, 169–176.

Sitarek, M., Grzyb, Z.S. and Olszewski, T. (1998) The mineral elements concentration in leaves of two sweet cherry cultivars grafted on different rootstocks. *Acta Horticulturae* 468, 373–376.

Sotirov, D. (2005) Growth and reproductive characteristics of sour cherry cultivars grown on own roots and grafted on IK-M9 mahaleb rootstock. *Scientific Works of National Center for Agrarian Sciences* 3, 67–71.

Sotirov, D. (2012) Growth characteristics of Van sweet cherry cultivar grafted on six rootstocks. *Plant Science (Bulgaria)* 49, 55–60.

Sotirov, D.K. (2015) Performance of the sweet cherry cultivars 'Van' and 'Kozerska' on clonal rotstocks. *Acta Horticulturae* 1099, 727–733.

Spornberger, A., Hajagos, A., Modl, P. and Végvári, G. (2015) Impact of rootstocks on growth, yield and fruit quality of sweet cherries (*Prunus avium* L.) cv. 'Regina' and 'Kordia' in a replanted cherry orchard in Eastern Austria. *Erwerbs-Obstbau* 57, 63–69.

Stachowiak, A., Świerczyński, S. and Kolasiński, M. (2014) Growth and yielding of sweet cherry trees grafted on new biotypes of *Prunus mahaleb* (L.). *Acta Scientiarum Polonorum Hortorum Cultus* 13, 131–143.

Stachowiak, A., Bosiacki, M., Świerczyński, S. and Kolasiński, M. (2015) Influence of rootstocks on different sweet cherry cultivars and accumulation of heavy metals in leaves and fruit. *Horticultural Science* 42, 193–202.

Stehr, R. (2005) Experiences with dwarfing sweet cherry rootstocks in Northern Germany. *Acta Horticulturae* 667, 173–178.

Stortzer, M. and Grossman, G. (1988) Neue Möglichkeiten für wuchshemmende Zwischenveredlungen bei Süsskirschen. *Gartenbau* 35, 241–243.

Szabó, V., Németh, Z., Sárvári, A., Végvári, G. and Hrotkó, K. (2014) Effects of biostimulator and leaf fertilizers on *Prunus mahaleb* L. stockplants and their cuttings. *Acta Scientiarum Polonorum Hortorum Cultus* 13, 113–125.

Tanasescu, N., Mladin, G.H., Sumedrea, D., Marin, F.C. and Popescu, C. (2013) Preliminary results on growth and fruiting of trees, grafted on the Romanian vegetative rootstock 'IP-C-7' and German vegetative rootstock 'Gisela 5', in an intensive sweet cherry orchard. *Acta Horticulturae* 981, 361–366.

Tatarinov, A.N. and Zuev, V.F. (1984) *Pitomnik Plodovyh y Yagodnych Kultur*. Rosselhosisdat, Moscow.

Thomas, M. and Sarger, J. (1965) Sélection de *Prunus mahaleb* porte greffe du cerisier. In: *Rapport General du Congrès Pomologique de Bordeaux*, France, pp. 175–201.

Trefois, R. (1980) New dwarfing rootstocks for cherry trees. *Acta Horticulturae* 114, 208–217.

Treutter, D., Feucht, W. and Liebster, G. (1993) *40 Jahre Wissenschaft für den Obstbau in Weihenstephan*. Obst- und Gartenbauverlag, Munich, Germany.

Tukey, R.B. (1964) *Dwarfed Fruit Trees*. Macmillan, New York.

Vercammen, J. (2004) Dwarfing rootstock for sweet cherries. *Acta Horticulturae* 658, 307–311.

Vogel, T. (1997) Süsskirschen-Unterlagenversuch in Buckenreuth-Fränkirsche Schweiz. *Obstbau* 22, 246–247.

Walther, E. and Franken-Bembenek, S. (1998) Evaluation of interspecific cherry hybrids as rootstocks for sweet cherries. *Acta Horticulturae* 468, 285–290.

Webster, A.D. (1980) Dwarfing rootstocks for plums and cherries. *Acta Horticulturae* 114, 201–207.

Webster, A.D. (1996) Propagation of sweet and sour cherries. In: Webster, A.D. and Looney, N.E. (eds) *Cherries: Crop Physiology, Production and Uses*. CAB International, Wallingford, UK, pp. 167–202.

Webster, A.D. (1998) Strategies for controlling the tree size of sweet cherry trees. *Acta Horticulturae* 410, 229–240.

Webster, A.D. and Lucas, A. (1997) Sweet cherry rootstock studies: comparisons of *Prunus cerasus* L. and *Prunus* hybrid clones as rootstocks for Van, Merton Glory and Merpet scions. *Journal of Horticultural Science* 72, 469–481.

Webster, A.D. and Schmidt, H. (1996) Rootstocks for sweet and sour cherries. In: Webster, A.D. and Looney, N.E. (eds) *Cherries: Crop Physiology, Production and Uses*. CAB International, Wallingford, UK, pp. 127–163.

Wertheim, S.J. (1998) *Rootstock Guide*. Fruit Research Station, Wilhelminadorp, The Netherlands.

Westwood, M.N. (1978) Mahaleb × Mazzard hybrid cherry stocks. *Fruit Varieties Journal* 32, 39–39.

Wharton, P.S., Iezzoni, A. and Jones, A.L. (2003) Screening cherry germplasm for resistance to leaf spot. *Plant Disease* 87, 471–477.

Whiting, M.D. and Lang, G.A. (2004) 'Bing' sweet cherry on the dwarfing rootstock 'Gisela 5': thinning affects fruit quality and vegetative growth but not net $CO_2$ exchange. *Journal of the American Society for Horticultural Science* 129, 407–415.

Wicks, T.J., Bumbieris, M., Warcup, J.H. and Wallace, H.R. (1984) *Phytophthora* in fruit orchards in South Australia. In: Biennial Report of the Waite Agricultural Research Institute, Adelaide, Australia, p. 147.

Wójcicki, J.J. (1991) *Prunus × stacei* (Rosaceae), a new spontaneous hybrid of *P. fruticosa, P. cerasus* and *P. avium*. *Fragmenta Floristica et Geobotanica* 35, 139–14.

Wolfram, B. (1971) Unterlagenzüchtung für Süsskirschen. *Der Neue Deutsche Obstbau* 17, 3–4.

Wolfram, B. (1996) Advantages and problems of some selected cherry rootstocks in Dresden-Pillnitz. *Acta Horticulturae* 410, 233–237.

Yoltuchovski, M.K. (1977) Proisvodstvenno-biologicheskaya characteristika raslychnych form podvoyev vishnyi. In: Andryushtchenko, D.P. (ed.) *Plodovoje Pitomnikovodstvo Moldavii*. Redact. Isdatelstvo Krtya Moldavenyanske, Kishinau, Moldavia, pp. 39–69.

Zepp, L. and Szczygiel, A. (1985) Pathogenicity of *Pratylenchus crenatus* and *Pratylenchus neglectus* to three fruit tree seedling rootstocks. *Fruit Science Reports* 12, 109–117.

# 7 Rain-Induced Cracking of Sweet Cherries

**Moritz Knoche* and Andreas Winkler**
*Institute for Horticultural Production Systems, Leibniz-University Hannover, Hannover, Germany*

## 7.1 Introduction

Rain-induced cracking is probably the most serious limitation to sweet cherry production in almost all regions where this high-value crop is grown. Cracking occurs during or after rainfall and usually shortly before harvest. Cracking may result in complete crop failure. As a general rule, if the canopy contains above about 25% cracked fruit, the harvest becomes uneconomic (Looney, 1985). This is due to the high labour cost associated with eliminating the cracked fruit, both during picking (in the orchard) and also during subsequent grading (in the packhouse). Furthermore, after rainfall, even the uncracked fruit has much decreased storage quality, despite its macroscopically intact surface. This is because surface wetness also causes the formation of numerous microscopic cracks or microcracks in the cuticle, which bypass its barrier function and result in any, or several, of the following: increased incidence of fruit rot (Børve *et al.*, 2000), increased water uptake during rainfall (Knoche and Peschel, 2006), increased transpiration both pre- and postharvest (Knoche *et al.*, 2002; Beyer *et al.*, 2005), and loss of firmness and impaired appearance (shrivelled and dull) and thus reduced market appeal and price.

Despite considerable research over many years, the exact mechanism of cracking of sweet cherries is unknown. Nevertheless, for at least two centuries (e.g. von Wetzhausen, 1819), the close relationship between the incidence of rainfall and the incidence of cracking has been well known. A comprehensive compilation of such research is given in a number of reviews that summarize the studies on fruit cracking (e.g. Sekse, 1995a, 1998, 2008; Christensen, 1996; Sekse *et al.*, 2005; Simon, 2006; Balbontín *et al.*, 2013; Khadivi-Khub, 2015). Most of the research cited in these reviews is descriptive or correlative.

In this chapter, we will focus on the research that concentrates on the mechanistic processes potentially associated with rain-induced cracking. In particular, we will look at quantitative studies of: (i) fruit water relationships (including surface water transfers through the fruit skin and vascular water transfers through the fruit pedicel, and also fruit and tissue water potentials and the components thereof); and (ii) the development of the fruit skin (including the skin's

* moritz.knoche@obst.uni-hannover.de

mechanical properties and the primary determinants of these properties). We will summarize this quantitative/mechanistic research and relate it to earlier and current hypotheses regarding the mechanistic causes of rain-induced cracking.

It is worth noting that, while the focus of this chapter is almost entirely on sweet cherries, the problem of rain-induced cracking is a large problem in general among the world's commercial fruit crop species. Although for some fruit crop species (e.g. kiwifruit, apples, pears) rain-induced cracking is a minor problem, or does not occur at all, for many others its importance varies from significant (tomatoes and grapes) to major (many stone fruit and berry fruit species). Hence, the physical/physiological mechanisms explored in this chapter have potential application to a much wider problem (biologically) and to a much more serious problem (economically) than occurs with sweet cherries. Cherries are a minor crop (in terms of both tonnage and value) compared with some other rain-cracking-susceptible fruit crop species such as wine grapes and tomatoes.

## 7.2 Types of Cracks

Most evaluations of fruit cracking rely on quantifying the percentage of 'cracked fruit' or 'split fruit' based on macroscopic inspection of a representative sample of fruit. In a few studies, attempts were made to distinguish between different categories of cracks and to relate the different categories and positions of cracks on the surface to their potential causes (Measham *et al.*, 2009, 2010, 2014) or to their genetic background (Quero-García *et al.*, 2014). We therefore briefly review the various categories (or types) of cracks.

### 7.2.1 Cracks by size

*Microscopic cracks (microcracks)*

The sweet cherry fruit skin comprises cuticle, epidermis and several hypodermal cell layers (Fig. 7.1A). Microcracks are cracks in the skin that are limited to the cuticle and do not extend into the underlying epidermal and hypodermal cell layer (Fig. 7.1B, C; Peschel and Knoche, 2005). Microcracks are usually not detectable by visual inspection with the naked eye. Only in severe cases is an altered light reflection of a concentric ring pattern around the stylar scar a reliable indicator of massive microcracks in that region. Proper and sensitive detection of microcracks requires incubating the fruit in, for example, aqueous solutions of fluorescent tracers such as acridine orange and subsequent inspection of the surface by fluorescence microscopy (Peschel and Knoche, 2005). Water and fluorescent tracer infiltrate openings in the cuticle such as cracks and, if not sealed, the stem/fruit junction (Fig. 7.1B, C; Peschel and Knoche, 2005). In the presence of Silwet L-77 (Knoche, 1994) or hydrostatic pressure (Knoche and Peschel, 2002), mass flow of dye solution through open stomata occurs (Knoche, 1994) and this may make observation of infiltrated microcracks more difficult (Fig. 7.1D).

Microcracking may be quantified either by counting the number of total cracks, or by measuring their total length or the number of small and large cracks (see rating schemes by Peschel and Knoche, 2005; Sekse, 1995b), or by determining the area of the infiltration zone around a crack as done by Grimm *et al.* (2012a) in apple. Care must be taken when using automatic detection by image analysis, because of fluorescence of stomata and/or dye or dust particles. In the absence of surfactants, fluorescence of stomata is probably due to uptake of the dye along polar pathways associated with the stomatal apparatus (Franke, 1964, 1967; Weichert and Knoche, 2006a) rather than to mass flow through the stomatal pore (Schönherr and Bukovac, 1972).

Fruit with microcracks is usually not detected by visual inspection in the marketing chain. Hence, microcracks appear to have no immediate consequence for the market value of crop. However, the impaired barrier function of a cuticle with microcracks has dramatic consequences for the fruit's shelf life. A number of negative influences result from microcracking. The incidence of fungal

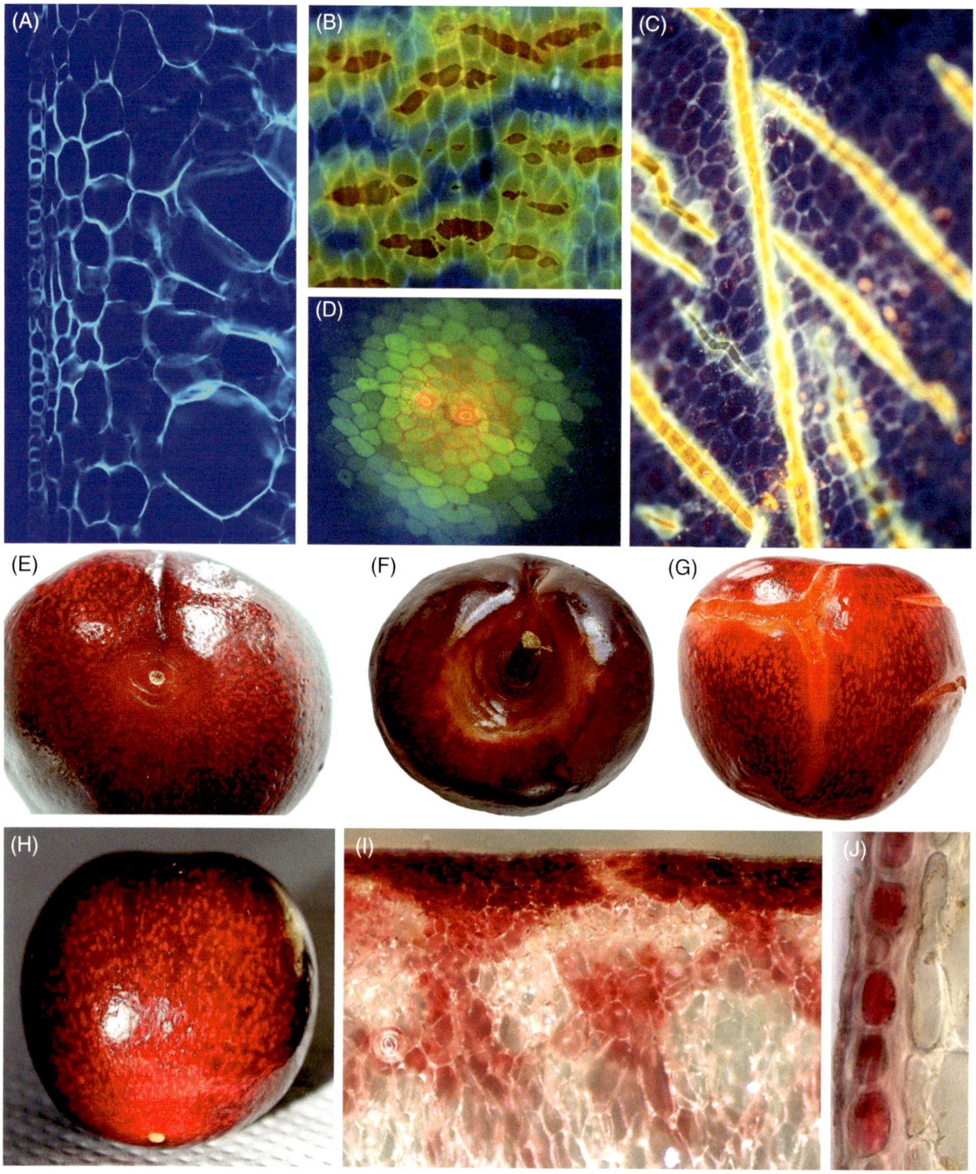


**Fig. 7.1.** (A) Micrograph of cross-section of the skin of a mature fruit after staining with calcofluor white. The skin comprises cuticle, epidermis and several hypodermal cell layers. (B, C) Microcracks in the cuticle after infiltration by acridine orange solution. (D) Stomata after infiltration by acridine orange solution. (E–G) Macroscopic cracks by position: in the stylar scar region (E), in the stem cavity region (F) and in the cheek region (G). (H–J) Strain spots on the surface of mature fruit: overview (H), cross-section through fruit skin and underlying flesh (I), and cross-section through the epidermis (with anthocyanin) and first layer of hypodermis (without anthocyanin) (J). (Reproduced with permission from Knoche, 2015 (C, D) and Grimm *et al.*, 2013 (H–J).)

infections increases (Børve *et al.*, 2000). In the packhouse, sweet cherries are commonly floated in water to cool them, wash them and transport them from the point of entry to the point of packing. Microcracking increases the rate of water uptake during this grading/packing process (Knoche and Peschel, 2006). Transpiration is also increased by microcracking so the fruit lose firmness and shine more quickly and suffer increased shrivel.

The frequency with which microcracks appear is positively correlated to the duration of surface wetness. Fruit grown under rain shelters or in greenhouses typically have a lower density of microcracks. Hence, in some regions, fruit grown under protected/semi-protected conditions receive a higher price at the market. It is interesting that cuticular microcracks do not significantly affect the fruit skin's mechanical properties (Brüggenwirth *et al.*, 2014).

#### *Macroscopic cracks (macrocracks)*

Macrocracks are cracks in the skin that traverse the cuticle and extend into the epidermal and hypodermal cell layers, possibly into the flesh and occasionally down to the pit. Macrocracks are visible with the naked eye (Fig. 7E–G). They 'gape' because the fruit skin in mature fruit is strained elastically and this strain is released when a crack occurs (Grimm *et al.*, 2012b). Macrocracks are thought to have their origins in microcracks (Glenn and Poovaiah, 1989). This idea is plausible, although direct experimental evidence is lacking. Microcracks impair the barrier function of the skin and thus increase water uptake (Knoche and Peschel, 2006). This may, in turn, cause the formation of a macrocrack at the site of water uptake. From a market perspective, fruit with small macrocracks around the stylar scar (Fig. 7E) or in the stem cavity (Fig. 7F) are tolerated as long as there is no fungal decay, but fruit with larger macrocracks on the cheek and/or suture (Fig. 7G) are usually rejected.

### 7.2.2 Cracks by position

Three different types of macrocracks are reported in the literature (Christensen, 1996): (i) cracks at the stylar scar (also known as apical cracks or nose cracks) (Fig. 7.1E); (ii) cracks in the stem cavity or around the rim of the stem cavity (also known as ring cracks) (Fig. 7.1F); and (iii) cracks on the cheek or suture side of the fruit (Fig. 7.1G).

Macrocracks around the stylar scar and in the stem cavity are usually the first visible cracks to appear on a fruit. Both of these positions also exhibit the first and most severe microcracking (Peschel and Knoche, 2005). Cracks on the cheek are often just an elongation of pre-existing apical or ring cracks (Verner and Blodgett, 1931; Glenn and Poovaiah, 1989).

Preferential cracking in the stylar-scar and pedicel cavity regions may be caused by one or several of the following factors:

**1.** Surface wetness induces microcracking and both regions exhibit extended wetness duration (Knoche and Peschel, 2006). During and after rainfall, a pendant water drop often collects at the bottom (stylar end) of the fruit, while at the top, a puddle collects in the pedicel cavity.
**2.** The pedicel/fruit junction (Beyer *et al.*, 2002b) and apex (Glenn and Poovaiah, 1989) are sites of preferential water uptake.
**3.** The stylar-scar and pedicel cavity regions exhibit marked curvatures. Small radii of curvature concentrate stresses and hence increase the likelihood of failure (Considine and Brown, 1981).
**4.** The stylar scar and the pedicel/fruit junction are stiffer than the rest of the skin. Stiffness focuses the stresses in the skin immediately adjacent to these structures. The resulting stress concentrations may cause failure as has been demonstrated for lenticels on the grape berry surface (Brown and Considine, 1982).

Measham *et al.* (2010) related the type of cracking to the dominant pathway of water entry into the fruit. Thus, irrigating the soil around trees while maintaining a dry canopy caused deep side-cracking, whereas the same amount of water deposited by overhead sprinkler resulted in small cracks in the stem cavity or at the stylar end of the fruit, but no side-cracking.

### 7.2.3 Mode of failure

To the best of our knowledge, there is limited published information on the morphology of fracture surfaces in sweet cherries (Glenn and Poovaiah, 1989; Weichert *et al.*, 2004). Considering the diagnostic importance of investigating fracture surfaces in materials science, this is surprising.

The first detectable microcracks most often form above the periclinal cell walls of the epidermal cells, rather than above the anticlinal ones (Peschel and Knoche, 2005). Microcracks are mostly orientated perpendicular to the longitudinal axis of the underlying epidermal cell. Epidermal cells below a microcrack do not differ in dimensions or orientation from neighbouring cells or cells at some distance from the crack. These observations suggest that microcracks in the cuticle do not release any strain from the cells below. This conclusion is consistent with the finding that it is epidermal and hypodermal cells, rather than the cuticle, that form the structural 'backbone' of a sweet cherry fruit (Brüggenwirth *et al.*, 2014).

To the best of our knowledge, there is no published information on the mode of failure of the epidermis and hypodermis. Unpublished results from our laboratory indicate that the dominant failure mode of fruit submerged in water is along cell walls, rather than across cell walls. When confirmed, this observation must be interpreted as indicating some failure of cell-to-cell adhesion because the pectin middle lamellae are likely to be the weakest links in a skin under tension (M. Brüggenwirth, unpublished data).

Glenn and Poovaiah (1989) observed a loosening of the cuticle from the underlying cell wall before a fracture occurred. Thus, the cuticle is only loosely attached to the underlying pectin layer, which would be consistent with a hypothetical failure of the pectin middle lamellae in mature, cracking-susceptible fruit.

## 7.3 Quantifying Cracking

Assessing the cracking susceptibility of a batch of fruit is often required to make cultivar comparisons for the purposes of breeding, consulting or research. Ideally, a standardized protocol would be used that allows a reproducible *in vitro* quantification of cracking susceptibility in the laboratory that perfectly reproduces the *in vivo* observations in the field. Unfortunately, such a protocol is not available. Until the mechanistic basis of cracking is understood, any assessment of cracking susceptibility using detached fruit in the laboratory can only approximate that occurring on the tree in the field, under naturally rainy conditions. In the following sections, we describe some of the tests in current use.

### 7.3.1 Quantifying cracking in the orchard

#### *Measuring cracking in the orchard after rainfall*

The easiest and most realistic way to quantify cracking is to determine the percentage of cracked fruit after a rain event. All fruit is picked from the tree and partitioned into cracked versus non-cracked fractions (Quero-García *et al.*, 2014). To avoid confounding with ripening stages, it is advisable to assess cracking percentages within different ripening stages as indexed by colour (A. Lang, Eastbourne, NZ, 2016, personal communication). The disadvantage of this procedure is the lack of control of the amount, distribution and duration of rainfall. Furthermore, the stage of maturity is difficult to define reproducibly for the non-climacteric sweet cherry, yet maturity has a significant effect on cracking susceptibility (Christensen, 1996). Finally, orchard factors such as crop load and environmental variables such as temperature are important confounding variables that can affect cracking and that usually are not standardized (Christensen, 1996; Measham *et al.*, 2012).

#### *Inducing cracking under artificial rain*

Cracking may also be induced by artificial 'rain' using overhead sprinklers and either orchard or potted trees (Quero-García *et al.*, 2014). As with field assessments, the fruit remains attached to the tree but, in contrast to the field, the timing, duration and intensity of the 'rain' are adjustable. Performing experiments using potted trees in a greenhouse or growth chamber also allows control of related environmental factors such as light, temperature and humidity. To obtain reproducible results, the use of deionized water, or at least rain water of neutral pH, is mandatory. Even low concentrations of calcium (<1 mm) may inhibit cracking (Christensen, 1972d). These concentrations are often reached or exceeded in tap water.

### 7.3.2 Laboratory-based assessments of cracking

Classical cracking tests are performed in the laboratory by immersing detached fruit in water and subsequent inspection for macroscopic cracks. Such assays are often used for head-to-head comparisons of cultivars or of treatment effects such as the effects of pH, organic acids (Winkler *et al.*, 2015), temperature (Bullock, 1952), minerals (Christensen, 1972d; Weichert *et al.*, 2004) and fruit size (Christensen, 1975).

#### *Cracking index (CI)*

The most popular test is the determination of the CI. The original test was established by Verner and Blodgett (1931) and modified by Christensen (1972b).

Briefly, 50 fruit, free of visual defects, are picked in the morning and brought to the laboratory within 1 h. Here, the fruit are immersed in distilled water at a constant temperature. After 2, 4 and 6 h, the fruit are inspected for macroscopic cracks. Those without cracks are re-incubated, and cracked ones are removed and counted. The CI is calculated as:

$$\text{CI} = \frac{(5a + 3b + c) * 100}{250}$$

In this equation, $a$, $b$ and $c$ represent the number of cracked cherries after 2, 4 and 6 h, respectively. The equation means that cultivars that crack sooner show a higher CI than those that crack more slowly (for the same overall percentage of fruit cracking by the end of the test).

Time courses of cracking usually follow a sigmoidal pattern. Using appropriate regression models, the time to half-maximum cracking ($T_{50}$, h) may be calculated by analogy to a half-time in radioactive decay. The CI carries information on both the kinetics and the percentage of cracking, while the $T_{50}$ only gives the time to reach half-maximum cracking. The CI and $T_{50}$ are thus both more informative than a simple assessment of the percentage of fruit cracked after a fixed time.

For a detailed description of how to determine the CI, see Christensen (1972b, 1996).

#### *Intrinsic cracking susceptibility*

Determination of intrinsic cracking susceptibility was introduced by Weichert *et al.* (2004). The underlying concept is that cracking susceptibility is determined by a combination of the fruit skin's water uptake characteristics and its mechanical characteristics (Winkler *et al.*, 2015). Thus, cracking may result from high water uptake and/or from a mechanically weak fruit skin. The intrinsic cracking susceptibility expresses cracking as a function of water uptake. It therefore focuses on the mechanical properties of the fruit skin, which are independent of the fruit's water uptake characteristics.

The test implies that uptake increases at a constant rate. This is usually the case for the time periods over which most of these tests are run (Beyer *et al.*, 2005). An intrinsic cracking assay requires measurements of: (i) the time course of cracking as is typically done for CI determinations; and (ii) the rate of water uptake in fruit of the same batch. For unbiased results, the pedicel end and the pedicel/fruit junction should be treated in the same way in both the cracking and the uptake assessments. Usually, two samples of 25 fruit each are incubated in distilled water and inspected for cracks until all fruit are cracked or beginning to rot. The $T_{50}$ (h) is calculated from the time course data of cumulative cracking versus time using appropriate regression models. For water uptake, 15 representative fruit (e.g. same mass, maturity) from the same batch are selected and incubated individually in deionized water. Fruit is removed from the solution at 0, 0.75 and 1.5 h, blotted dry using tissue paper, weighed and then re-incubated. The rate of water uptake is calculated as the slope of a linear regression fitted through a plot of cumulative mass versus time (Fig. 7.2A). Multiplying the $T_{50}$ by the mean rate of water uptake ($R$; mg h$^{-1}$) yields the amount of water taken up at 50% fruit cracking ($WU_{50}$, mg):

$$WU_{50} = R * T_{50}$$

Because the $WU_{50}$ is standardized for the amount of water taken up, the $WU_{50}$ is an indirect measure of the extensibility of the fruit skin on a whole-fruit basis.

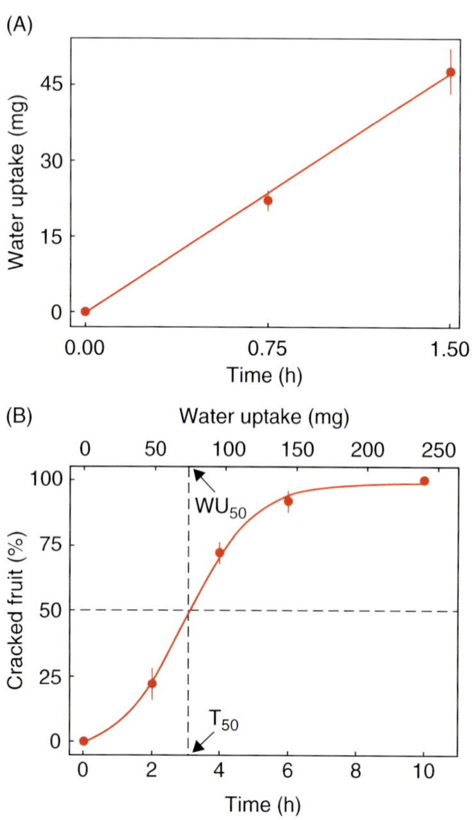


**Fig. 7.2.** (A) Time course of water uptake. (B) Percentage of fruit cracked as a function of time and as a function of water uptake. The dashed horizontal line indicates 50% cracking. Arrows indicate the time to 50% cracking ($T_{50}$) and the amount of water taken up at the $T_{50}$ ($WU_{50}$). The $WU_{50}$ is inversely related to the cracking susceptibility of the respective batch of fruit.

### 7.3.3 Opportunities and limitations of laboratory-based cracking assays

To the best of our knowledge, the only direct comparison of different methods for assessing the cracking susceptibility of sweet cherry was performed by Quero-García *et al.* (2014). Cracking susceptibility of different cultivars was compared for field assessments, artificial rain application in a tunnel and CI determinations in the laboratory in two seasons. Cracking was assessed by position. The highest year-to-year coefficients of correlation among the different methods were obtained for cracking at the stylar-scar end ($r = 0.49$ and $r = 0.44$ for artificial rain and field assessment, $P<0.01$ for both), followed by cracking in the pedicel cavity region ($r = 0.32$ and $r = 0.26$ for artificial rain and field assessment, $P<0.01$ for both). All three methods were closely correlated with coefficients of correlation ranging from 0.41 to 0.53 ($P<0.01$ for both) for cracking in the stylar-scar region. Coefficients of correlation were lower for cracking in the stem cavity and significant only for field/CI ($r = 0.25$, $P<0.01$) and artificial rain/CI assessments ($r = 0.32$, $P<0.01$). None of these relationships was significant for the cheek (Quero-García *et al.*, 2014). From these data, it may be concluded that surface wetness duration plays an important role in 'on-tree cracking' (field assessments, artificial rain) and that the CI is most suited for simulating this effect.

Unfortunately, cracking susceptibility of cultivars at different sites as indexed by the CI is not consistently correlated. For example, the CI values determined in Denmark (Christensen, 1996) were significantly correlated with those from Norway ($n = 13$, $r = 0.66$, $P<0.05$) or with ratings scores from field observations by the Federal Fruit Variety Office in Germany ($n = 38$, $r = 0.40$, $P<0.05$). However, there was no correlation with the CI determined in Oregon, USA ($n = 18$, $r = 0.37$; data from Zielinski (1964), cited in Christensen (1996)) or Spain ($n = 13$, $r = 0.24$; data from Tabuenca and Cambra

(1982), cited in Christensen (1996)). Low *r* values and lack of significance reflect large variability, possibly resulting from environmental effects (e.g. temperature, precipitation, duration of surface wetness) and the difficulty of standardizing maturity in a reproducible way. The environment must play some role in this, as may be inferred from the significance of correlations between sites having similar climates, such as Denmark, Norway and Germany. Reliable and reproducible CI assessments of cultivars probably require: (i) repeated CI determinations during maturation, possibly in two to three seasons (Christensen, 1996); (ii) proper experimental designs for head-to-head comparisons using appropriate controls; and (iii) tests of fruit grown under representative climatic conditions.

## 7.4 Factors Affecting Cracking

Numerous studies have been published that investigate the effects of a range of factors on cracking. The results obtained are summarized in Table 7.1. For more detailed information, the reader is referred to the original references or to the reviews cited in the introduction.

## 7.5 Cracking from a Mechanistic Perspective

Cracking is assumed to be caused by a net influx of water into the fruit. For cracking to occur, two conditions must be met. First, the fruit must increase in volume and hence surface area, thereby subjecting the skin to marked strain. Second, the strained fruit skin must rupture. Thus, two groups of mechanistically unrelated factors affect cracking: (i) factors that affect the mechanical properties of the fruit skin; and (ii) factors that affect fruit water inflows and outflows.

In the following sections, we review the recent literature on the anatomy and development of the fruit skin, its mechanical architecture, the fruit water potential and its components, and water transport in the pedicel vascular system and across the fruit surface.

### 7.5.1 Morphology and development of fruit skin

#### *Fruit skin and flesh*

The fruit skin is a complex material composite comprising a polymeric layer, i.e. the cuticular membrane (CM), and cellular layers, i.e. the epidermis and hypodermis (Fig. 7.1A).

The CM is a lipophilic composite of polyesters deposited on the outer cell wall of the epidermis. It comprises the cutin matrix, soluble cuticular lipids (referred to as wax) and polysaccharides in its inner surface. Compared with the CM of other fruit crops, the CM of a sweet cherry is very thin. Across 31 sweet cherry cultivars, the CM mass per unit area averaged 1.28±0.01 g $m^{-2}$ and ranged from 0.85±0.04 in 'Rainier' to 1.58±0.06 g $m^{-2}$ in 'Rube' at maturity (Peschel and Knoche, 2012).

The cutin matrix is a natural biopolyester of mostly C16 (69.5%) and C18 (19.4%) alkanoic, ω-hydroxyacids, α,ω-dicarboxylic and midchain hydroxylated acids (Peschel *et al.*, 2007). The two most abundant constituents of sweet cherry cutin are 9(10),16-dihydroxyhexadecanoic acid (53.6%) and 9,10,18-trihydroxyoctadecanoic acid (7.8%). There was no qualitative difference in cutin or wax composition among 'Hedelfinger', 'Kordia', 'Sam' and 'Van' (Peschel *et al.*, 2007). Also, there was no relationship between the cracking susceptibility of the fruit and the qualitative or quantitative composition of the cutin or wax fractions of the cuticle.

Within the wax fraction, triterpenes (75.6%) are the most abundant, followed by alkanes (19.1%) and alcohols (1.2%). Ursolic and oleanolic acid dominate the triterpene fraction and nonacosane and heptacosane the alkane fraction. The most abundant alcohol is nonacosanol (Peschel *et al.*, 2007). Wax occurs as embedded cuticular wax (25.6% of total CM mass; M. Hinz, unpublished data) impregnating the cutin matrix and as epicuticular wax (8.6% of total CM mass; M. Hinz, unpublished data) deposited as an amorphous film on the fruit surface. Wax mass per unit area averaged 0.33±0.00 g $m^{-2}$ (range 0.21±0.01 g $m^{-2}$ in 'Bing' to 0.42±0.04 g $m^{-2}$ in 'Zeppelin', all at maturity; Peschel and Knoche, 2012).

**Table 7.1.** Factors affecting the cracking susceptibility of sweet cherry fruit (data compiled from the literature).

| Factor | Levels | Susceptibility to cracking | Reference(s) |
|---|---|---|---|
| Fruit size | 2.8–10 g | Increased susceptibility in large fruited cultivars, high variability between cultivars | Tucker (1934); Christensen (1975); Yamaguchi *et al.* (2002) |
| Water uptake | | No, or only weak correlation with susceptibility | Kertesz and Nebel (1935); Christensen (1972a) |
| | | Positive correlation between rate of uptake and cracking | Belmans and Keulemans (1996); Yamaguchi *et al.* (2002) |
| Firmness | Fruit firmness (kg) 1–3.6, grade 3–10 | No correlation | Tucker (1934); Christensen (1975) |
| | Flesh firmness (g) 29.1–148.1 | Positive correlation between firmness and rate of cracking | Yamaguchi *et al.* (2002) |
| Temperature | 1–48°C | Higher temperature, faster cracking | Bullock (1952); Richardson (1998) |
| Osmolarity | 10.1–20% sugar | Positive correlation with susceptibility | Verner and Blodgett (1931) |
| | 12.8–26.4% soluble solids | Lack of or a weak correlation | Tucker (1934); Christensen (1972c); Moing *et al.* (2004) |
| Skin | Skin $g^{-1}$ of soluble solids | Higher mass of cell walls in less susceptible cultivars | Tucker (1934) |
| | Thickness of inner epidermal cell wall | Positive correlation between thickness of inner epidermal cell wall and susceptibility | Kertesz and Nebel (1935) |
| | Thickness of cuticle and epidermal cell wall (μm) 7.5–12.5 | Thicker cuticle and thicker epidermal cell wall in less susceptible cultivars | Belmans *et al.* (1990) |
| | Cuticle thickness (μm) 0.9–4.02 | Thicker cuticle in less susceptible cultivars | Belmans and Keulemans (1996); Demirsoy and Demirsoy (2004) |
| Cell size | Size of subepidermal cells | Smaller cell size in more susceptible cultivars; higher variability of cell size in flesh of susceptible cultivars | Kertesz and Nebel (1935) |
| | | Larger cells (longitudinally and latitudinally) in susceptible cultivars | Yamaguchi *et al.* (2002) |
| Stomata | Density of stomata | No correlation | Christensen (1972c); Glenn and Poovaiah (1989) |
| Rootstock | 'Colt', 'MaxMa 14', 'MaxMa 97' | Differences in cracking rates between cherries of different rootstocks after 4 and 24 h immersion in water | Simon *et al.* (2004) |
| | 'Colt', 'F 12/1' | More cracking on 'Colt' than on 'F 12/1' | Cline *et al.* (1995a,b) |
| Calcium salts | Immersion test, overhead sprinkler | Decreased susceptibility | Verner (1937); Christensen (1972d); Glenn and Poovaiah (1989); Meheriuk *et al.* (1991); Brown *et al.* (1995); Fernandez and Flore (1998); Lang *et al.* (1998); Wójcik *et al.* (2013); Erogul (2014) |

*Continued*

**Table 7.1.** Continued.

| Factor | Levels | Susceptibility to cracking | Reference(s) |
|---|---|---|---|
| | | No or minor effect | Looney (1985); Koffmann *et al.* (1996) |
| Aluminium, iron and copper salts | $Al_2(SO_4)_3$, $Al_3(PO_4)_2$, $CuSO_4$, $AlCl_3$, $FeCl_3$, $Fe(NO_3)_3$, $Fe_2(SO_4)_3$ | Decreased water uptake and less cracking, $WU_{50}$ differed from control | Bullock (1952); Christensen (1972d); Beyer *et al.* (2002a) |
| Growth regulators | 4 × 10 p.p.m. $GA_3$, 1 × 10 p.p.m. $GA_3$ | Increased cracking | Cline and Trought (2007) |
| | $GA_3$ at 20 p.p.m. | Decreased cracking | Demirsoy and Bilgener (1998) |
| | NAA at 0.5, 1 and 2 p.p.m. | Decreased cracking | Yamamoto *et al.* (1992); Demirsoy and Bilgener (1998) |
| Fungicides | Borax, captan, maneb | No effect | Christensen (1972d) |
| Coatings | Vaporgard™ at 2% 14 or 7 days before harvest | No or negative effect of tested coatings | Richardson (1998) |
| | Mobileaf, SureSeal, RainGard®, mixture of T1+T2[a] | Decreased susceptibility | Davenport *et al.* (1972); Kaiser *et al.* (2014); Meland *et al.* (2014); Torres *et al.* (2014); Dumitru *et al.* (2015) |

[a]T1: 1% calcium chloride, 1% zinc sulphate, 0.1% polyphenols extracted from *Vitis vinifera* seeds and 0.1% humic acid extracted from lignite; T2: solution of 1% galactomannan extracted from seeds of *Gleditsia triacanthos*; 1% calcium chloride; 1% zinc sulphate; 0.1% polyphenols extracted from *Vitis vinifera* seeds and 0.1% humic acid extracted from lignite. For details, see Dumitru *et al.* (2015).

The polysaccharide content of sweet cherry fruit cuticles has not been quantified. Typically, the polysaccharide content of cuticles is in the order of 18–26% (Schreiber and Schönherr, 1990).

The sweet cherry epidermis is formed by a single layer of small collenchyma-type cells with thick cell walls (Fig. 7.1A). In stage II fruit, the cell shape is more or less isodiametric but cells extend during fruit maturation such that their lengths (latitudinal and longitudinal) to width (radial) ratio increases. At the cheek, the diameters of cells in the longitudinal and latitudinal directions averaged 44.1±1.0 μm and 63.2±0.9 μm, respectively (mean of 38 cultivars at maturity; Yamaguchi *et al.*, 2002). Epidermal cells are preferentially orientated on the fruit surface, the orientation depending on position. Cells in the stem cavity are elongated longitudinally in the direction of the pedicel/stylar-scar axis, while those on the cheek are elongated latitudinally in the direction of the equator.

There are no trichomes or hairs. The surface of sweet cherry is stomatous, but stomatal density is low compared with the leaves. Stomatal density increases from 0.00 $mm^{-2}$ in the stem cavity to 1.71 $mm^{-2}$ in the stylar-scar region of mature 'Hedelfinger' fruit (Peschel *et al.*, 2003), compared with 458.2 and 542.7 $mm^{-2}$ in leaves from 'Summit' and 'Burlat', respectively (Gonçalves *et al.*, 2008). The number of stomata depends on cultivar and ranges from a low of 143±26 per fruit in 'Adriana' to a maximum of 2124±142 per fruit in 'Hedelfinger'. Stomata lose functionality during stage III (Peschel *et al.*, 2003). Bukovac *et al.* (1999) observed wax occlusions and plugging of stomatal pores. Unlike in plum, stomata of sweet cherry are not sites of preferential microcracking (Knoche and Peschel, 2007).

The hypodermis is formed by several layers of collenchymatous cells (Fig. 7.1A). Typically, cell walls are thick and the hypodermal cells are larger than the epidermal cells. Cell size increases with depth from those immediately beneath the epidermis to those adjacent to the flesh (Brüggenwirth and Knoche, 2016b).

The flesh comprises large, thin-walled and approximately isodiametric parenchymatous cells. Diameters average about 227.4±2.9 μm (mean of 53 cultivars in three seasons; Yamaguchi *et al.*, 2004).

### *Fruit growth, skin development and cuticle deposition*

The growth pattern of sweet cherry follows the classical double-sigmoidal pattern of stone fruit development (Lilleland and Newsome, 1934; Tukey, 1934). During stage I, cell division in the pericarp accounts for a small increase in mass to about 1.5–2.5 g per fruit. In stage II, the mass remains essentially constant, the endocarp lignifies and the embryo develops. Stage III ('final swell') represents the final phase of development characterized by a rapid increase in mass, primarily as a result of cell enlargement in the flesh (Fig. 7.3). Pit hardening and the onset of colour change mark the stage II/stage III transition. The beginning of stage III also coincides with a rapid decrease (greater concentration) in osmotic potential from about −0.7 MPa to less than −3 MPa at maturity (Knoche *et al.*, 2004). Maximum rates of growth during stage III averaged 0.54 g day$^{-1}$ and 0.96 mm$^2$ day$^{-1}$ (Knoche *et al.*, 2001). This is very high considering the small size of the fruit at mid-stage III.

During stage III, the fruit surface area increases and the skin becomes markedly strained. Evidence for elastic skin strain is based on the following observations:

**1.** Cutting into the fruit results in 'gaping' of the cut (Grimm *et al.*, 2012b).
**2.** Exocarp segments excised by cutting tangentially underneath the surface rapidly decrease in area (Fig. 7.4; Grimm *et al.*, 2012b).
**3.** The fruit surface has a mottled appearance, probably resulting from tensional failure

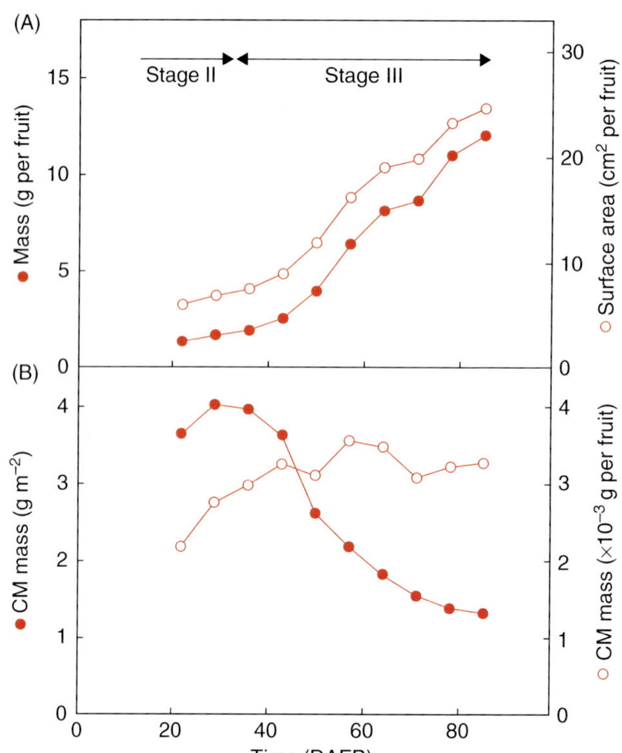


**Fig. 7.3.** Developmental time course of changes in fruit mass and surface area (A) and mass of the cuticular membrane (CM) (B) on the basis of unit fruit surface area and also the whole fruit. Stages II and stage III of fruit development are indicated. DAFB, days after full bloom. (Redrawn with permission from Peschel *et al.*, 2007.)

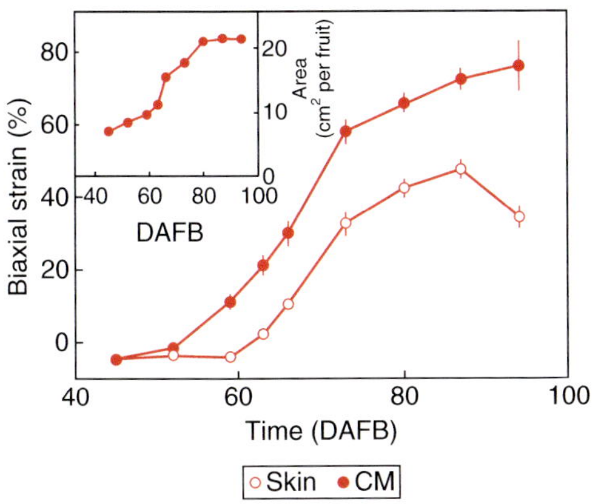


Fig. 7.4. Developmental time course of change in biaxial elastic strain of the fruit skin and the cuticular membrane (CM) of 'Hedelfinger' sweet cherry. Inset: time course of change in fruit surface area during stage III development. DAFB, days after full bloom. (Redrawn with permission from Grimm *et al.*, 2012b.)

during stage III. The hypodermal cell layer tears and separates from the epidermis in a manner analogous to the 'stretch marks' occurring in human skin in puberty, obesity and pregnancy (Fig. 7.1H–J; Grimm *et al.*, 2013).

**4.** Microcracks in the cuticle are orientated perpendicular to the longest dimension of the underlying epidermal cell, suggesting a cause-and-effect relationship (Peschel and Knoche, 2005).

**5.** The length-to-width ratio of epidermal and hypodermal cells increases from stage II to maturity, which is indicative of strain.

There is also considerable elastic strain in the CM. This increases from close to zero at the end of stage II to 80% at maturity. Over the same time period, the elastic biaxial strain in the skin composite (epidermis including cuticle plus hypodermis) increases to about 40% (Fig. 7.4; Grimm *et al.*, 2012b). Thus, the CM contains more elastic strain than the epidermis/hypodermis layer that undergoes cell division to accommodate the increase in fruit surface area. The rapid increase in CM strain is accompanied by a marked increase in the frequency and severity of microcracks in the cuticle (Peschel and Knoche, 2005). Recently, we discovered that extracting wax from the isolated CM resulted in significant further shrinkage. This observation suggests that: (i) as in the CMs of other crops, the wax 'fixes' the strain in the sweet cherry CM (Khanal *et al.*, 2013); and (ii) total biaxial elastic strain in the isolated and dewaxed CM of sweet cherry fruit may be as high as 159% (Lai *et al.*, 2016).

Cell division, cell enlargement and a continuous increase in the planar length/width ratio of skin cells accompany the increase in fruit volume and hence skin area during stage III (Knoche *et al.*, 2004). In contrast, the CM must follow the increase in area by strain only. From stage II development onwards, the mass of cutin and wax on a whole-fruit basis remains nearly constant, indicating the absence of significant deposition of new cutin or wax material (Knoche *et al.*, 2004; Peschel and Knoche, 2005; Peschel *et al.*, 2007). Thus, the increase in CM area during stage III redistributes a near constant amount of CM material over an enlarging surface. That the CM in sweet cherry fruit is markedly strained is based on the following observations: (i) the marked decrease in CM area following excision and isolation (Knoche *et al.*, 2004) – the decrease in CM area upon isolation exceeds that of the isolated fruit skin (Fig. 7.4; Grimm *et al.*, 2012b); and (ii) the formation of highly orientated cuticle microcracks, and the positive relationship between the surface area increase and the frequency and severity of microcracks in the CM (Peschel and Knoche, 2005).

The cessation of CM deposition during stage II and subsequent stage III development results from a downregulation of genes involved in cutin monomer and wax synthesis (Alkio *et al.*, 2012, 2014). There appears to be no genetic variability in the essential shutdown of cutin and wax deposition during stage II. Across 32 cultivars, a linear relationship (slope 1.14±0.10, $r^2$ = 0.90, $P$<0.001) was reported between the amount of CM per fruit during stage III and that during stage II (Peschel and Knoche, 2012).

### 7.5.2 Mechanical properties of fruit skin and cuticle

There are only a limited number of studies that quantify the mechanical properties of skins and/or cuticles in sweet cherry (Bargel *et al.*, 2004; Brüggenwirth *et al.*, 2014; Brüggenwirth and Knoche, 2016a,b). The behaviour of the sweet cherry fruit skin under test suggests that the fruit may be considered like a fluid-filled balloon held under slight pressure by the elastic strain in the skin. In this sense a sweet cherry is similar to a grape berry (Considine and Kriedemann, 1972; Considine and Brown, 1981).

In principle, excised skins and isolated cuticles may be tested mechanically in uniaxial or biaxial tensile tests. In uniaxial tensile tests, a force is applied in one direction, whereas in biaxial tests, specimens are loaded in multiple directions. For sweet cherry fruit skin, biaxial testing is essential for two reasons. First, the near-spherical shape of the fruit results from multiaxial strains requiring biaxial testing if natural strain caused by growth is to be mimicked. Second, due to its high Poisson ratio, uniaxial testing leads to a marked overestimation of (fracture) strains of the skin caused by the narrowing of the test specimen during extension, just like when stretching a knitted woollen garment (Brüggenwirth *et al.*, 2014).

The first biaxial test of a sweet cherry fruit skin was described by Bargel *et al.* (2004). In this test, an excised skin segment is pressurized from the inner side using water. As a consequence, the skin segment bulges. The pressure and extent of bulging are monitored. Brüggenwirth *et al.* (2014) modified the elastometer such that: (i) the *in vivo* strain of the excised fruit skin is preserved after excision; and (ii) any contact of water with the flesh at the inner side of the skin segment is avoided. The strain is maintained by first mounting a washer on the fruit surface *before* the skin segment is excised by a tangential cut beneath. The skin segment fixed in the washer is then mounted in the elastometer. Bursting of flesh and skin cells due to water uptake (Simon, 1977) is avoided by pressurizing the skin segment using silicone oil. The pressure and extent of bulging are monitored. A modulus of elasticity ($E$) may be calculated from the following equation (Brüggenwirth *et al.*, 2014):

$$E = \frac{p * r^2 * (r^2 + h^2)}{h^3 * t * 2}$$

In this equation, $r$ (mm) is the radius of the washer orifice, $p$ (MPa) is the pressure applied to the skin segment, $h$ (mm) is the height of the bulging skin segment and $t$ is the thickness of the load-bearing layer ($t = 0.1$ mm). The last value is obtained from direct observation (light microscope) or from the literature (Glenn and Poovaiah, 1989). A high $E$ value implies a stiff skin, while a low $E$ implies a skin offering limited resistance to extension.

The following conclusions were derived from biaxial tensile tests (Brüggenwirth *et al.*, 2014; Brüggenwirth and Knoche, 2016a,b):

**1.** The epidermis and hypodermis, and *not* the cuticle, represent the principal mechanical member of a sweet cherry skin. The contribution of the cuticle to the mechanical properties of the skin is negligible.
**2.** The skin is isotropic in the planar axes because strains of a bulging skin segment do not differ between longitudinal and latitudinal directions.
**3.** Strain relaxation upon releasing the pressure is complete and time dependent, suggesting the skin exhibits both elastic and viscoelastic behaviour.
**4.** There is little difference in the skin's mechanical properties when these are measured in samples from the cheek, shoulder, suture and stylar-scar regions of the fruit surface.
**5.** Water uptake (up to the point of fruit cracking) has surprisingly little effect on the mechanical properties of the skin.
**6.** Destroying cell turgor decreases the stiffness of the fruit skin.
**7.** The mechanical properties of the skin are only slightly affected by temperature.
**8.** The biaxial tensile test detected differences in cracking susceptibility between cultivars. The less cracking-susceptible 'Regina' skin is stiffer, as indexed by a higher $E$ value, and also has a higher pressure at fracture than 'Burlat'. The differences between 'Regina' and 'Burlat' are most likely accounted for by differences in physical and chemical properties of the epidermal/hypodermal cell walls and not by properties related to cell turgor.

The elastometer is a useful technique for mechanical testing of excised fruit skins in a defined and reproducible manner. The $E$ value, and possibly the fracture threshold (i.e. pressure at fracture and/or strain at fracture) determined can then be related to the physical and chemical properties of the cell wall. However, some limitations of biaxial tensile testing are worth noting. First, routine testing is restricted to regions of the surface having uniform radius of curvature, such as the fruit shoulders on either side of the cheek. Second, the technique is laborious, which limits its application for extensive screening of, say, the large numbers of progeny emerging from a breeding programme. Third, the equipment may need to be modified to accommodate low rates of loading to simulate natural rates of fruit growth and water uptake. Finally, the results obtained using this technique estimate fracture strains that are markedly larger than those obtained in classical immersion assays. The reason for this deviation is unknown. The pressures at fracture are of the same order of magnitude as those reported for fruit turgor (Brüggenwirth *et al.*, 2014; Knoche *et al.*, 2014).

### 7.5.3 Water potential, osmotic potential and turgor

The water potential ($\Psi$, MPa) of sweet cherry fruit comprises two principal components, the hydrostatic pressure (P or $\Psi_P$, MPa) in the cells of the flesh and their osmotic or solute potential ($\Psi_\Pi$ or $\Pi$, MPa). The osmotic properties of the flesh symplast will probably be close to those of the expressed juice. The gravitational and matric water potential components are expected to be insignificant.

Water potentials are important because gradients in the hydrostatic pressure component represent the driving force for water transport in the xylem. As with other tissues, cell turgor is also an important component of the fruit's mechanical integrity.

Water potentials ($\Psi_{fruit}$) reported in the literature have mostly been determined using pressure bombs. This technique assumes the vascular system to be fully functional in the pedicel and fruit and the pedicel/fruit junction to be hydraulically conductive. Whether these conditions are met in mature sweet cherries is unknown. The values for $\Psi_{fruit}$ determined for sweet cherries using a pressure bomb range from –1.2 to almost –2 MPa (Andersen and Richardson, 1982; Measham *et al.*, 2014). Measham *et al.* (2014) reported an increase in $\Psi_{fruit}$ from –2 MPa to a less negative $\Psi_{fruit}$ of about –0.7 MPa during simulated rain. The corresponding increase for leaf water potential ($\Psi_{leaf}$) was from –1.4 to –0.7 MPa (Measham *et al.*, 2014).

Schumann *et al.* (2014) quantified water potentials of excised tissue discs and strips ($\Psi_{tissue}$) using a water uptake/release assay. In this assay, the fruit and/or tissue is incubated in hypertonic and hypotonic solutions of known osmolarity, and the change in mass, volume and/or dimensions is monitored. By regression analysis, the osmolarity of a hypothetical solution yielding zero change in mass or curvature is calculated. This osmolarity corresponds to the water potential of the tissue. Using this approach, $\Psi_{tissue}$ was found to decrease in a sigmoid pattern in developing fruit. The decrease in $\Psi_{tissue}$ closely paralleled the decrease in $\Psi_\Pi$ (Schumann *et al.*, 2014).

The value of $\Psi_\Pi$ in developing sweet cherry fruit has been quantified by water vapour pressure osmometry. This technique is useful in analysing solutions containing residual particles, such as that extracted from the flesh mechanically, e.g. by using a garlic or spaghetti press. The value of $\Psi_\Pi$ decreases rapidly (becomes more negative) beginning in stage II and throughout stage III development. At maturity, $\Psi_\Pi$ ranging from –2.2 to –4.1 MPa has been reported (Andersen and Richardson, 1982; Moing *et al.*, 2004; Knoche *et al.*, 2014; Measham *et al.*, 2014). Interestingly, the $\Psi_\Pi$ value differs between skin and flesh. Compared with the flesh, the skin has a markedly less negative $\Psi_\Pi$ (by about 1.1 MPa) (Grimm and Knoche, 2015).

Few turgor data are available on a whole-fruit basis (flesh turgor) or on an individual cell basis (cell turgor). Often, turgor has been calculated by subtracting $\Psi_\Pi$ from $\Psi_{fruit}$. The $\Psi_P$ values thus obtained are unrealistically high, exceeding the pressure of a fully inflated car tyre by tenfold or more ($\Psi_P$ range 2.0–3.8 MPa; Measham *et al.*, 2009). A fruit having this pressure would feel like steel! Direct approaches for quantifying cell and fruit turgor include pressure-probe techniques, compression plates, and incubation assays and corresponding measurements of osmolarities (Knoche *et al.*, 2014; Schumann *et al.*, 2014). These techniques yield pressures very much lower than those calculated by the difference between $\Psi_{\Pi}$ and $\Psi_{fruit}$. For mature fruit, pressures range from 8 to 64 kPa (Knoche *et al.*, 2014). Only in late stage II was the turgor markedly higher (mean 200 kPa with individual cells occasionally up to 1000 kPa using a cell pressure probe) (Fig. 7.5; Schumann *et al.*, 2014). At maturity, turgor pressures on a whole-fruit basis and an individual cell basis are negligibly low compared with the osmotic potentials of juice extracted from the same batch of fruit. The reason for the very low turgor pressures are unknown but may be related to either one or several of the following: (i) the *E* value of the skin indicates low stiffness and hence low resistance to extension (Brüggenwirth *et al.*, 2014); and (ii) in grape berries, osmolytes accumulate in the apoplast, thereby decreasing the gradient in osmotic potential between symplast and apoplast and, hence turgor (Lang and Düring, 1991; Tilbrook and Tyerman, 2008; Wada *et al.*, 2008, 2009). Whether this also applies to sweet cherries is unknown. It is remarkable that neither osmotic water uptake (wet fruit) nor transpiration (dry air) has a significant effect on fruit turgor (Knoche *et al.*, 2014).

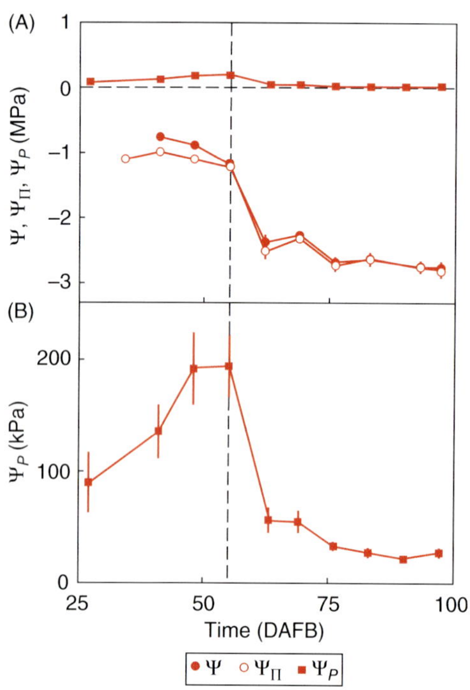


**Fig. 7.5.** (A) Developmental time course of change in tissue water potential (Ψ), the osmotic potential of the fruit ($\Psi_{\Pi}$), and the cell turgor ($\Psi_P$) of developing 'Regina' sweet cherry. (B) Same data for $\Psi_P$ as in (A), but redrawn on an expanded *y*-axis scale. The $\Psi_P$ was quantified using a pressure probe, and Ψ was calculated from $\Psi = \Psi_{\Pi} + \Psi_P$. The vertical dashed line indicates the transition from stage II to stage III at 55 days after full bloom (DAFB). (Reproduced with permission from Schumann *et al.*, 2014.)

From these data, we infer that, for mature sweet cherry: (i) the fruit water potential and osmotic potential are essentially equal; (ii) fruit and cell turgor are negligibly low relative to the osmotic potentials; and (iii) the sweet cherry fruit skin has a less negative osmotic potential, and perhaps a less negative water potential, than the flesh.

### 7.5.4 Water transfer

Water transfer into and out of the fruit occurs as vascular flow through the stem but also through the fruit surface.

#### *Vascular flow*

Vascular flow through the pedicel has received relatively little attention compared with that across the fruit surface. Measham *et al.* (2010, 2014) published flow rates and oscillations of fruit diameter for fruit that remained attached to the tree. Flow rates were determined using heat pulse sensors affixed to the pedicel. In this technique, a heat pulse is applied to a section of the pedicel and its propagation along the pedicel axis is monitored. The data obtained represent net water flows, but the technique does not differentiate between flows through xylem and phloem (during the night, these are likely to be co-directional in a fruit pedicel but could well be in opposite directions on a sunny day). The flow rates reported were very low, averaging

0.8 and 1.5 μl $h^{-1}$ before and after rain (Measham *et al.*, 2010). Alternatively, changes in fruit diameter may be quantified using linear variable displacement transducers (Measham *et al.*, 2014). From diameter changes, net flows may be calculated, provided that rates of water loss via the fruit surface due to transpiration are known (Lang, 1990). Using this approach, Brüggenwirth *et al.* (2016) determined xylem, phloem and transpiration flows in developing sweet cherry. Important findings were a continuous decrease in xylem flow in the course of stage III development from about 85% (equivalent to 11.6 μl $h^{-1}$) of total sap (12.4 μl $h^{-1}$) inflow to essentially zero at maturity (0.6 μl $h^{-1}$ of a total of 11.9 μl $h^{-1}$). In the same interval, the phloem flow continuously increased from 0.8 to 11.3 μl $h^{-1}$ at maturity (Brüggenwirth *et al.*, 2016). Furthermore, the phloem flow closely paralleled the rate of increase in dry matter of the fruit, suggesting that phloem sap concentration remains essentially constant at about 18% (weight/volume) throughout development.

Hovland and Sekse (2004a,b) and Winkler *et al.* (2016) used a potometric approach to quantify pedicel flow rates of detached fruit. In this technique, the fruit is cut from the tree under water to avoid air embolism in the xylem. Subsequently, a water-filled capillary is affixed to the pedicel. Flow rates are quantified by monitoring the movement of a meniscus along the capillary. Hovland and Sekse (2004a,b) reported flow rates in the range of 3.0–11.6 μl $h^{-1}$ for mature fruit. In a developmental time course determined by Winkler *et al.* (2016), flow rates from fruit held at 0% relative humidity increased from 12.2 μl $h^{-1}$ during stage II to a maximum of 24.9 μl $h^{-1}$ at the early stage III, but then continuously decreased to 5.2 μl $h^{-1}$ at harvest (Winkler *et al.*, 2016). For fruit held at 100% relative humidity, corresponding flow rates were 7.1, 18.8 and 5.0 μl $h^{-1}$. It is interesting to note that flow rates during stage II depended on the relative humidity of the atmosphere surrounding the fruit. In contrast, for mature fruit, flow rates were essentially independent of humidity. The flows determined by potometry reflect xylem flows at a hypothetical tree water potential of 0 MPa. Under orchard conditions, the tree water potential will depend on the tree's water supply, and even for a well-watered tree, the water potential will be significantly lower (0 to −1 MPa). Thus, potometric xylem flows offer conservative estimates. Under orchard conditions, xylem flows to the fruit are likely to be lower and possibly even negative during the daytime (i.e. from fruit to tree, as tree water potential reaches its most negative values during the early afternoon; Brüggenwirth *et al.*, 2016).

Brüggenwirth and Knoche (2015) quantified conductances of detached pedicels using a modified root pressure probe. Because the flow was induced by pressurizing the pedicel end, flow must have occurred via the xylem. Conductance of the xylem in the pedicel decreased somewhat during the stage II/III transition, but remained constant throughout stage III. Conductance estimates were lower than those calculated from cross-sectional areas of the xylem vessels using Hagen–Poiseuille's law. Attempts to quantify the conductance of the vascular system within the fruit were not successful, probably due to the low conductance (i.e. high resistance) of the xylem within the fruit (M. Brüggenwirth, unpublished data).

### *Transport across the fruit surface*

Water transfer across the fruit surface has been reviewed recently (Knoche, 2015). Briefly, due to the coincidence of rain and fruit cracking, water transfer through the surface is usually considered the dominant factor in cracking (Christensen, 1996). Thus, many studies have focused on osmotic water uptake through the fruit surface and on fruit transpiration.

Water transfer through the fruit skin is usually quantified gravimetrically by repeated weighing of detached fruit incubated in water (uptake) or in a non-saturated atmosphere (transpiration). Flow rates ($F$) are calculated from slopes of plots of the cumulative change in mass versus time. From flow rates, surface areas and driving forces, the hydraulic conductance (osmotic water uptake) and permeance (transpiration) of the fruit skin may be calculated. These coefficients represent the 'material constants' of the rate-limiting barrier presented by the fruit skin. Conductances and permeances are useful for comparing these properties in different cultivars, following different treatments or in different seasons, locations, etc. However, for head-to-head comparisons of fruit of the same size and the same batch (and hence the same water potential), the conversion of flow rates or fluxes into hydraulic conductances or permeances usually offers little or no additional insight.

### *Pathways of transport*

Uptake and transpiration through the surface occur along a number of parallel pathways. These are through the cuticle, microcracks, stomata, stem/fruit junction, stylar scar and (for detached fruit) the cut end of the pedicel.

CUTICLE. The cuticle represents the primary barrier to water transfer. Abrading the cuticle increased rates of water uptake 33-fold from 19.4±2.8 to 641.7±26.0 mg $h^{-1}$ and rates of transpiration fivefold from 39.4±1.3 to 198.4±8.8 mg $h^{-1}$ (A. Winkler, unpublished data).

Cuticles isolated from mature fruit are very fragile and markedly strained. Due to the presence of stomata and strain, it is difficult to study the permeability of isolated cuticles *in vitro* (Beyer *et al.*, 2005). However, robust estimates of permeability of the entire fruit skin composite (including epidermis and hypodermis) are obtained by quantifying water transfer on a whole-fruit basis.

Water uptake through a lipophilic cuticle occurs along a continuum of polar domains in the cutin matrix, which results from the hydration and orientation of polar functional groups. These polar domains are referred to as aqueous pores or polar pathways (Franke, 1964; Schönherr, 2006; Weichert and Knoche, 2006a). Polar pathways provide an aqueous continuum across the lipophilic cuticle that allows rapid transport by viscous flow. Polar pathways occur at higher frequency above anticlinal cell walls, cuticular ledges and the guard cells of the stomatal apparatus (Franke, 1964, 1967; Schönherr, 2006). Polar pathways may account for the high permeability of the sweet cherry fruit cuticle as compared with CMs from other crops (Becker *et al.*, 1986).

MICROCRACKS. Microcracks impair the barrier function of the cuticle and hence increase the permeability of the fruit skin, particularly to water uptake and, to a lesser extent, to transpiration (Knoche and Peschel, 2006).

STOMATA. The sweet cherry fruit surface is stomatous, and stomata represent openings in the cuticle envelope. However, a mass flow of water through open stomata is unlikely. The critical surface tension of the fruit surface (24.9 mN $m^{-1}$; Peschel *et al.*, 2003) is well below that of water (72 mN $m^{-1}$) making mass flow through open stomata unlikely (Schönherr and Bukovac, 1972).

Among sweet cherry cultivars, the permeability of the surface in transpiration and in osmotic water uptake is only weakly related to stomatal density (19 cultivars, 5 years, $n = 918$, $r^2 = 0.095$, $P<0.001$ for transpiration; 15 cultivars, 2 years, $n = 236$, $r^2 = 0.042$, $P<0.01$ for water uptake; Peschel and Knoche, 2012).

STEM/FRUIT JUNCTION. The stem/fruit junction represents a site of preferential water uptake into the fruit (Beyer *et al.*, 2002b). Averaged across eight cultivars, uptake along the junction amounted to about 46% of total surface uptake (Weichert *et al.*, 2004). In the course of development, penetration along the junction increases from an initial 30% of total uptake to a maximum of 70% of the total uptake of a submerged mature fruit, indicating that the junction becomes leakier as maturity progressed. In 'Sam', the amount of water penetrating at the junction was negatively related to the fruit removal force (Beyer *et al.*, 2002b).

The mechanistic basis for the high uptake in this region is not entirely clear. Several factors are involved. First, the stem/fruit junction represents the attachment zone between carpel and receptacle and the cuticle appears to be discontinuous in this region. Second, abscission zones are present between the fruit and stem in at least some but not all cultivars (Stösser *et al.*, 1969) and also at the edge of the receptacle where stamina, petals and sepals attached during anthesis. The permeability of these zones to water is unknown. In grape berries, abscission zones on the receptacle are highly permeable (Becker *et al.*, 2012). Third, the stem/fruit junction has a high density of microcracks in the cuticle (Peschel and Knoche, 2005). This is because the curvature of the surface in the stem cavity region is high, causing stress concentration and hence failure of the skin (Considine and Brown, 1981; Yamamoto *et al.*, 1990).

From a practical point of view, penetration along the junction is extremely important. The shape of the fruit, with a marked depression around the junction, results in extended periods of surface wetness and hence continuing water uptake long after a rain event. The pedicel/fruit junction does not play a significant role in transpiration.

STYLAR SCAR. The permeability of the stylar scar in transpiration is higher than that of the surrounding cuticle. However, due to the small area of the scar relative to the remaining fruit surface, the effect of the scar on whole-fruit transpiration is small (Knoche *et al.*, 2000, 2002). There is no evidence for increased uptake through the stylar scar in mature sweet cherry (Beyer *et al.*, 2002b). However, the region around the stylar scar has a high density of microcracks (Peschel and Knoche, 2005).

PEDICEL END. When detaching fruit from the tree, the proximal end of the pedicel is exposed to atmospheric pressure. Because the fruit water potential is negative, an air embolism develops instantaneously. The embolism interrupts the water column in the xylem and virtually blocks water entry through the pedicel end. Phloem transport is also expected to cease instantaneously because of callose formation within minutes after wounding. These factors make water uptake by detached fruit through the pedicel end unlikely.

### 7.5.5 Whole-fruit water balance

From rates of water in- and outflow through the fruit surface and pedicel vasculature, a water balance may be established (Knoche and Measham, 2017). Calculations reveal that (daytime) transpiration is one of the largest of the various flows contributing to fruit water balance. Here, it is useful to consider a number of simplified scenarios to better appreciate the effects of weather on fruit water balance. Thus, on a sunny day, the rate of transpiration outflow exceeds the sum of the vascular (xylem and phloem) inflow rates, resulting in a net loss of fruit water (Table 7.2). Meanwhile, on a humid cloudy day, and if the stem cavity contains a few drops of residual water from an overnight shower or heavy dew, the osmotic and vascular inflows can be more than sufficient to replace the water lost by transpiration through an otherwise dry skin. The result will be a net gain in water by the fruit. Of course, on a rainy day, with little or no significant transpiration outflow, a number of significant inflows (xylem and phloem vascular inflows and the roughly similar osmotic uptake through the wetted fruit surface) combine to give a strong gain in fruit water content.

Against this background, the effectiveness of a rain shelter may also be assessed. Under a rain shelter, the osmotic inflows through the fruit surface and at the stem/fruit junction will be eliminated. However, the net flow into the fruit could well remain positive, because the associated high humidity and reduced radiation levels under a rain shelter during rain will greatly reduce transpiration outflow, while vascular inflows will continue. This net inflow may well be sufficient to result in fruit cracking even under a rain shelter (Cline *et al.*, 1995b).

## 7.6 Prevention of Cracking

Here, we focus on studies reporting successful strategies to reduce cracking at the field scale or in the laboratory.

### 7.6.1 Rain shelters

The use of rain shelters effectively prevents contact between liquid water and the fruit

**Table 7.2.** Estimates of sweet cherry water balance under different weather scenarios as indexed by different combinations of temperature and relative humidity. Hypothetical flow rates in transpiration and water uptake through the fruit surface including the pedicel/fruit junction and the overall mean of published flow rates through the pedicel were combined and the net flow rates calculated.

| Parameter | Weather scenario | | | |
|---|---|---|---|---|
| | Sunny | Partly cloudy | Cloudy | Rainy |
| Temperature (°C) | 25 | 25 | 25 | 25 |
| Relative humidity (%) | 60 | 80 | 90 | 98 |
| Stem cavity | Dry | Dry | Wet | Wet |
| $F_{skin\ transpiration}$ (mg $h^{-1}$) | –21.1 | –12.6 | –6.9 | 0 |
| $F_{skin\ uptake}$ (mg $h^{-1}$) | 0 | 0 | 0 | 9.7 |
| $F_{juncture\ uptake}$ (mg $h^{-1}$) | 0 | 0 | 9.7 | 9.7 |
| $F_{vascular\ uptake}$ (mg $h^{-1}$) | 8.3 | 8.3 | 8.3 | 8.3 |
| $F_{net}$ (mg $h^{-1}$) | –12.8 | –4.3 | 11.1 | 27.7 |

Assumptions: 12 g per fruit, surface area 25 $cm^2$, density 1 kg $dm^{-3}$, leakiness of pedicel fruit junction about 50% in mature fruit, permeance (*P*) in transpiration ($P_{transpiration}$) $1.7 \times 10^{-4}$ m $s^{-1}$ (equivalent to liquid-based $P_{transpiration}$ of $3.9 \times 10^{-9}$ m $s^{-1}$) at 0% humidity, $P_{uptake}$ $67.1 \times 10^{-9}$ m $s^{-1}$, $\Psi_{fruit}$ –2.18 MPa, 100% surface wetness during rain, no boundary layer resistance.

surface, thereby markedly reducing cracking (Cline *et al.*, 1995b; Balmer, 1998; Børve and Meland, 1998; Børve and Stensvand, 2003; Børve *et al.*, 2003; Usenik *et al.*, 2009; Thomidis and Exadaktylou, 2013). Occasionally, a low percentage (<5%) of the fruit crack under a shelter (Cline *et al.*, 1995b). This cracking may be due to uptake via the vascular system (Measham *et al.*, 2010), the lack of transpiration and possibly uptake from the vapour phase (Beyer *et al.*, 2005). Because the surface of sheltered fruit stays dry, microcracking is markedly reduced (Knoche and Peschel, 2006).

### 7.6.2 Spray application of calcium salts

The effects of calcium ions on cracking have been the subject of numerous studies (see Table 7.1). Calcium has been applied in submersion assays (Verner, 1937; Glenn and Poovaiah, 1989; Weichert *et al.*, 2004), as foliar sprays in the field (Verner, 1937; Meheriuk *et al.*, 1991; Wójcik *et al.*, 2013; Erogul, 2014) or by overhead sprinklers during rain (Fernandez and Flore, 1998; Lang *et al.*, 1998). The effects on cracking have not been consistent. Calcium reduced cracking in a significant number of studies but had no effect in others. Unfortunately, in the latter group, it remains mostly unclear whether the lack of effectiveness was due to a lack of penetration or a lack of action of the calcium taken up. Due to its charge, cuticular penetration of calcium ions is poor (Schlegel and Schönherr, 2002a,b; Schlegel *et al.*, 2005).

The mechanism of calcium in reducing cracking has been related to: (i) effects on the mechanical properties of the cell walls; and (ii) reduced water uptake. Improved cross-linking of cell wall constituents and particularly those of pectins in the middle lamellae is well established in postharvest physiology (Demarty *et al.*, 1984; Aghdam *et al.*, 2012). Indeed, $CaCl_2$ increased the $WU_{50}$ in cracking assays (Weichert *et al.*, 2004). The second argument for reduced cracking is a reduction in water uptake due to an osmotic effect caused by the decrease in osmotic potential of the calcium solution. A 0.5–1% solution of $CaCl_2 \times 2H_2O$ has an osmotic potential of –0.25 to –0.5 MPa. For fruit with a water potential of –4.1 MPa (Knoche *et al.*, 2014), this would translate into a 6–12% reduction in water uptake rate through the wetted portion of the fruit surface. This is a conservative estimate. Most likely, the benefit would be lower because the solution would be much diluted when applied during rainfall. These effects are probably too small to be detectable under field conditions.

### 7.6.3 Use of other mineral salts

Some studies have reported reduced cracking and decreased water uptake when incubating fruit in solutions of salts containing iron ($Fe^{3+}$), aluminium ($Al^{3+}$) or mercury ($Hg^{2+}$) ions (Bullock, 1952; Beyer *et al.*, 2002a; Weichert *et al.*, 2004). The mechanism is twofold: (i) improved mechanical properties of the fruit skin, as indicated by an increased $WU_{50}$ (Weichert *et al.*, 2004); and (ii) a marked reduction in the water permeability of the fruit surface (Weichert and Knoche, 2006b). The latter is caused by a pH-dependent precipitation reaction where the precipitates selectively block polar pathways across the sweet cherry fruit surface (Weichert and Knoche, 2006b; Weichert *et al.*, 2010). Because polar pathways are not involved in water transport in the vapour phase, transpiration and gas exchange are unaffected. The effect obtained at fairly low concentrations (lower limit 2.5–10 mm) is impressive in laboratory-based immersion assays but unfortunately has been useless to date in the field. Solutions of effective ferric salts are very acidic and corrosive, while those of mercury and aluminium are toxic (Chang, 1977; Boegman and Bates, 1984). In addition, the ferrous precipitates forming in the fruit skin discolour the fruit and cover the tree with unacceptable spray residues.

### 7.6.4 Other methods

Three other strategies are worth mentioning: the use of antitranspirants, the removal of surface water using blowers or helicopters, and the use of less susceptible cultivars.

It has often been suggested that rain-induced cracking might be averted by coating the fruit with a waterproof layer or applying an antitranspirant. Significant reductions in cracking are occasionally reported with such products (Davenport *et al.*, 1972; Kaiser *et al.*, 2014; Meland *et al.*, 2014; Torres *et al.*, 2014). To the best of our knowledge, no mechanistic studies have identified the modes of action of these substances. Provided that the spray deposit created on the fruit surface is sufficiently thick, osmotic uptake at the stem/fruit junction may be lowered. In addition, a spray solution applied at high volume collects as a pendant drop in the stylar-scar area and as a puddle in the stem cavity. As the solution dries off, it leaves behind a thick deposit. In these areas, the fruit surface is more permeable due to a higher density of cracks, making coatings more likely to be effective in slowing water uptake. However, a number of limitations of these coating strategies must be kept in mind:

**1.** Any spray application is not selective and so would affect the leaves and the fruit equally. Gas exchange of leaves and fruit must not be impaired.
**2.** Antitranspirants have a contact mode of action and hence are effective only in the wetted portion of the fruit surface. For water without a surfactant, the wetted fraction of a sweet cherry fruit surface after spray application averages only about 18% of the surface (M. Knoche, unpublished data).
**3.** A sweet cherry cuticle represents a very significant barrier to water transfer. Hence, to lower its permeability requires the application of a film having a permeance as low as, or lower than, the cuticle. These effects will limit the success of all coating strategies.

In some orchards, growers use helicopters or air blast blowers (without solution) to blow adhering moisture off the fruit. To the best of our knowledge, no experimental studies evaluating the benefit of such efforts have been published. While the removal of surface moisture would have a positive effect, the agitation of the fruit and its pedicel may increase the leakiness of the stem/fruit junction (Beyer *et al.*, 2002b).

Lastly, it should be mentioned that an effective, economic and environmentally friendly way to reduce cracking is to grow less cracking-susceptible cultivars. However, none of the cultivars currently available is completely resistant (see Chapter 4, this volume).

## 7.7 Conclusions

Good progress has been made recently in a number of areas relevant to rain-induced cracking. In particular, the molecular background of cuticle deposition in sweet cherry, important chemical and physical characteristics of the cuticle and the fruit skin, and the mechanisms, pathways and driving forces for water uptake through the surface have largely been identified. In addition, data on xylem flow, the fruit's water potential, osmotic potential and turgor have also recently been reported. A remaining gap resides with the vascular flow via the phloem. There is also no published information on the mode of fracture of the sweet cherry fruit skin.

There has been little progress in developing models that explain rain-induced cracking on a whole-fruit basis. The prevailing hypothesis is based on a two-compartment fruit model – an elastic skin enveloping a sugary flesh. Water uptake into the flesh increases the fruit's volume, surface area and turgor. When a critical turgor and/or critical strain of the skin is exceeded, the skin will rupture and the fruit will crack (Considine and Kriedemann, 1972; Sekse, 1995a; Sekse *et al.*, 2005; Measham *et al.*, 2009). However, the absence of significant turgor, the lack of any response of turgor to water uptake or transpiration (Knoche *et al.*, 2014) and the observation of cracking despite a net loss in mass (Knoche and Peschel, 2006) must bring this hypothesis into serious question.

An alternative hypothesis is to consider cracking as the result of a localized phenomenon, i.e. a local defect. This now appears more likely. A local defect would then cause a zipper-type propagation of the defect to form a crack, in much the same way as a 'ladder' will 'run' in a fine, knitted fabric. It would seem likely that microcracks in the cuticle may very well represent the initial defect, allowing: (i) a high rate of localized water uptake (Peschel and Knoche, 2005; Knoche and Peschel, 2006); (ii) the consequent bursting of individual cells in the vicinity of the crack (Simon, 1977) with leakage of cell contents into the apoplast; and (iii) a concentration of stress in the elastically strained skin at the point of this defect. Malic acid, a major osmolyte, present in high concentrations in the fruit symplast, leaks into the apoplast where it weakens the cell walls that form the structural member of the fruit skin (Winkler *et al.*, 2015). Furthermore, membrane permeability increases, causing the local defect to spread, the skin to 'unzip' and eventually the fruit to crack. This hypothesis merits further research. Identifying the mechanism of cracking is also a prerequisite for the development of efficient and high-throughput phenotyping methods to characterize cultivars and hybrids for cracking tolerance.

## References


Aghdam, M.S., Hassanpouraghdam, M.B., Paliyath, G. and Farmani, B. (2012) The language of calcium in postharvest life of fruits, vegetables and flowers. *Scientia Horticulturae* 144, 102–115.

Alkio, M., Jonas, U., Sprink, T., van Nocker, S. and Knoche, M. (2012) Identification of putative candidate genes involved in cuticle formation in *Prunus avium* (sweet cherry) fruit. *Annals of Botany* 110, 101–112.

Alkio, M., Jonas, U., Declercq, M., van Nocker, S. and Knoche, M. (2014) Transcriptional dynamics of the developing sweet cherry (*Prunus avium* L.) fruit: sequencing, annotation and expression profiling of exocarp-associated genes. *Horticulture Research* 1, 11. DOI: 10.1038/hortres.2014.11.

Andersen, P.C. and Richardson, D.G. (1982) A rapid method to estimate fruit water status with special reference to rain cracking of sweet cherries. *Journal of the American Society for Horticultural Science* 107, 441–444.

Balbontín, C., Ayala, H., Bastías, R.M., Tapia, G., Ellena, M., Torres, C., Yuri, J.A., Quero-García, J., Ríos, J.C. and Silva, H. (2013) Cracking in sweet cherries: a comprehensive review from a physiological, molecular, and genomic perspective. *Chilean Journal of Agricultural Research* 73, 66–72.

Balmer, M. (1998) Preliminary results on planting densities and rain covering for sweet cherry on dwarfing rootstocks. *Acta Horticulturae* 468, 433–439.

Bargel, H., Spatz, H.C., Speck, T. and Neinhuis, C. (2004) Two-dimensional tension test in plant biomechanics – sweet cherry fruit skin as a model system. *Plant Biology* 6, 432–439.

Becker, M., Kerstiens, G. and Schönherr, J. (1986) Water permeability of plant cuticles: permeance, diffusion and partition coefficients. *Trees* 1, 54–60.

Becker, T., Grimm, E. and Knoche, M. (2012) Substantial water uptake into detached grape berries occurs through the stem surface. *Australian Journal of Grape and Wine Research* 18, 109–114.

Belmans, K. and Keulemans, J. (1996) A study of some fruit skin characteristics in relation to the susceptibility of cherry fruit to cracking. *Acta Horticulturae* 410, 547–550.

Belmans, K., Keulemans, J., Debarsy, T. and Bronchart, R. (1990) Influence of sweet cherry epidermal characters on the susceptibility to fruit cracking. In: *Proceedings of the XXIII International Horticulture Congress* XXIII, Firenze, Italy, 27 August–1 September, p. 637.

Beyer, M., Peschel, S., Weichert, H. and Knoche, M. (2002a) Studies on water transport through the sweet cherry fruit surface: VII. $Fe^{3+}$ and $Al^{3+}$ reduce conductance for water uptake. *Journal of Agricultural and Food Chemistry* 50, 7600–7608.

Beyer, M., Peschel, S., Knoche, M. and Knörgen, M. (2002b) Studies on water transport through the sweet cherry fruit surface: IV. Regions of preferential uptake. *HortScience* 37, 637–641.

Beyer, M., Lau, S. and Knoche, M. (2005) Studies on water transport through the sweet cherry fruit surface: IX. Comparing permeability in water uptake and transpiration. *Planta* 220, 474–485.

Boegman, R.J. and Bates, L.A. (1984) Neurotoxity of aluminum. *Canadian Journal of Physiology and Pharmacology* 62, 1010–1014.

Børve, J. and Meland, M. (1998) Rain cover protection against cracking of sweet cherries. I. The effects on marketable yield. *Acta Horticulturae* 468, 449–453.

Børve, J. and Stensvand, A. (2003) Use of a plastic rain shield reduces fruit decay and need for fungicides in sweet cherry. *Plant Disease* 87, 523–528.

Børve, J., Sekse, L. and Stensvand, A. (2000) Cuticular fractures promote postharvest fruit rot in sweet cherries. *Plant Disease* 84, 1180–1184.

Børve, J., Skaar, E., Sekse, L., Meland, M. and Vangdal, E. (2003) Rain protective covering of sweet cherry trees – effects of different covering methods on fruit quality and microclimate. *HortTechnology* 13, 143–148.

Brown, G., Wilson, S., Boucher, W., Graham, B. and McGlasson, B. (1995) Effects of copper–calcium sprays on fruit cracking in sweet cherry (*Prunus avium*). *Scientia Horticulturae* 62, 75–80.

Brown, K. and Considine, J. (1982) Physical aspects of fruit growth. *Plant Physiology* 69, 585–590.

Brüggenwirth, M. and Knoche, M. (2015) Xylem conductance of sweet cherry pedicels. *Trees* 29, 1851–1860.

Brüggenwirth, M. and Knoche, M. (2016a) Factors affecting mechanical properties of the skin of sweet cherry fruit. *Journal of the American Society for Horticultural Science* 141, 45–53.

Brüggenwirth, M. and Knoche, M. (2016b) Mechanical properties of skins of sweet cherry fruit of differing susceptibilities to cracking. *Journal of the American Society for Horticultural Science* 141, 162–168.

Brüggenwirth, M., Fricke, H. and Knoche, M. (2014) Biaxial tensile tests identify epidermis and hypodermis as the main structural elements of sweet cherry skin. *AoB Plants* 6, plu019.

Brüggenwirth, M., Winkler, A. and Knoche, M. (2016) Xylem, phloem, and transpiration flows in developing sweet cherry fruit. *Trees* 30, 1822–1830.

Bukovac, M.J., Knoche, M., Pastor, A. and Fader, R.G. (1999) The cuticular membrane: a critical factor in rain-induced cracking of sweet cherry fruit. *HortScience* 34, 549.

Bullock, R.M. (1952) A study of some inorganic compounds and growth promoting chemicals in relation to fruit cracking of Bing cherries at maturity. *Proceedings of the American Society for Horticultural Science* 59, 243–253.

Chang, L.W. (1977) Neurotoxic effects of mercury – a review. *Environmental Research* 14, 329–373.

Christensen, J.V. (1972a) Cracking in cherries I. Fluctuation and rate of water absorption in relation to cracking susceptibility. *Danish Journal of Plant and Soil Science* 76, 1–5.

Christensen, J.V. (1972b) Cracking in cherries III. Determination of cracking susceptibility. *Acta Agriculturae Scandinavica* 22, 128–136.

Christensen, J.V. (1972c) Cracking in cherries IV. Determination of cracking susceptibility. *Acta Agriculturae Scandinavica* 22, 153–162.

Christensen, J.V. (1972d) Cracking in cherries V. The influence of some salts and chemicals on cracking. *Frukt og Baer Oslo* 1, 37–47.

Christensen, J.V. (1975) Cracking in cherries VII. Cracking susceptibility in relation to fruit size and firmness. *Acta Agriculturae Scandinavica* 25, 11–13.

Christensen, J.V. (1996) Rain-induced cracking of sweet cherries: its causes and prevention. In: Webster, A.D. and Looney, N.E. (eds) *Cherries: Crop Physiology, Production and Uses*. CAB International, Wallingford, UK, pp. 297–327.

Cline, J.A. and Trought, M. (2007) Effect of gibberellic acid on fruit cracking and quality of Bing and Sam sweet cherries. *Canadian Journal of Plant Science* 87, 545–550.

Cline, J.A., Sekse, L., Meland, M. and Webster, A.D. (1995a) Rain cracking of sweet cherries: I. Influence of cultivar and rootstock on fruit water absorption, cracking and quality. *Acta Agriculturae Scandinavica, Section B — Soil and Plant Science* 45, 213–223.

Cline, J.A., Meland, M., Sekse, L. and Webster, A.D. (1995b) Rain cracking of sweet cherries: II. Influence of rain covers and rootstocks on cracking and fruit quality. *Acta Agriculturae Scandinavica, Section B — Soil and Plant Science* 45, 224–230.

Considine, J. and Brown, K. (1981) Physical aspects of fruit growth. Theoretical analysis of distribution of surface growth forces in fruit in relation to fruit cracking and splitting. *Plant Physiology* 68, 371–376.

Considine, J.A. and Kriedemann, P.E. (1972) Fruit splitting in grapes. Determination of the critical turgor pressure. *Australian Journal of Agricultural Research* 23, 17–24.

Davenport, D.C., Uriu, K. and Hagan, R.M. (1972) Antitranspirant film: curtailing intake of external water by cherry fruit to reduce cracking. *HortScience* 7, 507–508.

Demarty, M., Morvan, C. and Thellier, M. (1984) Calcium and the cell wall. *Plant, Cell and Environment* 7, 441–448.

Demirsoy, L. and Demirsoy, H. (2004) The epidermal characteristics of fruit skin of some sweet cherry cultivars in relation to fruit cracking. *Pakistan Journal of Botany* 36, 725–731.

Demirsoy, L.K. and Bilgener, S. (1998) The effects of preharvest chemical applications on cracking and fruit quality in 0900 'Ziraat', 'Lambert' and 'Van' sweet cherry varieties. *Acta Horticulturae* 468, 663–670.

Dumitru, M.G., Vasile, N.I., Baciu, A.A. (2015) The use of natural biopolymer derived from *Gleditsia triacanthos* in reducing the cracking process of cherries. *Revista de Chimie* 66, 97–100.

Erogul, D. (2014) Effect of preharvest calcium treatments on sweet cherry fruit quality. *Notulae Botanicae Horti Agrobotanici* 42, 150–153.

Fernandez, R.T. and Flore, J.A. (1998) Intermittent application of $CaCl_2$ to control rain cracking of sweet cherry. *Acta Horticulturae* 468, 683–689.

Franke, W. (1964) Role of guard cells in foliar absorption. *Nature* 202, 1236–1237.

Franke, W. (1967) Mechanisms of foliar penetration of solutions. *Annual Review of Plant Physiology* 18, 281–300.

Glenn, G.M. and Poovaiah, B.W. (1989) Cuticular properties and postharvest calcium applications influence cracking of sweet cherries. *Journal of the American Society for Horticultural Science* 114, 781–788.

Gonçalves, B., Moutinho-Pereira, J., Bacelar, E., Correia, C., Silva, A.P. and Santos, A. (2008) Relationships among sweet cherry leaf gas exchange, morphology and chemical composition. *Acta Horticulturae* 795, 633–637.

Grimm, E. and Knoche, M. (2015) Sweet cherry skin has a less negative osmotic potential than the flesh. *Journal of the American Society for Horticultural Science* 140, 472–479.

Grimm, E., Khanal, B.P., Winkler, A., Knoche, M. and Köpcke, D. (2012a) Structural and physiological changes associated with the skin spot disorder in apple. *Postharvest Biology and Technology* 64, 111–118.

Grimm, E., Peschel, S., Becker, T. and Knoche, M. (2012b) Stress and strain in the sweet cherry skin. *Journal of the American Society for Horticultural Science* 137, 383–390.

Grimm, E., Peschel, S. and Knoche, M. (2013) Mottling on sweet cherry fruit is caused by exocarp strain. *Journal of the American Society for Horticultural Science* 138, 18–23.

Hovland, K.L. and Sekse, L. (2004a) Water uptake through sweet cherry (*Prunus avium* L.) fruit pedicels: influence of fruit surface water status and intact fruit skin. *Acta Agriculturae Scandinavica, Section B — Soil and Plant Science* 54, 91–96.

Hovland, K.L. and Sekse, L. (2004b) Water uptake through sweet cherry (*Prunus avium* L.) fruit pedicels in relation to fruit development. *Acta Agriculturae Scandinavica, Section B — Soil and Plant Science* 54, 264–266.

Kaiser, C., Fallahi, E., Meland, M., Long, L.E. and Christensen, J.M. (2014) Prevention of sweet cherry fruit cracking using SureSeal, an organic biofilm. *Acta Horticulturae* 1020, 477–488.

Kertesz, Z.I. and Nebel, B.R. (1935) Observations on the cracking of cherries. *Plant Physiology* 10, 763–772.

Khadivi-Khub, A. (2015) Physiological and genetic factors influencing fruit cracking. *Acta Physiologiae Plantarum* 37, 1718.

Khanal, B.P., Grimm, E., Finger, S., Blume, A. and Knoche, M. (2013) Intracuticular wax fixes and restricts strain in leaf and fruit cuticles. *New Phytologist* 200, 134–143.

Knoche, M. (1994) Organosilicone surfactants: performance in agricultural spray application. A review. *Weed Research* 34, 221–239.

Knoche, M. (2015) Water uptake through the surface of fleshy soft fruit: barriers, mechanism, factors, and potential role in cracking. In: Kanayama, Y. and Kochetov, A. (eds) *Abiotic Stress Biology in Horticultural Plants*. Springer, Tokyo, pp. 147–166.

Knoche, M. and Measham, P. (2017) The permeability concept: a useful tool in analyzing water transport through the sweet cherry fruit surface. *Acta Horticulturae* (in press).

Knoche, M. and Peschel, S. (2002) Studies on water transport through the sweet cherry fruit surface: VI. Effect of hydrostatic pressure on water uptake. *Journal for Horticultural Science and Biotechnology* 77, 609–614.

Knoche, M. and Peschel, S. (2006) Water on the surface aggravates microscopic cracking of the sweet cherry fruit cuticle. *Journal of the American Society for Horticultural Science* 131, 192–200.

Knoche, M. and Peschel, S. (2007) Deposition and strain of the cuticle of developing European plum fruit. *Journal of the American Society for Horticultural Science* 132, 597–602.

Knoche, M., Peschel, S., Hinz, M. and Bukovac, M.J. (2000) Studies on water transport through the sweet cherry fruit surface: characterizing conductance of the cuticular membrane using pericarp segments. *Planta* 212, 127–135.

Knoche, M., Peschel, S., Hinz, M. and Bukovac, M.J. (2001) Studies on water transport through the sweet cherry fruit surface: II. Conductance of the cuticle in relation to fruit development. *Planta* 213, 927–936.

Knoche, M., Peschel, S. and Hinz, M. (2002) Studies on water transport through the sweet cherry fruit surface: III. Conductance of the cuticle in relation to fruit size. *Physiologia Plantarum* 114, 414–421.

Knoche, M., Beyer, M., Peschel, S., Parlakov, B. and Bukovac, M.J. (2004) Changes in strain and deposition of cuticle in developing sweet cherry fruit. *Physiologia Plantarum* 120, 667–677.

Knoche, M., Grimm, E. and Schlegel, H.J. (2014) Mature sweet cherries have low turgor. *Journal of the American Society for Horticultural Science* 139, 3–12.

Koffmann, W., Wade, N.L. and Nicol, H. (1996) Tree sprays and root pruning fail to control rain induced cracking of sweet cherries. *Plant Protection Quarterly* 11, 126–130.

Lai, X., Khanal, B.P. and Knoche, M. (2016) Mismatch between cuticle deposition and area expansion in fruit skins allows potentially catastrophic buildup of elastic strain. *Planta* 244, 1145–1156.

Lang, A. (1990) Xylem, phloem and transpiration flows in developing apple fruits. *Journal of Experimental Botany* 41, 645–651.

Lang, A. and Düring, H. (1991) Partitioning control by water potential gradient: evidence for compartmentation breakdown in grape berries. *Journal of Experimental Botany* 42, 1117–1122.

Lang, G., Flore, J.A., Guimond, C., Southwick, S., Facteau, T., Kappel, F. and Azarenko, A. (1998) Performance of calcium/sprinkler based strategies to reduce sweet cherry rain-cracking. *Acta Horticulturae* 468, 649–656.

Lilleland, O. and Newsome, L. (1934) A growth study of the cherry fruit. *Proceedings of the American Society for Horticultural Science* 32, 291–299.

Looney, N.E. (1985) Benefits of calcium sprays below expectations in BC tests. *Goodfruit Grower* 36, 7–8.

Measham, P.F., Bound, S.A., Gracie, A.J. and Wilson, S.J. (2009) Incidence and type of cracking in sweet cherry (*Prunus avium* L.) are affected by genotype and season. *Crop and Pasture Science* 60, 1002–1008.

Measham, P.F., Gracie, A.J., Wilson, S.J. and Bound, S.A. (2010) Vascular flow of water induces side cracking in sweet cherry (*Prunus avium* L.). *Advances in Horticultural Science* 24, 243–248.

Measham, P.F., Bound, S.A., Gracie, A.J. and Wilson, S.J. (2012) Crop load manipulation and fruit cracking in sweet cherry (*Prunus avium* L.). *Advances in Horticultural Science* 26, 25–31.

Measham, P.F., Wilson, S.J, Gracie, A.J. and Bound, S.A. (2014) Tree water relations: flow and fruit. *Agricultural Water Management* 137, 59–67.

Meheriuk, M., Neilsen, G.H. and McKenzie, D.-L. (1991) Incidence of rain splitting in sweet cherries treated with calcium or coating materials. *Canadian Journal of Plant Science* 71, 231–234.

Meland, M., Kaiser, C. and Christensen, M.J. (2014) Physical and chemical methods to avoid fruit cracking in cherry. *AgroLife Scientific Journal* 3, 177–183.

Moing, A., Renaud, C., Christmann, H., Fouilhaux, L., Tauzin, Y. and Zanetto, A. (2004) Is there a relation between changes in osmolarity of cherry fruit flesh or skin and fruit cracking susceptibility? *Journal of the American Society for Horticultural Science* 129, 635–641.

Peschel, S. and Knoche, M. (2005) Characterization of microcracks in the cuticle of developing sweet cherry fruit. *Journal of the American Society for Horticultural Science* 130, 487–495.

Peschel, S. and Knoche, M. (2012) Studies on water transport through the sweet cherry fruit surface: XII. Variation in cuticle properties among cultivars. *Journal of the American Society for Horticultural Science* 137, 367–375.

Peschel, S., Beyer, M. and Knoche, M. (2003) Surface characteristics of sweet cherry fruit: stomata number, distribution, functionality and surface wetting. *Scientia Horticulturae* 97, 265–278.

Peschel, S., Franke, R., Schreiber, L. and Knoche, M. (2007) Composition of the cuticle of developing sweet cherry fruit. *Phytochemistry* 68, 1017–1025.

Quero-García, J., Fodor, A., Reignier, A., Capdeville, G., Joly, J., Tauzin, Y., Fouilhaux, L. and Dirlewanger, E. (2014) QTL detection of important agronomic traits for sweet cherry breeding. *Acta Horticulturae* 1020, 57–64.

Richardson, D.G. (1998) Rain-cracking of 'Royal Ann' sweet cherries: fruit physiological relationships, water temperature, orchard treatments, and cracking index. *Acta Horticulturae* 468, 677–682.

Schlegel, T.K. and Schönherr, J. (2002a) Selective permeability of cuticles over stomata and trichomes to calcium chloride. *Acta Horticulturae* 594, 91–96.

Schlegel, T.K. and Schönherr, J. (2002b) Penetration of calcium chloride into apple fruits as affected by stage of fruit development. *Acta Horticulturae* 594, 527–533.

Schlegel, T.K., Schönherr, J. and Schreiber, L. (2005) Size selectivity of aqueous pores in stomatous cuticles of *Vicia faba* leaves. *Planta* 221, 648–655.

Schönherr, J. (2006) Characterization of aqueous pores in plant cuticles and permeation of ionic solutes. *Journal of Experimental Botany* 57, 2471–2491.

Schönherr, J. and Bukovac, M.J. (1972) Penetration of stomata by liquids. Dependence on surface tension, wettability, and stomatal morphology. *Plant Physiology* 49, 813–819.

Schreiber, L. and Schönherr, J. (1990) Phase transitions and thermal expansion coefficients of plant cuticles: the effects of temperature on structure and function. *Planta* 182, 186–193.

Schumann, C., Schlegel, H.J., Grimm, E., Knoche, M. and Lang, A. (2014) Water potential and its components in developing sweet cherry. *Journal of the American Society for Horticultural Science* 139, 349–355.

Sekse, L. (1995a) Fruit cracking in sweet cherries (*Prunus avium* L.). Some physiological aspects – a mini review. *Scientia Horticulturae* 63, 135–141.

Sekse, L. (1995b) Cuticular fracturing in fruits of sweet cherry (*Prunus avium* L.) resulting from changing soil water contents. *Journal of Horticultural Science* 70, 631–635.

Sekse, L. (1998) Fruit cracking mechanisms in sweet cherries (*Prunus avium* L.) – a review. *Acta Horticulturae* 468, 637–648.

Sekse, L. (2008) Fruit cracking in sweet cherries – some recent advances. *Acta Horticulturae* 795, 615–623.

Sekse, L., Bjerke, K.L. and Vangdal, E. (2005) Fruit cracking in sweet cherries – an integrated approach. *Acta Horticulturae* 667, 471–474.

Simon, E.W. (1977) Leakage from fruit cells in water. *Journal of Experimental Botany* 28, 1147–1152.

Simon, G. (2006) Review on rain induced fruit cracking of sweet cherries (*Prunus avium* L.), its causes and the possibilities of prevention. *International Journal of Horticultural Science* 12, 27–35.

Simon, G., Hrotkó, K. and Magyar, L. (2004) Fruit quality of sweet cherry cultivars grafted on four different rootstocks. *Acta Horticulturae* 658, 365–370.

Stösser, R., Rasmussen, H.P. and Bukovac, M.J. (1969) A histological study of abscission layer formation in cherry fruits during maturation. *Journal of the American Society for Horticultural Science* 94, 239–243.

Tabuenca, M.C. and Cambra, M. (1982) Susceptibilidad al agrietamiento de los frutos de distintas variedades de cerezo. *An Aula Dei* 16, 95–99.

Thomidis, T. and Exadaktylou, E. (2013) Effect of a plastic rain shield on fruit cracking and cherry diseases in Greek orchards. *Crop Protection* 52, 125–129.

Tilbrook, J. and Tyerman, S.D. (2008) Cell death in grape berries: varietal differences linked to xylem pressure and berry weight loss. *Functional Plant Biology* 35, 173–184.

Torres, C.A., Yuri, A., Venegas, A. and Lepe, V. (2014) Use of a lipophilic coating pre-harvest to reduce sweet cherry (*Prunus avium* L.) rain-cracking. *Acta Horticulturae* 1020, 537–543.

Tucker, R. (1934) A varietal study of the susceptibility of sweet cherries to cracking. *University of Idaho Agriculture Experimental Station Bulletin* 211, 1–15.

Tukey, H.B. (1934) Growth of the embryo, seed, and pericarp of the sour cherry (*Prunus cerasus*) in relation to season of fruit ripening. *Proceedings of the American Society for Horticultural Science* 31, 125–144.

Usenik, V., Zadravec, P. and Štampar, F. (2009) Influence of rain protective tree covering on sweet cherry fruit quality. *European Journal of Horticultural Science* 74, 49–53.

Verner, L. (1937) Reduction of cracking in sweet cherries follow the use of calcium sprays. *Proceedings of the American Society for Horticultural Science* 36, 271–274.

Verner, L. and Blodgett, E.C. (1931) Physiological studies of the cracking of sweet cherries. *Bulletin Agricultural Experiment Station University of Idaho* 184, 1–5.

von Wetzhausen, C. (1819) *Systematische Classification und Beschreibung der Kirschensorten*. Cottaische Buchhandlung, Stuttgart, Germany.

Wada, H., Shackel, K.A. and Matthews, M.A. (2008) Fruit ripening in *Vitis vinifera*: apoplastic solute accumulation accounts for pre-veraison turgor loss in berries. *Planta* 227, 1351–1361.

Wada, H., Matthews, M.A. and Shackel, K.A. (2009) Seasonal pattern of apoplastic solute accumulation and loss of cell turgor during ripening of *Vitis vinifera* fruit under field conditions. *Journal of Experimental Botany* 60, 1773–1781.

Weichert, H. and Knoche, M. (2006a) Studies on water transport through the sweet cherry fruit surface: 10. Evidence for polar pathways across the exocarp. *Journal of Agricultural and Food Chemistry* 54, 3951–3958.

Weichert, H. and Knoche, M. (2006b) Studies on water transport through the sweet cherry fruit surface: 11. $FeCl_3$ decreases water permeability of polar pathways. *Journal of Agricultural and Food Chemistry* 54, 6294–6302.

Weichert, H., von Jagemann, C., Peschel, S., Knoche, M., Neumann, D. and Erfurth, W. (2004) Studies on water transport through the sweet cherry fruit surface: VIII. Effect of selected cations on water uptake and fruit cracking. *Journal of the American Society for Horticultural Science* 129, 781–788.

Weichert, H., Peschel, S., Knoche, M. and Neumann, D. (2010) Effect of receiver pH on infinite dose diffusion of $^{55}FeCl_3$ across the sweet cherry fruit exocarp. *Journal of the American Society for Horticultural Science* 135, 95–101.

Winkler, A., Ossenbrink, M. and Knoche, M. (2015) Malic acid promotes cracking of sweet cherry fruit. *Journal of the American Society for Horticultural Science* 140, 280–287.

Winkler, A., Brüggenwirth, M., Ngo, N.S. and Knoche, M. (2016) Fruit apoplast tension draws xylem water into mature sweet cherries. *Scientia Horticulturae* 209, 270–278.

Wójcik, P., Akgül, H., Demirtas, I., Sarisu, C., Aksu, M. and Gubbuk, H. (2013) Effect of preharvest sprays of calcium chloride and sucrose on cracking and quality of 'Burlat' sweet cherry fruit. *Journal of Plant Nutrition* 36, 1453–1465.

Yamaguchi, M., Sato, I. and Ishiguro, M. (2002) Influences of epidermal cell sizes and flesh firmness on cracking susceptibility in sweet cherry (*Prunus avium* L.) cultivars and selections. *Journal of the Japanese Society for Horticultural Science* 71, 738–746.

Yamaguchi, M., Sato, I., Takase, K., Watanabe, A. and Ishiguro, M. (2004) Differences and yearly variation in number and size of mesocarp cells in sweet cherry (*Prunus avium* L.) cultivars and related species. *Journal of the Japanese Society for Horticultural Science* 73, 12–18.

Yamamoto, T., Hosoi, K. and Watanabe, S. (1990) Relationship between the degree of fruit cracking of sweet cherries and the distribution of surface stress of the fruit analyzed by a newly developed system. *Journal of the Japanese Society for Horticultural Science* 59, 509–517.

Yamamoto, T., Satoh, H. and Watanabe, S. (1992) The effects of calcium and naphthalene acetic acid sprays on cracking index and natural rain cracking in sweet cherry fruits. *Journal of the Japanese Society for Horticultural Science* 61, 507–511.

Zielinski, B.Q. (1964) Resistance of sweet cherry varieties to fruit cracking in relation to fruit and pit size and fruit colour. *Proceedings of the American Society for Horticultural Science* 84, 98–102.

# 8 Climatic Limiting Factors: Temperature

**Bénédicte Wenden,[1]* José Antonio Campoy,[1] Martin Jensen[2] and Gregorio López-Ortega[3]**

[1]*UMR 1332 Biologie du Fruit et Pathologie, INRA et Université de Bordeaux, Villenave d'Ornon, France;* [2]*Department of Food Science, Aarhus University, Aarslev, Denmark;* [3]*Murcia Institute of Agri-Food Research and Development (IMIDA), Murcia, Spain*

## 8.1 Introduction

Survival and production of woody and perennial plants in temperate and boreal zones depend on precise timing of growth and rest periods in synchrony with seasonal changes in temperature (Olsen, 2010). In cherry trees, several stages of growth are subject to strict temperature control or are at risk from temperature extremes, including dormancy, flowering and fruit development. In order to survive the freezing temperatures of winter, cherry trees and other perennials have developed adaptive mechanisms that include cessation of meristem activity and bud set and an acquired tolerance to cold. In most woody plants in temperate climates, these processes are induced by decreasing photoperiod and temperature, which leads to a greater tolerance to cold and leaf fall (Allona *et al.*, 2008). In spring and summer, warm and hot temperatures can also affect flower and fruit quality. These constraints have a high impact on production and, in the context of global warming, it is essential to have a better understanding of temperature-based limitations in order to anticipate future production changes and research needs. In particular, temperature responses are partly determined genetically, and it is important to take this into account for future breeding strategies adapted to specific climates, especially in regions exposed to warmer or more variable winters.

## 8.2 Temperature Control of Dormancy

For trees in general and cherry in particular, one of the strategies to survive freezing winter temperatures is a period of dormancy, which is strongly influenced by variations in temperature. Fruiting cherry trees typically set terminal buds (ceasing active shoot growth) in mid- to late summer, as the photoperiod decreases. During autumn, the transition to short days followed by a drop in temperature increases the depth of dormancy and begins the process of cold acclimatization. This period can be separated into two main phases (Lang *et al.*, 1987): (i) endodormancy, mainly under the control of cold temperatures and defined as the inability to initiate growth from meristems under favourable conditions (Rohde and Bhalerao, 2007), followed by (ii) ecodormancy, corresponding to the period during which meristems can resume growth if temperatures are optimal. Because

* benedicte.wenden@inra.fr

the quality of flowering and fruit production are directly dependent on optimal conditions during dormancy, an exhaustive understanding of endogenous and environmental control of the processes involved is critical.

### 8.2.1 Control of dormancy stages by temperature and photoperiod

As dormancy is a mechanism that has evolved to enable plants to survive the adverse conditions of winter in temperate and cold climates, the onset of endodormancy prior to the arrival of extreme temperatures is essential (Campoy *et al.*, 2011). Cessation of apical elongation growth and bud set are initial processes demarcating the dormancy cycle (Cooke *et al.*, 2012). After bud scale initiation, organogenic activity continues for several weeks. In sweet cherry, floral organ differentiation continues during endodormancy establishment (Guimond *et al.*, 1998; Wang *et al.*, 2004). Photoperiod and temperature are the primary environmental cues that control bud set, with the relative contribution of these cues varying among species (Cooke *et al.*, 2012). In pear and apple, growth cessation and endodormancy induction and release are induced by low temperature (Heide and Prestrud, 2005). However, in several *Prunus* spp. and interspecific hybrids, an interaction between photoperiod and temperature controls growth cessation and endodormancy (Heide, 2008). Using non-bearing potted juvenile seedlings or plants derived from tissue culture, it was demonstrated that after 8 weeks of low-temperature treatment, a relatively deep state of endodormancy is attained (Heide, 2008). In *Prunus cerasus* and *Prunus avium*, endodormancy onset was insensitive to photoperiod at high temperatures, maintaining continuous growth. However, at lower temperatures, growth was controlled by the interaction of photoperiod and temperature. This interaction was stronger in *P. cerasus* and wild sweet cherry seedlings than in cultivated sweet cherry clones. *P. cerasus* and wild sweet cherry required the combination of low temperature and short days for growth cessation and formation of winter buds, whereas *P. avium* maintained active growth in short-day conditions up to moderately low temperatures (9°C). This interaction of photoperiod and temperature suggests that *Prunus* spp. may have a dual endodormancy induction control system, securing timely growth cessation and endodormancy induction in response to the progressive decrease in day length and temperature in the autumn (Heide, 2008). It is also likely that this induction system is modulated further by juvenility and maturity (G.A. Lang, Michigan, USA, 2016, personal communication). In cherry orchard trees, which are composite grafted plants usually of one species on to another or on to an interspecific hybrid, the shifts in hormonal dynamics and in nutrient and carbohydrate partitioning due to rootstock–scion interactions and the transition from non-reproductive trees to flowering and fruit production clearly modify tree response to photoperiod and temperature, resulting in growth cessation earlier during mid- to late summer than Heide's (2008) controlled environment experiments would predict.

Once endodormancy has been induced, chilling temperatures are required to initiate growth and flowering in spring (Saure, 1985). In Rosaceae species, endodormancy induction and release are driven by similar temperature conditions (Heide, 2003; Heide and Prestrud, 2005). The chilling temperatures needed to transition from endodormancy to ecodormancy are known as the chilling requirement (CR). As with endodormancy induction, the CR is not an absolute constant for a given cultivar and may vary depending on many factors such as climatic conditions, juvenile period and stress conditions. In apple, Jonkers (1979) demonstrated a tendency for endodormancy to deepen when the temperature at which the buds had been formed was higher. In other species, high temperatures induced endodormancy faster and deeper (increased requirement for chilling) than low temperatures (e.g. Westergaard and Eriksen, 1997; Heide, 2003; Junttila *et al.*, 2003). In addition, long periods of time during endodormancy with warm temperatures above 16°C may reverse accumulated chilling units and increase the CR necessary for the endodormancy-to-ecodormancy transition (Longstroth and Perry, 1996).

Due to the importance of winter chilling for temperate-zone fruit production, a number of efforts have been made to model this agroclimatic factor (reviewed by Luedeling, 2012). The most-used models in horticulture are the chilling hours model (Bennet, 1949; Weinberger, 1950), Utah model (Richardson *et al.*, 1974) and dynamic model (Fishman *et al.*, 1987a,b). In the chilling hours model, temperatures between 0 and 7.2°C are assumed to have a chilling effect for alleviation of endodormancy, with each hour at temperatures between these thresholds contributing 1 chilling hour (Bennet, 1949; Weinberger, 1950). The Utah model proposes differential temperature weights, including negative weights for temperatures above 15.9°C (Richardson *et al.*, 1974). The dynamic model assumes that chill accumulates by a two-step process and that chill can be enhanced by moderate temperatures (Erez and Couvillon, 1987). These models are not proportional, and CRs determined in a given location may not be valid elsewhere (Luedeling, 2012). In addition, the CR of a given cultivar may vary depending on the climatic conditions, either by yearly variation or when cultivated in a different region, as shown in peach (Balandier *et al.*, 1993) and apricot (Campoy *et al.*, 2012). Also, methods for estimating the date of endodormancy release (Samish and Lavee, 1962; Dennis, 2003) and for calculating CRs (Luedeling, 2012) are not standardized and may bias the values available for a given cultivar. To avoid the standardization problem and the time-consuming protocols needed to assess the CR, partial least-squares regression can be used upon availability of long-term phenology data sets (Luedeling *et al.*, 2013), although experimental data will always be necessary to validate the estimations.

Cherry CRs vary widely depending on cultivar and climatic conditions (Table 8.1) (Cortés and Gratacós, 2008; Tersoglio *et al.*, 2012). Alburquerque *et al.* (2008) calculated the CR of a group of seven cultivars in southern Spain, which ranged from 30.4 ('Cristobalina') to 57.6 ('Marvin') chill portions using the dynamic model (Table 8.1) (Fishman *et al.*, 1987a,b). They estimated the altitude above sea level needed for each cultivar to satisfy the CR in a warm winter area. Castède *et al.* (2014) found a full range of 40–85 chill portions in south-western France in a mapping population derived from 'Regina' and 'Garnet' evaluated for three consecutive years.

The main symptoms of inadequate chilling fulfilment in temperate fruit trees are delayed bud burst, reduced bud burst, and uneven bud burst and flowering (reviewed by Erez, 2000). In sweet cherry, taking the chilling duration into account is crucial for assuring profitable production. The low productivity of 'Burlat' in mild winters is due largely to the lack of chilling, leading to a low rate of and sporadic bud break in addition to various anatomical abnormalities in flower buds, such as the absence of pistils, atrophied ovules and immature pollen (Oukabli and Mahhou, 2007). In warm winter regions in southern China, poor fruit set, abnormal floral organs and irregular fruit growth have also been described in sweet cherry (Xu *et al.*, 2014). The poor fruit set is related to inadequate ovule and embryo sac development in mild winter areas (Wang *et al.*, 2004).

Once endodormancy has been released, warm temperatures are needed to resume growth and achieve flowering. This is known as the heat requirement (HR) for the transition from ecodormancy to active growth. It is still not clear whether cultivars have specific HRs for flowering (Overcash, 1965; Rom and Arrington, 1966; Gianfagna and Mehlenbacher, 1985; Alonso *et al.*, 2005) or whether flowering date is determined basically by CR (Brown, 1957; Swartz and Powell, 1981; Couvillon and Erez, 1985; Pawasut *et al.*, 2004; Ruiz *et al.*, 2007; Campoy *et al.*, 2010; Okie and Blackburn, 2011). Several studies have shown that CRs have much stronger effects on flowering date than HRs in sweet cherry (Alburquerque *et al.*, 2008; Castède *et al.*, 2014), as well as in other *Prunus* spp., such as apricot (Campoy *et al.*, 2012), almond (Sánchez-Pérez *et al.*, 2012) and peach (Okie and Blackburn, 2011). In sweet cherry, a multi-year quantitative trait locus (QTL) analysis of flowering time, CR and HR showed that the highly heritable

**Table 8.1.** Chilling requirements calculated for different cultivars at various locations.

| Location | Altitude above sea level (m) | Species | Cultivar | Hours below 45°F (7.2°C) | Chill units[a] | Dynamic model[b] | Growing degree hours[a] | Flowering date | Methodology to assess chill requirement | Source |
|---|---|---|---|---|---|---|---|---|---|---|
| Pozanti, Turkey | 1100 | *Prunus avium* | 'Kordia' | 700–750 | 150 | – | 14,000 | – | 50% green tip in cuttings after 21 days forcing at 24°C | Kuden *et al.* (2012) |
| | | | 'Lapins' | 400–450 | 94 | – | 15,500–16,000 | – | | |
| | | | 'Larian' | 450 | 94 | – | 15,500–16,000 | – | | |
| | | | 'Nafrina' | 500–550 | 120 | – | 15,000–15,500 | – | | |
| | | | 'Noir de Guben' | 600–650 | 110 | – | 14,000–14,500 | – | | |
| | | | 'Summit' | 650 | 125 | – | 15,000 | – | | |
| | | | 'Sunburst' | 650–700 | 141 | – | 14,000–14,500 | – | | |
| | | | '0900-Ziraat' | 600–650 | 134 | – | 15,500–16,000 | – | | |
| Abarán, Spain | 270 | *Prunus avium* | 'Cristobalina' | 176 | 397 | 30.4 | 9195 | 14 March | 50% green tip in branches after 10 days forcing at 24°C | Alburquerque *et al.* (2008) |
| Jumilla, Spain | 360 | *Prunus avium* | 'Brooks' | 411.5 | 556 | 36.7 | 7863.2 | 27 March | 50% green tip in branches after 10 days forcing at 24°C | Alburquerque *et al.* (2008) |
| | | | 'Ruby' | 618 | 806 | 48 | 7326.2 | 29 March | | |
| | | | 'Somerset' | 618 | 806 | 48 | 8625.2 | 3 April | | |
| | | | 'Burlat' | 618 | 806 | 48 | 8750.2 | 4 April | | |
| | | | 'New Star' | 709.5 | 909.3 | 53.5 | 8257 | 4 April | | |
| | | | 'Marvin' | 788 | 1001.5 | 57.6 | 9449.7 | 9 April | | |
| | | | 'Bing' | 900 | 900 | | | | | Adapted from Longstroth and Perry (1996) |
| | | | 'Emperor Francis' | 1300 | 1100 | | | | | |
| | | | 'Early Burlat' | 1300 | 1100 | | | | | |
| | | | 'Van' | 1350 | 1150 | | | | | |
| | | | 'Hedelfingen' | 1400 | 1200 | | | | | |
| | | *Prunus cerasus* | 'Montmorency' | | 950 | | | | | Adapted from Longstroth and Perry (1996) |

[a]Richardson *et al.* (1974).
[b]Fishman *et al.* (1987a,b).

component of flowering time is CR, and that HR has a high genotype × environment interaction (Castède *et al.*, 2014). Consequently, it is essential to correctly evaluate the response to warm temperature and anticipate the flowering period. The effect of temperature on plant development rate can be modelled and is often described using a thermal time concept. Several methods proposed include growing degree days (McMaster and Wilhelm, 1997; Zavalloni *et al.*, 2006; Ruml *et al.*, 2010), growing degree hours (Alburquerque *et al.*, 2008; Ruml *et al.*, 2011; Guo *et al.*, 2014) and photothermal units (Blümel and Chmielewski, 2012; Chmielewski and Götz, 2016). These models assume that heat accumulates when daily or hourly temperatures occur above a base temperature, starting from a fixed date, 1 January for example, or following satisfaction of the CR. More recently, several models have been proposed for the prediction of sour and sweet cherry flowering based on CR and HR. Neilsen *et al.* (2015) developed a multi-stage model in sweet cherry to predict flowering time in a temperate climate involving separate estimates of CR fulfilment, a lag period following endodormancy release with no activation occurring, and a final growth activation stage towards flowering. The model demonstrated a mean absolute error at 1.6–2 days for estimating full bloom. Furthermore, phenological models for sour and sweet cherry flowering validated the idea that the addition of a day-length term in the forcing equations improves the predictive capacity of the models (Matzneller *et al.*, 2014; Chmielewski and Götz, 2016).

### 8.2.2 Molecular control of dormancy and flowering

Inheritance studies of the genetic control of flowering time in cherry have shown that dormancy release and flowering time are polygenic traits (Wang *et al.*, 2000; Castède *et al.*, 2014). The flowering process in perennial plants is different from that of annual plants, with bud dormancy release and reset being specific to perennials. In trees, and fruit trees in particular, flower buds are differentiated in the year preceding flowering, but the precise timing for flowering is determined by the response to temperature during dormancy. In the model plant *Arabidopsis*, as well as in many other annual plants such as wheat and rice, analyses of genes involved in flowering processes have led to the identification of four major pathways involved in transduction of environmental and endogenous signals: vernalization, photoperiod, gibberellic acid and the autonomous pathways; and a large number of genes were identified to regulate flowering time (Amasino, 2010). In woody perennials, genomic studies conducted on model species, including *Populus* and peach, have led to the identification of candidate genes potentially involved in the response to temperature, controlling the timing of dormancy and flowering. Overall, genetic and molecular approaches have revealed a remarkable conservation of the genetic pathways regulating phenology and the pathway controlling flowering time in *Arabidopsis* (reviewed by Ding and Nilsson, 2016).

In *Arabidopsis*, *FLOWERING LOCUS T* (*FT*) acts as a central activator of flowering, being one of the main points of convergence of all signalling pathways controlling flowering time. Several analyses in perennials have led to the discovery that *FT* homologues are key players in growth cessation, bud set and probably dormancy onset (Hsu *et al.*, 2011; Srinivasan *et al.*, 2012). In sweet cherry, as in peach (Zhang *et al.*, 2015), only one *FT* homologue was identified. This gene co-localized with a QTL for flowering date located on linkage group (LG) 6 (Castède *et al.*, 2015). Another key gene, *TERMINAL FLOWER 1* (*TFL1*) is expressed in *Arabidopsis* during the vegetative phase and acts antagonistically to *FT* for floral determination (Ahn *et al.*, 2006). In woody perennials, *TFL1*-like genes or other members of the *TFL1/CENTRORADIALIS* (*CEN*) family appear to be involved in shoot phenology including growth cessation and dormancy (Ruonala *et al.*, 2008; Mimida *et al.*, 2009; Mohamed *et al.*, 2010). Several members of the *TFL1/CEN* family were isolated from sweet cherry but their function was not precisely deciphered (Mimida *et al.*, 2012).

Gibberellins (GAs) are key mediators between the perception of environmental signals and the resulting growth responses, including stem elongation and flowering time. In transgenic hybrid aspen trees overexpressing *GA20ox* genes, cessation of apical growth was delayed, thus suggesting that GAs are also regulators of growth cessation in trees (Eriksson *et al.*, 2000, 2015). Transcriptome analysis of Japanese pear (*Pyrus pyrifolia* Nakai) flower buds revealed that genes involved in GA metabolism were also linked to dormancy stages, since transcripts of *GA20ox* were less abundant during ecodormancy, whereas *GA2ox* was upregulated after dormancy release (Bai *et al.*, 2013). Both *GA2ox* and *GA20ox* homologues are located in the confidence interval of the flowering date QTL on LG4 in sweet cherry. They therefore appear to be serious candidate genes for flowering time and dormancy control (Castède *et al.*, 2014, 2015).

Much research has characterized 'Evergreen', a non-dormant peach tree genotype that fails to both cease growth and enter dormancy under dormancy-inducing conditions (Rodriguez *et al.*, 1994). Genomic studies revealed that the mutant was affected in a cluster of six *DORMANCY ASSOCIATED MADS*-box (*DAM*) genes (Bielenberg *et al.*, 2008). These genes belong to the *SHORT VEGETATIVE PHASE/AGAMOUS-LIKE 24* (*SVP/AGL24*) subfamily of MADS-box genes in *Arabidopsis*, which are known to act in the flowering response to environmental signals (Hartmann *et al.*, 2000; Michaels *et al.*, 2003). *DAM* genes are seasonally regulated and their expression seems to be correlated to dormancy stages. In particular, peach *DAM5* and *DAM6* genes are upregulated during growth cessation and their expression is subsequently inhibited by chilling temperatures during winter (Jiménez *et al.*, 2010), thus suggesting a role in establishing and maintaining endodormancy (Li *et al.*, 2009; Saito *et al.*, 2013). These expression patterns were confirmed in Chinese cherry (*Prunus pseudocerasus*) (Zhu *et al.*, 2015). In sweet cherry, *DAM5* and *DAM6* co-localize with a major QTL for flowering time in LG1 (Castède *et al.*, 2015) but it is probable that other genes also play a crucial role in the regulation of dormancy, since in that study the QTL with the highest effect is on LG4.

Both in endodormancy and vernalization systems, processes leading to flowering and flower induction, respectively, are triggered after plants have been exposed to a certain amount of cold temperature, suggesting similar response mechanisms (Chouard, 1960; Horvath, 2009). The signalling pathways controlling the vernalization response in *Arabidopsis* have been investigated in detail and involve the chromatin remodelling of a central flowering repressor, the MADS-box transcription factor *FLOWERING LOCUS C* (*FLC*) (He, 2012). In sweet cherry, several candidate genes identified as homologues of *Arabidopsis* genes involved in chromatin remodelling or modification complexes were located in major QTLs for flowering and dormancy, such as *EMBRYONIC FLOWER2* (*EMF2*), *PHOTOPERIOD-INDEPENDENT EARLY FLOWERING 1* (*PIE1*) and *ACTIN-RELATED PROTEIN 4-LIKE* (*ARP4*) (Castède *et al.*, 2014, 2015). At present, no *FLC* orthologue has been found in either the peach genomic sequence or the available sweet cherry transcribed sequences (Castède *et al.*, 2014), so the search for the key player in the dormancy response to prolonged cold is ongoing. However, interestingly, as is the case for the regulation of *FLC*, chromatin remodelling and histone modifications have been observed in *DAM* genes as a consequence of prolonged exposure to low temperatures, suggesting another similarity between the signalling pathways controlling flowering in *Arabidopsis* and the regulation of growth and dormancy cycles in trees (Leida *et al.*, 2012; Ríos *et al.*, 2014; de la Fuente *et al.*, 2015; Saito *et al.*, 2015).

## 8.3 Cold Resistance and Spring Frost Damage

### 8.3.1 Molecular control of cold hardiness

Cold hardiness is the ability of plants to adapt to and withstand freezing temperatures. Cold-hardy woody plants can develop freezing tolerance to temperatures below

−30°C and can even acclimate to extremely low temperatures down to −196°C, as shown by Rinne *et al.* (1998) for birch (*Betula pubescens*). The major stress encountered by cells during freezing is severe freeze-induced dehydration resulting from the presence of extracellular ice, which leads to membrane structural damage, protein denaturation and oxidative stress (Pearce, 1999). Cold hardiness is triggered by environmental factors that usually precede freeze periods such as short photoperiod and low non-freezing temperature, but also endogenous cues including abscisic acid (ABA) (Junttila *et al.*, 2002). There are various associated tolerance mechanisms and these involve biophysical processes including deep supercooling of xylem tissues and the existence of preferred sites for ice formation, in addition to the production of specific metabolites and proteins and changes in membrane structure (Gusta and Wisniewski, 2013; Wisniewski *et al.*, 2014). Changes in plasma membrane composition are a key feature of cold acclimatization, including an increase in fatty acid desaturation in membrane lipids, which maintains functional membrane fluidity under low temperatures. Water content and osmotic potential of the buds and stem have been shown to decrease in autumn and in response to a short photoperiod. This programmed dehydration contributes to extreme hardiness of the tissues by preventing ice formation and reducing the effects of freeze-induced cellular dehydration. Sugars also play an essential role in cold acclimatization, as shown by the correlation observed between sugar content and freezing tolerance in woody plants, associated with various processes including the formation of a metastable cell solution and osmoregulation (Wisniewski *et al.*, 2003). By studying the content of soluble sugars and sucrose-metabolizing enzymes in sweet cherry, Turhan and Ergin (2012) confirmed that soluble sugar, reducing sugars and sucrose contents were higher in cold-acclimatized stages than in non-acclimatized stages.

Plant hormones, especially ABA and ethylene, have been shown to play an essential role in plant stress signalling. ABA levels increase in woody plants under conditions that lead to cold acclimatization, and application of exogenous ABA can increase freezing tolerance without low-temperature exposure (Rinne *et al.*, 1998). As well as its demonstrated role in dormancy regulation, ethylene levels increase during cold acclimatization and ethylene acts as a positive regulator of plant freezing tolerance through the activation of cold-induced gene expression and the production of antifreeze proteins (Yu *et al.*, 2001; Catalá *et al.*, 2014). Several studies on diverse plants have identified common proteins associated with low temperatures such as the cold-regulated (*COR*) genes and the cold-inducible *C-repeat-binding* (*CBF*)/*Dehydration-responsive element-binding1* (*DREB1*) factors (Welling and Palva, 2006). The *CBF* genes are induced within 15 min of plant exposure to cold, followed by the induction of CBF target genes, the 'CBF regulon', which all share the LTRE/DRE/CRT (low-temperature response element, drought responsive element, or c-repeat) element in their promoter (Medina *et al.*, 2011). The CBF protein family appears to be highly conserved among plant species. However, regulation of *CBF* genes in woody plants seems to be more complex than in herbaceous species, with specific expression patterns for gene, tissue and age (Wisniewski *et al.*, 2014). For example, woody-plant *CBF* genes are induced in various conditions including low and freezing temperatures after short-day exposure and growing season, suggesting that they participate not only in seasonal cold acclimatization but also in acclimatization to periodic freeze episodes during the growing season (Welling and Palva, 2006).

At least three *CBF*/*DREB1* homologues have been identified in sweet cherry, and the conservation of function has been confirmed by the overexpression of one of the sweet cherry *CBF*/*DREB1* genes in *Arabidopsis* (Kitashiba *et al.*, 2002, 2004). In one of the transgenic plant lines, the CBF/DREB1 target gene *cor15a* was induced in the absence of stress treatment and the plant displayed higher freezing tolerance (Kitashiba *et al.*, 2004). One putative sour cherry orthologue of *CBF1* was also shown to be upregulated in leaves after exposure to cold temperature

(Owens *et al.*, 2002), thus supporting the hypothesis that the *CBF* gene system is widely conserved. Some peach *DAM* genes have CRT/DREB response elements in their promoters, and one apple *DAM* gene promoter has a CRT-like element, suggesting that cold resistance and dormancy pathways are intertwined (Wisniewski *et al.*, 2011). In *Arabidopsis*, many *COR* genes are characterized by the presence of a CRT or DRE element in their promoter and their expression is increased under non-acclimating conditions in transgenic plants overexpressing *CBF* genes (Polashock *et al.*, 2010). Zalunskaitė *et al.* (2008) showed that the expression of sweet and sour cherry homologues of the *Arabidopsis COR47* gene increased during cold acclimatization. Among the *COR* genes, dehydrins are induced by stresses that cause cellular dehydration, including low and freezing temperatures, and results from a wide range of species suggest that dehydrins operate as protective proteins (Welling and Palva, 2006; Wisniewski *et al.*, 2014). In particular, dehydrins were characterized in peach, induced by low temperature or water stress (Artlip *et al.*, 1997; Wisniewski *et al.*, 2006). At present, knowledge on dehydrins in cherry is very limited, but one study seems to have identified cold-responsive proteins in cherry microshoots (Lukoševičiūtė *et al.*, 2009).

The processes of de-acclimatization and re-acclimatization remain less understood. De-acclimatization is often defined as a reduction in the levels of cold hardiness due to diverse factors such as environmental stimuli (warmer temperatures and long days), phenological changes and reactivation of growth. In woody plants, de-acclimatization occurs more rapidly than cold acclimatization (Howell and Weiser, 1970; Arora *et al.*, 1992). This process, together with growth renewal, is associated with tissue and cellular rehydration, and activation of enzymes involved in sugar metabolism (Andrews and Proebsting, 1987; Kalberer *et al.*, 2006; Turhan and Ergin, 2012). De-acclimatization and re-acclimatization capacity play a significant role in determining plant hardiness during bud burst and flowering when plants are particularly vulnerable to cold injury, and can vary widely among species and cultivars (Mathers, 2004).

### 8.3.2 Physiological effect of freezing temperatures on buds

Although low mid-winter freezing temperatures are an important limiting parameter in high-latitude temperate regions marginal for cherry cultivation, freezing temperatures during acclimatization in autumn, de-acclimatization in spring and flowering are also strongly associated with loss of sweet and sour cherry crops (Mathers, 2004; Caprio and Quamme, 2006). Freeze damage may be caused by different climatic situations, such as radiative, advective and evaporative cooling, and the damage will depend on the intensity, rate of temperature decrease and duration of the freezing event (Rodrigo, 2000). The actual damage is caused mainly by intracellular ice formation, whereas extracellular ice formation may not cause damage. A slow extracellular freezing event will produce a water potential gradient that will pull water out of the cells, thus lowering the water content and lowering the risk of intracellular ice formation. Sudden rapid freezing events do not allow time to lower intracellular water content and therefore often produce greater damage to buds. Severe desiccation during long exposure to very low temperatures may cause cell membranes to coagulate and kill the cells, but most freezing damage is incurred by the formation of ice crystals that disrupt cell membranes and intracellular components. Different flower bud tissues may have different sensitivities to ice formation, potentially caused by physical discontinuity between tissues, differences in water potential of tissues, or the presence or absence of ice nucleators in the tissue. During flowering, such differences seem much smaller than in mid-winter when buds are still endodormant. The long-term consequence of ice formation depends on the tissue type affected and how many cells are involved. Damage to the ovule itself and style is usually detrimental, whereas local surface damage to the ovary or integuments may be repaired partially but results in abnormal fruit development.

Ice nucleation normally spreads rapidly into nearby tissue, and the phase transition of water to ice produces an exotherm in buds that can be measured by differential thermal analysis using thermocouples, infrared thermography or differential scanning calorimetry coupled with an infrared thermal video. A high-temperature exotherm and a low-temperature exotherm typically are measured. The ability to supercool water in buds is extremely important for avoiding ice formation and damage, and this ability depends on water content (Mathers, 2004).

Oxidative browning of tissue is a very common symptom of damage following severe frost damage, and topographical evaluation of the extent and intensity of this in buds is important as a method to evaluate tissue damage. Membrane disruption also results in leakage of electrolytes from tissue and this can be measured in standardized tests to quantify the extent of damage, albeit as an average for the tissue tested (Jensen and Kristiansen, 2014). Damage may be evaluated immediately after freezing by dissection of flower buds for symptoms (Longstroth, 2013). Alternatively, branches may be grown at controlled temperatures or excised buds cultivated on agar medium for a period to allow normal budburst and development of flowers or deterioration of floral tissue.

Flower buds of cherry may be damaged by freezing temperatures from autumn to after flowering and during early fruit development in spring. Sour cherry is generally considered to be slightly more resistant to frost than sweet cherry (Szabó *et al.*, 1996; Mathers, 2004), but there is a large overlap depending on variation among cultivars and interaction with regional climates.

In autumn, damage often occurs when a warm period is followed by a rapid drop to freezing temperatures. In sour cherry in the northern hemisphere, this would typically be in November (Mathers, 2004). Generally, in mid-winter, cherries are normally hardy to between −20 and −25°C, but this varies significantly with cultivar and regional climate. In sour cherry, significant bud damage occurred at winter temperatures of −12°C (Dencker and Toldam-Andersen, 2005).

Andrews *et al.* (1983) suggested four stages of changing frost tolerance in cherry during de-acclimatization in spring: (i) during dormancy when flower buds have the ability to supercool; (ii) a transition period when buds begin to swell and supercooling is progressively lost: (iii) before petal tip emergence; and (iv) following petal emergence, during which flowers and fruit are very sensitive to frost. Spring frost damage also may occur after flowering during early fruit development.

Early warm spells occurring just after endodormancy release in winter or early spring may lead to transient de-acclimatization and bud activation; when followed by low temperatures and rapid freezing, buds may be damaged (Mathers, 2004). Cultivars with a low CR may begin de-acclimatization earlier if temperatures are high. Cultivars with a low base temperature for de-acclimatization and/or a short HR are often more prone to frost damage.

During flowering, bud hardiness is very limited and even shallow freezing (−2°C) may cause damage to some buds. Flowering time thus has a large impact on the risk of frost damage in different cultivars. The most cited and still widely used studies on low-temperature resistance to frost in sweet and sour cherry buds are those by Proebsting and Mills (1978) and Ballard *et al.* (1982), which show temperatures that can cause damage to 10% and 90% of buds for each developmental stage. These, together with revised numbers, are shown on Michigan State University's Cherry website (Anon., 2014).

In sour cherry, Stepulaitienė *et al.* (2013) investigated the onset and duration of different phenological phases and the frost tolerance in each of these periods in Lithuania. In early phases before and during meiosis (swollen bud, side green, green tip and tight cluster), no damage was seen at −8°C. After these stages, however, testing at −4°C revealed first injury to ovaries and styles during the developmental stages of first white, full bloom and fruit germ. Szabó *et al.* (1996) found almost 100% damage in fully open flowers at –2.5°C in 'Érdi Bőtermő', 'Érdi Nagygyümölcsu' and 'Meteor Korai'.

Recently, Salazar-Gutiérrez *et al.* (2014) used differential thermal analysis and freezing tests with microscope dissection of buds to update this knowledge and demonstrated changes in the lethal temperatures for killing 10 ($LT_{10}$), 50 ($LT_{50}$) or 90% ($LT_{90}$) of a population of flower buds in three cultivars of sweet cherry, during the period from autumn to flowering in spring. $LT_{10}$ values for dormant buds were about –20°C, whereas at the tight cluster stage $LT_{10}$ was only –6°C. De-hardening was especially fast from the end of February in Washington, USA, following endodormancy release. De-acclimatization and growth activation depend strongly on temperature, and at constant low temperatures, ecodormancy is retained, thus delaying further bud development. From green tip to full flowering in March, $LT_{10}$ was only –5 to –2°C. For most of the period from November to March, the difference between $LT_{10}$ and $LT_{90}$ was about 7–10°C, suggesting a fairly large difference in hardiness of individual buds. This difference decreased to only 1–3°C in April near flowering, demonstrating that the risk of damaging a large proportion of buds is much higher during spring frost than during winter. Similar results with 'Burlat' and 'Sunburst' were obtained by Miranda *et al.* (2005), who showed that the temperature differential for 10% and 90% damage was negligible during flowering. Szabó *et al.* (1996) found 82–100% damaged flowers in sweet cherry cultivars at –2.5°C during full bloom, whereas almost no damage occurred at the same temperature in the balloon stage. Knowledge on current LT values in buds according to tests may, together with local weather forecasts, give information on frost damage risks for use to initiate frost protection activities in orchards.

### 8.3.3 Variations in cultivar resistance to frost damage

Generally, damage to a small percentage of buds does not lower marketable yield due to increased compensation growth by non-damaged fruit. In sweet cherry, damage to less than 20–30% of flower buds may not reduce marketable yield. In sour cherry, around 15% non-damaged buds may result in a normal yield. The exact impact on yield depends on the cultivar, cultivation conditions, marketable fruit quality and potential for compensation growth.

Genetic control of spring bud hardiness may be expressed through differences in flowering time, variation in flower bud stage, the density of flower buds, critical temperatures for damage and the ability to supercool (Rodrigo, 2000). The most recent multi-stage model developed by Neilsen *et al.* (2015) may be used to identify both regional areas for growing cherries and the most adapted cultivars for local regions. Cittadini *et al.* (2006) developed models for risk of frost damage for 'Bing', 'Burlat', 'Lapins', 'Stella', 'Sunburst' and 'Van', and found only small cultivar differences, with 'Sunburst' having the lowest risk. Kadir and Proebsting (1994) compared the hardiness of 'Bing' with three breeding selections and found about a 2°C difference in lethal temperatures between the most and least hardy of these genotypes, and that this difference was evident during the entire dormant period. Similarly, Liu *et al.* (2012) and Jensen and Kristiansen (2014) found a persistent difference in hardiness between the sour cherry cultivars 'Stevnsbaer Birgitte' and 'Kelleris 16'.

A number of studies have evaluated cultivar performance after severe frost events or after artificial freezing tests (Table 8.2). Salazar-Gutiérrez *et al.* (2014) only found small differences between 'Bing', 'Chelan' and 'Sumtare' (Sweetheart™) in freezing exotherms for $LTE_{10}$ (10% low-temperature exotherms), $LTE_{50}$ and $LTE_{90}$. Buds of frost-sensitive sweet cherry cultivars were found to begin bud activation earlier and also flowered earlier (Kadir and Proebsting, 1994). Stepulaitienė *et al.* (2013) found that sour cherry cultivars differed both in onset time and duration of each phenological phase, and this was related to the risk of frost events in the most sensitive periods. Iezzoni and Mulinix (1992) showed that sour cherry flowering time and heat unit accumulation (base temperature 10°C) from pollen mother

**Table 8.2.** Overview of investigations on frost hardiness in sweet and sour cherries.

| Cherry type | No. cultivars/ genotypes compared | Country | Year of experiments | Experimental protocol[a] | Test result parameters | Reference |
|---|---|---|---|---|---|---|
| Sweet | 4 | USA | 1991–1992 | Artificial | DTA, LTE, damage to flower buds | Kadir and Proebsting (1994) |
| Sweet | 17 | Germany | 1992/1993 | Artificial | Flower bud damage, wood damage | Fischer and Hohlfeld (1998a) |
| Sweet | 131 (38) | Germany | 1995/1996/1997 | Natural (131) and artificial (38) | Damage to vegetative buds, flower buds, wood | Fischer and Hohlfeld (1998a) |
| Sweet | 2 | Spain | 2000–2001 | Artificial | % damage to flower buds | Miranda *et al.* (2005) |
| Sweet | 6 | Argentina | 1997, 1999, 2000, 2001 | Natural | Frost damage frequency, indirect | Cittadini *et al.* (2006) |
| Sweet | 10 | Poland | 2005/2006 | Natural | Flower bud frost damage index, fruit set | Szewczuk *et al.* (2007) |
| Sweet | 23 | Romania | 2011/2012 | Natural | % damage to flower buds, | Asănică *et al.* (2012) |
| Sweet | 3 | USA | 2012–2013 | Artificial, DTA | $LT_{10}$, $LT_{50}$, $LT_{90}$, % damage to flower buds, HTE and LTE | Salazar-Gutiérrez *et al.* (2014) |
| Sweet | 5 | Romania | 2013/2014 | Artificial | % damage to flower buds | Asănică *et al.* (2014) |
| Sour | 1 cultivar on 4 rootstocks | Denmark | 1998–1999 | Natural | % damage to flower buds | Kühn and Callesen (2001) |
| Sour | 1 cultivar on 2 rootstocks | Denmark | 1998–2000 | Natural and artificial | % damage to flower buds | Dencker and Toldam-Andersen (2005) |
| Sour | 14 | Hungary | 2005/2006, 2006/2007 | Artificial | % damage to flower buds | Pedryc *et al.* (2008) |
| Sour | 2 | Denmark | 2010–2011 | Natural and artificial | % bud damage, | Liu *et al.* (2012) |
| Sour | 8 | Denmark | 2009–2010 | Artificial | % damage to flower buds | Clausen *et al.* (2012) |
| Sour | 7 | Lithuania | 2010–2012 | Artificial | Damage to flower buds | Stepulaitienė *et al.* (2013) |
| Sour | 12 | Denmark | 2008 | Artificial | REL %, % bud damage, bud water content | Jensen and Kristiansen (2014) |
| Sour, sweet | 23 sour, 16 sweet | Hungary | Sweet: 1986, 1987, 1991; sour: 1987, 1991 | Natural | % damage to flower buds | Szabó *et al.* (1996) |
| Sour, sweet | 12 sour, 2 sweet, 1 *P. fruticosa* | USA | 1990–1991 | Artificial | LTE, $LT_{50}$ of flower buds | Mathers (2004) |

DTA, differential thermal analysis; LTE, low-temperature exotherm; HTE, high-temperature exotherm; $LT_{10/50/90}$, lethal temperature for 10/50/90% of population.
[a]Natural frost event or artificial frost testing.

cell formation (beginning of April) until bloom is genetically controlled and varied by more than 2 weeks between individual seedlings (from 8 to 24 May).

In Romania, Asănică *et al.* (2012) evaluated winter freeze damage to flower buds of 23 sweet cherry cultivars at three locations in early February and found the highest percentage damage in 'Germersdorf' (48%) and 'Giant Red' (45%), but less than 10% damage in 'Regina' and 'Van'. In some cases, bud damage was up to 28% higher in buds at the lower positions in crowns compared with the top of the crown, probably due to colder air lower in the canopy. Asănică *et al.* (2014) compared spring freeze damage to flower buds of five sweet cherry cultivars in the field with artificial freezing at different durations and temperatures from −1.5 to −7°C and found that 'Rivan', 'Burlat' and 'Kordia' had about 40–50% bud damage at −1.5°C for 1 h, whereas 'Katalin' and 'Regina' had only 17–20% damage. 'Rivan' and 'Burlat' also had lower HRs from swelling to budburst and bloom, and flowered earlier than the other cultivars.

Szewczuk *et al.* (2007) evaluated bud damage and fruit set in ten sweet cherry cultivars after a late January −25°C freeze event in Poland. 'Regina' and 'Karina' were damaged the most, while damage was less in 'Summit', 'Bianca' and 'Viola'. Fischer and Hohlfeld (1998a) tested 17 sweet cherry cultivars in Germany by artificial freezing from −18 to −28°C and found that 'Naresa', 'Pi-Na481', 'Van' and 'Nalina' were the most hardy, whereas 'Nadino', 'Vinka', 'Hedelfinger', 'Altenburger' and 'Querfurter' were the least hardy. In a follow-up study, Fischer and Hohlfeld (1998b) evaluated 131 sweet cherry cultivars in Dresden, Germany, after field freezes down to −25°C and by laboratory testing with artificial freezing. Vegetative buds, flower buds and wood damage were evaluated. Although the correlation between field and laboratory results was not very high, they were able to classify cultivars into nine groups of frost tolerance. 'Bianca', 'Namare', 'Namati', 'Naresa', 'Rivan', 'Valera', 'Drogens Gelbe', 'Meckenheimer Fruhe' and 'Rainier' were very hardy, whereas 'Cristobalina', 'Sparkle', 'Napoleon', 'Bigarreau Reverchon' and 'Velvet' were very susceptible to low-temperature damage.

Szabó *et al.* (1996) in Hungary compared freeze injury in 16 sweet cherry and 23 sour cherry cultivars over several years and in different growing regions after various frost events. In general, early-flowering cultivars such as 'Meteor Korai' were much more damaged than late-flowering cultivars such as 'Újfehértói Fürtös'. 'Cigány Meggy 7', 'Paraszt Meggy', 'Pándy Meggy' and 'Újfehértói Fürtös' were the most freeze resistant, whereas 'Érdi Bőtermő', 'Erdi Nagygyümölczu' and 'Meteror Korai' were very freeze sensitive. In sweet cherry, 'Schneiders Späte Knorpelkirsche', 'Bigarreau Marmotte', 'Van' and 'Ulster' were freeze damaged, whereas 'Tardif de Vignola', 'Sam', 'Lambert' and 'Burlat' were damaged less. Pedryc *et al.* (2008) compared freeze tolerance in 14 sour cherry cultivars at different freezing temperatures depending on the time of year. They classified 'Cigány Meggy 59', 'Erdi Jubileum', 'Pándy 279' and 'Erdi Nagygyümölczu' as the most tolerant and 'Érdi Bőtermő', 'Maliga Emleke', 'Piramis' and 'Meteor Korai' as the most sensitive cultivars.

Liu *et al.* (2012) showed that 'Stevnsbaer Birgitte' was significantly less hardy and had a much higher degree of damage to buds than 'Kelleris 16', as early as October in Denmark at −15°C, when 80–90% buds were damaged, but also in mid-winter and during de-acclimatization. Similar damage profiles in 'Stevnsbaer' were described by Kühn and Callesen (2001). Dencker and Toldam-Andersen (2005) compared young 'Birgitte' trees on 'Colt' or 'Weiroot 10' rootstocks exposed to a mild winter with minimum temperatures of −12°C in January. Bud mortality in the field was 62% for trees on 'Colt', but only 29% on 'Weiroot 10'. Clausen *et al.* (2012) further investigated different selections of 'Stevnsbaer' compared with 'Birgitte' at artificial freezing of dormant buds to 9°C in December and found that 'Birgitte' was the least hardy of all eight tested cultivars and was less hardy than 'Stevnsbaer Viki'.

Jensen and Kristiansen (2014) found that bud endodormancy was released in all 12 sour cherry cultivars tested in Denmark in week 5 (beginning of February). Bud water content rose significantly between weeks 5 and 8, suggesting activation was initiated. Freezing tolerance declined significantly between weeks 5 and 8, after which buds were damaged at −10°C. 'Stevnsbaer Verner Skov' and 'Kelleris 16' were the most frost tolerant to −15°C in week 2, and 'Stevnsbaer Birgitte' was less tolerant than 'Kelleris 16' in all periods from week 49 to week 2. Stepulaitienė *et al.* (2013) investigated freezing of flower buds to −4°C during spring in seven sour cherry cultivars that differed in their freeze resistance during first white, full bloom and fruit germ stages.

Mathers (2004) compared flower bud $LT_{50}$ values after artificial freezing of 12 sour cherry and two sweet cherry cultivars. $LT_{50}$ values were about 5–7°C lower in November and in March in 'Oblačinska' and 'Meteor' than in 'Pándy 114' and 'Cigány Meggy', suggesting a higher freeze tolerance. Measuring LTEs after a short standardized hardening period was a more reliable method for comparing hardiness of cultivars than multi-year field trials. Delayed de-acclimatization was suggested to be important for future development of hardy sour cherries.

## 8.4 Effects of Warm Temperatures on Flower and Fruit Development

Traditionally, sweet cherries have been produced in cool areas, but in order to harvest earlier when market prices are higher, sweet cherry producers have been moving to warm regions all over the world (Micke *et al.*, 1983; Southwick *et al.*, 1991; García-Montiel *et al.*, 2010; Li *et al.*, 2010; Beppu and Kataoka, 2011; James and Measham, 2011). However, production stability in these areas is limited by poor fruit set, the occurrence of double fruit (Fig. 8.1A), malformed flowers containing pistil-like appendages in lieu of anthers (Philp, 1933; Beppu and Kataoka, 1999; Roversi, 2001; Martin, 2008) (Fig. 8.1B) and fruit with deep sutures, because the carpel margins do not fuse at the base and remain open (Southwick *et al.*, 1991; Engin *et al.*, 2009). Among these problems, poor fruit set and the frequent occurrence of double fruit are the most serious ones affecting orchard profitability in warm regions. In addition, global warming has produced a recent increase in the incidence of double fruit (Martin, 2008; Imrak *et al.*, 2014), even in new areas previously not familiar with the problem (Roversi *et al.*, 2008).

### 8.4.1 Formation of double pistils and double fruit

High summer temperatures during flower bud differentiation are believed to cause double pistil formation, resulting in double fruit the following year (Micke *et al.*, 1983). Double pistil formation is due to abnormal differentiation of pistil primordia (Philp, 1933; Tucker, 1934). Flower bud induction and the initial stages of bud development can occur before harvest (Tufts and Morrow, 1925; Westwood, 1993), and development continues throughout the rest of the season (Guimond *et al.*, 1998). In cherries, the interval between initial flower formation and

Fig. 8.1. Double fruit (A) and abnormal flower (B) caused by high temperatures.

final reproductive development can vary from 86 to 112 days, depending on climate and cultivar (Faust, 1989). Flower bud induction occurs via a biochemical signal to change from a vegetative to a reproductive state (Faust, 1989), as result of the balance of gibberellins, auxin, cytokinins and ethylene (Westwood, 1993). Under natural conditions, floral initiation of sweet cherry starts in July, and, sepals, petals, stamens and pistils then differentiate sequentially (Guimond *et al.*, 1998; Engin and Ünal, 2007; see Chapter 2, this volume).

Temperatures above 30°C are critical for the formation of double pistils (Beppu and Kataoka, 1999). Hot temperatures cause double pistils most severely in buds with differentiated sepal and petal primordia compared with buds at earlier stages of flower bud differentiation or buds that already have stamen and pistil primordia formed (Beppu and Kataoka, 2011), suggesting that buds are most sensitive to the induction of abnormal flower primordia at the transition stage from sepal to petal differentiation. Considerable cultivar variation in the percentage of double fruit between years has been reported (Tucker, 1934, 1935; Micke *et al.*, 1983; Beppu, 2000; Engin and Ünal, 2008; Roversi *et al.*, 2008; García-Montiel *et al.*, 2010). Coastal areas exposed to cooling breezes tend to have fewer double fruit than regions without a coastal influence (Southwick *et al.*, 1991). In the northern hemisphere, doubling tends to occur more severely in the south and in the top parts of the canopy than in the north and in the bottom parts because bud temperature may be increased by higher exposure to solar radiation (Philp, 1933; Tucker, 1934, 1935; Micke *et al.*, 1983; Southwick *et al.*, 1991; Beppu, 2000).

### 8.4.2 Variations in doublings among cultivars

In areas with hot summers, the high risk of producing double fruit (and consequent reduction in marketable fruit) can limit the planting of sweet cherries. Hand thinning can be used to remove double fruit selectively, but in some cultivars this cost may be considerable (Patten *et al.*, 1989). Various strategies have been used to modify the orchard microclimate and minimize doubling in cherries (see Chapter 11, this volume). The high variation in doubling susceptibility among cultivars has been described previously (Table 8.3). These findings suggest that cultivar susceptibility for double-fruit formation has a strong genetic influence. Therefore, it may be possible to breed new cultivars with a low occurrence of double fruit.

## 8.5 Global Warming Consequences

Seasonal timing of phenological events is crucial not only for plant survival but also for maintaining high production in fruit trees. According to some models, as global temperatures rise, harvests in some areas are expected to fall by as much as 6–10% for every 1°C increase in temperature. Recent studies have highlighted substantial changes in phenology already observed with the current global climatic changes. Notably, advances in spring phenology resulting from warming in most northern-hemisphere regions have been documented widely (e.g. Badeck *et al.*, 2004; Menzel *et al.*, 2006; Fu *et al.*, 2015). This trend has also been observed with sweet cherry phenology in Europe (Chmielewski *et al.*, 2004; Dose and Menzel, 2004; Estrella *et al.*, 2007). In Germany, cherry bloom of early-maturing cultivars has advanced up to 4.7 days per °C (Chmielewski *et al.*, 2004), and in south-western France, warmer winters have dramatically affected sweet cherry production (G. Charlot, Balandran, France, 2016, personal communication). In addition to increased risk of spring freeze damage, earlier flowering dates can have an impact on flowering synchronization with pollinator activity and impair growth and fruit development. This will be especially true for areas with a cold winter where HR accumulation is limiting (Guo *et al.*, 2015). Warmer winters can also be associated with delayed spring phenology for species and cultivars with high CRs, resulting in abnormal flowering phenology and reduced productivity. Models for chill availability predict an

**Table 8.3.** Potential of sweet cherry cultivars to produce double fruit in open-field conditions.

| Place | High potential | Low potential | Reference |
|---|---|---|---|
| California | 'Chinook', 'Stella', 'Bing', 'Royal Ann', 'Early Burlat', 'Corum', 'Compact Van', 'Van' | 'Vernon', 'Sam', 'Sue', 'Black Republican', 'Black Tartarian', 'Rainier', 'Jubilee', '24A-9A' | Micke *et al.* (1983) |
| Italy | 'Moreau', 'Karabodur', 'Colafemmina' | 'Stella', 'Sweetheart', 'Ferrovia', 'Durone nero I', 'Stark glorious gold', 'Droganova', 'Denisenova', 'Cherie' | Roversi *et al.* (2008) |
| California | 'Tieton' | | Whiting and Martin (2008) |
| Spain | 'Somerset', 'Burlat', 'Ruby' | 'Marvin', 'Newstar' | García-Montiel *et al.* (2010) |
| Japan | 'Satohnishiki', 'Napoleon' | 'Takasago' | Beppu and Kataoka (2011) |
| Turkey | 'Napoleon', 'Early Burlat', 'Na-1', 'Early Van Compact', 'Bing Spur', 'Lapins', 'Cristobalina' | | Imrak *et al.* (2014) |
| Spain | 'Ruby', 'New Moon', 'Bing', 'Prime Giant', 'Tieton', 'Somerset', 'Lala Star', 'Sylvia', 'Mister Early', 'Utah Giant', 'Sumgita' (Canada Giant™), 'Crystal Champaing' | 'Brooks', 'Cashmere', '4-84', 'Skeena', 'Lapins', 'Larian', 'Newstar', 'Jarandilla' | López *et al.* (2014) |
| France | 'Bellise®Bedel', 'Burlat', 'Feroni', 'Folfer', 'Giant Red', 'Mariant', 'Napoleon', 'New Moon', 'Sweet Early', 'Tieton' | 'Kordia', 'SPC207' (Starblush™), 'Sweetheart', 'Bigalise', 'Black Star', 'Brooks', 'Cambrina', 'Coralise', 'Earlise®Rivedel', 'Early Red', 'Early Star', 'Ferdiva', 'Ferdouce', 'Fermina', 'Fernier', 'Fertard', 'Grace Star', 'Impériale', 'Lapins', 'Penny', 'Poisdel', 'Rainier', 'Regina', 'Rosie', 'Rosilam', 'Rubilam', 'Rubin', 'Sabrina', 'Sumste' (Samba™), 'Sumele' (Satin™), '13N0770' (Stardust™), 'Stark Hardy Giant', 'Summit', 'Staccato', 'Techlovan', 'Van' | G. Charlot (Balandran, France, 2016, personal communication) |

increase in the delaying effect of mild winters as the temperature increase becomes more pronounced (Vitasse *et al.*, 2011; Luedeling, 2012), especially in warmer locations (Guo *et al.*, 2015).

Precise phenological modelling of climate change impact is needed for production area risk assessment, including lack of chilling temperatures (Webb *et al.*, 2007; Luedeling *et al.*, 2009; Luedeling, 2012) or freeze exposure (Eccel *et al.*, 2009; Molitor *et al.*, 2014; Mosedale *et al.*, 2015), to predict the regions that will become unfavourable or favourable for growth and production (e.g. Darbyshire *et al.*, 2013; Ford *et al.*, 2016). Such an approach has been reported for choosing sweet cherry cultivars adapted to anticipated climatic conditions in threatened production areas (Measham *et al.*, 2014).

As a consequence, the response to temperature should represent a priority target for cherry breeding strategies. Increasing bud endodormancy and CR to delay cold de-acclimatization and onset of growth in spring could be a limited approach, since future warm winters may not allow the full release of dormancy in some regions. Alternatively, cultivars that have a higher HR and/or base temperature for activation would be another approach to produce later bud activation and flowering for such regions. Additionally, attention should be focused on breeding cultivars with lower risks of double fruit where summer temperatures are increasing. Predictive modelling approaches could be powerful tools to assist the breeding strategies in overcoming the complexity of all mechanisms involved in temperature responses of cherries. Phenology models are valuable for assessing the impact of temperature, but at the moment they are limited by their lack of transferability between sites and cultivars (Linkosalo *et al.*, 2008; Luedeling and Brown, 2011). Therefore, future approaches for predictive models should be based on more precise measurable parameters, such as biochemical or molecular markers (Satake *et al.*, 2013) or phenological observations for endodormancy induction and release, as well as other environmental parameters such as photoperiod (Chmielewski and Götz, 2016).

## References


Ahn, J.H., Miller, D., Winter, V.J., Banfield, M.J., Lee, J.H., Yoo, S.Y., Henz, R.S., Brady, R.L. and Weigel, D. (2006) A divergent external loop confers antagonistic activity on floral regulators FT and TFL1. *EMBO Journal* 25, 605–614.

Alburquerque, N., García-Montiel, F., Carrillo, A. and Burgos, L. (2008) Chilling and heat requirements of sweet cherry cultivars and the relationship between altitude and the probability of satisfying the chill requirements. *Environmental and Experimental Botany* 64, 162–170.

Allona, I., Ramos, A., Ibañez, C., Contreras, A., Casado, R. and Aragoncillo, C. (2008) Molecular control of winter dormancy establishment in trees. *Spanish Journal of Agricultural Research* 6, 201–210.

Alonso, J.M., Anson, J.M., Espiau, M.T. and Socias I Company, R. (2005) Determination of endodormancy break in almond flower buds by a correlation model using the average temperature of different day intervals and its application to the estimation of chill and heat requirements and blooming date. *Journal of the American Society for Horticultural Science* 130, 308–318.

Amasino, R. (2010) Seasonal and developmental timing of flowering. *Plant Journal* 61, 1001–1013.

Andrews, P. and Proebsting, E.L. (1987) Effects of temperature on the deep supercooling characteristics of dormant and deacclimating sweet cherry flower buds. *Journal of the American Society for Horticultural Science* 112, 334–340.

Andrews, P., Proebsting, E.L. and Gross, D. (1983) Differential thermal analysis and freezing injury of deacclimating peach and sweet cherry reproductive organs. *Journal of the American Society for Horticultural Science* 108, 755–759.

Anon. (2014) Critical spring temperatures for cherry bud development stages. MSU Cherries website based on WSU EB 1128 and MSU Research Report 220. Available at: http://cherries.msu.edu/weather/critical_spring_temperatures_for_cherry_bud_development_stages (accessed 16 November 2015).

Arora, R., Wisniewski, M.E. and Scorza, R. (1992) Cold acclimation in genetically related (sibling) deciduous and evergreen peach (*Prunus persica* (L.) Batsch). *Plant Physiology* 99, 1562–1568.

Artlip, T.S., Callahan, A.M., Bassett, C.L. and Wisniewski, M.E. (1997) Seasonal expression of a dehydrin gene in sibling deciduous and evergreen genotypes of peach (*Prunus persica* [L.] Batsch). *Plant Molecular Biology* 33, 61–70.

Asănică, A., Hoza, D., Tudor, V. and Temocico, G. (2012) Evaluation of some sweet cherry cultivars to winter freeze in different areas of Romania. *Scientific Papers, Series B, Horticulture* LVI, 23–28.

Asănică, A., Tudor, V. and Tiu, J. (2014) Frost resistance of some sweet cherry cultivars in the Bucharest area. *Scientific Papers, Series B, Horticulture* LVIII, 19–24.

Badeck, F.-W., Bondeau, A., Bottcher, K., Doktor, D., Lucht, W., Schaber, J. and Sitch, S. (2004) Responses of spring phenology to climate change. *New Phytologist* 162, 295–309.

Bai, S., Saito, T., Sakamoto, D., Ito, A., Fujii, H. and Moriguchi, T. (2013) Transcriptome analysis of Japanese pear (*Pyrus pyrifolia* Nakai.) flower buds transitioning through endodormancy. *Plant and Cell Physiology* 54, 1132–1151.

Balandier, P., Gendraud, M., Rageau, R., Bonhomme, M., Richard, J.P. and Parisot, E. (1993) Bud break delay on single node cuttings and bud capacity for nucleotide accumulation as parameters for endo- and paradormancy in peach trees in a tropical climate. *Scientia Horticulturae* 55, 249–261.

Ballard, J., Proebsting, E.L. and Tukey, R. (1982) Critical temperatures for blossom buds: cherries. Washington State University Cooperative Extension, COOP EXT. Bulletin 1128, 2. Available at: http://extension.wsu.edu/benton-franklin/wp-content/uploads/sites/27/2013/12/critical-temp-bud-cherries-eb1128e.pdf (accessed 17 January 2017).

Bennet, J. (1949) Temperature and bud rest period. *California Agriculture* 3, 9–12.

Beppu, K. (2000) Morphological and physiological studies on reproduction of sweet cherry in warm climate. PhD thesis. Kyoto University, Kyoto, Japan.

Beppu, K. and Kataoka, I. (1999) High temperature rather than drought stress is responsible for the occurrence of double pistils in 'Satohnishiki' sweet cherry. *Scientia Horticulturae* 81, 125–134.

Beppu, K. and Kataoka, I. (2011) Studies on pistil doubling and fruit set of sweet cherry in warm climate. *Journal of the Japanese Society of Horticultural Science* 80, 1–13.

Bielenberg, D.G., Wang, Y., Li, Z., Zhebentyayeva, T., Fan, S., Reighard, G.L., Scorza, R. and Abbot, A.G. (2008) Sequencing and annotation of the evergrowing locus in peach [*Prunus persica* (L.) Batsch] reveals a cluster of six MADS-box transcription factors as candidate genes for regulation of terminal bud formation. *Tree Genetics and Genomes* 4, 495–507.

Blümel, K. and Chmielewski, F.M. (2012) Shortcomings of classical phenological forcing models and a way to overcome them. *Agricultural and Forest Meteorology* 164, 10–19.

Brown, D. (1957) The rest period of apricot flower buds as described by a regression of time of bloom on temperature. *Plant Physiology* 32, 75–85.

Campoy, J.A., Ruiz, D. and Egea, J. (2010) Effects of shading and thidiazuron + oil treatment on dormancy breaking, blooming and fruit set in apricot in a warm-winter climate. *Scientia Horticulturae* 125, 203–210.

Campoy, J.A., Ruiz, D. and Egea, J. (2011) Dormancy in temperate fruit trees in a global warming context: a review. *Scientia Horticulturae* 130, 357–372.

Campoy, J.A., Ruiz, D., Allderman, L., Cook, N. and Egea, J. (2012) The fulfilment of chilling requirements and the adaptation of apricot (*Prunus armeniaca* L.) in warm winter climates: an approach in Murcia (Spain) and the Western Cape (South Africa). *European Journal of Agronomy* 37, 43–55.

Caprio, J.M. and Quamme, H.A. (2006) Influence of weather on apricot, peach and sweet cherry production in the Okanagan Valley of British Columbia. *Canadian Journal of Plant Science* 86, 259–267.

Castède, S., Campoy, J.A., Quero-García, J., Le Dantec, L., Lafargue, M., Barreneche, T., Wenden, B. and Dirlewanger, E. (2014) Genetic determinism of phenological traits highly affected by climate change in *Prunus avium*: flowering date dissected into chilling and heat requirements. *New Phytologist* 202, 703–715.

Castède, S., Campoy, J.A., Le Dantec, L., Quero-García, J., Barreneche, T., Wenden, B. and Dirlewanger, E. (2015) Mapping of candidate genes involved in bud dormancy and flowering time in sweet cherry (*Prunus avium*). *PloS One* 10, e0143250.

Catalá, R., López-Cobollo, R., Mar Castellano, M., Angosto, T., Alonso, J.M., Ecker, J.R. and Salina, J. (2014) The *Arabidopsis* 14-3-3 protein RARE COLD INDUCIBLE 1A links low-temperature response and ethylene biosynthesis to regulate freezing tolerance and cold acclimation. *Plant Cell* 26, 1–18.

Chmielewski, F.M. and Götz, K.P. (2016) Performance of models for the beginning of sweet cherry blossom under current and changed climate conditions. *Agricultural and Forest Meteorology* 218–219, 85–91.

Chmielewski, F.M., Müller, A. and Bruns, E. (2004) Climate changes and trends in phenology of fruit trees and field crops in Germany, 1961–2000. *Agricultural and Forest Meteorology* 121, 69–78.

Chouard, P. (1960) Vernalization and its relations to dormancy. *Annual Review of Plant Physiology* 11, 191–238.

Cittadini, E.D., de Ridder, N., Peri, P.L. and Keulen, H. van. (2006) A method for assessing frost damage risk in sweet cherry orchards of South Patagonia. *Agricultural and Forest Meteorology* 141, 235–243.

Clausen, S., Grout, B., Toldam-Andersen, T. and Pederson, O. (2012) Enhanced winter-bud frost resistance and improved production in selected clones of sour cherry. *Acta Horticulturae* 935, 161–166.

Cooke, J.E., Eriksson, M.E. and Junttila, O. (2012) The dynamic nature of bud dormancy in trees: environmental control and molecular mechanisms. *Plant, Cell and Environment* 35, 1707–1728.

Cortés, A. and Gratacós, E. (2008) Chilling requirements of ten sweet cherry cultivars in a mild winter location in Chile. *Acta Horticulturae* 795, 457–462.

Couvillon, G.A. and Erez, A. (1985) Influence of prolonged exposure to chilling temperatures on bud break and heat requirement for bloom of several fruit species. *Journal of the American Society for Horticultural Science* 110, 47–50.

Darbyshire, R., Webb, L., Goodwin, I. and Barlow, E.W.R. (2013) Impact of future warming on winter chilling in Australia. *International Journal of Biometeorology* 57, 355–366.

de la Fuente, L., Conesa, A., Lloret, A., Badenes, M.L. and Ríos, G. (2015) Genome-wide changes in histone H3 lysine 27 trimethylation associated with bud dormancy release in peach. *Tree Genetics and Genomes* 11, 45.

Dencker, I. and Toldam-Andersen, T. (2005) Effects of rootstock, winter temperature, and potassium fertilization on yield components of young sour cherries. *Acta Horticulturae* 667, 409–414.

Dennis, F. (2003) Problems in standardizing methods for evaluating the chilling requirements for the breaking of dormancy in buds of woody plants. *HortScience* 38, 347–350.

Ding, J. and Nilsson, O. (2016) Molecular regulation of phenology in trees – because the seasons they are a-changin'. *Current Opinion in Plant Biology* 29, 73–79.

Dose, V. and Menzel, A. (2004) Bayesian analysis of climate change impacts in phenology. *Global Change Biology* 10, 259–272.

Eccel, E., Rea, R., Caffarra, A. and Crisci, A. (2009) Risk of spring frost to apple production under future climate scenarios: the role of phenological acclimation. *International Journal of Biometeorology* 53, 273–286.

Engin, H. and Ünal, A. (2007) Examination of flower bud initiation and differentiation in sweet cherry and peach by scanning electron microscope. *Turkish Journal of Agriculture and Forestry* 31, 373–379.

Engin, H. and Ünal, A. (2008) The effect of irrigation, gibberellic acid and nitrogen on the occurrence of double fruit in 'Van' sweet cherry. *Acta Horticulturae* 795, 645–649.

Engin, H., Şen, F., Pamuk, G. and Gökbayrak, Z. (2009) Investigation of physiological disorders and fruit quality of sweet cherry. *European Journal of Horticultural Science* 74, 118–123.

Erez, A. (2000) Bud dormancy; phenomenon, problems and solutions in the tropics and subtropics. In: Erez, A. (ed.) *Temperate Fruit Crops in Warm Climates*. Springer, The Netherlands, pp. 17–48.

Erez, A. and Couvillon, G.A. (1987) Characterization of the moderate temperature effect on peach bud rest. *Journal of the American Society for Horticultural Science* 112, 677–680.

Eriksson, M.E., Israelsson, M., Olsson, O. and Moritz, T. (2000) Increased gibberellin biosynthesis in transgenic trees promotes growth, biomass production and xylem fiber length. *Nature Biotechnology* 18, 784–788.

Eriksson, M.E., Hoffman, D., Kaduk, M., Mauriat, M. and Moritz, T. (2015) Transgenic hybrid aspen trees with increased gibberellin (GA) concentrations suggest that GA acts in parallel with FLOWERING LOCUS T2 to control shoot elongation. *New Phytologist* 205, 1288–1295.

Estrella, N., Sparks, T.H. and Menzel, A. (2007) Trends and temperature response in the phenology of crops in Germany. *Global Change Biology* 13, 1737–1747.

Faust, M. (1989) *Physiology of Temperate Zone Fruit Trees*. Wiley, New York.

Fischer, M. and Hohlfeld, B. (1998a) Resistance tests in sweet cherries. *Acta Horticulturae* 468, 87–96.

Fischer, M. and Hohlfeld, B. (1998b) Resistenszprüfungen an Süsskirschen (*Prunus avium* L.) Teil 5: evaluierung des Süsskirchensortiments der genbank Obst Dresden Pilnitz auf winterfrost resistens. *Erwerbsobstbau* 40, 42–51.

Fishman, S., Erez, A. and Couvillon, G.A. (1987a) The temperature dependence of dormancy breaking in plants: mathematical analysis of a two-step model involving a cooperative transition. *Journal of Theoretical Biology* 124, 473–483.

Fishman, S., Erez, A. and Couvillon, G.A. (1987b) The temperature dependence of dormancy breaking in plants: computer simulation of processes studied under controlled temperatures. *Journal of Theoretical Biology* 126, 309–321.

Ford, K.R., Harrington, C.A., Bansal, S., Gould, P.J. and St Clair, J.B. (2016) Will changes in phenology track climate change? A study of growth initiation timing in coast Douglas-fir. *Global Change Biology* 22, 3712–3723.

Fu, Y.H., Zhao, H., Piao, S., Peaucelle, M., Peng, S., Zhou, G., Ciais, P., Huang, M., Menzel, A., Peñuelas, J., Song, Y., Vitasse, Y., Zeng, Z. and Janssens, I.A. (2015) Declining global warming effects on the phenology of spring leaf unfolding. *Nature* 523, 104–107.

García-Montiel, F., Serrano, M., Martínez-Romero, D. and Alburquerque, N. (2010) Factors influencing fruit set and quality in different sweet cherry cultivars. *Spanish Journal of Agricultural Research* 8, 1118–1128.

Gianfagna, T. and Mehlenbacher, S. (1985) Importance of heat requirement for bud break and time of flowering in apple. *HortScience* 20, 909–911.

Guimond, C.M., Andrews, P.K. and Lang, G.A. (1998) Scanning electron microscopy of floral initiation in sweet cherry. *Journal of the American Society for Horticultural Science* 123, 509–512.

Guo, L., Dai, J., Ranjitkar, S., Yu, H., Xu, J. and Luedeling, E. (2014) Chilling and heat requirements for flowering in temperate fruit trees. *International Journal of Biometeorology* 58, 1195–1206.

Guo, L., Dai, J., Wang, M., Xu, J. and Luedeling, E. (2015) Responses of spring phenology in temperate zone trees to climate warming: a case study of apricot flowering in China. *Agricultural and Forest Meteorology* 201, 1–7.

Gusta, L.V. and Wisniewski, M. (2013) Understanding plant cold hardiness: an opinion. *Physiologia Plantarum* 147, 4–14.

Hartmann, U., Höhmann, S., Nettesheim, K., Wisman, E., Saedler, H. and Huijser, P. (2000) Molecular cloning of SVP: a negative regulator of the floral transition in *Arabidopsis*. *Plant Journal* 21, 351–360.

He, Y. (2012) Chromatin regulation of flowering. *Trends in Plant Science* 17, 556–562.

Heide, O.M. (2003) High autumn temperature delays spring bud burst in boreal trees, counterbalancing the effect of climatic warming. *Tree Physiology* 23, 931–936.

Heide, O.M. (2008) Interaction of photoperiod and temperature in the control of growth and dormancy of *Prunus* species. *Scientia Horticulturae* 115, 309–314.

Heide, O.M. and Prestrud, A.K. (2005) Low temperature, but not photoperiod, controls growth cessation and dormancy induction and release in apple and pear. *Tree Physiology* 25, 109–114.

Horvath, D. (2009) Common mechanisms regulate flowering and dormancy. *Plant Science* 177, 523–531.

Howell, G. and Weiser, C. (1970) The environmental control of cold acclimation in apple. *Plant Physiology* 45, 390–394.

Hsu, C.-Y., Adams, J.P., Kim, H., No, K., Ma, C., Strauss, S.H., Drnevich, J., Vandervelde, L., Ellis, J.D., Rice, B.M., Wickett, N., Gunter, L.E., Tuskan, G.A., Brunner, A.M., Page, G.P., Barakat, A., Carlson, J.E., DePamphilis, C.W., Luthe, D.S. and Yuceer, C. (2011) *FLOWERING LOCUS T* duplication coordinates reproductive and vegetative growth in perennial poplar. *Proceedings of the National Academy of Sciences USA* 108, 10756–10761.

Iezzoni, A.F. and Mulinix, C.A. (1992) Variation in bloom time in a sour cherry germplasm collection. *HortScience* 27, 1113–1114.

Imrak, B., Sarier, A., Kuden, A., Kuden, A.B., Comlekcioglu, S. and Tutuncu, M. (2014) Studies on shading system in sweet cherries (*Prunus avium* L.) to prevent double fruit formation under subtropical climatic conditions. *Acta Horticulturae* 1059, 171–176.

James, P. and Measham, P. (2011). Rain and its impacts. In: *Australian Cherry Production Guide*. Cherry Growers of Australia, Hobart, Australia, pp. 196–209.

Jensen, M. and Kristiansen, K. (2014) Aspects of freezing tolerance in sour cherry. In: *Proceedings of EU COST Action FA1104: Sustainable production of high-quality cherries for the European market*. COST meeting, Plovdiv, Bulgaria, May 2014. Available at: www.bordeaux.inra.fr/cherry/docs/dossiers/Activities/Meetings/2014%2005%2026-27%20WG1-WG3%20Meeting_Plovdiv/Presentations/Jensen_Plovdiv2014.pdf (accessed 16 November 2015).

Jiménez, S., Reighard, G.L. and Bielenberg, D.G. (2010) Gene expression of *DAM5* and *DAM6* is suppressed by chilling temperatures and inversely correlated with bud break rate. *Plant Molecular Biology* 73, 157–167.

Jonkers, H. (1979) Bud dormancy of apple and pear in relation to the temperature during the growth period. *Scientia Horticulturae* 10, 149–154.

Junttila, O., Welling, A., Chunyang, L., Tsegay, B.A. and Palva, E.T. (2002) Physiological aspects of cold hardiness in Northern deciduous tree species. In: Li, P.H. and Palva, E.T. (eds) *Plant Cold Hardiness: Gene Regulation and Genetic Engineering*. Kluwer Academic/Plenum Publishers, New York, pp. 65–76.

Junttila, O., Nilse, J. and Igeland, B. (2003) Effect of temperature on the induction of bud dormancy in ecotypes of *Betula pubescens* and *Betula pendula*. *Scandinavian Journal of Forest Research* 18, 209–217.

Kadir, S.A. and Proebsting, E.L. (1994) Screening sweet cherry selections for dormant floral bud hardiness. *HortScience* 29, 104–106.

Kalberer, S.R., Wisniewski, M. and Arora, R. (2006) Deacclimation and reacclimation of cold-hardy plants: current understanding and emerging concepts. *Plant Science* 171, 3–16.

Kitashiba, H., Matsuda, N., Ishizaka, T., Nakano, H. and Suzuki, T. (2002) Isolation of genes similar to *DREB1/CBF* from sweet cherry (*Prunus avium* L.). *Journal of the Japanese Society of Horticultural Science* 71, 651–657.

Kitashiba, H., Ishizaka, T., Isuzugawa, K., Nishimura, K. and Suzuki, T. (2004) Expression of a sweet cherry *DREB1/CBF* ortholog in *Arabidopsis* confers salt and freezing tolerance. *Journal of Plant Physiology* 161, 1171–1176.

Kuden, A., Imrak, B., Bayazit, S., Comlekcioglu, S. and Kuden, A. (2012) Chilling requirements of cherries grown under subtropical conditions of Adana. *Middle East Journal of Scientific Research* 12, 1497–1501.

Kühn, B. and Callesen, O. (2001) Morphologic differentiation of flower buds and development of dead flowers in autumn and winter 1998/99: flowering and fruit set in 1999 in the sour cherry cv. Stevnsbaer on four rootstocks. *Gartenbauwissenschaft* 66, 39–45.

Lang, G., Early, J., Martin, G. and Darnell, R. (1987) Endo-, para-, and ecodormancy: physiological terminology and classification for dormancy research. *HortScience* 22, 371–377.

Leida, C., Conesa, A., Llácer, G., Badenes, M.L. and Ríos, G. (2012) Histone modifications and expression of *DAM6* gene in peach are modulated during bud dormancy release in a cultivar-dependent manner. *New Phytologist* 193, 67–80.

Li, B., Xie, Z., Zhang, A., Xu, W., Zhang, C., Liu, Q., Liu, C. and Wang, S. (2010) Tree growth characteristics and flower bud differentiation of sweet cherry (*Prunus avium* L.) under different climate conditions in China. *Horticultural Science – UZEI (Czech Republic)* 37, 6–13.

Li, Z., Reighard, G.L., Abbott, A.G. and Bielenberg, D.G. (2009) Dormancy-associated MADS genes from the *EVG* locus of peach [*Prunus persica* (L.) Batsch] have distinct seasonal and photoperiodic expression patterns. *Journal of Experimental Botany* 60, 3521–3530.

Linkosalo, T., Lappalainen, H. and Hari, P. (2008) A comparison of phenological models of leaf bud burst and flowering of boreal trees using independent observations. *Tree Physiology* 28, 1873–1882.

Liu, G., Pagter, M. and Andersen, L. (2012) Preliminary results on seasonal changes in flower bud cold hardiness of sour cherry. *European Journal of Horticultural Science* 77, 109–114.

Longstroth, M. (2013) Assessing frost and freeze damage to flowers and buds of fruit trees. Michigan State University Exension, Michigan, USA. Available at: http://msue.anr.msu.edu/news/assessing_frost_and_freeze_damage_to_flowers_and_buds_of_fruit_trees (accessed 16 November 2015).

Longstroth, M. and Perry, R.L. (1996) Selecting the orchard site, orchard planning and establishment. In: Webster, A.D. and Looney, N.E. (eds) *Cherries: Crop Physiology, Production and Uses*. CAB International, Wallingford, UK, pp. 203–221.

López, G., García, F., Bayo, A., Frutos, C. and Frutos, D. (2014) Sweet cherry cultivar evaluation in the Region of Murcia, Spain. Proceedings of the EU COST Action FA1104: Sustainable production of high-quality cherries for the European market. COST Meeting, Bordeaux, France, November 2014. Available at: www.bordeaux.inra.fr/cherry/docs/dossiers/Activities/Meetings/2014 10 13-15_4th MC and all WG Meeting_Bordeaux/Posters/Lopez-Ortega_Bordeaux2014.pdf (accessed 30 September 2015).

Luedeling, E. (2012) Climate change impacts on winter chill for temperate fruit and nut production: a review. *Scientia Horticulturae* 144, 218–229.

Luedeling, E. and Brown, P.H. (2011) A global analysis of the comparability of winter chill models for fruit and nut trees. *International Journal of Biometeorology* 55, 411–421.

Luedeling, E., Zhang, M. and Girvetz, E.H. (2009) Climatic changes lead to declining winter chill for fruit and nut trees in California during 1950–2099. *PLoS One* 4, e6166.

Luedeling, E., Kunz, A. and Blanke, M.M. (2013) Identification of chilling and heat requirements of cherry trees – a statistical approach. *International Journal of Biometeorology* 57, 679–689.

Lukoševičiūtė, V., Rugienius, R., Stanienė, G., Stanys, V. and Baniulis, D. (2009) Thermostable protein expression during hardening of sweet and sour cherry microshoots *in vitro*. *Sodininkystė ir Daržininkystė* 28, 35–43.

Martin, R. (2008) Causes of and practical strategies for reducing sweet cherry polycarpy. MS thesis, Washington State University, Washington, DC.

Mathers, H.M. (2004) Supercooling and cold hardiness in sour cherry germplasm : vegetative tissue. *Journal of the American Society for Horticultural Science* 129, 682–689.

Matzneller, P., Blümel, K. and Chmielewski, F.-M. (2014) Models for the beginning of sour cherry blossom. *International Journal of Biometeorology* 58, 703–715.

McMaster, G. and Wilhelm, W. (1997) Growing degree-days: one equation, two interpretations. *Agricultural and Forest Meteorology* 87, 291–300.

Measham, P.F., Quentin, A.G. and MacNair, N. (2014) Climate, winter chill, and decision-making in sweet cherry production. *HortScience* 49, 254–259.

Medina, J., Catalá, R. and Salinas, J. (2011) The CBFs: three *Arabidopsis* transcription factors to cold acclimate. *Plant Science* 180, 3–11.

Menzel, A., Sparks, T.H., Estrella, N., Koch, E., Aasa, A., Ahas, R., Alm-Kübler, K., Bissoli, P., Braslavslá, O., Briede, A., Chmielewski, F.M., Crepinsek, Z., Curnel, Y., Dahl, A., Defila, C., Donnelly, A., Fifella, Y., Jatczak, K., Mage, F., Mestrem A., Nordli, O., Penuelas, J., Pirinen, P., Remisova, V., Scheifinger, H., Striz, M., Susnik, A., van Liet, A.J.H., Wielgolaski, F.-E., Zach, S. and Zust, A. (2006) European phenological response to climate change matches the warming pattern. *Global Change Biology* 12, 1969–1976.

Michaels, S.D., Ditta, G., Gustafson-Brown, C., Pelaz, S., Yanofsky, M. and Amasino, R.M. (2003) *AGL24* acts as a promoter of flowering in *Arabidopsis* and is positively regulated by vernalization. *Plant Journal* 33, 867–874.

Micke, W.C., Doyle, J.F. and Yeager, J.T. (1983) Doubling potential of sweet cherry cultivars. *California Agriculture* March–April, 24–25.

Mimida, N., Kotoda, N., Ueda, T., Igarashi, M., Hatsuyama, Y., Iwanami, H., Moriya, S. and Abe, K. (2009) Four *TFL1/CEN*-like genes on distinct linkage groups show different expression patterns to regulate vegetative and reproductive development in apple (*Malus* × *domestica* Borkh.). *Plant and Cell Physiology* 50, 394–412.

Mimida, N., Li, J., Zhang, C., Moriya, S., Moriya-Tanaka, Y., Iwanami, H., Honda, C., Oshino, H., Takagishi, K., Suzuki, A., Komori, S. and Wada, M. (2012) Divergence of *TERMINAL FLOWER1*-like genes in Rosaceae. *Biologia Plantarum* 56, 465–472.

Miranda, C., Santesteban, L.G. and Royo, J.B. (2005) Variability in the relationship between frost temperature and injury level for some cultivated *Prunus* species. *HortScience* 40, 357–361.

Mohamed, R., Wang, C.-T., Ma, C., Shevchenko, O., Dye, S.J., Puzey, J.R., Etherington, E., Sheng, X., Meilan, R., Strauss, S.H. and Brunner, A.M. (2010) Populus *CEN/TFL1* regulates first onset of flowering, axillary meristem identity and dormancy release in *Populus*. *Plant Journal* 62, 674–688.

Molitor, D., Caffarra, A., Sinigoj, P., Pertot, I., Hoffmann, L. and Junk, J. (2014) Late frost damage risk for viticulture under future climate conditions: a case study for the Luxembourgish winegrowing region. *Australian Journal of Grape and Wine Research* 20, 160–168.

Mosedale, J.R., Wilson, R.J. and Maclean, I.M.D. (2015) Climate change and crop exposure to adverse weather: changes to frost risk and grapevine flowering conditions. *PLoS One* 10, e0141218.

Neilsen, D., Losso, I., Neilsen, G. and Guak, S. (2015) Development of chilling and forcing relationships for modeling spring phenology of apple and sweet cherry. *Acta Horticulturae* 1068, 125–132.

Okie, W.R. and Blackburn, B. (2011) Increasing chilling reduces heat requirement for floral budbreak in peach. *HortScience* 46, 245–252.

Olsen, J.E. (2010) Light and temperature sensing and signaling in induction of bud dormancy in woody plants. *Plant Molecular Biology* 73, 37–47.

Oukabli, A. and Mahhou, A. (2007) Dormancy in sweet cherry (*Prunus avium* L.) under Mediterranean climatic conditions. *Biotechnologie Agronomie Société et Environnement* 11, 133–139.

Overcash, J. (1965) *Heat Required for Pear Varieties to Bloom*. Association of Southern Agricultural Workers 1962 Convention, Texas.

Owens, C.L., Thomashow, M.F., Hancock, J.F. and Iezzoni, A.F. (2002) *CBF1* orthologs in sour cherry and strawberry and the heterologous expression of *CBF1* in strawberry. *Journal of the American Society for Horticultural Science* 127, 489–494.

Patten, K., Nimr, G. and Neuendorff, E. (1989) Fruit doubling of peaches as affected by water stress. *Acta Horticulturae* 254, 319–321.

Pawasut, A., Fujishige, N., Yamane, K., Yamaki, Y. and Honjo, H. (2004) Relationships between chilling and heat requirement in ornamental peaches. *Journal of the Japanese Society of Horticultural Science* 73, 519–523.

Pearce, R. (1999) Molecular analysis of acclimation to cold. *Plant Growth Regulation* 29, 47–76.

Pedryc, A., Hermán, R., Szabó, T., Szabó, Z. and Nyéki, J. (2008) Determination of the cold tolerance of sour cherry cultivars with frost treatments in climatic chamber. *International Journal of Horticultural Science* 14, 49–54.

Philp, G. (1933) Abnormality in sweet cherry blossoms and fruit. *Botanical Gazette* 94, 815–820.

Polashock, J.J., Arora, R., Peng, Y., Naik, D. and Rowland, L.J. (2010) Functional identification of a C-repeat binding factor transcriptional activator from blueberry associated with cold acclimation and freezing tolerance. *Journal of the American Society for Horticultural Science* 135, 40–48.

Proebsting, E.L. and Mills, H. (1978) Low temperature resistance of developing flower buds of six deciduous fruit species. *Journal of the American Society for Horticultural Science* 103, 192–198.

Richardson, E., Seeley, S.D. and Walker, D. (1974) A model for estimating the completion of rest for Redhaven and Elberta peach trees. *HortScience* 9, 331–332.

Rinne, P., Welling, A. and Kaikuranta, P. (1998) Onset of freezing tolerance in birch (*Betula pubescens* Ehrh.) involves LEA proteins and osmoregulation and is impaired in an ABA-deficient genotype. *Plant, Cell and Environment* 21, 601–611.

Ríos, G., Leida, C., Conejero, A. and Badenes, M.L. (2014) Epigenetic regulation of bud dormancy events in perennial plants. *Frontiers in Plant Science* 5, 247.

Rodrigo, J. (2000) Spring frosts in deciduous fruit trees. Morphological damage and flower hardiness. *Scientia Horticulturae* 85, 155–173.

Rodriguez, A., Sherman, W., Scorza, R., Wisniewski, M. and Okie, W.R. (1994) 'Evergreen' peach, its inheritance and dormant behavior. *Journal of the American Society for Horticultural Science*. 119, 789–792.

Rohde, A. and Bhalerao, R.P. (2007) Plant dormancy in the perennial context. *Trends in Plant Science* 12, 217–223.

Rom, R. and Arrington, E. (1966) Effect of varying temperature regimes on degree-days to bloom in Elberta peach. *Proceedings of the American Society for Horticultural Science* 88, 239–244.
Roversi, A. (2001) Osservazioni sulla comparsa di frutti gemellati nel ciliegio dolce. *Frutticoltura* 3, 33–37.
Roversi, A., Monteforte, A., Panelli, D. and Folini, L. (2008) Observations on the occurrence of sweet cherry double-fruits in Italy and Slovenia. *Acta Horticulturae* 795, 849–854.
Ruiz, D., Campoy, J.A. and Egea, J. (2007) Chilling and heat requirements of apricot cultivars for flowering. *Environmental and Experimental Botany* 61, 254–263.
Ruml, M., Vukovic, A. and Milatovic, D. (2010) Evaluation of different methods for determining growing degree-day thresholds in apricot cultivars. *International Journal of Biometeorology* 54, 411–422.
Ruml, M., Milatovic, D., Vulic, T. and Vukovic, A. (2011) Predicting apricot phenology using meteorological data. *International Journal of Biometeorology* 55, 723–732.
Ruonala, R., Rinne, P.L.H., Kangasjärvi, J. and van der Schoot, C. (2008) *CENL1* expression in the rib meristem affects stem elongation and the transition to dormancy in populus. *Plant Cell* 20, 59–74.
Saito, T., Bai, S., Ito, A., Sakamoto, D., Saito, T., Ubi, B.E., Imai, T. and Moriguchi, T. (2013) Expression and genomic structure of the dormancy-associated MADS box genes *MADS13* in Japanese pears (*Pyrus pyrifolia* Nakai) that differ in their chilling requirement for endodormancy release. *Tree Physiology* 33, 654–667.
Saito, T., Bai, S., Imai, T., Ito, A., Nakajima, I. and Moriguchi, T. (2015) Histone modification and signalling cascade of the dormancy-associated MADS-box gene, *PpMADS13-1*, in Japanese pear (*Pyrus pyrifolia*) during endodormancy. *Plant, Cell and Environment* 38, 1157–1166.
Salazar-Gutiérrez, M.R., Chaves, B., Anothai, J., Whiting, M. and Hoogenboom, G. (2014) Variation in cold hardiness of sweet cherry flower buds through different phenological stages. *Scientia Horticulturae* 172, 161–167.
Samish, R. and Lavee, S. (1962) The chilling requirement of fruit trees. *Proceedings of the 16th International Horticultural Congress* 5, 372–388.
Sánchez-Pérez, R., Dicenta, F. and Martínez-Gómez, P. (2012) Inheritance of chilling and heat requirements for flowering in almond and QTL analysis. *Tree Genetics and Genomes* 8, 379–389.
Satake, A., Kawagoe, T., Saburi, Y., Chiba, Y., Sakurai, G. and Kudoh, H. (2013) Forecasting flowering phenology under climate warming by modelling the regulatory dynamics of flowering-time genes. *Nature Communications* 4, 2303.
Saure, M. (1985) Dormancy release in deciduous fruit trees. *Horticultural Reviews* 7, 239–299.
Southwick, S., Shackel, K., Yeager, J., Asai, W. and Katacich, M. Jr (1991) Over-tree sprinkling reduces abnormal shapes in 'Bing' sweet cherries. *California Agriculture* 45, 24–26.
Srinivasan, C., Dardick, C., Callahan, A. and Scorza, R. (2012) Plum (*Prunus domestica*) trees transformed with poplar FT1 result in altered architecture, dormancy requirement, and continuous flowering. *PLoS One* 7, e40715.
Stepulaitienė, I., Žebrauskienė, A. and Stanys, V. (2013) Frost resistance is associated with development of sour cherry (*Prunus cerasus* L.) generative buds. *Žemdirbystė-Agriculture* 100, 175–178.
Swartz, H. and Powell, L. Jr (1981) The effect of long chilling requirement on time of bud break in apple. *Acta Horticulturae* 120, 173–178.
Szabó, Z., Nyéki, J. and Soltész, M. (1996) Frost injury to flower buds and flowers of cherry varieties. *Acta Horticulturae* 410, 315–321.
Szewczuk, A., Gudarowska, E. and Dereń, D. (2007) The estimation of frost damage of some peach and sweet cherry cultivars after winter 2005/2006. *Journal of Fruit and Ornamental Plant Research* 15, 55–63.
Tersoglio, E., Naranjo, G., Quiroga, O. and Setien, N. (2012) Identification of start conditions of sweet cherry ecodormancy varieties: Brooks, New Star, Garnet and Stella. *ITEA* 108, 131–147.
Tucker, L. (1934) Notes on sweet cherry doubling. *Proceedings of the American Society of Horticultural Science* 32, 300–302.
Tucker, L. (1935) Additional notes on sweet cherry doubling. *Proceedings of the American Society of Horticultural Science* 33, 237–239.
Tufts, W. and Morrow, E. (1925) Fruit-bud differentiation in deciduous fruits. *Hilgardia* 1, 1–14.
Turhan, E. and Ergin, S. (2012) Soluble sugars and sucrose-metabolizing enzymes related to cold acclimation of sweet cherry cultivars grafted on different rootstocks. *Scientific World Journal* 2012, 979682.
Vitasse, Y., François, C., Delpierre, N., Dufrêne, E., Kremer, A., Chuine, I. and Delzon, S. (2011) Assessing the effects of climate change on the phenology of European temperate trees. *Agricultural and Forest Meteorology* 151, 969–980.
Wang, D., Karle, R. and Iezzoni, A.F. (2000) QTL analysis of flower and fruit traits in sour cherry. *Theoretical and Applied Genetics* 100, 535–544.

Wang, S., Yuan, C., Dai, Y., Shu, H., Yang, T. and Zhang, C. (2004) Development of flower organs in sweet cherry in Shanghai area. *Acta Horticulturae Sinica* 31, 357–359.

Webb, L.B., Whetton, P.H. and Barlow, E.W.R. (2007) Modelled impact of future climate change on the phenology of winegrapes in Australia. *Australian Journal of Grape and Wine Research* 13, 165–175.

Weinberger, J. (1950) Chilling requirements of peach varieties. *Proceedings of the American Society of Horticultural Science* 56, 122–128.

Welling, A. and Palva, E.T. (2006) Molecular control of cold acclimation in trees. *Physiologia Plantarum* 127, 167–181.

Westergaard, L. and Eriksen, E. (1997) Autumn temperature affects the induction of dormancy in first-year seedlings of *Acer platanoides* L. *Scandinavian Journal of Forest Research* 12, 11–16.

Westwood, M. (1993) *Temperate-zone Pomology: Physiology and Culture*, 3rd edn. Timber Press, Portland, Oregan.

Whiting, M. and Martin, R. (2008) Reducing sweet cherry doubling. *Good Fruit Grower* 59, 24–26.

Wisniewski, M., Bassett, C. and Gusta, L. (2003) An overview of cold hardiness in woody plants: seeing the forest through the trees. *HortScience* 38, 952–959.

Wisniewski, M.E., Bassett, C.L., Renaut, J., Farrell, R., Tworkoski, T. and Artlip, T.S. (2006) Differential regulation of two dehydrin genes from peach (*Prunus persica*) by photoperiod, low temperature and water deficit. *Tree Physiology* 26, 575–584.

Wisniewski, M., Norelli, J., Bassett, C., Artlip, T. and Macarisin, D. (2011) Ectopic expression of a novel peach (*Prunus persica*) *CBF* transcription factor in apple (*Malus* × *domestica*) results in short-day induced dormancy and increased cold hardiness. *Planta* 233, 971–983.

Wisniewski, M., Nassuth, A., Teulières, C., Marque, C., Rowland, J., Cao, P.B. and Brown, A. (2014) Genomics of cold hardiness in woody plants. *Critical Reviews in Plant Sciences* 33, 92–124.

Xu, F., Zhang, X. and Luo, J. (2014) Major limiting factors and improvements measures on sweet cherry cultivation in southern China. *Acta Agriculturae Shanghai* 30, 106–112.

Yu, X.M., Griffith, M. and Wiseman, S.B. (2001) Ethylene induces antifreeze activity in winter rye leaves. *Plant Physiology* 126, 1232–1240.

Zalunskaitė, I., Rugienius, R., Vinskienė, J., Bendokas, V. and Gelvonauskienė, D. (2008) Expression of *COR* gene homologues in different plants during cold acclimation. *Biologia* 54, 33–35.

Zavalloni, C., Andresen, J.A. and Flore, J.A. (2006) Phenological models of flower bud stages and fruit growth of 'Montmorency' sour cherry based on growing degree-day accumulation. *Journal of the American Society for Horticultural Science* 131, 601–607.

Zhang, X., An, L., Nguyen, T.H., Liang, H., Wang, R., Liu, X., Li, T., Qi, Y. and Yu, F. (2015) The cloning and functional characterization of peach *CONSTANS* and *FLOWERING LOCUS T* homologous genes *PpCO* and *PpFT*. *PLoS One* 10, e0124108.

Zhu, Y., Li, Y., Xin, D., Chen, W., Shao, X., Wang, Y. and Guo, W. (2015) RNA-Seq-based transcriptome analysis of dormant flower buds of Chinese cherry (*Prunus pseudocerasus*). *Gene* 555, 362–376.

# 9 Environmental Limiting Factors for Cherry Production

G.H. Neilsen,* D. Neilsen and T. Forge
*Summerland Research and Development Centre, Agriculture and AgriFood Canada, Summerland, British Columbia, Canada*

## 9.1 Introduction

Cherry production, especially of sweet cherries, has increased worldwide in the past 20 years as a consequence of increased consumer demand for fresh market products. This has stimulated a demand for improved understanding of environmental limitations affecting cherry as production has expanded to regions and soils not previously planted to cherry. At the same time, growers have begun to experiment with high-density plantings on dwarfing rootstocks following economic strategies pioneered previously for apple orchards. It is anticipated that higher and earlier yields per unit area will alter nutrient and water-management strategies if economically desirable maximum fruit size is to be achieved.

Despite increases in production area, cherry remains a specialty fruit crop with considerably less research conducted than for other deciduous fruit crops such as apple and peach. Thus, earlier reviews of the nutrient (Westwood and Wann, 1966) and water (Hanson and Proebsting, 1996) requirements of cherry have acknowledged the necessity to adapt research from other tree fruits while summarizing cherry-specific research. This chapter will review the major potential limitations to optimizing cherry production as imposed by abiotic and biotic soil characteristics, summarizing strategies to overcome or avoid such limitations.

## 9.2 Abiotic Soil Factors Influencing Cherry Production

### 9.2.1 Soil organic matter

The benefits of organic matter (OM) in enhancing soil quality have long been recognized and described in standard soil texts (Brady and Weil, 1996). Organic production has a core philosophy of soil OM augmentation by application of organic fertilizers (Neilsen *et al.*, 2009a). Although generally containing lower nutrient contents than inorganic chemical fertilizers, organic amendments release multiple plant nutrients upon decomposition, in contrast to chemical fertilizers that supply one to several nutrients. The organic decomposition process also generates humus and humic and fulvic acid, providing an abundance of chemically reactive compounds, increasing the soil's nutrient exchange capacity and chemically

* gerry.neilsen@agr.gc.ca

buffering the soil against adverse changes in pH and salinity. Improved soil physical structure resulting from the interaction between organic and mineral components can increase water-holding capacity and aeration, reducing susceptibility to excessive compaction and erosion. Soils with higher OM generally are expected to have increased soil biological activity, resulting in enhanced soil and root health. Cherry has, however, been grown successfully in soils with OM contents ranging from <1% in sandy soils in warm, semi-arid regions to as high as 10% in cooler temperate regions where soils have a history of high organic inputs from previous land management and vegetative cover. Although the association between improved cherry orchard performance and increased soil OM status is not always direct, it is good management strategy to enhance the OM content of soils that are below 2%. This can be achieved by the application of organic mulches or amendments (see Chapter 10, this volume).

### 9.2.2 Soil pH

It is generally recommended that soil pH is maintained at pH 6.0–7.0 for optimum growth of cherry. This slightly acid to neutral pH range optimizes the potential availability of most plant nutrients (Fig. 9.1). Cherry can exhibit micronutrient deficiencies of manganese (Mn), zinc (Zn) and iron (Fe), which are much more likely when the soil pH exceeds 7.0, especially in soil containing free calcium carbonate of pH 8.2 and higher. Soils with high pH are also susceptible to toxic accumulations of sodium (Na) and chloride (Cl). Excessively acid soil pH can inhibit cherry growth. This situation can develop in coarse-textured loam soils as a consequence of acidification associated with leaching of soil bases and is also associated with the release of large quantities of plant-available aluminium (Al) and Mn. Toxic soil and leaf levels of Al can accumulate when the soil pH falls below 5.5, which can result in the death of sweet

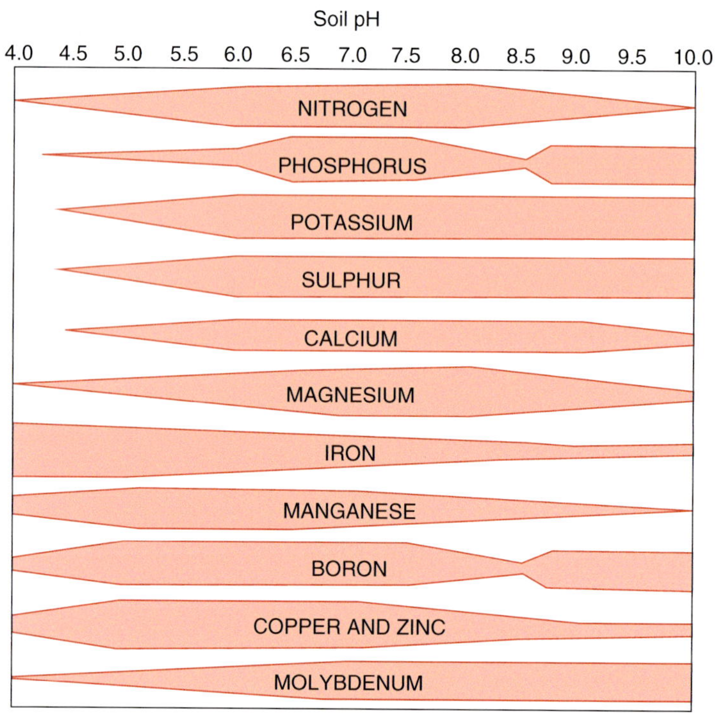


**Fig. 9.1.** General relations between soil pH and nutrient availability. A wider bar width indicates greater relative availability.

cherry seedlings (Melakeberhan *et al.*, 1995) and inhibition of initial root and top growth (Neilsen *et al.*, 1990). The consequences of low soil pH may be observed more readily when replanting previous cherry orchard sites that have experienced management-induced soil pH decline. It is possible to adjust the soil pH to the optimum range for cherry trees, particularly prior to establishment of an orchard when amendments can readily be incorporated deeper in the soil profile without damaging existing roots (see Chapter 10, this volume).

### 9.2.3 Soil salinity

Little detailed information is available concerning the salt tolerance of cherry trees, although, in common with most deciduous fruit crops, they are considered sensitive to excessive soil salinity. Reduced vegetative growth probably occurs at a relatively low salinity threshold of 1.5–1.7 $dSm^{-1}$, as measured in saturation extracts of soils for other related *Prunus* spp. including peach and plum (Maas, 1987). Intolerance to excess salinity may be associated, in part, with sensitivity to specific ion toxicities, including those for Cl, Na and boron (B). There is no information available concerning salt tolerance among cherry rootstocks.

### 9.2.4 Soil fertility

Most research to assess soil fertility by calibrating various nutrient extracts with plant nutrient uptake and growth has been conducted on annual rather than perennial crops. Thus, in most production areas, growers seeking to characterize their orchard nutrient status receive soil test results that were developed for crops other than cherry. Therefore, extractible measures of key nutrients can vary among regions, with their utility for cherry often dependent on local experience rather than any formal soil calibration work with cherry. For North America, for example, common individual nutrient tests are summarized in Table 9.1. Recently, a universal chemical extractant (Mehlich-3, Table 9.1) has been developed that is capable of simultaneously measuring multiple plant nutrients in a single extract by an inductively coupled plasma emission spectrophotometer. So far, there has been limited calibration of the values generated with cherry tree performance. Despite these limitations, various extractable nutrient tests can be useful for identifying orchards with extremely high or low availability of specific nutrients.

#### *Nitrogen (N)*

Orchard soils need to supply sufficient N for annual leaf and fruit production, with N requiring the highest concentration of all essential nutrients. The bulk of soil N is taken up as nitrate ($NO_3$-N) or ammonium ($NH_4$-N) via cherry roots (Fig. 9.2). The availability of these forms is dependent on

**Table 9.1.** Common chemical extractants used to measure plant-available nutrients, acidity (pH), salinity (electrical conductivity) and lime requirement.[a]

| Nutrients | Extractant |
|---|---|
| $NH_4$-N + $NO_3$-N | Potassium chloride (KCl) |
| P (Olsen-P) | Sodium bicarbonate ($NaHCO_3$) |
| Exchangeable Ca, Mg, K, Na | Ammonium acetate pH 7.0 ($NH_4CH_3COOH$) |
| B | Hot water |
| Trace elements (Zn, Mn, Fe, Cu) | Calcium or magnesium chloride ($CaCl_2$, $MgCl_2$), EDTA and diethylene triamine penta-acetic acid (DTPA) |
| Universal extractant (P, K, Ca, Mg, Na, Cu, Zn, Mn, B, Fe) | Mehlich-3[b] |
| Soil pH | Water, $CaCl_2$ |
| Soil electrical conductivity | Water |
| Soil lime requirement | Shoemaker–McLean–Pratt (SMP) buffer |

[a]For details, see local soil test laboratory procedures or standard analysis methods texts such as Carter and Gregorich (2008).
[b]Universal extracting solution comprised of acetic acid ($CH_3COOH$), ammonium nitrate ($NH_4NO_3$), ammonium fluoride ($NH_4F$), nitric acid ($HNO_3$) and ethylene diamine tetra-acetic acid (EDTA).

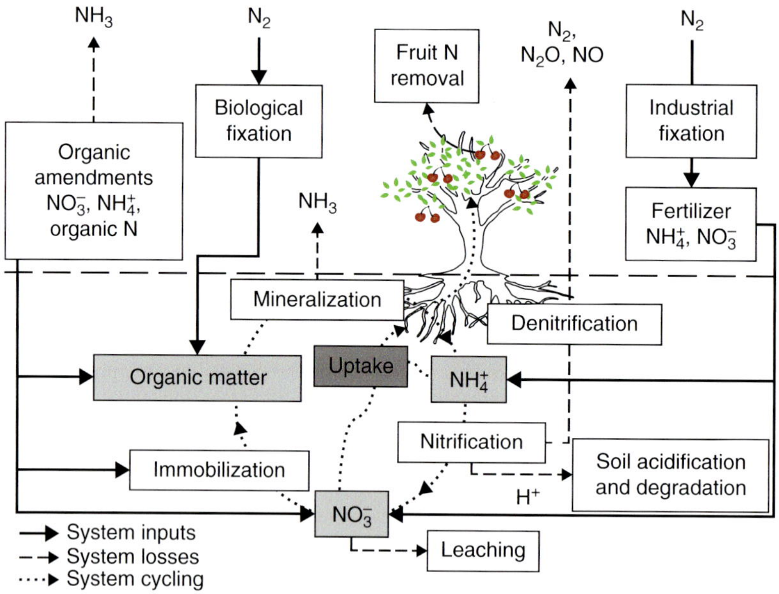


**Fig. 9.2.** Potential flow of nitrogen (N) within a typical cherry orchard.

microbiologically mediated processes in the soil involving mineralization of N from OM to generate $NH_4$-N and subsequent nitrification of the $NH_4$-N to $NO_3$-N. Soil N reserves can be augmented by the addition of chemical fertilizers capable of supplying N in either ammonium or nitrate form, depending on their chemical composition. Organic amendments usually contain much lower N contents than chemical fertilizers, with N supplied in organic forms that are converted to readily available ammonium or nitrate prior to root uptake. Nitrate forms of N are very soluble and subject to leaching loss when large quantities of water pass through the orchard root zone. This might occur during spring snowmelt or at times of high rainfall intensity or application of irrigation in excess of tree water requirements. Nitrogen also can be lost in gaseous form from orchard soils as ammonia ($NH_3$) and nitrous oxide ($N_2O$). $NH_3$ gas evolution can largely be reduced by avoiding application of ammonium-containing fertilizer or organic amendments to the surface of high-pH soils. $N_2O$, a potent greenhouse gas in the atmosphere, can be generated through denitrification when soil moisture content is high.

Determination of available soil N is made difficult by the dynamic nature of N transformations and additions and losses from orchard soils. Nevertheless, high soil $NO_3$-N or $NH_4$-N values, as measured in a commonly used extractant such as potassium chloride (KCl, Table 9.1), are indicators of excessive N fertilization when measured at the end of the growing season.

### *Phosphorus (P)*

P has limited solubility relative to N in cherry orchard soils due to the formation of insoluble phosphate precipitation products with Ca in high-pH soils and with Al and Fe in low-pH soils (Fig. 9.1). Commonly used P extracts (Table 9.1) have been of limited use for predicting the response of sweet cherry to P fertilization. At best, extractable soil P values can be useful to alert cherry growers to extremely low or high P status in specific orchards.

### *Potassium (K), magnesium (Mg) and calcium (Ca)*

These three soil nutrients, often referred to as base cations, are considered to have intermediate mobility in soils due to potential interaction with clay minerals and soil OM. This interaction, which can be measured as a soil's cation exchange capacity, can act to buffer the supply of these nutrients against leaching losses. However, the availability of K to plants can be decreased by irreversible adsorption on specific clay minerals such as illite or vermiculite. Such soils, with high K-fixation capacity, require K fertilizer application rates to be increased relative to non K-fixing soils to ensure sufficient K uptake by roots. The availability of soluble K can be further decreased by low soil moisture content. Low K uptake from soils with otherwise adequate soil K levels may be an indication of water stress.

In general, the availability of base cations is Ca > Mg > K, as measured after extraction in a solution containing a more strongly adsorbed cation such as ammonium (Table 9.1). The relative availability of each depends on the composition of the mineral material contained within the soil, raising the possibility of soils that are exceptionally enriched or depleted in individual nutrients. Soil testing can identify soils with proportionally high concentrations of individual cations that may adversely affect plant uptake of the others. High ratios of soil Ca/Mg, Ca/K and K/Mg may be particularly problematic for the nutrient of low availability. Coarse-textured sands, sandy loams and loamy sands of low pH and OM content are most prone to inadequate Mg and K supplies. Most soils are characterized by high Ca availability, which is rarely a limitation to tree growth. Since Ca availability is closely associated with pH, liming with calcitic materials to avoid the detrimental effects of low pH also would suffice to maintain adequate soil Ca supplies.

#### *Micronutrients: boron (B), zinc (Zn), manganese (Mn), iron (Fe) and copper (Cu)*

Total soil B contents normally are modest, ranging from 2 to 200 p.p.m. with minor proportions (e.g. 5%) readily available to plants. As the soluble boric acid form, B is readily absorbed but is also readily leached from sandy soils when high quantities of low-B irrigation water are applied. Soil OM can be an important potential source of plant-available B, as can precipitation in humid regions. Soil B availability increases at high soil pH (Fig. 9.1). However, cherry growth in such high-pH soils is frequently inhibited by excessive salinity and Na or Cl toxicity. As a result of its ready solubility, toxic B levels can accumulate in landscape positions where leachate accumulates or when irrigating with water containing a high B concentration. The narrow range between deficiency and toxicity for B and the known sensitivity of fruit trees to altered B supply indicate that it is useful to determine B levels in cherry orchard soils. Guidelines have been developed for a range of crops using hot-water-extractable B values as indicators of B availability (Table 9.1).

The availability of the metal micronutrients Zn, Mn, Fe and Cu is decreased when soil pH values exceed 7.0. Diethylenetriamine penta-acetic acid (DTPA) extraction of Zn, Mn, Fe and Cu has been used widely for establishing threshold values for non-orchard crops (Table 9.1) and may be useful, in combination with tissue analysis, for identifying deficient or toxic conditions for cherry.

### 9.2.5 Soil texture, porosity and water-holding capacity

In addition to being the primary source of nutrients for growth, the physical arrangement of soil particles has a strong effect on essential soil properties, including porosity and hence soil aeration and moisture-holding capacity. Soil texture, which can be determined by standard analysis in soil-testing laboratories, indicates the percentage size distribution of individual soil particles as sand (0.05–2.0 mm diameter), silt (0.002–0.05 mm) and clay (<0.002 mm). Reference to a standard texture triangle (Fig. 9.3) allows classification of orchard soils into groupings based on whether fine-, medium- or coarse-textured particles predominate. There are important management differences among soil textural groupings since particle size affects critical soil properties, including water-holding capacity, bulk density, porosity and hence aeration capacity (Table 9.2).

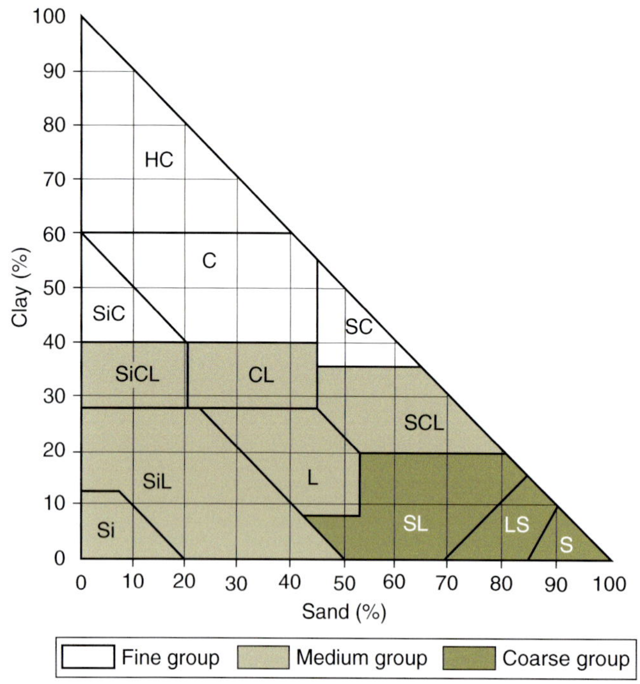


**Fig. 9.3.** Soil texture triangle representing the relative percentages of sand, silt and clay in each soil textural class. H, heavy; C, clay; Si, silt; L, loam; S, sand.

**Table 9.2.** Typical available water-storage capacity, bulk density and estimated porosity as influenced by soil texture.

| Soil texture | Available water-storage capacity (cm m$^{-1}$ of soil) | Bulk density (g cm$^{-3}$) | Porosity (% of total volume) |
|---|---|---|---|
| Coarse | | | |
| Sand | 8.3 | >1.6 | 60 |
| Loamy sand | 10.0 | 1.6 | |
| Sandy loam | 12.5 | 1.4 | |
| Fine sandy loam | 14.2 | | |
| Medium | | | |
| Loam | 17.5 | | |
| Silt loam | 20.8 | 1.3 | |
| Clay loam | 20.0 | 1.2 | |
| Fine | | | |
| Clay | 20.0 | 1.1 | 30 |
| Organic soils | >25.0 | <1.0 | |

The water-holding capacity of soils is defined by characterizing moisture retention over a range of externally imposed energy levels calculated to simulate soil drainage and drying (Fig. 9.4). Water content between field capacity and the permanent wilting point (the soil moisture content at which most plants suffer irrecoverable damage) is considered a soil's storage capacity for available water (Table 9.2). Saturation represents complete occupancy of pores by water, whereas field capacity (equivalent to 10 kPa externally imposed pressure) represents soil moisture content after soils cease draining freely. The soil moisture content at the permanent wilting point is assumed to be equivalent to 1500 kPa. However, for perennial fruit crops such as cherry, which have

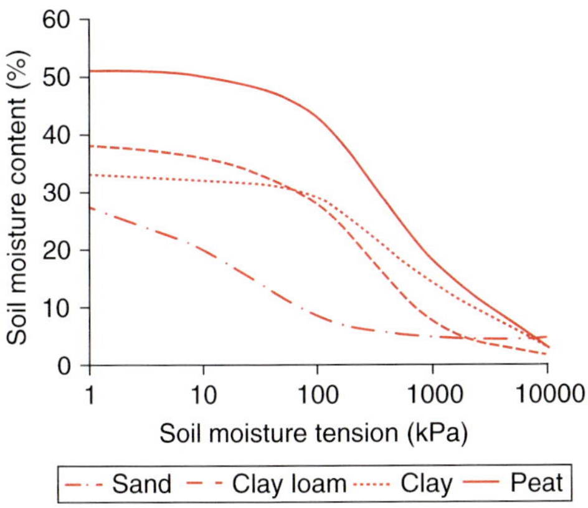


Fig. 9.4. Generalized soil moisture retention curves for different soil textures indicating the amount of soil moisture at different soil moisture tensions.

low root densities, significant water stress can occur prior to reaching 1500 kPa in bulk soil due to high evaporative demand that creates dry soil conditions immediately surrounding the roots. Nevertheless, defining plant-available water content as the difference between water content at 10 and 1500 kPa usefully indicates the important influence soil texture can exert on water-management strategies. Coarse-textured soils retain less plant-available water, drain rapidly and thus are more prone to drought, particularly if they are shallow and stony (Table 9.2). Fine-textured soils retain more water but are susceptible to aeration problems at high water contents.

## 9.3 Biotic Soil Factors Influencing Cherry Production

### 9.3.1 Replant disease complex and root-lesion nematodes (*Pratylenchus penetrans*)

Young fruit trees replanted into old orchard soil are usually affected by a replant disease syndrome that can have significant impacts on productivity throughout the life of the planting. Root-lesion nematodes (genus *Pratylenchus*) have been associated with poor replant establishment of cherry (Mai and Abawi, 1978; Mai *et al.*, 1994). The predominant species affecting fruit trees grown in temperate regions is *P. penetrans*, although other species of *Pratylenchus* are parasites of fruit trees and have been linked to poor replant growth of apple (e.g. Nyczepir and Halbrendt, 1993; Dullahide *et al.*, 1994). In Mediterranean regions, *Pratylenchus vulnus* is particularly damaging to a wide variety of fruit and nut tree species including *Prunus* spp. (Pinochet *et al.*, 1991, 1996a). Conversely, some species of *Pratylenchus* do not reproduce on *Prunus* spp. (Pinochet *et al.*, 1991).

It is widely recognized that poor replant growth also can occur in soil without root-lesion nematodes or with low population densities, indicating that other soil-borne pathogens are important causes of poor replant establishment (Mai and Abawi, 1978; Traquair, 1984; Mazzola and Manici, 2012). A wide variety of fungi in the genera *Cylindrocarpon/Ilyonectria*, *Rhizoctonia* and *Pythium* that have been isolated from fruit trees suffering from replant disease can cause disease when re-inoculated into apple seedlings, and are potential causes of replant disease of apple (Traquair, 1984; Mazzola, 1998; Mazzola and Manici, 2012; Manici *et al.*, 2013). Some researchers have speculated that the fungi causing replant disease on apple and cherry might be specific to those plant species, respectively, which led to the concept of 'specific replant disease' (Traquair, 1984). Research to date has not provided clear evidence to support the specific replant disease idea, with non-pasteurized or non-fumigated soil from old apple orchards generally suppressing growth of cherry seedlings relative to pasteurized or fumigated soil, and vice versa (Mai and Abawi, 1978). In addition, the complement of fungi that have been isolated from cherry and other *Prunus* spp. displaying symptoms of replant disease are broadly similar to those commonly isolated from apple (Fliegel *et al.*, 1963; Browne *et al.*, 2006; Urbez-Torres *et al.*, 2016), although such investigations of cherry and other *Prunus* spp. are few relative to the body of research on apple.

While root-lesion nematodes are most often associated with poor replant establishment, they can have important influences on other aspects of cherry tree physiology and growth at maturity. Root-lesion nematodes were observed to decrease the cold hardiness of cherry seedlings in one study (Edgerton and Parker, 1958). The mechanisms of this effect are unknown and worthy of additional research.

Management of root-lesion nematodes and the replant disease complex has historically relied on preplant fumigation. Since the phase-out of methyl bromide, fumigants of choice include broad spectrum isothiocyanate-liberating formulations (e.g. Vapam, AMVAC; Basamid®, Certis USA), chloropicrin and the fumigant nematicide 1,3-dichloropropene (Telone® II, Dow AgroSciences). Cultural and biological management practices that can be deployed to reduce root-lesion nematode populations and the effects of the replant disease complex include biofumigation with *Brassica* green manure cover crops, *Brassica* seed meals, or some manures and manure slurries (reviewed by Forge *et al.*, 2016b). Preplant growth of nematode-suppressive cover crops such as marigold (*Tagetes* spp.) also can be effective (Forge *et al.*, 2016b). Incorporation of mature compost with a neutral carbon (C)/N ratio (~12) at relatively high rates (i.e. >50 t $ha^{-1}$) can improve replant establishment of raspberry and cherry via root-lesion nematode population suppression and improvements in multiple aspects of soil quality (Forge *et al.*, 2016a,c). Similarly, recent research suggests that anaerobic soil disinfestation, which has been tested in field vegetable production systems, is a strategy worthy of consideration; it has not yet been evaluated for replant of tree fruit crops (Lamers *et al.*, 2010; Butler *et al.*, 2012).

Some recently released apple rootstocks are more tolerant of replant disease than industry standards such as M9 (Robinson *et al.*, 2012). While a greater variety of cherry rootstocks have become available in recent years, little is known of their differential susceptibility to root-lesion nematodes or fungal components of the replant disease complex. Preliminary research has found consistently higher *P. penetrans* colonization of 'Skeena' sweet cherry trees on 'GiSelA 3' rootstock than on adjacent 'GiSelA 5' or 'GiSelA 6' in a high-density training systems trial (Neilsen *et al.*, 2016).

### 9.3.2 Other nematodes

#### *Ring nematode (*Mesocriconema xenoplax*)*

The ectoparasitic ring nematode is recognized as a serious pathogen of peach and plum wherever these crops are grown. Studies on the host status of a wide range of *Prunus* genotypes, including *Prunus avium*, indicate that there is little if any difference among *Prunus* genotypes with respect to the ability of ring nematode to feed and reproduce, suggesting susceptibility of cherry to this nematode (Westcott and Zehr, 1991; Westcott *et al.*, 1994). The effects of ring nematode on peach physiology have been studied extensively and include increased predisposition to winter injury and subsequent susceptibility to bacterial canker, resulting in a disease complex known as peach tree short life (Olien *et al.*, 1995; Cao *et al.*, 2006; Browne *et al.*, 2013). Little research has explored specific host–parasite interactions for cherry. Considering that fecundity of ring nematode on cherry is similar to peach, it seems likely that the nematode has significant negative impacts on cherry.

#### *Dagger nematodes (*Xiphinema *spp.)*

Dagger nematodes are ectoparasites capable of directly reducing the vigour and yield of trees when they are present in soil at high population densities. These nematodes are particularly important, however, as vectors of viruses. Most species populations in the *Xiphinema americanum* group can vector cherry rasp leaf virus and tomato ringspot virus, both of which can be particularly damaging to cherry. If *X. americanum*-group nematodes are found in the root zone of virus-infected trees, it is necessary to fumigate the soil in an area around the infected trees in addition to removal of the trees themselves.

#### *Root-knot nematodes (*Meloidogyne *spp.)*

Root-knot nematodes are sedentary endoparasites. Infective juveniles present in the soil invade root tips and set up permanent feeding sites that result in the formation of swellings or 'knots', impaired root functioning, and other changes in host physiology that ultimately lead to reduced vigour. The three most damaging species on cherry and other *Prunus* spp. are *Meloidogyne incognita*, *Meloidogyne arenaria* and *Meloidogyne javanica*, all of which are confined to regions where the soil does not freeze more than a few centimetres deep. The main species in northern temperate growing regions, *Meloidogyne hapla*, has been recorded parasitizing *Prunus cerasus* (Nyczepir and Halbrendt, 1993). Its relationships with *P. avium* and *Prunus mahaleb* have not been studied, but it is generally not recognized as a concern on cherry in northern temperate growing regions.

There is considerable variation within *Prunus* spp. germplasm with respect to root-knot nematode (*M. incognita*, *M. arenaria* and *M. javanica*) resistance (Marull and Pinochet, 1991; Fernández *et al.*, 1994; Saucet *et al.*, 2016), with resistance genes being identified in peach, plum and almond rootstock material. The root-knot nematode-resistant peach rootstocks 'Nemaguard', Guardian' and 'Nemared' are widely used in commercial orchards. Root-knot nematode resistance does not appear to have been a major criterion in selection of cherry rootstocks, and no commercially available cherry rootstocks currently carry resistance to root-knot nematodes. Ongoing research to stack root-knot resistance genes from peach, plum and almond into interspecific *Prunus* rootstock materials (Saucet *et al.*, 2016) may lead to the development of resistant rootstocks for cherry.

### 9.3.3 Crown gall

Crown gall is a severe disease of the roots and crowns of cherry trees. It is caused by the bacterium *Rhizobacterium radiobacter*, previously known as *Agrobacterium tumefaciens*. The bacterium is widespread in agricultural soils and infects roots through wounds, often during propagation processes such as digging and handling of trees from nurseries, and during planting into infested soil. Crown gall can have devastating impacts on productivity in severely infested orchards. Mazzard (*P. avium*) and 'Colt' rootstocks are very susceptible to crown gall, whereas 'GiSelA 5' and 'GiSelA 6' rootstocks appear to be more resistant.

Preplant fumigation effectively controls crown gall in most soils, and the historical widespread reliance on preplant fumigation for control of parasitic nematodes and replant disease has masked the potential impacts of crown gall on cherry production. As with plant-parasitic nematodes, the prominence of crown gall is likely to increase, as recent regulatory changes in the acceptability of fumigation in Europe and North America have led to reduced reliance on preplant fumigation. Specific *R. radiobacter* strains (e.g. K84, K1026) that do not cause disease but compete with pathogenic strains of the bacterium at infection sites are commercially available as preventative root dip treatments, such as Nogall™ (BASF) and Galltrol® (AgBioChem). Chemical control products are also available (e.g. Gallex®, AgBioChem).

### 9.3.4 Rhizosphere symbionts

Several groups of microorganisms that form symbioses with plant roots or colonize the rhizosphere, including arbuscular mycorrhizal fungi (AMF), rhizosphere bacteria and root-associated fungi, have the potential to suppress pathogens (such as those associated with replant disease), enhance P uptake, and otherwise enhance root growth and functioning. With respect to fruit trees, interactions with AMF have been studied more extensively than interactions with other groups of rhizosphere organisms. Enhanced acquisition of P, Zn and Cu appears to be the principal benefit of AMF to most plant species, but increased resistance to root pathogens is another important benefit of AMF for fruit trees (Pinochet *et al.*, 1995, 1996b). Pinochet *et al.* (1996b) reviewed ten controlled studies of the effectiveness of AMF for reducing nematode infection and improving growth of a range of fruit tree rootstocks, including *Prunus* spp. Inoculation with AMF reduced infection by root-lesion nematodes (*P. vulnus*) in approximately half of the interactions reviewed, including some involving *Prunus* spp. In nearly all experiments, increased host tolerance of nematode infection was observed (i.e. enhanced capacity for growth and nutrient uptake despite the nematodes).

While the benefits of AMF are well documented for *Prunus* rootstocks under experimental conditions, when orchard sites possess inadequate inoculum, it is not clear whether commercial AMF inoculation is worthwhile, especially after fumigation. One study with 'Ottawa 3' apple rootstocks found that AMF inoculation reduced infection by root-lesion nematodes, and improved growth and P, Cu and Zn uptake in fumigated but not in non-fumigated soil (Forge *et al.*, 2001). Improved understanding of inoculum potential in replant soils is needed before the practical benefits of commercial AMF inoculants can be fully assessed.

Non-arbuscular mycorrhizal microbes that beneficially colonize the rhizosphere include species and strains of rhizosphere-colonizing bacterial genera such as *Bacillus*, *Burkholderia*, *Pseudomonas*, *Arthrobacter*, *Rhizobium*, *Actinomyces* and *Streptomyces*, and fungal genera such as *Aspergillus*, *Trichoderma* and *Penicillium* (Whipps, 2001; Harman *et al.*, 2004; Haas and Défago, 2005; Owen *et al.*, 2015). Like AMF, many of these potential bioinoculants appear to have multiple benefits, including enhanced resistance to pathogens, tolerance to abiotic stress and P mobilization in the rhizosphere. Despite the considerable body of research and commercial interest in some of these organisms as bioinoculants, remarkably few published studies have documented their utility for improving the replant success of tree fruit crops, especially cherry.

### 9.3.5 General soil health

Soil health refers to 'the capacity of a living soil to function, within natural or managed ecosystem boundaries, to sustain plant and animal productivity, maintain or enhance water and air quality, and promote plant and animal health' (Doran, 2002). The attributes that distinguish a healthy soil from an unhealthy soil of the same parent material, texture and climate are primarily the result of biological activity. These include aggregation (improved soil structure) and enhanced water-holding capacity, effective litter decomposition and the modulation of nutrient immobilization and mineralization processes, and the promotion of rhizosphere interactions that directly enhance the growth and healthy functioning of roots.

The development of stable pools of soil nutrients is the result of decomposition of added OM and incorporation of the nutrients into the microbial biomass, from which subsequent release of the nutrients is stimulated by trophic interactions in the soil food web involving a diversity of microbivorous soil micro- and mesofauna. Such trophic interactions can also reduce the extent or duration of severe nutrient immobilization that can occur after addition of OM with high C/N ratios, such as when certain crop residues are incorporated into soil (Ferris *et al.*, 1998; Chen and Ferris, 1999). The presence of a large microbial biomass can also help temporarily immobilize excess mineral N and P added in fertilizer or nutrient-rich amendments such as poultry manure, thus reducing the likelihood of nutrient leaching.

The rhizosphere interactions that directly affect the health, functioning and growth of roots are more complex and less well understood than bulk soil processes. They include the beneficial symbioses described above, such as rhizosphere colonization by plant growth-promoting rhizobacteria and similar beneficial microbes such as *Trichoderma* spp. Recent research suggests that rhizosphere microbial-grazing protists and nematodes can directly stimulate root growth via mechanisms other than enhanced nutrient turnover; the production of growth-stimulating factors such as plant hormones has been postulated (Cheng *et al.*, 2011). Biologically based suppression of the build-up of populations of parasitic nematodes and fungal pathogens, and the suppression of their ability to attack crop roots, are also attributes of healthy soil as the greater overall microbial activity and diversity of soil fauna associated with healthy soil generally contributes to suppression of root pathogens and populations of parasitic nematodes (Janvier *et al.*, 2007; Timper, 2014). Thus, the expression of orchard replant disease can be viewed as a result of poor soil ecosystem health. Management practices that enhance overall soil health, promoting multiple beneficial soil functions, will encourage vigorous root growth and may allow economically acceptable replant establishment and the development of uniformly productive orchards despite the presence of soil-borne pathogens such as root-lesion nematodes.

## 9.4 Seasonal Nutrient Limitations

There have been few well-documented estimations of annual nutrient requirements for perennial cherry orchards. Nutrient requirements per unit area for nutrients removed with the harvested fruit can, however, be readily calculated if yield, tree density and fruit nutrient concentration on a fresh weight basis are measured. It is more difficult to estimate annual nutrient requirements associated with incremental growth of above-ground wood organs and buds and below-ground roots without approximations associated with successive, destructive sampling of representative trees. Furthermore, it can be a challenge to determine the extent to which nutrients are recycled from shed flowers, fruit and fruit stalks, as well as summer and dormant prunings and dropped leaves, all of which can vary in the extent to which they are retained and decomposed within an individual orchard. When estimates of these gross annual nutrient requirements have been attempted, values tend to be low when expressed in kg ha$^{-1}$ (Table 9.3), relative to many annual crops.

Although it is useful to be aware of whole-tree nutrient requirements, most orchards are managed to avoid the development of nutrient stresses during the growing season. The chemical composition of mid-terminal, new extension shoots of representative vigour in mid-summer is standard for monitoring the nutritional status of cherry orchards. Comparisons can be made to recommended ranges for most nutrients (Table 9.4). Relative to apple, much less information has been compiled concerning fresh weight nutrient concentrations of cherry fruit, and deficiency and excess thresholds have not been established. Typical fresh weight values, reported from a range of experimental studies, are shown in Table 9.4.

Cultivar and rootstock genotype affect leaf and fruit nutrient concentrations, but there is insufficient information to allow the development of cultivar- or rootstock-specific nutrient standards. 'Bing' sweet cherry trees on dwarfing rootstocks had higher yield efficiency and lower leaf Mg and K than on Mazzard, suggesting a different inherent capacity to take up nutrients (Neilsen and Kappel, 1996). Trees on 'Colt' had lower leaf N, P and K relative to 'F 12/1' (Ystaas and Frøynes, 1995b). Macro- and micronutrient analysis of dormant spurs of 'Bing' and 'Hedelfinger' sweet cherry and 'Montmorency' sour cherry on up to 18 different rootstocks at three sites showed few consistent rootstock trends, varying by scion, site and soil type, but not revealing any general correlations with rootstock vigour

**Table 9.3.** Estimated gross annual nutrient requirements in cherry orchards.

| Cultivar | Rootstock | Tree density (trees ha$^{-1}$) | Age (years) | Nutrient (kg ha$^{-1}$) N | K | Ca | Mg | P |
|---|---|---|---|---|---|---|---|---|
| 'Schattenmorelle'[a] | *P. avium* | 667 | 8–10 | 95 | 57 | 42 | 10 | 4.5 |
| Six sweet cherry cultivars[b] | *P. avium* | 333 | 13 | 39–65 | 16–47 | 26–55 | 5–9 | 6–11 |

[a]Data adapted from Baghdadi and Sadowski (1998) for nutrients contained in harvested fruit, summer prunings (wood and leaves) and dropped leaves.
[b]Data adapted from Roversi and Monteforte (2006) for nutrients contained in harvested fruit, summer and dormant prunings, and dropped leaves, shown as a range of values for six cultivars.

**Table 9.4.** Nutrient concentration ranges for leaves[a] and fruit[b] of sweet and sour cherry.

| Nutrient | Deficiency | Normal range | Potential toxicity |
|---|---|---|---|
| Leaf (mid-terminal shoot, mid-summer, dry-weight basis) | | | |
| Nitrogen (%) | <1.9 | 1.9–3.0 | >3.4 |
| Phosphorus (%) | <0.1 | 0.16–0.40 | NE |
| Potassium (%) | <1.0 | 1.3–3.0 | NE |
| Calcium (%) | NE | 1.0–3.0 | NE |
| Magnesium (%) | <0.24 | 0.3–0.6 | NE |
| Sulfur (%) | NE | 0.13–0.8 | NE |
| Boron (p.p.m.) | <20 | 25–60 | >80 |
| Zinc (p.p.m.) | <10 | 15–70 | NE |
| Manganese (p.p.m.) | <20 | 20–200 | NE |
| Iron (p.p.m.) | NE | 20–500 | NE |
| Copper (p.p.m.) | NE | 5–20 | NE |
| Fruit (at harvest, whole fruit minus seeds and stem, fresh weight basis) | | | |
| Nitrogen (mg per 100 g) | NE | 110–190 | NE |
| Phosphorus (mg per 100 g) | NE | 15–25 | NE |
| Potassium (mg per 100 g) | NE | 120–220 | NE |
| Calcium (mg per 100 g) | NE | 7–14 | NE |
| Magnesium (mg per 100 g) | NE | 7–12 | NE |
| Boron (mg per 100 g) | NE | 0.2–0.5 | NE |

[a]Adapted from Hanson and Proebsting (1996).
[b]Typical values compiled from published literature for healthy fruiting sweet cherry.
NE, Not established.

(Lang *et al.*, 2011). Levels of K, P, Mg and sulfur (S) were most consistent across rootstock and scion genotypes. Variations in crop load also can affect annual cherry leaf nutrient concentration. 'Lapins' sweet cherry leaf P and K concentrations on a 'GiSelA 5' rootstock were lower in a high-crop year (Neilsen *et al.*, 2007). This implies that inadequate P and K nutrition for an individual orchard might be more readily apparent after leaf analysis in high-crop years.

### 9.4.1 Nitrogen

N is the nutrient most likely to be inadequate in cherry orchards, as a consequence of relatively high demand (Table 9.3) and low N availability in soils, especially those with coarse texture and low OM content. Cherry, like most fruit trees, also does not compete well for N relative to sod and weed species that have root lengths and densities that are orders of magnitude greater than that of cherry (Buwalda, 1993). N deficiency is characterized by reduced shoot length and small, pale-green leaves that may abscise prematurely in autumn. These factors contribute to lower yields and smaller fruit.

In cherry orchards, fertilizer N is generally applied annually to the soil surface. For example, in France, applications of 30–40 kg N $ha^{-1}$ are recommended for newly planted cherry trees, increasing incrementally to 80–100 kg N $ha^{-1}$ for mature fully fruiting orchards (Lichou *et al.*, 1990). Application rates should be reduced for fertile soils. In a recent experiment in Chile, N supply from a clay loam soil containing 2% OM was sufficient to meet N demands for the first 3 years for newly planted 'Bing' sweet cherry trees on 'GiSelA 6' (Bonomelli and Artacho, 2013). Application of N at rates as high as 120 kg N $ha^{-1}$ increased tree N content but did not increase growth relative to no supplemental N. Detrimental effects of high N application on sweet cherry yield and fruit size have been reported in a 6-year N application trial with 'Lapins' on 'GiSelA 5' in which N was applied with irrigation for 8 weeks post-bloom (Neilsen *et al.*, 2004, 2007). Yield (Fig. 9.5) and average fruit size (Table 9.5) were highest

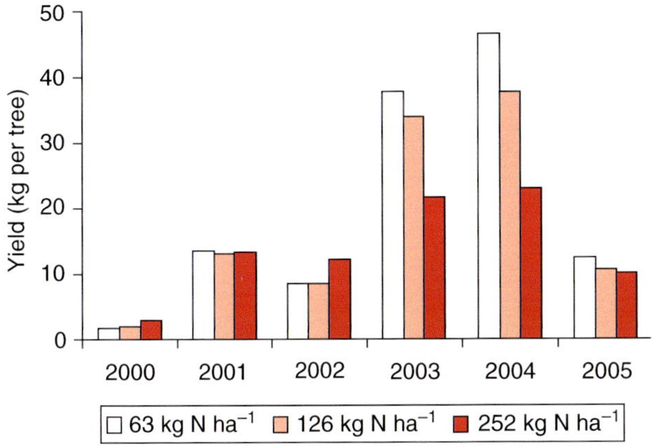


**Fig. 9.5.** Yield of 'Lapins' on 'GiSelA 5' rootstock as affected by rate of N applied via fertigation over the first 6 fruiting years. (Adapted from Neilsen *et al.*, 2004, 2007.)

**Table 9.5.** Average fruit size (g) of 'Lapins' sweet cherry grown on 'GiSelA 5' rootstock as affected by rate of N fertilization over the first 6 years of fruiting. (Adapted from Neilsen *et al.*, 2007.)

| | N fertilization rate | | | |
|---|---|---|---|---|
| Year | Low (63 kg N $ha^{-1}$) | Medium (126 kg N $ha^{-1}$) | High (254 kg N $ha^{-1}$) | *P* value |
| 2000 | 12.6 | 12.0 | 12.3 | NS |
| 2001 | 11.0 | 10.0 | 9.6 | <0.05 |
| 2002 | 10.0 | 9.0 | 9.0 | <0.05 |
| 2003 | 11.2 | 9.9 | 8.5 | <0.01 |
| 2004 | 9.7 | 10.1 | 9.4 | NS |
| 2005 | 14.9 | 14.4 | 13.8 | <0.01 |

NS, Not significant.

at low (about 60 kg N $ha^{-1}$) relative to high (about 250 kg N $ha^{-1}$) annual N application rates, suggesting N fertilizer should be applied in moderation. More research is needed to determine the relations between N fertilization rate and other cherry quality characteristics such as fruit firmness.

The timing of N application is influenced by the growth pattern of cherry, which flowers prior to leaf emergence and is the earliest deciduous fruit to ripen in summer. Uptake of soil-applied N very early in the growing season is minimal due to cool soils and lack of significant evapotranspirational movement of soil water and dissolved nutrients into the plant until leaf emergence and expansion. Flowering and fruit set therefore depend largely on remobilization of N from storage reserves for about 3 weeks after bud break (Grassi *et al.*, 2002, 2003). Fertigation strategies have been developed to deliver incremental spring N applications with irrigation water for a 4–6-week period commencing at bloom and coinciding with rapid shoot growth and fruit cell division and expansion. The short fruit development period also creates opportunities for postharvest N applications, although several studies have indicated inefficient uptake of N when these applications are made to the soil (Neilsen *et al.*, 2004; Azarenko *et al.*, 2008). In contrast, late summer postharvest urea spray applications in the 2–5% (w/v) concentration range have been effective at increasing N stored in flowering spurs, shoot tips and buds, improving availability for remobilization the following spring (Ouzounis and Lang, 2011; Thielemann *et al.*, 2014). The effectiveness of postharvest foliar urea sprays for improving subsequent cherry performance decreases for trees with high N status at the season end (Wójcik and Morgaś, 2015).

### 9.4.2 Phosphorus

Many fertilizer trials have indicated no measurable response of cherry to P fertilization, even when soil P levels were low (Westwood and Wann, 1966). Absolute cherry P requirements are low relative to other major plant nutrients (Table 9.3). For other plant species grown in solution culture under known P deficiency, symptoms of P deficiency have included red or purple coloration on young shoot tip leaves. However, for P-deficient young 'Golden Delicious' apple trees growing in gravel culture, the only symptom was small leaf size (Benson and Covey, 1979). Thus, the identification of P deficiency in cherry under field conditions is difficult.

Nevertheless, there have been some reports of cherry response to P. First-year canopy and root growth was increased at high soil P in acid soil, implying a role for P in improved cherry establishment (Neilsen *et al.*, 1990). This is consistent with the benefits of P fertilization and fumigation observed for improving apple establishment and initial growth in replant disease-affected orchard soils (Neilsen and Yorston, 1991). In a long-term P fertilization trial on a low-P, acid soil of pH 4.4, broadcast soil-applied P fertilizer increased the vigour and yield of 'Kristin' sweet cherry over a 13-year period (Ystaas and Frøynes, 1995a). Annual fertigation of 20 g of P as ammonium polyphosphate at bloom improved the growth and yield of 'Sumnue' (Cristalina™) and 'Skeena' sweet cherry in the first 4 years on 'GiSelA 6' (Neilsen *et al.*, 2010). After the fifth growing season, this P fertigation was associated with delayed fruit harvest (Neilsen *et al.*, 2014). While foliar P applications of dilute (0.75%) soluble P compounds such as $Ca(H_2PO_4)_2$ have been associated with increased P uptake and improved fruit quality of apple (Johnson and Yogaratnam, 1978), there have been no documented benefits of such sprays for cherry.

### 9.4.3 Potassium, calcium and magnesium

Total cherry requirements for K are second to those of N (Table 9.3) and generally represent the highest mineral concentration in fruit (Table 9.4). K deficiency has been associated with chlorosis, leaf curl, and marginal leaf scorch and necrosis, especially apparent on basal leaves of new-year extension growth. Deficiency reduces tree vigour and fruit size. Leaf K concentrations below 1% on a dry-weight basis are frequently associated with deficiency and are indicative of a need for supplemental K fertilization. Due to the relatively high requirements for K, soil broadcast applications in the 100–200 kg $ha^{-1}$ range, applied early in the growing season, have been effective at overcoming K deficiency. Fertigation with any soluble K fertilizer can also be timed to coincide with the period of maximum fruit growth and K demand. Inadequate K uptake has been associated with moisture stress, occurring more commonly during periods of drought in non-irrigated regions. This reflects the dependence of K uptake by roots on diffusion through soil water films that decrease rapidly in volume and continuity in dry soils (Neilsen and Neilsen, 2003). In a recent study in a 10-year-old irrigated sweet cherry orchard with high tree-to-tree variation in vigour and yield, trees that were smaller with lower yields had lower leaf K concentrations. These trees were in warmer locations within the block and probably reflected the consequences of greater cumulative water stress, which also reduced K uptake (Neilsen *et al.*, 2009b). Furthermore, in a drip versus micro-jet sprinkler irrigation study in sandy soils, drip irrigation resulted in a smaller wetted soil volume and reduced leaf and fruit K uptake for 'Lapins' sweet cherry on 'GiSelA 5' (Table 9.6). As a consequence, 5 years after planting, deficient leaf K concentrations requiring supplemental K fertilization were measured for the drip-irrigated trees. Years of high crop load can also depress annual leaf K concentration due to the high concentration of K in cherry fruit.

Ca requirements in cherry orchards are high, often approaching that for N and K (Table 9.3). Much of the required Ca is contained within wood comprising the supporting framework of the trees. As a consequence of this and the high availability of Ca in most soils, Ca leaf deficiency symptoms

Table 9.6. Leaf and fruit K of 'Lapins' on 'GiSelA 5' rootstock as affected by type of irrigation (microsprinkler or drip) over the first 6 years of fruiting. Adapted from Neilsen *et al.* (2007).

| | Leaf K (% DW) | | | Fruit K (mg per 100 g FW) | | |
|---|---|---|---|---|---|---|
| Year | Microsprinkler | Drip | *P* value[a] | Microsprinkler | Drip | *P* value[a] |
| 2000 | 2.08 | 1.75 | <0.01 | 231 | 213 | <0.01 |
| 2001 | 2.00 | 1.48 | <0.001 | 160 | 139 | <0.001 |
| 2002 | 1.55 | 1.23 | <0.01 | NM | NM | – |
| 2003 | 1.48 | 1.19 | <0.001 | 190 | 181 | NS |
| 2004 | 1.35 | 0.89 | <0.0001 | 172 | 146 | <0.0001 |
| 2005 | 1.92 | 1.42 | <0.0001 | 175 | 154 | <0.0001 |

DW, dry weight; FW, fresh weight; NM, not measured; NS, not significant.
[a]Averages within years significantly different at the indicated *P* values.

in the field have not been reported. However, for horticultural crops, deficient concentrations of Ca in fruit have long been associated with major fruit quality disorders including bitter pit of apple and blossom end rot of tomato, with continued uncertainty as to the ultimate causes of these localized Ca deficiencies (Vang-Petersen, 1980; Saure, 2005).

The relationship between cherry fruit quality and Ca concentration has not been consistently positive, with reports of preharvest and postharvest calcium chloride ($CaCl_2$) applications increasing fruit firmness and reducing surface pitting (Lidster *et al.*, 1979), and other reports indicating no association between fruit Ca and firmness (Facteau, 1982). Similarly, foliar applications of calcium hydroxide ($Ca(OH)_2$) (Callan, 1986), $CaCl_2$ (Meheriuk *et al.*, 1991) or combinations of copper and calcium sprays (Brown *et al.*, 1995) have been reported to reduce cracking in sweet cherry fruit, although not consistently across years and crop loads (see Chapter 7, this volume). Thus, it has been difficult to develop target fruit Ca concentrations for optimizing cherry quality (Table 9.4). The preferred method of augmenting fruit Ca concentration has been via multiple dilute Ca salt sprays applied during the growing season. Soil Ca applications generally are not effective at increasing cherry Ca concentration but have been applied to increase soil pH.

Cherry requirements for Mg are lower than those for K and Ca. Field symptoms of Mg deficiency on cherry are not common (Westwood and Wann, 1966), but when observed include interveinal leaf chlorosis progressing to necrosis on basal leaves of new extension growth. Mg deficiency does not affect leaf size but, if severe, can be associated with early leaf senescence and abscission (Mulder, 1950). Such leaves commonly exhibit Mg concentrations around 0.2% dry weight or lower (Ashby and Stewart, 1969). The cherry rootstock 'F 12/1' is more susceptible to Mg deficiency than 'Colt', which has been attributed to its higher shoot/root dry-weight ratio (Troyanos *et al.*, 1997). It has been difficult to increase cherry leaf Mg concentrations via soil applications of magnesium sulfate ($MgSO_4$), whereas soil K applications readily decrease leaf Mg concentrations (Ashby and Stewart, 1969). Multiple foliar sprays of soluble Mg salts, including 2% (w/v) $MgSO_4$ sprays, applied during the period of rapid shoot growth have ameliorated Mg deficiency symptoms and maintained leaf Mg concentrations (Swietlik and Faust, 1984). Application of Mg-containing dolomite lime is an option to improve long-term soil Mg status for acidic soils.

### 9.4.4 Micronutrients: boron, zinc, manganese, iron and copper

Micronutrient requirements are low in cherry orchards, usually less than 1 kg $ha^{-1}$, and are required at p.p.m. concentrations in cherry leaves and fruit (Table 9.4). Nevertheless, with the exception of Cu, deficiencies of micronutrients have been identified for cherry.

Sand culture experiments with 'Bing' sweet cherry indicated that B deficiency was associated with reduced terminal growth, bud death, defoliation of shoot terminals after growth started, leaves with irregular margins and failure of blossoms to develop (Woodbridge, 1955). B toxicity also readily occurred in solutions containing 10 p.p.m. and could result in shoot dieback and gumming. Node development and leaf size and shape were normal, but necrotic areas along the leaf main vein and within the leaf were observed.

Modest soil B applications of 1 kg $ha^{-1}$ have been sufficient to alleviate B deficiency and increase leaf B concentrations above 20 p.p.m. dry weight, below which B deficiency has been observed (Table 9.4). Foliar B sprays at dilute concentrations (0.5%, w/v) are often applied as annual maintenance sprays in regions susceptible to low leaf B, with early season applications often recommended due to the perceived importance of B during flowering and pollination. However, a comparison of soil-applied B at bud break, foliar-applied B from bud break to petal fall and foliar-applied B before leaf drop showed no differences among application methods or timings in sweet cherry; all improved B levels in plant tissues and improved soluble solids levels in fruit but had no effect on tree growth, fruit yield or fruit size (Wójcik and Wójcik, 2006). Controlled trace $^{10}B$ studies have indicated rapid uptake and export of foliar-applied B throughout the growing season by most fruit trees, including sweet cherry (Picchioni *et al.*, 1995). Hanson (1991) reported that foliar B sprays applied postharvest to sour cherry increased fruit set and yield the following season, especially for trees of low leaf B status (<25 p.p.m., dry weight). Overapplication of fertilizer B in orchards has resulted in toxicity symptoms in other *Prunus* spp., such as peach and nectarine, when dry-weight concentrations of leaf B were between 38 and 198 p.p.m. and fruit B was between 90 and 311 p.p.m. (Dye *et al.*, 1984). The narrow range between B deficiency and toxicity for most plants suggests that B applications to sweet cherry should be moderate once adequate B levels are achieved.

Zn is widely considered the most common micronutrient deficiency, especially for high-pH soils that have limited Zn solubility (Broadley *et al.*, 2007). Severe Zn deficiency in cherry causes shoot tip rosetting (Woodbridge, 1954). A cluster of leaves with shortened internodes develops with some leaves of normal size and colour while others are quite small (little leaf) with interveinal chlorosis. Affected branches are stunted with undeveloped blind buds. Mild Zn deficiency symptoms are restricted to terminal leaves, which are somewhat elongated and yellow–green in colour. Trees are considered to be Zn deficient at leaf concentrations below 15 p.p.m. (Table 9.4).

Dormant sprays of zinc sulfate ($ZnSO_4$) are recommended to manage Zn deficiency (Swietlik and Faust, 1984). Soil applications are often ineffective, since the underlying causes of the deficiency are soil properties, such as high pH, which reduce the availability of Zn to tree roots. However, Benson *et al.* (1957) reported short-term success in increasing cherry leaf Zn concentration with two soil applications of 2.2 kg of Zn chelate, including Zn-EDTA. Wójcik and Morgaś (2015) reported that autumn foliar Zn sprays applied to sour cherry trees increased Zn concentrations in the sprayed leaves but did not improve Zn concentrations in growth the next spring, indicating a lack of remobilization during autumn leaf senescence.

Mn deficiency in sweet cherry is characterized by interveinal leaf chlorosis, especially on older, basal leaves of new growth (Westwood and Wann, 1966). Under severe deficiency conditions, leaves can become completely chlorotic and appear similar to lime-induced Fe chlorosis. Temporary chlorosis associated with Mn deficiency has been observed during cold springs but disappears as soil temperature warms. Leaf tissue analysis is a good indicator of Mn deficiency, with concentrations less than 20 p.p.m. indicative of possible problems (Table 9.4).

Mn deficiency can readily be corrected by early season manganese sulfate sprays at petal fall and 4–6 weeks later if required. Soil Mn applications are generally ineffective. Extremely high leaf Mn concentrations (exceeding deficiency concentrations by 10–20-fold) may occur on acid orchard soils resulting from long-term N fertilization. This can be confirmed by soil pH analysis and may require corrective liming, especially for replanting.

Fe deficiency, frequently referred to as lime-induced chlorosis, is common for fruit trees, including cherry, growing in semi-arid regions with calcareous or alkaline soils. Fe availability in these soils is reduced due to their high pH and bicarbonate content. Although considerable research related to overcoming Fe deficiency has been reported for other susceptible tree fruits such as peach, pear and apple, little information has been reported specifically for cherry. It is likely that much of the information pertinent to these other fruits is relevant to cherry and thus will be used to guide the following discussion. Inadequate Fe inhibits the normal development of chlorophyll and is initially apparent as interveinal chlorosis on young (shoot tip) leaves, resulting in prominent green veins. Symptoms can progress from younger to older leaves, with chlorosis eventually extending to complete yellowing of leaves that can be susceptible to leaf scorch late in the growing season. Leaf and fruit size can be severely decreased. These symptoms are a more reliable indicator of deficiency than leaf Fe concentration (dry-weight basis), which can frequently be high in deficient leaves (Morales *et al.*, 1998). Determination of 'active Fe' ($Fe^{2+}$) in leaves after extraction in 0.5–1.0 M hydrochloric acid has been proposed as an improved measure of Fe deficiency (Neilsen and Neilsen, 2003). Flower petal Fe concentrations have been correlated positively with the degree of late-season Fe chlorosis for peach (Belkhodja *et al.*, 1998).

Overcoming Fe deficiency for fruit trees has frequently been problematic. Multiple sprays of iron sulfates or chelates to foliage can be moderately effective but need annual reapplication. Temporary regreening of kiwifruit leaves has been reported after foliar sprays of acidic compounds, including citric, sulfuric and ascorbic acid, which may solubilize Fe already present in chlorotic leaves (Tagliavini *et al.*, 2000). Application of iron sulfate ($FeSO_4$) in association with OM and acidifying fertilizers, particularly if banded, may provide temporary relief. Broadcast applications of Fe chelates have limited long-term effectiveness and high costs. In general, soil effectiveness is curtailed, especially if the lime ($CaCO_3$) content of soil is high. Trunk injection of 0.5–2.0% ferrous sulfate solutions have been temporarily effective but can cause trunk damage. Use of graminaceous cover crops, fertilized with $FeSO_4$ or application of Fe-enriched organic amendments, have been proposed as agronomic methods to improve Fe nutrition (Tagliavini *et al.*, 2000).

## 9.5 Specific Nutrient Management Strategies

Nutrient management in traditional orchards has been reviewed thoroughly (Hanson and Proebsting, 1996). The advent of high-density orchards and dwarfing rootstocks over the past 20 years, as well as a greater focus on orchard management impacts on environmental sustainability (soil health, nutrient leaching, nutrient runoff in surface water), have been the impetus for new techniques to provide nutrients precisely and efficiently. The broadcasting of seasonal nutrient needs in a single or split application is being replaced by multiple, more precise applications via irrigation (fertigation) and/or foliage sprays and/or utilization of organic sources. Foliar applications are largely utilized to remedy specific deficiencies or address specific localized needs (Table 9.7). Strategies and sources for fertigation and organic fertility are discussed below.

### 9.5.1 Fertigation

Application of soluble nutrients directly with irrigation, termed fertigation, can be particularly effective in irrigated horticultural production systems as well as beneficial in more humid climates characterized by variation in natural precipitation patterns.

**Table 9.7.** Selected foliar spray applications of major and minor nutrients for cherry. (Adapted from Swietlik and Faust, 1984.)

| Nutrient | Compound | Timing | Concentration[a] | Comment |
|---|---|---|---|---|
| Nitrogen | Urea | Postharvest | 2–5 kg $l^{-1}$ | To enhance tree N reserves; deficiency usually overcome by soil applications |
| Calcium | Calcium chloride (food grade) | Prior to harvest after pit hardening | 0.5 kg $l^{-1}$ | Apply by sprayer to reduce splitting |
| | | Prior to harvest after pit hardening | 0.75 kg $l^{-1}$ | Apply by overhead sprinklers to reduce splitting |
| Magnesium | Magnesium sulfate | Apply to foliage preharvest | 2 kg $l^{-1}$ | As maintenance, to overcome deficiency |
| Zinc | Zinc sulfate | Late dormant | 1.25 kg $l^{-1}$ | To overcome deficiency or as annual maintenance |
| | | Postharvest (within 2 weeks of harvest) | 1.0 kg $l^{-1}$ | |
| | Liquid zinc sulfate | Late dormant/ postharvest | 2.5 l per 100 l | |
| | Zinc chelate or organic complex | Throughout growing season | Label rates | |
| Boron | Borospray or Solubor® | Throughout growing season | 0.1 kg $l^{-1}$ | To overcome deficiency or as annual maintenance |
| Manganese | Manganese sulfate | Throughout growing season | 0.2 kg $l^{-1}$ | To overcome deficiency |
| Iron | Ferrous sulfate | To chlorotic foliage during growing season | | To temporarily overcome deficiency |
| | Iron chelate | | Label rates | |

[a]Application rate depends on the water applied and number of applications.

The topic has been the subject of several reviews (Bar-Yosef, 1999; Neilsen *et al.*, 1999) including with specific reference to sweet cherry (Neilsen and Neilsen, 2008). Fertigation provides frequent small doses of fertilizer instead of a typical high broadcast or banded dose, which is particularly effective in high-density plantings with compact root systems and limited wetted soil volumes under drip emitters (Neilsen and Neilsen, 2008). This can reduce fertilizer inputs by more precisely matching nutrient application and timing to plant requirements. A method of injecting nutrients into the irrigation system is required (Fig. 9.6), as are suitable soluble nutrient sources (Table 9.8) (Burt *et al.*, 1995; Waterman, 2001). Some combinations of soluble fertilizers are incompatible when mixed because they can precipitate, potentially plugging irrigation lines and emitters. Similarly, local irrigation water may have chemical compositions, especially those containing high concentrations of Ca and Mg, which similarly interact with fertigation constituents to cause precipitation. It is often useful to determine the nutrient content of the irrigation source prior to fertigation. Mixing samples of irrigation water and potential fertigation solutions in the proportion with which they will be mixed in irrigation lines can provide a visual indication of whether precipitation problems are likely to occur.

In irrigated plantings, with minimal growing-season precipitation, a high degree of control of root-zone water application can be achieved. This can be particularly effective for root-zone retention of N that is susceptible to leaching due to its high solubility and rapid transformation to the

1. Backflow preventer
2. Filter
3. Solenoid valve
4. Flow meter
5. Pressure regulator
6. Injection port
   i. Venturi – inline passive
   ii. Pumps – active injection
7. Pressure-compensating emitters
8. Automated soil moisture or ET data

**Fig. 9.6.** Components of an automated, regulated microirrigation system. ET, evapotranspiration.

**Table 9.8.** Commonly fertigated nutrient sources for cherry.

| Nutrient | Compound | Nutrient content | Comment |
|---|---|---|---|
| Nitrogen | Ammonium nitrate | 33–34% N | Acidifying |
| | Ammonium sulfate | 21% N, 24% S | Very acidifying |
| | Urea | 45–46% N | Acidifying |
| | Calcium nitrate | 15.5% N | |
| | Potassium nitrate | 13–14% N, 44–46% $K_2O$ | |
| | Urea solutions – various | 20–23% N | |
| | Urea/ammonium solutions – various | 28–32% N | |
| Phosphorus | Phosphoric acids | 52–75% $P_2O_5$ | Acidifying |
| | Ammonium polyphosphates | 8–11% N, 34% P205 | Acidifying |
| | Monoammonium phosphate | 11% N, 50% $P_2O_5$ | Acidifying |
| | Diammonium phosphate | 18% N, 46% $P_2O_5$ | Acidifying |
| Potassium | Potassium chloride | 60–62% $K_2O$ | |
| | Potassium sulfate | 50% $K_2O$, 17% S | Ultrafine grind |
| | Potassium/magnesium sulfate | 22% $K_2O$, 11% MgO | |
| | Potassium thiosulfate | 22% $K_2O$, 17% S | |
| Boron | Sodium borate | 20% B | |

mobile $NO_3$-N form. Control of within-season N leaching can thus be achieved by controlled water applications. An evapotranspiration (ET)-based irrigation method, in which daily water applications vary in response to seasonal canopy development and atmometer-based daily ET (Parchomchuk *et al*., 1996), is discussed in section 9.6.

Fertigation strategies need to take into account the fertility (nutrient-supplying capacity) of the orchard soil and the tree's seasonal requirement for important nutrients. Experiences with fertigating individual nutrients for cherry have been described previously regarding methods for achieving seasonal nutrient sufficiency. An example of a multi-nutrient fertigation regime for sweet cherry on dwarfing rootstocks in a relatively infertile sandy loam soil involves annual fertigation of N, P, K and B augmented with dormant application of $ZnSO_4$ (Table 9.9). This resulted in successful establishment, good growth and yield, and a healthy range of leaf nutrient concentrations over the first five growing seasons for 'Skeena' sweet cherry (Neilsen *et al.*, 2016).

The cumulative evidence from fertigation research is that nutrients can be applied at lower rates than one-time broadcast applications across the whole orchard floor (Neilsen *et al.*, 1998, 1999; Stoilov *et al.*, 1999; Koumanov *et al.*, 2017). The possibility of excessive nutrient enrichment of the soil via fertigation can be monitored by periodic electroconductivity measurements of the soil, especially where soil profile drainage is limited. Continued fertigation of ammonium-containing N fertilizers into restricted soil volumes can result in soil acidification due to nitrification of the applied $NH_4$-N. The sensitivity of orchard soils to this type of acidification varies with soil properties, and the buffering capacity of individual soils can be estimated by an acidification resistance index (Neilsen *et al.*, 1995). Soils with minimal resistance to acidification should receive N fertigated in nitrate rather than ammonium form.

### 9.5.2 Organic and integrated nutrient management

In most cherry-growing regions, organic production is limited due to the difficulty of achieving organically acceptable control of insect pests and/or diseases. However, central to the concept of organic production is the enhancement of soil OM and soil biological activity (Treadwell *et al.*, 2003). Thus, for organic production systems, seasonal nutrient demands are met by the use of cover crops and/or application of composts or organic amendments (Table 9.10) as surface mulches or incorporated within the soil profile instead of relying on application of manufactured inorganic chemical fertilizers. Achieving adequate N can be particularly challenging if relying solely on organic sources due to their relatively low N content and variability in rate of mineralization to plant-available $NO_3$-N and $NH_4$-N. A number of soil test laboratories offer analyses to estimate the N mineralization potential of specific organic materials and to measure their C/N ratio, which is a useful indicator of available N (Gale *et al.*, 2006). Furthermore, as a consequence of the narrow N/P ratios of many organic sources, particularly composts, relative to the ratio of N/P required for cherry production (Table 9.3), long-term compost application can result in relative enrichment in orchard soil P, thereby increasing the risk of P contamination of water sources (Nelson and Janke, 2007). A range of liquid organics, including compost teas and various humate materials and fulvic acids, has been proposed as acceptable soil amendments, but there has been little verification of their effectiveness in cherry production.

Integrated nutrient management involving the application of both organic and biological products, in combination with reduced rates of inorganic fertilizers, has been proposed for intensive horticultural production systems, although field assessment for cherry

**Table 9.9.** Annual fertigation strategy applied during the first five growing seasons to 'Skeena' sweet cherry on dwarfing rootstocks on a sandy loam soil. (Adapted from Neilsen *et al.*, 2016.)

| Nutrient | Fertilizer form | Application duration | Annual application rate |
|---|---|---|---|
| Nitrogen | Calcium nitrate (15.5-0-0)[a] | Daily 6 weeks after bloom | 16.5 g N per tree (27.5 kg N $ha^{-1}$) |
| Phosphorus | Ammonium polyphosphate (10-34-0)[a] | One day after bloom | 20 g P+13.5 g N per tree (33 kg P+22.5 kg N $ha^{-1}$) |
| Potassium | Potassium chloride (0-0-60)[a] | Daily for 6 weeks starting 4 weeks after bloom | 20 g K per tree (33 kg K $ha^{-1}$) |
| Boron | Solubor® (20.3% B) | Daily for 6 weeks starting 4 weeks after bloom | 0.17 g B per tree (0.28 kg B $ha^{-1}$) |

[a]Percentages by weight of N-$P_2O_5$-$K_2O$.

**Table 9.10.** Selected soil amendment strategies suitable for organic cherry production. (Adapted from Neilsen *et al.*, 2009a.)

| Amendment | Qualifications/limitations | Reference |
|---|---|---|
| Cover crops/green manures | No genetically engineered crops allowed | Blackshaw *et al.* (2005) |
| Uncomposted manure | Potential pathogenic bacteria problem if there is contact with fruit | |
| Composts | When produced, requires a thermophilic digestion stage at 55–60°C for several days to kill weed seeds and pathogens; can be comprised of manures and yard, fish, seaweed and food wastes, but no prohibited materials (e.g. biosolids); caution should be taken with the use of high-salinity composts | |
| Meals | Can comprise blood, fish, feather, bone and soybean containing unadulterated components | |
| Liquid organics | Little documented research; can be comprised of compost teas (leachate of acceptable composts), liquid fish (without synthetic preservatives), or humate and fulvic acid (unfortified with synthetic fertilizers) | |
| Surface mulches | | |
| Wood chips, bark | Untreated wood; may require supplemental N | Yao *et al.* (2005) |
| Shredded or waste paper, spray-on mulch | Containing no metal-based or other chemical contaminants | Cline *et al.* (2011) |
| Straw (cereals, wheat) | | Yin *et al.* (2012) |
| Hay (meadow, lucerne) | | |

is limited to date (Flores *et al.*, 2015). Nevertheless, the adoption of management practices relevant to organic production by conventional cherry growers is of increasing interest. Pertinent research has indicated that in-row application of straw or polypropylene mulches in cherry orchards can enhance fruit yields and reduce water consumption (Yin *et al.*, 2007, 2012). An important component of future estimates of the C footprint of cherry orchards will be the contribution of orchard floor management and the C balance of orchard soils. Increased use of in-row organic amendments, mulches and beneficial inter-row cover crops can enhance the C sequestration potential of cherry orchards (Tozzini *et al.*, 2015).

## 9.6 Seasonal Water Limitations

### 9.6.1 Water use

Compared with other tree fruits (e.g. apple and peach) or grapes, there has been relatively little research on the water requirements of cherry. Although cherries are grown under both rain-fed and irrigated conditions, the best estimates of cherry water requirements come from irrigated production. In regions with low summer precipitation and high vapour pressure deficit (VPD), such as Washington State, USA, sweet cherry trees may require between 750 and 1000 mm of irrigation during the growing season, if watered to 100% ET (Hanson and Proebsting, 1996). In more humid regions, requirements are lower, for example 600–650 mm for 'Summit' on 'SL-64' (*P. mahaleb*) in Lleida, Spain (Marsal *et al.*, 2010) and 550–700 mm for a range of cultivars/rootstocks in British Columbia, Canada (D. Neilsen, unpublished data). Oyarzún *et al.* (2010) estimated transpiration rates for mature 'Bing' on 'GiSelA 5' from in-field cuvettes that enclosed the whole tree. Maximum whole-canopy transpiration rates were around 75 mmol $s^{-1}$ per tree, with a daily average of approximately 50 mmol $s^{-1}$ per tree. Assuming a 10 h day during which transpiration was occurring, daily whole-tree transpiration can be estimated at around 32 l during hot dry weather in a semi-arid climate where daily reference crop ET ($ET_0$) is around 7 mm.

Whole-tree transpiration under both rain-fed and rain-fed plus supplemental irrigation conditions also was estimated from sap flow measured over the growing season for 'Rita' grafted on 'GiSelA 6' using heat-balance probes (Juhász *et al.*, 2013). Transpiration rates were directly related to estimates of VPD, global radiation and air temperature, and seasonal requirements were estimated to be 700–800 l year$^{-1}$. Water uptake was strongly affected by crop phenology, with a decrease in water use and demand after harvest, which rebounded when shoot growth resumed later in the season. This is supported by observations of reduced post-harvest stomatal conductance and photosynthesis (Whiting and Lang, 2004).

### 9.6.2 Excess water

A major concern for cherry production under both rain-fed and irrigated conditions is the impact of excess water on fruit quality. Sweet cherry fruit cracking due to rain on fruit and/or excess water in the soil is a major disorder that seriously affects economic returns. This is discussed mechanistically in Chapter 7 (this volume).

### 9.6.3 Tree water status

#### *Water potential*

Measurements of tree water status are often used to determine the degree of stress in response to environmental conditions. Water potentials of leaves, stems and fruit, taken at different times during the day, have all been used to describe tree water status. Consequently, a range of values exists as 'thresholds' of damaging stress. The diurnal time course of water potential in cherry tissues depends on temperature and VPD. Diurnal leaf water potential varied between more than –1 MPa and –2.6 MPa for 'Corum' and 'Napoleon' ('Royal Ann') sweet cherries, with the minimum occurring around 2 p.m. and the maximum at predawn (Tvergyak and Richardson, 1979). A similar diurnal time course was observed for the water potential of sunlit leaves (minimum –2.4 MPa) in 'Bing' grafted on 'GiSelA 6', but this was moderated for shaded leaves (minimum –1.6 MPa) and for stem water potential assessed from covered leaves (–1.0 MPa) (Oyarzún *et al.*, 2010). In all cases, predawn water potential was around –0.5 MPa. However, transpiration rates were not reduced at minimum leaf potentials or estimated whole-canopy water potentials of –1.8 MPa. In contrast, under high VPD, recovery of leaf water potential was associated with stomatal closure, with a threshold for damaging stress around –1.5 to –1.8 MPa in 'Simone' on 'F 12/1' (Measham *et al.*, 2014). Similarly, Proebsting *et al.* (1981) found severe tree damage when the leaf water potential fell below –1.5 MPa in response to imposed water deficits of <50% ET replacement in 'Bing', 'Chinook' and 'Rainier' trees on *P. mahaleb*. As a consequence, measurements of plant water potential are often used as triggers for irrigation scheduling. In an experiment with 'Brooks' on 'MaxMa 14', a predawn leaf water potential of <0.5 MPa was indicative of plant water deficits (Livellara *et al.*, 2011), and in deficit irrigation experiments with 'New Star' and 'Summit' on 'SL-64', stem water potential was not allowed to fall below –1.5 MPa (Marsal *et al.*, 2009, 2010).

#### *Hydraulic conductance and water storage*

There have been few estimates of cherry tree hydraulic conductance and water storage. Measurements of leaf and stem water potential and whole-canopy transpiration were used to assess tree hydraulic conductance in 'Bing' on 'GiSelA 5' (Oyarzún *et al.*, 2010). Average hydraulic conductance across the soil–plant–atmosphere continuum was estimated at ~60 ± 6 mmol s$^{-1}$ MPa$^{-1}$. Hydraulic conductance was higher on the stem–leaf–atmosphere pathway (150 ± 50 mmol s$^{-1}$ MPa$^{-1}$) than the soil–root–stem pathway (100 ± 20 mmol s$^{-1}$ MPa$^{-1}$), indicating that the latter is the limiting step for water transport, as has been seen in other grafted fruit tree species

(Moreshet *et al.*, 1990; Alarcón *et al.*, 2003). The rootstock–scion graft union may play a role in these differential hydraulic conductivities, as discussed in section 9.6.5.

A weak hysteresis in diurnal whole-canopy transpiration and plant water potential was suggested as evidence that tree water storage plays a small part in cherry tree water balance (Oyarzún *et al.*, 2010). Similar results have been reported for other fruit trees, such as apricot (Alarcón *et al.*, 2003) and olive (Tognetti *et al.*, 2004).

#### *Diagnosing water stress*

In addition to measurements of plant water potential, other methods have been used for the diagnosis of plant water stress, ranging from in-plant sensors to remote sensing imagery. Compared with other woody perennial horticultural species, there are relatively few studies with plant sensors and imagery for cherry. Plant sensors for continuous monitoring of water stress in other species of fruit trees include those measuring shrinkage and expansion of the trunk (e.g. Goldhamer and Fereres, 2001), plant water tension (e.g. Pagay, 2014) and sap flow (e.g. Dragoni *et al.*, 2009).

Using automated systems, measured daily trunk shrinkage and trunk growth rate were found to be related to stem water potential for 'Brooks' on 'MaxMa 14' in response to season-long 50/100/150% potential ET replacement irrigation (Livellara *et al.*, 2011). Based on reduced tree productivity and growth, the water deficit threshold was assessed to be between 50 and 100% ET replacement in this study. Correlations between stem water potential and automated trunk diameter measurements indicated that a critical predawn stem water potential value around –0.5 MPa was associated with reference values of 165 μm for maximum daily trunk shrinkage and 83 μm day$^{-1}$ trunk growth rate, which could potentially be used to control irrigation applications automatically. Some new approaches have yet to reach the commercialization stage but hold great promise. For example, a microtensiometer that can be embedded in the stems of woody plants has been developed for continuous measurement of water potential (Pagay, 2014).

Remote sensing techniques have been employed to determine plant water stress, including thermal (Jones *et al.*, 2009) and leaf reflectance imagery accessed through a range of technologies (satellites, ultralight aircraft, unmanned aerial vehicles, and in-orchard static or mobile platforms). Infrared spectrum thermal imagery has been used to determine a range of indices including the water deficit index (Moran *et al.*, 1994) and crop water-stress index based on temperature differences between the canopy and surrounding air (e.g. Bellvert *et al.*, 2016) for peach. By comparison with a well-watered control and other plant water relations measurements such as stem potential and stomatal conductance, thermal imagery can be used to examine the spatial distribution of water stress and potentially to schedule irrigation. There are no reported studies for cherry. Other types of spectral imaging have been used to determine canopy development, such as the normalized distance vegetation index using the visible and near-infrared spectra. The use of spectral reflectance for diagnosing water stress was examined for mature 'Bing' on Mazzard (Antúnez *et al.*, 2008). Individual leaf reflectance in the range 540–710 nm correlated well with the range of measured stem water potential and was considered to have good potential for rapid screening of plant water status across an orchard.

### 9.6.4 Mitigation of water stress

Water stress is usually mitigated by irrigation. Cherries can be grown with many different types of irrigation systems ranging from flood and furrow to microirrigation such as drip irrigation (Hanson and Proebsting, 1996; see Chapter 10, this volume). The threat of water shortages and changes in cherry production systems have led to increased use of microirrigation (Neilsen *et al.*, 2010, 2014) and in some cases conversion from less efficient systems (Proebsting *et al.*, 1981; Yin *et al.*, 2012).

### Conversion to efficient irrigation systems

Proebsting *et al.* (1981) converted furrow-irrigated 'Bing', 'Chinook' and 'Rainier' on *P. mahaleb* trees to drip irrigation and imposed a range of water stresses to mimic drought. The conversion to daily drip from furrow irrigation applied just twice before harvest did not reduce yield or fruit size when water was supplied to meet the 100% ET calculated from class A evaporation pans (860 mm). Conversion to drip irrigation from a microsprinkler was combined with different types of mulches to determine water savings from reduced evaporation from the soil surface in 'Lapins' on Mazzard sweet cherry (Yin *et al.*, 2012; Long *et al.*, 2014). There was up to 70% reduction in water use with drip irrigation compared with use of a microsprinkler, with no corresponding loss in production, fruit size or firmness, and the use of straw mulch reduced water input by 9%. Drip irrigation increased the amount of marketable fruit by reducing surface pitting and bruising.

### Irrigation scheduling

The timing and amount of applied irrigation water needed are influenced by prevailing climatic conditions, the stage of growth, soil properties and the extent of root development. Historically, two methods have been used for scheduling irrigation applications to meet plant requirements: (i) estimates of water loss through ET derived from class A evaporation pans or climate-based measurements (Allen *et al.*, 1998); and (ii) soil moisture measurements (USDA-NRCS, 1993). In the past 20–25 years, plant-based water-stress measurements have gained popularity, including leaf and stem water potential (Shackel *et al.*, 1997), permanently installed sensors such as dendrometers for trunk expansion (Goldhamer and Fereres, 2001; Livellara *et al.*, 2011), sap flow (Green *et al.*, 2003; Dragoni *et al.*, 2009; Juhász *et al.*, 2013) and, more recently, plant water tension (Pagay, 2014). In addition, remote sensing and direct measurement of canopy temperatures and spectral reflectance are being developed for determining field-scale plant stress and water deficits, with the aim of refining irrigation scheduling techniques (Barria, 2006; Buyukcangaz *et al.*, 2007; Jones *et al.*, 2009; Köksal *et al.*, 2010; Barton, 2012). Where irrigation scheduling is used, often a range of methods and sensors are employed.

ET-based methods require the use of a canopy development factor to modify the reference $ET_0$ estimates made from climate or class A evaporation pan data. Average canopy area estimates were used to modify class A evaporation pan data in a sweet cherry drought trial (Proebsting *et al.*, 1981), and crop and pan coefficients (Allen *et al.*, 1998) and a shade coefficient were used for a trial in which all coefficients combined to give a coefficient of 0.63 (Buyukcangaz *et al.*, 2007). Crop coefficients ($K_c$) were used to modify reference $ET_0$ estimates from weather data (Allen *et al.*, 1998) based on photosynthetically active radiation interception by sweet cherry canopies measured with a ceptometer (Marsal *et al.*, 2009, 2010). At mid-season (peak canopy development), $K_c$ was around 0.94, which is within the range of 0.9–1.20 suggested by Allen *et al.* (1998) for sweet cherry grown with or without ground cover. More recently, remote sensing techniques have been used to assess $K_c$ and the basal crop coefficient ($K_{cb}$), which modifies the transpiration component of ET in a dual-crop coefficient system (Pereira *et al.*, 2015). Vegetation indices such as the normalized difference vegetation index have been used to estimate the fraction of vegetation cover, and are applied to ET estimates from either weather station variables or from other remotely sensed imagery used in energy or water-balance calculations. Crop ET for high-density apples grown with drip irrigation was evaluated using latent heat fluxes in an eddy covariance system and compared with a remote sensing-based soil water balance with reasonably good results (Odi-Lara *et al.*, 2016). To date, there are no such studies for cherry.

Soil moisture-based methods for scheduling irrigation use either soil moisture content or soil moisture potential measurements. These could be discrete measurements on a predetermined time step or continuous recording methods that may or may not be linked to automated irrigation systems. Soil moisture measurements also may be components of water-balance calculations. Yin *et al.* (2012) used a neutron probe to determine weekly soil moisture content in a study of 'Lapins' on Mazzard. The irrigation amount was based on weekly soil moisture depletion and estimates of crop ET ($ET_c$), precipitation and deep drainage. The importance of soil for water storage becomes less important as irrigation frequency increases, particularly if applied on a daily or subdaily basis (Hillel, 2004). Plant growth (trunk cross-sectional area) and yield were greater for 'Skeena' and 'Cristalina' on 'GiSelA 6' receiving high-frequency (four times daily) irrigation than for trees receiving irrigation every second day. In both cases, water was added to replace 100% $ET_c$ (Neilsen *et al.*, 2010).

Multiple sensors are often used in scheduling irrigation. A combination of automatic weather station data used to calculate $ET_0$ and automated soil moisture monitoring was used to schedule supplemental irrigation based on maintenance of soil moisture between predefined trigger levels (Juhász *et al.*, 2013). A combination of $ET_0$ estimated from an electronic atmometer (ETGage Co., Loveland, Colorado) calibrated against weather station estimates of $ET_0$ has been used in conjunction with $K_c$ values for sweet cherry to automatically schedule and control daily or subdaily irrigation in sweet cherry (Neilsen *et al.*, 2010, 2014, 2016).

#### *Targeted water management (water conservation)*

As the water supply for irrigation becomes more uncertain, there is increased interest in conserving water by reducing water inputs through regulated deficit irrigation (RDI). Associated with the need to accommodate water shortages are the supplemental benefits of potentially improving fruit quality and controlling floral bud development. In sweet cherry production, the relative earliness of the harvest allows postharvest applications of water deficits without affecting the current season crop (Barria, 2006). 'New Star' grafted on *P. mahaleb* trees were subjected to 100, 80 and 50% $ET_c$ irrigation postharvest, but trees were not allowed to fall below –1.5 MPa of stem water potential (Marsal *et al.*, 2009). The second-year yield and crop load were unaffected by irrigation treatments, but 50% RDI advanced crop maturity and slightly reduced fruit firmness and soluble solids content after storage for the subsequent crop. Postharvest RDI (100, 80 and 50% $ET_c$) was also tested as a means of reducing biennial bearing in 'Summit' on 'SL-64' (Marsal *et al.*, 2010). RDI in year 1 reduced the winter starch content of roots and fruit set in year 2, but did not affect subsequent yield, as all trees required crop thinning in the 'on' year. In year 3, the 'off' year, 50% RDI increased fruit set, suggesting a potential role for RDI in correcting alternate bearing.

### 9.6.5 Rootstocks and cherry water relations

The role of rootstocks in controlling plant water uptake, transport and stress has long been of interest for woody perennial crops. The issue of rootstock–scion compatibility and its effect on sweet cherry water relations was examined for 'Sam' grafted on *P. cerasus* ('Weiroot'), *Prunus acida* or *P. avium* ('F 12/1') and compared with *P. cerasus* seedlings (Schmitt *et al.*, 1989). The lowest leaf water potentials (–2.4 MPa) in response to high VPD were associated with *P. cerasus* seedlings and 'Sam' on 'F 12/1'. Closed stomata, low transpiration rates and low leaf water potential were associated with wilted leaves indicative of graft incompatibility on *P. cerasus* and *P. acida*. However, there was no resistance to flow found across the graft union. A comparison of the graft unions of 'Lapins' on vigorous 'Colt' and dwarfing 'GiSelA 5' revealed reduced xylem flow in the dwarfing graft union, probably resulting from smaller vessel diameters and more lignified tissue in the union (Olmstead *et al.*, 2006).

The effects of size-controlling traits on water relations and gas exchange were examined for 'Burlat', 'Summit' and 'Van' cultivars grown on five rootstocks (Gonçalves *et al.*, 2003, 2005). More vigorous rootstocks had higher stem water potential, net $CO_2$ assimilation, stomatal conductance and intercellular $CO_2$ concentration than less vigorous rootstocks (*P. avium* > 'CAB11E' > 'MaxMa 14' > 'GiSelA 5' > 'Edabriz'), but there was no effect of the scion. Photosynthesis was strongly correlated with stomatal conductance, with the effect increasing with rootstock vigour control, indicating that water relations and photosynthesis of sweet cherry are largely influenced by rootstock genotype rather than by the scion.

Similar effects were found when 'Skeena' trees on three dwarfing rootstocks, 'GiSelA 6', 'GiSelA 5' and 'GiSelA 3', were compared (Neilsen *et al.*, 2016). 'GiSelA 3' (the most dwarfing) trees had a consistently lower midday stem water potential and stomatal conductance (Fig. 9.7) than 'GiSelA 6' (semi-vigorous) trees, particularly up to harvest and during the hottest part of the growing season, with 'GiSelA 5' trees intermediate in size and response. After harvest, these effects largely disappeared, consistent with the postharvest reduction in photosynthesis reported by Whiting and Lang (2004).

### 9.6.6 Summary

Water requirements for cherry are influenced by interactions among environmental conditions, management practices and genetics. In high VPD environments, cherry may require up to 800–1000 mm of irrigation per year when watered at full ET replacement. Under the same environmental conditions, trees may display a range of responses to water inputs depending on irrigation method (type, frequency and application rates), crop load, rootstock (size-controlling characteristics and compatibility with scion) and a range of other management practices. However, from a number of studies, detrimental stress appears to occur when the midday stem water potential approaches –1.5 MPa or when the predawn leaf water potential approaches –0.5 MPa.

Mitigation of water stress is usually through irrigation, and a wide range of practices has been developed to supply water to cherries. Frequently, the methods used for assessing plant water stress (water potential measurements, canopy temperature, remote sensing of vegetation cover or spectral reflectance, sap flow, and stem and fruit linear displacement gauges) are also used in irrigation management to determine the timing and duration of water applications. Other methods include soil moisture sensing and estimates of $ET_c$ from weather station or evaporation data.

As cherry production systems adopt size-controlling rootstocks and higher

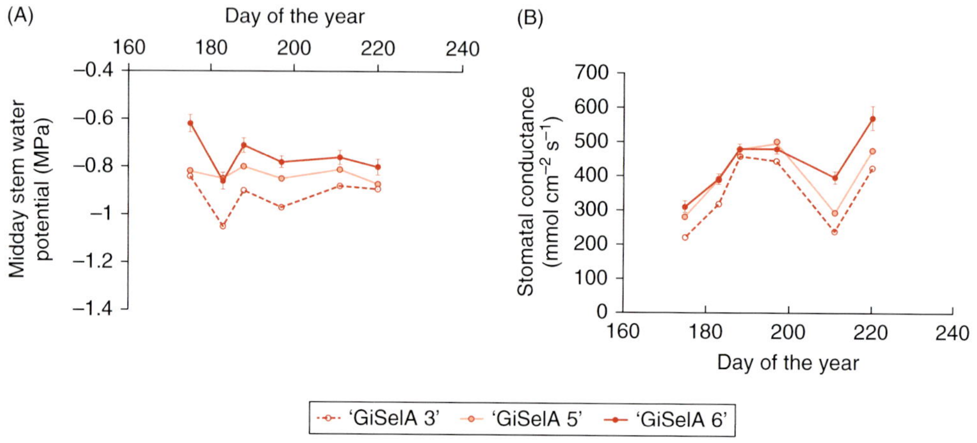


**Fig. 9.7.** Midday stem water potential (A) and stomatal conductance (B) over the growing season for 'Skeena' sweet cherry on 'GiSelA 3', 'GiSelA 5' and 'GiSelA 6' rootstocks. Harvest date was day 202. (Adapted from Neilsen *et al.*, 2016.)

density plantings, there is increasing utilization of high-efficiency microirrigation technologies, particularly drip irrigation. This also helps address the growing concerns about available and reliable water supplies. In addition to efficient irrigation systems and irrigation scheduling to meet plant water demand, other techniques have been adopted that reduce annual water use, including the use of mulches to reduce evaporation and RDI. As cherries are frequently harvested several months before leaf senescence, there is an opportunity to reduce water inputs postharvest without affecting current season fruit quality. Some benefits may also accrue, including improved subsequent season fruit quality and a reduction in biennial bearing tendencies.

## 9.7 Future Challenges

New environmental challenges can be expected to arise for cherry growers. Among these will be variability and change in current climates, which over time may alter the suitability of existing planting areas while increasing the production potential of land at higher altitudes and latitudes. Not all newly suitable sites will have optimum soil chemical and physical properties for cherry production, necessitating increased attention to overcoming soil limitations. Increased ET resulting from higher growing-season temperatures, combined with potentially increasingly variable precipitation, will stimulate the adoption of irrigation to supplement orchard water supplies and fertigation to supply nutrients with irrigation waters. Increased development of technologies that measure plant and soil moisture stresses will be useful, especially if these measurements can be automated to provide scheduling of irrigation to prevent excessive water applications and to optimize cherry fruit quality, especially fruit size. Orchard production will be increasingly scrutinized for its role in mitigating the production of major greenhouse gases including $CO_2$, methane and $N_2O$. Modification of major orchard management strategies, including irrigation frequency and timing, use of cover crops and soil C amendments, may occur to improve the climate mitigation effects of cherry production.

The unique nutrient and water requirements of new cultivar and rootstock genotypes will always have to be assessed. The nutrient profiles of cherry for optimization of postharvest storage quality need to be determined. It would be desirable if rootstocks could be developed with enhanced nutrient acquisition capabilities or drought resistance, provided such improved genetic capabilities did not detrimentally affect the production of high-quality fruit. Increased plantings of high-density cherry on dwarfing rootstocks to increase precocity and unit area yields will increase the demand for nutrients such as K that occur at high concentrations in harvested fruit. The water relations of dwarfing rootstocks will have to be better understood to maximize their efficient irrigation. The high cost of orchard establishment and replanting will necessitate the development of effective strategies to overcome replant problems, especially with the disappearance of many standard fumigants for environmental reasons. Therefore, it is desirable to develop a better understanding of the biological elements of a healthy soil that will sustain cherry production.

Augmenting orchard soil OM content may be the single most important factor preventing long-term degradation of soil quality in cherry orchards. In addition to increasing soil biological diversity and improving soil structure and soil water-holding capacity, OM acts as a source of multiple plant nutrients. The cost of conventional fertilizers is currently low relative to other management inputs but is likely to rise in the future due to the close association between natural gas energy costs, N fertilizer production and diminished supplies of non-renewable P fertilizer. Furthermore, it will be desirable to better understand the suitability and effectiveness of complete or partial use of organic fertilizers as nutrient sources. Similarly, more information is required to judge the effectiveness of various new and proposed biological amendments.

## References


Alarcón, J.J., Domingo, R., Green, S.R., Nicolás, E. and Torrecillas, A. (2003) Estimation of hydraulic conductance within field-grown apricot using sap flow measurements. *Plant Soil* 251, 125–135.

Allen, R.G., Pereira, L.S., Raes, D. and Smith, M. (1998) *Crop Evapotranspiration Guidelines for Computing Crop Water Requirements*. FAO Irrigation and Drainage Paper 56. United Nations Food and Agriculture Organization, Rome.

Antúnez, A., Whiting, M.D., Pierce, F. and Stockle, C. (2008) Estimation of sweet cherry tree water status by spectral reflectance. *Acta Horticulturae* 795, 711–716.

Ashby, D.L. and Stewart, J.A. (1969) Magnesium deficiency in tree fruits in relation to leaf concentration of magnesium, potassium and calcium. *Journal of the American Society for Horticultural Science* 94, 310–313.

Azarenko, A.N., Chozinski, A. and Brutcher, L. (2008) Nitrogen uptake efficiency and partitioning in sweet cherry is influenced by time of application. *Acta Horticulturae* 795, 717–721.

Baghdadi, M. and Sadowski, A. (1998) Estimation of nutrient requirements of sour cherry. *Acta Horticulturae* 468, 515–521.

Barria, A.J.A. (2006) The impact of deficit irrigation strategies on sweet cherry (*Prunus avium*, L.) physiology and spectral reflectance. PhD thesis, Washington State University, Pullman, Washington.

Barton, C.V.M. (2012) Advances in remote sensing of plant stress. *Plant and Soil* 354, 41–44.

Bar-Yosef, B. (1999) Advances in fertigation. *Advances in Agronomy* 65, 1–76.

Belkhodja, R., Morales, F., Sanz, M., Abadia, A. and Abadia, J. (1998) Iron deficiency in peach trees: effects on leaf chlorophyll and nutrient concentrations in flowers and leaves. *Plant Soil* 203, 257–268.

Bellvert, J., Marsal, J., Girona, J., González-Dugo, V., Fereres, E., Ustin, S.L. and Zarco-Tejada, P.J. (2016) Airborne thermal imagery to detect the seasonal evolution of crop water status in peach, nectarine and Saturn peach orchards. *Remote Sensing* 8, 39.

Benson, N.R. and Covey, R.P. Jr (1979) Phosphorus nutrition of young 'Golden Delicious' trees growing in gravel culture. *Journal of the American Society for Horticultural Science* 104, 682–685.

Benson, N.R., Batjer, L.P. and Chmelir, I.C. (1957) Response of some deciduous fruit trees to zinc chelates. *Soil Science* 84, 63–75.

Blackshaw, R.E., Moyer, J.R. and Huang, H.C. (2005) Beneficial effects of cover crops on soil health and crop management. *Recent Research Developments in Soil Science* 1, 15–35.

Bonomelli, C. and Artacho, P. (2013) Nitrogen application to non-bearing 'Bing' sweet cherry trees on Gisela 6 rootstock: effects on accumulation and partitioning of biomass and nitrogen. *Scientia Horticulturae* 162, 293–304.

Brady, N.C. and Weil, R.R. (1996) *The Nature and Properties of Soils*, 11th edn. Prentice Hall, Upper Saddle River, New Jersey.

Broadley, M.R., White, P.J., Hammond, J.P., Zelko, I. and Lux, A. (2007) Zinc in plants. *New Phytologist* 173, 677–702.

Brown, G., Wilson, S., Boucher, W., Graham, B. and McGlasson, B. (1995) Effects of copper-calcium sprays on fruit cracking in sweet cherry. *Scientia Horticulturae* 62, 75–80.

Browne, G.T., Connell, J.H. and Schneider, S.M. (2006) Almond replant disease and its management with alternative pre-plant soil fumigation treatments and rootstocks. *Plant Disease* 90, 869–876.

Browne, G.T., Lampinen, G.D., Holtz, B.A., Doll, D.A., Upadhyaya, S.K., Schmidt, L.S., Bhat, R.G., Udompetaikul, V., Coates, R.W., Hanson, B.D., Klonsky, K.M., Gao, S., Wang, D., Gillis, M., Gerik, J.S. and Johnson, R.S. (2013) Managing the almond and stone fruit replant disease complex with less soil fumigant. *California Agriculture* 67, 128–138.

Burt, C.M., O'Connor, K. and Ruehr, T. (1995) *Fertigation*. Irrigation Training and Research Center, California Polytechnic State University, California.

Butler, D.M., Kokalis-Burelle, N., Muramoto, J., Shennan, C., McCollum, T.G. and Roskopf, E.N. (2012) Impact of anaerobic soil disinfestation combined with soil solarisation on plant-parasitic nematodes and introduced inoculum of soil-borne plant pathogens in raised-bed vegetable production. *Crop Protection* 39, 33–40.

Buwalda, J.G. (1993) The carbon costs of root systems of perennial fruit crops. *Environmental Experimental Botany* 33, 131–140.

Buyukcangaz, H., Koksal, E.S. and Yazgagn, S. (2007) Diurnal variation of remotely sensed ET and some indices on sweet cherry trees in sub-humid climate conditions. *Pakistan Journal of Biological Sciences* 10, 1380–1389.

Callan, N.W. (1986) Calcium hydroxide reduces splitting of 'Lambert' sweet cherry. *Journal of the American Society for Horticultural Science* 111, 173–175.

Cao, T., McKenry, M.V., Duncan, R.A., DeJong, T.M., Kirkpatrick, B.C. and Shackel, K.A. (2006) Influence of ring nematode infestation and calcium, nitrogen and indoleacetic acid applications on peach susceptibility to *Pseudomonas syringae* pv. *syringae*. *Phytopathology* 96, 608–615.

Carter, M.R. and Gregorich, E.G. (2008) *Soil Sampling and Methods of Analysis*, 2nd edn. CRC Press, Boca Raton, Florida.

Chen, J. and Ferris, H. (1999) The effects of nematode grazing on nitrogen mineralization during fungal decomposition of organic matter. *Soil Biology and Biochemistry* 31, 1265–1279.

Cheng, Y., Jiang, Y., Griffiths, B.S., Li, D., Hu, F. and Li, H. (2011) Stimulatory effects of bacterial-feeding nematodes on plant growth vary with nematode species. *Nematology* 13, 369–372.

Cline, J., Neilsen, G., Hogue, E., Kuchta, S. and Neilsen, D. (2011) Spray-on-mulch technology for intensively grown irrigated apple orchards: influence on tree establishment, early yields, and soil physical properties. *HortTechnology* 21, 398–411.

Doran, J.W. (2002) Soil health and global sustainability: translating science into practice. *Agriculture, Ecosystems and Environment* 88, 119–127.

Dragoni, D., Lakso, A.N. and Piccioni, R.M. (2009) Transpiration of apple trees in a humid climate using heat pulse sap flow gauges calibrated with whole-canopy gas exchange chambers. *Agricultural and Forest Meteorology* 130, 85–94.

Dullahide, S.R., Stirling, G.R., Nikulin, A. and Stirling, A.M. (1994) The role of nematodes, fungi, bacteria and abiotic factors in the etiology of apple replant problems in the Granite Belt of Queensland. *Australian Journal of Experimental Agriculture* 34, 1177–1182.

Dye, M.H., Buchanan, L., Dorofaeff, F.D. and Beecroft, F.G. (1984) Boron toxicity in peach and nectarine trees in Otago. *New Zealand Journal of Experimental Agriculture* 12, 303–313.

Edgerton, L.J. and Parker, K.G. (1958) Effect of nematode infestation and rootstock on cold hardiness of Montmorency cherry trees. *Journal of the American Society for Horticultural Science* 72, 134–138.

Facteau, T.J. (1982) Relationship of soluble solids, alcohol-insoluble solids, fruit calcium, and pectin levels to firmness and surface pitting in 'Lambert' and 'Bing' sweet cherry fruit. *Journal of the American Society for Horticultural Science* 107, 151–154.

Fernández, C., Pinochet, J., Esmenjaud, D., Salesses, G. and Felipe, A. (1994) Resistance among new *Prunus* rootstocks and selections to root-knot nematodes in Spain and France. *HortScience* 29, 1064–1067.

Ferris, H., Venette, R.C., van der Meulen, H.R. and Lau, S.S. (1998) Nitrogen mineralization by bacterial-feeding nematodes: verification and measurement. *Plant Soil* 203, 159–171.

Fliegel, P., Parker, K.G. and Mai, W.F. (1963) The fungous flora of non-suberized roots of poorly growing cherry trees. *Phytopathology* 53, 1368–1369.

Flores, L., Martínez, M.M. and Ortega, R. (2015) Integrated nutrition program in cherry (*Prunus avium* L.) 'Lapins' in the VI region of Chile, based on soil bio-inoculants and organic matter. *Acta Horticulturae* 1076, 187–191.

Forge, T., Muehlchen, A., Hackenberg, C., Neilsen, G. and Vrain, T. (2001) Effects of preplant inoculation of apple (*Malus domestica* Borkh.) with arbuscular mycorrhizal fungi on population growth of the root-lesion nematode, *Pratylenchus penetrans*. *Plant Soil* 236, 185–196.

Forge, T., Kenney, E., Hashimoto, N., Neilsen, D. and Zebarth, B. (2016a) Compost and poultry manure as preplant soil amendments for red raspberry: comparative effects on root lesion nematodes, soil quality and risk of nitrate leaching. *Agriculture, Ecosystems and Environment* 223, 48–58.

Forge, T.A., Neilsen, G.H. and Neilsen, D. (2016b) Organically acceptable practices to improve replant success of temperate tree-fruit crops. *Scientia Horticulturae* 200, 205–241.

Forge, T.A., Neilsen, D., Neilsen, G. and Watson, T. (2016c) Using compost amendments to enhance soil health and replant establishment of tree-fruit crops. *Acta Horticulturae* 1146, 103–108.

Gale, E.S., Sullivan, D.M., Cogger, C.G., Bary, A.I., Hemphill, D.D. and Myhre, E.A. (2006) Estimating plant available nitrogen release from manures, composts and specialty products. *Journal of Environmental Quality* 35, 2321–2332.

Goldhamer, D.A. and Fereres, E. (2001) Irrigation scheduling protocols using continuously recorded trunk diameter measurements. *Irrigation Science* 20, 115–125.

Gonçalves, B., Santos, A., Silva, A.P., Moutinho-Pereira, J. and Torres-Pereira, J.M.G. (2003) Effect of pruning and plant spacing on the growth of cherry rootstocks and their influence on stem water potential of sweet cherry trees. *Journal of Horticultural Science and Biotechnology* 78, 667–678.

Gonçalves, B., Moutinho-Pereira, J., Santos, A., Silva, A.P., Bacelar, E., Correia, C. and Rosa, E. (2005) Scion–rootstock interaction affects the physiology and fruit quality of sweet cherry. *Tree Physiology* 26, 93–104.

Grassi, G., Millard, P., Wendler, R., Minnotta, G. and Tagliavini, M. (2002) Measurement of xylem sap amino acid concentrations in conjunction with whole tree transpiration estimates spring N remobilization by cherry (*Prunus avium* L.) trees. *Plant, Cell and Environment* 25, 1689–1699.

Grassi, G., Millard, P., Gioaccchini, P. and Tagliavini, M. (2003) Recycling of nitrogen in the xylem of *Prunus avium* trees starts when spring remobilization of internal reserves declines. *Tree Physiology* 23, 1061–1068.

Green, S.R., Clothier, B.E. and Jardine, B. (2003) Theory and practical application of heat-pulse to measure sap flow. *Agronomy Journal* 95, 1371–1379.

Haas, D. and Défago, G. (2005) Biological control of soil-borne pathogens by fluorescent pseudomonads. *Nature Reviews Microbiology* 3, 307–319.

Hanson, E.J. (1991) Sour cherry trees respond to foliar boron applications. *HortScience* 26, 1142–1145.

Hanson, E.J. and Proebsting, E.L. (1996) Cherry nutrient requirements and water relations. In: Webster, A.D. and Looney, N.E. (eds) *Cherries: Crop Physiology, Production and Uses*. CAB International, Wallingford, UK, pp. 243–257.

Harman, G.E, Howell, C.R., Viterbo, A., Chet, I. and Lorito, M. (2004) *Trichoderma* species: opportunistic, avirulent plant symbionts. *Nature Reviews Microbiology* 2, 43–56.

Hillel, D. (2004) *Introduction to Environmental Soil Physics*. Elsevier Academic Press, New York.

Janvier, C., Villeneuve, F., Alabouvette, C., Edel-Hermann, V., Mateille, T. and Steinberg, C. (2007) Soil health through soil disease suppression: which strategy from descriptors to indicators? *Soil Biology and Biochemistry* 39, 1–23.

Johnson, D.S. and Yogaratnam, N. (1978) The effects of phosphorus sprays on the mineral composition and storage quality of Cox's orange pippin apples. *Journal of Horticultural Science* 53, 171–178.

Jones, H.G., Serraj, R., Loveys, B.R., Xiong, L.Z., Wheaton, A. and Price, A.H. (2009) Thermal infrared imaging of crop canopies for the remote diagnosis and quantification of plant responses to water stress in the field. *Functional Plant Biology* 36, 978–989.

Juhász, A., Sepsi, P., Nagy, Z., Tokei, L. and Hrotkóba, K. (2013) Water consumption of sweet cherry trees estimated by sap flow measurement. *Scientia Horticulturae* 164, 41–49.

Köksal, E.S., Candoğan, B.N., Yildirim, Y.E. and Yazgan, S. (2010) Determination of water use and water stress of cherry trees based on canopy temperature, leaf water potential and resistance. *Žemdirbystė-Agriculture* 97, 57–64.

Koumanov, K.S., Tsareva, I.N. and Kornov, G.D. (2017) Fertigation: content of mineral nutrients in the soil and in the leaves of sweet cherry trees between two applications. *Acta Horticulturae* (in press).

Lamers, J.G., Runia, W.T., Molendijk, L.P.G. and Bleeker, P.O. (2010) Perspectives of anaerobic soil disinfestation. *Acta Horticulturae* 883, 277–283.

Lang, G., Valentino, T., Robinson, T., Freer, J., Larsen, H. and Pokharel, R. (2011) Differences in mineral nutrient contents of dormant cherry spurs as affected by rootstock, scion, and orchard site. *Acta Horticulturae* 903, 963–971.

Lichou, J., Edin, M., Tronel, C. and Saunier, R. (1990) *Le Cerisier*. Centre Technique Interprofessionnel des Fruits et Légumes (CTIFL), Paris, France.

Lidster, P.D., Tung, M.A. and Yada, R.G. (1979) Effects of pre-harvest and postharvest calcium treatments on fruit calcium content and the susceptibility of 'Van' cherry to impact damage. *Journal of the American Society for Horticultural Science* 104, 790–793.

Livellaria, N., Saavedra, E. and Salgado, E. (2011) Plant based indicators for irrigation scheduling in young cherry trees. *Agricultural Water Management* 98, 684–690.

Long, L.E., Yin, X., Huang, X.L. and Jaja, N. (2014) Responses of sweet cherry water use and productivity and soil quality to alternate groundcover and irrigation systems. *Acta Horticulturae* 1020, 131–138.

Maas, E.V. (1987) Salt tolerance of plants. In: Christie, B.R. (ed.) *CRC Handbook of Plant Science in Agriculture*, Vol. II. CRC Press, Boca Raton, Florida, pp. 57–75.

Mai, W.F. and Abawi, G.S. (1978) Determining the cause and extent of apple, cherry, and pear replant diseases under controlled conditions. *Phytopathology* 68, 1540–1544.

Mai, W.F., Merwin, I.A. and Abawi, G.S. (1994) Diagnosis, etiology and management of replant disorders in New York cherry and apple orchards. *Acta Horticulturae* 363, 33–41.

Manici, L.M., Kelderer, M., Franke-Whittle, I.H., Ruhmer, T., Baab, G., Nicoletti, F., Caputo, F., Topp, A., Insam, H. and Naef, A. (2013) Relationship between root-endophytic microbial communities and replant disease in specialized apple growing areas in Europe. *Applied Soil Ecology* 72, 207–214.

Marsal, J., López, G., Arbones, A., Mata, M., Vallverdú, X. and Girona, J. (2009) Influence of post-harvest deficit irrigation and pre-harvest fruit thinning on sweet cherry (cv. New Star) fruit firmness and quality. *Journal of Horticultural Science and Biotechnology* 84, 273–278.

Marsal, J., López, G., del Campo, J., Mata, M., Arbones, A. and Girona, J. (2010) Postharvest regulated deficit irrigation in 'Summit' sweet cherry: fruit yield and quality in the following season. *Irrigation Science* 28, 181–189.

Marull, J. and Pinochet, J. (1991) Host suitability of *Prunus* rootstocks to four *Meloidogyne* species and *Pratylenchus vulnus* in Spain. *Nematropica* 21, 185–195.

Mazzola, M. (1998) Elucidation of the microbial complex having a causal role in the development of apple replant disease in Washington. *Phytopathology* 88, 930–938.

Mazzola, M. and Manici, L.M. (2012) Apple replant disease: role of microbial ecology in cause and control. *Annual Review of Phytopathology* 50, 45–65.

Measham, P.F., Wilson, S.J., Gracie, A.J. and Bound, S.A. (2014) Tree water relations: flow and fruit. *Agricultural Water Management* 137, 59–67.

Meheriuk, M., Neilsen, G.H. and McKenzie, D.-L. (1991) Incidence of rain splitting in sweet cherries treated with calcium or coating materials. *Canadian Journal of Plant Science* 71, 231–234.

Melakeberhan, H., Jones, A.L., Hanson, E. and Bird, G.W. (1995) Effect of low soil pH on aluminum availability and on mortality of cherry seedlings. *Plant Disease* 79, 886–892.

Morales, F., Grass, R., Abadia, A. and Abadia, J. (1998) Iron chlorosis paradox in fruit trees. *Journal of Plant Nutrition* 21, 815–825.

Moran, M.S., Clarke, T.R., Inooue, Y. and Vidal, A. (1994) Estimating crop water deficit using the relation between surface-air temperature and spectral vegetation index. *Remote Sensing of Environment* 49, 246–263.

Moreshet, S., Cohen, Y., Green, D.C. and Fuchs, M. (1990) The partitioning of hydraulic conductance within mature orange trees. *Journal of Experimental Botany* 41, 833–839.

Mulder, D. (1950) Magnesium deficiency in fruit trees on sandy soils and clay soils in Holland. *Plant Soil* 2, 145–157.

Neilsen, D. and Neilsen, G. (2008) Fertigation of deciduous fruit trees: apple and sweet cherry. In: Imus, P. and Price, M.R. (eds) *Fertigation: Optimizing the Utilization of Water and Nutrition*. International Potash Institute, Horgen, Switzerland, pp. 76–88.

Neilsen, D., Hoyt, P.B., Parchomchuk, P., Neilsen, G.H. and Hogue, E.J. (1995) Measurement of the sensitivity of orchard soils to acidification. *Canadian Journal of Soil Science* 75, 391–395.

Neilsen, D., Parchomchuk, P., Neilsen, G.H. and Hogue, E.J. (1998) Using soil solution monitoring to determine the effect of irrigation management and fertigation on nitrogen availability in high-density apple orchards. *Journal of the American Society for Horticultural Science* 123, 706–713.

Neilsen, D., Neilsen, G.H., Forge, T. and Lang, G. (2016) Dwarfing rootstocks and training systems affect initial growth, cropping and nutrition in 'Skeena' sweet cherry. *Acta Horticulturae* 1130, 199–206.

Neilsen, G. and Kappel, F. (1996) 'Bing' sweet cherry leaf nutrition is affected by rootstock. *HortScience* 31, 1169–1172.

Neilsen, G.H. and Neilsen, D. (2003) Nutritional requirements for apple. In: Ferree, D. and Warrington, I.J. (eds) *Apples, Botany, Production and Uses*. CABI publishing, Wallingford, UK, pp. 267–302.

Neilsen, G.H. and Yorston, J. (1991) Soil disinfection and monoammonium phosphate fertilization increase precocity of apples on replant problem soils. *Journal of the American Society for Horticultural Science* 116, 651–654.

Neilsen, G.H., Neilsen, D. and Atkinson, D. (1990) Top and root growth and nutrient absorption on *Prunus avium* L. at two soil pH levels. *Plant Soil* 121, 137–144.

Neilsen, G.H., Neilsen, D. and Peryea, F. (1999) Response of soil and irrigated fruit trees to fertigation or broadcast application of nitrogen, phosphorus, and potassium. *HortTechnology* 9, 393–401.

Neilsen, G., Kappel, F. and Neilsen, D. (2004) Fertigation method affects performance of 'Lapins' sweet cherry on Gisela 5 rootstock. *HortScience* 39, 1716–1721.

Neilsen, G., Kappel, F. and Neilsen, D. (2007) Fertigation and crop load affect yield, nutrition and fruit quality of 'Lapins' sweet cherry on Gisela 5 rootstock. *HortScience* 42, 1456–1462.

Neilsen, G.H., Lowery, D.T., Forge, T.A. and Neilsen, D. (2009a) Organic fruit production in British Columbia. *Canadian Journal of Soil Science* 89, 677–692.

Neilsen, G.H., Neilsen, D., Herbert, L., Losso, I. and Rabie, B. (2009b) Factors affecting within orchard variability of nutrition, yield and quality of sweet cherry (*Prunus avium* L.). Paper posted at eScholarship, University of California. Available at: http://repositories.cdlib.org/ipnc/xvi/1160 (accessed 13 January 2017).

Neilsen, G.H., Neilsen, D., Kappel, F., Toivonen, P. and Herbert, L. (2010) Factors affecting establishment of sweet cherry on Gisela 6 rootstock. *HortScience* 45, 939–945.

Neilsen, G.H., Neilsen, D., Kappel, F. and Forge, T. (2014) Interaction of irrigation and soil management on sweet cherry productivity and fruit quality at different crop loads that simulate those occurring by environmental extremes. *HortScience* 49, 215–220.

Nelson, N.O. and Janke, R.R. (2007) Phosphorus sources and management in organic production systems. *HortTechnology* 17, 442–454.

Nyczepir, A.P. and Halbrendt, J.M. (1993) Nematode pests of deciduous fruit and nut trees. In: Evans, K., Trudgill, D.L. and Webster, J.M. (eds) *Plant-parasitic Nematodes in Temperate Agriculture*. CAB International, Wallingford, UK, pp. 381–425.

Odi-Lara, M., Campos, I., Neale, C.M.U., Ortega-Farias, S., Poblete-Echeverría, C., Balbontín, C. and Calera, A. (2016) Estimating evapotranspiration of an apple orchard using a remote sensing-based soil water balance. *Remote Sensing* 8, 253; doi:10.3390/rs8030253.

Olien, W.C., Graham, C.J., Hardin, M.E. and Bridges, W.C. Jr (1995) Peach rootstock differences in ring nematode tolerance related to effects on tree dry weight, carbohydrate and prunasin contents. *Physiologia Plantarum* 94, 117–123.

Olmstead, M.A., Lang, N.S., Lang, G.A., Ewers, F.W. and Owens, S.A. (2006) Examining the vascular pathway of sweet cherries grafted onto dwarfing rootstocks. *HortScience* 41, 674–679.

Ouzounis, T. and Lang, G.A. (2011) Foliar applications of urea affect nitrogen reserves and cold acclimation of sweet cherries (*Prunus avium L.*) on dwarfing rootstocks. *HortScience* 46, 1015–1021.

Owen, D., Williams, A.P., Griffith, G.W. and Withers, P.J.A. (2015) Use of commercial bio-inoculants to increase agricultural production through improved phosphorus acquisition. *Applied Soil Ecology* 86, 41–54.

Oyarzún, R., Stöckle, C. and Whiting, M. (2010) Analysis of hydraulic conductance components in field grown, mature sweet cherry trees. *Chilean Journal of Agricultural Research* 70, 58–66.

Pagay, V. (2014) Physiological responses of grapevine shoots to water stress and the development of a microtensiometer to continuously measure water potential. PhD thesis, Cornell University, Ithaca, New York.

Parchomchuk, P., Berard, R.C. and Van der Gulik, T.W. (1996) Automatic irrigation scheduling using an electronic atmometer. In: Camp, C.R., Sadler, E.J. and Yoder, R.E. (eds) *Evapotranspiration and Irrigation Scheduling*. Proceedings of the International Conference of the American Society of Agricultural Engineers, San Antonio, Texas, pp. 1099–1104.

Pereira, L.S., Allen, R.G., Smith, M. and Raes, D. (2015) Crop evapotranspiration with FAO 56: past and future. *Agricultural Water Management* 147, 4–20.

Picchioni, G.A., Weinbaum, S.A. and Brown, P.H. (1995) Retention and kinetics of uptake and export of foliage-applied labelled boron by apple, pear, prune, and sweet cherry leaves. *Journal of the American Society for Horticultural Science* 120, 28–35.

Pinochet, J., Verdejo-Lucas, S. and Marull, J. (1991) Host suitability of eight *Prunus* spp. and one *Pyrus communis* rootstocks to *Pratylenchus vulnus*, *P. neglectus*, and *P. thornei*. *Journal of Nematology* 23, 570–575.

Pinochet, J., Calvet, C., Camprubi, A. and Fernández, C. (1995) Interaction between the root-lesion nematode *Pratylenchus vulnus* and the mycorrhizal association of *Glomus intraradices* and Santa Lucia 64 cherry rootstock. *Plant Soil* 170, 323–329.

Pinochet, J., Fernández, C. and Alcaniz, E. (1996a) Damage by a lesion nematode, *Pratylenchus vulnus*, to *Prunus* rootstocks. *Plant Disease* 80, 754–757.

Pinochet, J., Calvet, C., Camprubi, A. and Fernández, C. (1996b) Interactions between migratory endoparasitic nematodes and arbuscular mycorrhizal fungi in perennial crops: a review. *Plant Soil* 185, 183–190.

Proebsting, E.L. Jr, Middleton, J.E. and Mahon, M.O. (1981) Performance of bearing cherry and prune trees under very low irrigation rates. *Journal of the American Society for Horticultural Science* 106, 243–346.

Robinson, T., Autio, W., Clements, J., Cowgill, W., Embree, C., González, V., Hoying, S., Kushad, M., Parker, M., Parra, R. and Schupp, J. (2012) Rootstock tolerance to apple replant disease for improved sustainability of apple production. *Acta Horticulturae* 940, 521–528.

Roversi, A. and Monteforte, A. (2006) Preliminary results on the mineral uptake of six sweet cherry varieties. *Acta Horticulturae* 721, 123–128.

Saucet, S.B., van Ghelder, C., Abad, P., Duval, H. and Esmenjaud, D. (2016) Resistance to root-knot nematodes *Meloidogyne* spp. in woody plants. *New Phytologist* 211, 41–56.

Saure, M.C. (2005) Calcium translocation to fleshy fruit: its endogenous control. *Scientia Horticulturae* 105, 65–89.

Schmitt, E.R., Duhme, F. and Schmid, P.P.S. (1989) Water relations in sweet cherries (*Prunus avium* L.) on sour cherry rootstocks (*Prunus cerasus* L.) of different compatibility. *Scientia Horticulturae* 39, 189–200.

Shackel, K.A., Ahmadi, H., Biasi, W., Buchner, R., Goldhamer, D., Gurusinghe, S., Hasey, J., Kester, D., Krueger, B. and Lampinen, B. (1997) Plant water status as an index of irrigation need in deciduous fruit trees. *HortTechnology* 7, 23–29.

Stoilov, G., Koumanov, K. and Dochev, D. (1999) Investigations on fertigation of peach on three soils. I. Migration and localization of nitrogen. *Bulgarian Journal of Agricultural Science* 5, 605–614.

Swietlik, D. and Faust, M. (1984) Foliar nutrition of fruit crops. *Horticultural Reviews* 6, 287–355.

Tagliavini, M., Abadia, J., Rombola, A.D., Abadia, A., Tsipouridis, C. and Marangoni, B. (2000) Agronomic means for the control of iron deficiency chlorosis in deciduous fruit crops. *Journal of Plant Nutrition* 23, 2007–2022.

Thielemann, M., Toro, R. and Ayala, M. (2014) Distribution and recycling of canopy nitrogen storage reserves in sweet cherry (*Prunus avium* L.) fruiting branches following $^{15}$N-urea foliar applications after harvest. *Acta Horticulturae* 1020, 353–361.

Timper, P. (2014) Conserving and enhancing biological control of nematodes. *Journal of Nematology* 46, 75–89.

Tognetti, R., d'Andria, R., Morelli, G., Calandrelli, D. and Fragnito, F. (2004) Irrigation effects on daily seasonal variations of trunk sap flow and leaf water relations in olive trees. *Plant Soil* 263, 249–264.

Tozzini, L., Lakso, A.N. and Flore, J.A. (2015) Estimating the carbon footprint of Michigan apple and cherry trees – lifetime dry matter accumulation. *Acta Horticulturae* 1068, 85–90.

Traquair, J.A. (1984) Etiology and control of orchard replant problems: a review. *Canadian Journal of Plant Pathology* 6, 54–62.

Treadwell, D.D., McKinney, D.E. and Creamer, N.G. (2003) From philosophy to science: a brief history of organic horticulture in the United States. *HortScience* 38, 1009–1014.

Troyanos, Y.E., Hipps, N.A., Moorby, J. and Ridout, M.S. (1997) The effects of external magnesium concentration on the growth and magnesium inflow rates of micropropagated cherry rootstocks 'F.12/1' (*Prunus avium L.)* and 'Colt' (*Prunus avium* L. × *Prunus pseudocerasus* L.). *Plant Soil* 197, 25–33.

Tvergyak, P.J. and Richardson, D.G. (1979) Diurnal changes of leaf and fruit water potentials of sweet cherries during the harvest period. *HortScience* 14, 520–521.

Urbez-Torres, J.R., Boulé, J., Haag, P., Hampson, C.R. and O'Gorman, D.T. (2016) First report of root and crown rot caused by *Fusarium oxysporum* Schltdl. on sweet cherry (*Prunus avium* L.) in British Columbia. *Plant Disease* 100, 855.

USDA-NRCS (1993) Chapter 2. Irrigation water requirements. In: *National Engineering Handbook Part 623*. US Department of Agriculture, Natural Resources Conservation Service, Washington, DC.

Vang-Petersen, O. (1980) Calcium nutrition of apple trees: a review. *Scientia Horticulturae* 12, 1–9.

Waterman, P.F. (2001) *Fertilization Guidelines in High Density Apples and Apple Nurseries in the Okanagan – Similkameen*. British Columbia Ministry of Agriculture, Food and Fisheries, Victoria, British Columbia, Canada.

Westcott, S.W. and Zehr, E.I. (1991) Evaluation of host suitability in *Prunus* for *Criconemella xenoplax*. *Journal of Nematology* 23, 393–401.

Westcott, S.W., Zehr, E.I., Newall, W.C. Jr and Cain, D.W. (1994) Suitability of *Prunus* selections as hosts for the ring nematode (*Criconemella xenoplax*). *Journal of the American Society for Horticultural Science* 119, 920–924.

Westwood, M.N. and Wann, F.B. (1966) Cherry nutrition. In: Childers, N.F. (ed.) *Temperate to Tropical Fruit Nutrition*. CABI Horticultural Publications, Rutgers University, New Brunswick, New Jersey, pp. 158–173.

Whipps, J.L. (2001) Microbial interactions and biocontrol in the rhizosphere. *Journal of Experimental Botany* 52, 487–511.

Whiting, M.D. and Lang, G.A. (2004) 'Bing' sweet cherry on the dwarfing rootstock 'Gisela5': thinning affects tree growth and fruit yield and quality but not net $CO_2$ exchange. *Journal of the American Society for Horticultural Science* 129, 407–415.

Wójcik, P. and Morgaś, H. (2015) Impact of postharvest sprays of nitrogen, boron and zinc on nutrition, reproductive response and fruit quality of 'Schattenmorelle' tart cherries. *Journal of Plant Nutrition* 38, 1456–1468.

Wójcik, P. and Wójcik, M. (2006) Effect of boron fertilization on sweet cherry tree yield and fruit quality. *Journal of Plant Nutrition* 29, 1755–1766.

Woodbridge, C.G. (1954) Zinc deficiency in fruit trees in the Okanagan valley in British Columbia. *Canadian Journal of Agricultural Science* 34, 545–551.

Woodbridge, C.G. (1955) The boron requirements of stone fruit trees. *Canadian Journal of Agricultural Science* 35, 282–285.

Yao, S., Merwin, I.A., Bird, G.W., Abawai, G.S. and Theis, J.A. (2005) Orchard floor management practices that maintain vegetative or biomass groundcover, stimulate soil microbial activity and alter soil microbial community composition. *Plant Soil* 271, 377–389.

Yin, X., Seavert, C.F., Turner, J., Núñez-Elisea, R. and Cahn, H. (2007) Effects of polypropylene groundcover on nutrient availability, sweet cherry nutrition, and cash costs and returns. *HortScience* 42, 147–151.

Yin, X., Long, L.E., Huang, X.-L., Jaja, N., Bai, J., Seavert, C.F. and le Roux, J. (2012) Transitional effects of double-lateral drip irrigation and straw mulch on irrigation water consumption, mineral nutrition, yield, and storability of sweet cherry. *HortTechnology* 22, 484–492.

Ystaas, J. and Frøynes, O. (1995a) Sweet cherry nutrition: effects of phosphorus and other major elements on vigour, productivity, fruit size and fruit quality of 'Kristin' sweet cherries grown on a virgin, acid soil. *Norwegian Journal of Agricultural Science* 9, 105–114.

Ystaas, J. and Frøynes, O. (1995b) The influence of Colt and F 12/1 rootstock on sweet cherry nutrition as demonstrated by the leaf content of major nutrients. *Acta Agriculturae Scandinavica, Section B – Soil and Plant Science*, 45, 292–296.

# 10 Site Preparation and Orchard Infrastructure

K.S. Koumanov[1]* and L.E. Long[2]
[1]Fruitgrowing Institute, Agricultural Academy, Plovdiv, Bulgaria;
[2]Oregon State University Extension, The Dalles, Oregon, USA

## 10.1 Introduction

Both establishment and management of a cherry orchard require significant investments, much of which will occur before planting the first tree. Proper planning and preparation are critical to success. Orchards should be productive for 20 years or more; therefore, initial poor decisions can have long-term negative effects on profit potential. There is a wide spectrum of issues that have to be analysed carefully to make good preplant decisions, such as site selection and preparation, pollenizers and pollinators, tree support, drainage and irrigation, soil mineral fertility and organic matter, and weed management.

## 10.2 Site Selection

One of the most important decisions an orchardist makes is site selection. Soil and water quality, potential for winter damage and spring frost, pressure for disease infection and rain cracking are all determined by orchard location. A poorly chosen site may reduce production or fruit quality, or increase the cost of disease control.

Ideally, a history of the site will be available for analysis, including climatic data that includes winter and spring cold patterns, average last frost date, potential for rain prior to and during the projected harvest, and wind speeds and patterns. Lacking official records, conversations with neighbours may provide some of this information.

### 10.2.1 Topography

Gentle slopes with a gradient of 4–8% are often the best locations for cherry orchards. Slopes allow for the movement of both cold air and excess water away from trees. Good air drainage helps reduce the potential for freeze damage in the winter and frost just prior to and during bloom. For adequate air movement away from the orchard, a clear and open path must be present so that cold air is capable of moving away from the trees. Hedgerows, windbreaks, road cuts and other obstructions can dam up cold air near trees and cause substantial damage.

In the same way, slopes can facilitate the movement of excess water away from trees. This can be helpful where soils are heavier. However, ravines and gullies can collect

* kskoumanov@hotmail.com

 *Cherries: Botany, Production and Uses* (J. Quero-García, A. Iezzoni, J. Puławska and G. Lang) water runoff from surrounding slopes, leading to root asphyxiation. In valleys or flat plains, winter freeze and frost control and the installation of drainage systems may be required, leading to increased costs.

Steep slopes, of 10% or greater, promote soil erosion and may hinder the safe use of orchard equipment. Due to wind and water erosion, soil on ridges and hilltops is often depleted, and winds can have detrimental effects on tree growth and fruit quality. Winds can also affect yield by reducing bee activity during bloom and limit the opportunities to apply critical sprays.

Slope exposure affects the potential for wind damage, time of bloom and harvest, and the potential for diseases, as well as summer and winter injury. In the northern hemisphere, south-facing slopes are prone to greater temperature fluctuation during the winter than north-facing slopes. North-facing slopes do not warm as quickly during the day, retain moisture longer on leaves and fruit, which may increase the potential for diseases, and delay bloom and harvest. Southern exposures bloom and harvest earlier but may be more prone to trunk injury in the winter or sunburn in the summer. Eastern slopes warm faster in the morning and thus dry more quickly due to sun exposure, which would decrease disease potential; they are cooler in the afternoon than western slopes. In the northern hemisphere, western slopes often are more exposed to prevailing winds.

### 10.2.2 Soil characteristics

Cherries require well-drained and aerated soils. The best soil is a light, well-drained silt loam, but cherries can tolerate soils ranging from sandy loam to clay loam as long as there is good drainage. Ideally, soil depth should be 1.0 m or more. Compacted layers must be broken up prior to planting or trees will suffer from poor anchorage, drought stress and poor health. Poorly drained subsoils are typically mottled in colour with prominent grey streaks and rusty brown spots indicative of poor aeration (Roper and Frank, 2004).

### 10.2.3 Irrigation water quality

Although cherry trees are tolerant to drought, in most parts of the world, premium-quality fruit can only be grown with supplemental irrigation. The irrigation water electrical conductivity, pH, sodium adsorption ratio, boron and bicarbonate levels should be analysed. Analysis for specific ions such as chloride, sulfate and nitrate-nitrogen, as well as potential impurities including heavy metals and microbial contaminants, also may be necessary (Ayers and Westcot, 1985).

In general, fruit trees are sensitive to salt, and cherries are no exception. High levels of salts and sodium may come from a number of different sources, including irrigation water. Salinity levels of 1.3–2.5 dS $m^{-1}$ in the water and 1.9–3.1 dS $m^{-1}$ in the soil can decrease the yield by 10–50% (Kotuby-Amacher *et al.*, 2000). On the other hand, irrigation water with low electrical conductivity values can aggravate soil sodicity problems. High levels of bicarbonates can break down the soil structure, which reduces irrigation water penetration (Long and Kaiser, 2013). High bicarbonate levels were found to be a primary cause of iron chlorosis in orchards in The Netherlands (Boxma, 1972).

### 10.2.4 Site history

Orchard site cropping history is important, since trees planted on old orchard sites (especially, but not limited to, cherry) can be affected by replant disease. Replant disease is not a specific disease, but may be caused by biotic factors such as pathogenic fungi and nematodes or by abiotic factors such as phytotoxins, persistent herbicides, nutrient deficiencies or excesses, and/or poor soil structure associated with previous orchard sites. Newly planted trees will be weak, more susceptible to disease, and more prone to reduced yields and fruit quality. Where available, soil fumigation can help reduce the effect of replant disease, although the list of approved traditional fumigants, such as methyl bromide, has decreased in recent years.

In some cases, 'Colt' rootstock has shown tolerance to replant symptoms (Webster and Schmidt, 1996).

Native stands of pine, oak and other species may have survived for years while infected with *Armillaria* root rot. Clearing these trees and planting cherries could cause infections that rapidly or slowly kill them. Currently, there are no known rootstocks resistant to *Armillaria*. It is also possible for soil pathogens to build up where susceptible agronomic crops have been grown previously. For example, potatoes, tomatoes, peppers, lucerne and a number of other crops are highly sensitive to *Verticillium* wilt, which also infects cherries. Residual herbicides used on agronomic crops and pastures may persist in the soil for years, causing phytotoxicity to newly planted cherry trees. Symptoms may include leaf burn, significantly reduced growth and high mortality rates.

## 10.3 Site Preparation

### 10.3.1 Designing the orchard

A topographical site map should be used or created, showing the location of different soil types. The map should also indicate roads, the planned irrigation system, sprayer tank filling area locations, an area for handling harvested fruit, and planned or existing windbreaks. To maximize light interception, rows should be oriented, wherever possible, in a north–south direction. However, on hills, rows should run across the slope, and cover crop vegetation strips should be planted to help to prevent erosion and slow runoff. Rows across the slope allow more uniform applications of irrigation water, as well as more precise sprayer application of pesticides. Where there is a potential for frost, open spaces should be left in the rows, or row orientation should be changed to run downhill, to allow for movement of cold air away from the orchard.

Planting density and orchard layout depend on the training system, rootstock and cultivar. Planting densities for modern systems on dwarfing rootstocks range from 0.5 × 3.0 m (between trees × between rows) for super-high-density systems to 1.8 × 4.5 m for multiple leader systems (Long *et al.*, 2015). Single- or multi-planar (V-shaped) canopy systems that necessitate a trellis will require additional planning and expense.

### 10.3.2 Soil preparation, analysis and modification

Preparation for planting includes the removal of any old root systems, as possible, from the orchard site. The physical conditions of the soil should be analysed by digging holes, by hand or backhoe, to determine soil texture, structure, depth and the presence of any hard layers that might inhibit water, air and roots from penetrating deeper layers. It is important to examine each soil type within a field. Soils should then be ripped as deeply as possible in at least two directions prior to planting to break up hardpans close to the surface. If drainage is a problem, tilling the site to remove excess water should be considered.

If the soil is a replant site, the presence of pathogenic nematodes, such as root-lesion nematodes (*Pratylenchus penetrans*), dagger nematodes (*Xiphinema* spp.) and ring nematodes (*Macroposthonia* spp.), should be determined (see Chapter 9, this volume). Other microorganisms can also contribute to replant disease. One of the most common methods to reduce replant disease symptoms is soil fumigation. Unfortunately, fumigation kills not only pathogenic microorganisms but also beneficial organisms. Inoculating roots with mycorrhizal fungi after fumigation can help reintroduce some beneficial microorganisms back into the rhizosphere. Alternatively, Brassicaceae seed meal can be used as a soil amendment to substitute for fumigation. According to Mazzola and Manici (2012), seed meal formulations of *Brassica juncea*/*Sinapis alba* or *B. juncea*/*Brassica napus* can outperform fumigation in replant situations involving both root-lesion nematodes and *Pythium* spp.

### 10.3.3 Surveying and staking the plot

There are many different ways to lay out an orchard in preparation for planting. When using a tree planter to plant the trees, only the rows need to be marked, as the machine will plant the tree at a prescribed spacing. Tree planters are a fast and easy way to plant an orchard, and significantly reduce labour needs for both planting and staking. However, the accuracy of tree placement is not as good as some other methods, and planting super-high-density blocks can be difficult. Using an auger or shovels to dig holes is much more time consuming, but trees can be placed more accurately in rows. It is common to stake the outside border of rows and simply run a string or wire marked at the desired tree spacing, so that stakes, marking the tree location, can be placed accurately. It can be difficult to line rows up properly on undulating hills. In these cases, or when orchard blocks are very large, laser surveying can provide nearly perfect row and tree placement. Increasingly, global positioning system (GPS) coordinates are being used for precision planting, providing highly accurate tree placement in even the most difficult terrain. The use of GPS coordinates at planting can provide the capacity for GPS mapping of yields, nutrient analyses/fertilizer needs and other parameters for precision tree performance management in future years.

It is important to leave adequate room (generally ~10 m) at the end of each row to turn a tractor and sprayer. The loading area should be planned near a road for easy access.

## 10.4 Pollenizers and Pollinators

### 10.4.1 Pollenizers

Many recently developed sweet cherry cultivars are self-fertile and do not require pollenizers, allowing solid blocks of a single genotype to be planted. However, the majority of older, and some newer, varieties are self-incompatible, requiring cross-pollination to set fruit (see Chapter 2, this volume). Compatible cultivars must be planted with the main cultivar in reasonably close proximity, and they must bloom at the same time. Typically, a pollenizer is planted at every third tree position in every third row. In most cases, this provides reasonable pollen transfer to the main cultivar and assures that every main crop tree is adjacent to a pollenizer. Since pollenizers are usually of lower market value than the main cultivar, the number of pollenizers should be kept to an effective minimum to keep the block profitable. For cultivars of relatively low productivity, such as 'Regina', it may be advantageous to plant a solid block of trees, plant a pollenizer between every fifth tree in the row and offset the pollenizers in adjacent rows. Since no additional space is allotted to these pollenizers, they should be pruned to a slender spindle or other architecture that reduces their canopy volume. This is an efficient system as bees typically work down the row, thus assuring that each tree has sufficient cross-compatible pollen. Planting multiple pollenizer genotypes may be valuable for improving the fruit set of cultivars having low productivity. When main and pollenizer cultivars have similar market value, alternating two solid rows of the main cultivar with two solid rows of the pollenizer places every tree in the block adjacent to a pollenizer. Pollen compatibility charts are available from many commercial tree fruit nurseries and depict the relative bloom times of the most popular cultivars.

### 10.4.2 Pollinators

The European honeybee (*Apis mellifera* L.) is the most common pollinator used in cherry orchards around the world (Free, 1993). In ideal conditions, a honeybee may visit 5000 flowers a day and transfer pollen from the anther to the stigma. Managed hives are placed in the orchard at 10% bloom. This timing assures that early-opening flowers are pollinated, while preventing bees from establishing their foraging habits on other species before the cherry blossoms open. It is important that hives contain healthy, active bees. Beekeepers should be willing to open their hives and show the grower the quality of the hive. In the USA, regulations in Washington State maintain that hives should consist of six frames, with two-thirds of each frame covered with bees at a temperature of 18°C. This means that growers should expect colonies to contain 14,400 bees (Sagili and Burgett, 2011). Generally, honeybee populations range between 10,000 and 30,000 worker bees, even at their nadir in late winter and early spring (NAS, 2007). Three to five hives per hectare is regarded as adequate to pollinate sweet cherries (James, 2010). The denser the planting and the longer the rows, the greater the number of bees required. To ensure maximum bee activity, hives should be located close to or in between trees, in an elevated position in a warm sunny area and protected from prevailing winds (Somerville, 1999). It is also possible to assess hive strength by observing flight activity. On a calm, warm day, with winds of less than 16 km $h^{-1}$ and temperatures above 18°C, there should be more than 100 incoming bees per minute in a healthy colony.

The availability of large colonies, and ease of rearing and transportation, make honeybees a popular pollinator (Delaplane and Mayer, 2000). However, honeybees are strongly dependent on suitable weather conditions to be effective. Other bee species are much less demanding. For example, bumblebees and bees from the genus *Osmia* will fly at cooler temperatures, lower light and in some cases even in light rain. Worldwide, there are approximately 250 species of bumblebees. Most, however, are not managed by commercial producers. For example, in North America, only one species, *Bombus impatiens*, currently is available commercially. Bumblebee colony size is much smaller than those of honeybees, with colonies of 300 rather than 30,000. However, bumblebees can be much more efficient as pollinators and tolerate weather extremes (cold, heat, rain, light levels and wind) better than other pollinators. In fact, honeybee activity is minimal around 10°C, whereas bumblebees continue to forage at temperatures as low as 7°C (Mader *et al.*, 2010; Pfiffner and Müller, 2014).

Besides bumblebees, mason bees are also excellent pollinators. In the USA, the most commonly managed mason bee is *Osmia lignaria*, the blue orchard bee (Bosch *et al.*, 2006). This species is native to most western states and the eastern seaboard, but it is not readily found in the midwest or the south. The blue orchard bee has a strong preference for tree fruit pollen, and the bees emerge in early spring, so they are present as cherries begin to bloom. Just 600–750 are needed to pollinate 1 ha of cherries. Like bumblebees, they forage at lower temperatures than honeybees, and begin and end their foraging earlier and later in the day than honeybees. The second most commonly managed mason bee is the hornfaced bee (*Osmia cornifrons*), originally native to Japan (Mader *et al.*, 2010).

Non-honeybee species can be used as the sole pollinator in an orchard or to supplement honeybees. The presence of multiple bee species in an orchard can have a synergistic effect on pollination (Holzschuh *et al.*, 2012; Garibaldi *et al.*, 2013). When pollenizers are not located in every row, bumblebees can enhance pollination due to their erratic flight pattern. In fact, it appears that the presence of other species also increases the probability of honeybees crossing rows (Mader *et al.*, 2010). Moreover, native bees can serve as a buffer to potential declines in honeybee colonies (Winfree *et al.*, 2007; Breeze *et al.*, 2011). Wild bee communities have also been declining over the last half century, and measures to support pollinator diversity are of vital importance. These include providing habitat suitable for nesting, nest-building materials and flowering plants for native bees to forage when the trees are not in bloom, as well as access to clean water, the use of bee-friendly pesticides, no-tillage farming and transition from flood- to microirrigation (Shepherd *et al.*, 2003; Isaacs and Tuell, 2007; Garibaldi *et al.*, 2014).

Under conditions in which natural pollination is limited, artificial (assisted, controlled or supplementary) pollination is a commercial practice in many tree crops. This is biological or mechanical application of supplementary compatible pollen that has been collected previously. Controlled pollination is especially useful when the pollenizer bloom is poor or does not overlap with the main crop cultivar, or when weather conditions are unfavourable for pollination. Pollen transfer by honeybees may be increased by placing pollen dispensers at the hive entrance to dispense pollen to bees as they leave the hive (Pinillos and Cuevas, 2008). This system is also being tested for the prevention of brown rot in cherry orchards (Lehnert, 2015). Instead of spraying a fungicide, bees are used to deliver a biological control agent (spores of the fungus *Trichoderma harzianum*, a parasite of *Monilinia* sp.) to flowers susceptible to infection (Mommaerts and Smagghe, 2011).

Various mechanized pollen-application methods have been developed, including diffusers or insufflators, mechanical sprayers, atomizers and electrostatic dusters. These devices can either be carried by workers or mounted on vehicles, aircraft or helicopters. However, the mechanized application methods use large amounts of pollen and many have shown low pollination efficiency on entomophilous species such as cherry (Vaknin *et al.*, 2001; Pinillos and Cuevas, 2008). Recently, however, low-volume electrostatic applications of pollen suspensions have shown promise for artificial pollination in commercial orchards. Such research at Washington State University has documented increases in fruit set of nearly 20% compared with open bee-mediated pollination (M. Whiting and P. Das, Prosser, Washington, USA, 2016, personal communication).

## 10.5 Tree Support

Many high-density planting systems (see Chapter 12, this volume) require installation of tree support structures (van Dalfsen, 1989; Dart, 2008; Craig, 2012; Hoying, 2012; OMAFRA, 2012) for canopy orientation or fruit load support. The bending and binding of branches on to trellis wires often promotes early yields, and economic analyses of orchard system profitability have shown that input costs, including support systems, are a secondary profitability factor compared with yield and fruit quality (Robinson *et al.*, 2007). Trellis systems must not only support the crop load but also withstand additional forces exerted by wind, snow and rain. Mistakes made at establishment are very costly, if not impossible, to correct. Building a good tree support system starts with consideration of the orchard design. The cultivar, rootstock, tree density, canopy architecture and final tree height will determine the strength and type of support required.

The simplest sweet cherry tree support systems minimize the use of posts or trellis wire. For trees on dwarfing rootstocks or rootstocks such as 'GiSelA 6' that tend to lean with some cultivars, a single standalone post at each tree is satisfactory, being most important for providing support during the establishment years (Fig. 10.1). Alternatively, the 'single stake, single high wire' system for tree support uses a 3 m high 1/2″ conduit or 25 mm bamboo stake. A single wire is attached to inline posts at approximately 2 m high. Each stake is attached to the wire with a trellis clip or wire tie. The tree is initially fastened to the stake and then, as the tree grows, additional ties are added to reach the top of the stake. This system facilitates easier movement among trees and rows for harvest crews.

For a multi-wire trellis, three or more wires are spaced evenly along inline posts (Fig. 10.2). Usually, the first wire is 0.5–1 m from the ground and often is used to support irrigation lines. From this point upwards, a wire at least every 0.5–1 m is required. Closer spacing is better, but this may limit the ability of workers to move through rows. Wires should be added as the trees grow to avoid the potential for growing tips rubbing on wires and being snapped off on windy days. As trees age, the bottom wires can be removed to facilitate easier movement around the tree for picking and pruning. In the case of three wires, a vertical soft wire can be used to bridge the gaps between the top, middle and bottom wires at each tree. It is used as a training aid and for fruit support. Many other materials also can serve the same purpose, for example a 1.8 m high 3/8″ conduit, very lightweight 12–16 mm bamboo stakes or 20×20 mm metal angles. Usually, these are fastened with trellis clips to the upper and middle wire without

Fig. 10.1. Single, stand-alone wooden posts for tree support in a high-density sweet cherry orchard in Oregon, USA. (Photo courtesy of L. Long.)

Fig. 10.2. A four-wire trellis for support of the planar upright fruiting offshoots sweet cherry training system in Washington state, USA. (Photo courtesy of G. Lang.)

reaching the ground. The rigidity of the stake allows two wires to be used instead of three. Trees are secured to support systems with plastic tubing, wire clips and hand-twisted wire ties. In most climates, the periodic rubbing of trees against steel posts or trellis wires can lead to bacterial canker infections (Fig. 10.3). The use of small pieces of plastic tubing or discarded drip irrigation lines to sleeve trellis wires (Fig. 10.4) or the use of high-tensile (HT) plastic trellis wire or plastic-coated trellis wire can significantly reduce the risk of bacterial canker in trellised sweet cherry orchards (Lillrose *et al.*, 2017).

**Fig. 10.3.** Bacterial canker (*Pseudomonas syringae*) infection from improperly tying the trellis wire directly against the sweet cherry tree trunk. Plastic ties should be twisted between the tree and the wire to prevent bark contact with the wire and then tied outside the wire. (Photo courtesy of G. Lang.)

Typically, trellis wires are 2.5 mm diameter galvanized HT steel with a minimum tensile strength of 630–1260 MPa. It is not recommended to splice wires within a row. The HT wire is rigid and difficult to splice. According to a study carried out by the Australian Wire Industries (van Dalfsen, 1989), the most effective knots failed at 60–66% of the HT wire breaking strength. If splicing is necessary, special mechanical splices should be used. The lines should be tightened to a tension force of 0.7–0.9 kN to carry the load without sagging. Wire tightening should be corrected periodically for temperature deformation. This can be done using one or several permanently installed inline tensioners. Wires should be fastened on the upwind side of the post with galvanized staples driven across the grain of the wood.

The wires are carried by inline posts (poles), 100–125 mm in diameter, which transfer the load forces into the ground. All posts should be pounded or driven to a depth below the frost line. Every one-third increase in depth doubles the post resistance to movement. Posts that are driven into the ground have 50% higher resistance to movement compared with posts placed in an oversized hole with the earth rammed back around it. Posts should be driven into the ground with the small end down. The driving of blunt, larger diameter and longer posts will be easier if

**Fig. 10.4.** Affixing sweet cherry trees to trellis wire using plastic tape for lateral branches and plastic tubing sleeves over trellis wire to reduce the potential for bacterial canker (*Pseudomonas syringae*) infection from trunk rubbing. (Photo courtesy of G. Lang.)

done in augered pilot holes 3–5 cm smaller in diameter than the post. When the posts are hand-set in oversized holes, they have to be placed with the large end down. When using wood, posts that have been pressure-treated with chromated copper arsenate reduce the chance of breakage due to rot. Post spacing depends on rootstock, row length, tree height, crop load at maturity and soil conditions. Clay soil provides more resistance than sandy soil, and dry soils offer more resistance than wet soils. Closer spacing may be required on rolling terrain to prevent the wire from pulling the posts out of the ground.

End posts and anchors should be 125–150 mm in diameter to securely anchor the wires. The strongest end-post assembly is an equilateral triangle created by the end post, the trellis wire and the distance between them along the ground. There are two main types of anchors: tie-back and brace. The tie-back anchor can be a wooden post or commercially manufactured earth screw, metal plate or other device of various materials and shapes such as arrowhead and duckbill. The end post should lean 15–18° from the vertical towards the anchor. To be effective, the tie-back post has to be driven 1.2–1.5 m deep in undisturbed firm soil at a 4/1 backwards slope ratio. It can then be cut off at 0.3 m above the ground. These posts resist pullout better than screw anchors, but are more difficult, time consuming and costly to install. If the loads are too great (e.g. a large number of wires) or the soils are weak, a brace assembly should be used. This consists of one horizontal and two vertical posts, as well as a diagonal brace wire. The brace anchor may be constructed in two versions: a single-span brace assembly or a 'stay' brace assembly. In extremely rocky soils, wooden posts are installed in backhoe-dug holes. Since disturbed soil provides weak support, thrust blocks should be placed on the row side of the post near the top and on the opposite side at the bottom of the hole. Pouring concrete in the hole will not strengthen the anchor. Unless the concrete anchor is supported by undisturbed soil, its mass will be insufficient to resist the wires' pull. Van Dalfsen (1989) provides more details concerning wooden-post trellises.

Screw anchors (augers) are used frequently for trellis systems. The anchor should be as far away from the end pole as the height of the pole, and the end post should lean 15–18° from the vertical, towards the anchor. Screw anchors come with shafts of various lengths depending on the load they have to resist. The diameter of the shaft can range from 16 to 22 mm. The plate at the base should be at least 150 mm wide to resist pullout and bending. The anchor shaft should point towards the point of attachment on the end post.

Alternatives to wood posts include metal posts or steel pipes, or combinations to develop complex structures such as multi-planar trellises (Fig. 10.5). Metal posts can be supplied with holes or notches positioned where the wires can be fastened, following customer specification. Steel pipes do not have holes, so attachment points may need to be welded. Flat-sided steel can be used for V- and Y-shaped trellis systems. Recently, a new trellis system has been developed based on hoop-house technology, which incorporates bent and swaged tubes coupled with other profiles to complete the structure (Hansen, 2013). This system is quick and inexpensive to assemble and can easily be adjusted to orchard specifics. Wire forms and wrap clips provide secure wire attachment to the metal posts.

## 10.6 Drainage Systems

Cherry tree growth is negatively affected by excessive wetting of the soil in the root zone or by ponded water on the surface. Different rootstocks (see Chapter 6, this volume) can have quite different tolerances for excess water volume and exposure time. Root zone saturation results in oxygen deficiency and the accumulation of toxic gases. A short period of oxygen deficiency can reduce water uptake, nutrient uptake and root respiration, and can result in the build-up of toxins, all of which can lead to the death of cells and roots, and, if prolonged, to death of the tree (Sutton *et al.*, 1971). Flooding of the soil surface for more than a week may be

**Fig. 10.5.** A five-wire multi-planar Y-shaped trellis structure created with wooden posts and angled iron steel braces at each wire connection in Washington State, USA. (Photo courtesy of G. Lang.)

detrimental for cherry trees. Even short-term flooding can have damaging long-term effects. Overwetting of the soil favours the development of *Phytophthora* root rots (James, 2010). In temperate climates, wet soils warm more slowly in the spring, which can delay vegetative growth and negatively affect crop yields (van der Molen *et al.*, 2007). Hence, it is critical to ensure that any site selected for cherries has good surface and subsurface drainage. Transient soil saturation during ripening also can exacerbate the incidence of sweet cherry fruit cracking (see Chapter 7, this volume).

Drainage systems collect and remove excess water from the orchard, as well as diverting water before it reaches the orchard boundaries. The sources of excess water may be precipitation, snowmelt, irrigation, overland flow or underground seepage from adjacent areas, artesian flow from deep aquifers, floodwater from channels or water applied for special purposes such as leaching salts from the soil or for temperature control (Sutton *et al.*, 1971). There are two types of drainage systems: surface and subsurface. The first step to proper water management in the orchard (drainage, but also irrigation) is land forming.

### 10.6.1 Land forming

Mechanically changing the land surface to modify the movement of surface water is referred to as land forming (Sutton *et al.*, 1971; van der Molen *et al.*, 2007). This may be done by smoothing, grading, bedding or levelling. Smoothing corrects minor differences in elevation without changing the general contours of the land. It seldom involves cuts and fills of more than 15 cm.

Land smoothing is also the finishing operation in land grading. Land grading is the shaping of the land surface by cutting, filling and smoothing to planned grades so that runoff will flow over the surface without ponding. Land grading for drainage does not require shaping of the land into plane surfaces with uniform slopes. Surface ridging can be established with cuts from the edge of an interrow ditch and fills towards the two adjacent tree rows. Thus, convex surfaces are developed from the ditch shoulders towards the rows and artificial ridges that are created along the tree rows. Row grades on permeable, but easily erodible, soils should not exceed 0.5%. On slowly permeable soils with limited row lengths, grades may reach a maximum of 2.0% (Sutton *et al.*, 1971). Bedding, also known as crowning, berming or ridging, is the process of mounding soil into broad ridges along the rows before trees are planted or where feasible in existing orchards. Planting trees on berms is recommended on flat lands and especially on soils with low infiltration rates. The system of raised beds also includes shallow furrows or ditches located in between the tree rows. Beds are shaped to a convex surface with a slope towards the furrows/ditches, which furthers the effective evacuation of excess water from the tree root zone. The height of the berms can range from 30 to 90 cm and the crown width from 40 to 180 cm, depending on the depth of the groundwater table and the drainage ability of the soil. Perry (1998), for example, reported the best performance of cherries on beds 30 cm high and 180 cm wide. In raised bed systems, it is essential to have a permanent sod cover on the sides of the beds and in the alleys to prevent soil erosion. Land levelling is a precise operation of modifying the land surface to planned grades to improve drainage.

### 10.6.2 Surface drainage

Surface drainage is useful primarily for flat land in which deep water percolation is prevented by low-permeable or restricting layers in the soil profile. Ditches, the collecting elements of a surface drainage system, are open waterways excavated in the earth to collect and/or convey drainage water, often both surface and subsurface. Their size and shape should allow the passage of agricultural implements. Diversion ditches can be graded channels constructed across the slope of the land at the orchard edge to intercept seepage, as well as surface flow, and divert water to a suitable outlet. Dikes are embankments constructed to protect land against overflow from streams, lakes and tidal influences, as well as from diffused surface waters.

### 10.6.3 Subsurface drainage

Subsurface drainage removes excess water below the ground surface and lowers high water tables, thus improving the soil profile for plants to develop their natural root pattern. For fruit trees, the design groundwater depth (the steady-state depth of the groundwater) varies with soil texture, and is generally at 1.0–1.2 m (van der Molen *et al.*, 2007). Where excessive soluble salts are present, good subsurface drainage re-establishes downward percolation of water in the soil profile and permits leaching of these salts. Subsurface drainage can be used to lower a high water table (relief drainage) or to intercept, reduce the flow, and lower the flowline of water in the orchard (interception drainage). Subsurface drainage may be horizontal, using a system of parallel drains or vertical, pumping water from collection wells. Parallel open ditches in the tractor alley middle are suitable for flat fields where lack of grade, soil characteristics or economic conditions do not favour buried drains. Usually, field drains consist of buried conduits with open joints or perforations.

Interception drains may be either open ditches or buried drains. On sloping lands, the aim of subsurface drainage is to intercept the surface and groundwater flow at the base of a slope when this is easier than correcting the excess water in the orchard. Combination systems provide both surface and subsurface drainage. For more details about drainage systems, see Sutton *et al.* (1971) and van der Molen *et al.* (2007).

## 10.7 Irrigation

### 10.7.1 Traditional irrigation techniques

Cherry crop water requirements vary during the season according to tree developmental stage (see Chapter 9, this volume). The quantitative and temporal variations in cherry tree water requirements and soil water availability determine the necessity of irrigation. There are three types of irrigation systems: surface, sprinkler and microirrigation.

Surface irrigation techniques, such as flood, furrow or basin irrigation, are the oldest and least capital-intensive methods for supplying water to orchards, but they are labour intensive, inefficient and/or wasteful compared with other methods, and their use has decreased significantly around the world.

Sprinkler irrigation (USDA-SCS, 1983; Rieul, 1990; Vaysse *et al.*, 1990; Solomon, 2013) distributes water relatively uniformly over the whole soil surface or a large part of it. The necessary pressure head (0.2–0.4 MPa) may be obtained by pumping or by gravity using a natural elevation of the water source. The water is conveyed through subsurface pipelines, thus allowing free movement of orchard machinery. The most widely used distribution systems are: (i) portable lateral with sprinklers, which is moved as a whole; (ii) a semi-solid set, where only the sprinklers are moved; (iii) a dragline, where only the sprinklers and hoses are moved; and (iv) permanent, comprising a solid set.

Sprinkler irrigation may be applied above or below the tree canopies, either of which can also be used to maintain perennial sod in the orchard alleys or for microclimate modification (see Chapter 11, this volume). However, the use of sprinklers overhead is limited by water quality. High evaporation rates may cause salt accumulation on leaf surfaces, with subsequent damage to plant organs (Evans, 2006a). Moreover, prolonged wetting of trees can stimulate the development of fungal and bacterial diseases (Barfield *et al.*, 1990; Evans, 2006a,b). Special care also must be taken to avoid damage to fruit, as irrigation water can cause ripening fruit to split.

### 10.7.2 Microirrigation

Microirrigation (drip and microsprinklers) has increasingly been used in fruit production because of higher water-use efficiencies and more precise control of water delivery (Lamm *et al.*, 2007; USDA-NRCS, 2013). It can be the most water-, energy-, nutrient- and pesticide-efficient of all irrigation systems due to its low working pressure, high distribution uniformity and high application efficiency. With microirrigation, the full or partial soil surface area may be wetted. Small and low-intensity irrigation rates are applied frequently and directly within the tree root zone, thus providing favourable water availability, nutrient updates and soil aeration. Soil moisture is maintained close to field capacity. Microirrigation laterals and emitters can be located aboveground (e.g. on a trellis), on the ground or below the soil surface. Generally, systems are solid set and relatively capital intensive, but with low labour and management costs. Microirrigation is suitable for sloping or irregularly shaped terrains, and water application can be fully automated via a computer. Since the emitters are prone to clogging by soil particles, algae, bacteria and solids suspended in the water, filtration and regular flushing of the system are critical. Brackish water can be used but tends to result in a build-up of salts at the periphery of the wetted soil volume. Hence, soil salinity issues must be addressed adequately.

With drip irrigation, water is applied at minimal pressure of around 0.1–0.2 MPa, (Schwankl and Hanson, 2007) and low discharge rates of 1–4 l $h^{-1}$ for point-source emitters and 8–12 l $m^{-1}$ $h^{-1}$ for laterals with porous walls. Some emitters are pressure compensating and deliver a nearly constant rate over a range of pressures. Usually, the necessary application rate is supplied by two to four drippers per adult tree, depending on their discharge. After planting of the trees, irrigation can begin using one to two drippers per tree, with more emitters added as the trees mature. The application efficiency of drip irrigation is strongly affected by the soil properties. On soils with a low infiltration rate, the loss to evaporation may increase dramatically (Koumanov *et al.*, 1998; Koumanov, 2007). If not taken into account, such a reduction of the actual water volume delivered to the root zone could have a significant negative effect on tree growth and yield. Evaporation losses can be substantially decreased and even eliminated if the laterals are buried (as deep as 60 cm) below the soil surface (Camp, 1998; Lamm and Camp, 2007; USDA-NRCS, 2013). Such subsurface drip irrigation (SDI) systems require additional flushing manifolds and vacuum relief valves to prevent the drawing of muddy water into the system. Pinching of the drip lateral lines and clogging of emitters penetrated by roots are potential problems in SDI systems. However, SDI systems create a dry soil surface that reduces weed growth.

With microsprinklers, irrigation water is spread over the soil surface as a mist or fine droplets via a wide variety of emitters referred to as misters, microsprinklers, microjets or sprayers (Boman, 2007; USDA-NRCS, 2013). Water is emitted through one or more small openings on a fixed or rotating distributor and spread into various patterns of surface wetting. Microsprinklers usually are operated under the tree canopy, keeping foliage dry and reducing disease occurrence. Microsprinklers can be coupled with polyethylene laterals directly, or with short or long spaghetti tubing, depending on the position of the lateral. Laterals may be mounted on stakes fixed in the ground, attached to a trellis or hung in tree branches. The last two approaches allow mechanical cultivation in the row strip. The emitter discharge can vary between 20 and 1000 $l\ h^{-1}$ at working pressures of 0.1–0.2 MPa. The irrigation water wets a larger soil surface area than that with drip irrigation and therefore is suitable for a wider variety of soils, although evaporation losses can be significant (Koumanov *et al.*, 1997, 2006; USDA-NRCS, 2013). The tree root system develops in a larger soil volume and can exploit soil fertility more efficiently, and a higher proportion of the roots develop close to the soil surface where mineral nutrient content is higher. Compared with drip irrigation, the larger emitter orifice diameter makes microsprinklers less prone to clogging and therefore water filtration is relatively easier.

### 10.7.3 Chemigation

As well as meeting irrigation needs, microirrigation systems can be used to apply fertilizers, herbicides, insecticides, fungicides, fumigants, nematocides, soil amendments, growth regulators, etc., collectively known as 'chemigation' (Trimmer *et al.*, 1992; Waterman, 2001; Burt, 2003; Storlie, 2004; van der Gulik *et al.*, 2007; Zhu *et al.*, 2011). Chemicals also may be injected into the irrigation system for maintenance purposes, such as algicides and chlorine (Evans and Waller, 2007). The potential advantages of chemigation include uniformity, flexibility, avoidance of heavy machinery, cost efficiency, independence of weather conditions and reduced environmental contamination. However, to date, most pesticides are not approved for application with irrigation water (US EPA, 2015). Only pesticides labelled for chemigation can be applied by injecting them into an irrigation system.

#### *Fertigation*

Fertigation, described in detail in Chapter 9, provides dynamic fertilizer applications with the highest potential for achieving an optimized nutritional regime, the best application efficiency, and the lowest potential for overdosing, leaching or toxicity (Kafkafi and Tarchitzky, 2011). Together with rootstock selection and proper carbon nutrition, it can be crucial for intensive cherry production (Koumanov and Tsareva, 2017).

#### *Herbigation*

Studies of herbigation in cherry orchards are rare and information is scarce, especially regarding herbicide efficiency, selectivity, durability of the effect, mobility and persistency in the soil. Promising results from cherry orchard herbigation experiments with the pre-emergence herbicide pendimethalin were reported by Koumanov *et al.* (2009) and Rankova *et al.* (2009): the post-treatment herbicide effect, effect on growth and yield of trees with different vigour ('Burlat' on Mazzard, and 'Lapins' on 'GiAelA'), and the soil microbial activity were similar to those under standard spray herbicide applications.

## 10.8 Mineral Fertility and Organic Matter

Soil fertility is a complex concept referring to the potential of a soil to supply nutrients in amounts, forms and proportions required for optimum plant growth (Roper, 2000). Soil fertility management has two requirements: (i) establishing the soil pH and providing essential plant nutrient and organic matter (OM) content for initial tree growth (as described in this chapter); and (ii) maintaining soil pH, nutrients and OM within their desired ranges (Jones, 2012) during tree maturation and cropping (see Chapter 9, this volume).

### 10.8.1 Preplant fertilization

Typically, plants derive mineral nutrients as ions from the soil solution. Nutrient deficiencies are alleviated by the application of fertilizers and/or adjustment of pH. The fertilizer type and application rates can be estimated from data about crop nutrient removal, nutritional deficiency symptoms, and plant and soil analyses. Reference values of these qualitative and quantitative indices and corresponding fertilization recommendations can be found in various orchard management guides (e.g. Hart, 1990; James, 2010; Stasiak, 2010; Horneck *et al.*, 2011; OMAF-MRA, 2013; UMCAFE, 2013; WSU-TFREC, 2015). Soil nutrient analysis (see Chapter 9, this volume) is important for determining how to amend the soil nutritional status before planting the orchard. Since some nutrients are relatively immobile in the soil, amendments for nutrients, pH correction or OM may need to be incorporated deeply into the soil layer. For example, phosphorus fertilizer should be incorporated into the soil at a depth of 40 cm or more. Because of the adsorption binding of phosphate ions on soil colloids, the fertilization rate has to exceed the net plant requirements. According to Soing (2004), no more than 25% of phosphorus (P) applied as $P_2O_5$ is absorbed by the tree. On the other hand, excessive P levels may block the absorption of some micronutrients (iron, copper, zinc, manganese), which can negatively affect cherry tree physiological functions, growth and yield.

Generally, surface application of moderate to high potassium (K) rates can raise the soil exchangeable-K content in the root zone (Boynton and Oberly, 1966). However, due to relatively low K mobility, cherry fertilization can be problematic on clay soils. When spread on the surface of such a soil, the K fertilizer is bound in the upper 10 cm of the soil profile. On heavy soils, therefore, preplant incorporation of K fertilizers into the soil is preferable. In that case, the application rate is determined as a difference between the desired and the actual K level in the upper 40 cm soil layer (UMCAFE, 2013). However, excess K can lead to magnesium (Mg) deficiency.

### 10.8.2 pH correction

Soil acidity (low pH) can be corrected by incorporating neutralizing substances into the soil, a practice known as liming. Liming is also used for fertilization of calcium (Ca) and Mg. A range of liming materials, including oxides, hydroxides or carbonates of Ca and Mg, are available to increase suboptimal soil pH. Common liming materials are limestone, dolomitic limestone and slags. Dolomitic limestone is recommended on soils with a Mg content less than 100 mg $kg^{-1}$ or with a high K level, which is an indication of Mg deficiency. The quality of the liming material is related to its degree of fineness (particle size), chemical composition (calcium carbonate equilibrium) and water content (Hart, 1990). The calcium carbonate equilibrium is an expression of the acid-neutralizing value of a liming material compared with pure calcium carbonate. The particle size is reciprocal to the rate at which a liming material dissolves in water, since lime does not neutralize acidity or release its nutrients until it has dissolved. In addition, the acid-neutralizing value of a liming material is reciprocal to its moisture content. Effectiveness of liming is increased by higher solubility and finer particle size of the liming material and its incorporation within the potential rooting zone of the orchard.

Under certain circumstances, the liming of an acid soil can result in a temporary reaction (lime shock) that can adversely affect cherry trees. Therefore, it is generally recommended that a liming material be applied at least 4 months prior to tree planting, thus allowing sufficient time for the lime–soil interaction to stabilize. Deep ploughing is used to incorporate the liming material within, if possible, or at the surface of the subsoil in order to provide good root development. The amount of lime to be added is the sum of topsoil and subsoil requirements. The lime should be thoroughly harrowed into the surface soil and then ploughed to work it as deeply as possible into the soil. If large amounts of lime are required (more than 3 t per acre), split application is recommended, working one-half or two-thirds of the total amount of lime into the soil, as indicated above, plus thoroughly harrowing the remainder into the topsoil after ploughing (OMAF-MRA, 2013; UMCAFE, 2013).

High soil pH can be reduced by similar incorporation of acidic materials within the soil profile. The most commonly used acidifiers are elemental sulfur (S) and ammonium sulfate. These should be incorporated throughout the soil 3 months to 1 year before planting (OMAFRA, 2006). Ammonium sulfate should not be applied at rates of nitrogen (N) higher than those recommended for fruiting cherry trees. The effect of soil acidification is reciprocal to the soil's buffering capacity and to the content of free lime. Hence, it is easier to acidify a sandy soil than a clay soil. On highly buffered soils, the reduction in soil pH may be short-lived. Soil acidification using aluminium and iron sulfate, sulfuric acid or S has been assessed for orchard soils in British Columbia, Canada, with finely ground and incorporated granulated S judged the most effective (Neilsen *et al.*, 1993). Some soils are resistant to pH decline as a result of very high contents of calcium carbonate, which can approach 20% by weight, requiring the addition of impractical quantities of acidifying amendments. Generally, the pH reduction that can be achieved at a reasonable cost is less than 2 pH units.

### 10.8.3 Soil OM

The term OM covers a wide variety of carbon-containing materials, comprised of living, dead or decomposing organisms. Soil OM helps maintain soil structure, enhances soil moisture-holding capacity, improves drainage and increases the ability of the soil to hold nutrients. Organic residues act as slow-release fertilizers, directly and indirectly providing nutrients to the plants. Adequate soil OM levels can help maintain crop yields and long-term plant health, especially in adverse weather conditions. The soil OM content varies between 1 and 5%, with most soils being in the 1–2% range. The preferable content for a range of soil types, however, is between 2 and 5%, with the lowest value for sandy soils and the highest for clay soils. Under cultivation and cropping, the soil OM content declines. The use of green manures can temporarily increase the OM content, but with a tendency to decrease and stabilize at certain levels specific for each combination of soil, climate and cultivation practice. Many of the soils for growing cherries are light-textured and frequently cultivated. The maintenance of OM levels in these soils is critical for maintaining productivity (Roper, 2000; Jones, 2012; OMAF-MRA, 2013; Tisdall and van den Ende, 2015).

In recent years, there has been a growing interest in what is referred to as the 'soil food web' (Ingham, 2000). This refers to the significant biological diversity that exists in the rhizosphere and includes bacteria, fungi, protozoa, arthropods, nematodes, earthworms and insects. Understanding how these organisms affect the health, vigour and yield of fruit trees is a relatively new science. Much of the evidence suggesting benefits from enhancing this soil food web is anecdotal. Nevertheless, there appears to be increasing evidence that practices to enhance the soil food web provide benefits to tree health, yield and fruit quality (Forge *et al.*, 2002, 2013; Neilsen *et al.*, 2003; Pokharel and Zimmerman, 2015).

The addition of organic residues is the only sure way to increase soil OM levels to enhance the soil food web. Common organic amendments include manure, crop residues, composts, biosolids and cover crops. Compost supplies relatively low amounts of readily available nutrients but releases nutrients over time. That is why it should be applied together with fresh residues. Determining the right OM level for a soil depends on the soil texture and the aggregate stability target. Details on OM, its nutrient composition and the levels for maximum aggregate stability can be found in the specialized literature (e.g. OMAFRA, 2006, 2009; Jones, 2012). Adding OM to soils before planting is more beneficial than surface applications with no incorporation after plant establishment.

Application of excessive amounts of OM to a mineral soil, however, may have an adverse effect. Since OM content and soil water-holding capacity are correlated positively, wetter soils may remain cool for a longer time in the spring, and cultivation may also be hindered. Usually, soils with more than 5% OM require increased herbicide rates, because the herbicides may become partly inactivated due to their adsorption on the OM particles (Tisdall and van den Ende, 2015). Nutrients are not well balanced in most organic substances, and this may result in imbalanced plant mineral nutrition. For instance, the high K content in poultry manure may eventually induce either Mg or Ca plant deficiency. Some organic waste products contain heavy metals that can become toxic to plants (Jones, 2012). Additions of OM will help minimize soil limitations due to texture but will not entirely alleviate severe problems (Roper, 2000).

## 10.9 Weed Management and Cover Cropping

Cherry trees, especially young trees, do not compete well with weeds. Therefore, most weeds should be controlled during site preparation for the new orchard. This is especially true if the field has been fumigated prior to planting. Good cultural practices during the year prior to planting of trees can reduce many weed problems (Majek, 2004; Roper, 2004; Tworkoski and Glenn, 2008; OMAFRA, 2009, 2015; PSU, 2012; OMAFMRA, 2013; UMCAFE, 2013). A successful weed control programme must integrate both cultural and chemical weed control practices. Ideally, the site to be planted should be either fallow, cover-cropped or farmed with row crops such as maize or wheat that do not support tomato ringspot virus. Primary tillage operations should be completed the summer before planting of the orchard. Persistent perennial weeds should be treated and removed using systemic herbicides during the preplanting year.

The establishment of a cover (green manure) crop will reduce weed growth due to competition and will add OM to the soil. Cover crops are non-economic crops grown either between orchard rows or over the entire soil surface 1 or 2 years before orchard establishment. The cover crop turns into green manure when it is ploughed under. Sorghum, Sudan grass, hybrid kale, sugarbeets, winter rye, clovers, field peas and hairy vetch all make good cover crops. The first green manure crop should be planted in the autumn after the land has been cleared of old orchard or native brush. It will be ploughed under in the spring and replanted, and this can be repeated until the soil is properly prepared. This should be done after or in conjunction with control of perennial weeds. Non-selective short-residual herbicides can be applied before seeding the cover/manure crop and before ploughing the green manure crop under. Residual herbicides are not recommended because of the risk of a carry-over effect on young cherry trees during the planting year. An application of N at 50 kg $ha^{-1}$ is recommended for cover crop establishment. An additional 50 kg $ha^{-1}$ is recommended when the cover crop is ploughed in.

Grasses and legumes are commonly used as cover crops in cherry orchards. Grasses have fine, fibrous and dense roots that pervade the entire surface layer of the soil and can scavenge large quantities of residual N. Grasses can be seeded in either spring or late summer. The late-summer establishment is preferred because the warm soil will stimulate rapid germination, and winter frost will kill annual weeds before they flower and produce seed. Legume cover crops can fix N from the air as well as take up both residual soil N and N from manure applications. However, excess nitrogen release late in the season could induce excessive vegetative growth in young cherry trees and delay cold acclimation.

Decisions about how the orchard floor will be managed should be made before an orchard is planted. The preferred orchard floor management system combines vegetation-free tree rows with a ground cover (sod) in the alleys between rows. Sod prevents soil erosion, provides traction for equipment, adds OM to the soil, improves soil moisture and structure, and can provide shelter for beneficial predatory insects. Perennial ryegrass, tall fescue and hard fescue are among the species used most frequently as ground covers in the alleys. Recently, slow-growing turfgrasses have gained popularity as cover crops, especially on low-fertility soils, in poor growing conditions and under conditions of heavy traffic. A perennial sod can be established in the orchard by sowing over the entire soil surface in the year or 2 years prior to tree planting and killing it in the tree row strips with a systemic non-residual herbicide several months in advance of (Roper and Frank, 2004; Tworkoski and Glenn, 2008) or immediately after (Majek, 2004) tree planting. Conversely, alleys can be sown after tree rows are planted, particularly when extensive soil modifications such as raised beds are required. Clover and other legumes are not recommended as sod constituents because of their N-release inconsistency. The sod should not include plants attractive to pests (e.g. insects, diseases, nematodes) that could attack the cherry tree or fruit.

## References


Ayers, R.S. and Westcot, D.W. (1985) *Water Quality for Agriculture*. FAO Irrigation and Drainage Paper 29, Food and Agriculture Organization of the United Nations, Rome.

Barfield, B.J., Perry, K.B., Martsolf, J.D. and Morrow, C.T. (1990) Modifying the aerial environment. In: Hoffman, G.J., Howell, T.A. and Solomon, K.H. (eds) *Management of Farm Irrigation Systems*. American Society of Agricultural Engineers, St Joseph, Michigan, pp. 827–869.

Boman, B.J. (2007) Microsprinkler irrigation. In: Lamm, F.R., Ayars, J.E. and Nakayama, F.S. (eds) *Microirrigation for Crop Production: Design, Operation and Management*. Elsevier, Amsterdam, pp. 575–608.

Bosch, J., Kemp, W. and Trostle, G. (2006) Bee population returns and cherry yields in an orchard pollinated with *Osmia lignaria* (Hymenoptera: Megachilidae). *Journal of Economic Entomology* 99, 408–413.

Boxma, R. (1972) Bicarbonate as the most important soil factor in lime-induced chlorosis in The Netherlands. *Plant and Soil* 37, 233–243.

Boynton, D. and Oberly, G.H. (1966) Apple nutrition. In: Childers, N.F. (ed.) *Nutrition of Fruit Crops, Temperate, Subtropical, Tropical*. Horticultural Publications, New Brunswick, New Jersey, pp. 1–50.

Breeze, T.D., Bailey, A.P., Balcombe, K.G. and Potts, S.G. (2011) Pollination services in the UK: how important are honeybees? *Agriculture, Ecosystems and Environment* 142, 137–143.

Burt, C. (2003) *Chemigation and Fertigation Basics for California*. Irrigation Training and Research Center, California Polytechnic State University, San Luis Obispo, California.

Camp, C.R. (1998) Subsurface drip irrigation: a review. *Transactions of the ASAE* 41, 1353–1367.

Craig, B. (2012) Building better trellis systems for Nova Scotia orchards. Available at: http://www.perennia.ca/fieldservices/fruit-crops/tree-fruits/building-better-trellis-systems-for-orchards/ (accessed 19 July 2016).

Dart, J. (2008) *Intensive Apple Orchard Systems*. NSW Department of Primary Industries, New South Wales, Australia.

Delaplane, K.S. and Mayer, D.E. (2000) *Crop Pollination by Bees*. CAB International, Wallingford, UK.
Evans, R.G. (2006a) *Overtree Evaporative Cooling System Design and Operation for Apples in the PNW*. US Department of Agriculture, Agricultural Research Service, Northern Plains Agricultural Research Laboratory, Sidney, Montana.
Evans, R.G. (2006b) *Frost Protection in Orchards and Vineyards*. US Department of Agriculture, Agricultural Research Service, Northern Plains Agricultural Research Laboratory, Sidney, Montana.
Evans, R.G. and Waller, P.M. (2007) Application of chemical materials. In: Lamm, F.R., Ayars, J.E. and Nakayama, F.S. (eds) *Microirrigation for Crop Production: Design, Operation and Management*. Elsevier, Amsterdam, pp. 285–327.
Forge, T.A., Hogueb, E., Neilsen, G. and Neilsen, D. (2002) Effects of organic mulches on soil microfauna in the root zone of apple: implications for nutrient fluxes and functional diversity of the soil food web. *Applied Soil Ecology* 22, 39–54.
Forge, T.A., Neilsen, G.H., Neilsen, D., Hogue, E. and Faubion, D. (2013) Composted dairy manure and alfalfa hay mulch affect soil ecology and early production of 'Braeburn' apple on M9 rootstock. *HortScience* 48, 645–651.
Free, J.B. (1993) *Insect Pollination of Crops*. Academic Press, London.
Garibaldi, L.A., Steffan-Dewenter, I., Winfree, R., Aizen, M.A., Bommarco, R., Cunningham, S.A., Kremen, C., Carvalheiro, L.G., Harder, L.D., Afik, O., Bartomeus, I., Benjamin, F., Boreux, V., Cariveau, D., Chacoff, N.P., Dudenhöffer, J.H., Freitas, B.M., Ghazoul, J., Greenleaf, S., Hipólito, J., Holzschuh, A., Howlett, B., Isaacs, R., Javorek, S.K., Kennedy, C.M., Krewenka, K.M., Krishnan, S., Mandelik, Y., Mayfield, M.M., Motzke, I., Munyuli, T., Nault, B.A., Otieno, M., Petersen, J., Pisanty, G., Potts, S.G., Rader, R., Ricketts, T.H., Rundlöf, M., Seymour, C.L., Schüepp, C., Szentgyörgyi, H., Taki, H., Tscharntke, T., Vergara, C.H., Viana, B.F., Wanger, T.C., Westphal, C., Williams, N. and Klein, A.M. (2013) Wild pollinators enhance fruit set of crops regardless of honey bee abundance. *Science* 339, 1608–1611.
Garibaldi, L.A., Carvalheiro, L.G., Leonhardt, S.D., Aizen, M.A., Blaauw, B.R., Isaacs, R., Kuhlmann, M., Kleijn, D., Klein, A.M., Kremen, C. et al. (2014) From research to action: enhancing crop yield through wild pollinators. *Frontiers in Ecology and the Environment* 12, 439–447.
Hansen, M. (2013) Growers try hooped trellises. *Good Fruit Grower* 64, 30–31.
Hart, J. (1990) *Fertilizer and Lime Materials*. Fertilizer Guide 52. Oregon State University, Corvallis, Oregon.
Holzschuh, A., Dudenhöffer, J.-H. and Tscharntke, T. (2012) Landscapes with wild bee habitats enhance pollination, fruit set and yield of sweet cherry. *Biological Conservation* 153, 101–107.
Horneck, D.A., Sullivan, D.M., Owen, J.S. and Hart, J.M. (2011) *Soil Test Interpretation Guide*. Oregon State University Extension Service 1478, Corvallis, Oregon.
Hoying, S.A. (2012) Experiences with support systems for the Tall Spindle apple planting system. *New York Fruit Quarterly* 20, 3–8.
Ingham, E.R. (2000) The soil food web. In: Tugel, A.J., Lewandowski, A.M. and Happe-von Arb, D. (eds) *Soil Biology Primer*. Soil and Water Conservation Society, Ankeny, Iowa, pp. 2–6.
Isaacs, R. and Tuell, J. (2007) *Conserving Native Bees on Farmland*. Michigan State University Extension, Michigan.
James, P. (2010) *Australian Cherry Production Guide*. Department of Agriculture, Fisheries and Forestry, Tasmanian Institute of Agricultural Research, Rural Solutions SA, Lenswood, South Australia.
Jones, J.B. (2012) *Plant Nutrition and Soil Fertility Manual*, 2nd edn. CRC Press, Boca Raton, Florida.
Kafkafi, U. and Tarchitzky, J. (2011) *Fertigation: A Tool for Efficient Fertilizer and Water Management*. International Fertilizer Industry Association and International Potash Institute, Paris, France.
Kotuby-Amacher, J., Koenig, R. and Kitchen, B. (2000) *Salinity and Plant Tolerance*. Utah State University Extension AG-SO-03, Logan, Utah.
Koumanov, K.S. (2007) On the necessity for further improvement in microirrigation scheduling. In: *Water Resources Management and Irrigation and Drainage Systems Development in the European Environment. Proceedings of the 22nd European Regional Conference of the International Commission on Irrigation and Drainage*, Pavia, Italy, 2–6 September.
Koumanov, K.S. and Tsareva, I.N. (2017) Intensive sweet cherry production: potential, "bottlenecks", perspectives. *Acta Horticulturae* (in press).
Koumanov, K., Hopmans, J.W., Schwankl, L.J., Andreu, L. and Tuli, A. (1997) Application efficiency of micro-sprinkler irrigation of almond trees. *Agricultural Water Management* 34, 247–263.
Koumanov, K., Dochev, D. and Stoilov, G. (1998) Investigations on fertigation of peach on three soil types – patterns of soil wetting. *Bulgarian Journal of Agricultural Science* 4, 745–753.

Koumanov, K., Hopmans, J.W. and Schwankl, L.J. (2006) Spatial and temporal distribution of root water uptake of an almond tree under microsprinkler irrigation. *Irrigation Science* 24, 267–278.

Koumanov, K.S., Rankova, Z., Kolev, K. and Shilev, S. (2009) Herbigation in a cherry orchard – translocation and persistency of pendimethalin in the soil. *Acta Horticulturae* 825, 305–312.

Lamm, F.R. and Camp, C.K. (2007) Subsurface drip irrigation. In: Lamm, F.R., Ayars, J.E. and Nakayama, F.S. (eds) *Microirrigation for Crop Production: Design, Operation and Management*. Elsevier, Amsterdam, pp. 473–551.

Lamm, F.R., Ayars, J.E. and Nakayama, F.S. (2007) *Microirrigation for Crop Production: Design, Operation and Management*. Elsevier, Amsterdam, The Netherlands.

Lehnert, R. (2015) Flying doctors: honeybees deliver brown rot control to sweet cherry orchards. *Good Fruit Grower* 66, 40–41.

Lillrose, T., Lang, G.A. and Sundin, S.W. (2017) Strategies to minimize bacterial canker in high density sweet cherry orchards. *Acta Horticulturae* (in press).

Long, L.E. and Kaiser, C. (2013) *Sweet Cherry Orchard Establishment in the Pacific Northwest, Important Considerations for Success*. Pacific Northwest Extension, Oregon State University, Corvallis, Oregon.

Long, L., Lang, G., Musacchi, S. and Whiting, M. (2015) *Cherry Training Systems*. Pacific Northwest Extension, Oregon State University, Corvallis, Oregon.

Mader, E., Spivak, M. and Evans, E. (2010) *Managing Alternative Pollinators: A Handbook for Beekeepers, Growers, and Conservationists*. Natural Resource, Agriculture, and Engineering Service Cooperative Extension, Ithaca, New York.

Majek, B. (2004) Weed management strategies for tree fruits. In: *Proceedings of 'Mapping Your Road to the Future', Great Lakes Fruit, Vegetable and Farm Market EXPO*, Grand Rapids, Michigan.

Mazzola, M. and Manici, L.M. (2012) Apple replant disease: role of microbial ecology in cause and control. *Annual Review of Phytopathology* 50, 45–65.

Mommaerts, V. and Smagghe, G. (2011) Entomovectoring in plant protection. *Arthropod–Plant Interactions* 5, 81–95.

NAS (2007) *Status of Pollinators in North America*. National Academy of Sciences, National Academies Press, Washington, DC.

Neilsen, D., Hogue, E.J., Hoyt, P.B. and Drought, B.G. (1993) Oxidation of elemental sulphur and acidulation of calcareous orchard soils in southern British Columbia. *Canadian Journal of Soil Science* 73, 103–114.

Neilsen, G.H., Hogue, E.J., Forge, T.A. and Neilsen, D. (2003) Mulches and biosolids affect vigor, yield and leaf nutrition of fertigated high density apple. *HortScience* 38, 41–45.

OMAF-MRA (2013) *Guide to Fruit Production*. Ontario Ministry of Agriculture and Food and the Ministry of Rural Affairs, Toronto, Ontario, Canada.

OMAFRA (2006) *Soil Fertility Handbook*. Ontario Ministry of Agriculture and Food and Rural Affairs, Toronto, Ontario, Canada.

OMAFRA (2009) Soil management. In: Brown, C. (ed.) *Agronomy Guide for Field Crops*. Ontario Ministry of Agriculture and Food and Rural Affairs, Toronto, Ontario, Canada, pp. 137–154.

OMAFRA (2012) Tall spindle orchard support systems. Ontario Ministry of Agriculture and Food and Rural Affairs, Toronto, Ontario, Canada. Available at: http://www.omafra.gov.on.ca/neworchard/english/apples/9trellis.html (accessed 3 September 2015).

OMAFRA (2015) *Guide to Weed Control 2016–2017*. Ontario Ministry of Agriculture and Food and Rural Affairs, Toronto, Ontario, Canada.

Perry, R. (1998) Tree fruit root systems and soils. In: *Proceedings of the Utah State Horticultural Association Annual Convention*, 27–28 January, Provo, Utah. Available at: http://www.hrt.msu.edu/uploads/535/78649/Tree-Root-Systems-and-Soils.pdf (accessed 6 September 2015).

Pfiffner, L. and Müller, A. (2014) *Wild Bees and Pollination*. Research Institute of Organic Agriculture, Frick, Switzerland.

Pinillos, V. and Cuevas, J. (2008) Artificial pollination in tree crop production. *Horticultural Reviews* 34, 239–276.

Pokharel, R.R. and Zimmerman, R. (2015) Impact of organic and conventional peach and apple production practices on soil microbial populations and plant nutrients. *Organic Agriculture* 6, 19–30.

PSU (2012) *Pennsylvania Tree Fruit Production Guide 2012–2013*. Pennsylvania State University, University Park, Pennsylvania.

Rankova, Z., Koumanov, K., Kolev, K. and Shilev, S. (2009) Herbigation in a cherry orchard – efficiency of pendimethalin. *Acta Horticulturae* 825, 459–464.

Rieul, L. (ed.) (1990) *Irrigation*, 1st edn. French Research Institute for Agricultural and Environmental Engineering (CEMAGREF)/Le Centre d'Études et de Prospective France Agricole (CEP-FA)/Réseau National Expérimentation Démonstration, Secteur Hydraulique Agricole (RNED-HA), Montpellier, France.

Robinson, T.L., DeMarree, A.M. and Hoying, S.A. (2007) An economic comparison of five high density apple planting systems. *Acta Horticulturae* 732, 481–490.

Roper, T.R. (2000) Mineral nutrition of fruit crops. Available at: https://soilsextension.triforce.cals.wisc.edu/wp-content/uploads/sites/68/2016/06/4C.roper_.pdf (accessed 26 September 2015).

Roper, T.R. (2004) *Orchard Floor Management for Fruit Trees*. A3562, University of Wisconsin Cooperative Extension, Madison, Wisconsin.

Roper, T.R. and Frank, G.G. (2004) *Planning and Establishing Commercial Apple Orchards in Wisconsin*. University of Wisconsin Cooperative Extension, Balsam Lake, Wisconsin.

Sagili, R.R. and Burgett, D.M. (2011) *Evaluating Honey Bees for Pollination: A Guide for Commercial Growers and Beekeepers*. Pacific Northwest Extension, Oregon State University, Corvallis, Oregon.

Schwankl, L.J. and Hanson, B.R. (2007) Surface drip irrigation. In: Lamm, F.R., Ayars, J.E. and Nakayama, F.S. (eds) *Microirrigation for Crop Production: Design, Operation and Management*. Elsevier, Amsterdam, pp. 431–472.

Shepherd, M., Buchmann, S.L., Vaughan, M. and Black, S.H. (2003) *Pollinator Conservation Handbook: A Guide to Understanding, Protecting, and Providing Habitat for Native Pollinator Insects*. Xerces Society/Bee Works, Portland, Oregon.

Soing, P. (2004) *Fertilization des Vergers: Environnement et Qualité*. Centre Technique Interprofessionnel des Fruits et Légumes (CTIFL), Paris, France.

Solomon, K.H. (2013) Irrigation systems. In: Goyal, M.R. (ed.) *Management of Drip/Trickle or Micro Irrigation*. Apple Academic Press, Oakville, Ontario, Canada, pp. 72–102.

Somerville, D. (1999) Honey bees in cherry and plum pollination. NSW Agriculture, New South Wales, Australia. Available at: http://www.dpi.nsw.gov.au/__data/assets/pdf_file/0019/117109/bee-cherry-plum-pollination.pdf (accessed 19 July 2016).

Stasiak, M. (2010) *Nutrition Guidelines for Wisconsin Apple and Cherry Orchards*. Peninsular Agricultural Research Station, University of Wisconsin, Madison, Wisconsin.

Storlie, C.A. (2004) *Treating Drip Irrigation Systems with Chlorine*. Rutgers Cooperative Research and Extension, The State University of New Jersey, Rutgers, New Jersey.

Sutton, J.G., Brownscombe, R.H., Fasken, G.B., Foreman, H.J., Gain, E.W., Herndon, L.W., Long, W.F. and Schlaudt, E.A. (1971) *Drainage of Agricultural Land, National Engineering Handbook Section 16*. Natural Resources Conservation Service, US Department of Agriculture, Washington, DC.

Tisdall, J. and Van den Ende, B. (2015) Understanding soil organic matter: don't overlook the soil if you want your orchard to thrive. *Good Fruit Grower* 66, 44–47.

Trimmer, W.L., Ley, T.W., Clough, G. and Larsen, D. (1992) *Chemigation in the Pacific Northwest*. Pacific Northwest Extension, Washington State University Digital Collection, Pullman, Washington.

Tworkoski, T.J. and Glenn, D.M. (2008) Orchard floor management systems. In: Layne, D.R. and Bassi, D. (eds) *The Peach: Botany, Production and Uses*. CAB International, Cambridge, Massachusetts, pp. 332–351.

UMCAFE (2013) Nutrient management of apple orchards. In: *2013 New England Tree Fruit Management Guide*. University of Massachusetts Center for Agriculture, Food and the Environment, Amherst, Massachusetts, pp. 100–108.

US EPA (2015) Restricted Use Product Summary Report (August 27, 2015). United States Environmental Protection Agency. Available at: http://www2.epa.gov/sites/production/files/2015-08/documents/rupreport-sec3-update_0.pdf (accessed 3 September 2015).

USDA-NRCS (2013) Microirrigation. In: *National Engineering Handbook Part 623*. US Department of Agriculture, Natural Resources Conservation Service, Washington, DC.

USDA-SCS (1983) Sprinkler irrigation, Section 15 irrigation. In: *National Engineering Handbook*. US Department of Agriculture, Soil Conservation Service, Washington, DC.

Vaknin, Y., Gan-Mor, S., Bechar, A., Ronen, B. and Eisikowitch, D. (2001) Improving pollination of almond (*Amigdalus communis* L., Rosaceae) using electrostatic techniques. *Journal of Horticultural Science and Biotechnology* 76, 208–212.

van Dalfsen, K.B. (1989) *Support Systems for High-density Orchards*. Province of British Columbia Ministry of Agriculture and Fisheries, Victoria, British Columbia, Canada.

van der Gulik, T.W., Evans, R.G. and Eisenhauer, D.E. (2007) Chemigation. In: Hoffman, G.J., Evans, R.G., Jensen, M.E., Martin, D.L. and Elliot, R.L. (eds) *Design and Operation of Farm Irrigation Systems*, 2nd edn. American Society of Agricultural and Biological Engineers, St Joseph, Michigan, pp. 725–753.

van der Molen, W.H., Beltrán, J.M. and Ochs, W.J. (2007) *Guidelines and Computer Programs for the Planning and Design of Land Drainage Systems*. Food and Agriculture Organization of the United Nations, Rome.

Vaysse, P., Soing, P. and Peyremorte, P. (1990) *L'Irrigation des Arbres Fruitiers*. Centre Technique Interprofessionnel des Fruits et Légumes (CTIFL), Paris, France.

Waterman, P.F. (2001) *Fertilization Guidelines in High Density Apples and Apple Nurseries in the Okanagan – Similkameen*. British Columbia Ministry of Agriculture, Food and Fisheries, Victoria, British Columbia, Canada.

Webster, A.D. and Schmidt, H. (1996) Rootstocks for sweet and sour cherries. In: Webster, A.D. and Looney, N.E. (eds) *Cherries: Crop Physiology, Production and Uses*. CAB International, Wallingford, UK, pp. 127–163.

Winfree, R., Williams, N.M., Dushoff, J. and Kremen, C. (2007) Native bees provide insurance against ongoing honey bee losses. *Ecology Letters* 10, 1105–1113.

WSU-TFREC (2015) Diagnosing nutrient problems in orchards. Washington State University, Tree Fruit Research and Extension Center, Washington. Available at: http://soils.tfrec.wsu.edu/webnutritiongood/TreeFruitStuff/04diagnosing.htm (accessed 27 September 2015).

Zhu, H., Wang, X., Reding, M. and Locke, J. (2011) Distribution of chemical and microbial pesticides through drip irrigation systems. In: Stoycheva, M. (ed.) *Pesticides – Formulations, Effects, Fate*. InTech, Rijeka, Croatia, pp. 155–180.

# 11 Orchard Microclimate Modification

**Michael M. Blanke,[1]* Gregory A. Lang[2] and Mekjell Meland[3]**
*[1]INRES Horticultural Science, University of Bonn, Bonn, Germany; [2]Department of Horticulture, Michigan State University, East Lansing, Michigan, USA; [3]Norwegian Institute of Bioeconomy Research – NIBIO Ullensvang, Lofthus, Norway*

## 11.1 Microclimates in Cherry Production and Climate Change

Microclimates comprise spatial variations in light, temperature, wind and humidity, which determine the extent to which cherry can be grown in a particular environment. Geological features such as bodies of water or mountain ranges modify regional climates in ways that enhance or limit the production of cherries, which are among the earliest flowering fruit tree species in spring and among the most sensitive to rain with respect to fruit disorders and diseases. Cherry cultivars have originated from a wide range of temperate climates and exhibit a subsequently wide growing potential – second only to apple – ranging in Europe from the moderate Mediterranean climates of Turkey and Macedonia in northern Greece at latitudes of 39–40°N to Slovenia at 46°N, through the continental climate of central Europe into northern Germany at 53°N, to the Fjord areas in western Norway at 61°N. North American microclimates influenced by the Pacific ocean and the Coastal, Cascade and Sierra mountain ranges allow excellent cherry production from 36°N in California to 47°N in Washington and 49°N in British Columbia, and the Great Lakes moderate the more continental climate in the midwest USA for cherry production in Michigan at 43–45°N. In South America, the Pacific Ocean and the Andes mountains influence cherry production in Argentina and Chile from 33 to 46°S. Unique and diverse microclimates across other significant cherry-producing areas range from Australia to China and Iran.

Cherry cultivars have been bred for acclimatization to local conditions and microclimates over decades. Global climate change during recent decades has brought about a range of changes in microclimate, which must be accommodated for successful cherry cultivation in the future. This includes not only warmer and drier summers, but also warmer winters (Luedeling *et al.*, 2013a). For example, the date for the green tip stage of flower bud development, after the endodormancy-to-ecodormancy transition, has advanced markedly over the past 75 years in Michigan, USA, averaging 10 days earlier by 2010 compared with 1935 and earlier (Fig. 11.1). This has significant consequences for the probability of spring frost damage to buds or open flowers. During that same period, the number of freeze events that occurred at or after the green bud stage has nearly doubled and become more extreme (Fig. 11.2), with fewer

* ulp304@uni-bonn.de

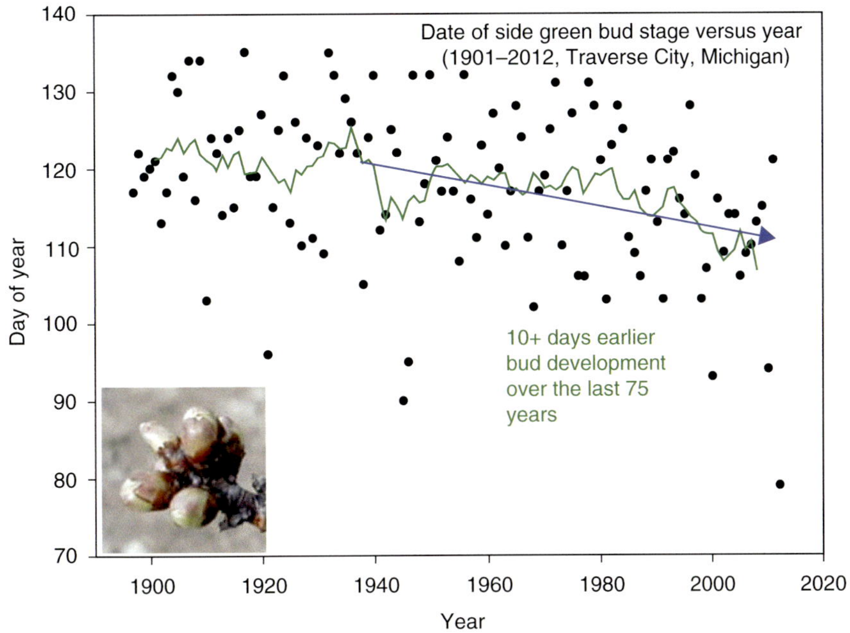


**Fig. 11.1.** Annual date of achieving the side green bud developmental stage of 'Montmorency' sour cherry over the past 110+ years in Traverse City, Michigan, USA. (Graph courtesy of J. Andresen.)

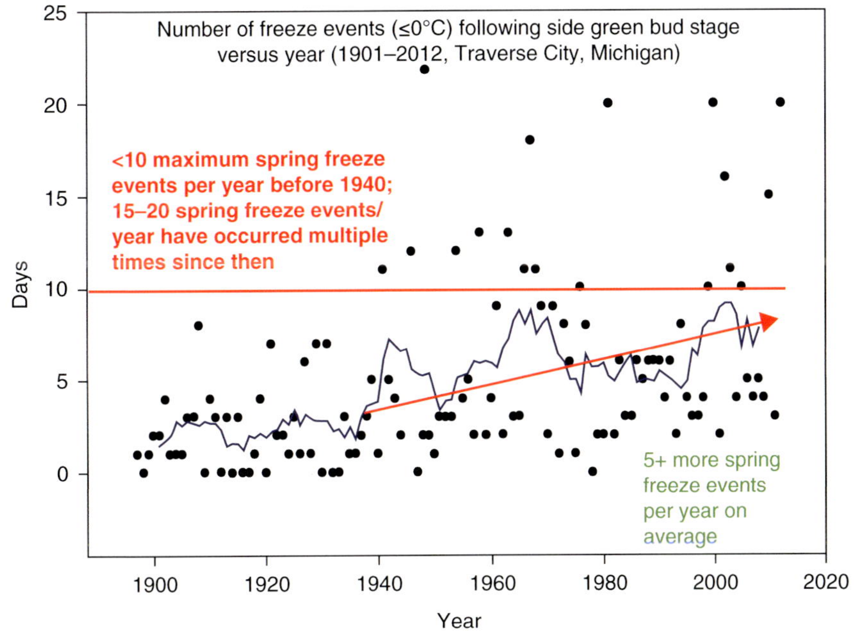


**Fig. 11.2.** Annual number of freeze events at or following the side green bud developmental stage over the past 110+ years in Traverse City, Michigan, USA. (Graph courtesy of J. Andresen.)

than ten in any season before 1940, but as many as 15 seasons with more than ten since 1940 and as many as 15–20 freeze events per season occurring seven times in that period.

If recent increases in orchard exposure to solar radiation and hours of sunshine (e.g. Fig. 11.3) continue, earlier bud development in temperate climates and heat-induced abnormalities in hot climates will continue to present challenges for cherry growers. In California's Central Valley, the typical periodic occurrence of 'Tule' fog during winter helped block solar radiation and warming of buds, facilitating satisfaction of chilling requirements and the natural transition from endodormancy to ecodormancy. As the Tule fog has become less common, winter bud chilling has decreased, prolonging or preventing the natural endodormancy-to-ecodormancy transition (Baldocchi and Waller, 2014). This seemingly minor shift in climate has had negative effects on completing pollen and pistil development late in floral meristem differentiation, resulting in a weaker bloom and lower fruit set. Typical sweet cherry fruit set in California is about half of that in Washington State, 1100 km to the north. Another consequence of climate change tied to floral differentiation is the increasing number of summer days with extremely high temperatures (e.g. >35°C) in locations such as California's Central Valley (e.g. Cal-adapt, 2017) and the Yakima and Columbia River Basins of Washington (e.g. EPA, 2016). The occurrence of such temperatures during July and August coincide with the increased incidence of double pistil formation, resulting in double or spur fruit during the subsequent fruiting season (see Chapter 8, this volume).

Global solar radiation and temperature extremes are not the only climatic changes that can have critical effects on cherry production, and such changes are not always negative. Earlier bloom generally leads to earlier ripening, which may shift production to a higher value return in traditionally early markets or to a lower value return (when peak supplies are still high) in traditionally late markets. Producers with late-maturing cultivars in cooler environments can benefit from warmer temperatures to reach full maturity and produce high-quality fruit. The incidence of rain is a major climatic factor that influences economically viable cherry production, from cracking in sweet cherries to prevalence of debilitating diseases such as brown rot (*Monilinia* spp.), cherry leaf spot (*Blumeriella jaapii*) and bacterial canker (*Pseudomonas syringae*) in sour or sweet cherry. In Michigan's historic sour and sweet cherry production areas, average rainfall has increased by 10% (Fig. 11.4) over the past

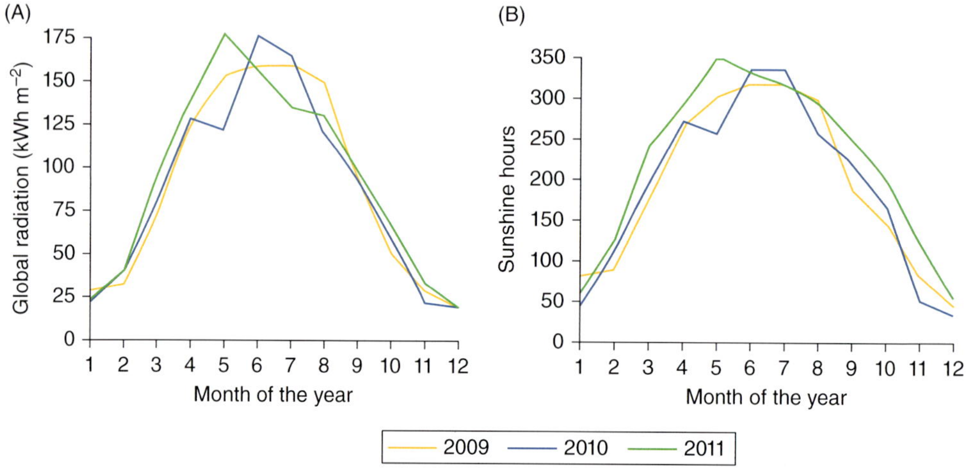


**Fig. 11.3.** (A) Monthly global radiation and (B) monthly sunshine hours in a changing climate at Klein-Altendorf, Germany, from 2009 to 2011.

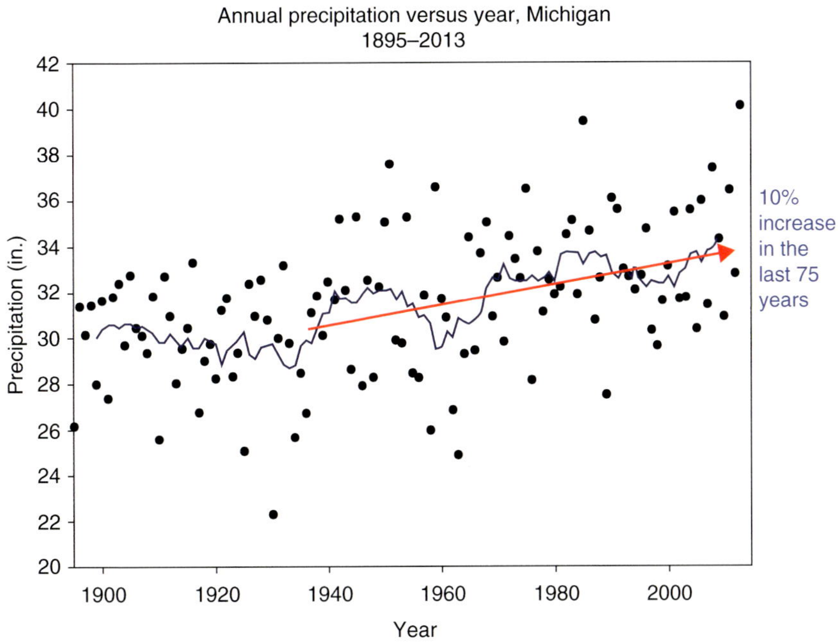


**Fig. 11.4.** Climate change in annual precipitation over the past 115+ years in Michigan, USA. (Graph courtesy of J. Andresen.)

75 years. Currently, there are few fully effective spray-based orchard control measures for preventing rain-induced cracking or canker, and ever-evolving resistance to fungicides for leaf spot and brown rot. Strategies for manipulation of orchard microclimates can provide several potential benefits to growers.

## 11.2 Microclimate Modification

### 11.2.1 Protection from low temperatures

Production of cherries is limited by low temperature in numerous ways (see Chapter 8, this volume). The occurrence of unusually early or extremely low temperatures in autumn, before trees have acquired an adequate level of cold hardiness or even before leaf abscission, can cause damage to flower buds and damage to xylem and phloem in vegetative tissues, or prevent the remobilization of carbon and nitrogen from leaves during senescence, which results in poor resources for winter hardiness and for strong bloom and growth in spring. Extremely low temperatures during winter can damage flower buds, shoots, tree trunks and even the root system. In cold locations on sunny winter days, warming of the bark in the afternoon on the south-west side of the tree in the northern hemisphere or the north-west side of the tree in the southern hemisphere, followed by plunging overnight low temperatures, can cause damage to conducting tissues that may not be apparent until spring. Low temperatures in spring after the endodormancy-to-ecodormancy transition can kill pistils in the bud or during bloom, or kill young fruit after pollination and fruit set has already occurred. Such spring freeze damage often then leads to bacterial canker infections that kill vegetative meristems, not only affecting the loss of the current season's crop but also reducing the cropping potential for several future seasons as well due to death of entire spurs (see Chapter 15, this volume).

Growers attempt to gain some level of control over these situations with microclimate modification strategies to add heat to the orchard, prevent heat from escaping the orchard or reduce localized heating of plant tissues that may be detrimental in certain situations. Traditional strategies include the open-air burning of fuel sources directly in the orchard during predicted winter or spring low-temperature events to attempt to prevent tissue damage, although these can be quite inefficient since heat quickly rises from the orchard. The burning of fuel sources with obviously negative by-products such as excessive smoke and other noxious gases, for example from wood, tyres, hay bales, kerosene or diesel fuel in heaters, has thankfully given way in most production regions to cleaner burning fuels such as propane heaters or other less environmentally questionable strategies. Another supplemental heat method is the use of tractor-mounted propane heaters to distribute added heat up and down orchard rows, although this is also relatively inefficient since the heated air can rise rapidly beyond the canopy. Under-tree sprinklers may also be used to increase air temperature by 1–2°C up to 2.0 m above the ground for spring frost protection. However, the effect is highly dependent on water temperature for the amount of potential heat to be added, and on the presence and strength of climatological air temperature inversions that can slow the rising of heat above the orchard (USDA-NRCS, 1993).

One of the most significant advances in the past 20 years has been the use of wind machines. These are comprised of an internal combustion engine (usually fuelled by propane or diesel) that drives a large propeller mounted at the top of a steel tower in the orchard, the purpose of which is to mix warmer upper layers of air with dangerously cold layers of air in the orchard during low-temperature inversions (e.g. Ribeiro *et al.*, 2006). The towers are generally 10–12 m tall, usually with two to four blades having an end-to-end diameter of 5–6 m. The propeller gearbox on the tower may tilt and rotate to spread its impact across 4–6 ha and uneven orchard terrains. Wind machines may be used alone during strong thermal inversion events or in combination with other supplemental heat strategies, such as under-tree sprinkler operation (to add heat from irrigation water to radiant heat from the soil) or open-air heaters. Auto-start features tied to thermostats make wind machines one of the most labour- and energy-efficient techniques for low-temperature microclimate manipulation.

With the increasing adoption of wind machines to modify low orchard temperatures, the use of over-tree sprinklers for microclimate modification has decreased. Rather than rely on adding heat directly from the water source to the air and soil as with under-tree sprinkler operation, the use of over-tree sprinklers relies on adding endothermic heat directly to buds through latent heat transfer as water freezes on the plant. Thus, such water application must be continuous until temperatures rise above freezing and all ice melts, otherwise heat transfer can become exothermic and result in plant damage. Over-tree sprinklers are not conducive to use with wind machines that could increase evaporative cooling. Therefore, over-tree sprinklers for frost protection can result in the use of large volumes of water that create saturated soil conditions in the orchard, as well as potentially spreading bacterial diseases such as canker.

Another significant advance in microclimate modification over the past 20 years is the increasing use of plastic solid or net orchard covers. Pole-and-cable single-row covers, high-tunnel multiple-row covers,- and retractable-roof greenhouse-like orchard covering structures can improve the conservation of radiant orchard heat from the soil, from under-tree sprinklers or from clean-burning fuel sources (e.g. Fig. 11.5). However, orchard covers do not provide complete frost protection, depending on the type of frost and the covering material. Common formulations of single-layer plastic orchard covers transmit radiant energy relatively quickly, resulting in inefficient trapping of heat in the orchard. Without supplemental heat, night temperatures inside poly-covered high tunnels will equilibrate relatively rapidly with those outside, including falling below freezing (Dekova and Blanke, 2007). The frost protection of orchard covers ranges from none to minimal. Research at Michigan State University showed that 13 propane heaters, each rated at up to 85,000 kJ output,

Fig. 11.5. Automated gas heater (with preset temperature range) and sweet cherry trees in a small gothic high tunnel at Klein-Altendorf, Germany. Without supplemental heat, the entire crop was lost due to spring frost in 2005.

were needed to raise and maintain the air temperature by 1°C ha$^{-1}$ in single-layer plastic-covered multi-bay high tunnels, or five heaters per hectare in a retractable-roof single-layer plastic greenhouse (G.A. Lang, East Lansing, Michigan, USA, 2015, personal communication). In fact, high tunnels without heat can get colder than outside air temperatures due to dry conditions lowering the dew point below that outside the tunnel. Thus, irrigating the soil in covered orchards can improve the soil's radiant heat capacity, raise the dew point and alleviate mild frost events to some extent.

To prevent south-west or north-west winter freeze damage to cherry trunks in the northern or southern hemisphere, respectively, painting the trunk white is common to protect it from direct or snow-reflected solar radiation, creating less of a differential between day and night temperatures and the loss of localized cold acclimatization. This is a very localized means of microclimate modification, affecting primarily the specific plant tissue in danger of low-temperature damage.

### 11.2.2 Enhancement of growth-promoting temperatures

Orchard microclimate temperatures can be modified to promote dormancy alleviation (cool temperatures) or to promote active growth, development and ripening (warm temperatures). The former can result in a stronger bloom and improved fruit set, while the latter can result in reduced fruit abnormalities, such as doubling and earlier or later ripening for improved harvest windows for marketing.

#### *Evaporative cooling*

Evaporative cooling occurs in all orchards having adequate soil moisture during warm days. The underlying physics are based on the exothermic energy transfer when fluid water is evaporated in the gas phase and cools the evaporative surface. The cooling of leaves through evapotranspirational loss of water through stomata, concomitant with photosynthetic gas exchange, is an indication of a well-functioning cherry tree. This natural cooling can be augmented by over-tree or under-tree microsprinkler applications of water.

Supplemental evaporative cooling can be promoted by over-tree impact sprinkler or microsprinkler application of water to the surface of plant buds, leaves or fruit. The efficacy of evaporative cooling is limited by the dew point, which is a function of temperature and humidity. Moisture will evaporate when the plant surface is warmer than the air, or will condense on the plant surface when it is cooler than the air. The temperature at which these two processes are in balance is the dew point, which limits the potential for evaporative cooling (Table 11.1). During evaporative cooling, the temperature can decrease to the dew point but not beyond. Evaporative cooling is most effective when the humidity is low and wind disperses the evaporated moisture from the orchard.

Evaporative cooling for microclimate modification in cherry production can be used during winter in warm (low-chill) climates to enhance chilling unit accumulation for the transition from endodormancy to ecodormancy. This improves the final stages of floral primordia development, resulting in a stronger and more uniform bloom and vegetative bud break, and can reduce or replace the need for dormancy-breaking chemicals such as hydrogen cyanamide. The most efficient use requires an over-tree microsprinkler system operated by a programmable datalogger/controller (e.g. CR1000, Campbell Scientific) with climatic inputs for temperature, humidity, wind speed and solar radiation (J. Flore, Michigan, USA, 2015, personal communication; similar to the controller described by Fernández and Flore, 1998, and Lang *et al.*, 1998). The controller can be programmed to pulse-operate the microsprinklers for less than 1 min on days with good climatic potential for evaporative cooling. The pulse application provides just enough water to wet the surface of the tree buds, and the controller applies the next pulse when its programme estimates that the previous pulse has finished evaporating, thereby using water very efficiently. The cooling is discontinued once the chilling requirement has been met for bud development and bloom to proceed normally. This is of particular interest in California as a way to compensate for the decreasing occurrence of the Tule fog during winter. One grower in southern California operates his orchard microsprinklers whenever the temperature exceeds 12.5°C from November to January, which is a less water-efficient but still effective way to improve chilling.

**Table 11.1.** Dew point (given as °C) as a function of relative humidity and air temperature, for determining evaporative cooling potential for cherry orchards. (Data courtesy of P. Buch.)

| Temperature (°C) | Humidity (%) 20 | 30 | 40 | 50 | 60 | 70 | 80 | 90 |
|---|---|---|---|---|---|---|---|---|
| 20 | −3 | 2 | 6 | 9 | 12 | 14 | 16 | 18 |
| 21 | −2 | 3 | 7 | 10 | 13 | 15 | 17 | 19 |
| 22 | −2 | 4 | 8 | 11 | 14 | 16 | 18 | 20 |
| 23 | −1 | 5 | 9 | 12 | 15 | 17 | 19 | 21 |
| 24 | −1 | 6 | 10 | 13 | 16 | 18 | 20 | NE |
| 25 | 0 | 7 | 11 | 14 | 17 | 19 | 21 | NE |
| 26 | 1 | 8 | 12 | 15 | 18 | 20 | NE | |
| 27 | 2 | 8 | 13 | 16 | 19 | 21 | NE | |
| 28 | 3 | 9 | 14 | 17 | 20 | NE | | |
| 29 | 4 | 9 | 15 | 18 | 21 | NE | | |
| 30 | 5 | 10 | 16 | 19 | 22 | NE | | |

NE, Use of evaporative cooling at high humidity is not effective.

The same microsprinkler-based evaporative cooling concept can be used in cold (high-chill) climates to delay cherry bud break for reducing the risk of spring frosts or to delay bloom to achieve a later ripening window. Instead of operating the microsprinklers during endodormancy for enhanced chilling, they are activated on sunny days once the chilling requirement has been met so as to retard the accumulation of heat units during ecodormancy. Although the concept was first tested with impact sprinklers several decades ago (Hewitt and Young, 1980), the massive quantity of water used in the days before microsprinklers and computer controllers was problematic. However, several years of testing the modern systems in cherries and apples have demonstrated bud break delays of 7–10 days with no detrimental orchard effects (J. Flore, Michigan, USA,

2015, personal communication). The potential extent of bud-break delay is greater than the potential extent of ripening delay, since heat-unit accumulation increases exponentially during the post-bud-break fruit development period compared with the pre-bud-break ecodormant period. Thus, a 10-day delay in bud-break date may only translate to a harvest delay of a few days.

In addition to evaporative cooling of cherry buds during endodormancy and ecodormancy, a third important period of potential use is during flower bud differentiation in the summer. Sweet cherry flower buds are initiated in the late spring, but floral organ differentiation does not begin until mid-summer (Guimond *et al.*, 1998). High (>30°C) summer temperatures at this time (i.e. July–August in the northern hemisphere, January–February in the southern hemisphere) induce abnormal pistil development, resulting in double or spur fruit formation during the next growing season (Beppu *et al.*, 2001). In growing locations where high summer temperatures are becoming more extreme with climate change, the incidence of such abnormalities for susceptible cultivars is likely to increase. This can be minimized by evaporative cooling during early flower bud differentiation (Southwick *et al.*, 1991; Whiting and Martin, 2008). The probable high frequency of applications needed during hot summer days suggest that the water-use efficiencies achieved using pulsed, programmable microsprinkler applications could be significant.

Other considerations for any of the potential uses of evaporative cooling in cherry production include a requirement for the use of good-quality water with minimal dissolved salts, since evaporation on the plant can concentrate such salts on the cooled tissues, creating phytotoxicities to buds, leaves and fruit. Although evaporative cooling might have potential for use on extremely hot days that could endanger cherry fruit quality during ripening, in at least some cultivars, it also may have the potential to cause fruit cracking if the fruit are wetted too much during the sensitive stage of fruit development (see Chapter 7, this volume).

Under-tree microsprinkling lacks significant evaporative cooling potential but can still affect the orchard microclimate. Using two microsprinkler emitter types (with fixed and rotating heads, respectively), Koumanov (2002) reported an average decrease in canopy air temperature ranging from 1.2 to 1.9°C at 2.50 m, 2.4 to 2.6°C at 1.75 m and 3.1 to 3.6°C at 1.0 m above ground. The maximum departures from the reference were 3.7–4.7°C at the canopy base. As a result, the critical threshold of ambient air temperatures at which photosynthesis could be maintained in an under-tree microsprinkler-cooled orchard was shifted up to 40°C. The average increase in relative humidity was 12–13% at 2.50 m, 14–15% at 1.75 m and 16–18% at 1.0 m, with maximum departures from the reference of 20–28%. The microclimatic effect of microsprinkling occurred only after the ambient air temperature exceeded 28°C and the relative humidity dropped to 45%. The effect on microclimate was essentially unaffected by microsprinkler type or by the daily application regime (continuous or intermittent). The effect was positively related to the magnitude of the meteorological factors and to the duration of operation. Drip irrigation had no direct effect on air temperature or humidity.

### *Heat accumulation*

The trapping of heat from daily solar radiation to enhance growing degree units and advance sweet cherry bloom and shoot, leaf and fruit growth is usually accomplished with orchard covers (see section 11.3). This can be particularly attractive economically for early-ripening (i.e. warmer) locations or for extending early offerings at on-farm or local markets. Trials to cover cherry trees for the purpose of advancing fruit development began in 2002 at Klein-Altendorf/Bonn (Balmer and Blanke, 2005b, 2008; Dekova and Blanke, 2007), with similar subsequent work in Norway (Meland *et al.*, 2017), Chile and the USA. Flowering under closed orchard covers was advanced by as much as 16–18 days, followed by an earlier harvest of 12–19 days, depending on cultivar. Lang (2013, 2014) used multiple dates before bloom to cover sweet cherries under high tunnels to stagger their bloom dates by up to 11 days. Shoot growth, leaf area development and fruit ripening were advanced proportionally to covering date. Although the trees covered latest, only 3 weeks before bloom, did not bloom earlier than the uncovered control trees, accumulation of greater heat units post-bloom resulted in new shoot growth that was 88% greater at the end of stage II fruit growth and fruit that was larger and sweeter at the same sampling stage during mid-stage III.

The successful forcing of earlier bloom is dependent on completion of the chilling requirement and endodormancy-to-ecodormancy transition before heat units begin accumulating. Earlier bud development and bloom lead to a higher probability of more severe frost events in many climates, so the severity of potential frosts and the efficacy of frost protection measurements must be considered. In cold, snowy climates, the potential for heavy snowfalls and the structural integrity of the covering system must also be considered when covers are used to advance bloom. Generally, plastic covers are removed during the winter where air temperatures can be low and the potential for mid-winter heat accumulation on sunny days would adversely affect cold acclimatization, or in locations with very mild winters where the accumulation of chilling to break endodormancy could be adversely affected. In a fully enclosed covering system, temperatures on a sunny day can rise dramatically, from –5°C at daybreak to 45°C by midday. Higher orchard covers, such as taller high tunnels, create a greater volume of modified air and therefore moderate potential heat accumulation better than shorter covers; taller covers also let hot air rise further above the tree canopy (Dekova and Blanke, 2007).

### 11.2.3 Protection from rain, hail and wind

The most common microclimate modification goal for cherry orchards is the use of orchard covers during stage III fruit development to reduce rain-induced fruit cracking (Schäfer, 2007; Lang, 2009; Usenik *et al.*, 2009; Lang *et al.*, 2011; Wallberg and Sagredo, 2014; Kafkaletou *et al.*, 2015). If covers are only being used for rain-cracking protection, timing for covering generally begins about 4–6 weeks before harvest. This also allows harvest in the orchard to occur even when it is raining (Meland *et al.*, 2014). Protection from rain also has implications for reduced disease incidence (Børve *et al.*, 2007; Lang, 2009), longer efficacy of some pesticides and improved potential for organic cherry production (Lang *et al.*, 2013). Several types of covers, ranging from pole-and-cable row covers to steel-hoop high tunnels, are commonly used for the purpose of rain exclusion (see section 11.3 for a more in-depth discussion). All types that exclude rain from the root zone must also, therefore, include irrigation systems. The more complete the orchard coverage and more impermeable the cover, the higher the potential for increased relative humidity, especially in the early part of the day.

Most orchard covers provide protection from hail as well as rain, although if rain protection is not a priority, netting covers that transmit rain but capture or divert hail can be installed. All orchard covering materials, from solid plastic to hail netting to bird or insect netting, have the potential to reduce wind exposure of the orchard to varying degrees. This can be particularly important for the production of high-value, yellow-fleshed sweet cherry varieties such as 'Rainier', which readily shows bruises from wind whipping. Reducing cherry orchard exposure to wind, via covering systems or simply via natural or man-made windbreaks (Brandle and Finch, 1991) placed strategically around or on the windward side of the orchard, can not only protect cherry fruit from bruising but also improve tree water relations by reducing evapotranspiration. Wind is also the greatest climatic danger for structural damage to most orchard covering systems. In general, orchard covers least susceptible to wind damage are those that divert wind over the top of the structure. If wind enters the orchard under the covering system, it can create a pressure differential and tremendous lift, much like an airplane wing, that can pull even well-anchored posts out of the ground. For such covering systems, gaps between covering panels or partial hoop venting of high tunnels can reduce the magnitude of potential lift, allowing wind to move with less impedance below the cover.

### 11.2.4 Light modification and other microclimatic considerations

The light microclimate can be increased or decreased in the orchard, with potential impacts on cherry growth and fruit development. Orchard covers alter light transmittance, both in overall quantity and in specific wavelengths, depending on the formulation of the plastic covering material. Many plastics also increase the quantity of dispersed light (Healey *et al.*, 1998). Synthetic orchard floor fabric mulches (e.g. interwoven white plastic, aluminium foil, industrial paper) or materials (e.g. paint, kaolin clay, titanium dioxide) can reflect light not initially intercepted by the tree back up into portions of the canopy, especially decreasing shade in lower portions of the tree.

As with some of the evaporative cooling microclimatic modification goals, the potential to decrease bud, leaf or fruit temperatures involves solar radiation. In low-chill growing areas, growers and researchers have just begun to explore the potential use of orchard covers that could be used to shade cherry trees on sunny winter days to improve chilling. Similarly, for sites at higher latitudes and/or altitudes typically planted with late-ripening varieties, the opportunity to delay fruit development by reducing growing-season heat-unit accumulation via partial shading could provide another strategy to achieve a wider harvest window for the end of the cherry marketing season. Likewise, covers that reduce radiation during flower bud differentiation could lower the maximum air temperatures below thresholds for abnormal pistil development (Beppu and Kataoka, 2000; Whiting and Martin, 2008; Imrak *et al.*, 2014). In addition, sprayable particles to reflect light and reduce bud temperature have been reported to reduce the proportion of double fruit (Whiting and Martin, 2008).

Reports of a reduction in orchard-level solar radiation by covering systems range from 15% in The Netherlands (Balkhoven-Baart and Groot, 2005) to 25% in Michigan (42°N; average April–September; Lang *et al.*, 2011) to 40% in Chile (Wallberg and Sagredo, 2014). Some loss of light, particularly from high light environments, may not have a negative effect on growth; light levels of 1000–1100 µmol photosynthetically active radiation (PAR) measured under orchard covers (Wallberg and Sagredo, 2014) are adequate for photosynthetic saturation of well-exposed leaves (Fig. 11.6) (Tartachnyk and Blanke, 2004). Orchard covers had no significant effect on leaf chlorophyll content (Balmer and Blanke, 2005a; Dekova and Blanke, 2007). Although light is less under orchard covering systems, plant water relations can be improved due to higher humidity and reduced wind. Vegetative growth is generally enhanced, as indicated by longer shoots and greater leaf area per tree (Balmer and Blanke, 2005a; Dekova and Blanke, 2007; Lang, 2009), 35% more shoot elongation and 25% larger leaves (Wallberg and Sagredo, 2014), and 35% greater trunk cross-sectional area (TCSA) (Lang *et al.*, 2011).

Management strategies to compensate for light loss under covers include: (i) north–south row planting; (ii) open-tree or narrow 'fruiting wall' training (Lang *et al.*, 2011); (iii) reflective mulches to increase canopy light distribution (Lang, 2014); and (iv) cover films with increased UV transmission and light dispersal, such as Luminance THB and Lumiso (BPI Visqueen). The recent concept of periodically pruning 'light windows' into tree fruit canopies to allow light penetration either vertically or horizontally (A. Engel, Bonn, Germany, 2015, personal communication) is another potential strategy to partially compensate for the reduction in light under orchard covers.

Reflective mulches (e.g. from Extenday, Auckland, New Zealand) have been tested in cherry orchards with mixed results compared with trials with apple, depending on the plant response examined, solar traits of the location, orchard training system and time of application. Greater fruit effects reported in apple, compared with cherry, may be due, in part, to differences in anthocyanin synthesis in both the skin and flesh of cherry fruit compared with skin only in apples (Schmitz-Eiberger and Blanke, 2012), and due to higher solar angles at the time of cherry harvest in early to mid-summer versus that of apples in the autumn (Funke and

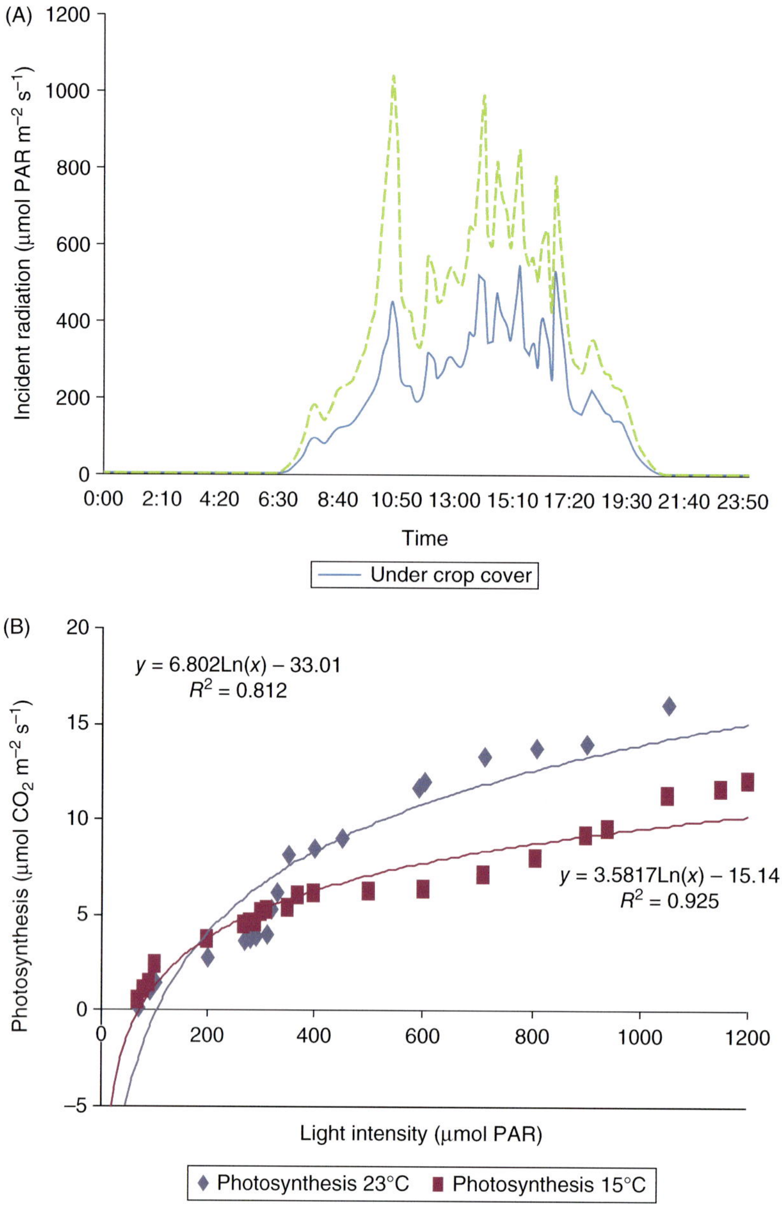


**Fig. 11.6.** Light (A) and the photosynthetic response of cherry leaves to light intensity and temperature (B) in a small high tunnel in April at Klein-Altendorf, Germany. Green and blue lines in (A) represent open and covered cherry orchards, respectively.

Blanke, 2005; Meinhold *et al.*, 2011). The high levels of anthocyanins in red cherry fruit may make it more difficult to detect potential effects of reflected light on cherry colour compared with quantifying red versus non-red skin colour in apple. The reflection of lower angle solar radiation, as in autumn, would be more dispersed within the tree canopy than higher-angle radiation in mid-summer. However, Whiting *et al.* (2008) reported that season-long use of reflective mulch in the high light environments of

Rancagua, Chile, and Washington State, USA, increased cherry shoot length by 32% and TCSA by 90%. The fruit ripened earlier and were firmer. Lang (2014) reported that reflective mulch use in cherry orchards in the lower light environment of Michigan, USA, increased TCSA by 24% in open orchards and by 34% under high tunnels.

In Germany, woven reflective mulches (Folitec and Propex, Agrarfolien Vertriebs GmbH, Westerburg, Germany; PhormiFlex, Bonar Technical Fabrics Co., Zele, Belgium) were spread in the tractor alleys of a high-density orchard (1100 trees ha$^{-1}$) to improve light conditions (Fig. 11.7A). At ripening, measurements indicated that light was increased several-fold in the otherwise most light-limited (i.e. shaded) canopy positions (Fig. 11.7B). In the cool, cloudy climate of Norway, similar reflective mulches (Extenday™) have been tested with the red-fruited 'Sumtare' (Sweetheart™) in high tunnels; no extra effects on fruit colour development and maturity levels were observed (M. Meland, Ullensvang, Norway, 2015, personal communication). Overall, it may be concluded for cherry that reflective mulches may not be as beneficial for improving the colour of red-fleshed cherries as for apples or blushed yellow-fleshed cherries (G.A. Lang, East Lansing, Michigan, USA, 2015, personal communication). However, where canopies are not fully closed over the tractor alley, as in high-density orchards, the use of reflective mulches can improve light penetration into the shaded parts of the canopy from below, thereby improving growth, flower initiation and bud development, particularly in light-limited environments, as in some orchard covering systems.

## 11.3 Orchard Covers

The benefits and limitations of microclimate-modifying orchard covers vary with the type of cover (Table 11.2). Therefore, for such a significant additional production cost, the anticipated market value(s) of the benefit(s) to be gained must be taken into account when considering whether to design a new orchard or retro-fit an existing orchard with covers. Similarly, the impacts of orchard covering systems on microclimate and plant biology should be considered to optimize their use.

### 11.3.1 Types of covers

Covering systems for sweet cherries range from relatively inexpensive plastic sheets or

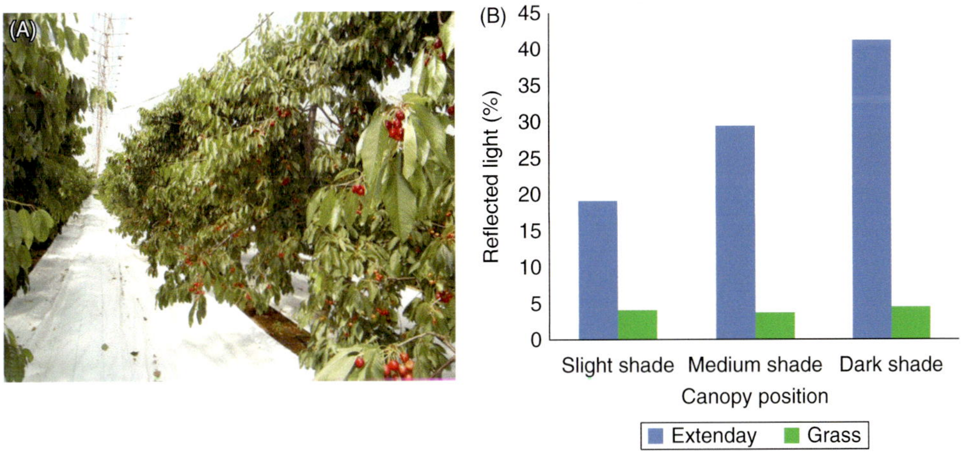


Fig. 11.7. Reflective mulches employed in a sweet cherry orchard near Bonn, Germany (A) and subsequent improvement in reflected light conditions in 'Brändlin Frustar' cherry tree canopies measured in July 2015 at 1 m height at 3:00 p.m. (100% = 1400 µmol PAR m$^{-2}$ s$^{-1}$) (B). (M. Blanke, unpublished data; photo courtesy of A. Engel and M. Blanke.)

**Table 11.2.** Advantages and challenges of using orchard covers in sweet cherry production.

| Potential advantages | Potential challenges |
|---|---|
| Protection from rain, hail and birds (with supplemental netting) | Fruit cracking can still occur if root zone becomes saturated during rain or by irrigation |
| Healthier trees; better growth; more rapid return on investment | High investment (structure) and maintenance (periodic replacement of plastic) costs; intensive annual labour for installation and removal of covers |
| Higher and more consistent marketable yields | Protection from frost requires supplemental heat; specialized pollinators may be needed |
| Ability to alter ripening window to increase or extend market value | Excessive heat before or during bloom can cause floral malformation and/or reduce fruit set |
| Larger fruit size; more glossy appearance; higher sugar content | Excessive heat during ripening may reduce fruit quality (lower firmness, soluble solids) |
| Improved potential for organic production; reduced fungicide use | Lower light transmission, especially as the plastic ages, and excessive vegetative growth may increase shade issues |
| Harvest and pruning is possible during rain; more efficient management | Structures must be managed to prevent damage to the structure or plastic from wind or snow |

nets temporarily clipped to post-and-wire support (or pole-and-cable) structures, to plastic-covered small (individual rows, e.g. Balmer, 1998; Meland *et al.*, 2014) or large (high tunnels, e.g. Haygrove Ltd, Ledbury, UK) steel-hoop structures, to expensive glasshouse or plastic greenhouse-like structures that can more precisely manipulate multiple environmental factors for potential off-season production. Material costs of orchard covers and the labour needed to install and remove them are discussed by Overbeck *et al.* (2013).

In their simplest form, pole-and-wire or pole-and-cable structures are used to cover many orchard rows at a time with plastic bird netting to protect the fruit from bird predation, or with more dense plastic netting to protect the fruit and trees from hail damage. Alternatively, this construction technique can be used to cover single rows with solid plastic tent-like sheeting (Fig. 11.8A) primarily to protect fruit from rain-induced cracking (Meland and Skjervheim, 1998). Such covers generally are installed each spring after fruit set and removed after harvest, thereby altering the microclimate for only 3–6 weeks or so. Posts are usually made of treated wood, steel or concrete. Row covers for rain protection are usually solid plastic, which keep trees dry but can capture hot air under the covers during ripening in warm climates. The Norwegian three-wire system has overhead plastic curtains that can slide back and forth to open and close manually. In case of rain or other climatic threats, the curtains can be closed. If the wind speed approaches 30 m $s^{-1}$, major damage can be expected on various covering systems, including tunnels (see below). The Norwegian system is constructed such that the plastic cover is the weakest point, which is easier and cheaper to replace than the main frames of poles, wires and transverse woodwork. More complex specialized covers constructed of netting with overlapping plastic panels attached at the upper side of the plastic also can be obtained (e.g. Voen Vöhringer GmbH & Co KG, Berg, Germany), which provide passive venting of rising hot air and some wind, while retaining protection from rain.

High tunnels can be stand-alone single-bay structures for covering a couple of rows in very small orchards or, more typically, multi-bay structures for covering any size of orchard (Fig. 11.8B). High-tunnel bays are usually comprised of steel legs augured into the soil and connected to the next row of legs by a steel hoop 7–10 m wide. Two or three rows of cherry trees are usually planted per bay. Each bay is covered with a solid plastic sheet, which can be sourced with different specifications for light spectra transmittance (e.g. varying in transmittance of PAR, UV light, infrared radiation, etc.). The plastic is usually held in place by ropes that

Fig. 11.8. A three-wire covering system (A) and multi-bay polyethylene tunnels (B) in Norway for prevention of fruit cracking.

criss-cross the plastic from one hoop to the next. The plastic is usually removed during winter in locations where accumulating snow loads would be a concern for structural limits or where daily heat trapped during the winter would preclude alleviation of endodormancy (Luedeling *et al.*, 2013b).

If the hoop connections are designed for it and the plastic is fixed to remain fully extended during the season, gutters can be installed where the hoops meet, providing a means for removing rainwater runoff and preventing it from entering the root zone of the nearest trees. If gutters are not installed, the plastic can be pushed up the hoop during the season to provide some venting and cross-air flow, but with the trade-off of allowing rain to run off into the soil between the tunnel legs and the root zone of the adjacent trees. In cool rainy climates like Norway, preventing root zone saturation might be more important than venting of excessive heat that can collect quite readily under the hoop. In hotter climates, the ability to vent heat daily during fruit ripening may be the higher priority, and soil drainage tiles can be installed to handle rainwater runoff. Venting in multi-bay tunnels is usually done manually by pushing the plastic some distance up the side of the hoop. Improved venting also can be achieved passively with the use of some specialty plastics that have a net panel sown in at the peak (top of the tunnel) or offset to one side. This allows heat to be vented passively from or near the highest point, with the trade-off that some rain will enter the tunnel under the net vent, which may alter tree row planting design for avoidance of canopy wetting. Also, while improved permanent venting is desirable during ripening, the loss of ability to capture heat units early in the spring would preclude the use of such hybrid net/plastic tunnels as an orchard covering system for significantly advancing bloom and fruit development. High tunnels also generally require the use of bird or insect exclusion netting (if desired) at all openings (doors, sides and vented troughs where the hoops meet) to achieve protection from birds or insects.

The greatest level of control over microclimate is achieved by growing cherries in glasshouses or greenhouse-like structures with automated (electronic) controls for venting and heating, usually controlled by a thermostat or inputs from climatic sensors (e.g. air temperature, relative humidity, wind speed, precipitation, solar radiation). Such an investment requires extremely high-value markets. For example, sweet cherries produced in roof-vented greenhouses in Lleida province, Spain, command a very high economic return by filling the market void in cherry availability between mid-March and late April (http://www.glamour-edoa.com/cherries). Small-scale greenhouse cherries have also been produced in New Zealand, The Netherlands and Michigan, USA (G. Lang, Michigan, USA, 2015, personal communication). In Norway, greenhouses used previously for tomatoes are under test for highly intensive (10,000 trees $ha^{-1}$) sweet cherry production that offers more programmable environmental control, larger fruit and improved tree yields (M. Meland, Ullensvang, Norway, 2015, personal communication). Greenhouse production has been accomplished both with trees planted in the ground and with trees in pots or bags, the latter of which requires a high level of precision in irrigation and fertigation for water and nutrient supply to the limited root systems. Greenhouse production in pots allows groups of trees to be held dormant in refrigeration and moved into the greenhouse to provide a progression of bloom and ripening to target high-value market windows.

Plastic-covered greenhouse-like structures with programmable automated retractable roofs and sides also have been studied since 2012 at Michigan State University (Cravo Equipment Ltd, Brantford, Canada, http://www.cravo.com/) and have recently been adopted by commercial specialty market cherry producers in South Africa and Australia (G.A. Lang, East Lansing, Michigan, USA, 2015, personal communication). The retractable roof provides opportunities for full sunlight in moderate light environments such as northern Europe or the midwest USA, partial shading in high light environments such as California or Spain, open skies for honeybee pollination (see section 11.5.1), automated daily opening and closing to optimize chilling unit accumulation, spring heat-unit accumulation, and rain or hail protection. While significantly more expensive in establishment costs than pole-and-cable covers or high tunnels, automated greenhouse-like structures tend to have lower labour and maintenance (e.g. plastic replacement) costs.

An area of future potential microclimate modification is covering systems that selectively alter orchard-level light spectra. Coloured netting (i.e. selective transmission of specific light spectra such as blue (400–500 nm), red (600–700 nm) or far-red (700–800 nm) radiation) has been studied in apple (e.g. Solomakhin and Blanke, 2010; Bastias *et al.*, 2012) and peach (e.g. Rapparini *et al.*, 1999), but few investigations have reported any potential benefits of coloured netting use in sweet cherry. Studies are more common in herbaceous crops like vegetables, and the few in tree fruits have reported possible positive effects on fruit size or tree vigour, but such studies are challenging and expensive to conduct, and no definitive conclusions have yet been drawn for consistent impacts at an orchard scale.

### 11.3.2 Varieties, rootstocks and training systems for covered orchards

The high capital costs of orchard covering systems require precision in orchard design for optimized space utilization and in irrigation/fertigation, since exclusion of rain can adversely impact nutrient availability. Tree canopies must be developed for optimized light interception and distribution, since covers may reduce overall available light. Orchard management must be innovative and efficient, including the potential for orchard task mechanization when possible to reduce labour costs.

The choice of cultivars for production under orchard covers is more a matter of what the market demands than any specific fruit trait. Covers may provide the opportunity to produce marketable fruit from cultivars not typically grown in a particular region if the cover minimizes a critical limiting factor, such as the ability to grow cultivars with high rain-cracking susceptibility in rainy climates (although some cultivars, such as 'Brooks', have been observed to crack even in the absence of direct contact with rain water), or to grow extremely early-blooming cultivars in a region with frequent spring frosts. Since the drier climate under tunnels may increase susceptibility to powdery mildew (see section 11.5.3), cultivars with genetic resistance to mildew (Olmstead *et al.*, 2001) would be desirable.

Since well-managed cherry orchards under covers generally increase tree health, vigour, and the soil nitrogen mineralization rate and availability in spring (Balmer and Blanke, 2005a; Dekova and Blanke, 2007; Lang *et al.*, 2011), the use of vigorous to semi-vigorous rootstocks such as 'GiSelA 6', 'GiSelA 12', 'Krymsk 5', 'Colt', 'CAB6P', Mazzard and Mahaleb seedling may result in excessively large trees that become difficult to contain under the covers. Except on poor soils, trees on the dwarfing to semi-dwarfing rootstocks 'GiSelA 3' and 'GiSelA 5' have been more suitable for developing small-statured trees. Unlike the equivalent situation for apple with 'Malling 9' rootstock, these cherry rootstocks generally need no support unless the training system requires it. Furthermore, these rootstocks are very precocious, providing early yields for a more rapid return on investment. Thus, the strategy of planting a new orchard under an established covering system, rather than constructing the covering system over an orchard once it begins fruiting, will result in better growth, the development of more fruit-bearing wood earlier and therefore a higher production earlier (Lang, 2013). Where soils or climate are not suitable for dwarfing to semi-dwarfing rootstocks, semi-vigorous rootstocks may be combined with multiple-leader training systems and precocity-inducing horticultural techniques (such as tying limbs below the horizontal and/or deficit irrigation) to reduce vigour and hasten fruiting (see Chapter 12, this volume). If trees planted under cover become too vigorous, additional strategies to reduce vigour include the use of growth inhibitors, such as paclobutrazol (Facteau and Chestnut, 1991) or prohexadione calcium (Elfving *et al.*, 2003), where legal, or root pruning (Ferree, 1992).

Typical orchard design for covers depends on the type of covering system, although all designs should be as space efficient as possible to optimize yield for returns on investment. Orchards using pole-and-cable row covers should match the cover and tree canopy width. Tree row spacing for systems that cover multiple rows must account for the spacing of tunnel legs or greenhouse support posts, equipment access, and uniformity of spray coverage and light distribution. Common high-tunnel planting designs include two tree rows with a middle tractor alley or a double tree row in two-thirds of the tunnel with the tractor alley to one side. In terms of three-dimensional tree canopies, single-leader spindle trees tend to have better light distribution and space efficiency than multiple-leader trees. As growers have increasingly adopted dwarfing rootstocks and mechanical hedging, three narrow rows of tall spindle-type trees are increasingly common (Fig. 11.9).

Furthermore, with the advent of fruiting wall-type tree canopy architectures (see Chapter 12, this volume) such as the upright fruiting offshoots (UFO) (Zhang *et al.*, 2015) and the super slender axe (Musacchi *et al.*, 2015) systems, covered space efficiency and light distribution uniformity can be further optimized to increase yield potential. For example, a greenhouse-like structure with support posts 8.0 m apart (or a high tunnel with legs 8.0 m apart) could accommodate three rows of narrow UFO trees and access for narrow-gauge tractors, sprayers and/or orchard platforms (Fig. 11.10) with uniform light

**Fig. 11.9.** A three-row sweet cherry orchard under 8.5 m high tunnel covers at bloom (A) and pruned with a mechanical hedger (pruned row on the right, unpruned centre row on the left) (B). (Photos courtesy of M. Blanke.)

distribution throughout the canopy. Assuming a continuous fruiting wall at maturity, the fruiting area/orchard area ratio (an estimate of cropping potential) would be 10 m/8 m = 1.25. By retaining the narrow fruiting wall canopy architecture but expanding its light interception via angled walls, the cropping potential for a similar 8.0 m wide high-tunnel or greenhouse-like structure could be increased to 14 m/8 m = 1.75 (Fig. 11.11).

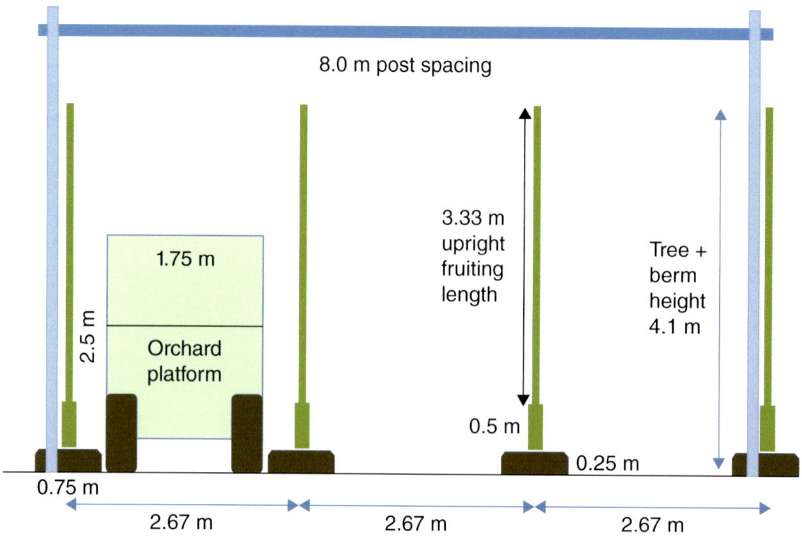


**Fig. 11.10.** Sweet cherry orchard design with trees trained to an upright fruiting offshoots (UFO) fruiting wall canopy architecture for a greenhouse-like covering structure having support posts 8 m apart (Figure courtesy of G.A. Lang.) The upright fruiting area/orchard area ratio is 10/8 = 1.25.

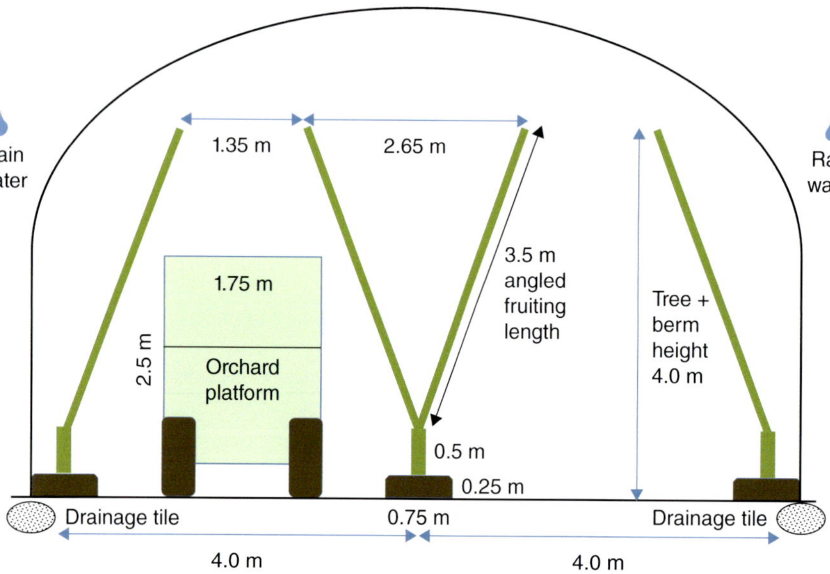


**Fig. 11.11.** Sweet cherry orchard design with trees trained to a V-shaped upright fruiting offshoots (V-UFO) angled fruiting wall canopy architecture for an 8.0 m wide high-tunnel covering structure (Figure courtesy of G.A. Lang.) The angled fruiting area/orchard area ratio is 14/8 = 1.75.

Note that both orchard designs utilize tree planting on raised beds, and the high-tunnel design includes installation of soil drainage tiles to improve drainage of rain runoff away from the root zone. Additional management efficiencies might be realized with the use of the covering structure for support of a solid-set canopy delivery spray system to replace the tractor sprayer (Lang, 2014).

## 11.4 Effects of Orchard Covers on Fruit Production

### 11.4.1 Flowering and fruit set

The early covering of orchards to increase heat-unit accumulation and advance bloom, fruit set and ripening was discussed in section 11.2.2. It should be noted, however, that when covers are used only to prevent rain-induced fruit cracking (i.e. deployed primarily during stage III fruit growth), there may be little effect on ripening date (Schäfer, 2007; Kafkaletou *et al.*, 2015). Covering at bloom must be managed carefully, since fully enclosed orchards can become very hot quickly on sunny days. When a 2008 high-tunnel cherry study in Michigan was first fully enclosed in mid-March, the temperature inside increased from –5°C at daybreak to 46°C within 6 h (G. Lang, Michigan, USA, 2015, personal communication). Subsequent flowering was abnormal and fruit set was poor. Hedhly *et al.* (2007) reported that heat stress at bloom of 5–7°C higher maximum (1–3°C higher average) temperatures reduced sweet cherry fruit set due to accelerated pollen tube growth, a reduced number of pollen tubes growing in the style, ovule degeneration and shorter stigmatic receptivity. Therefore, the goal of advancing bloom and ripening requires knowledgeable management to (i) successfully begin accumulating growing degree units only after the chilling requirement has been met; (ii) protect from possibly higher risks of spring frost; and (iii) prevent excessive heat during final floral organ differentiation, bloom and fertilization.

### 11.4.2 Yield and fruit size

The effect of orchard covers on sweet cherry yields and fruit size is dependent on many factors, and reports have varied widely. Typically, fruit size is related inversely to crop load, so any large effects on yield are likely to inversely affect fruit size. A high-density (1250 trees $ha^{-1}$) high-tunnel planting of 'Sweetheart' on 'Colt' in Norway yielded 11 and 24 t in the fourth and fifth leaf, respectively; 31% of the fourth leaf harvest was >34 mm, but only 4% of the fifth-leaf harvest was >34 mm (Meland *et al.*, 2017). Factors such as climate, variety, exposure to spring frosts, pollination conditions and pruning can have significant and variable effects on crop yields exclusive of orchard covers. In the majority of reports (Schäfer, 2007; Lang, 2009; Usenik *et al.*, 2009; Lang *et al.*, 2011; Schmitz-Eiberger and Blanke, 2012; Overbeck *et al.*, 2013; Wallberg and Sagredo, 2014; Meland *et al.*, 2017), cherry fruit size increased under orchard covers (e.g. Table 11.3). Lang (2009) connected initially lower yields under high tunnels with poor pollination by honeybees; the following year, with bumblebee pollinators, yields were equal or slightly higher under covers, and fruit size was slightly larger even with the higher yields. The following year, yields were higher in the tunnels due to frost damage outside (Lang *et al.*, 2011). Similar yield variability has been seen more recently under retractable-roof greenhouse-like structures and pole-and-cable row covers, which are attributable to variations in late winter freezes, spring pollination conditions, spring frosts, and the effect of postbloom rains on disease and cracking incidence (G.A. Lang, Michigan, USA, 2015, personal communication).

Yields over several years in a small research tunnel in Germany have tended to be lower (Balmer and Blanke, 2005a, 2008; Dekova and Blanke, 2007) or comparable to outside yields (Kafkaletou *et al.*, 2015). Others have reported that yield was unaffected by covers in a warm and sunny climate (Usenik *et al.*, 2009), decreased under a temporary orchard cover (Schäfer, 2007) or increased in a large tunnel (Overbeck *et al.*, 2013). It is clear that, in most situations,

**Table 11.3.** Quality of sweet cherry fruit grown under small orchard cover versus in the field. Results are given as mean ± standard error. (Modified from Dekova and Blanke, 2007.)

| | Diameter (mm) | | Fruit colour (°hue value) | |
|---|---|---|---|---|
| Cultivar | Under cover | Open field | Under cover | Open field |
| 'Earlise®Rivedel' | 26.4±0.5 | 26.8±0.4 | 16±0.4 | 29±0.5 |
| 'Burlat' | 28.9±0.3 | 25.6±0.6 | 12±0.5 | 12±0.4 |
| 'Souvenir' | 29.3±0.7 | 24.9±0.8 | 9±0.3 | 23±0.8 |
| 'Sumste' (Samba™) | 27.0±0.8 | 25.4±0.7 | 16±0.6 | 21±0.7 |
| 'Prime Giant' | 29.2±0.4 | 29.9±0.5 | 9±0.7 | 20±0.6 |

yields under orchard covers can be maintained or increased relative to uncovered orchards, provided the grower can anticipate the factors likely to potentially be limiting, and can manage the covered orchard accordingly (e.g. by temperature management, pollinator choice, frost protection).

The proportion of the fruit development period under cover also can affect fruit quality. In Norway, fruit grown in tunnels covered from flowering to harvest were significantly larger on all harvest dates compared with those from tunnels covered from straw colour to harvest (Table 11.4; Meland *et al.*, 2017). In contrast, fruit were significantly firmer when covered for the shorter duration.

### 11.4.3 Fruit cracking and shelf-life

The goal of preventing cherry fruit cracking requires management of orchard covers to exclude precipitation from contact with the fruit, divert precipitation from saturating the root zone, and supply adequate and consistent water for growth via irrigation, preferably drip irrigation (Balmer and Blanke, 2005a). Such management can lead to impressive reductions in cracking, from double to single digits (Børve *et al.*, 2003; Dekova and Blanke, 2007; Usenik *et al.*, 2009; Lang *et al.*, 2011; Kafkaletou *et al.*, 2015). However, covered orchard trials in various countries also have shown that cracking can still occur under orchard covers when water intrudes via the soil, for example when pole-and-cable row covers shed water into the alley between tree rows, or when high tunnels lack gutters, or the gutters or orchard soil drainage systems cannot remove water fast enough and overflow occurs. It is even possible in some situations for covered fruit to have a higher incidence of cracking than uncovered fruit, which Wallberg and Sagredo (2014) attributed to a higher frequency of condensation on the fruit under covers.

In severe storm conditions or frequent rain events, cherry cracking losses can be devastating, even under orchard covers. Lang (2013) reported unprotected cherry cracking levels of 89 and 91% for 'Rainier' and 'Lapins', respectively, while under adjacent high tunnels with no gutters, the respective rates of cracking were 60 and 32%, even though the fruit was never wet, due to root zone saturation from rainwater shed by the covers directly to the soil between the tunnel arches. For most operations, cherries with >35% cracking are not worth picking for marketing. In such instances, even fruit that have no visible cracks are likely to have a high proportion of microcracks, which can be exacerbated by high relative humidities (Peschel and Knoche, 2005) and can lead to poor postharvest performance or cracking during packing, storage and/or shipment. In general, however, as found by Kafkaletou *et al.* (2015), orchard covers have a positive effect on cherry fruit postharvest performance in terms of less fresh weight loss, less cracking and lower respiration.

### 11.4.4 Fruit sugar, organic acids, acidity, firmness and stem quality

Fruit from trees under orchard covers have been reported to have higher total soluble

**Table 11.4.** Effect of time of high-tunnel covering on fifth-leaf 'Sweetheart' cherry fruit weight and firmness at four harvest dates at Lofthus, Norway (2009). (Data from Meland *et al.*, 2017.)

| Fruit characteristic | Harvest date in August | Time of covering: Flowering to harvest | Time of covering: Straw colour to harvest | *F*-test |
|---|---|---|---|---|
| Fruit weight (g) | 14 | 11.8 | 10.9 | $P<0.05$ |
| | 19 | 12.6 | 11.4 | $P<0.001$ |
| | 24 | 11.8 | 10.7 | $P<0.05$ |
| | 28 | 12.5 | 10.9 | $P<0.01$ |
| Fruit firmness | 14 | 65 | 72 | $P<0.001$ |
| (Durofel units) | 19 | 66 | 74 | $P<0.01$ |
| | 24 | 62 | 70 | $P<0.001$ |
| | 28 | 66 | 75 | $P<0.001$ |

solids (TSS) levels (Børve *et al.*, 2003; Schäfer, 2007; Overbeck *et al.*, 2013), although the effects of covers on ripening can make direct comparisons with uncovered fruit challenging due to differences in ripeness. In an early comparison of different types of covers applied during ripening, fruit from trees under covers that retained higher temperatures and relative humidity had lower TSS and colour than those from under other covers (Børve *et al.*, 2003). Fruit from covered orchards contained more sugars (glucose, fructose and sorbitol) and more organic acids, but the difference values were not statistically significant (Usenik *et al.*, 2009). Schäfer (2007) found less titratable acid (TA) in covered fruit, and Kafkaletou *et al.* (2015) reported comparable TSS/TA ratios >25/1, a value suggested by Schmitz-Eiberger and Blanke (2012) for good eating quality and attractive consumer appeal in terms of taste, in both types of orchards. As with fruit size, microclimate effects on fruit firmness are similarly influenced by crop load and climate. The firmness of fruit under orchard cover was unaffected in a hot climate (Kafkaletou *et al.*, 2015). Firmer fruit of 'Regina' were reported under a temporary orchard cover (Schäfer, 2007), as was the fruit of five cultivars that were covered early to advance bloom and ripening (Schmitz-Eiberger and Blanke, 2012; Overbeck *et al.*, 2013). However, softer fruit under covers has been reported by Balmer and Blanke (2005a) and Dekova and Blanke (2007) in Bonn, Germany, as well as for early-covered 'Lapins' in Chile (Wallberg and Sagredo, 2014). G.A. Lang (East Lansing, Michigan, USA, 2015, personal communication) found that sugar and acid levels and fruit firmness varied widely from year to year between open orchards and those with various types and timings of covers (high tunnels, vented row covers and retractable-roof greenhouse-like structures). More studies are needed to understand and best utilize the various techniques for microclimate modification, including irrigation scheduling, for consistently optimized fruit quality parameters such as sugars, acids and firmness. To date, only Kafkaletou *et al.* (2015) has examined potential cover effects on green pedicel colour, finding none of consequence.

Dekova and Blanke (2007) reported that, under orchard covers managed to advance bloom and ripening, early-ripening cherry cultivars produced fruit of the same firmness with and without covers, whereas two late-ripening varieties developed firmer fruit without covers; this was attributed to the interaction between microclimate and fruit development period. Meland *et al.* (2017) reported that fruit from trees covered from straw colour until harvest were firmer (Table 11.4) and had higher TSS than fruit covered from flowering until harvest.

### 11.4.5 Fruit colour and human health compounds

Studies of the impact of orchard covers on sweet cherry fruit colour have reported lower (Børve *et al.*, 2003; Dekova and

Blanke, 2007; Schäfer, 2007; Wallberg and Sagredo, 2014) and higher (Schmitz-Eiberger and Blanke, 2012; Overbeck *et al.*, 2013; Kafkaletou *et al.*, 2015) colour values, but for the majority of cultivars that are dark red to dark purple in colour, the differences detectable with laboratory equipment are relatively meaningless to consumer preferences. Indeed, a shiny, glossy skin that tends to be obtained more often under protective covers is likely to rate higher for consumers than minor differences in analytical laboratory colour values. Orchard cover impact on fruit colour mainly is significant when blushed yellow-fleshed cherries, such as 'Rainier', are produced for premium markets (Lang, 2009). Depending on the formulation of the cover plastic, the premium reddish blush that forms in the skin of most yellow-fleshed cherries can be reduced (Mulabagal *et al.*, 2009), particularly by plastics with reduced transmission in the UV spectrum. However, many cover plastics also increase light diffusion, which can improve the proportion of the fruit with blush development.

Orchard covers had no effect on the concentration of the four predominant anthocyanins (cyanidin 3-glucoside, cyanidin 3-rutinoside, pelargonidin 3-rutinoside and peonidin 3-rutinoside) in sweet cherry fruit (Usenik *et al.*, 2009; Kafkaletou *et al.*, 2015). 'Regina' fruit contained higher concentrations of these four anthocyanins under orchard cover, whereas the opposite was found for 'Kordia' (Usenik *et al.*, 2009). However, antioxidative potential was unaffected (Kafkaletou *et al.*, 2015). The yellow-fleshed 'Rainier' normally develops a red blush due to cyanidin 3-*O*-rutinoside, but did not under high-tunnel covers (Mulabagal *et al.*, 2009). However, these tunnel-grown fruit had the highest activity against lipoxygenase and cyclooxygenase enzymes due to higher levels of ursolic, coumaric, ferulic and caffeic acids as well as β-carotene. In another study with tunnels, vitamin C (ascorbic acid), total phenolics, antioxidative potential and flavonoids, as well as anthocyanins, were enhanced under covers (Schmitz-Eiberger and Blanke, 2012; Overbeck *et al.*, 2013).

Sweet and sour cherries can elicit an allergic response in some people, caused by one or more cherry allergens such as Pru av 1, Pru av 2, Pru av 4 and cherry lipid transfer protein (LTP). To determine whether cultivation under covers would stimulate Pru av 1 in sweet cherry fruit, Schmitz-Eiberger and Blanke (2012) examined Pru av 1 levels in five cultivars grown with or without covers. The allergen was below the detection level with or without covers, possibly due to a lack of water stress or lack of pathogen attack preventing significant induction of the allergen.

## 11.5 Effects of Orchard Covers on Insects and Diseases

While orchard covers most often are used for protection from rain-induced fruit cracking or to manipulate ripening, in some climates a significant secondary benefit is their use throughout the growing season, which can reduce the incidence of several diseases that are promoted by rain, such as cherry leaf spot (*B. jaapii*) (see Chapter 14, this volume) and bacterial canker (*P. syringae*) (see Chapter 15, this volume). Microclimatic modifications that alter light, temperature and rainfall most certainly can affect greater orchard ecology with regard to insects (both good and bad), as well as diseases.

### 11.5.1 Beneficial insects

Excessive heat during bloom under covers has already been discussed with regard to the effects on floral organ development and effective pollination period longevity. If covers create hot, humid conditions during bloom, pollen grains can stick together, making collection and transport by insects less successful. Excessively dry conditions under covers may reduce floral nectar production and make flowers less attractive to pollinators (Wittmann *et al.*, 2005; Hamm *et al.*, 2007). Since honeybees (*Apis mellifera*) utilize UV and polarized UV light to navigate between the hive and their food sources (flowers), the alteration of light spectra by plastic covers can disorient honeybee foraging behaviour (Lang, 2009). This can be a greater problem in fully covered orchard systems, such as high tunnels, than in row covering systems that have significant gaps providing access to the open sky between the covers. Other potential supplemental pollinators for covered sweet cherries include bumblebees (*Bombus* spp.), solitary bees such as *Osmia* spp. and *Andrena* spp., and other native insects such as syrphid flies (family Syrphidae). Bumblebees are available commercially and have the advantages of foraging at cooler temperatures as well as having a higher foraging efficiency than honeybees.

Covering the orchard during bloom can provide a warmer microclimate and protection from the wind for improved pollinator activity where spring weather is often cool and windy. When using supplemental hives of honeybees or bumblebees, the hives should be introduced more than 48 h prior to bloom under the covers to allow time for pollinator orientation. Red solitary bees (*Osmia rufa*) normally are overwintered dry and frost free in aged timber or stones with ~8 mm boreholes, warmed up 2–3 weeks prior to flowering and then brought in and located under the orchard covers before flowering (Wittmann *et al.*, 2005).

### 11.5.2 Insect and other arthropod pests

Historically, the predominant insect pest of cherry has been the European fruit fly (*Rhagoletis cerasi*) and American fruit fly (*Rhagoletis cingulata*), and more recently, the spotted wing fruit fly (*Drosophila suzukii*) (see Chapter 13, this volume). These pests thrive under orchard covers, although the covering structures can also serve as supports for dense fruit fly exclusion netting (0.9 × 0.9 mm mesh size) to reduce infestations (Schäfer, 2007; C. Daniel, Frick, Switzerland, 2015, personal communication). These have the added advantage of providing protection from bird damage, although the density of the mesh makes management of excessive heat challenging. Secondary infection from soil-borne larvae emergence can be partially controlled or reduced by using drip irrigation in the tree row and leaving alleys unirrigated. This was found to be as effective as chemical sprays, and therefore may be a key management tool for organic production. The use of woven weed barrier fabric in the tree row can provide further exclusion of fruit fly (and other) fruit-originating larvae from completing their life cycle in the moist soil strip (G.A. Lang, East Lansing, Michigan, USA, 2015, personal communication).

In the eastern USA, the incidence of Japanese beetle (*Popillia japonica*) damage to sweet cherry trees was reduced significantly (>90%) under high tunnels, while some pests such as plum curculio (*Conotrachelus nenuphar*), oblique-banded leaf roller (*Choristoneura rosaceana*), red-banded leaf roller (*Argyrotaenia velutinana*) and tent caterpillars (*Malacosoma americanum*), remained unaffected (Lang *et al.*, 2011, 2013). Other arthropod pests, such as black cherry aphid (*Myzus cerasi*) and spider mites (e.g. *Tetranychus urticae*), can thrive in the dry hot microclimate under orchard covers and may become significant problems if integrated pest management (see Chapter 13, this volume) is not practised. Biological control measures such as the introduction of lady bird beetles (family Coccinellidae) and lacewings (family Chrysopidae) for aphids, and predatory mites (e.g. *Typhlodromus* spp.) for pest mites, can be utilized successfully under covers.

### 11.5.3 Diseases

Depending on the combination of orchard cover type, irrigation type (drip versus microsprinkler) and venting management, the covered orchard microclimate can significantly reduce some disease pressures as well as prolong efficacy of protective fungicide residues (Børve *et al.*, 2007). In the USA, the incidence of cherry leaf spot was essentially eliminated under all covering systems that minimized leaf wetness and exposure to rain (Lang, 2009; Lang *et al.*, 2011). The incidence of bacterial canker also was greatly reduced by eliminating rain-dissemination of *Pseudomonas* bacteria during key infection points, such as bloom in spring, leaf drop in autumn and during pruning (Lang, 2014), although uncovered orchards during winter remain a potential time for infection.

Børve *et al.* (2003) reported a fivefold reduction in fruit decay incidence under covers. While lower infection rates for European brown rot and blossom blight (*Monilinia* spp.) have been observed under isolated high tunnels (M.M. Blanke, Bonn, Germany, 2015, personal communication), Lang *et al.* (2011) reported that American brown rot (*Monilinia fructicola*) remains a common and pernicious disease of sweet cherry under multi-bay tunnels, requiring chemical controls and serving as a limitation for organic production in the eastern and midwestern USA. Additionally, powdery mildew (*Podosphaera clandestina*) is typically a minor disease in the midwestern USA, but can become more significant in the dry environment of high-tunnel cherries (Lang *et al.*, 2011). Cultivars with genetic resistance to powdery mildew (Olmstead *et al.*, 2001; Olmstead and Lang, 2002) have performed well under covers and are particularly useful for potential organic production (G.A. Lang, East Lansing, Michigan, USA, 2015, personal communication).

## 11.6 Research Needs, Trends and Outlook

The manipulation of one component of the orchard microclimate, such as a cover to provide protection from rain, has many secondary effects across other microclimatic and biological factors. Therefore, many potential studies remain to be conducted for many orchard microclimate modification technologies such as the myriad types of covering systems or seasonal uses of evaporative cooling described in this chapter. Physiological responses include tree and fruit water diurnal status and turgor, leaf photosynthesis, tree growth and flower bud initiation due to altered light wavelengths, as well as interactive impacts on localized orchard pests, diseases, beneficial insects and pesticide efficacy/residues. All microclimate modification technologies add to production expenses, so integration of space-efficient orchard designs and labour-saving technologies with microclimate modification techniques is critical to their economically successful adoption.

## References


Baldocchi, D. and Waller, E. (2014) Winter fog is decreasing in the fruit growing region of the central valley of California. *Geophysical Research Letters* 41, 3251–3256.

Balkhoven-Baart, J.M.T. and Groot, M.J. (2005) Evaluation of 'Lapins' sweet cherry on dwarfing rootstocks in high density plantings, with or without plastic covers. *Acta Horticulturae* 667, 345–351.

Balmer, M. (1998) Preliminary results on planting densities and rain covering for sweet cherry on dwarfing rootstock. *Acta Horticulturae* 468, 433–439.

Balmer, M. and Blanke, M.M. (2005a) Developments in high density cherries in Germany. *Acta Horticulturae* 667, 273–277.

Balmer, M. and Blanke, M.M. (2005b) Verfrühung von Süßkirschen unter geschlossener Folie [Forcing sweet cherry under closed cover]. *Erwerbs-Obstbau* 47, 78–86.

Balmer, M. and Blanke, M.M. (2008) Cultivation of sweet cherry under cover. *Acta Horticulturae* 795, 479–484.

Bastias, R.M., Manfrini, L. and Corelli Grappadelli, L. (2012) Exploring the potential use of photoselective nets for fruit growth regulation in apple. *Chilean Journal of Agricultural Research* 72, 224–231.

Beppu, K. and Kataoka, I. (2000) Artificial shading reduces the occurrence of double pistils in 'Satohnishiki' sweet cherry. *Scientia Horticulturae* 83, 241–247.

Beppu, K., Ikeda, T. and Kataoka, I. (2001) Effect of high temperature exposure time during flower bud formation on the occurrence of double pistils in 'Satohnishiki' sweet cherry. *Scientia Horticulturae* 87, 77–84.

Børve, J., Skaar, E., Sekse, L., Meland, M. and Vangdal, E. (2003) Rain protective covering of sweet cherry trees – effects of different covering methods on fruit quality and microclimate. *HortTechnology* 13, 143–148.

Børve, J., Meland, M. and Stensvand, A. (2007) The effect of combining rain protective covering and fungicide sprays against fruit decay in sweet cherry. *Crop Protection* 26, 1226–1233.

Brandle, J.R. and Finch, S. (1991) *How Windbreaks Work*. Papers in Natural Resources, Paper 121. DigitalCommons@University of Nebraska – Lincoln. Available at: http://digitalcommons.unl.edu/natrespapers/121 (accessed 17 January 2017).

Cal-Adapt (2017) Temperature: extreme heat tool. California Energy Commission, California. Available at: http://cal-adapt.org/temperature/heat/ (accessed 16 January 2017).

Dekova, O. and Blanke, M.M. (2007) Verfrühung von Süßkirschen im Folienhaus [Forcing sweet cherry in a polytunnel]. *Erwerbs-Obstbau* 49, 10–17.

Elfving, D.C., Lang, G.A. and Visser, D.B. (2003) Prohexadione-Ca and ethephon reduce shoot growth and increase flowering in young, vigorous sweet cherry trees. *HortScience* 38, 293–298.

EPA (2016) Climate change indicators: weather and climate. US Environmental Protection Agency, Washington, DC. Available at: www3.epa.gov/climatechange/science/indicators/weather-climate/index.html (accessed 16 January 2017).

Facteau, T.J. and Chestnut, N.E. (1991) Growth, fruiting, flowering, and fruit quality of sweet cherries treated with paclobutrazol. *HortScience* 26, 276–278.

Fernández, R.T. and Flore, J.A. (1998) Intermittent application of $CaCl_2$ to control cracking of sweet cherry. *Acta Horticulturae* 468, 683–689.

Ferree, D.C. (1992) Time of root pruning influences vegetative growth, fruit size, biennial bearing, and yield of 'Jonathan' apple. *Journal of the American Society for Horticultural Science* 117, 198–202.

Funke, K. and Blanke, M. (2005) Can reflective ground cover enhance fruit quality and colouration? *Journal of Food, Agriculture, and Environment* 3, 203–206.

Guimond, C.M, Andrews, P.K. and Lang, G.A. (1998) Scanning electron microscopy of floral initiation in sweet cherry. *Journal of the American Society for Horticultural Science* 123, 509–512.

Hamm, A., Lorenz, J., Papendieck, P., Dekova, O. and Blanke, M.M. (2007) Honigbienen als Bestäuber für verfrühte Süßkirschen im geschützten Anbau [Honey bee as pollinators for forced sweet cherries under protected cultivation]. *Erwerbs-Obstbau* 49, 85–92.

Healey, K.D., Hammer, G.L., Rickert, K.G. and Bange, M.P. (1998) Radiation use efficiency increases when the diffuse component of incident radiation is enhanced under shade. *Australian Journal of Agricultural Research* 49, 665–672.

Hedhly, A., Hormaza, J.J. and Herrero, M. (2007) Warm temperatures at bloom reduce fruit set in sweet cherry. *Journal of Applied Botany* 83, 158–184.

Hewitt, E.W. and Young, K. (1980) Water sprinkling to delay bloom in fruit trees. *New Zealand Journal of Agricultural Research* 23, 523–528.

Imrak, B., Sarier, A., Kuden, A., Kuden, A.B., Comlekcioglu, S. and Tutuncu, M. (2014) Studies on shading system in sweet cherries (*Prunus avium* L.) to prevent double fruit formation under subtropical climatic conditions. *Acta Horticulturae* 1059, 171–176.

Kafkaletou, M., Ktistaki, M.-E., Sotiropoulos, T. and Tsantili, E. (2015) Influence of rain cover on respiration, quality attributes and storage of cherries (*Prunus avium* L.). *Journal of Applied Botany* 88, 87–96.

Koumanov, K.S. (2002) Drought mitigation effects of microirrigation in orchards. In: *ICID International Conference on Drought Mitigation and Prevention of Land Desertification*, Bled, Slovenia, 20–26 April.

Lang, G.A. (2009) High tunnel tree fruit production – the final frontier? *HortTechnology* 19, 50–55.

Lang, G.A. (2013) Tree fruit production in high tunnels: current status and case study of sweet cherries. *Acta Horticulturae* 987, 73–81.

Lang, G.A. (2014) Growing sweet cherries under plastic covers and tunnels: physiological aspects and practical considerations. *Acta Horticulturae* 1020, 303–312.

Lang, G., Guimond, C., Flore, J., Southwick, S., Facteau, T., Kappel, F. and Azarenko, A. (1998) Performance of calcium/sprinkler-based strategies to reduce sweet cherry rain-cracking. *Acta Horticulturae* 468, 649–656.

Lang, G., Valentino, T., Demirsoy, H. and Demirsoy, L. (2011) High tunnel sweet cherry studies: innovative integration of precision canopies, precocious rootstocks, and environmental physiology. *Acta Horticulturae* 903, 717–723.

Lang, G., Hanson, E., Biernbaum, J., Brainard, D., Grieshop, M., Isaacs, R., Montri, A., Morrone, V., Schilder, A., Conner, D. and Koan, J. (2013) Holistic integration of organic strategies and high tunnels for Midwest/Great Lakes fruit production. *Acta Horticulturae* 1001, 47–55.

Luedeling, E., Guo, L., Dai, J., Leslie, C. and Blanke, M.M. (2013a) Differential responses of trees to temperature variation during the chilling and forcing phases. *Agricultural Forest Meteorology* 181, 33–42.

Luedeling, E., Kunz, A. and Blanke, M.M. (2013b) Effect of recent climate change on cherry phenology. *International Journal of Biometeorology* 57, 679–689.

Meinhold, T., Damerow, L. and Blanke, M.M. (2011) Reflective materials under hailnet improve fruit quality and particularly fruit colouration. *Scientia Horticulturae* 127, 447–451.

Meland, M. and Skjervheim, K. (1998) Rain cover protection against cracking for sweet cherry orchards. *Acta Horticulturae* 468, 441–447.

Meland, M., Kaiser, C. and Christensen, J.M. (2014) Physical and chemical methods to avoid fruit cracking in cherry. *AgroLife Scientific Journal* 3, 177–183.

Meland, M., Frøynes, O. and Kaiser, C. (2017) High tunnel production systems improve yields and fruit size of sweet cherry. *Acta Horticulturae* (in press).

Mulabagal, V., Lang, G.A., DeWitt, D.L., Dalavoy, S.S. and Nair, M.G. (2009) Anthocyanin content, lipid peroxidation and cyclooxygenase enzyme inhibitory activities of sweet and sour cherries. *Journal of Agricultural and Food Chemistry* 57, 1239–1246.

Musacchi, S., Gagliardi, F. and Serra, S. (2015) New training systems for high-density planting of sweet cherry. *HortScience* 50, 59–67.

Olmstead, J.W. and Lang, G.A. (2002) *Pmr1*, a gene for resistance to powdery mildew in sweet cherry. *HortScience* 37, 1098–1099.

Olmstead, J.W., Lang, G.A. and Grove, G.G. (2001) Assessment of severity of powdery mildew infection of sweet cherry leaves by digital image analysis. *HortScience* 36, 107–111.

Overbeck, M., Schmitz-Eiberger, M. and Blanke, M.M. (2013) Reflective mulch enhances ripening and health compounds in fruit. *Journal of the Science of Food and Agriculture* 93, 2575–2579.

Peschel, S. and Knoche, M. (2005) Characterization of microcracks in the cuticle of developing sweet cherry fruit. *Journal of the American Society for Horticultural Science* 130, 487–495.

Rapparini, R., Rotondi, A. and Baraldi, R. (1999) Blue light regulation of the growth of *Prunus persica* plants in a long term experiment: morphological and histological observations. *Trees, Structure, and Function* 14, 169–176.

Ribeiro, A.C., De Melo-Abreu, J.P and Snyder, R.L. (2006) Apple orchard frost protection with wind machine operation. *Agricultural Forest Meteorology* 141, 71–81.

Schäfer, S. (2007) Überdachungssysteme im Obstbau – Auswirkungen auf Mikro-klima, Baumwachstum, Fruchtqualität sowie den Krankheits- und Schädlingsbefall von Süßkirschen. PhD thesis, Horticultural and Agricultural Faculty, Humboldt University, Berlin, Germany.

Schmitz-Eiberger, M. and Blanke, M.M. (2012) Bioactive compounds in forced sweet cherries, anti-oxidative capacity and allergenic potential. *LWT – Food and Science Technology* 46, 388–392.

Solomakhin, A. and Blanke, M.M. (2010) Microclimate under coloured hailnets affects leaf and fruit temperature, transpirational cooling, leaf anatomy and vegetative and reproductive growth as well as fruit colouration in apple. *Annals of Applied Biology* 156, 121–136.

Southwick, S., Shackel, K., Yeager, J., Asai, W. and Katacich, M. Jr (1991) Over-tree sprinkling reduces abnormal shapes in 'Bing' sweet cherries. *California Agriculture* 45, 24–26.

Tartachnyk, I. and Blanke, M.M. (2004) Effect of delayed fruit harvest on photosynthesis, transpiration and nutrient remobilization of apple leaves. *New Phytologist* 164, 442–450.

USDA-NRCS (1993) Chapter 2. Irrigation water requirements. In: *National Engineering Handbook Part 623*. US Department of Agriculture, Natural Resources Conservation Service, Washington, DC.

Usenik, V., Zadravec, P. and Stampar, F. (2009) Influence of rain protective tree covering on sweet cherry fruit quality. *European Journal of Horticultural Science* 74, 49–53.

Wallberg, B.N. and Sagredo, K.X. (2014) Vegetative and reproductive development of 'Lapins' sweet cherry under rain protective cover. *Acta Horticulturae* 1058, 411–419.

Whiting, M. and Martin, R. (2008) Reducing sweet cherry doubling. *Good Fruit Grower* 59, 24–26.

Whiting, M.D., Rodriguez, C. and Toye, J. (2008) Preliminary testing of a reflective ground cover: sweet cherry growth, yield and fruit quality. *Acta Horticulturae* 795, 557–560.

Wittmann, D., Klein, D., Sieg, V., Schindler, M. and Blanke, M.M. (2005) Wildbienen für die Bestäubung von Obstanlagen [Solitary bees for cherry pollination]. *Erwerbs-Obstbau* 47, 27–36.

Zhang, J., Whiting, M.D. and Zhang, Q. (2015) Diurnal pattern in canopy light interception for tree fruit orchard trained to an upright fruiting offshoots (UFO) architecture. *Biosystems Engineering* 129, 1–10.

# 12 Morphology, Cropping Physiology and Canopy Training

**Marlene Ayala[1]* and Gregory A. Lang[2]**
[1]*Pontificia Universidad Católica, Santiago, Chile;* [2]*Michigan State University, East Lansing, Michigan, USA*

## 12.1 Introduction

The most significant development in cherry production during the past 20 years is the commercial availability of precocious, highly productive rootstocks that impart a range of levels of vigour control. These have impacted reproductive morphology in terms of temporal and positional development of flower buds (Maguylo *et al.*, 2004), as well as canopy physiology in terms of altering source–sink relations and root–shoot physiology with respect to acquisition of water and nutrients. For example, Olmstead *et al.* (2004, 2006) found that vessel diameter in the graft union of dwarfing rootstocks tended to be smaller than in more vigorous rootstocks. This suggests that water transport capacity might be limiting on a diurnal basis and therefore trees on dwarfing rootstocks may be subjected to transient daily water stress with potentially reduced photosynthesis and nutrient uptake, thereby reducing growth, as has been reported in peach (Basile *et al.*, 2003; Tombesi *et al.*, 2010). Indeed, Gonçalves *et al.* (2005) reported that midday stem water potentials and carbon (C) assimilation in sweet cherry trees on rootstocks of varying vigour decreased proportionally to the level of vigour imparted. Reduced shoot growth, as a result of rootstock genotype (Costas *et al.*, 2009), environmental water stress, nitrogen (N) deficiency and canopy manipulations such as limb bending, are often associated with increased flower bud formation. Consequently, these relatively recent vigour-controlling, precocity-inducing rootstocks have been areas of active research activity to understand how best to utilize these traits and develop or adopt canopy training systems and orchard practices to optimize their use in new intensive production systems.

While low-density sweet cherry orchards using vigorous seedling rootstocks still exist, the inefficiencies of such orchards are driving worldwide trends towards planting higher-density orchards and controlling tree vigour with dwarfing rootstocks and/or training systems (Lang, 2000, 2008; Balmer and Blanke, 2005; Lauri and Claverie, 2008; Calabro *et al.*, 2009). Large non-precocious trees delay return on investment (West *et al.*, 2012), have less uniform fruit ripening and are difficult to protect from biotic (insect pests, diseases, birds) and abiotic (rain, hail, frost, solar radiation) stresses (see Chapter 11, this volume). Producers worldwide are

* mayalaz@uc.cl

adopting simplified tree structures that are more conducive to partial mechanization of labour tasks (e.g. pruning, harvest) or smaller trees more conducive to pedestrian orchards and precision management (Robinson, 2005; Santos *et al.*, 2006; Vercammen and Vanrykel, 2009; Long *et al.*, 2014). Such trees have less permanent structure and a greater emphasis on optimizing leaf area-to-fruit (LA/F) ratios and yield per area, improving orchard economics, particularly in the early years (West *et al.*, 2012). High-density (1200–4000 trees $ha^{-1}$) and simplified canopy structures better facilitate hand labour or partial mechanization for such tasks as crop load regulation (e.g. pruning and thinning of buds, spurs, flowers and immature fruit) and harvest.

However, high-density orchards also present unique challenges. Smaller trees are more susceptible to a high proportion of the crop being exposed to the coldest air during spring frosts. Precocious cropping requires precise canopy development to fill the allotted orchard space quickly and efficiently. Imprecise tree structure during the early establishment years may not only lead to lower initial yields or excessive cropping, but may also be difficult to correct due to competition for photoassimilates between early vegetative growth and precocious crop loads. When sources of photoassimilates (i.e. LA and storage reserves) are insufficient to supply the various competing sinks (i.e. fruit, buds, spurs, extension shoot growth, wood and roots), fruit quality and vegetative vigour will be less than optimal. To develop best management practices and avoid such situations, understanding source–sink relations in cherries on precocious, and often vigour-limiting, rootstocks is useful for implementing strategies to obtain premium fruit and maintain an appropriate C balance and distribution in the tree.

## 12.2 Canopy Growth and Fruiting Habit

With non-precocious rootstocks such as *P. avium* or *P. mahaleb* seedlings, or non-precocious clonal rootstocks such as 'Colt' or Mazzard 'F 12/1', the orchard establishment period typically comprised 5 or 6 years during which an extensive root and canopy system developed with little fruit production. Thus, the tree's growth and energy was directed largely towards developing its 'source' structure for acquisition of water, nutrients and C prior to development of a significant reproductive 'demand' for resources by flowering and fruiting. Pruning and training practices that had evolved over decades of such traditional vegetative–reproductive relationships had to be completely reconsidered with the introduction in the 1990s of rootstocks that conferred an almost instantaneous shift to fruiting capacity. On trees grafted on these rootstocks, the new shoots that grow during the year of planting often form basal flower buds, and nodes on the central leader that did not elongate into extension shoots during the year of planting may then form flowering spurs, leading to cropping as early as 1 year after planting.

Thus, before the advent of precocious rootstocks, by the time significant fruiting began on standard trees, a large canopy and root system would typically be developed to support the eventual cropping capacity. Standard orchard canopy light distribution, LA per fruit and horticultural manipulations to reduce the non-precocious period, such as shoot bending, have been discussed by Flore *et al.* (1996). In arid growing regions, deficit irrigation could also be used to alter the hormonal physiology of standard trees to shift development away from vegetative growth and towards increased flower bud formation. Precocious rootstocks such as the 'GiSelA' series therefore resolved this developmental lag in fruiting capacity and required a shift in orchard management towards achieving a better reproductive–vegetative balance by increasing LA and reducing cropping potential.

### 12.2.1 Precocious canopy structure, leaf, flower bud and fruit development

Sweet and sour cherry trees grow and flower from simple buds that are either vegetative or reproductive. These develop into canopies that contain several distinct populations of leaves and fruit (Fig. 12.1). Growth of a vegetative bud produces multiple nodes with single leaves: these may have minimal internode lengths and only form a rosette of about five to eight preformed (primordial) leaves, resulting in a spur, or have significant internode lengths and continue new leaf and node formation beyond the preformed leaves, resulting in extension of the primary shoot or formation of a new lateral shoot. Vegetative or reproductive meristems are then initiated in the axil of each spur or shoot leaf. If the newly formed meristem also elongates in the same season that it forms, it becomes a sylleptic lateral extension shoot. This usually occurs when growth is extremely vigorous.

If newly formed axillary meristems remain primordial in the season of formation, with no further differentiation, they become latent buds arrested in further development (i.e. paradormant) and are unlikely to grow the following year unless the terminal shoot structure is damaged or pruned. Alternatively, such meristems may differentiate into a new vegetative or reproductive bud, transitioning from paradormancy to endodormancy as the season progresses but ready to grow the following spring. If the extension shoot is pruned during the first half of the growing season (thereby releasing apical dominance), one or two of the most terminal axillary buds will elongate into a replacement terminal shoot. For cherry trees on precocious rootstocks, the most basal axillary buds of the extension shoot often become reproductive. The rest of the axillary buds along the length of the extension shoot remain vegetative, each containing a single primordial shoot meristem subtended by several lateral leaf primordia.

Meristems that differentiate into single reproductive buds (see Chapter 2, this volume) may occur in the axils of shoot leaves at the base of that shoot's extension growth, thereby becoming non-spur fruit buds (single flower buds with no adjacent vegetative buds containing a shoot meristem), or they may occur in the axils of spur leaves, thereby becoming one of the spur fruit buds that can form a cluster in the spur rosette with a single terminal vegetative bud. Therefore, the fruiting potential for cherry is comprised of two types of fruit bud populations, each associated with different local leaf populations (Fig. 12.1). The non-spur fruit buds generally make up only a small proportion of the overall population of fruit within the tree canopy, but the fruit at these nodes tend to have the greatest potential size and quality. However, because they lack a vegetative meristem, after fruiting the node becomes 'blind', incapable of forming additional leaves, fruit or new shoots.

Cherry reproductive buds usually contain one to six flowers, and fruiting spurs usually have one to six buds, although more is possible. The propensity for forming basal flower buds on extension shoots and fruiting spurs varies by scion genotype, and these traits as well as flower numbers per bud can be significantly modulated further by rootstock genotype. For example, Maguylo *et al.*

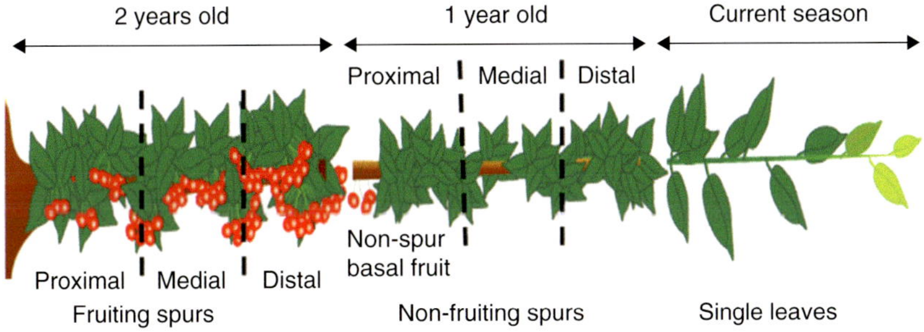


Fig. 12.1. The fundamental fruiting unit of a sweet cherry, comprised of three populations of leaves (fruiting spur, non-fruiting spur and new shoot) and two populations of fruiting sites (fruiting spurs and non-spur basal fruit on the 1-year-old shoot segment) (Lang, 2005).

(2004) found that 'Hedelfinger' sweet cherry on 'GiSelA 3' ('Gi 209/1') had about one flower per spur on the proximal segment of a 2-year-old shoot (Fig. 12.2), although flower numbers averaged around eight to 11 for the medial and distal segments, respectively. Trees on both 'GiSelA 5' and 'GiSelA 6' averaged about four flowers per spur on the proximal segment and six to 13 for the medial and distal segments, respectively, while trees on Mazzard averaged fewer than one flower per spur in any segment.

Non-spur fruit buds bloom, usually bear one to three fruit if fertilized, and then become blind nodes since they lack vegetative meristems. Spur fruit buds bloom, bear fruit if fertilized, and the terminal vegetative bud either gives rise to a continued rosette of preformed leaf nodes, thereby continuing as a fruiting spur for the next year, or the terminal vegetative bud elongates to create a new (lateral) extension shoot.

Cherry canopies can be deconstructed into three main leaf populations: extension shoot leaves, non-fruiting spur leaves and fruiting spur leaves (Fig. 12.1):

- ***Extension shoot leaves*** occur one per node on new seasonal shoot growth and include the initial five to eight leaves preformed in the bud plus an indeterminate number of neoformed leaves that ultimately may number 15–75 or more by the time seasonal shoot extension terminates (usually in mid-summer, i.e. July or January in the northern or southern hemisphere, respectively). Preformed leaves and initial shoot expansion constitute sinks that rely on growth resources (C and N) translocated from storage reserves in spring (Kappel, 1991; Thielemann *et al.*, 2014; Ayala and Lang, 2015). However, the ever-increasing number of mature leaves in this population constitutes an important C source for distant fruit, particularly by mid-stage III of fruit growth (Ayala, 2004; Ayala and Lang, 2008;

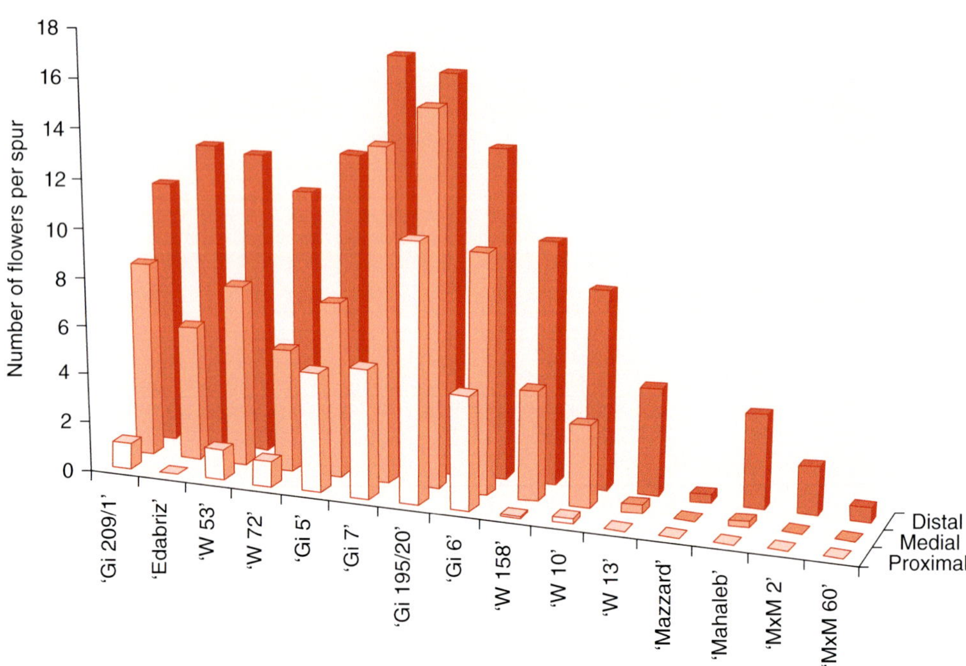


Fig. 12.2. The total number of flowers per spur in 'Hedelfinger' sweet cherry 2-year-old fruiting branch sections as influenced by rootstock, listed in order of increasing vigour (left to right), and segment within the branch section, depicted from proximal (pink bars) to distal (dark red bars) (Maguylo *et al.*, 2004).

Correa, 2008). The extension shoot leaf population can become a net exporter of C with only ten leaves (Ayala and Lang, 2008). Leaf size on extension shoots can vary from relatively small (≥20–50 $cm^2$) for the preformed leaves to very large (≥110–150 $cm^2$) for the neoformed leaves (although terminal leaves may be smaller), with the size of the neoformed leaves directly proportional to tree vigour, as their development is a function of genotype and acquisition of growth resources from the environment – C from current photosynthesis and N from root uptake of nutrients largely through evapotranspiration. The stimulation of new shoots by pruning, and the associated altered distribution of C and N (from reserves and current acquisition) to fewer actively growing meristems, often results in an increase in neoformed leaf size. Consequently, pruning can be an important tool for increasing localized source LA on highly productive trees.

- ***Non-fruiting spur leaves*** occur on the 1-year-old section of the branch, usually with six to eight leaves per spur rather than the single leaf per node during the first year of growth. Leaf size within a spur can vary from small (≤20 $cm^2$) to moderately large (75–90 $cm^2$). Thus, the total LA per spur can range from about 200 to 450 $cm^2$, in comparison with the previous year's LA per node at the same location ranging from about 50 to 200 $cm^2$. Spur leaf size is unrelated to current-season C and N acquisition, but rather is proportional to storage reserve availability and thus the previous year's growth. Some preliminary studies have suggested that early-season foliar applications of nutrients (e.g. N) or growth regulators (e.g. brassinosteroids) may be able to improve spur leaf size (M. Ayala, Santiago, Chile, 2016, personal communication). All spur leaf expansion is completed within 3–4 weeks of bud break, before current-season C and N acquisition by roots becomes significant. This leaf population is a key supplemental carbohydrate source for growth of distant fruit on older wood, and for terminal or lateral shoot extension growth (Ayala and Lang, 2008; Correa, 2008). The proximity of this significant LA to the small non-spur fruit population at its base contributes to the factors that result in these fruit usually being the largest and highest-quality fruit in the canopy. During the course of the season, the primordial meristems in the axil of each non-fruiting spur leaf may remain undifferentiated or begin differentiation into flower buds for the following season.
- ***Fruiting spur leaves*** occur on the 2-year-old and older sections of the branch; not all spurs on these older sections may be reproductive. If a spur had no axillary meristems differentiated the previous season, it will again serve as a non-fruiting spur. The proportion of spurs that become reproductive can be influenced by rootstock (more on precocious rootstocks; Maguylo *et al.*, 2004), by shoot orientation (more on horizontal than on vertical shoots) and growth rate (more on weaker than on vigorous shoots), as well as by light exposure (fewer in heavily shaded areas of the canopy). There are usually six to nine leaves per spur. Leaf size within 2-year-old and older spurs varies similarly to that of non-fruiting spurs, from small (≤20 $cm^2$) to moderately large (75–90 $cm^2$), with leaves on reproductive spurs tending to be slightly smaller compared with non-fruiting spurs. The rapid expansion and maturation of fruiting and non-fruiting spur leaves provides the primary source of C during the transition from storage reserves to current-season acquisition (Roper and Loescher, 1987; Ayala *et al.*, 2014; Ayala and Lang, 2015), as well as the evapotranspirational driving force for current-season acquisition of N from root uptake (Ayala *et al.*, 2014).

Since the number of buds per spur and flower spurs per 2-year-old and older shoots typically exceeds the number of basal flower buds on 1-year-old shoots, spur fruit tend to make up a greater proportion of the fruit population in most sweet cherry tree canopies. Deconstructing the tree into its component leaf and fruit populations facilitates the modelling of canopy dynamics (Lang *et al.*, 2004; Lang and Lang, 2009) and the potential impacts of training system, pruning and crop regulation decisions on their relative populations, as will be described later in this chapter.

### 12.2.2 Seasonal growth and fruit developmental timeline

It is valuable to understand the timing, acquisition and distribution of growth resources among the various tissues to optimize cherry orchard management strategies for desired fruit yields and quality. As noted in Chapter 2, sweet and sour cherry flower bud induction begins a little more than a year before those buds will eventually yield ripe fruit. The major reproductive and vegetative developmental events and physiological processes during this approximately 15-month period are outlined in Fig. 12.3 for sweet cherry. During the year preceding fruit harvest, flower buds are induced, and floral organs differentiate within each bud. Current photosynthates and exposure to at least moderate levels of light are requisite for developing strong flower buds. Spurs in significant shade may defoliate prematurely, with abortion of induced flower buds, and the terminal vegetative meristem may die, resulting in a lost spur ('blind' node). It is not uncommon for fruitful nodes with good light exposure at bud break to become heavily shaded by the end of summer when the primordial floral organs, and primordial spur leaves, are forming for the next year. The acrotonic growth habit of sweet cherry, often vigorous shoot growth and potentially large shoot leaves can cause significant interior shading within the canopy in a single season, which can be alleviated or exacerbated by orchard management decisions. Where flower buds are well positioned, differentiation is arrested prior to autumn leaf senescence, and buds become dormant in advance of winter. During this period, storage forms of N and C increase by 50% or more in spurs (Ouzounis and Lang, 2011; Thielemann *et al.*, 2014), as well as in other storage tissues such as shoots, trunk and roots (Zavalloni, 2004; Azarenko *et al.*, 2008).

Remobilization of growth resources occurs concomitant with bud swell (Grassi *et al.*, 2002; Ouzounis and Lang, 2011). Remobilized N, transported primarily as glutamine, predominates for about 3 weeks after bud break, at which point current-season N uptake by roots, transported as asparagine, begins to occur (Grassi *et al.*, 2003). Following bloom, fertilization and fruit set (see Chapter 2, this volume), the double-sigmoidal curve of fruit growth (Choi *et al.*, 2002) occurs concomitantly with the sigmoidal curve of new and extension shoot growth. The duration of stage I is relatively similar among most genotypes regardless of differences in fruit harvest date (Choi *et al.*, 2002; Gibeaut *et al.*, 2016). Genotypic differences in ripening date have been reported to be associated with differential durations in stage II (Choi *et al.*, 2002) or during the second exponential increase in fruit growth, in stage III (Gibeaut *et al.*, 2017). The fruit development period for sweet and sour cherries (averaging about 55–70 days from bloom to ripening for mid-season genotypes) is one of the most rapid for temperate zone tree fruits, generally being shorter than that for apricot, peach, plum, pear and apple. This is probably a limiting factor for cherry yields per hectare, which tend to be significantly less than for tree fruits such as apple with longer fruit development periods (and thus with longer timeframes for acquiring current-season photosynthates for fruit growth).

Initial shoot growth (comprised of preformed nodes and the first neoformed nodes) is exponential, followed by a linear portion of continued extension (neoformed node) growth, and then decreasing shoot node formation. Terminal bud set often coincides with the final phase of fruit maturation and harvest. On less vigorous rootstocks or orchard soils, bud set may occur much earlier, and on more vigorous rootstocks or sites, bud set may occur well after harvest, or a

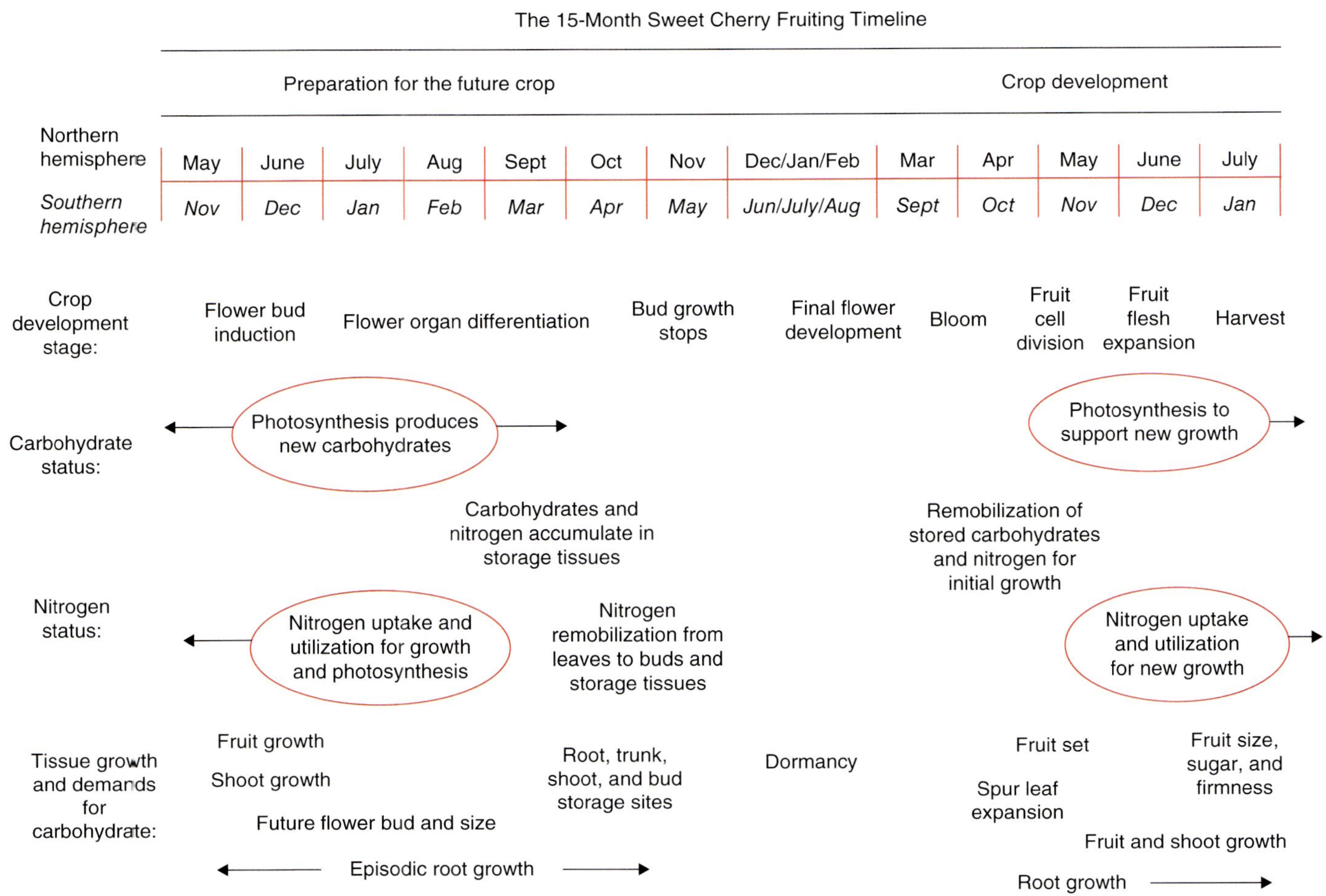


**Fig. 12.3.** Developmental timeline for sweet cherry cropping and growth, from flower bud induction to fruit harvest (Lang, 2005).

second flush of growth may even occur in late summer after apparent bud set. When mature, neither cherry species tends to exhibit additional significant shoot growth after harvest (Fig. 12.3). Pruning during the period of fruit and shoot growth usually results in reiterative regrowth from the most terminal one or two nodes below the pruning cut. Depending on tree vigour, pruning just after harvest might stimulate a small amount of regrowth, but later in the summer (August or February in the northern or southern hemispheres, respectively), pruning generally does not stimulate shoot regrowth. At this point, photosynthates are used mainly to support radial growth, and presumably that of roots, followed by an eventual shift in distribution to storage tissues rather than to actively growing shoot, root or cambial meristems. Foliar applications of urea after harvest are partitioned to storage tissues such as flower spurs (Thielemann *et al.*, 2014). While flower buds are induced during fruit growth, floral organs generally begin differentiating after fruit and shoot growth has ceased (Guimond *et al.*, 1998a).

This short fruit development period, its overlap with concurrent shoot growth and the fact that bloom precedes LA development have important consequences for strategies to optimize fruit quality. Remobilized storage reserves in spring (Fig. 12.3) represent a significant portion of growth resources for fruit set and early development, as well as the sole resources for growth of the spur leaves and initial extension shoot growth and shoot leaf development. Spur leaf size is directly impacted by storage reserves (Ouzounis and Lang, 2011), and thus postharvest leaf health is critical for building storage reserves that influence a significant proportion of the next spring's LA. Spur leaf photosynthesis then supports early growth of both fruit and new shoots, as will be discussed in greater detail in section 12.3.

## 12.3 Canopy Photosynthesis and Carbon Distribution

In fruit trees, photoassimilate availability and distribution are affected by numerous factors, including: canopy LA, architecture and light interception; C supply from photosynthesis and storage reserves; crop load and organ development; respiration and environmental conditions; rootstock; and various cultural practices (McCamant, 1988; Keller and Loescher, 1989; Flore and Layne, 1999). The tree is a collection of multiple independent C sources and sink organs between which C moves as a function of source supply, sink demand, and distance between the sources and sinks (DeJong and Grossman, 1995). The competitive strength of sink organs as net importers of assimilates changes with growth (Ho, 1988; Flore and Layne, 1999). C distribution studies in cherry grown on traditional vigorous rootstocks have provided insight for orchard management regarding pruning and balancing crop load and LA (McCamant, 1988; Keller and Loescher, 1989; Flore and Layne, 1999). However, the advent of precocious, vigour-controlling rootstocks in the 1990s changed temporal and proportional developmental relationships between cherry sources and sinks, with reduced aerial woody structures, smaller root systems and higher harvest indices in the earlier years of orchard development. During the first decade of dwarfing rootstock adoption in sweet cherries, higher yields on smaller trees often resulted in sink demands that exceeded photoassimilate capacity, resulting in decreased fruit quality and stunted growth as early as 5 years after planting (Lang, 2001a; Whiting *et al.*, 2005; Correa, 2008; Ayala and Andrade, 2009).

Early attempts to adopt dwarfing rootstocks for sweet cherry led many growers to conclude that the rootstocks limit not only tree size but also fruit size. Thus, one of the most important areas of sweet cherry physiological study during the past two decades has been to examine the relationships governing C distribution among vegetative and reproductive sinks. Sweet cherry source and sink organ interactions on precocious rootstocks include the following: (i) increasing crop loads during the early years of canopy development can lead to earlier termination of extension shoot growth, reduced LA development and therefore less C storage for subsequent year initial growth; (ii) earlier termination of shoot growth can lead to greater flower bud formation, thereby increasing the potential crop load for the subsequent year even more; (iii) lower storage C availability for subsequent spring growth can result in smaller spur leaves and therefore reduced early-season LA for initial shoot and fruit growth; and (iv) fruit consequently fail to achieve their full genetic growth potential due to disproportionately reduced C sources and increased competition between the developing fruit. Other considerations for sweet cherry source–sink relationships include: (i) spatial and temporal changes in reproductive and vegetative organ sink strength throughout the season; (ii) temporal changes in the direction of carbohydrate translocation; (iii) distance between source leaves and sinks; and (iv) rootstock vigour-mediated sink dynamics between fruit and vegetative growth (Kappes and Flore, 1986, 1989; Kappel, 1991; Flore and Layne, 1999; Ayala, 2004; Correa, 2008; Mora, 2008; Ayala and Andrade, 2009).

### 12.3.1 Canopy and fruit photosynthesis

Photosynthesis and carbohydrates play an important role in sweet cherry fruit quality since 20–25% of the fruit dry weight is dry matter, of which ~90% is carbohydrate (Ayala, 2004). Photosynthetic potential is controlled by the environment and the sink demand of various organs (Flore, 1994). The presence of fruit and/or increased vegetative growth has been associated with an increase in photosynthetic assimilation rate (*A*) (Flore and Lakso, 1989). The effect of crop load on *A* has been studied by comparing fruiting and non-fruiting trees. Increases in *A* during fruit development have been reported for several species including sweet cherry (Gucci *et al.*, 1991). Partial defoliation also increases *A* in sour cherry due to photosynthetic compensation (Layne and Flore, 1993). However, some studies in sweet and sour cherry have not found a sink effect of fruit on *A* (Sams and Flore, 1983; Flore and Layne, 1999). Whiting and Lang (2001, 2004b) used whole-canopy cuvettes to estimate photosynthesis of 5-year-old fruiting sweet cherry trees on the dwarfing precocious rootstock 'GiSelA 5', and found a lack of sink effect on *A* when crop loads were varied experimentally.

Sweet cherry fruit are photosynthetically competent during the early stages of development, but their contribution to fruit and/or vegetative growth is minor. Ayala (2004) reported that fruits exposed to $^{13}CO_2$ fixed variable amounts of $^{13}C$ depending on developmental stage (Table 12.1). At 25 days after full bloom (DAFB; stage I), fruit fixed the highest amounts of $^{13}C$. At 44 DAFB (stage II), fruit continued fixing $^{13}C$ but in lower quantities. In sour cherry, fruit gross photosynthesis contributed ~19, 30 and 1.5% of the carbohydrate used during stages I, II and III of fruit development, respectively (Flore and Layne, 1999). Overall, 70% of the fixed C was incorporated into fruit dry matter, while the rest was used in dark respiration.

### 12.3.2 Source–sink relations: storage reserves, leaves and fruit

*Importance of storage reserves*

Storage reserves are important for multiple life-cycle processes in perennial fruit trees, including winter survival, metabolism, respiration, defence, and periods of vegetative and reproductive growth. Carbohydrates, in the form of starch and soluble sugars, are

**Table 12.1.** Total $^{13}C$ content in sweet cherry fruit sampled immediately (0 h) after labelling of the fruiting spur leaves at 25, 40, 44, 56 and 75 days after full bloom. Results are shown as means ± standard error ($n$ = 10). Means followed by the same lower-case letter are not significantly different ($\alpha$ = 0.05). DW, dry weight. (Data from Ayala, 2004.)

| Stage of fruit development | Days after full bloom | Total $^{13}C$ content ($\mu g\ ^{13}C\ g^{-1}$ DW) |
|---|---|---|
| I | 25 | 188.3±64.8 a |
| II | 40 | 69.3±12.5 b |
| III | 44 | 98.0±17.6 b |
| | 56 | 8.4±2.7 c |
| | 75 | 11.6±3.7 c |

the major component of storage reserves in sweet cherry (Loescher *et al.*, 1990). After harvest and terminal bud set, sink demand for carbohydrates decreases significantly, being comprised mainly of growth of secondary woody structure, roots and reproductive buds. At this time, photoassimilate supply is in excess and storage reserves, mainly starch, accumulate and reach a maximum concentration at leaf abscission (McCamant, 1988; Keller and Loescher, 1989).

The whole perennial structure of the cherry tree can serve as a storage organ (Loescher *et al.*, 1990). C reserves accumulate predominantly in living ray and axial parenchyma cells of woody structures (i.e. branches and trunk) and roots (Oliveira and Priestley, 1988). In sweet cherry, carbohydrate reserves accumulate preferentially in the roots compared with other storage organs such as trunk and shoots (Keller and Loescher, 1989; Loescher *et al.*, 1990; Grassi *et al.*, 2003). Ayala and Lang (2015), using $^{13}C$ enrichment during autumn, reported higher $^{13}C$ levels stored in older wood of the trunk, branches and coarse roots of 'Regina' on 'GiSelA 6' trees. $^{13}C$ was also detected in the phloem, buds and extension shoots.

Since sweet cherry flowering usually occurs before leaves are fully expanded, during bud break and early growth in the spring, fruit, spur leaves and extension shoots compete for C storage reserves (Loescher *et al.*, 1990; Kappel, 1991; Ayala and Lang, 2015). Keller and Loescher (1989) demonstrated that the carbohydrate in sweet cherry roots, wood and bark declines rapidly during full bloom. Ayala and Lang (2015) found the highest levels of $^{13}C$ (from that fixed the previous season) at bud break in fruiting and non-fruiting spur buds, as well as in roots. Following bud break, 3–11% of the $^{13}C$ was translocated to new aerial organs (flowers, spur leaves and developing fruit) up to 14 DAFB. During mid-to-late stage I (after 21 DAFB), $^{13}C$ values from storage reserves no longer changed appreciably in growing organs. Within several weeks of bud break, leaves become net C exporters and provide developing fruit and elongating shoots with photoassimilates (Roper and Loescher, 1987; Ayala and Lang, 2008).

### *Importance of leaf area and C translocation patterns*

The amount (number and size) and quality (exposure to solar radiation and photosynthetic capacity) of fruiting spur, non-fruiting spur and extension shoot leaves have a direct impact on sweet cherry fruit quality, since LA constitutes the primary C source for fruit growth during stages II and III (Ayala, 2004). Each leaf population provides carbohydrates for fruit and shoot development. Spur LA can vary greatly due to the wide range of leaf sizes, and leaf exposure to solar radiation varies with canopy architecture and position within the canopy. More vigorous trees are more likely to develop shade leaves, which are larger, thinner and positioned more horizontally within the canopy.

In the late 1990s, in response to grower concerns about small fruit size on the new 'GiSelA' rootstocks, G.A. Lang and M. Whiting (unpublished data) tested the hypothesis that dwarfing rootstocks do not restrict fruit size, but rather that sweet cherry LA/F ratio limits fruit size. Entire young 'Bing' on 'GiSelA 5' trees were bud-thinned to one, two or three flower buds per spur, or left unthinned. At harvest, more than half (53%) of the fruit on the control trees was considered to be too small for the fresh market, a result typical of many growers at that time. However, changing the LA:F ratio by thinning flower buds per spur to three, two or one shifted the proportion of unmarketable small fruit to only 37, 21 and 17% of the crop, respectively (as well as increasing soluble solids proportionally). Many studies and grower experiences since then have demonstrated that large fruit can indeed be grown on small trees through appropriate horticultural management techniques (Whiting and Lang, 2004a,b; Whiting *et al.*, 2005; Ayala and Lang, 2008; Correa, 2008; Ayala and Andrade, 2009). A hierarchy of developmental sensitivity to insufficient LA for aerial organs of sweet cherry trees on dwarfing rootstocks was proposed by Whiting and Lang (2004b) as (from most to least sensitive): trunk > fruit soluble solids (stage III) > fruit growth (stage III) > LA/spur > shoot elongation > fruit growth (stages I and II) > LA/shoot. That is, when LA (or other photosynthetic factors such as daily solar radiation) is limiting during the fruit development period, trunk growth would be reduced first. Fruit soluble solids would be impacted before fruit size, which would decrease before shoot length, and spur LA would be impacted before shoot LA. Olmstead *et al.* (2007) reported that sweet cherry mesocarp cell number is genetically stable and therefore cell size is the main determinant of fruit size, which is consistent with the proposed hierarchy above that a limited LA/F ratio can impact stage III fruit growth (cell expansion) more than growth in stage I (cell division).

In Ayala's (2004) study of $^{13}C$ distribution to fruit by pulsing the three populations of leaves from fruit set to ripening, fruiting spur leaves provided the highest proportion (57–79% at mid-stage III) and most consistent source of $^{13}C$ to developing fruit (Table 12.2). The contributions of non-fruiting spur and extension shoot leaves varied to a greater extent with the stage of fruit and shoot growth. The exception to this general trend was at mid-stage III (56 DAFB), during rapid cell enlargement and dry matter accumulation, when the proportions of $^{13}C$-photoassimilate translocation from all leaf populations were significantly higher to fruit than to any other sink. The lowest $^{13}C$ contents in all leaf populations occurred at this time, indicating rapid export of fixed $^{13}C$ to meet the high fruit sink demands. Roper *et al.* (1987, 1988) similarly found decreased carbohydrate levels in sweet cherry leaves during the period of most active demand for fruit growth.

**Table 12.2.** Relative $^{13}C$ distribution among sweet cherry organs on 2-year-old branches of 'Ulster' on 'GiSelA 6' trees 48 h after $^{13}CO_2$ pulsing of fruiting spur (FS), non-fruiting spur (NFS) or extension shoot (ES) leaf populations. Mean percentage calculations are based on absolute amounts of $^{13}C$ recovered for each organ from each $^{13}CO_2$ pulse-labelling source and date (25, 40, 44, 56 and 75 days after full bloom (DAFB)) ($n = 7$). (Data from Ayala, 2004.)

| Pulsed source leaf population | Sampled organ | Relative $^{13}C$ distribution (%) recovered 48 h after pulsing each source leaf population at the following DAFB: | | | | |
|---|---|---|---|---|---|---|
| | | 25 | 40 | 44 | 56 | 75 |
| FS leaves | Fruit | 63.2[a] | 59.9[a] | 58.9[a] | 79.1[a] | 57.3[a] |
| | FS leaves | 32.5[a] | 30.5[a] | 34.1[a] | 17.9[a] | 36.4[a] |
| | NFS leaves | <1 | <1 | <1 | <1 | <1 |
| | ES leaves | <1 | <1 | <1 | <1 | <1 |
| | FS wood | 3.0 | 8.8 | 7.0 | 2.6 | 5.4 |
| | NFS wood | <1 | <1 | <1 | <1 | <1 |
| | ES wood | <1 | <1 | <1 | <1 | <1 |
| NFS leaves | Fruit | 45.8 | 31.7 | 31.3 | 70.9 | 32.7 |
| | FS leaves | <1 | <1 | <1 | <1 | <1 |
| | NFS leaves | 41.2[a] | 42.7[a] | 46.1[a] | 19.9[a] | 49.3[a] |
| | ES leaves | <1 | <1 | <1 | <1 | <1 |
| | FS wood | 8.4 | 16.9 | 14.5 | 5.1 | 12.1 |
| | NFS wood | 4.2 | 7.8 | 7.6 | 3.3 | 5.0 |
| | ES wood | <1 | <1 | <1 | <1 | <1 |
| ES leaves | Fruit | 27.2 | 22.3 | 17.5 | 59.2 | 28.3 |
| | FS leaves | <1 | <1 | <1 | <1 | <1 |
| | NFS leaves | <1 | <1 | <1 | <1 | <1 |
| | ES leaves | 50.4[a] | 46.3[a] | 59.8[a] | 28.1[a] | 45.0[a] |
| | FS wood | 8.8 | 10.1 | 4.8 | 5.0 | 10.7 |
| | NFS wood | 8.1 | 15.7 | 9.1 | 3.7 | 10.0 |
| | ES wood | 4.9 | 5.5 | 8.7 | 3.1 | 5.8 |

[a]Organ was labelled with $^{13}C$ directly.

When either non-fruiting spur or extension shoot leaves were pulsed with $^{13}CO_2$, the wood located basipetally from these sources was highly enriched in $^{13}C$ (Ayala, 2004). While photosynthetically competent leaves on extension shoots supported terminal extension growth, $^{13}C$ was translocated basipetally to the fruiting and non-fruiting spur sections of the branch, with the highest proportion being found in the fruit, ranging from approximately 18% at early stage III to 59% at mid-stage III (Table 12.2). The greatest retention of $^{13}C$ in extension shoots, presumably for active shoot growth, occurred in early stage III. Non-fruiting spur leaves exported $^{13}C$ predominantly to fruit as well, ranging from 31 to 71% at mid-stage III. This pattern of $^{13}C$ distribution occurred both for 'Ulster' on semi-vigorous 'GiSelA 6' rootstock and for 'Sam' on semi-dwarfing 'GiSelA 5' rootstock. In the latter experiment, during stage III, crop load affected the distribution of C from non-fruiting spur leaves among the spur fruit locations. $^{13}C$ translocation to fruit among the fruiting spurs was relatively uniform at higher LA/F ratios (i.e. low crop loads); however, at lower LA/F ratios (i.e. high crop loads), less $^{13}C$ was translocated to fruit on basipetal fruiting spurs and more was recovered in fruit on the upper fruiting spurs nearest the non-fruiting spur source leaves. Directional transport of $^{13}C$ among sweet cherry reproductive and vegetative sinks also can be affected by winter pruning (Mora, 2008) (Fig. 12.4) and spur removal (Correa, 2008) (also known as spur 'extinction'; Lauri and Claverie, 2005). Unidirectional and bidirectional C transport from different leaf populations has also been reported for sour cherry (Kappes and Flore, 1986; Toldam-Andersen, 1998). Based on recovery of $^{13}C$, some also was exported out of the branch or used for respiration; Loescher *et al.* (1986) estimated that 16–23% of the total carbohydrate requirements for sweet cherry fruit growth are used in respiration.

### Fruit as a sink

During the double-sigmoidal curve of dry-weight accumulation in cherry fruit (Ayala, 2004), 50–80% of fruit growth occurs during stage III (Flore, 1994) and 90% of the dry-weight accumulation is comprised of carbohydrate (Flore, 1985; Whiting *et al.*, 2005). Relative to many other temperate tree fruits, the sweet cherry fruit development period is short (60–70 days) and concomitant with shoot growth, which Roper

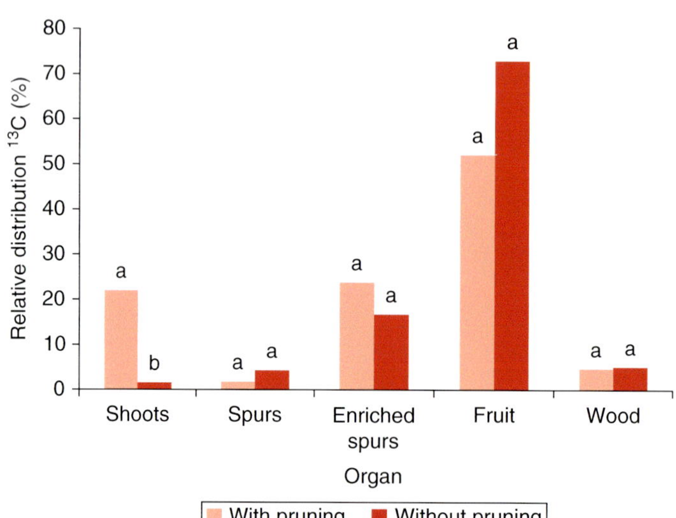


**Fig. 12.4.** The influence of pruning on relative distribution of $^{13}C$ to various sweet cherry organs, with and without pruning.

*et al.* (1987) suggested indicates a highly prioritized fruit sink demand. This competition for carbohydrate was studied using $^{13}CO_2$ to label different source leaf populations (i.e. fruiting spur, non-fruiting spur and extension shoot leaves) at various times during the sweet cherry fruit growth curve (Ayala and Lang, 2008) as well as to label storage reserves to examine their contributions to early fruit growth the following spring (Ayala and Lang, 2015). Fruit were a high priority sink for storage reserves from bud swell to 2 weeks after full bloom, as well as during the entire period of the leaf population study, from mid-stage I (cell division) to the end of stage III. With regard to translocation from reserves, the highest sink activity was detected in flowers and fruit; however, considering total dry weight, vegetative structures (spurs and extension shoots) had the greatest sink strength for reserves.

$^{13}C$ derived from photoassimilation by each leaf population accumulated in fruit during development (Table 12.3; Ayala and Lang, 2008). The fruiting spur leaves were the greatest source of C throughout fruit development, although in mid-stage III (56 DAFB) when fruit were growing most rapidly, the non-fruiting spur leaves were a comparable C source. Extension shoots become net exporters soon after fruit set. The highest fruit sink activity on a per-unit basis was observed during stage I, although the highest fruit sink strength relative to each leaf population occurred during mid-stage III. At that time, even the extension shoots provided a significant amount of C, indicating that C for optimal fruit growth during the late to final stages of ripening cannot be met solely by spur leaves.

**Table 12.3.** Total $^{13}C$ content in fruit on 2-year-old branches of 'Ulster' on 'GiSelA 6' sweet cherry trees 48 h after $^{13}CO_2$ pulsing of leaves of fruiting spurs (FS), non-fruiting spurs (NFS) and extension shoots (ES). Mean calculations are based on absolute amounts of $^{13}C$ recovered in fruit at each pulse-labelling date (25, 40, 44, 56 and 75 days after full bloom) ($n = 7$). Means within a row followed by the same lower-case letter are not significantly different ($\alpha = 0.05$). (Data from Ayala and Lang, 2008.)

| Days after full bloom | Total $^{13}C$ content (μg $^{13}C$) | | |
|---|---|---|---|
| | FS leaves | NFS leaves | ES leaves |
| 25 | 13,253 a | 10,977 b | 2,622 c |
| 40 | 10,422 a | 6,274 b | 3,952 c |
| 44 | 12,506 a | 7,242 b | 3,476 c |
| 56 | 28,531 a | 24,45 a | 12,073 c |
| 75 | 14,685 a | 8,950 b | 7,916 b |

### *Shoot extension growth as sink and source*

In sweet cherry trees on vigorous rootstocks, the presence of fruit reduced shoot extension growth, but overall, vegetative growth comprised a greater C sink than fruit (Kappel, 1991). In sour cherry on vigorous rootstocks, extension shoots became net carbohydrate exporters at 27% expansion, about 17 days after leaf emergence (Kappes and Flore, 1989; Flore and Layne, 1999). Extension shoot growth of sweet cherry trees on dwarfing or semi-dwarfing rootstocks provides a temporally changing C source for its own growth as well as other sinks such as fruit (Ayala, 2004; Ayala and Lang, 2008; Correa, 2008). In Ayala's (2004) study with 'Ulster' on 'GiSelA 6', extension shoot growth became a C-exporting source for fruit development early in the season (by 25 DAFB; Table 12.2), when only ten leaves had formed. The lowest $^{13}C$ export from this leaf population was detected in early stage III, when the shoot was rapidly elongating and developing additional leaves. However, by mid-stage III when fruit were rapidly accumulating dry matter and terminal shoot growth rate was decreasing (having reached 30 cm in length and 20 leaves), the majority (59%) of the C fixed by the extension shoot leaf population was exported to fruit. These results are consistent with the observations of Roper *et al.* (1987) for vigorous 'Bing' on Mazzard trees as well as those of Correa (2008) for semi-vigorous 'Bing' on 'GiSelA 6' trees, wherein a portion of C for fruit growth was thought to come from shoot leaves when spur leaf photoassimilation was insufficient to meet fruit demand during stage III. As will be seen later in this chapter, this is particularly important for the Super Slender Axe (SSA) and other training systems that are mostly dependent on shoot leaves for fruit growth.

This strong, dynamic competition between shoot extension growth and fruit is the crux of the problem for high crop loads, which can lead to a cascade of negative events, such as reduced extension shoot growth, yield and fruit quality in the current season, plus reduced spur leaf size in the subsequent season (Whiting and Lang, 2004b; Correa, 2008; Villasante *et al.*, 2012). Spur LA is often insufficient to support high-quality fruit development of the spur crop load, so optimal fruit growth often requires shoot LA to supplement that of the spurs. When the LA/F ratio is not optimized (i.e. the crop load is not reduced or additional shoot LA is not promoted by pruning), not only will the current-season fruit size and quality be less than optimal, but the shoot growth and LA will be reduced and flower bud formation for the next year will tend to increase. This creates an even more unbalanced LA/F ratio situation for subsequent years, leading to a progressively weaker tree. The more dwarfing the rootstock, the more quickly this imbalance can occur and to a greater degree (Correa, 2008; Mora, 2008; Villasante *et al.*, 2012). When very young trees are allowed to fruit too heavily, resulting in negligible extension shoot growth, it can be difficult to stimulate adequate new growth for recovery. Once vigour is lost, it is difficult for this type of tree to ever become suitably productive with good fruit quality.

### Source and sink limitations

Storage reserves and photoassimilates from various leaf populations supply C for sink organ growth and maintenance (Farrar and Williams, 1991; Grossman and DeJong, 1995; Flore and Layne, 1999; Basile *et al.*, 2002). Their distribution varies among sink organs and with developmental stage, as suggested above by Whiting and Lang (2004b). Sink growth and development can be restricted due to limited C availability (a 'source limitation') or by the inherent ability of the organ to utilize assimilates (a 'sink limitation') (Basile *et al.*, 2002). Lang (2001a) proposed that sweet cherry carbohydrate reserves are critical for final flower differentiation, bloom and fruit set, and therefore management practices should promote the acquisition of optimized reserves after harvest and into the autumn. From bud swell through bloom and shortly thereafter, cells divide rapidly in young leaves and fruit, ultimately influencing final fruit size and spur LA. The detrimental effects of C source limitations on fruit quality and vegetative growth have been described (Whiting, 2001; Ayala and Lang, 2004, 2008; Whiting and Ophardt, 2005; Correa, 2008). In most sweet cherry varieties, reproductive and vegetative growth occur simultaneously during fruit development (Roper *et al.*, 1988; Ayala, 2004). This generates competition between aerial sinks (i.e. flowers, fruit, spur leaves and shoot extension) for the available C provided by storage reserves and various leaf populations (i.e. fruiting spurs, non-fruiting spurs and extension shoots).

### Sink and source manipulation

Source–sink ratios can be modified in sweet cherry by increasing or decreasing sink strength (e.g. crop load demand for C) or source strength (e.g. LA to provide C). The experimental manipulation of source and sink relationships is useful for developing information about the optimal LA/F ratio for scion/rootstock/environment combinations, since scions and rootstocks vary in precocity, productivity, vigour and growth habit, while orchard environments vary in daily solar radiation, evapotranspiration, vapour pressure deficit, chilling and heat unit accumulation, and soil type. For example, the daily solar radiation during fruit development (primarily May and June in the northern hemisphere) is 35% higher in the cherry-growing region of eastern Washington (e.g. 24.7 MJ $day^{-1}$ at Prosser) compared with the cherry-growing region of western Michigan (e.g. 18.2 MJ $day^{-1}$ at Clarksville). This is one key factor when making orchard decisions to assure an adequate balance between vegetative growth, fruit yield and fruit quality in cherry (Whiting, 2001; Ayala, 2004; Correa, 2008; Ayala and Andrade, 2009). Regions with lower solar energy may require lower crop loads (higher LA/F ratios) to achieve comparable fruit quality and/or shoot growth, or greater light interception per orchard area to achieve comparable fruit yields. With lower incident radiation, source leaves in the interior of the tree canopy photosynthesize less, so canopy architectures that distribute light better (narrower or more-open canopies) may be required to achieve functionally optimal LA/F ratios.

In mature sweet cherry trees, spurs become fruitful in the third year after the shoot forms, and can remain fruitful for several years if light exposure is adequate for continued flower bud formation, although fruit quality tends to decline as spurs age. Thus, trees grown on precocious dwarfing rootstocks can begin yielding significantly in the third year of canopy development, and cumulative fruiting spur populations can increase rapidly, with a correspondingly rapid decrease in LA/F ratio if no intervening tree management measures are taken (Lang, 2001a,b). Therefore, source–sink manipulation, such as selective (or 'precision') pruning to reduce future spur number and increase shoot LA (Lang, 2001a, 2005; Mora, 2008), spur extinction (Correa, 2008; Ayala and Andrade, 2009), or flower or fruit thinning (Whiting and Ophardt, 2005; Whiting *et al.*, 2006) is required to optimize fruit quality.

Whiting (2001) proposed that a typical balanced sweet cherry canopy should have about 5.5 leaves per fruit to achieve optimal fruit size and quality. However, sweet cherry leaf size can be quite variable, generally ranging from 30 to 150 $cm^2$. Current standards for high-value fresh market sweet cherries demand a fruit size of 10–12 g, which in turn is thought to require a LA/F ratio of around 200 $cm^2$ per fruit as an optimized source–sink ratio (Whiting, 2001). This is the equivalent of two to three average shoot leaves or nearly an entire set of average spur leaves. This is about two to three times the recommended leaf number per fruit for sour cherries (Flore and Lakso, 1989), which are about 40–50% the size of fresh market sweet cherries. Consequently, source–sink ratios can be manipulated directly by fruit number reduction or leaf number increase, as well as indirectly by nutrient (Ouzounis and Lang, 2011) or bioregulator (M. Ayala, Santiago, Chile, 2016, personal communication) application strategies that increase leaf size, or pest management practices that maintain leaf health throughout the season to optimize storage reserves for spur leaf growth in early spring (Ayala and Lang, 2015).

Of the various source–sink manipulation strategies, precision pruning is a relatively cheap, fast and effective way to regulate crop load and change C distribution (Lang, 2001b; Villasante *et al.*, 2012). Anticipated outcomes, however, must consider integration of the scion/rootstock/site factors noted above for precision pruning decisions to be most successful. Thinning cuts that remove reproductive meristems and/or improve light distribution in the canopy can improve source–sink relationships, which in turn can improve fruit quality. Selective tipping or heading cuts to previous season extension shoots remove the portion of the shoot upon which future fruiting spurs tend to form most densely (where internodes become shorter as terminal growth slows). This has the effect of reducing future sink demand, while stimulating the formation of one or more new lateral extension shoots, thereby increasing current-season source LA. Less effective is the use of selective heading cuts into older wood of the branch, which directly removes current fruiting spurs to reduce sink demand and stimulates new extension shoot growth to increase source LA, but also removes the section of the shoot that formed the previous year, which would provide non-fruiting spur LA, a significant C source (Ayala and Lang, 2008). Structural renewal pruning (the periodic removal of older fruiting wood structure) improves light distribution within the canopy and 're-sets' source–sink relationships by invigorating the formation of new LA, eliminating old fruiting spurs and initiating new fruiting sites. Dormant pruning also tends to increase spur leaf size in spring (Villasante *et al.*, 2012) since storage reserves for leaf expansion are partitioned to fewer spurs and shoots.

While spur extinction (Claverie and Lauri, 2005; Whiting and Ophardt, 2005; Neilsen *et al.*, 2007) and hand thinning of blossoms or fruit (Whiting and Ophardt, 2005; Lenahan and Whiting, 2006b) are effective crop regulation techniques, they tend to be quite labour intensive and are employed most economically as a complement to precision pruning for cultivars that tend to crop heavily. The challenges for source–sink manipulation limited only to spur extinction include the following: (i) while crop load is reduced selectively, spur removal also eliminates LA since the vegetative bud of the spur is also removed; (ii) the elimination of growing points along the limb can create a situation similar to 'blind nodes'; and (iii) the use of spur extinction in lieu of additional pruning does not promote new extension shoot formation and corresponding source LA (Correa, 2008; Ayala and Andrade, 2009). Whiting and Ophardt (2005) found that 50% spur extinction reduced yield more than 50% blossom thinning, but did not increase fruit size comparably, even though yields were reduced, which was probably due to the loss of spur LA with extinction. Like spur extinction, mechanical blossom thinning can effectively reduce sink demand, but often with some indiscriminate removal of vegetative meristems (source LA) as well. In contrast to spur extinction and mechanical flower thinning, blossom and fruit thinning by hand, while extremely labour intensive, selectively removes only reproductive sinks, improving LA/F ratios without reducing fruiting or non-fruiting spur LA. As with other tree fruits, the earlier that flower or fruit thinning can be imposed, the more rapid the improvement in source–sink ratio and the greater the positive effect on fruit quality. However, growers must consider climatic risks such as postbloom frosts when deciding how soon to impose thinning measures.

## 12.4 Canopy Management

Canopy establishment, tree training and annual pruning (in either summer or winter) are key tree fruit management tools to increase uniformity of light interception and distribution, renew fruiting wood and adjust LA/F ratios. Efficient canopy management begins with designing the cherry orchard for a specific canopy architecture and training system (see section 12.5), using an appropriate rootstock and optimized tree spacing, to rapidly fill the three-dimensional orchard space allotted for each tree and then shift distribution of growth resources to promote fruit production rather than further structural development. This strategy emphasizes having an understanding of the fruit and leaf populations that ultimately will comprise the mature orchard, and then developing the tree architecture accordingly to bring the maximum number of these fruiting sites into production concomitant with canopy fulfilment in the allotted orchard space. Reaching high fruit production before the space is filled may slow tree growth too much to finish filling the space. Not reaching high fruit production by the time the space is filled may result in excessive growth, shading and senescence of potential fruiting sites, or require remedial pruning to prevent excessive shading, which then further invigorates vegetative growth at the potential expense of a continued delay in flower bud formation.

### 12.4.1 Structural establishment

Most modern cherry tree training systems focus on simplifying the tasks for establishing and maintaining a specific cherry tree architecture. However, there are two schools of thought regarding efficiency in canopy structural development: (i) minimal, generalized training techniques that are labour efficient; and (ii) intensive, precise training techniques that may require more initial labour but result in a more uniform tree structure for which maintenance may be more labour efficient or potentially semi-mechanized in future years. The first factor in establishing a suitable tree structure is deciding on whether to plant nursery trees with no or few lateral branches ('whips') or well-branched ('feathered') nursery trees (where available). Typically, well-branched knip (2-year-old) nursery trees are available in Europe, but whip or minimally branched (1-year-old) nursery trees are more common in North and South America. When the training system requires many lateral branches on a single leader, a well-feathered nursery tree can hasten the filling of space and precocious production. Nursery tree 'feather' lateral branches are sylleptic shoots and therefore will not form basal flower buds as would typically occur on a primary lateral branch; however, the nodes on sylleptic branches can potentially form fruitful spurs in 2 years as would normally occur on a primary lateral branch. Therefore, the key for efficient tree establishment with a well-feathered nursery tree is how many uniform and well-distributed sylleptic lateral branches are present; those that are retained provide important future fruiting structure, while those that are too vigorous, too weak or in the wrong location are essentially worthless and must be removed or headed back to a basal vegetative bud for regrowth.

When whip nursery trees are planted and three to five strong laterals are required to form structural scaffolds (as for an open-vase canopy), simply heading the tree at a height just above where the scaffolds are desired is often sufficient and quick. This is also the technique for establishing multiple-leader bush canopy architectures, for which the resulting multiple shoots may be headed a second time during the growing season to double the upright shoot number. When many laterals on a central leader are required, the application of growth regulators, such as mixtures of gibberellins and cytokinins (e.g. Promalin® or Cytolin®), at the green-tip bud stage in spring can promote good lateral formation, as described by Looney (1996). However, growth regulators may have a relatively narrow window for successful application and often exhibit poor activity where spring temperatures are cool or on older wood. Non-chemical means of promoting feather-like lateral shoots on the leader include scoring above individual buds (Fig. 12.5A) or selecting specific buds for retention (around 15–20%) and removing the others (80–85%) (Fig. 12.5B), often called 'bud removal' or 'bud selection' (Lang, 2005; Long *et al.*, 2015). These techniques also can be used to form planar upright leaders on cordon-like training systems, such as the Upright Fruiting Offshoots (UFO) system (Long *et al.*, 2015; Law and Lang, 2016). Bud selection should not be used where winter chilling accumulation is inadequate for good bud development and strong bud break in spring, nor where severe low temperatures may damage buds during winter or early spring.

Recent nursery tree innovations include: (i) potted trees, which transplant with less shock and thus exhibit stronger first-year establishment growth; and (ii) dual-leader (bi-axis) trees, developed by double-budding on the rootstock or by early-season heading of the scion leader in the nursery, for specific training systems that require two leaders (e.g. Musacchi *et al.*, 2015). For the double-budding technique, two scion buds are usually placed directly opposite each other on the rootstock, ostensibly to encourage equal growth that is less common when one growing point is located above the other (as can result from outgrowth of the two most terminal buds on a headed single leader or offset double-budding). Potted nursery trees are more expensive, but have the advantage of an intact, strong root system. They are especially useful for establishing orchards later in spring or for replanting missing trees in an existing orchard, as long as irrigation and leaf protection measures are timely.

Following initial tree training interventions at planting to promote branching, subsequent early training techniques may be desirable, such as the use of clothes pegs to adjust lateral shoot crotch angle, pinching or tying down of stronger new shoots to slow growth, and tying of upright or horizontal shoots for precise positioning. Summer pruning timing and severity can modify reiterative growth and flower bud formation (Guimond *et al.*, 1998b). These and pruning manipulations for older trees, such as sectorial double pruning to adjust resulting shoot angles and numbers, have been discussed in detail by Flore *et al.* (1996) and Long *et al.* (2015).

Fig. 12.5. (A) Sweet cherry lateral branch formation resulting from scoring (note non-spur cluster of flower buds at the base of each lateral shoot, typical of young trees on precocious rootstocks). (B) Selective bud removal to promote lateral branching without scoring or growth regulator use. (Photos courtesy of S. Musacchi (A) and G.A. Lang (B).)

### 12.4.2 Structural maintenance

Sweet cherry vigour is influenced by the scion/rootstock combination, soil type and fertility, climatic conditions and orchard management inputs (mineral nutrition, irrigation, pruning and training system, growth regulators and plant protection). The goal for all orchard canopy architectures is to maintain a structure that balances vegetative growth with reproductive potential (Lang, 2005). Imbalanced LA/F ratios resulting in low vigour can be caused by various factors, such as: (i) inadequate crop load regulation; (ii) improper timing and intensity of pruning (e.g. excessive summer pruning, especially on dwarfing rootstocks); (iii) insufficient irrigation to maintain daily turgor and photosynthesis (especially for trees on dwarfing rootstocks due to their small root systems); (iv) insufficient mineral nutrition, especially N; (v) excessive bending of branches and/or minimal pruning, which can promote the formation of too many fruiting spurs; (vi) improper use of growth regulators that reduce seasonal vegetative growth (e.g. paclobutrazol); and (vii) adverse weather conditions such as high temperatures, low relative humidity and/or high solar radiation. Imbalanced LA/F ratios resulting from high vigour can be caused by: (i) spring frosts that kill flower buds, flowers and/or young fruit, thus reducing reproductive sink demand; (ii) overly vigorous rootstock/scion combinations for the orchard spacing and/or canopy training system; (iii) improper timing and intensity of pruning (e.g. excessive dormant pruning, especially on vigorous rootstocks); and (iv) excessive mineral nutrition, especially N.

Low-vigour sweet cherry trees have fewer spurs (fruiting and non-fruiting), small spur leaves (<30–35 $cm^2$), and greatly decreased lateral and terminal extension shoot growth (<15–20 cm in length with fewer than 10–12 modest-sized leaves) that terminates in late spring or early summer, before harvest. Low-vigour trees are characterized by low LA/F ratios and insufficient LA to supply adequate photoassimilates to reproductive and vegetative growth in spring and early summer, and to storage organs in late summer and autumn. Consequently, fruit quality (i.e. size, weight, soluble solids, firmness) is reduced, and the fruit bruise more easily and have a shorter postharvest life (Facteau and Rowe, 1979; Zoffoli *et al.*, 2008). In high-light environments, sunburn damage to the tree structure (i.e. bark/cambium) and even to fruit can occur. On precocious (especially dwarfing) rootstocks, remediating insufficient vigour can be challenging, generally requiring multiple management inputs such as increased mineral nutrition, regular irrigation, aggressive dormant pruning and significantly reducing crop loads. Any single input is unlikely to improve vigour adequately since the shoot growth–crop load–LA–nutrient availability cycle is strongly interconnected. Vigour-limited trees should be pruned either in winter or early spring to achieve a greater distribution of storage reserves to a reduced number of growing points, thereby promoting a more vigorous vegetative response. Heading cuts stimulate new shoots with larger leaves; thinning cuts remove weak shoots or branches without stimulating weak replacements. After pruning (the first stage of LA/F ratio adjustment), additional thinning of buds, spurs, flowers and/or fruit can further reduce excess fruit sink demand and improve the LA/F ratio (Villasante *et al.*, 2012). On fruiting spurs, thinning of flower buds or flowers is the most effective timing and strategy to increase fruit quality.

Conversely, excessive vigour reduces precocity and promotes greater wood and shoot formation. In vigorous trees with a high LA/F ratio, LA is not a limiting factor for fruit growth, as long as leaves are well illuminated and photosynthesis is optimal. However, excessive vegetative growth tends to create shade, which in turn reduces productivity, and increases pruning costs. Highly vigorous trees develop strong lateral shoots and water sprouts (>1 m in length) in the upper part of the canopy; if these are not eliminated with thinning cuts or managed, shading of reproductive wood and leaves of spurs and shoots can occur within the canopy. As an example, in sweet cherry trees trained as a central leader, some growers do not eliminate water sprouts at the top of the leader but instead bend them horizontally to promote precocious spur formation. After a couple of years, those vigorous shoots (4–5 units initially) become strong limbs full of fruiting spurs and together form a ‘heavy umbrella-type structure’ at the top. This ‘umbrella’ is undesirable since it reduces light interception by lower leaves, and reproductive spurs at the base of scaffolds die, creating blind wood. After a few years, fruiting occurs mostly in the outer part of the canopy, and interior and exterior fruit mature unevenly. Branch renewal becomes less successful due to blind wood at the base of the scaffolds and few latent vegetative meristems on the central axis to initiate new shoots after pruning, and excessive shade within the canopy.

Several diagnostic factors can be used to identify orchards with excessive vigour: the presence of dead spurs at the base of the branches, very large and thin leaves in horizontal positions inside the canopy, and senescent leaves within the canopy in midsummer. To reduce vigour, summer pruning (after terminal bud set) using thinning cuts will reduce canopy LA that would contribute to the building of storage reserves, without stimulating vigorous replacement-shoot development. The following spring, the only pruning to be done should utilize thinning cuts and focus on improved light interception to promote uniform fruit maturation. When these steps are not enough to control vigour, and deficit irrigation is not an option (e.g. in non-arid temperate climates), root pruning in the spring can significantly reduce shoot growth (Flore *et al.*, 1996), as can branch bending towards or below the horizontal to reduce apical dominance and induce spur formation. Similarly, the use of growth inhibitors such as paclobutrazol (Looney, 1996) or prohexadione calcium (Elfving *et al.*, 2003, 2004,

2005) during active shoot growth can provide temporary to significant (depending on the number of applications) growth control, as well as increase flower bud density and size (Guak *et al.*, 2005; Manriquez *et al.*, 2005; Zhang and Whiting, 2011b; Cares *et al.*, 2014).

### 12.4.3 Crop load management

Management of excessive sweet cherry crop loads by selective pruning and/or thinning of reproductive structures is used widely by growers around the world. Thus, regulation of fruiting has become a key labour input for genotypes with high fruit set, having the aim of lowering the number of fruit and adjust the LA/F ratio to achieve a balance among major sinks and sources prior to harvest. In most cases, winter pruning and, depending on the scion/rootstock combination, subsequent adjustment of the final fruit number by thinning of spurs (spur extinction), individual buds, flowers and/or fruit are used by growers. The strategy used to regulate crop load varies depending on the productivity of the scion/rootstock combination, tree vigour, tree age, climatic conditions (frost risk), labour availability, training system and market destination, among others. It is not a recipe to repeat every year, but must be analysed and planned accurately, from season to season, considering the characteristics and condition of the orchard. Growers must integrate a series of productive and physiological concepts to make pruning and/or subsequent thinning decisions each season. The most important factors that growers should consider prior to deciding on the crop load regulation strategy are discussed below.

#### *Scion/rootstock productivity potential*

Since fruit set can vary from low to high, depending on the interaction between the cultivar, rootstock and climate, each orchard requires a specific crop load regulation strategy. Traditionally, sweet cherry orchards were established with vigorous trees at low density, resulting in many years of low yields until the trees completely filled their orchard space (Flore *et al.*, 1996). However, the use of dwarfing precocious rootstocks has increased yield in low-fruit-set cultivars (e.g. 'Regina', 'Santina®', 'Benton'), which consequently require more intensive pruning and thinning strategies. In contrast, highly productive or self-fertile cultivars (e.g. 'Lapins', 'Sumtare' (Sweetheart™), 'Royal Dawn®') often are grafted on more vigorous and less precocious rootstocks to promote moderate fruit sets and more balanced crop loads. For example, growing the highly productive 'Sweetheart' on vigorous 'Colt' or Mazzard rootstock, in an appropriate training system, may only require annual pruning to achieve a good balance between yield and fruit quality (Einhorn *et al.*, 2011).

#### *Climatic conditions*

Sweet cherry production can be influenced significantly by winter chilling accumulation, spring heat accumulation, frost and rain during bloom and fruit set, hail during fruit development, rain near harvest and high summer temperatures. Each cultivar has specific genetic requirements for chilling and heat, as well as susceptibility to rain-induced cracking and heat-induced fruit doubling (see Chapters 7 and 8, this volume). Therefore, site selection is critical to guarantee cultivar adaptation and profitability.

When growers establish cultivars in suboptimal production areas, it makes crop load management more complex. Recent projections for a changing global climate may impact an increasing number of sour and sweet cherry growers (e.g. Zavalloni *et al.*, 2008; Measham *et al.*, 2014). For example, if a sweet cherry cultivar, with a high chilling requirement (e.g. 'Bing') on a highly productive rootstock (e.g. 'GiSelA 6') is established in a location with mild winters (i.e. low chilling unit accumulation), in particularly warm winters (e.g. El Niño climatic events) bloom is likely to be reduced and delayed, with lower fruit set and yield. Thus, in addition to considering the use of dormancy-breaking chemicals such as hydrogen cyanamide or calcium–nitrogen fertilizers (CAN-17, Erger), growers must adopt changes in the timing and intensity of pruning and thinning. When chilling is insufficient, pruning should be delayed (e.g. late winter or early spring) and reduced in intensity (e.g. 500 versus 400 fruiting spurs per tree), and thus thinning of buds or flowers might be unnecessary.

### *Crop load management during orchard development and maturation*

During orchard establishment, crop regulation is not a major management issue; precise pruning and training are the main focus. However, with precocious rootstocks, anticipation of significant cropping can begin as early as the second or third year after planting (Lang, 2000). Fruiting precocity requires growers to be more precise in tree structural development and crop load management, since growth in the year of planting has the potential to provide some cropping sites (basal non-spur flowers) in the second year and significant cropping sites (fruiting spur flowers) in the third. Therefore, mistakes made in tree development in years 1 and 2 that require later correction can reduce early yield potential; similarly, mistakes made in crop load management in years 2 and 3 can lead to potential overcropping and poor fruit quality in years 4 and beyond (Lang, 2001a). Lang (2005) projected that, without intervention by pruning, trees on precocious rootstocks would require about a 25% reduction in potential fruiting sites in year 4 to achieve a LA/F ratio suitable for achieving full genetic potential in fruit size and quality, and this proportional reduction in cropping potential would increase to as much as 45% in subsequent years if the tree was left unpruned. Law and Lang (2016) demonstrated how training decisions made in the year of planting could promote initially higher yields in the second and third years at the expense of lower cumulative yields by the fifth year. Elimination of nursery tree nodes that otherwise would form fruiting spurs, while promoting greater structural development of future fruiting sites during year 1, lowered early yields while the canopy was being developed, but advanced overall canopy growth, resulting in a higher annual yield by year 4 and a higher cumulative yield by year 5. Understanding the developmental progression of leaf and fruit populations in each year of canopy growth provides the knowledge necessary to anticipate future cropping potential and implement strategies to maintain favourable LA/F ratios (Lang, 2005). This is particularly important for trees on dwarfing and semi-dwarfing rootstocks, which are difficult to reinvigorate once they have been allowed to overcrop heavily enough to stunt vegetative growth.

Generally, there is an inverse relationship between sweet cherry vegetative vigour and reproductive capacity (Flore and Layne, 1999). Flower bud induction tends to be suppressed by greater shoot growth and promoted by weak or minimal shoot growth. Thus, when vigour is too strong, few flower buds form and excessive vegetative growth may occur, including some new shoot nodes that form sylleptic shoots concomitant with primary shoot extension. Conversely, when growth is stunted, flower buds may form not only at the basal nodes of current-season growth, but acropetally at almost every node along the shoot, even to the terminal in extreme cases. Trees in the former condition yield poorly, and trees in the latter condition produce small fruit of poor quality, store minimal reserves for the next year, and become extremely difficult to invigorate to achieve a healthy balance of growth and cropping.

In high-density orchards on precocious scion/rootstock combinations, crop management strategies should be initiated prior to significant orchard fruiting, that is, at least from the second year after planting onwards (Lang, 2005). Structural pruning should be focused on promoting a balance between reproductive bud formation and elongation of extension shoots; this balance will vary according to training system. Selection of primary canopy structure or scaffolds, and light heading (tipping) cuts on future fruit-bearing shoots when they are only 1 year old, can provide appropriate initial crop load regulation by efficiently eliminating anticipated fruiting capacity before it becomes excessive. Heading cuts usually generate new extension shoots, increasing the shoot LA, which is an important C source for the rapid phase of fruit growth in stage III (Ayala and Lang, 2008). This tipping of 1-year-old growth to regulate the future crop load should begin no later than the year before the tree begins significant production on spurs, thereby maintaining balanced LA/F ratios and contributing to the regulation of the anticipated crop loads two seasons ahead. This is a key technique for the Central Axis (Villasante *et al.*, 2012) and Tall Spindle Axe (TSA) (Long *et al.*, 2015) canopy training systems to moderate future crop loads while promoting lateral shoot formation. Heading cuts or tipping of 1-year-old branches also avoids the formation of long branches with no lateral ramification and excessive spur number, which can become too dense to maintain an optimal LA/F ratio. If the excessive crop load has not been anticipated, pruning into the 2-year-old portion of the shoot (removing up to 50% of the excessive spurs) to reduce crop load can improve fruit size somewhat, with the best results from pruning pre-bloom to shuck-fall (Gutzwiler and Lang, 2001).

This strategy for high-density orchard crop load regulation is essentially the opposite of previous training techniques for trees on traditional vigorous, non-precocious rootstocks such as 'Colt' or Mazzard; since heading cuts delay flower bud induction and promote vigour, non-precocious vigorous trees are pruned mainly with thinning cuts in summer after terminal bud set. The resulting branches often are bent to a horizontal or below horizontal position to reduce vigour and increase fruiting spur formation. Typically, most lateral shoots are removed so that only fruiting and non-fruiting spurs are borne on long, narrow branches. This is a key technique for the Solaxe canopy training system (Claverie *et al.*, 1997). Somewhat similarly, the Kym Green Bush (KGB) (Green, 2005) and UFO (Ampatzidis and Whiting, 2013) training systems promote long narrow fruiting sections comprised of spurs, although these are oriented vertically rather than being bent or tied to horizontal or below horizontal orientations. For any of these three or other similar canopy fruiting unit architectures, a consequence of the primarily spur-fruiting population is that fruit demand for carbohydrate is supplied by spur LA only. This has the potential for an unbalanced LA/F ratio due to a lack of supplemental lateral shoot leaves (Ayala and Lang, 2008). Thus, optimizing spur leaf size, reducing spur number where they occur densely (spur extinction), and/or thinning of flowers or fruit can be critical to optimizing LA/F ratios for good fruit quality.

### *Plant growth regulator use for crop load management*

Since sweet cherry crop load management generally was not an issue prior to the advent of precocious, productive, vigour-controlling rootstocks, studies of chemical thinning agents have not been extensive to date. Therefore, unlike apples, for which a range of chemical thinning agents for blossoms and fruitlets are used routinely to adjust annual crop loads, sweet cherry growers have few such options. Most chemicals studied to date are caustic agents for blossom thinning, such as ammonium thiosulfate, fish oil plus lime sulfur, and surfactants (vegetable oil, Tergitol™) (Whiting *et al.*, 2006). In addition to damaging floral organs, these have been shown to damage leaf photosynthesis as well, for up to 23 days when applied at bloom (Lenahan and Whiting, 2006b) or a week when applied as a postbloom thinner (Lenahan and Whiting, 2006a, 2008). However, as with many chemical thinning agents tested for use in stone fruits, results thus far have been inconsistent and unreliable (Lenahan and Whiting, 2006b; Whiting *et al.*, 2006). Early spring applications of gibberellins (GAs) to inhibit flower bud initiation for the following year have also been studied (Lenahan *et al.*, 2006). Both $GA_3$ and $GA_{4+7}$, applied from stage I to beginning of stage II, have been shown to effectively and reliably reduce flower bud formation in a negatively linear fashion relative to concentration (Lenahan *et al.*, 2008). The reduction in fruit number resulted in larger fruit with higher soluble solids, although higher rates reduced flower bud formation too much and resulted in some floral organ abnormalities. All strategies that reduce crop loads at bloom (blossom thinners or flower bud inhibitors) provide both the greatest potential for increasing fruit size and quality as well as the highest risk of additional crop reduction from postbloom frost events.

During the past 20 years, preharvest use of $GA_3$, applied at the beginning of stage III fruit growth (straw colour), has become routine in North and South American cherry production to enhance fruit firmness and size, and can also be used to delay ripening to spread out harvest in large blocks of the same variety (Kappel and MacDonald, 2002, 2007; Lenahan *et al.*, 2006, 2008). Use of $GA_3$ does not affect fruit soluble solids accumulation consistently, but has been associated with higher titratable acidity (Choi *et al.*, 2002; Cline and Trought, 2007; Zhang and Whiting, 2011a,b). Treated fruit also generally maintain better postharvest quality (Horvitz *et al.*, 2003; Özkaya *et al.*, 2006; Einhorn *et al.*, 2013). Commercial formulations of this natural growth regulator are even available for use in organic cherry production. While the use of $GA_3$ can enhance fruit quality, it also has been associated with increased susceptibility to rain-induced fruit cracking (Cline and Trought, 2007). Foliar applications of $GA_3$ also have been used experimentally to promote lateral extension shoot formation in young trees trained with a central axis (M. Ayala, Santiago, Chile, 2016, personal communication). Application of prohexadione calcium with $GA_3$ at the beginning of stage II had a synergistic effect on delay of ripening, by up to 7 days, as well as on increased size and improved postharvest performance (Zhang and Whiting, 2011a,b). Prohexadione calcium alone had less of an effect on fruit size and firmness than $GA_3$ alone. Initial studies have shown that the ethylene synthesis inhibitor, aminovinylglycine (AVG), can increase fruit set when applied at bloom, although results have varied by cultivar (Bound *et al.*, 2014). Nevertheless, it is promising that AVG may provide a potential crop load management tool for cherry cultivars that set poorly.

### *Precision crop load management and orchard cropping records*

Smaller trees with more simplified canopy structures make precision management decisions more feasible (Lang, 2005). Detailed orchard crop load and pruning records provide valuable information for making annual decisions about pruning and thinning, since both have a direct effect not only on current-season production (i.e. yield and quality), but also on cropping factors for the next two growing seasons (i.e. reproductive wood formation and LA/F ratios). Thus, growing seasons are interdependent, albeit with a strong annual climatic variable, and therefore crop regulation decisions must consider impacts across years. Annual variables to monitor (in representative trees within the orchard) for precise crop load and canopy management decisions include:

- fruiting spur number per tree after pruning;
- extension shoot number per tree after pruning;
- pruning date and intensity (weight of prunings per tree);
- average flower bud density (bud number per spur);
- average flower density (flower number per bud);
- annual yield (kg per tree and t $ha^{-1}$);
- fruit size distribution (and percentage premium or export grade, if applicable);
- average shoot leaf number and size;
- average spur leaf number and size; and
- leaf mineral nutrient content

This information will allow estimation of the potential crop load for a given year and facilitate accurate planning for pruning intensity, mineral nutrition requirements and subsequent potential adjustment of fruit number by thinning. Inferences can be made about annual fruit set, the proportions of the spur and shoot fruit populations, the adequacy of vigour for renewal shoot growth, the adequacy of nutrients for leaf and fruit size optimization, etc. Evaluation of such records over several years will provide insight into each orchard's optimized fruit capacity for trees at maturity, and the pruning and mineral nutrition programmes required to maintain optimal LA:F ratios year in and year out.

## 12.5 Canopy Architectures and Training Systems

Prior to the widespread availability of precocious, vigour-controlling rootstocks, traditional sweet and sour cherry training and production systems were reviewed by Flore *et al.* (1996). In the 20 years since then, high-density orchards have become more commonplace, with many innovative ideas for tree training and management. A compilation of sweet cherry canopy development and management guidelines for several contemporary orchard training systems has recently been published online and as an application for tablet computers (Long *et al.*, 2015). While little has changed in sour cherry production systems over the past two decades, innovations are beginning to take place in the adaptation of new technologies and tree architectures for mechanical harvest; these are reviewed in Chapter 18 (this volume).

As with all tree fruit, yields and fruit quality are directly related to light interception efficiency and distribution. A common question among cherry growers is whether high-density orchards, which become productive earlier than traditional lower-density orchards, will remain productive as long as the more traditional orchards. The productive life of these orchards, when good tree health is maintained, could be 35 years or longer. Since most high-density orchards to date are younger than 20 years old, this remains an open question until those orchards have reached similar ages. However, the tree management strategy inherent in most high-density training systems provides a logical basis for the likelihood that optimal productivity can be equal to that of traditional orchards. This strategy is the minimization of permanent tree structure, coupled with regular renewal of temporary fruit-bearing canopy structure. This shifts the distribution of tree growth resources (i.e. mainly C and N) away from the previous strategy of building an extensive trunk and scaffold structure that would take many years to achieve and results in a mantle of well-exposed LA and fruit-bearing surface overlaying a more shaded interior structure with lower productivity and fruit quality. Instead, contemporary high-density canopy training strategies direct growth resources towards maintaining a proportionally greater young, well-exposed LA and fruit-bearing surface, often achieving an optimum production potential within 3–5 years of planting. This is accomplished, in part, by increasing the number of trees per hectare concomitant to the decrease in permanent tree structure and canopy volume. The key to sustained productivity after reaching earlier optimum yields is a management plan to periodically remove the oldest portion of the canopy and promote its renewal. The periodic cycling of new growth of moderate vigour replaces ageing fruiting structure that would be increasingly shaded and of lower productivity. This maintains a relatively young, well-exposed fruit-bearing canopy that is expected to sustain yields and fruit quality as long as new shoots can be generated, conceivably for many decades. How this recycling and renewal is achieved can vary by training system and is an area of active physiological research.

### 12.5.1 Multi-dimensional/self-supported systems

Traditional cherry canopy architectures are three-dimensional, generally creating self-supported trees that 'stand alone' as symmetrical canopies of some depth that are harvested by pickers moving around the perimeter of the tree, usually with tall ladders to access the upper regions of the canopy. These architectures can be adapted to higher densities, usually by reducing tree height and canopy volume, particularly when grown on vigour-controlled rootstocks.

#### *Single-leader canopies*

At moderate densities on rootstocks that impart moderate levels of vigour, single central-leader canopy architectures can be maintained as stand-alone trees. Alternatively, at high densities on semi-dwarfing to dwarfing rootstocks, single-leader trees can be maintained as tall, narrow, conical canopies or adapted to near-continuous fruiting walls centred on the single leader. Examples of such canopy training systems include variations of the traditional spindle (conical or pyramidal) tree architecture, such as Zahn (Flore *et al.*, 1996), Vogel (Long *et al.*, 2015), Solaxe (Claverie *et al.*, 1997; Lauri *et al.*, 1998) and TSA (Long *et al.*, 2015). The Solaxe system, developed originally for apples, has been adopted in sweet cherries for trees on more vigorous rootstocks, since its governing principles of limb bending and fruiting primarily on spurs on long pendant shoots improves precocity and productivity. However, these traits make the Solaxe generally unsuitable for cherries on dwarfing or semi-dwarfing rootstocks, due to reduced LA/F ratios that can lead to overcropping and small fruit size. The concept of spur extinction (Lauri and Claverie, 2005), the permanent removal of a number of spurs to reduce the crop load (but also removing some spur LA), was developed to address these challenges, although it has not been widely effective. Similarly, the TSA system was derived from the Tall Spindle in apple, with differences including beginning with a whip (rather than a well-feathered) nursery tree, and annual dormant heading of 1-year-old lateral shoot growth to promote lateral branching and pre-emptive removal of the future dense spur section that forms below the annual growth juncture. This latter technique is similar to spur extinction in that spurs are permanently removed, but unlike extinction, the reduction in fruiting site density is accompanied by increased LA from new shoot growth in spring after pruning.

### *Multiple-leader canopies*

Multiple-leader canopies generally are suitable for trees on rootstocks of moderate to high vigour at wider spacing, creating orchards of moderate density (i.e. 1000–1250 trees $ha^{-1}$). The greater the number of leaders, the more the typically vigorous upright growth of sweet cherry can be 'diluted' or distributed among the leaders to maintain a moderate tree stature. Therefore, multi-leader canopy architectures such as Open Vase or Goblet ('Gobelet' in France) have long been traditional techniques for growing sweet cherries and will not be discussed here. Variations have been developed over the years for specific regions, such as the Steep Leader in Washington State and the Spanish Bush in Spain (Robinson and Domínguez, 2014; Long *et al.*, 2015), and the Aussie Bush and KGB in Australia (Green, 2005). The innovative contributions of the Steep Leader include the hybridizing of open-vase and spindle canopy elements: the multiple leaders help dilute vigour but are grown more vertically and closely ('steep') together, and the lateral scaffolds and branches are developed only towards the outside of the narrow leaders, in a partial conical shape for improved light distribution from top to bottom and with a narrow open middle between the scaffolds. This maintains greater productivity lower in the Steep Leader canopy than for a traditional Open Vase or Goblet tree.

The innovative contributions of the KGB include very simple training and annual pruning rules (Long *et al.*, 2015), with the goal of forming 20–25 upright leaders during the first 2–3 years of development. This not only dilutes vigour very effectively, but also provides the opportunity to retain the upright leaders having most uniform vigour as the majority for bearing fruit and remove the overly vigorous (least fruitful) and weakest (poorest fruit quality) during development. Furthermore, the many moderate-vigour fruiting leaders can be momentarily pulled down to easily pick the crop entirely from the ground, establishing a truly pedestrian orchard (Fig. 12.6). Finally, the largest one or two upright leaders are removed annually, thereby renewing fruiting units to maintain a relatively young spur population in the canopy that results in large fruit of uniform quality when the LA/F ratio is adequate. Since fruiting on KGB leaders is preferentially on spurs, some fruit thinning or spur extinction may be

Fig. 12.6. Multiple-leader sweet cherry trees trained as Kym Green Bush (KGB) canopies with many spur-bearing upright fruiting units, in Chile.

required on highly productive cultivars. Cultivars with upright growth habits and that do not readily produce lateral shoots (e.g. 'Lapins') are best for developing KGB canopies. Some rootstocks, such as the 'GiSelA' series, that tend to promote greater lateral branching, can make maintenance of narrow upright leaders in KGB canopies more problematic. On less vigorous sites and/or on vigour-limiting rootstocks, fewer upright leaders should be developed, proportional to the level of rootstock-induced vigour control. Factors that reduce vigour also increase the difficulty of annual renewal of the largest upright leader after the trees reach full production, due to C competition with the crop load at maturity.

### 12.5.2 Planar/trellised systems

With orchard labour becoming more expensive and difficult to source throughout most cherry-growing regions, there is significant interest in increased labour efficiency and mechanization of orchard operations, from pruning (hedging) to crop thinning to harvesting (pickers on self-propelled mechanized platforms or even mechanical or robotic harvesters; see Chapter 18, this volume). Orchards amenable to increased mechanization must be comprised of trees with simplified, uniformly structured canopies, ideally as a continuous wall of fruit-bearing surface in which fruit is readily accessible to pickers or machine harvesters. Contemporary cherry canopy architectures that are trained as relatively narrow (planar) continuous fruiting walls generally require tree support, such as one or more trellis wires, to maintain the precision of canopy orientation, efficiency of light interception and uniformity of fruiting surface development in the allotted narrow canopy space. The goal is to facilitate the unidirectional movement of labourers (and/or machines) parallel to the tree row and not around individual trees. These architectures must be developed with higher densities, but depending on training system, do not necessarily require vigour-controlling rootstocks. Precocity, however, is a critical trait for quickly shifting the distribution of growth resources from filling the allotted canopy space to producing full crop loads, which then helps to maintain a moderate vigour that achieves a balanced leaf-to-fruit ratio.

The narrower the planar canopy, the closer the tree rows can be spaced, increasing tree density and yield potential per hectare. Zhang *et al.* (2015) found no reduction in light interception for narrow planar sweet cherry canopies at tree height-to-row spacing ratios of 1.25 (e.g. a mature tree height of 3.75 m and a row spacing of 3.0 m) compared with those at height-to-row spacings of 1.0. Vertical planar canopies also help facilitate the use of narrow row covers for rain or hail protection while minimizing adverse heat accumulation due to the narrow profile of the cover (Fig. 12.7A, B). Likewise, such canopies can facilitate individual row netting systems for exclusion of insects (such as spotted wing drosophila, *Drosophila suzukii* and other fruit flies) and birds (Fig. 12.7C).

### *Single-leader planar canopies*

Single-leader planar tree architectures can be developed with precocious semi-dwarfing to dwarfing rootstocks for very-high-density orchards (e.g. 2000–4800 trees ha$^{-1}$), such as the SSA system described by Long *et al.* (2015) (Fig. 12.7B, C). Trees are planted about 0.5 m apart, which provides some vigour control due to root competition for soil resources, and are subject to drastic annual pruning to promote fruiting primarily on non-spur buds at the base of the previous season's shoot growth. This system is best for cultivars that readily form lateral branches, are not too upright in growth habit and are very productive on basal fruit buds. Root pruning may be required to maintain the modest vigour necessary for balanced fruitfulness, as excessive growth in such a high-density system can lead to shading and overly vigorous shoots that form few basal flower buds. Such tree training can also be used for very-high-density SSA V-trellis orchards in which alternate trees are planted in opposite directions at approximately 60–70° angles to form dual inclined planes on a multi-wire trellis for greater light interception per orchard area. Pruning to promote basal flower bud fruiting is similar to that for the vertical SSA.

The French fruiting wall ('Mur Fruitier') is based on a similar concept for apples (Masseron, 2002) developed at the Centre Technique Interprofessionel des Fruits et Légumes (CTIFL). The basic canopy architecture is that of a central-leader tree planted relatively closely (e.g. 1.5 m) with 3.5–4.0 m between rows and 2.7–4.0 m mature tree height (Charlot and Pinczon du Sel, 2016). As with the SSA system, multiple lateral branches should be developed and distributed along the leader during the first 3 years. However, these are not pruned back annually to basal buds, but are allowed to form spurs. Beginning after the third leaf, trees are summer-pruned mechanically (hedged) at 40–50 cm from the leader, generally about 3 weeks prior to harvest to enhance light penetration to the fruiting sites. This system is best for cultivars that are very productive, branch readily and have a spreading to willowy growth habit (versus an upright habit). The summer hedging is supplemented by hand-pruning during dormancy every year or every other year, depending on vigour.

Alternatively, with semi-dwarfing to semi-vigorous rootstocks, single-leader planar architectures can be developed along multi-wire trellises as horizontal Espalier canopies (Fig. 12.7A). The development of more extensive, permanent lateral shoot structure along each trellis wire requires more vigour than for SSA canopies, and consequently allows planting at moderate to high densities, depending on rootstock and site vigour. Fruiting populations are comprised primarily of spurs on each horizontal branch, creating a narrower planar canopy than the SSA or Mur Fruitier system at maturity. To intercept greater light and increase orchard productivity, single-leader Espalier trees can also be used to develop Espalier V-trellis orchards by alternate tree planting in opposite directions at approximately 60–70° angles to form dual inclined planes of horizontal spur-fruiting branches.

**Fig. 12.7.** (A–C) Vertical planar sweet cherry orchards with narrow single-row covers for rain protection in New Zealand (A), for hail protection in Italy (B) and for rain protection plus insect exclusion side-netting in Italy (C). (D–F) The 1999 planar bilateral cordon sweet cherry canopy architecture trial at Washington State University, Washington, USA (D, E) that gave rise to the Upright Fruiting Offshoots (UFO) canopy training system (F).

### Multiple-leader planar canopies

Multiple-leader planar tree architectures can be developed with precocious vigorous to dwarfing rootstocks for moderate- to high-density orchards. As with three-dimensional canopies, the use of multiple leaders provides a strategy for the proportional partitioning of vigour into more than one vertical leader to achieve moderate annual growth, balanced productivity, decreased annual pruning costs and more efficient use of photosynthates. The greater the number the leaders, the more successfully vigorous upright growth can be 'diffused' among the leaders to maintain a planar fruiting wall of moderate stature. Various multiple-leader planar canopy architectures are feasible.

The Mur Fruitier system described above for a single leader can also be developed with multiple leaders aligned within the tree row (e.g. bi-axis, tri-axis, or a four- to six-leader palmette, depending on scion/rootstock/site vigour). The UFO canopy structure creates the narrowest architecture (Fig. 12.8), with fruiting primarily on spurs comparable to the Espalier but with the fruit-bearing leaders oriented vertically rather than horizontally (described by Long *et al.*, 2015). The ideas behind this architecture, which is the oldest of the modern planar canopy designs (developed in 1999 by G.A. Lang, East Lansing, Michigan, USA, 2001, personal communication) include: (i) utilization of the natural strongly acrotonic growth habit of sweet cherry; (ii) deconstruction of the canopy into simplified fruiting units for ease of estimating and fine-tuning crop loads and LA/F ratios; and (iii) optimizing uniform light distribution throughout the canopy. The architecture of multi-leader upright fruiting units arising from a bilateral cordon (Fig. 12.7D, E) subsequently evolved into a single-cordon version achieved by planting the nursery tree at a 45° angle (Fig. 12.7F) to begin developing the upright fruit-bearing leaders at planting rather than growing two leaders-to-be-cordons during the first year in the orchard (M. Whiting, Prosser, Washington, USA, 2008, personal communication; Law and Lang, 2016), much like a Drapeau Marchand (Moreno *et al.*, 1998) training system but with a mostly horizontal cordon and vertical leaders rather than a 45° cordon and 45° reverse-angled leaders. The concept of selective annual leader renewal

**Fig. 12.8.** The spread within the tree row of the permanent fruiting zone of Tall Spindle Axe (TSA) (A), Super Slender Axe (SSA) (B) and Upright Fruiting Offshoots (UFO) (C) sweet cherry canopies at Michigan State University, Michigan, USA.

was adopted from the KGB system in 2007 and, more recently, the use of bi-axis nursery trees, like those developed for the BiBaum® dual vertical leader training system in Italy, has now been applied to UFO-like cherry training in New Zealand to begin developing the upright fruit-bearing leaders on bilateral cordons during the planting year (S. Tustin, Cromwell, New Zealand, 2014, personal communication). Like single-leader planar trees, the multiple upright leader planar architecture can be adopted for UFO V- or Y-trellis orchards to increase light interception and potential yields. UFO trees planted alternately such that entire cordons and their upright leaders fill one side of the trellis, and then the other, become UFO V-trellis orchards; UFO trees planted in the row middle with the upright leaders alternating in orientation to each side of the trellis become UFO Y-trellis orchards (Fig. 12.9A). The latter are best developed with semi-vigorous to vigorous rootstocks so that vigour is adequate to fill both sides of the trellis.

The aforementioned dual vertical leader (bi-axis) nursery trees can be utilized for vertical planar or dual-plane training systems. When both leaders are oriented parallel within the row (usually north–south), they can be trained as two SSA canopies with half the number of trees required for single-leader SSAs. The dual-leader SSA trees should be planted on somewhat more vigorous rootstocks than for single-leader SSA trees. When both leaders are oriented perpendicular to the row (usually east–west), they can be trained as an SSA Y-trellis orchard with half the number of trees as for a single-leader SSA V-trellis, or as Espalier Y-trellis trees that fruit primarily on spurs on the horizontal branches (Fig. 12.9B). Similar training decisions can be made to develop multi-leader planar vertical or dual-plane canopies using tri-axis nursery trees or, by heading at planting, four or more leader trees, creating narrow Trident, Candelabra or Palmette (Moreno *et al.*, 1998) canopy architectures.

### 12.5.3 Cultivar and rootstock genotype influence on training system

As canopy architectures become more uniform and precise with regard to the management of leaf and fruit populations, the influence of the cultivar and rootstock on genetic architectural traits increases in importance. Training systems such as KGB and UFO, which focus on spur-fruiting sites and multiple upright leaders for production, perform better when matched to cultivars having strongly upright spur-bearing growth habits (e.g. 'Lapins'). Cultivars that readily produce lateral branches instead of retaining

**Fig. 12.9.** Dual planar Y-trellis sweet cherry canopies that fruit primarily on spurs developed as alternating vertically angled fruiting units of the Upright Fruiting Offshoots (UFO) architecture (A) compared with horizontal fruiting units of the Espalier architecture (B), both in Chile.

spurs (e.g. 'Hedelfinger' and 'Santina'), that bear a high proportion of non-spur fruit on 1-year-old shoots instead of spurs (e.g. 'Kordia' and 'Regina'), or rootstocks that promote less upright (more horizontal) shoot growth (e.g. 'GiSelA 5') are less well-suited for such training systems.

Similarly, training systems such as the SSA, which focus on high numbers of weak lateral shoots to form basal non-spur-fruiting sites, can be poorly suited to cultivars that initiate too many basal flower buds (thereby creating large segments of blind wood annually) or that fail to form enough lateral shoots. In a trial carried out by Musacchi *et al.* (2015), comparing production of 11 cultivars in three very-high-density (1905–5714 trees $ha^{-1}$) training systems, only four cultivars ('Ferrovia', 'Giorgia', 'Grace Star' and 'Sylvia') produced impressive cumulative yields over the first 7 years. For the SSA and V-system, which consistently had higher yields than the spindle system, these ranged from about 50 t $ha^{-1}$ ('Ferrovia'/'GiSelA 5' and 'Giorgia'/'GiSelA 6') to about 45 t $ha^{-1}$ ('Grace Star'/'GiSelA 5', 'Grace Star'/'GiSelA 6') to around 40 t $ha^{-1}$ ('Sylvia'/'GiSelA 5'). The remaining cultivars (which included 'Regina'/'GiSelA 5', 'Kordia'/'GiSelA 5', 'Summit'/'GiSelA 5' and 'Black Star'/'GiSelA 5') generally had cumulative yields of less than 20 t $ha^{-1}$ over the same period.

## 12.6 Future Research Trends and Needs

The future of sweet cherry production systems undoubtedly will continue to move towards simplified, narrow tree architectures that facilitate increasingly efficient use of labour and partial mechanization of tasks, while improving fruit quality and ripening uniformity, precision in optimizing LA/F balance and ease of implementing covering strategies. As orchards become easier to cover, research on covering technologies (e.g. plastics that transmit or reflect selective light wavelengths) and operational strategies for manipulation of bloom and ripening and/or protection from rain, frost, disease and pests will be of increasing value. Desirable research to support these goals includes: (i) developing a better understanding of the relationship(s) between vegetative growth rates and reproductive meristem formation; (ii) node-by-node canopy planning and improved techniques for promoting precise meristem determination and development; and (iii) seasonal timings for providing specific growth resources to optimize shoot and fruit development. Better knowledge in these areas will facilitate improved tree and crop growth modelling, which will advance precision in determining and achieving specific fruiting unit and LA growth targets. Knowledge of optimized growth rates for balanced LA/F ratio development, combined with knowledge of rootstock effects on scion vigour, scion growth habit and training system strategies for diluting vigour, will contribute to better adaptation of specific rootstock–scion–system combinations for varying edaphoclimatic conditions. As canopy architectures are optimized for precise structural placement of leaf and fruit populations, orchard management will be simplified, but a greater emphasis will need to be placed on preplanting decisions to integrate scion and rootstock genetic traits with the modifying influence of prevailing environmental factors (climate and soil).

Additionally, research to improve insight into optimization of storage reserve levels, spur leaf size potential, flower number per bud and floral bud number per spur will improve precision in orchard management decisions that advance crop load balancing at the primordial leaf-to-fruit ratio and cell number level. Future advances in genetic manipulation of these traits could have a significant impact. More precise control of meristem activation and differentiation includes not only the promotion of lateral shoots for structural development and basal and spur reproductive buds, but also the prevention of sylleptic shoot formation (and thus loss of potential spurs) when undesirable, and the activation of latent (epicormic) buds on visibly blind sections of leader or branch growth to fill in or renew fruiting units as needed for an optimized canopy structure.

The future of sour cherry production systems is also moving towards smaller, more simplified tree structures, albeit for hedgerow-type orchards that can be mechanically harvested by over-the-row machinery (see Chapter 18, this volume). Thus, important research questions that must be pursued include optimization of tree structure and fruit-bearing habit (spur versus basal fruiting sites on 1-year-old shoots) most suitable for interfacing with the harvester mechanisms for minimal fruit damage and maximum fruit removal. Strategies for renewal of fruiting sites and prevention of excessive growth are near-term canopy management research objectives that have yet to be elucidated for sour cherries.

Finally, the various components of these potential advances in orchard production (e.g. higher tree densities, investments in trellising and mechanization, requirements for more labour during tree establishment to reduce labour for later maintenance and harvest, covering systems) must be evaluated economically for returns on investment, years to break-even cash flow and profitable production sustainability.

## References

Ampatzidis, Y.G. and Whiting, M.D. (2013) Training system affects sweet cherry harvest efficiency. *HortScience* 48, 547–555.

Ayala, M. (2004) Carbon partitioning in sweet cherry (*Prunus avium* L.) on dwarfing precocious rootstocks during fruit development. PhD thesis, Michigan State University, East Lansing, Michigan.

Ayala, M. and Andrade, M.P. (2009) Effects of fruiting spur thinning on fruit quality and vegetative growth of sweet cherry (*Prunus avium*). *Ciencia e Investigacion Agraria* 36, 443–450.

Ayala, M. and Lang, G. (2004) Examining the influence of different leaf populations on sweet cherry fruit quality. *Acta Horticulturae* 636, 481–488.

Ayala, M. and Lang, G.A. (2008) $^{13}C$-Photoassimilate partitioning in sweet cherry on dwarfing rootstocks during fruit development. *Acta Horticulturae* 795, 625–632.

Ayala, M. and Lang, G.A. (2015) $^{13}C$-Photoassimilate partitioning in sweet cherry (*Prunus avium* L.) during early spring. *Ciencia e Investigacion Agraria* 42,191–203.

Ayala, M., Bañados, P., Thielemann, M. and Toro, R. (2014) Distribution and recycling of canopy nitrogen storage reserves in sweet cherry (*Prunus avium*) fruiting branches following $^{15}N$-urea foliar applications after harvest. *Ciencia e Investigacion Agraria* 41, 71–80.

Azarenko, A.N., Chozinski, A. and Brutcher, L. (2008) Nitrogen uptake efficiency and partitioning in sweet cherry is influenced by time of application. *Acta Horticulturae* 795, 717–721.

Balmer, M. and Blanke, M. (2005) Developments in high density cherries in Germany. *Acta Horticulturae* 667, 273–277.

Basile, B., Mariscal, M.J., Day, K.R., Johnson, R.S. and DeJong, T.M. (2002) Japanese plum (*Prunus salicina* L.) fruit growth: seasonal pattern of source/sink limitations. *Journal of the American Pomological Society* 56, 86–93.

Basile, B., Marsal, J. and DeJong, T.M. (2003) Daily shoot extension growth of peach trees growing on rootstocks that reduce scion growth is related to daily dynamics of stem water potential. *Tree Physiology* 23, 695–704.

Bound, S.A., Close, D.C., Jones, J.E. and Whiting, M.D. (2014) Improving fruit set of Kordia and Regina sweet cherry with AVG. *Acta Horticulturae* 1042, 285–292.

Calabro, J.M., Spotts, R.A. and Grove, G.G. (2009) Effect of training system, rootstock, and cultivar on sweet cherry powdery mildew foliar infections. *HortScience* 44, 481–482.

Cares, J., Sagredo, K.X., Cooper, T. and Retamales, J. (2014) Effect of prohexadione calcium on vegetative and reproductive development in sweet cherry trees. *Acta Horticulturae* 1058, 357–363.

Charlot, G. and Pinczon du Sel, S. (2016) High-density training systems. Proceedings of the EU COST Action FA 1104: Sustainable production of high-quality cherries for the European market. Cost Training School, Balandran, France, February 2016. Available at: http://www.bordeaux.inra.fr/cherry/docs/dossiers/Activities/Meetings/2016%2002%2002-04%20Training%20School_Balandran/Presentations/Charlot2_Balandran2015.pdf (accessed 17 January 2017).

Choi, C., Wiersma, P.A., Toivonen, P. and Kappel, F. (2002) Fruit growth, firmness and cell wall hydrolytic enzyme activity during development of sweet cherry fruit treated with gibberellic acid (GA3). *Journal of Horticultural Science and Biotechnology* 77, 615–621.

Claverie, J. and Lauri, P.E. (2005) Extinction training of sweet cherries in France – appraisal after 6 years. *Acta Horticulturae* 667, 376–372.

Claverie, J., Lespinasse, J.M. and Lauri, P.E. (1997) Arcina® Fercer, Tabel® Edabriz and Solaxe. Pour un nouveau type de verger. *Réussir Fruits et Légumes* 155, 4.

Cline, J.A. and Trought, M. (2007) Effect of gibberellic acid on fruit cracking and quality of Bing and Sam sweet cherries. *Canadian Journal of Plant Science* 87, 545–550.

Correa, J.E. (2008) Effect of spur thinning on the photoassimilate translocation and the morphological characteristics of sweet cherry fruit (*Prunus avium* L.) in the combination 'Bing'/'Gisela 6'. MSc thesis, Pontificia Universidad Católica de Chile, Santiago, Chile.

Costas, P., Ko, J.-H., Lang, G.A., Iezzoni, A.F. and Han, K.-H. (2009) Rootstock-induced dwarfing in cherries is caused by differential cessation of terminal meristem growth and is triggered by rootstock specific gene regulation. *Tree Physiology* 29, 927–936.

DeJong, T.M. and Grossman, Y.L. (1995) Quantifying sink and source limitations on dry matter partitioning to fruit growth in peach trees. *Physiologia Plantarum* 95, 437–443.

Einhorn, T., Laraway, D. and Turner, J. (2011) Crop load management does not consistently improve crop value of 'Sweetheart'/'Mazzard' sweet cherry trees. *HortTechnology* 21, 546–553.

Einhorn, T., Wang, Y. and Turner, J. (2013) Pre-harvest applications of gibberellic acid ($GA_3$) improve fruit firmness and postharvest fruit quality of late-season sweet cherry cultivars. *HortScience* 48, 1010–1017.

Elfving, D.C., Lang, G.A. and Visser, D.B. (2003) Prohexadione-Ca and ethephon reduce shoot growth and increase flowering in young, vigorous sweet cherry trees. *HortScience* 38, 293–298.

Elfving, D.C., Whiting, M.D., Lang, G.A. and Visser, D.B. (2004) Growth and flowering responses of sweet cherry cultivars to prohexadione-Ca and ethephon. *Acta Horticulturae* 636, 122–129.

Elfving, D.C., Visser, D.B. and Lang, G.A. (2005) Effects of prohexadione-calcium and ethephon on growth and flowering of 'Bing' sweet cherry. *Acta Horticulturae* 667, 439–446.

Facteau, T.J. and Rowe, K.E. (1979) Factors associated with surface pitting of sweet cherry. *Journal of the American Society for Horticultural Science* 104, 707–710.

Farrar, J.F. and Williams, M.L. (1991) The effects of increased atmospheric carbon dioxide and temperature on carbon partitioning, source-sink relations and respiration. *Plant, Cell and Environment* 14, 819–830.

Flore, J.A. (1985) The effect of water stress and vapor pressure gradient in stomatal conductance, water use efficiency and photosynthesis of fruit crops. *Acta Horticulturae* 171, 207–218.

Flore, J.A. (1994) Stone fruit. In: Schaffer, B. and Anderson, P.C. (eds) *Handbook of Environmental Physiology of Fruit Crops. Vol. 1: Temperate Crops*. CRC Press, Boca Raton, Florida, pp. 233–270.

Flore, J.A. and Lakso, A.N. (1989) Environmental and physiological regulation of photosynthesis in fruit crops. *Horticultural Reviews* 11, 111–157.

Flore, J.A. and Layne, D.R. (1999) Photoassimilate production and distribution in cherry. *HortScience* 34, 1015–1019.

Flore, J.A., Kesner, C.D. and Webster, A.D. (1996) Tree canopy management and the orchard environment: principles and practices of pruning and training. In: Webster, A.D. and Looney, N.E. (eds) *Cherries: Crop Physiology, Production and Uses*. CAB International, Wallingford, UK, pp. 259–277.

Gibeaut, D.M., Whiting, M.D. and Einhorn, T. (2017) Time indices of multiphasic development in genotypes of sweet cherry are similar from dormancy to cessation of pit growth. *Annals of Botany* 119, 465–475.

Gonçalves, B., Moutinho-Pereira, J., Santos, A., Silva, A.P., Bacelar, E., Correia, C. and Rosa, E. (2005) Scion–rootstock interaction affects the physiology and fruit quality of sweet cherry. *Tree Physiology* 26, 93–104.

Grassi, G., Millard, P., Wendler, R., Minnotta, G. and Tagliavini, M. (2002) Measurement of xylem sap amino acid concentrations in conjunction with whole tree transpiration estimates spring N remobilization by cherry (*Prunus avium* L.) trees. *Plant, Cell and Environment* 25, 1689–1699.

Grassi, G., Millard, P., Gioaccchini, P. and Tagliavini, M. (2003) Recycling of nitrogen in the xylem of *Prunus avium* trees starts when spring remobilization of internal reserves declines. *Tree Physiology* 23, 1061–1068.

Green, K. (2005) High density cherry systems in Australia. *Acta Horticulturae* 667, 319–324.

Grossman, Y.L. and DeJong, T.M. (1995) Maximum fruit growth potential and seasonal patterns of resource dynamics during peach growth. *Annals of Botany* 75, 553–560.

Guak, S., Beulah, M. and Looney, N. (2005) Controlling growth of sweet cherry trees with prohexadione-calcium: its effect on cropping and fruit quality. *Acta Horticulturae* 667, 433–438.

Gucci, R., Petracek, P.D. and Flore, J.A. (1991) The effect of fruit harvest on photosynthetic rate, starch content, and chloroplast ultrastructure in leaves of *Prunus avium* L. *Advances in Horticultural Science* 5, 19–22.

Guimond, C.M., Andrews, P.K. and Lang, G.A. (1998a) Scanning electron microscopy of floral initiation in sweet cherry. *Journal of the American Society for Horticultural Science* 123, 509–512.

Guimond, C.M, Lang, G.A. and Andrews, P.K. (1998b) Timing and severity of summer pruning affects flower initiation and shoot regrowth in sweet cherry. *HortScience* 33, 647–649.

Gutzwiler, J. and Lang, G.A. (2001) Sweet cherry crop load and vigor management on Gisela rootstocks. *Acta Horticulturae* 557, 321–325.

Ho, L.C. (1988) Metabolism and compartmentation of imported sugars in sink organs in relation to sink strength. *Annual Review of Plant Physiology and Plant Molecular Biology* 39, 355–378.

Horvitz, S., Lopez Camelo, A.F., Yommi, A. and Godoy, C. (2003) Application of gibberellic acid to 'Sweetheart' sweet cherries: effects on fruit quality at harvest and during cold storage. *Acta Horticulturae* 628, 311–316.

Kappel, F. (1991) Partitioning of above-ground dry matter in 'Lambert' sweet cherry trees with or without fruit. *Journal of the American Society for Horticultural Science* 116, 201–205.

Kappel, F. and MacDonald, R.A. (2002) Gibberellic acid increases fruit firmness, fruit size and delays maturity of 'Sweetheart' sweet cherry. *Journal of the American Pomological Society* 56, 219–222.

Kappel, F. and MacDonald, R.A. (2007) Early gibberellic acid sprays increase firmness and fruit size of 'Sweetheart' sweet cherry. *Journal of the American Pomological Society* 61, 38–43.

Kappes, E.M. and Flore, J.A. (1986) Carbohydrate balance models for 'Montmorency' sour cherry leaves, shoots and fruits during development. *Acta Horticulturae* 184, 123–128.

Kappes, E.M. and Flore, J.A. (1989) Phyllotaxy and stage of leaf and fruit development influence initiation and direction of carbohydrate export from sour cherry leaves. *Journal of the American Society for Horticultural Science* 114, 642–648.

Keller, J.D. and Loescher, W.H. (1989) Nonstructural carbohydrate partitioning in perennial parts of sweet cherry. *Journal of the American Society for Horticultural Science* 114, 969–975.

Lang, G.A. (2000) Precocious, dwarfing, and productive – how will new cherry rootstocks impact the sweet cherry industry? *HortTechnology* 10, 719–725.

Lang, G.A. (2001a) Intensive sweet cherry orchard systems – rootstocks, vigor, precocity, productivity and management. *Compact Fruit Tree* 34, 23–26.

Lang, G.A. (2001b) Critical concepts for sweet cherry training systems. *Compact Fruit Tree* 34, 70–73.

Lang, G.A. (2005) Underlying principles of high density sweet cherry production. *Acta Horticulturae* 667, 325–335.

Lang, G.A. (2008) Sweet cherry orchard management: from shifting paradigms to computer modeling. *Acta Horticulturae* 795, 597–604.

Lang, G.A. and Lang, R.J. (2009) VCHERRY – an interactive growth, training, and fruiting model to simulate sweet cherry tree development, yield and fruit size. *Acta Horticulturae* 803, 235–242.

Lang, G.A., Olmstead, J.W. and Whiting, M.D. (2004) Sweet cherry fruit distribution and leaf populations: modeling canopy dynamics and management strategies. *Acta Horticulturae* 636, 591–599.

Lauri, P.E. and Claverie, J. (2005) Sweet cherry training to improve fruit size and quality – an overview of some recent concepts and practical aspects. *Acta Horticulturae* 667, 361–366.

Lauri, P.E. and Claverie, J. (2008) Sweet cherry tree architecture, physiology, and management; towards an integrated view. *Acta Horticulturae* 795, 605–614.

Lauri, P.E., Claverie, J. and Lespinasse, J.M. (1998) The effect of bending on the growth and fruit production of INRA Fercer® sweet cherry. *Acta Horticulturae* 468, 411–417.

Law, T.L. and Lang, G.A. (2016) Planting angle and meristem management influence sweet cherry canopy development in the "Upright Fruiting Offshoots" training system. *HortScience* 51, 1010–1015.

Layne, D.R. and Flore, J.A. (1993) Physiological responses of *Prunus cerasus* to whole-plant source manipulation. Leaf gas exchange, chlorophyll fluorescence, water relations and carbohydrate concentrations. *Physiologia Plantarum* 88, 45–51.

Lenahan, O. and Whiting, M.D. (2006a) Fish oil plus lime sulphur shows potential as a sweet cherry post-bloom thinning agent. *HortScience* 41, 860–861.

Lenahan, O. and Whiting, M.D. (2006b) Physiological and horticultural effects of sweet cherry chemical blossom thinners. *HortScience* 41, 1547–1551.

Lenahan, O.M. and Whiting, M.D. (2008) Chemical thinners reduce sweet cherry net $CO_2$ exchange, stomatal conductance and chlorophyll fluorescence. *Acta Horticulturae* 795, 681–684.

Lenahan, O., Whiting, M.D. and Elfving, D.C. (2006) Gibberellic acid inhibits floral bud induction and improves 'Bing' sweet cherry fruit quality. *HortScience* 41, 654–659.

Lenahan, O., Whiting, M.D. and Elfving, D.C. (2008) Gibberellic acid is a potential sweet cherry crop management tool. *Acta Horticulturae* 795, 513–516.

Loescher, W., Roper, T. and Keller, J. (1986) Carbohydrate partitioning in sweet cherries. *Proceedings of the Washington State Horticultural Society* 81, 240–248.

Loescher, W., McCamant, T. and Keller, J. (1990) Carbohydrate reserves, translocation, and storage in woody plant roots. *HortScience* 25, 274–281.
Long, L., Brewer, L. and Kaiser, C. (2014) Cherry rootstocks for the modern orchard. *Compact Fruit Tree* 47, 24–28.
Long, L., Lang, G., Musacchi, S. and Whiting, M. (2015) Cherry training systems. Pacific Northwest Extension Publication PNW667, Oregon State University, Washington State University and the University of Idaho in cooperation with Michigan State University.
Looney, N.E. (1996) Principles and practice of plant bioregulator usage in cherry production. In: Webster, A.D. and Looney, N.E. (eds) *Cherries: Crop Physiology, Production and Uses*. CAB International, Wallingford, UK, pp. 279–295.
Maguylo, K., Lang, G.A. and Perry, R.L. (2004) Rootstock genotype affects flower distribution and density of 'Hedelfinger' sweet cherry and 'Montmorency' sour cherry. *Acta Horticulturae* 636, 259–266.
Manriquez, D., Defilippi, B. and Retamales, J. (2005) Prohexadione-calcium, a gibberellin biosynthesis inhibitor, can reduce vegetative growth in 'Bing' sweet cherry trees. *Acta Horticulturae* 667, 447–452.
Masseron, A. (2002) *Pommier, le Mur Fruitier*. Centre Technique Interprofessionel des Fruits et Légumes (CTIFL), Paris, France.
McCamant, T. (1988) Utilization and transport of storage carbohydrates in sweet cherry. MSc thesis, Washington State University, Pullman, Washington.
Measham, P.F., Quentin, A.G. and MacNair, N. (2014) Climate, winter chill, and decision-making in sweet cherry production. *HortScience* 49, 254–259.
Mora, L.C. (2008) El efecto de la poda en la distribución de $^{13}C$ y $^{15}N$ durante la fase III de crecimiento del fruto en cerezo dulce (*Prunus avium* L) (project title). Pontificia Universidad Católica de Chile, Santiago, Chile.
Moreno, J., Toribio, F. and Manzano, M.A. (1998) Evaluation of palmette, Marchand, and vase training systems in sweet cherry varieties. *Acta Horticulturae* 468, 485–489.
Musacchi, S.M., Gagliardi, F. and Serra, S. (2015) New training systems for high-density planting of sweet cherry. *HortScience* 50, 59–67.
Neilsen, G., Kappel, F. and Neilsen, D. (2007) Fertigation and crop load affect yield, nutrition and fruit quality of 'Lapins' sweet cherry on Gisela 5 rootstock. *HortScience* 42, 1456–1462.
Oliveira, C.M. and Priestley, C.A. (1988) Carbohydrate reserves in deciduous fruit trees. *Horticultural Reviews* 10, 403–430.
Olmstead, J.W., Iezzoni, A. and Whiting, M.D. (2007) Genotypic differences in sweet cherry (*Prunus avium* L.) fruit size are primarily a function of cell number. *Journal of the American Society for Horticultural Science* 132, 697–703.
Olmstead, M.A., Lang, N.S., Lang, G.A., Ewers, F.W. and Owens, S.A. (2004) Characterization of xylem vessels in sweet cherries (*Prunus avium* L.) on dwarfing rootstocks. *Acta Hortiulturae* 636, 129–135.
Olmstead, M.A., Lang, N.S., Lang, G.A., Ewers, F.W. and Owens, S.A. (2006) Examining the vascular pathway of sweet cherries grafted onto dwarfing rootstocks. *HortScience* 41, 674–679.
Ouzounis, T. and Lang, G.A. (2011) Foliar applications of urea affect nitrogen reserves and cold acclimation of sweet cherries (*Prunus avium* L.) on dwarfing rootstocks. *HortScience* 46, 1015–1021.
Özkaya, O., Dündar, Ö. and Küden, A. (2006) Effect of preharvest gibberellic acid treatments on postharvest quality of sweet cherry. *Journal of Food, Agriculture, and Environment* 4, 189–191.
Robinson, T.L. (2005) Developments in high density sweet cherry pruning and training systems around the world. *Acta Horticulturae* 667, 269–272.
Robinson, T.L. and Domínguez, L.I. (2014) Comparison of the modified Spanish Bush and the Tall Spindle cherry production systems. *Acta Horticulturae* 1058, 45–53.
Roper, T.R. and Loescher, W.H. (1987) Relationships between leaf area per fruit and fruit quality in Bing sweet cherry. *HortScience* 22, 1273–1276.
Roper, T.R., Keller, J.D., Loescher, W.H. and Rom, C.R. (1988) Photosynthesis and carbohydrate partitioning in sweet cherry: fruiting effects. *Physiologia Plantarum* 72, 42–47.
Roper, T.R., Loescher, W.H., Keller, J. and Rom, C.R. (1987) Sources of photosynthate for fruit growth in Bing sweet cherry. *Journal of the American Society for Horticultural Science* 112, 808–812.
Sams, C.E. and Flore, J.A. (1983) Net photosynthetic rate of sour cherry (*Prunus cerasus* L. 'Montmorency') during the growing season with particular reference to fruiting. *Photosynthesis Research* 4, 307–316.
Santos, A., Santos-Ribeiro, R., Cavalheiro, J., Cordeiro, V. and Lousada, J. (2006) Initial growth and fruiting of 'Summit' sweet cherry (*Prunus avium*) on five rootstocks. *New Zealand Journal of Crop and Horticultural Science* 34, 269–277.

Thielemann, M., Toro, R. and Ayala, M. (2014) Distribution and recycling of canopy nitrogen storage reserves in sweet cherry (*Prunus avium* L.) fruiting branches following [15] N-urea foliar applications after harvest. *Acta Horticulturae* 1020, 353–361.

Toldam-Andersen, T.B. (1998) The seasonal distribution of $^{14}$C-labelled photosynthates in sour cherry (*Prunus cerasus*). *Acta Horticulturae* 468, 531–540.

Tombesi, S., Johnson, R.S., Day, K.R. and DeJong, T.M. (2010) Relationships between xylem vessel characteristics, calculated axial hydraulic conductance and size-controlling capacity of peach rootstocks. *Annals of Botany* 105, 327–331.

Vercammen, J. and Vanrykel, T. (2009) Use of Gisela 5 for sweet cherries. *Acta Horticulturae* 1020, 395–400.

Villasante, M., Godoy, S., Zoffoli, J.P. and Ayala, M. (2012) Pruning effects on vegetative growth and fruit quality of 'Bing'/'Gisela 5' and 'Bing'/'Gisela 6' sweet cherry trees (*Prunus avium*). *Ciencia e Investigacion Agraria* 39, 117–126.

West, T., Sullivan, R., Seavert, C. and Long, L. (2012) Orchard economics: the costs and returns of establishing and producing high-density sweet cherries in Wasco county. Oregon State University Extension Bulletin AEB 0032, Corvallis, Oregon.

Whiting, M.D. (2001) Whole canopy source-sink relation and fruit quality in 'Bing' sweet cherry trees on a dwarfing precocious rootstock. PhD thesis, Washington State University, Pullman, Washington.

Whiting, M.D. and Lang, G.A. (2001) Flow rate, air delivery pattern, and canopy architecture influence temperature and whole-canopy net $CO_2$ exchange of sweet cherry. *HortScience* 36, 691–698.

Whiting, M.D. and Lang, G.A. (2004a) Effects of leaf area removal on sweet cherry vegetative growth and fruit quality. *Acta Horticulturae* 636, 467–472.

Whiting, M.D. and Lang, G.A. (2004b) 'Bing' sweet cherry on the dwarfing rootstock Gisela 5: crop load effects on fruit quality, vegetative growth, and carbon assimilation. *Journal of the American Society for Horticultural Science* 129, 407–415.

Whiting, M.D. and Ophardt, D. (2005) Comparing novel sweet cherry crop load management strategies. *HortScience* 40, 1271–1275.

Whiting, M.D., Lang, G. and Ophardt, D. (2005) Rootstock and training system affect sweet cherry growth, yield and fruit quality. *HortScience* 40, 582–586.

Whiting, M.D., Ophardt, D. and McFerson, J.R. (2006) Chemical blossom thinners vary in their effect on sweet cherry fruit set, yield, fruit quality, and crop value. *HortTechnology* 16, 66–70.

Zavalloni, C. (2004) Evaluation of nitrogen-fertilizer uptake, nitrogen-use and water-use efficiency in sweet cherry (*Prunus avium* L.) on dwarfing and standard rootstocks. PhD thesis, Michigan State University, East Lansing, Michigan.

Zavalloni, C., Andresen, J.A., Black, J.R., Winkler, J.A., Guentchev, K., Piromsopa, K., Pollyea, A. and Bisanz, J.M. (2008) A preliminary analysis of the impacts of past and projected future climate on sour cherry production in the Great Lakes region of the USA. *Acta Horticulturae* 803, 123–130.

Zhang, C. and Whiting, M.D. (2011a) Improving 'Bing' sweet cherry fruit quality with plant growth regulators. *Scientia Horticulturae* 127, 341–346.

Zhang, C. and Whiting, M.D. (2011b) Pre-harvest foliar application of Prohexadione-Ca and gibberellins modify canopy source–sink relations and improve quality and shelf-life of 'Bing' sweet cherry. *Plant Growth Regulation* 65, 145–156.

Zhang, J., Whiting, M.D. and Zhang, Q. (2015) Diurnal pattern in canopy light interception for tree fruit orchard trained to an upright fruiting offshoots (UFO) architecture. *Biosystems Engineering* 129, 1–10.

Zoffoli, J.P., Muñoz, S., Valenzuela, L., Reyes, M. and Barros, F. (2008) Manipulation of 'Van' sweet cherry crop load influences fruit quality and susceptibility to impact bruising. *Acta Horticulturae* 795, 877–882.

# 13 Invertebrate and Vertebrate Pests: Biology and Management

**Nikolaos T. Papadopoulos,[1]* Sławomir A. Lux,[2] Kirsten Köppler[3] and Tim Beliën[4]**
[1]*University of Thessaly, Volos, Greece;* [2]*inSilico-IPM, Konstancin-Jeziorna, Poland;* [3]*Center for Agricultural Technology Augustenberg (LTZ), Karlsruhe Germany;* [4]*Proefcentrum Fruitteelt VZW, Sint-Truiden, Belgium*

## 13.1 Introduction

Sour and sweet cherry trees co-evolved with a complex of associated indigenous organisms, such as insects, mites, birds and mammals. Many of them thrive in the contemporary cherry-production regions and environments, and some attain the status of a pest of economic concern. Although mammal (rodent) and avian pests may cause significant damage, the most notorious belong to a wide variety of insect families such as Tephritidae, Drosophilidae, Torticidae, Sessiidae, Cecidomyidae, Diaspididae, Coccidae, Aphididae, Tenthredinidae, Tingidae, Curculionidae, Cerambycidae, Scarabaeidae, Scolytidae, and Buprestidae (Table 13.1). Among these, frugivorous insects infesting ripening fruit pose a special challenge, because the bulk of the sweet cherry crop is consumed immediately after harvest, fresh and unprocessed, which largely restricts management options. The universal key pest of cherries all over Europe and western Asia is the European cherry fruit fly, *Rhagoletis cerasi*. Recent invasions of alien fruit-infesting pests, such as *Drosophila suzukii* and *Rhagoletis cingulata*, add to the challenge and complexity of pest management. In addition, over the last few decades, banning of effective insecticides, such as dimethoate, has established a new more challenging environment to protect cherries from major pests.

This chapter provides an overview of the most important cherry pests, taking into account their destructiveness, distribution and regularity of occurrence, and discusses the main approaches and trends in their management.

## 13.2 Description, Biology, Significance and Management of Cherry Pests

### 13.2.1 European cherry fruit fly, *Rhagoletis cerasi* (L.)

#### *Distribution*

The European cherry fruit fly, *R. cerasi*, belongs to the family Tephritidae (order Diptera) (known as true fruit flies), which includes a large number of damaging pests of fruit and vegetables, many of regional or global concern, including several present in the Mediterranean region, such as the Mediterranean fruit fly, *Ceratitis capitata* (Wiedemann), the olive fly, *Bactrocera oleae* (Rossi), the peach fruit

* nikopap@uth.gr

**Table 13.1.** List of key cherry pests of general and local importance

| Pest species | Order | Family | Tree part infested | Pest importance |
|---|---|---|---|---|
| *Rhagoletis cerasi* (L.) | Diptera | Tephritidae | Maturing–ripening fruit | Key, ubiquitous, regular |
| *Drosophila suzukii* (Matsumura) | Diptera | Drosophilidae | Ripe fruit | Key, ubiquitous, regular |
| *Sphaerolecanium prunastri* (Fonscolombe) | Hemiptera: Homoptera | Coccidae | Twigs, branches, rarely leaves and fruit | Low, local, occasional |
| *Myzus cerasi* Fabricius | Hemiptera: Homoptera | Aphididae | Leaves, buds, young twigs | Moderate, local, occasional |
| *Adoxophyes orana* (Fischer von Röslerstamm) | Lepidoptera | Torticidae | Leaves, fruit | Low, local, occasional |
| *Archips podana* Scopoli | Lepidoptera | Torticidae | Leaves, fruit | Low, local, occasional |
| *Archips rosana* (L.) | Lepidoptera | Torticidae | Leaves, fruit | Low, local, occasional |
| *Spilonota ocellana* (Denis & Schiffermüller) | Lepidoptera | Torticidae | Buds, leaves | Low, local, occasional |
| *Operophtera brumata* (L.) | Lepidoptera | Geometridae | Leaves, small fruit | Low, local, occasional |
| *Panonychus ulmi* (Koch) | Trombidiformes | Tetranychidae | Leaves, young shoots | Low, local, occasional |
| *Lyonetia clerkella* (L.) | Lepidoptera | Lyonetiidae | Leaves | Low, local, occasional |
| *Anthonomus rectirostris* (L.) | Coleoptera | Curculionidae | Buds, flowers | Low, local |
| *Dasineura tortrix* (Loew) | Diptera | Cecidomyiidae | Leaves, young shoots | Low, local, occasional |
| *Caliroa cerasi* (L.) | Hymenoptera | Tenthredinidae | Leaves | Low, local, occasional |
| *Rhagoletis cingulata* (Loew) | Diptera | Tephritidae | Maturing–ripening fruit | Low, regional, regular |
| *Pseudaulacaspis pentagona* (Targioni-Tozzetti) | Hemiptera: Homoptera | Diaspididae | Twigs, branches, fruit, rarely on leaves | Low, local, occasional |
| *Quadraspidiotus perniciosus* (Comstock) | Hemiptera: Homoptera | Diaspididae | Twigs, branches, fruit, rarely on leaves | Low, local, occasional |
| *Tetranychus viennensis* (Zacher) | Trombidiformes | Tetranychidae | Leaves, young shoots | Minor ubiquitous, regular |
| *Halyomorpha halys* (Stål) | Hemiptera | Pentatomidae | | Undetermined, local |

fly, *Bactrocera zonata* (Saunders), and the Ethiopian or lesser pumpkin fly, *Dacus ciliatus* Loew. In contrast to most tropical and subtropical members of the tephritid family, which are multivoltine and notorious for a broad spectrum of their fruit hosts, members of the temperate genus *Rhagoletis* are univoltine and stenophagous (Bush, 1966). Several species occur in North America, but out of the few native to Europe, only *R. cerasi* is of major importance for cherry production. Its current geographical distribution spans from western Asia (Caspian and Caucasus regions, Asia Minor and western Siberia) to western Europe (Portugal), spreading from Norway and Sweden in the north down to Crete and Sicily in the south.

### Host range

The European cherry fruit fly infests mainly sweet and occasionally tart cherries and fruits of several other *Prunus* and *Lonicera* spp., especially *Lonicera xylosteum* L.

### Life cycle

*Rhagoletis cerasi* completes one generation per year, or rarely one every second year. Exceptionally, some of the pupae may emerge during the same season, but these do not seem to reproduce (Köppler, 2008). Late in spring, adults emerge from puparia residing in the soil beneath the canopy of the host tree. Adult emergence starts 10–40 days after cherry bloom, and usually is well synchronized with the fruit growth and expansion phase, which precedes ripening. The process of emergence can be extended in time to 30–50 days (although 60–80% of adults emerge within 2 weeks), depending on the local temperatures, farm topography, slope exposure, soil moisture and cover. *R. cerasi* adults are shiny black (Fig. 13.1A), with transparent wings bearing four distinct black zones. The dorsal part of the metathorax (scutellum) is bright yellow–orange in colour, while the eyes are metallic brown–green. Females are substantially larger (~4.1 mm long) than males (~3.5 mm long). The newly emerged flies are less mobile, and move up to the nearest canopy and seek sugar and protein food sources to feed and become reproductively mature. Depending on local conditions, it takes approximately 5–15 days for adults to reach maturity and mate. Both males and females are extremely polygamous and mating takes place on oviposition sites, where males exercise fruit guarding (Katsoyannos, 1979; Jaastad, 1998a,b). The existence of male sexual pheromone, of short-range attraction and mostly having an aphrodisiac effect, has been demonstrated (Katsoyannos, 1979). Mated females search for ripe or ripening fruit on which to oviposit. Fecundity may vary widely, from one to ten eggs daily up to 80–300 eggs per female over a lifetime, depending on food, mating, hosts and weather conditions, while adult longevity ranges from 1 to 2 months and spans the cherry fruiting season (Moraiti *et al.*, 2012).

The fruit becomes attractive for oviposition and suitable for egg and larval development when its mesocarp is 2–3 mm thick, which is marked by a fruit hue change from dark green to yellowish- or reddish-green. A single elongated whitish egg (0.75 mm long, 0.25 mm wide) is deposited in the mesocarp, and usually a single larva occurs in each infested fruit (Fig. 13.1C). After oviposition, the female deposits a strong oviposition-deterrent pheromone on the newly infested fruit to prevent additional ovipositions (Katsoyannos, 1975) and thus intraspecific larval competition within the fruit. Multiple fruit infestations occur seldom, only when the ratio of *R. cerasi* females to the available fruit is high.

The first-instar larvae hatch 3–7 days postoviposition and immediately initiate feeding in the mesocarp. After completing three instars within the fruit, fully grown larvae (third instar; ~15 mm long) leave the fruit, drop to the ground and pupate 3–7 cm deep in the soil underneath the canopy of the host trees, enclosed in a pale yellow puparium, which serves as an effective buffer to environmental stresses and predators.

Pupae enter into an obligate summer–winter diapause, terminated in the middle of winter (Papanastasiou *et al.*, 2011; Moraiti *et al.*, 2014). From that point on, pupae remain in a state of quiescence, becoming responsive to thermal accumulation, which

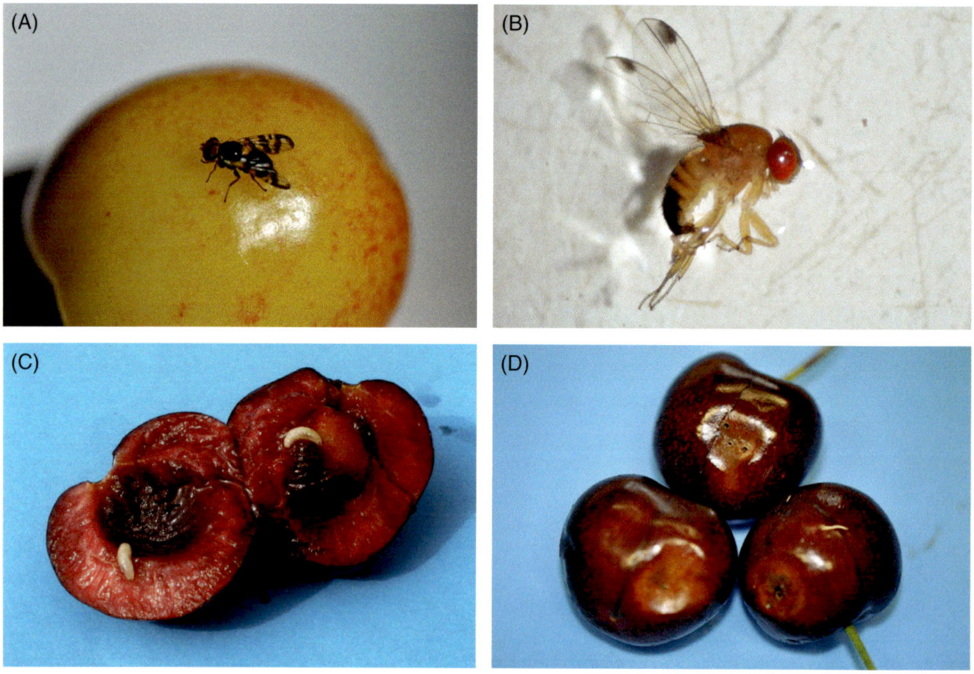


**Fig. 13.1.** Adult *Rhagoletis cerasi* (A) and *Drosophila suzukii* (B) and the respective infested fruit (C, D). (Photos courtesy of N.T. Papadopoulos and Proefcentrum Fruitteelt VZW, Zoology Department and DAT (C. de Schaetzen).)

gradually promotes development and leads to adult emergence (Fig. 13.2). Soil temperatures regulate both diapause termination and postdiapause development. An unusual duration of chilling conditions during winter (either shorter or longer than normal) may lead to life-cycle extension for one additional year (Moraiti *et al.*, 2014). Interestingly, annual diapause is genetically determined, while its prolongation to the next season constitutes an adaptive plastic response to prevailing local temperature conditions, and serves as a bet-hedging life-history strategy to enhance survival chances in unpredictable habitat.

## Damage

*Rhagoletis cerasi* is considered the key pest of cherry cultivation all over Europe and western Asia. The damage is caused by larval feeding in the mesocarp of the infested fruit, usually followed by secondary bacterial and fungal infections. Late fruit infestation is not always easy to detect. Oviposition stings inflicted soon before harvest are easily overlooked in the field or in the packinghouse, and frequently the infestation only becomes apparent on grocery shelves and consumers' tables.

The risk of fruit infestation is determined by the degree of alignment in the timing of the pest's peak fecundity and the period of fruit susceptibility to infestation (suitability for pest development). In very early-ripening cultivars, the alignment is imperfect, because fruit largely ripen and are harvested before females reach maturity and full fecundity potential, and consequently the infestation is usually minimal or negligible, even in the absence of pest control. Medium- and late-ripening cultivars bear the highest infestation risk, and the damage of unprotected fruit frequently exceeds 50%, occasionally reaching 100%. A recent survey by Moraiti (2013), covering many cherry-producing areas of Greece, focused on biological orchards and unmanaged, abandoned cherry groves in woods, revealed wide dispersion of *R. cerasi*

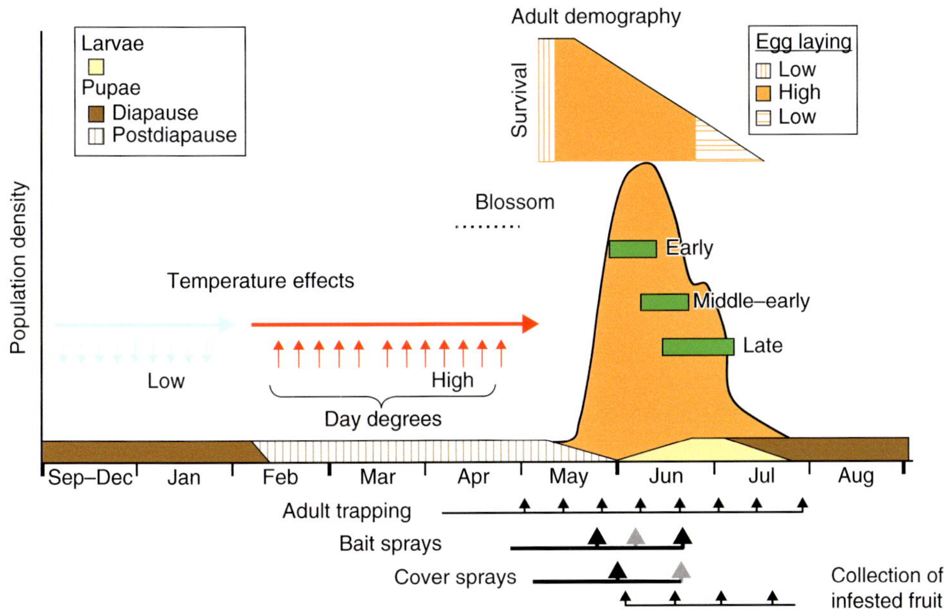


**Fig. 13.2.** Population model of the European cherry fruit fly (*Rhagoletis cerasi*) in a highland area of Greece, adult trapping and management practices considering the cherry tree phenology, and lifespan and reproduction of *R. cerasi*. (Based on Papanastasiou *et al.*, 2011; Moraiti *et al.*, 2014; N.T. Papadopoulos, unpublished data.)

and variable rates of infestation ranging from 15 to 100%.

### *Ecology and management*

Cherry fruit, and in particular sweet cherries, are a high-value commodity, in both the European and export markets, and thus any damage to the fruit translates into substantial economic loss. The European cherry fruit fly infests mature and ripening fruit, and the proximity of the infestation and harvest time greatly complicates the use of pesticides for its control, especially the systemic ones, which are the most effective to protect yield. For these reasons, and because of its ubiquitous occurrence, *R. cerasi* is the most important pest of sweet and to a lesser extent also tart cherries across its whole geographical distribution range, although the recent invasion by the spotted-wing drosophila, *D. suzukii*, might locally change the list of key cherry pests.

Currently, control of the European cherry fruit fly relies mainly on the application of contact insecticides (e.g. pyrethroids) targeting adult stages (ovipositing females) or systemic pesticides (e.g. organophosphates, neonicotinoids) that target immature stages developing inside the infested fruit (eggs and young larvae). Specific challenges associated with each tactic of the pesticide application are discussed in more detail in section 13.3.2.

Correct timing of the pesticide application is critical for the success of pest control and compliance with the residue standards. Time of pesticide application can be based on: (i) pest activity in the field; or (ii) fruit susceptibility for infestation. The first tactic is focused on the pest and its emergence in spring. Monitoring of adult populations with the yellow sticky Rebell® traps is the most reliable tool for both adult population monitoring and, under certain conditions, direct control through mass trapping (Remund and Boller, 1975; Daniel and Grunder, 2012). Efficacy of the Rebell traps can be slightly improved by including an ammonia dispenser (Katsoyannos *et al.*, 2000; Toth *et al.*, 2004). There are additional types of yellow sticky traps that have been developed recently, and several others in local markets, which could be used in both population monitoring and control programmes (Daniel *et al.*, 2014). Alternatively, adult emergence may be forecasted by the use of day-degree models (Łęski, 1963; Kovanci and Kovanci, 2006). However, due to the presence of locally adapted pest strains, the accuracy of these models is not universal; for example, they largely fail in coastal areas of Greece (N.T. Papadopoulos, 2009, unpublished data). Variability in diapause termination among *R. cerasi* populations, which has recently been demonstrated (Papanastasiou *et al.*, 2011; Moraiti *et al.*, 2014), that is regulated by high rates of gene flow and the adaptive developmental response of pupae to the local thermal environment highlights the importance of local studies to provide thermal summation models with more reliable and locally relevant biological and temperature inputs.

The second tactic is based on the host, and more specifically on the developmental status of the cherry fruit. The fruit becomes suitable for pest development and thus susceptible to infestation from the stage of its rapid growth, maturity and ripeness up to harvest. The beginning of this critical period is marked by a fruit hue change from green to yellowish- to reddish-green, and its end is marked by the harvest time. With the exception of the earliest cultivars, all others have to be protected during this critical period, and any 'protection gap' will result in fruit infestation and loss.

The two tactics of timing pesticide application, when correctly applied, offer similar results (Belien *et al.*, 2013), with the second method having obvious simplicity advantages. However, in areas of low or erratic pest prevalence, over-reliance on the second tactic may lead to executing pest control without the pest actually being present on the farm at densities above the required economic threshold.

A range of non-chemical alternative control approaches have been tried in the past or are currently under development, which are based on exploitation of the peculiarities of pest ecology and behaviour (e.g. sterile insect technique, application of oviposition-deterrent pheromone), application of natural enemies or pathogens (entomopathogenic nematodes), mechanical pest exclusion (ground and/or tree netting), supportive application of a 'virtual farm' concept and a range of modelling techniques emulating pest development and its control, location awareness systems and spatial decision algorithms. Several examples of these approaches are discussed in section 13.3.

### 13.2.2 Spotted-wing drosophila, *Drosophila suzukii* (Matsumura)

#### *Distribution*

The spotted-wing drosophila, *D. suzukii* (Diptera: Drosophilidae), is native to South-east Asia (Walsh *et al.*, 2011) and is closely related to the well-known and ubiquitous *Drosophila melanogaster*. In Europe, it was first recorded in 2008 in Spain (Calabria *et al.*, 2012) and within a few successive years in most European countries. According to the EPPO PQR database on quarantine pests and the CABI Invasive Species Compendium (EPPO, 2105a; CABI, 2016a), *D. suzukii* was officially recorded in 18 out of the 50 European countries with widespread distribution in the Mediterranean, western and central European regions (details of its invasion history are given in section 13.3). Long-distance invasion events are largely attributed to trading of infested fruit (EPPO, 2015b).

#### *Host range*

The spotted-wing drosophila (Fig. 13.1B) is a highly polyphagous species infesting fruits from many different plant genera and families including highly commercial crops, as well as ornamental and wild-growing species, especially small berries and other soft and stone fruits including sweet and sour cherries (Baroffio and Fischer, 2011; Asplen *et al.*, 2015; Rauleder and Köppler, 2015). Wild-growing berries and cherries (*Prunus avium*) as well as fruit of *Prunus cerasifera*, *Prunus serotina*, *Prunus laurocerasus* and *Prunus spinosa* are among the important hosts of *D. suzukii*. For a complete list of *D. suzukii* hosts, see, for example, CABI (2016a). The host status of the above plant species depends on *D. suzukii* phenology, especially on the occurrence of fertile adults early in spring

(Zerulla *et al.*, 2015; K. Köppler, 2015, unpublished data).

### Life cycle

Adult *D. suzukii* are 2–3 mm long, with red eyes, a pale yellowish–brown body and black bands at the posterior part of the abdomen. Male wings bare characteristic dark spot at the terminal end of the first vein of the wing (Fig. 13.1B), in contrast to the transparent wings of females. In addition, the front of the first and second segments of the first pair of male legs bear combs with three to six teeth (e.g. Hauser, 2011; Walsh *et al.*, 2011; Cini *et al.*, 2012). In contrast to other drosophilids, *D. suzukii* bare a large sclerotinized, serrated ovipositor, which facilitates oviposition in the mesocarp of healthy fruit. Eggs, which are laid just under the fruit skin, are equipped with two, visible, long respiratory stalks that are usually used as an early symptom of fruit infestation. Third-instar maggots (3–4 mm long; Kanzawa, 1939) are pale white, and pupate on or within the infested fruit. The reddish-brown cylindrical puparium bears two characteristic respiratory tubes at one end. The developmental cycle may last 9–14 days (1–2, 4–5 and 4–7 days for eggs, larvae and pupae, respectively) (Kanzawa, 1939). In optimal laboratory conditions (20–25°C), *D. suzukii* may complete 13 generations per year (Kanzawa, 1939; Tochen *et al.*, 2014), with age-specific egg laying reaching 25 eggs per female (Kinjo *et al.*, 2014). However, in suboptimal climatic conditions (e.g. the Netherlands) in the wild (temperatures <20 or >30°C), fecundity may be dramatically reduced and developmental duration increased reaching approximately six generations per year (H.H.M. Helsen, Wageningen, the Netherlands, 2016, personal communication).

*Drosophila suzukii* overwinters as an adult (male and female) in reproductive dormancy (Kanzawa, 1939; Sasaki and Sato, 1995). There are two adult morphotypes, the spring–summer and the overwintering one, which differ in body colour (darker in the winter morph) (P.W. Shearer, Oregon, USA, 2015, personal communication). In the northern hemisphere, winter adults appear in October/November, following a temperature drop and reduced day length, and exhibit a prolonged lifespan and cold hardiness (Stephens *et al.*, 2015; P.W. Shearer, Oregon, USA, 2015, personal communication). At the end of winter/beginning of spring, the gradual increase in temperature terminates the reproductive diapause (Zerulla *et al.*, 2015; Briem *et al.*, 2016). The population dynamics of *D. suzukii* during the summer fruiting season depends on the prevailing temperature and humidity, and the availability of suitable host fruits (EPPO, 2008; Tochen *et al.*, 2014, 2015).

### Damage

Sweet and sour cherries are heavily infested by *D. suzukii* (Sasaki and Sato, 1995; Lee *et al.*, 2011; Harzer and Köppler, 2015) compared with blueberries (Tochen *et al.*, 2014). Interestingly, however, *D. suzukii* is not the dominant pest of cherries in China (Guo, 2007; X. Chun, Kunming, China, 2015, personal communication), where it only occasionally causes economic damage (Guo, 2007; Zhang *et al.*, 2011). Oviposition of *D. suzukii*, and therefore infestation of cherries, occurs at the end of the ripening season, later than that of *Rhagoletis* spp. (K. Köppler, 2015, unpublished data). Larval feeding activity (followed by secondary fungal and bacterial infections resulting in fast fruit decay) occurs within the entire fruit pulp (Fig. 13.1D), unlike *Rhagoletis* spp., which feed close to the cherry pit. Under high population densities, the whole cherry crop may be infested, and under certain conditions *D. suzukii* may become a more severe pest for cherries than the European or American *Rhagoletis* spp. because of very fast population build-up and the timing of fruit infestation (just before harvest).

### Ecology and management

*Drosophila suzukii* adults can be monitored with various trap types containing acetic acid (e.g. cider vinegar), ethanol (e.g. red wine), sugar, fruit juice water, baker's yeast or other components in different combinations (Cha *et al.*, 2012; Landolt *et al.*, 2012a,b; Cha *et al.*, 2013; Grassi *et al.*, 2014), which form the basis of various commercial trapping systems (e.g. RIGA-Becherfalle, RIGA AG, Switzerland; Suzukii Trap®, Bioibérica, Spain; Pherocon, Trécé Inc., USA; Dros'Attract, Biobest, Belgium). Although monitoring is a very important tool to get information about the regional distribution and activity of the flies throughout and after the fruiting season, the currently available attractants are not fully reliable, since there is no correlation between trap catches and infestation level as well as population size. Ripening and ripe fruits are still more attractive than different multi-component bait blends, and the attractiveness of traps varies depending on availability of host fruits and food.

Management of the spotted-wing drosophila relies on the application of insecticides, mainly organophosphates, pyrethroids, neonicotinoids and spinosyns (Bruck *et al.*, 2011; Cuthbertson *et al.*, 2014; Wise *et al.*, 2014). In general, *D. suzukii* prefers higher humidity and moderate temperatures (see description of biology above). Thus, agronomic measures establishing 'uncomfortable' conditions for the flies, such as pruning to reduce density and compactness of the tree crown, continuous mulching or herbicide application, might reduce survival and thus pest population size in the orchard. The net effect of such agronomic measures on the actual fruit infestation depends heavily on the prevailing weather, and is more pronounced under adverse than under optimal climatic conditions.

Physical crop exclusion with net covers might be an option to prevent infestation of cherry orchards, although *D. suzukii* tends to be present in cherry orchards long before and after fruit ripening (K. Köppler, 2014, unpublished data; S. Kuske, Wädenswil, Switzerland, personal communication; P.W. Shearer, Oregon, USA, personal communication). Therefore, additional measures (traps, insecticides) inside the net cover (<1 $mm^2$ mesh size) might be necessary (Kaswase and Uchino, 2005; Weydert *et al.*, 2014; Rogers *et al.*, 2016). The overall economic merits of complete netting are not yet certain and remain to be evaluated. Apart from the heavy economic investment required, netting affects the development of other pests, beneficial organisms and diseases, as well as fruit growth and ripening patterns.

Biological control of *D. suzukii* is not easy, although its natural enemies (parasitoids of the families Figitidae, Braconidae, Pteromalidae and Diapriidae; predators of the family Chrysopiidae; entomopathogenic fungi) are known from various parts of the world. However, to date, there is no evidence for successful biological control (Asplen *et al.*, 2015; Englert and Herz, 2016; F. Chang, Beijing, China, 2015, personal communication).

### 13.2.3 Plum scale, *Sphaerolecanium prunastri* (Fonscolombe)

#### *Distribution*

Plum scale, *S. prunastri* (Homoptera: Coccidae), is cosmopolitan throughout the northern hemisphere and Australia (Ben-Dov, 1968). In Europe, its distribution covers the entire continent, from the coast of the Mediterranean Sea to Scandinavia and from the Iberian Peninsula to Caucasus. It is reported also from Japan, China, Central and East Asia, the Middle East, Russia, countries of the Black Sea and north-eastern states of the USA.

#### *Host range*

*Sphaerolecanium prunastri* (Fig. 13.3A) is the most important soft scale (Hemiptera: Coccidae) infesting cherry trees. It also infests several other *Prunus* spp., including peaches, prunes, plums and almonds, and to a lesser extent other Rosaceae such as apples, pears, quinces, related ornamental trees and shrubs, and occasionally grapes.

#### *Life cycle*

The plum scale is a univoltine pest. Early in spring the overwintering second-instar larvae develop to the last third instar and, within a few weeks, reach the adult stage. Compared with females, third-instar male larvae are smaller, elongated (1.5 mm long) and covered by a whitish transparent wax sheath. Adult males bear one pair of wings, are mobile and short lived, and die soon after mating. The dome-like (3.0–3.5 mm in diameter) wingless adult females are sedentary, and

Fig. 13.3. *Sphaerolecanium prunastri* (A), *Myzus cerasi* (B), *Pseudaulacaspis pentagona* (C) and *Caliroa cerasi* (D). (Photos courtesy of N.T. Papadopoulos and Proefcentrum Fruitteelt VZW, Zoology Department and DAT (C. de Schaetzen).)

become shiny brown–black when mature and producing offspring. The females are extremely fecund, each producing up to 1000 eggs (Tremblay, 1981; N.T. Papadopoulos, unpublished data). Young crawlers, which hatch within and emerge from the female's body, are deep red in colour, and disperse for a couple of days until they find a suitable feeding location, which becomes permanent for the rest of their adult life.

### Damage

All stages except adult males reside on twigs and branches, and seldom on leaves and fruit. During feeding, the sedentary stages remove large quantities of plant sap, usually causing twig death and a general 'weakening' of the tree (Argyriou and Paloukis, 1976). Like other sap-feeding insects, *S. prunastri* produces large quantities of honeydew, which promotes the development of sooty mould, further impairing the photosynthetic performance of the infested tree and reducing the quality of the produced cherries. Usually it is of local importance, but it has the potential to inflict significant damage, if left uncontrolled.

### Ecology and management

In undisturbed orchards, populations of the plum scale develop a dynamic equilibrium with a broad range of native, on-farm-dwelling natural enemies, such as parasitoids (e.g. Aphelinidae, Braconidae, Encyrtidae) and predators (e.g. Coccinellidae), which usually maintain the pest populations below the levels of economic concern. However, disturbance to the natural enemies, caused by pesticide abuse and/or plant stress due to poor cultivation practices (e.g. pruning, fertilization, irrigation), may promote a dramatic increase in the *S. prunastri* population, way above the acceptable level.

Similar to other Coccoidea, insecticide sprays are more effective when directed at crawlers and the first-instar larvae (Fig. 13.4). Because the oviposition period and emergence of crawlers span several weeks, often a second insecticidal application against crawlers is required for *S. prunastri*. In cases of severe infestation, dormant oils can be applied late in winter, and/or spring applications of insect growth regulators or juvenile hormone analogues may be considered (Fig. 13.4) (Paloukis *et al*., 1991). Laborious random sampling of twigs and branches and careful examination under a stereomicroscope in the laboratory is required to determine the optimal timing of intervention.

Ultimately, protecting the native fauna of natural enemies should be the target of long-term management of the plum scale, supported by occasional inoculative releases of parasitoids and predators.

### 13.2.4 Black cherry aphid (cherry blackfly), *Myzus cerasi* (Fabricius)

#### *Distribution*

The black cherry aphid, *M. cerasi* (Hemiptera: Aphididae) is a globally ubiquitous and serious pest of cherry trees, common throughout Europe, the Middle East and Asia, and has recently expanded to Australia, New Zealand and North America (Blackman and Eastop, 1994). Initially, the black shiny aphids (Fig. 13.3B) inhabiting cherry trees were described as a single species, *Myzus cerasi* Fabricius, 1775, but later, the European populations from *P. avium* were separated as *Myzus pruniavium* Börner, 1926. Recently, however, based on the gene sequence of the cytochrome oxidase I (COI) gene, both species, infesting sweet and sour cherries, were classified together again, and *M. cerasi* and *M. pruniavium* should now be regarded as synonyms.

#### *Host range*

Sour and sweet cherry are the main host plants of the black cherry aphid, although occasionally other *Prunus* spp. are also infested. A range of secondary hosts is known, such as Rubiaceae (*Galium* spp.), Scrophulariaceae (*Veronica* spp.) and Cruciferae (*Capsella* spp.), and, less frequently, Caprifoliaceae and Compositae, which vary in their importance in various parts of the world (Gilmore, 1960; Blackman and Eastop, 1984).

#### *Life cycle*

Black cherry aphids overwinter as eggs laid at the base of buds and in bud axils, on the spurs and young shoots of cherry trees. Larvae hatch coincides with the development/swelling of the buds, and all eggs have hatched before the white-bud stage. Similar to other aphids, *M. cerasi* is small and soft-bodied, occurring in both winged (alate) and wingless (apterous) forms. The larvae are dark brownish-purple, while the fundatrices (stem mothers) are black shiny with long black siphunculi.

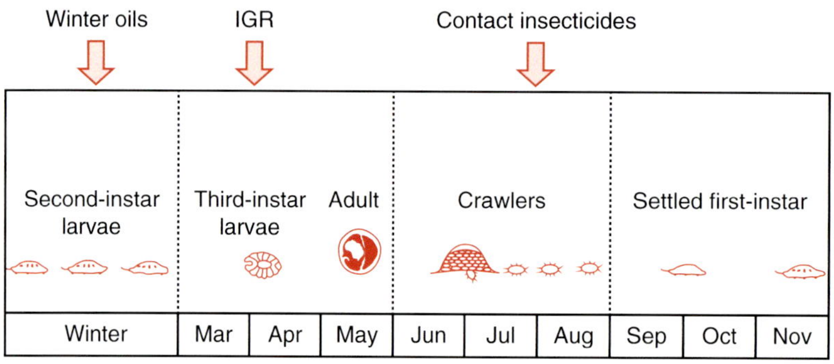


**Fig. 13.4.** Seasonal development of plum scale and optimum timing for insecticide application. Three different categories of insecticides have been considered for winter, spring and summer applications. IGR, insect growth regulator. (Based on Paloukis *et al.*, 1991, and N.T. Papadopoulos unpublished data.)

The fundatrice generation produce apterous, virginoparae females, which quickly form dense, black colonies on the underside of young cherry leaves, causing intense leaf curling. Following a number of generations on the primary cherry host, alate forms are produced and migrate to common summer weed host plants, such as *Galium* spp. or *Veronica* spp. Breeding colonies of wingless individuals may persist on cherry throughout the summer, but eventually die out. In autumn, winged virginoparous aphids migrate back to cherry trees where they produce sexual females, which mate with returning males and deposit overwintering eggs on the cherry trees. In general, aphids seem to thrive on larger trees, which provide more protection from direct sunlight (Gilmore, 1960). Ants, such as *Lasius niger* and *Myrmica laevinoides*, frequently visit aphid colonies on cherry trees, protect them from predators and thus promote their development (Gruppe, 1990).

#### *Damage*

Direct feeding on cherry tissues causes leaf curling and often premature leaf drop, shoot deformation, pseudogall (open gall) formation and the development of sooty moulds on excreted honeydews, which reduces the photosynthetic efficiency and eventually fruit set. Fruit are also contaminated by honeydews and sooty mould, which cause qualitative damage on harvested cherries.

#### *Ecology and management*

In undisturbed orchards, *M. cerasi* populations can be reduced by predators, such as lady beetles, lacewings and several species of parasitic wasps. In more severe infestations, spraying with contact or systemic insecticides during the pre- or postflowering period may be necessary. Visual inspection of curled, young shoots of cherry trees should be conducted to determine the economic injury level, which is set to two to five colonies per 100 shoot tips in several central European countries.

In organic orchards, banding tree bases with glue may be applied to prevent ants invading trees to 'protect' aphid colonies. Such simple treatment, if applied a few times during the season (to ensure continued glue function) may substantially increase the presence of natural enemies on the banded trees (Fontanari *et al.*, 1993), and its effectiveness in controlling *M. cerasi* may be comparable to pesticide application (Pérez *et al.*, 1995).

### 13.2.5 Summer fruit tortrix moth (reticulated tortrix), *Adoxophyes orana* (Fischer von Röslerstamm)

#### *Distribution*

The summer fruit tortrix moth, *A. orana* (Lepidoptera: Tortricidae) is widely distributed throughout the Palearctic ecozone (Pehlevan and Kovanci, 2014).

#### *Host range*

The summer fruit tortrix moth is a highly polyphagous species that may cause significant damage on cherry trees (Fig. 13.5A) and various other pome and stone fruit trees, mainly Rosaceae, including cherry, apple and pear. It is also recorded in species of the families Anacardiaceae, Betulaceae, Cannabaceae, Caprifoliaceae, Ebenaceae, Ericaceae, Fabaceae, Fagaceae, Grossulariaceae, Malvaceae and Pinaceae, among others.

#### *Life cycle*

Second- or third-instar larvae overwinter in silken hibernacula formed in the crook of a forking twig, bud axil, between dead leaves or mummified fruit, or in a crack in the bark. The larvae resume activity and begin to feed at the opening of buds early in spring. They eat the small outer leaves and floral parts, sometimes attacking a number of blossom trusses. After completing four instars, the larvae pupate in a shelter of webbed leaves. The pupae are 10–11 mm long and dark brown, while adult moths are greyish-brown to orange–brown with dark brown markings (~1 cm long, 1.5–2.2 cm wingspan). Following mating, females lay on leaves, producing shield-like, lemon-yellow egg batches of

**Fig. 13.5.** Infestation of cherries by *Adoxophyes orana* (A) and leaves by *Lyonetia clerkella* (B). (Photos courtesy of pcfruit vzw, Zoology Department and DAT (C. de Schaetzen).)

approximately 100 eggs. Yellowish-green to olive/dark green caterpillars of the first summer generation appear and feed beneath silken webs spun on the underside of leaves, usually removing the lower epidermal tissue close to a midrib, forming a shelter within webbed foliage, especially at the tips of shoots. The larvae often feed on the surface of fruit when infested leaves touch fruit. Fully grown larvae (~2 cm long) pupate in the larval shelter. Two to three generations (the third is incomplete in several cases) are completed before autumn larvae (second to third instar) seek overwintering refuges.

#### *Damage*

Damage is mainly due to loss of buds, while damage to foliage is usually relatively less important. Damage on the surface of developing fruit renders them blemished and unmarketable (Fig. 13.5A).

#### *Ecology and management*

Monitoring of the adult moth population is conducted with pheromone traps of various types, while that of larvae is by visual inspection of flower clusters and shoots. High trap catches (>30 moths per trap per week) and/or the presence of dense populations of caterpillars in blossom trusses indicate the need for insecticide application. The economic injury level is set to one caterpillar per 100 flower trusses. The occurrence of 5–10% damaged shoots in summer can be also considered an economic threshold. When considerable fruit damage occurs in the previous season (>1% fruit loss), a treatment in subsequent spring should also be considered.

Commercially available mating-disruption techniques can be successfully used against *A. orana* (Porcel *et al.*, 2015). Wherever the economic injury threshold is met, corrective (control) sprayings in the preflowering and postflowering period with chemical or biological (granulosis virus) crop protection agents might be considered. In addition, natural enemies such as *Trichogramma* spp. parasitic wasps can help keep *A. orana* populations at an endemic configuration.

### 13.2.6 Fruit-tree tortrix moth (great brown twist moth), *Archips podana* (Scopoli)

#### *Distribution*

The fruit-tree tortrix moth, *A. podana* (Lepidoptera: Tortricidae), is native to Europe and Anatolia, and has been introduced to North America.

#### *Host range*

*Archips podana* is a common pest of apples, but infests a wide variety of plant species, both commercial and ornamental trees and shrubs, including cherry trees (Alford, 2014).

### Life cycle

*Archips podana* hibernates as second-instar (occasionally) or third-instar (mostly) larvae, protected in dense, silken hibernacula spun under a twig, bud scale, dead leaf or some other kind of shelter. The larvae, emerging from the overwintering shelters, burrow into opening buds and develop further into the fourth and fifth instars, feeding on flowers and young fruitlets. Sixth- and seventh-instar larvae (up to 2.2 cm long, light green to greyish-green) feed on young leaves and pupate within freshly spun leaves or the larval habitation. Adult moths (~1 cm long, 19–28 mm wingspan; purplish-brown, purplish-ochreous to chestnut brown with yellowish and dark brown markings) (Bland *et al.*, 2014) appear from June to September. Eggs are deposited, in batches of approximately 50 eggs, on foliage or fruit, and second- to third-instar larvae prepare a shelter for overwintering. *A. podana* concludes one generation in northern and central Europe, and a partially incomplete second generation may occur only under extremely favourable climatic conditions. However, in warmer southern areas, two generations can be completed.

### Damage

Economic damage may arise by larvae feeding on fruit, which decreases cherry quality and facilitates the development of fungal pathogens. Feeding on buds and webbing affects the development of fruitlets and leaflets in spring.

### Ecology and management

Pheromone traps and visual inspection of bud scales, flower clusters, shoots and webbed leaves/flowers are used for monitoring adult and larval populations, respectively. The need to control the pest should be determined by high pheromone trap catches (>30 moths per trap) or by the presence of large populations of caterpillars in blossom trusses.

Commercially available mating-disruption techniques, and wherever threshold values are exceeded, corrective insecticide sprayings in the postflowering period can efficiently keep populations of *A. podana* at acceptable levels (Porcel *et al.*, 2015).

## 13.2.7 Rose tortrix moth (European leaf roller), *Archips rosana* (L.)

### Distribution

The rose tortrix moth, *A. rosana* (Lepidoptera: Tortricidae), is present in the Palearctic and Nearctic ecozones.

### Host range

*Archips rosana* is a highly polyphagous pest that, as well as cherries, infests apples, pears, plums, raspberries, cultivated roses and others (Alford, 2014).

### Life cycle

*Archips rosana* hibernates as eggs, deposited in batches of around 50–150 on the bark. The larvae (up to 22 mm long, light green body and brown/black head) pupate within curled, rolled leaves or other larval habitation. Adult moths (~1 cm, 15–24 mm wingspan; reddish-brown to grey–brown with darker markings; Bland *et al.*, 2014) appear from late June to September. Overwintering eggs are deposited in batches on the bark in a green liquid that quickly hardens, offering both camouflage and protection.

### Damage

The larvae feed on buds, and develop in flower-bud trusses, young leaves, blossoms and young fruitlets, causing symptoms similar to the other two leaf rollers described.

### Ecology and management

Pheromone traps and visual inspection of plant parts for possible infection are used to monitor adult and larval populations, respectively. The presence of egg batches can be inspected visually on the bark during winter/early spring. As well as high numbers of adult captures, and larvae on plant tissues, the number of egg batches (economic injury level of one egg batch per ten trees) early in spring can be used to make management decisions. Control methods are similar to those reported for the other leaf rollers (Sjöberg *et al.*, 2015). In addition, mineral oils targeting overwintering eggs can be applied early in the year and well before flowering. Other applications, mainly insecticides, should target larvae around the flowering period.

### 13.2.8 Eye-spotted bud moth (apple bud moth), *Spilonota ocellana* (Denis & Schiffermüller)

#### Distribution

The eye-spotted bud moth, *S. ocellana* (Lepidoptera: Tortricidae), is present in the Palearctic and Nearctic ecozones.

#### Host range

This is another polyphagous leaf roller that is quite a common pest of apples and pears, but is less frequent in stone fruits, such as cherries (Alford, 2014).

#### Life cycle

*Spilonota ocellana* overwinters as larvae within a hibernaculum in crotches of small shoots. The pink, reddish-brown, fully grown larvae may reach around 1.2 cm length before pupation. Adult moths (~0.6 cm long, 1.2–1.6 cm wingspan; reddish-brown to grey–brown with a thick white stripe and a dark, triangular dorsal spot on each wing) appear from mid-June to August. Eggs are deposited singly on leaves where young larvae initiate feeding before moving to overwintering refugia.

#### Damage

The larvae infest buds, flowers and later leaves.

#### Ecology and management

Only spring infestation can be considered important since *S. ocellana* can dramatically reduce blossom. Sampling of overwintering refugia, young shoots in spring and pheromone traps are used for population monitoring. Insecticide sprays can be applied before bloom, targeting active, overwintering larvae. More recently, *S. ocellana* has also been controlled by a commercially available mating-disruption technique (Porcel *et al.*, 2015).

### 13.2.9 European winter moth (common winter moth), *Operophtera brumata* (L.)

#### Distribution

The European winter moth, *O. brumata* (Lepidoptera: Geometridae), is distributed within Europe and the Middle East, and has invaded North America (Gwiazdowski *et al.*, 2013).

#### Host range

The European winter moth is a polyphagous pest with pome fruit trees (apple and pear) and stone fruit trees (plum and cherry) being important host plants (Alford, 2014).

#### Life cycle

Adult winter moths emerge from October to January (peak in November–December). Female adults are almost wingless (having only stubs), greyish-brown, approximately 5–8 mm long and are usually detected at the base of trees but also on other plant parts. Male winter moths bear fully developed, brownish-grey wings (wingspan 2.2–2.8 mm) and actively seek mates. After mating, females deposit an egg cluster on trunks and branches, in bark crevices, under bark scales, on loose lichen and elsewhere. The larvae hatch from bud break to green cluster (Salis *et al.*, 2016), crawl up the trunks and produce a long silken strand to passively disperse by 'ballooning'. The young larvae feed preferentially on fruit buds, while older ones commonly attack fruitlets, flower trusses and leaves. Fully grown larvae (up to 2.5 cm long, light green with a dark green dorsal stripe and several whitish stripes along the back and sides, looping gait moving) voraciously feeding on leaves may cause extensive defoliation until mid-June. They pupate in flimsy cocoons in the ground, entering into an aestivation dormancy state that lasts until autumn.

*Damage*

The young larvae damage fruit buds, while older ones attack fruitlets, flower trusses and leaves.

*Ecology and management*

Adult male moths can be monitored with species-specific pheromone trap, while females and egg presence can be monitored by visual inspection. Sampling of young shoots, buds and flowers can be employed to determine larval populations. A simple method to prevent winter moth infestation is the application of sticky bands around trees in October before the females start to climb up the trees, although it is practically difficult to apply this method in intensive cherry orchards. The best time to suppress the winter moth with (chemical) control sprayings is before bloom, targeting active larvae from the bud-bursting phenological stage onwards.

### 13.2.10 Fruit-tree red spider mite (European red mite), *Panonychus ulmi* (Koch)

*Distribution*

The fruit-tree red spider mite, *P. ulmi* (Trombidiformes: Tetranychidae), has a cosmopolitan distribution.

*Host range*

*Panonychus ulmi* is a notorious pest of many plant species (a wide range of plant families) including cherries (Alford, 2014).

*Life cycle*

*Panonychus ulmi* (Fig. 13.6A) overwinters as red eggs (0.17 mm in diameter, spherical onion-shaped), deposited from mid-August to September on bark crevices and smaller branches (Fig. 13.6B). Depending on the area and climatic conditions, larvae hatch from the overwintering eggs from April/May to mid-June. Adult females (0.5 mm long, dark red with long setae (hairs) arising from light-tipped tubercles) and males (smaller than females, pear-shaped and yellowish-green to bright red), as well as larvae, feed usually on the underside of leaves, where summer (paler brownish-red) eggs are deposited as well. Summer ovipositions can also take place on the upper side of the leaves along the midribs. Unfertilized eggs produce males, while fertilized eggs produce both males and females. *Panonychus ulmi* completes four to eight overlapping generations per year. Under high population densities, dispersal by 'ballooning' is common.

*Damage*

Red mite feeding activity causes pale spotting on the leaves that soon becomes brown–bronze in colour. As well as a dramatic reduction of photosynthesis, high infestation rates may lead to extensive leaf dropping, and a decrease in annual growth and bud formation for the next year. Similar damage can be caused by *Tetranychus viennensis* (Zacher), which can also be found in cherry orchards (Fig. 13.6C, D).

*Ecology and management*

Monitoring is usually conducted by visual inspection of winter eggs on branches prebloom (checking five 'searching sites', each covering 20 cm of branch) and repeated assessment of the percentage of heavily infested leaves later in the season. Economic injury levels have been set to 30–50 overwintering eggs per 'searching site' and at 50% heavily infested leaves later on.

The red spider mite is a secondary pest with a resurgence due to pesticide abuse; thus, for its effective control, application of integrated pest management (IPM) strategies, depending mainly on predatory mites (e.g. *Typhlodromus* spp.) is crucial (Thistlewood *et al.*, 2013). If high numbers of overwintering eggs are present, applying a delayed dormant

**Fig. 13.6.** *Panonychus ulmi* adult (A) and overwintering eggs (B), and *Tetranychus viennensis* adult, nymphs and eggs (C) and infestation on leaves (D). (Photos courtesy of Proefcentrum Fruitteelt VZW, Zoology Department and DAT (C. de Schaetzen).)

mineral oil spray is an effective way to suppress the population. Eggs are most vulnerable to control just before hatching. In the case of moderate infestation pressure, an acaricide spraying application is recommended (van Leeuwen *et al.*, 2015). The presence of the European red mite in cherry orchards later in the season may be managed either with an acaricide application or by relying on natural predators.

### 13.2.11 Cherry leaf miner (apple leaf miner), *Lyonetia clerkella* (L.)

#### *Distribution*

The cherry leaf miner, *L. clerkella*, (Lepidoptera: Lyonetiidae) is a small (0.8–0.9 cm wingspan, shiny white), slender moth, with narrow folding wings bordered by long bristles, folding backwards covering the hindwings and abdomen, and often having pointed apices and long hair fringes. It is distributed throughout Europe, northern Africa, the Middle East, Turkey, north-western Siberia, the Far East, India and Japan.

#### *Host range*

The cherry leaf miner is a polyphagous pest infesting mainly apples and cherries (and other Rosaceae), as well as species of the family Betulaceae (Alford, 2014).

#### *Life cycle*

Adults emerge from overwintering refugia in March–April and initiate egg laying on the underside of leaves. The larvae (8–9 mm long, green with a brown head and legs) mine elongated and narrow tunnels in leaves following a characteristic pattern that can be used as a diagnostic feature (Fig. 13.5B). It is regularly detected in cherry orchards, causing damage on fruit that may reach economic levels. Three to four generations are completed per year, depending on the local climatic conditions.

### *Damage*

Heavy infestation may cause extensive defoliation, which can be crucial for young trees and nurseries.

### *Ecology and management*

Pheromone traps are predominately used for population monitoring (Nakano *et al.*, 2014), accompanied by visual inspection of infested leaves (one to two mines per young leaf). Sprayings should be executed after the appearance of adult moths. Several hymenopterous parasitoids, as well as entomopathogenic fungi, may keep populations under economic injury levels.

## 13.2.12 Cherry-stone weevil, *Anthonomus rectirostris* (L.)

### *Distribution*

The cherry-stone weevil, *A. rectirostris* (Coleoptera: Curculionidae) is a univoltine species distributed throughout the Palearctic ecozone. The adults are 4–5 mm long and reddish-brown with yellowish hairs forming two pale crossbands on one elytra.

### *Host range*

*Anthonomus rectirostris* infests cherry and occasionally plum trees (Alford, 2014).

### *Life cycle*

The adults emerge from overwintering refugia (litter/leaves, soil and bark cracks) in April at 'buds breaking', and gnaw irregular-shaped holes on leaves, leaf stalks and apical sprouts (forming deep cavities). Single eggs are laid in the developing fruitlets, and larvae (~6 mm long, cylindrical, whitish with a reddish-brown head) feed on the developing (stone) seed and pupate inside the pit. Young adults emerge from infested cherries from late July onwards and reach the overwintering sites in autumn.

### *Damage*

Egg-laying females probe developing fruitlets with their rostrum, forming distinctive necrotic spots. Infested cherries stay small and do not ripen. The infested stones are filled with brown frass and have a small, round exit hole in the wall.

### *Ecology and management*

Monitoring of adult weevils can be executed by beating tray samples. Infested cherries can be observed on the tree and later fallen on the ground beneath trees (presence of stones with exit holes on the soil). The economic threshold is set at 5% infested cherries. Control measures in cherry plantations include treatments with insecticides in the postflowering period.

## 13.2.13 Plum leaf-curling midge, *Dasineura tortrix* (Loew)

### *Distribution*

The plum leaf-curling midge, *D. tortrix* (Diptera: Cecidomyiidae) is a univoltine species and is widespread in Europe.

### *Host range*

*Dasineura tortrix* is associated with various Rosaceae, such as fruiting species of *Prunus*, including cherry (Alford, 2014).

### *Life cycle*

Mature, fully grown larvae overwinter in a cocoon in the soil. Adult females (1.5 mm, females with red abdomen) lay eggs in swelling buds, and the larvae (~2.5 mm long, white) feed on the growing tip, causing crooked shoots, before pupating in the soil. There are two to three generations per year. Larvae of the summer generations infest leaves, which are rolled in a characteristic manner (fusiform bunches of distorted leaves and shortened internodes at the tip of shoots) (Alford, 2014).

#### Damage

Plum leaf-curling midge, except in nurseries, is not considered an important pest of cherry orchards. Infestations disrupt the growth of terminal shoots and shorten internodes, which causes the leaves to bunch together. Infested leaves also become interlocked, with their leaf margins curling upwards. Such damage is usually transient; however, new growth arising from infested buds may turn black and die. The pest can be especially harmful in nurseries.

#### Ecology and management

The presence of plum leaf-curling midges can be monitored by visual inspection of leaf galls (and the presence of feeding larvae inside). Control treatments can be executed during the bud-bursting stage, targeting the first generation.

### 13.2.14 Cherry slug sawfly (cherry slug, pear sawfly, pear slug, pear and cherry slug sawfly), *Caliroa cerasi* (L.)

#### Distribution

The cherry slug sawfly, *C. cerasi* (Hymenoptera: Tenthredinidae) is a cosmopolitan pest recorded in Asia, Africa, North America, South America, Antarctica, Europe and Australia.

#### Host range

The cherry slug sawfly (Fig. 13.3D) is associated with several rosaceous trees and shrubs, with cherry and pear as major hosts (Alford, 2014).

#### Life cycle

*Caliroa cerasi* overwinters as fully developed larvae in the soil. The glossy black adult sawflies (0.4–0.6 cm long) deposit two to five eggs per leaf. Larvae (up to 10 mm long, whitish initially but soon greenish-yellow to orange–yellow, covered with shiny olive-black slime, pear-shaped and broadest at the head side) feed on the upper epidermis of leaves. There are usually two generations per year. A partial third generation may occur during autumn. In heavy infestations, leaves turn brown, wither and drop, and fruit maturation may be delayed.

#### Damage

The cherry slug sawfly occasionally can cause significant damage in commercial cherry orchards; its larvae damage the leaves of cherry, pear and plum trees, leaving behind a 'skeleton' of veins (Fig. 13.3D).

#### Ecology and management

Visual inspections are performed to monitor the slug-like larvae in summer and autumn (Bartoloni *et al.*, 2012). Insecticide applications for other pests usually keep the cherry sawfly at low population levels.

### 13.2.15 North American cherry fruit fly (eastern cherry fruit fly), *Rhagoletis cingulata* (Loew)

#### Distribution

The North American cherry fruit fly, *R. cingulata* (Diptera: Tephritidae), is a North American ecological equivalent and close relative of *R. cerasi*. The adults of the two species look similar, but the thorax of *R. cingulata* adults is predominantly black, and the abdomen and scutellum are black as well. The apical band of the wing is forked, or an upper arm of the fork is separated by a clear area, leaving an isolated dark spot at the wing tip. *Rhagoletis cingulata* invaded Europe in the late 1980s from North America, where it is indigenous, and since then has been spreading at a very low rate in central European countries and the northern Balkan peninsula (Johannesen *et al.*, 2013).

#### Host range

The host range is similar to that of *R. cerasi* (see section 13.2.1).

#### Life cycle

The biology of *R. cingulata* is similar to that of *R. cerasi*, except that *R. cingulata* adults emerge 2–3 weeks later, and adult occurrence peaks from mid-June to September (Vogt *et al.*, 2010a).

#### Damage

The North American cherry fruit fly infests late-ripening cherry cultivars and, in contrast to *R. cerasi*, more severely sour cherries.

#### Ecology and management

Yellow sticky traps are used for adult population monitoring, and management is generally accomplished with methods similar to those used for *R. cerasi*.

### 13.2.16 White peach scale (peach scale, white scale, West India peach scale), *Pseudaulacaspis pentagona* (Targioni-Tozzetti)

#### Distribution

The white peach scale, *P. pentagona* (Homoptera: Diaspididae), is a cosmopolitan pest that occurs in Asia, Africa, North America, South America, Antarctica, Europe and Australia. In Europe, it is distributed throughout southern Europe and locally also occurs north of the Alps.

#### Host range

The white peach scale is a rather polyphagous species, infesting stone fruits (mainly peach, but also cherries), mulberry and kiwi trees, wild-growing deciduous and ornamental trees and shrubs of importance such as *Styphnolobium japonica*, *Aesculus* spp., *Catalpa bignonioides* and *Juglans* spp.

#### Life cycle

Adult females (1–1.5 mm wide) are yellow–orange, oval with five characteristic angles in the perimeter of the broader pygidium areas and covered by a whitish armour shield (2 mm wide) (Fig. 13.3C). Male armour shields are white elongated, and males, similar to other scale insects, are short lived and unable to feed. Male and female eggs are red–orange and white, respectively, while crawlers are rather deep red. Larvae armour shields are similar in shape to those of adults of the same sex.

#### Damage

Immature and adult females remove large quantities of plant sap and can develop dense colonies on twigs and branches that, on rare occasions, may lead to the death of stressed cherry trees. Infestation of leaves and fruit is not particularly common, but can substantially diminish the quality of the latter.

#### Ecology and management

Like other scale insects, *P. pentagona* is mainly a secondary pest, usually triggered by pesticide abuse and the resultant suppression of the beneficial native fauna of parasitoids and predators. Sticky pheromone traps are used to monitor adult male activity, while sticky traps determine the dispersion of crawlers and population densities. Targeting young larvae, the timing of insecticide applications is usually executed a few weeks after the peak of male captures. There are three generations per year, with mated adult females consisting of the overwintering stage in several Mediterranean countries (Pedata *et al.*, 1995; Tzanakakis and Katsoyannos, 2003). Similar to other scale insects, special attention should be placed on protecting natural enemies and restoring the native fauna of parasitoids and predators, which usually keep the pest population densities below economic injury levels.

### 13.2.17 Additional minor pests

The brown marmorated stink bug, *Halyomorpha halys* (Stål) (Hemiptera: Pentatomidae), native to Asia (Hoebeke and Carter, 2003; Lee *et al.*, 2013), invaded North America in 1998 (Hoebeke and Carter, 2003) and Europe (Switzerland) in 2007 (Wermelinger *et al.*, 2008). Since 2007, *H. halys* has been detected in Germany (Heckmann, 2012), France (Callot and Brua, 2013) and Italy

(Haye and Wyniger, 2013). Being extremely polyphagous with over 100 reported host plants (CABI, 2016b), including a large number of cultivated fruits, mainly apples, peaches and nectarines, but also cherries (Maistrello *et al.*, 2014), *H. halys* is a threat to cherry cultivation as it can infest both fruit and leaves.

Wood-boring insects of the family Scolytidae, such as *Ruguloscolytus rugulosus* (Müller) (Coleoptera: Scolytidae) may occasionally cause substantial damage (especially on trees weakened by drought, diseases and other pests) on cherries by infesting young buds, shoots and twigs. *Capnodis tenebrionis* L. (Coleoptera: Buprestidae) infestation may be severe in young plantations, while both *Zeuzera pyrina* (L.) (Lepidoptera: Cossidae) and *Synanthedon myopaeformis* (Borkhausen) may infest the trunks and branches of cherry trees, leading to the death of entire plants on rare occasions.

In the USA and Canada, the plum curculio, *Conotrachelus nenuphar* (Coleoptera: Curculionidae) is a serious pest of fruits such as plums, apples, pears and stone fruits, including cherries. In the wild, it infests wild plum, hawthorn and crab apple. It is indigenous to the regions east of the Rocky Mountains. It infests young fruit soon after bloom, and larvae develop inside the infested fruit and the fruit drops prematurely. It can be controlled with pesticides, applied at the stage of petal fall. Regular destruction of the prematurely fallen fruit conducted before the weevil develops can reduce the pest population.

Also in the USA, the Japanese beetle, *Popillia japonica*, can affect cherry trees, causing severe leaf damage. The beetle is native to Japan, where it is normally regulated by its co-evolved natural enemies and rarely causes economic problems. In the USA, it feeds on a wide range of over 200 wild and cultivated plant species, including cherry. The adults consume leaf tissue between veins, skeletonizing the leaves of the infested plant. Grasslands are required for larval development, eggs are deposited underground and the emerging grubs feed on grass roots. In the USA, typically, one generation develops annually, although in cooler regions, as well as in its native country of Japan, development can last up to 2 years. Immature stages can be controlled in lawns or grasslands with the bacterial biopesticide milky spore, *Paenibacillus* (formerly *Bacillus*) *popilliae*. Adults can be trapped in large numbers in pheromone-baited traps. Therefore, trapping action can be effective if applied more or less uniformly over a larger area. However, the traps have to be used with some caution: although they can attract large numbers of beetles to their vicinity, only a fraction of the attracted individuals finally ends up in the traps. Therefore, the traps, if deployed locally in a patchy fashion, are likely to attract more beetles than they can catch, which may actually worsen the situation and increase the damage locally.

## 13.3 Critical Overview of Current Pest Management Approaches: Advantages and Limitations

In contrast to the typically industrial scale of modern agriculture, in Europe, most cherry produce originates from mid- and small-scale orchards, where most of the pest control decisions are made and executed at the farm level. Consequently, the typical cherry-production landscape is highly heterogeneous at both spatial and temporal levels, and is challenging or virtually intractable for coordinated regional or area-wide pest management. Decades of reliance on preventative application of broad-spectrum contact and systemic pesticides, especially organophosphates with prolonged residual activity (FAO, 1985), allowed easy and efficient control of most insect pests, including the most troublesome frugivorous species. The gradual build-up of public awareness about the environmental costs and consumer health risks resulted in pressure to ban successive classes of pesticides, such as dimethoate, and the adoption of market-driven, low pesticide-residue standards for cherries. Rapid progression of these trends has steadily eliminated the convenient 'chemical' control agents from the pool of options available to cherry producers, and has created new and challenging scenarios for cherry protection and pest management.

Sweet and sour cherries are produced for high-quality table consumption, brandy (so-called 'Kirschwasser'), and tinned cherries or jam. For all these commodities, pest- and disease-free immaculate fruit have to be guaranteed. To produce cherries in an economically feasible way and of the right quality, both plants and fruit have to be maintained as healthy and free of pests and diseases. Only in backyard fruit-growing without economic pressure can the quality standards be lower. Management approaches differ among the two main production scenarios: (i) intense production (mainly pesticide based); and (ii) ecologically oriented production (organic, ecological, low pesticide levels, mechanical (netting) and biological control, such as exploitation of natural enemies and predators, etc.).

### 13.3.1 Intense production: conventional management

Modern and intense cherry production tries to cover a long period of the season, depending on the climatic region, from May/June with early varieties until July. A small proportion of cultivars (e.g. the Canadian 'Staccato'), which have limited economic importance, may be harvested as late as August/September. Most cherry pests build up dense populations in spring before, during and after cherry bloom (e.g. *M. cerasi*, *O. brumata*, *P. ulmi*, *T. viennensis*, *Tetranychus urticae*, *P. pentagona*, *Quadraspidiotus perniciosus*) or at fruit colour break, from green/yellow to reddish (*R. cerasi*). However, the infestation risk of the *D. suzukii* depends greatly on climatic conditions and may vary among years. In some regions, *A. orana* causes damage on leaves and blossoms (first generation), as well as on leaves and ripening or ripe fruit (second generation).

Control of these pests is insecticide based in integrated cherry production in Europe. Pesticide registration in Europe is regulated by Regulation (EC) No. 1107/2009, which introduces comparative assessment and substitution to the regulatory process for plant protection products (EC DG SANCO, 2013) that results in reduction of the availability of effective active ingredients against cherry pests. In the following sections, the control of some major and widely spread cherry pests is described.

#### *Fruit flies*

To control fruit flies, such as *R. cerasi* and *D. suzukii*, broad-spectrum insecticides, such as pyrethroids (e.g. lambda-cyhalothrin, deltamethrin), organophosphates (e.g. dimethoate, phosmet, chlorpyrifos/chlorpyrifos-methyl) and spinosyns (e.g. spinosad, spinetoram), as well as neonicotinoids (e.g. acetamiprid, thiacloprid) are employed. However, registration of insecticides differs among European countries, and effective insecticides may not be registered for use in cherries or can be applied after annual exceptional permits by member states. Furthermore, additional compounds such as dimethoate and lambda-cyhalothrin are candidates for substitution, by 2016 and 2018, respectively. In some countries such as in Germany and the Netherlands, only acetamiprid is registered against *R. cerasi*, while in others, such as Belgium, Hungary, Italy and France, the longer list (thiacloprid, acetamiprid, dimethoate, phosmet, spirotetramat, lambda-cyhalothrin) may soon be shortened by a substitution process, rendering control of the European cherry fruit fly challenging. Similar issues and even more challenging ones regard *D. suzukii*, which has become a pest of cherries in the European Union (EU) recently, and in many European countries its control is based on 'side-effects' of the control of *R. cerasi*, and on annual exemptions of insecticides, such as phosmet, dimethoate, lambda-cyhalothrin, cyantraniliprole and spinosad. It seems that the control of *D. suzukii* and probably *R. cerasi* in some EU countries will depend on spinosad and the related spinetoram in years to come. However, the effect on predatory mites and other beneficials should be evaluated for both. Furthermore, sufficient efficacies of insecticide treatments against *D. suzukii* infestation are not given in many cases.

### Aphids

Pirimicarb (known to have minor effects on beneficials) and thiacloprid can be applied against *M. cerasi* in commercial cherry orchards in some EU countries until 2018 and 2017, respectively. Acetamiprid is also registered in several countries for aphid control in cherries. The neonicotinoids have been suggested to have severe side-effects on honeybees, although both thiacloprid and acetamiprid are listed as low toxicity compounds for bees (Laurino *et al.*, 2011). Broad-spectrum insecticides belonging to the pyrethroids (e.g. lambda-cyhalothrin, deltamethrin) and organophosphates (e.g. dimethoate) are still registered for use against aphids in cherries in various EU countries, but in practice are used almost exclusively for the control of fruit flies (see below). Quite recently the new, two-way-systemic insecticide spirotetramat has been registered for controlling aphids and scale insects in cherries in some EU countries. Plant-derived oils have also been registered against aphids in cherries, although they have a lower efficacy compared with the abovementioned synthetic insecticides.

### Scale insects

Registered insecticides also differ among countries for use against scale insects. In the Netherlands and Germany, only plant oils can be used against these pests. Other EU-member states provide regular registrations for oils, organophosphates (e.g. chlorpyrifos, phosmet), pyrethroids (e.g. lambda-cyhalothrin, deltamethrin), pyriproxyfen and acetamiprid. Intense local problems caused by scale insect infestation may require additional effort to be solved in the near future.

### Leaf rollers and caterpillars

Caterpillars can be controlled with the insecticides indoxacarb, spinosad, tebufenocid, lambda-cyhalothrin, fenoxycarb, chlorpyrifos-methyl and *Bacillus thuringiensis* (less effective below 15°C), but there is no registration in any of the member states. Recently, a granulosis virus formulation was registered as a biological control agent against *A. orana* in cherries, in Belgium and elsewhere.

### Spider mites

Spirodiclofen and clofentezine are currently registered against spider mites in several EU countries. Abamectin is authorized for mite control on cherries only after harvest. In addition, certain mineral and plant oils can be used against spider mites on cherry trees.

Because the availability of pesticides against cherry pests is likely to become a major issue in years to come, research on developing new insecticidal compounds for sustainable cherry production, and alternative control strategies, must be intensified in the EU. Research on the biology, behaviour and epidemiology of important cherry pests as a basis for biological, biotechnical and other environmentally sound control strategies should also be intensified.

## 13.3.2 Ecologically oriented production

In the ecologically oriented (but still commercial) production of cherries, a distinction can be made between organic production and advanced integrated production systems such as low-pesticide strategies, for example the Ecophyto 2018 programme in France.

In organic cherry production, only biological compounds and control strategies of pests are allowed. Examples are natural pyrethrum for controlling aphids and fruit flies and *Bacillus thuringiensis* for caterpillars. For the control of caterpillars, mating-disruption products also exist and are registered in some EU countries (e.g. Belgium). For example, a granulosis virus product is authorized in several countries against leaf rollers including *A. orana*. Recently, there has also been increased interest in growing cherries under nets for pest protection. The advantage of netting is the total exclusion of pests larger than the mesh size. A disadvantage, however, is the creation of a microclimate, which might be more suitable for fast population build-up of other pests, such as spider mites and aphids. In practice, commercial organic production of cherries is very difficult and hence rather rare throughout Europe.

In advanced integrated and low-pesticide production systems, most of the abovementioned control strategies are also applied. However, under such a production system, the possibility of applying 'corrective' chemical insecticide or acaricide if required still exists. This type of production strategy has gained increased attention, with, for example, the introduction of mating disruption against moths and protective net covering of trees, and even complete packages against fruit flies, but is still minor compared with intense (completely pesticide based) production systems.

## 13.4 Trends, Challenges and New Directions in Cherry Pest Management

### 13.4.1 Global warming and impacts on cherry pest dislocations and management

Intensified by global warming, human mobility and globalization in trading, biological invasions have increasingly become one of the most important issues in pest management (Papadopoulos, 2014). Similar to other crops, cherries are under the pressure of such changes. The invasion of the American eastern cherry fruit fly, *R. cingulata*, and the more recent invasion of the spotted-wing drosophila, *D. suzukii*, are good examples that illustrate the challenges cherry cultivation is facing because of invasive pests. *Rhagoletis cingulata* is endemic to North America, including Mexico (Bush, 1966; Smith and Bush, 1997). It was reported in Europe for the first time in 1983 in Switzerland, erroneously as *Rhagoletis indifferens* Curran, and has dispersed since then to several central European countries: Italy, Slovenia, Hungary, Croatia, Germany and Poland. Populations of *R. cingulata* become active relatively late in the cherry fruiting period and therefore infest late-ripening cultivars of sweet and mainly sour cherries, extending the infestation period by *Rhagoletis* spp. Additional predictive and population monitoring tools need to be developed for *R. cingulata* and applied for accurate assessment of population dynamics, economic injury thresholds and timing of insecticide application.

*Drosophila suzukii* is of much greater importance since it has been dispersed in record time throughout Europe and has aggressively invaded cherry orchards, causing damage on ripening cherries. With its capacity to disperse by transport of infested fruit, its presence almost year-round, completing multiple generations per year, and the wide host range including, as well as cultivated fruits, numerous wild host plants (small fruits and berries), it seems perfectly adapted to our modern globalized world (Poyet *et al.*, 2015). Known from the beginning of the 20th century in Japan, and in later decades in Korea and China, *D. suzukii* emerged as a major pest of small fruits and berries worldwide in 2008 following its detection in both western America and Europe (for a comprehensive review, see Asplen *et al.*, 2015). It was first detected in the coastal area of California in 2008, and severe cherry infestation was recorded as early as 2009. Over the next few years, it was detected in Oregon, and in British Columbia in Canada, and also in several other states, such as Florida, North and South Carolina, Michigan, New Jersey, New York and Utah. Dispersion of *D. suzukii* was similarly impressive in Europe. Following its first detection in Spain and probably Italy in 2008, it was detected in France and Croatia in 2010, in Germany, Austria, Switzerland and Belgium in 2011, in the UK, the Netherlands, Hungary, Poland and Portugal in 2012, in Greece, Romania, Bosnia and Herzegovina, and Montenegro in 2013, and in the rest of Balkans and eastern European countries in 2014 (Asplen *et al.*, 2015). Dispersion of *D. suzukii* is not only impressive among the different countries and states but also within countries. For example, the first detection of a few individuals in north-west (Epirus, Greece) and central Macedonia (Thessaloniki, Greece) in the autumn of 2013 was followed by country-wide detection in 2014, including the island of Crete, the southernmost point of *D. suzukii* occurrence in Europe. Although cherries have been reported to suffer high infestation rates in North America, and they are generally considered a preferred host for *D. suzukii*, infestation rates in Europe are rather low. It seems that population densities during the ripening season of cherries are low, and early-maturing cultivars may 'escape' high infestation rates. The relatively recent invasion of *D. suzukii* in Europe does not currently allow a confident assessment of its importance in cherry cultivation.

In recent times, therefore, European cherry growers have had to deal with three instead of one fruit-infesting fly. This complicates management practices, increasing the cost of control activities and demanding high levels of coordination and scientific input at a local scale. Interspecific interactions among the two *Rhagoletis* spp. and *D. suzukii* should be addressed in years to come with the aim of developing sound control strategies against the fruit-infesting Diptera.

### 13.4.2 Trends, challenges and IPM approaches

The European sweet cherry domain is shaped by divergent trends, driven by consumers, markets, regulators and producers. Consumers demand unblemished, yet pesticide-free fruit. Markets prefer large, sweet, brightly coloured and unblemished fruit of consistent quality, compliant with the pesticide residue standards (EU, 2005). Regulators enforce integrated pest management (EU, 2009) despite the scarcity of sound IPM methods and dwindling lists of approved pesticides. Producers endeavour to comply with these demands and maximize their gains through adjustments to the spectrum of cherry cultivars grown, tree structure, orchard and pest management practices. These adjustments affect biological functions of the cherry orchard and influence the behaviour and development of the locally resident pests and beneficial insects. Each species responds in a unique fashion, modifying the local ecological equilibrium and the overall pest management scenario. Such effects are seldom anticipated and are difficult to predict, but invariably affect crop infestation risks and the results of the pest management efforts. The most acute effects occur in the case of pests such as *R. cerasi*, *R. cingulata* and *D. suzukii*, which infest the fruit shortly before harvest. Hence, the following illustration of such interactions is focused on *R. cerasi*.

#### *Strategy of pesticide use versus pest biology and pesticide residue compliance*

In spite of the progress made in non-chemical fruit fly management (Lux *et al.*, 2003; Daniel and Grunder, 2012), in the foreseeable future, pesticide use will remain the cornerstone in large-scale cherry production and a 'last resort option' for ecologically conscious systems. The producers are left with two options: (i) target the adult flies emerging from the soil in spring; or (ii) target the immature stages (eggs and larvae) developing inside the infested fruit. The third option – control of the pupae overwintering in the soil – has proved unfeasible in both the economic and environmental sense.

The first option – targeting the adult flies with non-systemic pesticides – bears substantial risk of non-target ecological side-effects and promotion of secondary pests. To protect the fruit, multiple pesticide applications are necessary, even if all the flies present on-farm during the treatment could be killed. This necessity is usually attributed to post-treatment pest immigration from the neighbourhood. However, in isolated plots surrounded by a 50–100 m wide non-host buffer, such immigration is of minor importance, because *R. cerasi* is intimately associated with its host tree and its mobility is limited (Daniel and Wyss, 2009; Daniel and Grunder, 2012; Daniel and Baker, 2013). Surprisingly, most of the 'post-treatment fly resurgence' originates from the soil within the treated plot. Typically, fly emergence lasts 4–7 weeks, with the bulk (60–80%) emerging over approximately 2 weeks (Vogt *et al.*, 2010b). This period may be extended by variation in farm topography and slope exposures, soil coverage, erratic weather spells, etc. Consequently, fractions of the 'overwintering' flies still emerge after the first and subsequent pesticide applications. The flies mature and attain capacity to lay eggs and infest fruits 5–14 days postemergence, which determines the frequency of pesticide reapplication and the duration of possible 'preharvest application pause'. Less frequent treatments, or incomplete pesticide efficacy, will inevitably result in fruit damage (infestation). Barring the inherent ecological costs, this approach – with modern, short-acting, non-persistent and non-systemic pesticides – offers good opportunities to comply with the stringent pesticide residue standards.

The second option – targeting immature stages inside the infested fruit with a systemic pesticide – is less cumbersome and is thus preferred. Such pesticides (e.g. neonicotinoids), in addition to the extended period of systemic killing activity against the immature stages, provide a short (1–3 days) protection against the adults. However, the flies immigrating or emerging on-farm 1–3 days post-treatment will continue unharmed and remain able to infest the fruit. Thus, effectively, this approach does not prevent fruit infestation, but eliminates its early stages in the infested fruit. Based on our estimations (Lux *et al.*, 2016), to provide the required fruit protection (infestation <1%), the effective daily in-fruit killing rate of the pesticide must stay at 50–70% until harvest; otherwise, the 'reduced protection gap' will result in substantial fruit infestation. This implies that production of 'maggot-free' and, at the same time, truly 'pesticide-free' fruit is practically impossible, which has already been recognized by the regulators, increasing the maximum residue level for acetamiprid residues in sweet cherries from the earlier 0.2 mg $kg^{-1}$ to 0.5 mg $kg^{-1}$ (EFSA, 2010). With this option, the challenge lies in precise timing of the pesticide application, to cover the whole period of fruit susceptibility to infestation (from green-yellowish stage to harvest) and, on the other hand, to ensure that prior to harvest, the pesticide residue present inside the fruit deteriorates below the stipulated maximum residue level. The latter is not easy to predict, since, apart from the legally defined 'preharvest interval' for each pesticide, the actual deterioration rate depends on the temperatures prevailing postapplication.

### *Evolution of spatial orchard structure and tree canopy size*

In recent decades, the structure of cherry orchards has evolved from the previously dominant high-growing trees with large canopies resembling 'dense cherry forest' towards 'sparse bush-land type' – rows of dwarf trees spaced by wide transects, with individual canopies reduced to just a few or even a single branch. For *R. cerasi*, the canopy of the host tree constitutes the primary environment, providing shelter, food and the fruit for its reproduction. The flies, emerging from the soil under the host tree, respond to visual signals from the canopy and adjust their within-canopy behaviour according to its size and structure (Prokopy, 1968; Prokopy *et al.*, 1987; Boller, 1969; Katsoyannos *et al.*, 1986; Stadler and Schoni, 1991; Senger *et al.*, 2009; Daniel and Grunder, 2012). Such a radical change in the canopy size and orchard macrostructure creates an entirely new environment, dissimilar to the natural habitat of *R. cerasi*, and substantially affects the flies' survival chances, behaviour and the patterns of its on-farm mobility.

The reduced and open canopy has much lower 'fly-holding capacity', which makes the flies more mobile, increasing their translocations both within and among plots, and promoting seasonal shifts among the plots containing cultivars of varying phenology. The latter translates to more uniform fruit infestation patterns and the need for spatially intense pest management. On the other hand, with 'shelter function' much compromised, the flies, especially the newly emerged ones, will be more vulnerable to unfavourable weather spells and more exposed to pesticides. Also yellow Rebel traps, relying on visual cues, when deployed in an open environment, will be more exposed and are likely to attract a greater fraction of the flies actually present on-farm. Hence, adjustment to the IPM action thresholds based on pest monitoring results might be necessary according to orchard structure. In the locations with diversified canopy structure, the flies will probably aggregate more on the larger, more prominent trees or on the plots with larger individual canopies, which may create opportunity for spatially focused IPM (Lux, 2014b; Lux *et al.*, 2014).

### *Evolution of fruit size, colour and sugar content*

Responding to consumers' predilection for attractive commodities, markets prefer larger, sweeter, brightly coloured fruits. This in turn sets the trend in development of new cultivars, and finally determines cultivar choice by the producers. Like human consumers, frugivorous insects also respond, in a species-specific manner, to the changes in fruit appearance and sugar content. Increase in sugar content will enhance larval development, and improve the survival and fecundity of the next pest generation (Daniel and Grunder, 2012). However, contrary to general expectations, enhancement of fruit coloration may be of lesser importance from the *R. cerasi* point of view. To gravid females, the fruit is the most attractive at earlier stages, when yellowish-green. Furthermore, on larger plots containing a single cultivar, the females are faced with no choice.

Quite unexpectedly, the increase in fruit size is more consequential. In contrast to *D. suzukii*, *R. cerasi* actively prevents overinfestation and in-fruit larval competition by depositing a single egg in each infested fruit and marking it with a pheromone (Katsoyannos, 1975). Thus, unlike most frugivorous pests, *R. cerasi* infestation is not determined by the volume of the available fruit, but by its number. This phenomenon has serious although seldom recognized consequences for the fruit producer. Two cherry cultivars differing only in fruit size, growing on the same farm under the same pest pressure (females $ha^{-1}$) and having the same net fruit productivity (t $ha^{-1}$) will be infested at dramatically different levels. A twofold increase in fruit size will result in an eightfold increase in the overall percentage of infested fruit (Lux, 2014b; Lux *et al.*, 2014). Thus, the trend to increase fruit size dramatically enhances the damaging potential of this pest. To maintain the same level of fruit infestation with large-fruiting cultivars, the management of *R. cerasi* must be much more effective (Lux 2014b; Lux *et al.*, 2014). This effect was largely concealed by the widespread use of highly effective persistent pesticides and a more liberal approach to pesticide residues. However, with the switch to IPM methods, which are inherently less effective and more erratic, and a more rigorous approach to residues, inevitably this problem will become apparent, and might even limit further increases in fruit size.

### *Phenology, dynamics of fruit growth and maturation*

Most of the sweet cherry cultivars flower and set fruit at approximately the same time. The process of fruit development, from the *R. cerasi* point of view, can be divided into two distinct periods: (i) unsuitable young fruitlets, with a thin layer of hard flesh around the pit; and (ii) rapidly expanding fruit with thicker and softer flesh, suitable for pest development and thus susceptible to infestation. The onset of the second period is marked by a fruit colour change from green to yellowish-green, and extends until harvest. In wild cherry (*P. avium*), the 'fruit susceptibility' lasts 40–45 days and perfectly fits the extended pest reproductive capacity. In cultivated cherries, the duration of the first period varies little among the cultivars (50–60 days), but the second period is very diverse, from 15–20 days in the earliest to 40–45 days in late cultivars (Schumann *et al.*, 2014; Lux *et al.*, 2016). It appears that development of the early cultivars was largely attained by acceleration of the fruit expansion process and maturity and ripeness. In extreme cases, this led to complete dis-alignment between the time of pest reproductive capacity and fruit suitability, resulting in practically no fruit infestation of the earliest cultivars, even in the absence of pest control.

An alternative strategy for development of late cultivars, focused on extension of the first period while retaining the short duration of the second one, has as yet unexploited potential. Such novel late cultivars with a shorter period of fruit susceptibility could make pest management easier and less expensive (Lux, 2014b).

### *IPM prospects*

Although approved pesticides will remain the key component of the cherry grower's IPM toolbox, their use will be more restrictive. The latter will revive interest in exploiting pest ecology and biological traits of the cherry-growing landscape for the development of less pesticide-reliant management approaches (Dominiak and Ekman, 2013). Typical medium-scale and scattered cherry production in Europe is not conducive to large-scale area-wide pest management approaches, but potentially the elements of structural landscape heterogeneity, such as the spatial arrangement of cultivars of varied phenology, diversification in canopy structure, establishment of buffer and/or pest interception zones, and other methods, could be exploited to design more 'pest-resilient production systems' (Lux *et al.*, 2016). However, effective exploitation of a purposefully created heterogeneity in agro-landscape requires extensive knowledge about pest biology and behaviour, and also meticulous long-term field experimentation, which will be much extended due to the perennial nature of the cherry crop. To alleviate this conundrum, novel approaches and tools have to be conceptualized, enabling the much required quantum boost in landscape design and site-tuned IPM.

### *'Virtual farm' concept and site-focused IPM modelling*

Extensive field experimentation could be facilitated and partially substituted by the development of comprehensive 'virtual farm' models, capable of emulating the key processes determining the performance of the target farming systems (Lux, 2014a; Lux *et al.*, 2016). Such models have the capacity to encapsulate large volumes of quantified knowledge about pest ecology and behaviour, crop phenology, IPM treatments, relevant on-farm processes and prevailing weather, and convert it into operable tools facilitating the development of pest-resilient landscapes and site-specific IPM. Admittedly, implementation of such tools will not eliminate field experiments, but they can radically shorten the usual 'development trajectory' by substituting major parts of long-term and expensive on-farm experiments by their 'virtual' emulations. However, in spite of the obvious benefits and numerous conceptual approaches possible, in the horticulture domain, the availability and utilization of such modelling tools remains limited.

The PESTonFARM model (Lux, 2014a; Lux *et al.*, 2016), operating as a 'virtual cherry farm', exemplifies such an approach. It is based on the assumption that the local pest development and IPM performance is determined by the local farm traits, which shape the outcome of interplay among concurrent processes, where the on-farm dwelling cohorts of individual, independently operating insects (*R. cerasi*) are the key actors. The latter was the reason for employing a 'bottom-up', individual-focused, 'ethological' approach, and the application of agent-based stochastic process emulation. Enacting the processes by its key 'virtual actors' (insects) permits evaluation of the net effects of multiple, concurrent and subtle modifications introduced into the local system, and offers insights into the mechanisms driving local pest development and IPM performance.

### *Automated pest surveillance and decision-making systems*

Recent advances in real-time data transfer, automated video analysis, electronic sensors, data-interpretation software and spatial decision-making algorithms (Cohen *et al.*, 2008) have fuelled the development of automated pest surveillance and decision-making systems, such as the 'location awareness pest management systems' (Pontikakos *et al.*, 2012; FruitFlyNet: http://fruitflynet.aua.gr). The system is based on: (i) automated traps for adult *R. cerasi* combined with soil and air temperature, precipitation, wind speed and humidity monitoring sensors; (ii) a wireless network to gather and transfer data in a real-time mode to cloud storage, accessible to designated users; (iii) a spatial decision support system that estimates the local adult emergence and development; and (iv) software that analyses the acquired information and generates local pest management decisions. The farmers are supported further by tools assisting in locally diversified spraying, following the 'path of spraying' and estimating pesticide use and environmental impact. The above system was tested in 2015 on commercial cherry orchards in the area of Thessaly, Greece, and resulted in a fourfold reduction in insecticide use, while keeping the fruit infestation level similar to that in control orchards following the standard IPM practice (N.T. Papadopoulos, 2015, unpublished data).

Combining systems approaches and site-focused pest management from different perspectives, such as the abovementioned location awareness system and the 'virtual farm' concept with the PESTonFARM simulation model, can introduce the concepts of advanced precision management on cherry farms, both at the individual farm and at the local/area-wide level.

## References

Alford, D.V. (2014) *Pests of Fruit Crops: A Colour Handbook*, 2nd edn. CRC Press, Boca Raton, Florida.

Argyriou, L.C. and. Paloukis, S.S. (1976) Some data on biology and parasitization of *Sphaerolecanium prunastri* Fonscolombe (Homoptera Coccidae) in Greece. *Annales de l'Institut Phytopathologique Benaki* 11, 230–240.

Asplen, M.K., Anfora, G., Biondi, A., Choi, D.-S., Chu, D., Daane, K.M,. Gibert, P., Gutierrez, A.P., Hoelmer, K.A., Hutchison, W.D., Isaacs, R., Jiang, Z.L., Kárpáti, Z., Kimura, M.T., Pascual, M., Philips, C.R., Plantamp, C., Ponti, L., Vétek, G., Vogt, H., Walton, V.M., Yu, Y., Zappalà, L. and Desneux, N. (2015) Invasion biology of spotted wing Drosophila (*Drosophila suzukii*): a global perspective and future priorities. *Journal of Pest Science* 88, 469–494.

Baroffio, C. and Fischer, S. (2011) Neue Bedrohung für Obstplantagen und Beerenpflanzen: die Kirschessigfliege [New threat to orchards and berry plants: the spotted wing drosophila]. *UFA-Revue* 11, 46–47.

Bartoloni, N., Gorosito, N., Laffaye, C. and Mareggiani, G. (2012) Flight detection of *Caliroa cerasi* L. (Hymenoptera: Tenthredinidae) adults in the Andean Region of Parallel 42, Argentina. *Boletín de Sanidad Vegetal – Plagas* 38, 233–238.

Belien, T., Bangels, E., Vercammen, J. and Bylemans, D. (2013) Integrated control of the European cherry fruit fly *Rhagoletis cerasi* in Belgian commercial cherry orchards. In: *Proceedings of the EU COST Action FA1104: Sustainable Production of High-quality Cherries for the European Market*. COST meeting, Pitesti, Romania, October 2013. Available at: https://www.bordeaux.inra.fr/cherry/docs/dossiers/Activities/Meetings/15-17%20 10%202013_3rd%20MC%20and%20all%20WG%20Meeting_Pitesti/Posters/Belien%20et%20al.pdf (accessed 8 May 2016).

Ben-Dov, Y. (1968) Occurrence of *Sphaerolecanium prunastri* (Fonscolombe) in Israel and description of its hitherto unknown third larval instar. *Annales des Epiphyties* 19, 615–621.

Blackman, R.L. and Eastop, V.F. (1984) *Aphids on the World's Crops. An Identification and Information Guide*. Wiley, Chichester, UK.

Blackman, R.L. and Eastop, V.F. (1994) *Aphids on the World's Trees: An Identification and Information Guide*. CAB International, Wallingford, UK.

Bland, K.P., Hancock, E.F. and Razowski, J. (2014) *Tortricidae, Part 1: Tortricinae & Chlidanotinae*, 1st edn. Brill Academic Publishers, Leiden, The Netherlands.

Boller, E. (1969) Neues über die Kirschenfliege: Freilandversuche im Jahr 1969. *Schweizerische Zeitschrift für Obst- und Weinbau* 105, 566–572.

Briem, F., Eben, A., Gross, J. and Vogt, H. (2016) An invader supported by a parasite: mistletoe berries as a host for food and reproduction of Spotted Wing Drosophila in early spring. *Journal of Pest Science* 89, 749.

Bruck, D.J., Bolda, M., Tanigoshi, L., Klick, J., Kleiber, J., DelFrancesco, J., Gerdeman, B. and Spitler, H. (2011) Laboratory and field comparisons of insecticides to reduce infestation of *Drosophila suzukii* in berry crops. *Pest Management Science* 67, 1375–1385.

Bush, G.L. (1966) The taxonomy, cytology, and evolution of the genus *Rhagoletis* in North America (Diptera, Tephritidae). *Bulletin of the Museum of Comparative Zoology at Harvard College* 134, 431–562.

CABI (2016a) *Drosphila suzukii* (spotted wing drosophila). CAB International, Wallingford, UK. Available at: http://www.cabi.org/isc/datasheet/109283#20123337590 (accessed 19 January 2017).

CABI (2016b) *Halyomorpha halys* (brown marmorated stink bug). CAB International, Wallingford, UK. Available at: http://www.cabi.org/isc/datasheet/27377 (accessed 19 January 2017).
Calabria, G., Maca, J., Bachli, G., Serra, L. and Pascual, M. (2012) First records of the potential pest species *Drosophila suzukii* (Diptera: Drosophilidae) in Europe. *Journal of Applied Entomology* 136, 139–147.
Callot, H. and Brua, C. (2013) *Halyomorpha halys* (Stål,1855), the marmorated stink bug, new species for the fauna of France (Heteroptera Pentatomidae). *Le Entomologiste* 69, 69–71.
Cha, D.H., Adams, T., Rogg, H. and Landolt, P.J. (2012) Identification and field evaluation of fermentation volatiles from wine and vinegar that mediate attraction of spotted wing Drosophila, *Drosophila suzukii*. *Journal of Chemical Ecology* 38, 1419–1431.
Cha, D.H., Hesler, S.P., Cowles, R.S., Vogt, H., Loeb, G.M. and Landolt, P.J. (2013) Comparison of a synthetic chemical lure and standard fermented baits for trapping *Drosophila suzukii* (Diptera: Drosophilidae). *Environmental Entomology* 42, 1052–1060.
Cini, A., Ioratti, C. and Anfora, G. (2012) A review of the invasion of *Drosophila suzukii* in Europe and a draft research agenda for integrated pest management. *Bulletin of Insectology* 65, 149–160.
Cohen, Y., Cohen, A., Hetzroni, A., Alchanatis, V., Broday, D., Gazit, Y. and Timar, D. (2008) Spatial decision support system for Medfly control in citrus. *Computers and Electronics in Agriculture* 62, 107–117.
Cuthbertson, A.G.S., Collins, D.A., Blackburn, L.F., Audsley, N. and Bell, H.A. (2014) Preliminary screening of potential products against *Drosophila suzukii*. *Insects* 5, 488–498.
Daniel, C. and Baker, B. (2013) Dispersal of *Rhagoletis cerasi* in commercial cherry orchards: efficacy of soil covering nets for cherry fruit fly control. *Insects* 4, 168–176.
Daniel, C. and Grunder, J. (2012) Integrated management of European cherry fruit fly *Rhagoletis cerasi* (L.): situation in Switzerland and Europe. *Insects* 3, 956–988.
Daniel, C. and Wyss, E. (2009) Migration und Ausbreitung der Kirschfruchtfliege Innerhalb von Obstanlagen – Möglichkeit der biologischen Bodenbehandlung. *Mitteilungen der Deutschen Gesellschaft für Allgemeine und Angewandte Entomologie* 17, 247–248.
Daniel, C., Mathis, S. and Feichtinger, G. (2014) A new visual trap for *Rhagoletis cerasi* (L.) (Diptera: Tephritidae). *Insects* 5, 564–576.
Dominiak, B.C. and Ekman, J.H. (2013) The rise and demise of control options for fruit fly in Australia. *Crop Protection* 51, 57–67.
EFSA (2010) Reasoned opinion. Modification of the existing MRL for acetamiprid in cherries. *EFSA Journal* 8, 1494.
Englert, C. and Herz, A. (2016) Native predators and parasitoids for biological regulation of *Drosophila suzukii* in Germany. In: Fördergemeinschaft Ökologischer Obstbau e.V. (FÖKO) (ed.) *Proceedings of the 17th International Conference on Organic Fruit-Growing*. FÖKO, 15–17 February, Weinsberg, Germany, pp. 284–285.
EPPO (2008) Pest risk analysis for: *Drosphila suzukii*. European and Mediterranean Plant Protection Organization. Available at: https://www.eppo.int/QUARANTINE/Pest_Risk_Analysis/PRAdocs_insects/11-17189_PRA_record_Drosophila_suzukii_final%20.pdf (accessed 19 January 2017).
EPPO (2015a) PQR – EPPO Plant Quarantine Data Retrieval system. Available at: https://www.eppo.int/DATABASES/pqr/pqr.htm (accessed 19 January 2017).
EPPO (2015b) *Drosophila suzukii* (Diptera: Drosophildae). European and Mediterranean Plant Protection Organization. Available at: https://www.eppo.int/QUARANTINE/Alert_List/insects/Drosophila_suzukii_factsheet_12-2010.pdf (accessed 19 January 2017).
EC DG SANCO (2013) Ad-hoc study to support the initial establishment of the list of candidates for substitution as required in Article 80(7) of Regulation (EC) No 1107/2009: Final report. European Commission Directorate General for Health and Consumer Affairs (DG SANCO), Brussels.
EU (2005) Regulation (EC) No 396/2005 of the European Parliament and of the Council of 23 February 2005 on maximum residue levels of pesticides in or on food and feed of plant and animal origin and amending Council Directive 91/414/EEC. Available at: http://eur-lex.europa.eu/legal-content/EN/TXT/PDF/?uri=-CONSLEG:2005R0396:20121026 (accessed 18 January 2017).
EU (2009) Directive 2009/128/EC of the European Parliament and of the Council of 21 October 2009 establishing a framework for Community action to achieve the sustainable use of pesticides. Official Journal of the European Union L309/71, 24.11.2009. Available at: http://www.eppo.int/PPPRODUCTS/information/2009_0128_EU-e.pdf (accessed 18 January 2017).
FAO (1985) Data and Recommendations of the Joint Meeting of the FAO Panel of Experts on Pesticide Residues in Food and the Environment and the WHO Expert Group on Pesticide Residues. Food and Agriculture Organization of the United Nations, Rome, 24 September–3 October 1984. Available at: http://www.inchem.org/documents/jmpr/jmpmono/v84pr19.htm (accessed 19 January 2017).

Fontanari, M., Sacco, M. and Girolami, V. (1993) Influence of ants on aphids and their predators in orchards. *Informatore Fitopatologico* 43, 47–55.

Gilmore, J.E. (1960) Biology of the black cherry aphid in the Willamette Valley, Oregon. *Journal of Economic Entomology* 53, 659–661.

Grassi, A., Anfora, G., Maistri, S., Maddalena, G., De Cristofaro, A., Savini, G. and Ioriatti, C. (2014) Development and efficacy of Droskidrink, a food bait for trapping *Drosophila suzukii*. In: *Book of Abstracts of the IOBC VIII Workshop on Integrated Soft Fruit Production*, 26–28 May, Pergine Valsugana, Italy, pp. 105–106. Available at: http://eventi.fmach.it/Iobc/Abstract-Manuscript-submission (accessed 19 February 2015).

Gruppe, A. (1990) Investigations on the significance of ants in the development and dispersal of the black cherry aphid *Myzus cerasi* F. (Hom., Aphididae). *Zeitschrift für Pflanzenkrankheiten und Pflanzenschutz* 97, 484–489.

Guo, J.M. (2007) Bionomics of fruit flies, *Drosophila melanogaster*, damage cherry in Tianshui. *Chinese Bulletin of Entomology* 44, 743–745.

Gwiazdowski, R.A., Elkinton, J.S., Dewaard, J.R. and Sremac, M. (2013) Phylogeographic diversity of the winter moths *Operophtera brumata* and *O. bruceata* (Lepidoptera: Geometridae) in Europe and North America. *Annals of the Entomological Society of America* 106, 143–151.

Harzer, U. and Köppler, K. (2015) Die Kirschessigfliege *Drosophila suzukii* – Befallssituation 2014 in Rheinland-Pfalz und Baden-Württemberg. *Obstbau* 4, 208–211.

Hauser, M. (2011) A historic account of the invasion of *Drosophila suzukii* (Matsumura) (Diptera: Drosophilidae) in the continental United States, with remarks on their identification. *Pest Management Science* 67, 1352–1357.

Haye, T. and Wyniger, D. (2013) Current distribution of *Halyomorpha halys* in Europe. Available at: http://www.halyomorphahalys.com/aktuelle-verbreitungskarte-current-distribution.html (accessed 19 January 2017).

Heckmann, R. (2012) First evidence of *Halyomorpha halys* (Stål, 1855) (Heteroptera: Pentatomidae) in Germany. *Heteropteron H* 36, 17–18.

Hoebeke, E.R. and Carter, M.E. (2003) *Halyomorpha halys* (Heteroptera: Pentatomidae): a polyphagous plant pest from Asia newly detected in North America. *Proceedings of the Entomological Society of Washington* 105, 225–237.

Jaastad, G. (1998a) Mating behavior and distribution of the European cherry fruit fly (*Rhagoletis cerasi* L.) in Norway; applied and evolutionary research aspects on an insect pest. DPhil thesis, Department of Zoology, University of Bergen, Norway.

Jaastad, G. (1998b) Male mating success and body size in the European cherry fruit fly, *Rhagoletis cerasi* L. (Dipt., Tephritidae). *Journal of Applied Entomology* 122, 1–5.

Johannesen, J., Keyghobadi, N., Schuler, H., Stauffer, C. and Vogt, H. (2013) Invasion genetics of American cherry fruit fly in Europe and signals of hybridization with the European cherry fruit fly. *Entomologia Experimentalis et Applicata* 147, 61–72.

Kanzawa, T. (1939) Studies on *Drosophila suzukii* Mats. *Review of Applied Entomology* 29, 622.

Kaswase, S. and Uchino, K. (2005) Effect of mesh size on *Drosophila suzukii* adults passing through the mesh. *Annual Report of the Kanto-Tosan Plant Protection Society* 52, 99–101.

Katsoyannos, B.I. (1975) Oviposition-deterring, male-arresting, fruit-marking pheromone in *Rhagoletis cerasi*. *Environmental Entomology* 4, 801–807.

Katsoyannos, B.I. (1979) Zum Reproduktions- und Wirtswahlverhalten der Kirschenßiege *Rhagoletis cerasi* L. (Diptera: Tephritidae). Dissertation, Eidgenössische Technische Hochschule, Zürich, No. 6409.

Katsoyannos, B.I., Boller, E.F. and Benz, G. (1986) Das verhalten der Kirschenfliege, *Rhagoletis cerasi* L., bei der Auswahl der Wirtspflanzen und ihre Dispersion. *Mitteilungen der Schweizerischen Entomologischen Gesellschaft* 59, 315–335.

Katsoyannos, B.I., Papadopoulos, N.T. and Stavridis, D. (2000) Evaluation of trap types and food attractants for *Rhagoletis cerasi*. *Journal of Economic Entomology* 93, 1005–1010.

Kinjo, H., Kunimi, Y. and Nakai, M. (2014) Effects of temperature on the reproduction and development of *Drosophila suzukii* (Diptera: Drosophilidae). *Applied Entomology and Zoology* 49, 297–303.

Köppler, K. (2008) Occurrence of non-diapausing pupae in *Rhagoletis cerasi* L. In: Papadopoulos, N. and Kouloussis, N. (eds) *6th TEAM newsletter*, Tephritid Workers of Europe, Africa and Middle East, December 2008, Thessaloniki, Greece, pp. 6–8.

Kovanci, O.B. and Kovanci, B. (2006) Effect of altitude on seasonal flight activity of *Rhagoletis cerasi* flies (Diptera: Tephritidae). *Bulletin of Entomological Research* 96, 345–351.

Landolt, P.J., Adams, T., Davis, T.S. and Rogg, H. (2012a) Trapping spotted wing drosophila, *Drosophila suzukii* (Matsumura) (Diptera: Drosophilidae), with combinations of vinegar and wine, and acetic acid and ethanol. *Journal of Applied Entomology* 136, 148–154.

Landolt, P.J., Adams, T., Davis, T.S. and Rogg, H. (2012b) Spotted wing drosophila, *Drosophila suzukii* (Diptera: Drosophilidae), trapped with combinations of wines and vinegars. *Florida Entomologist* 95, 326–332.

Laurino, D., Porporato, M., Patetta, A. and Manino, A. (2011) Toxicity of neonicotinoid insecticides to honey bees: laboratory tests. *Bulletin of Insectology* 64, 107–113.

Lee, D.H., Short, B.D., Joseph, S.V., Bergh, J.C. and Leskey, T.C. (2013) Review of the biology, ecology, and management of *Halyomorpha halys* (Hemiptera: Pentatomidae) in China, Japan, and the Republic of Korea. *Environmental Entomology* 42, 627–641.

Lee, J.C., Bruck, D., Curry, H., Edwards, D., Haviland, D.R. van Steenwyk, R.A. and Yorgey, B.M. (2011) The susceptibility of small fruits and cherries to the spotted-wing drosophila, *Drosophila suzukii*. *Pest Management Science* 67, 1358–1367.

Łęski, R. (1963) Studies on the biology and ecology of the cherry fruit fly, *Rhagoletis cerasi* L. (Diptera, Trypetidae). *Polskie Pismo Entomologiczne B*, 153–240.

Lux, S.A. (2014a) PESTonFARM – stochastic model of on-farm insect behaviour and their response to IPM interventions. *Journal of Applied Entomololgy* 138, 458–467.

Lux, S.A. (2014b) Influence of grower and consumer preferences on on-farm behaviour of *Rhagoletis cerasi*. In: *Proceedings of the Symposium on 'Opportunities for Enhancement of Integrated Pest Management'*, 1–3 April, Warsaw, Poland. Available at: https://www.bordeaux.inra.fr/cherry/docs/dossiers/Activities/Meetings/2014%2004%2001-03%20WG3%20Meeting_Warsaw/BOOKofABSTRACT.pdf (accessed 19 January 2017).

Lux, S.A., Ekesi, S., Dimbi, S., Mohamed, S. and Billah, M. (2003) Mango-infesting fruit flies in Africa: perspectives and limitations of biological approaches to their management. In: Neuenschwander, P., Borgemeister, C. and Langewald, J. (eds) *Biological Control in Integrated Pest Management Systems in Africa*. CABI International, Wallingford, UK, pp. 277–293.

Lux, S.A., Vogt, H. and Wnuk, A. (2014) Impacts of the cherry-market trends on *Rhagoletis cerasi* on-farm behaviour and IPM. In: *Proceedings of the EU COST Action FA1104: Sustainable Production of High-quality Cherries for the European market*. COST meeting, Bordeaux, France, October 2014. Available at: https://www.bordeaux.inra.fr/cherry/docs/dossiers/Activities/Meetings/2014%2010%2013-15_4th%20MC%20and%20all%20WG%20Meeting_Bordeaux/Presentations/Lux_Bordeaux2014.pdf (accessed 25 May 2016).

Lux, S.A., Wnuk, A., Vogt, H., Belien, T., Spornberger, A., Studnicki, M. (2016) Validation of individual-based Markov-like stochastic process model of insect behaviour and a 'virtual farm' concept for enhancement of site-specific IPM. *Frontiers in Physiology* 7, 363; doi: 10.3389/fphys.2016.00363. http://journal.frontiersin.org/article/10.3389/fphys.2016.00363/full

Maistrello, L., Costi, E., Caruso, S., Vaccari, G., Bortolotti, P.,Nannini, R., Casoli, L., Montermini, A., Bariselli, M. and Guidetti, R. (2014) *Halyomorpha halys* in Italy: first results of field monitoring in fruit orchards. In: Escudero-Colmar, L.A. and Damos, P. (eds) *Book of Abstracts of the Joint Meeting of the Sub-groups "Pome Fruit Arthropods" and "Stone Fruits", IOBC/WPRS Working Group "Integrated Protection in Fruit Crops"*, Vienna, Austria, p. 17.

Moraiti, C.A. (2013) Study on the biology, ecology and behavior of different populations of the European cherry fruit fly, *Rhagoletis cerasi* (Diptera: Tephritidae). PhD thesis, University of Thessaly, Volos, Greece.

Moraiti, C.A., Nakas, C.T., Köppler, K. and Papadopoulos, N.T. (2012) Geographical variation in adult life-history traits of the European cherry fruit fly, *Rhagoletis cerasi* (Diptera: Tephritidae). *Biological Journal of the Linnean Society* 107, 137–152.

Moraiti, C.A., Nakas, C.T. and Papadopoulos, N.T. (2014) Diapause termination of *Rhagoletis cerasi* pupae is regulated by local adaptation and phenotypic plasticity: escape in time through bet-hedging strategies. *Journal of Evolutionary Biology* 27, 43–54.

Nakano, R., Ihara, F., Mishiro, K., Toyama, M. and Toda, S. (2014) Confuser® mm can be used as an attractant to enable monitoring of the peach leafminer moth, *Lyonetia clerkella* (Lepidoptera: Lyonetiidae), in peach orchards treated with a mating disrupter. *Applied Entomology and Zoology* 49, 505–510.

Paloukis, S.S., Papadopoulos, N.T. and Katsoyannos, B.I. (1991) Observations on the bio-ecology and control of *Sphaerolecanium prunastri* (Fonsc.) (Coccidae) pest of fruit trees in northern Greece (in Greek). In: *Proceedings of the 3rd Symposium of the Greek Entomological Society*, 9–11 October 1989, Thessaloniki, Greece, pp. 85–93.

Papadopoulos, N.T. (2014) Fruit fly invasion: historical, biological, economic aspects and management. In: Shelly, T., Vargas, R. and Epsky, N. (eds) *Trapping and Detection, Control, and Regulation of Tephritid Fruit Flies*. Springer, Dordrecht, the Netherlands, pp. 219–252.

Papanastasiou, S.A., Nestel, D., Diamantidis, A.D., Nakas, C.T. and Papadopoulos, N.T. (2011) Physiological and biological patterns of a highland and a coastal population of the European cherry fruit fly during diapause. *Journal of Insect Physiology* 57, 83–93.

Pedata, P.A., Hunter, M.S., Godfray, H.C.J. and Viggiani, G. (1995) The population dynamics of the white peach scale and its parasitoids in a mulberry orchard in Campania, Italy. *Bulletin of Entomological Research* 85, 531–539.

Pehlevan, B. and Kovanci, O.B. (2014) First report of *Adoxophyes orana* in northwestern Turkey: population fluctuation and damage on different host plants. *Turkish Journal of Agriculture and Forestry* 38, 847–856.

Pérez, J.A., García, T., Arias, A. and Martínez de Velasco, D. (1995) La cola entomológica, un método alternativo a la lucha con insecticidas contra el pulgón negro del cerezo (*Myzus cerasi* F.). *Boletin de Sanidad Vegetal, Plagas* 21, 213–222.

Pontikakos, C.M., Tsiligiridis, T.A., Yialouris, C.P. and Kontodimas, D.C. (2012) Pest management control of olive fruit fly (*Bactrocera oleae*) based on a location-aware agro-environmental system. *Computers and Electronics in Agriculture* 87, 39–50.

Porcel, M., Sjöberg, P., Swiergiel, W., Dinwiddie, R., Rämert, B. and Tasin, M. (2015) Mating disruption of *Spilonota ocellana* and other apple orchard tortricids using a multispecies reservoir dispenser. *Pest Management Science* 71, 562–570.

Poyet, M., Le Roux, V., Gibert, P., Meirland, A., Prevost, G., Eslin, P. and Chabrerie, O. (2015) The wide potential trophic niche of the Asiatic fruit fly *Drosophila suzukii*: the key of its invasion success in temperate Europe? *PLoS One* 10, e0142785.

Prokopy, R.J. (1968) Orientation of the apple maggot flies *Rhagoletis pomonella* (Walsh) and European cherry fruit flies *R. cerasi* L. (Diptera: Tephritidae) to visual stimuli. In: *Proceedings of the 13 International Congress of Entomology*, 2–9 August, Moscow, Russia, pp. 34–35.

Prokopy, R.J., Papaj, D.R., Opp, S.B. and Wong, T.T.Y. (1987) Intra-tree foraging behavior of *Ceratitis capitata* flies in relation to host fruit density and quality. *Entomologia Experimentalis et Applicata* 45, 251–258.

Rauleder, H. and Köppler, K. (2015) Welche Früchte werden von Kirschessigfliegen befallen? *Obstbau* 4, 220–225.

Remund, U. and Boller, E. (1975) Qualitätskontrolle von Insekten: Die Messung von Flugparametern. *Zeitschrift für Angewandte Entomologie* 78, 113–126.

Rogers, M.A., Burknessn, E.C. and Hutchison, W.D. (2016) Evaluation of high tunnels for management of *Drosophila suzukii* in fall-bearing red raspberries: potential for reducing insecticide use. *Journal of Pest Science* 89, 815–821.

Salis, L., Lof, M., van Asch, M. and Visser, M.E. (2016) Modeling winter moth *Operophtera brumata* egg phenology: nonlinear effects of temperature and developmental stage on developmental rate. *Oikos* 125, 1772–1781.

Sasaki, M. and Sato, R. (1995) Bionomics of the cherry Drosophila, *Drosophila suzukii* Matsumura (Diptera: Drsosphilidae) in Fukushima prefecture 2. Overwintering and number of generations. *Annual Report of the Society of Plant Protection of North Japan* 46, 167–169.

Schumann, C., Schlegel, H.J., Grimm, E. and Knoche, M. (2014) Water potential and its components in developing sweet cherry. *Journal of the American Society for Horticultural Science* 139, 349–355.

Senger, S.E., Tyson, R.C., Roitberg, B.D., Thistlewood, H.M.A., Harestad, A.S., Chandler, M.T. (2009) Influence of habitat structure and resource availability on the movements of *Rhagoletis indifferens*. *Environmental Entomology* 38, 823–835.

Sjöberg, P., Rämert, B., Thierfelder, T. and Hillbur, Y. (2015) Ban of a broad-spectrum insecticide in apple orchards: effects on tortricid populations, management strategies, and fruit damage. *Journal of Pest Science* 88, 767–775.

Smith, J.J. and Bush, G.L. (1997) Phylogeny of the genus *Rhagoletis* (Diptera: Tephritidae) inferred from DNA sequences of mitochondrial cytochrome oxidase II. *Molecular Phylogenetics and Evolution* 7, 33–43.

Stadler, E. and Schoni, R. (1991) High sensitivity to sodium in the sugar chemoreceptor of the cherry fruit fly after emergence. *Physiological Entomology* 16, 117–129.

Stephens, A.R., Asplen, M.K., Hutchison, W.D. and Venette, R.C. (2015) Cold hardiness of winter-acclimated *Drosophila suzukii* (Diptera: Drosophilidae) adults. *Environmental Entomology* 44, 1619–1626.

Thistlewood, H.M., Bostanian, N.J. and Hardman, J.M. (2013) *Panonychus ulmi* (Koch) European red mite (Trombidiformes: Tetranychidae). In: Mason, P. and Gillespie, D.R. *Biological Control Programmes in Canada 2001–2012*. CAB International, Wallingford, UK, p. 238–243.

Tochen, S., Dalton, D.T., Wimann, N.G., Hamm, C., Shearer, P.W., Walton, V.M. (2014) Temperature-related development and population parameters for *Drosophila suzukii* (Diptera: Drosophilidae) on cherry and blueberry. *Environmental Entomology* 43, 501–510.

Tochen, S., Woltz, J.M., Dalton, D.T., Lee, J.C., Wiman, N.G., Walton, V.M. (2015) Humidity affects populations of *Drosophila suzukii* (Diptera: Drosophilidae) in blueberry. *Journal of Applied Entomology* 140, 47–57.

Toth, M., Szarukan, I., Voigt, E. and Kozar, F. (2004) Importance of visual and chemical stimuli in the development of an efficient trap for the European cherry fruit fly (*Rhagoletis cerasi* L.) Diptera, Tephritidae. *Növenyvedelem* 40, 229–236.

Tremblay, E. (1981) *Entomologia applicata*, Vol. II (1). Liguori, Napoli, Italy.

Tzanakakis, M.E. and Katsoyannos, B.I. (2003) *Insects of Fruit Trees and Grapevine*. Agrotypos, Marousi, Greece.

van Leeuwen, T., Tirry, L., Yamamoto, A., Nauen, R. and Dermauw, W. (2015) The economic importance of acaricides in the control of phytophagous mites and an update on recent acaricide mode of action research. *Pesticide Biochemistry and Physiology* 121, 12–21.

Vogt, H., Köppler, K., Dahlbender, W. and Hensel, G. (2010a) Observations of *Rhagoletis cingulata*, an invasive species from North America, on cherry in Germany. *IOBC-WPRS Bulletin* 54, 273–277.

Vogt., H., Kaffer, T., Just, J., Herz, A., Féjoz, B. and Köppler, K. (2010b) Key biological and ecological characteristics of European cherry fruit fly *Rhagoletis cerasi* with relevance to management. In: *IOBC-WPRS WG "Integrated Protection of Fruit Crops"*. Joint Meeting of the Sub-groups "Pome fruit arthropods" and "Stone fruits", 12–17 September, Vico del Gargano, Italy. Abstract, p. 54.

Walsh, D.B., Bolda, M.P., Goodhue, R.E., Dreves, A.J., Lee, J., Bruck, D.J., Walton, V.M., O'Neal, S.D. and Zalom, F.G. (2011) *Drosophila suzukii* (Diptera: Drosophilidae): invasive pest of ripening soft fruit expanding its geographic range and damage potential. *Journal of Integrated Pest Management* 2, 1–7.

Wermelinger, B., Wyniger, D. and Forster, B. (2008) First records of an invasive bug in Europe: *Halyomorpha halys* Stål (Heteroptera: Pentatomidae), a new pest on woody ornamentals and fruit trees? *Mitteilungen der Schweizerischen Entomologischen Gesellschaft* 81, 1–8.

Weydert, C., Charlot, G. and Mandrin, J.-F. (2014) Insect-proof nets to protect cherry trees from *Drosophila suzukii* and *Rhagoletis cerasi*. In: Escudero-Colmar, L.A. and Damos, P. (eds) *Book of Abstracts of the Joint Meeting of the sub-groups "Pome fruit arthropods" and "Stone fruits"*. IOBC-WPRS Working Group "Integrated Protection in Fruit Crops", Vienna, Austria, p. 32.

Wise, J.C., Vanderpoppen, R., Vandervoort, C., O'Donnell, C. and Isaacs, R. (2014) Curative activity contributes to control of spotted-wing drosophila (Diptera: Drosophilidae) and blueberry maggot (Diptera: Tephritidae) in highbush blueberry. *The Canadian Entomologist* 147, 109–117.

Zerulla, F.N., Schmidt, S., Streitberger, M., Zebitz, C.P. and Zelger, R. (2015) On the overwintering ability of *Drosophila suzukii* in South Tyrol. *Journal of Berry Research* 5, 41–48.

Zhang, C.-L., Zhang, Y.-H., He, W.-Y., Huang, J.-B., Zhou, L.-Y., Ning, H., Wu, Z.-P., Wang, D.-N. and Ke, D.-C. (2011) Integrated control technology of cherry fruit flies and experimental demonstration in Aba, Sichuan province. *China Plant Protection* 31, 26–28.

# 14 Fungal Diseases

**Jorunn Børve,[1]* Antonio Ippolito,[2] Brankica Tanović,[3] Monika Michalecka,[4] Simona Marianna Sanzani,[2] Anna Poniatowska,[4] Marta Mari[5] and Jovana Hrustić[3]**

[1]*Norwegian Institute of Bioeconomy Research, Ullensvang, Norway;* [2]*Università degli Studi Aldo Moro, Bari, Italy;* [3]*Institute of Pesticides and Environmental Protection, Belgrade, Serbia;* [4]*Research Institute of Horticulture, Skierniewice, Poland;* [5]*University of Bologna, CRIOF-DipSA, Bologna, Italy*

## 14.1 Introduction

In general, three factors need to be in place for a fungal disease to occur: (i) presence of the pathogen; (ii) suitable climatic conditions; and (iii) a suitable host tissue. Pathogen presence is avoidable for diseases with a narrow host spectrum and for which the main inoculum sources is in the trees. Mummified fruit are a typical example, since it is easy to avoid their formation by picking the whole yield, and if present, they can be removed during pruning and training of the trees. However, this is realistic only in intensive growing systems. Inoculum of pathogens with broader host spectrums is not as easy to avoid, but it is possible to reduce inoculum levels by removal of possible hosts in the surroundings. Climatic conditions can be optimized to avoid disease risk. On a local scale, the location of the orchard is essential, for example choosing a place with good air circulation and drainage. On a microclimatic scale, tree planting distance, tree vigour, training and pruning influence the openness of the canopy and thus the risk of wetness. In addition, covering systems reduce the wetness on the trees. Suitable host tissue differs among cultivars, but no cultivar is resistant to all diseases. In cherries, trees are more susceptible to infections during bloom and near harvest. Indeed, near harvest, the fruit becomes increasingly susceptible because of the tissue softening, which makes nutrients more easily available for the pathogen. Moreover, during the latter part of fruit development with the rapid increase in size, the cuticle can become fractured. Wounds and weakened tissues (e.g. doubles and aborted fruit) show a higher susceptibility to pathogens. Fungal spores can be transported to the wound by birds or insects such as wasps. Moreover, insect and bird damage provides very suitable openings even for weak pathogens. Bird nets or other tools to drive them out are commercially available. Preharvest preventative or curative sprays with fungicides are a possible way of avoiding or controlling infections. A common schedule for spraying in sweet cherry is: frequently during blossom time, less during the green fruit stage and then again frequently prior to harvest. In sour cherry, sprays are more focused on protecting leaves, and it is common to spray regularly throughout the whole season.

It is extremely important to avoid postharvest rots of sour and sweet cherries because

* jorunn.borve@nibio.no

of the high commercial value of the produce. Pathogen presence can be both as latent infections in the fruit or as fungal inoculum on the surface, introduced either before harvest or during postharvest handling. Fruit-to-fruit contamination is important. Sanitation of machines and the use of chlorinated water in hydrocoolers and grading machines reduce inoculum and the risk of possible contaminations. Alternative additives to chlorine are being investigated. At postharvest, climatic conditions can be better controlled than in the field. Temperature is essential to control fungal pathogens. Immediately cooling at harvest, cold storage and a continuous cold chain restrict the development of most pathogens. Since sweet cherries tolerate a high $CO_2$ level, controlled atmosphere storage or modified atmosphere packaging with high $CO_2$ and low $O_2$ also restrict fungal development. During storage, the host tissues become gradually more suitable for fungal growth. In addition, wounds from handling and tissue degradation increase susceptibility. Postharvest treatments can also include fungicides; however, they are not allowed in all countries.

## 14.2 Diseases on Fruit

### 14.2.1 Brown rot

Brown rot, the major disease of sweet and sour cherries throughout the world, is caused by four species of the genus *Monilinia*: *Monilinia laxa* (Aderh. and Ruhland) Honey, *Monilinia fructicola* (G. Wint.) Honey, *Monilinia fructigena* Honey in Whetzel, and *Monilinia polystroma* (G.C.M. Leeuwen) Kohn. There are differences in the geographical distributions of these species. In general, *M. laxa* is present worldwide and *M. fructicola* is important in North America, Australia and New Zealand, while *M. fructigena* is more common in Europe and the Middle East. *M. polystroma* on cherry fruit has been reported only in Poland (Poniatowska *et al.*, 2016). *Monilinia* spp. are known as polyphagous, but their main host range is fruit crops (Byrde and Willetts, 1977).

Infection of *Monilinia* spp. can cause serious losses to sour and sweet cherry fruit, especially in seasons with very wet weather during flowering or immediately before harvest. Losses are associated mainly with blossom blight reducing fruit set and potential yield, and with brown rot on maturing fruit. Blossom blight (Fig. 14.1A) is caused mainly by *M. laxa* and *M. fructicola* (Byrde and Willetts, 1977). *M. laxa* is well adapted to the relatively low temperatures during spring, causing infections at temperatures as low as 5°C with a very short period of wetness (Tamm *et al.*, 1995). In contrast, *M. fructicola* shows higher aggressiveness and infection ability at higher temperatures such as 20–25°C (Papavasileiou *et al.*, 2015). Blossoms are infected mainly by conidia (anamorph), which can be spread by rain or wind. When the spores of *M. laxa* land on susceptible tissue, they germinate in 2 h at favourable moisture and temperature (Keitt, 1948). However, for *M. fructicola*, the wetness duration plays an important role in the infection pathway. Without moisture, infection is nearly absent, even in the presence of a large inoculum, while with a wetness period of 15 h, over 80% of cherries are infected by the pathogen (Biggs and Northover, 1988). The teleomorph, which is rarely seen in *M. laxa*, plays a significant role in the life cycle of *M. fructicola*. Apothecia formed on fallen mummified fruit produce ascospores, which are an additional source of primary inoculum (Holtz *et al.*, 1998). Ascospores have not yet been found in European orchards (Villarino *et al.*, 2010).

Latent infections are important links between blossom blight and fruit rot. This type of infection is a result of conidia that have germinated but stopped growing, to resume only when the fruit are ripe (Jenkins, 1968). The incidence of latent infections of *M. fructicola* on immature sweet cherry fruit was studied by Adaskaveg *et al.* (2000). This species was isolated more frequently than *Botrytis cinerea* on cherry, and infections took place after 6–12 h of wetness rather than after 18–24 h, when active decay developed.

The brown rot of fruits can be caused by all four *Monilinia* spp., and even all present on the same fruit, relating to their presence

**Fig. 14.1.** Blossom blight on sweet cherry (A), brown rot on sour cherry (B), a sweet cherry cluster with aborted fruit (C) and grey mould on sweet cherry postharvest (D).

in the orchard and their potential competition in host colonization (Papavasileiou *et al.*, 2015). Interactions between species may occur during fruit formation and ripening (Villarino *et al.*, 2012). In the case of sour cherries, fruit infections are rarer (Fig. 14.1B). The phenological stage has a significant effect on fruit susceptibility to infections. Cherries are initially resistant to infection of *M. laxa* conidia before the fruit starts to get its red colour (Xu *et al.*, 2007). The infected fruit are covered by putrefactive spots (Fig. 14.1B, C), from which sporodochia (hyphae) with conidia appear. Conidia dispersed by insects, rain and wind penetrate through the stomata, lenticels and microfissures in the fruit skin (Fourie and Holz, 2003). The conidial concentration affects the appearance of lesions on sweet cherry, reducing the incubation time from 5 to 2 days (Northover and Biggs, 1995). With time, the fruit mummifies and the mycelium growing on such mummies gradually aggregates into pseudosclerotia, which are a source of primary inoculum for blossom infections in the next season. Similarly, the teleomorph of *M. fructigena* and *M. polystroma* does not play a significant role in their life cycle.

The incidence of disease in the postharvest phase is related to factors such as rapid cooling, proper refrigeration and packaging, since, more so than other stone fruits, sweet cherries are characterized by a short postharvest life (4–7 days in conventional cold storage) and, in this phase, they are very susceptible to *Monilinia* rots. In particular, the presence of cuticular fractures or similar minute injuries is positively correlated with the appearance of brown rot (Børve *et al.*, 2000). Preharvest colonization is reported to occur from petal fall to harvest, determining latent infections on maturing cherry fruit (Adaskaveg *et al.*, 2000). When fruit resistance declines, the pathogen growth occurs, resulting in brown rot symptoms.

Traditionally, the identification of *Monilinia* spp. was based on morphological features of fungal cultures grown on artificial nutrient media, such as colour, margin and growth rate of the colony, sporulation and size of spores, or germ-tube growth pattern (Hrustić *et al.*, 2015). However, the existence of atypical isolates within species and the need for early detection of quiescent infections necessitated the development of more accurate and less time-consuming molecular methods for identification of *Monilinia* spp. Such diagnostic methods include a multiplex PCR assay developed by Hu *et al.* (2011) and a PCR assay based on the presence and differences in size of a cytochrome *b* gene intron (Hily *et al.*, 2011).

Cultural practices, such as removal of rotten fruit or mummies and pruning of infected twigs, reduce inoculum levels but do not eliminate the disease. Insect control is essential for effective brown rot management, along with protective fungicide treatments (Ogawa *et al.*, 1995). Fungicides available for control of brown rot belong to at least 12 different chemical classes, but the most effective belong to the demethylation-inhibiting (DMI) triazole, phenylpyrrole and anilinopyrimidine classes. Timing of fungicide application is critical for blossom blight control, because flowers must be protected before the occurrence of the prolonged wetness and mild temperatures that favour the infection. In the case of fruits, because host resistance increases with the pit hardening stage and decreases around 3 weeks before harvest, the critical phase is immediately before harvest (Northover and Biggs, 1990). However, fungicide application is recommended at shuck fall and before harvest for both sweet and sour cherry, with an additional mid-season spray for sweet cherries. In order to reduce the use of fungicides, risk analysis and support systems for decision making have been developed. Parameters such as temperature, moisture conditions, inoculum potential, latent infections and fruit phenological stages, such as bloom and fruit ripening, may provide a risk prediction for an anticipated disease severity evaluation, aiding in making decisions related to the fungicide spraying programme (Tamm *et al.*, 1995; Xu *et al.*, 2007). Biological control of brown rot of cherries has been obtained with epiphytic fungi such as *Aureobasidium pullulans* and *Epicoccum purpurascens* (Wittig *et al.*, 1997), but is not yet used commercially. Picking and handling fruit carefully to avoid injuries, cooling fruit promptly after harvest by hydro- or forced air cooling, use of clean containers to hold the fruit and timely fruit harvest all help to reduce postharvest brown rot (Ritchie, 2000). In order to avoid spreading of *Monilinia* spores in the hydrocooling water, the addition of chlorine or chlorine dioxide is routine; however, this can be insufficient because of the instability of chlorine in water containing organic matter. The addition of peracetic acid in the hydrocooling water appears to be a valid alternative to chlorine (Mari *et al.*, 2004).

### 14.2.2 Grey mould

Grey mould, caused by the phytopathogenic fungus *Botrytis cinerea* (perfect stage *Botryotinia fuckeliana* Whetzel), is one of the most important pre- and postharvest diseases of cherry fruit (Adaskaveg *et al.*, 2000). The pathogen is capable of infecting all parts of cherry flowers. Depending on weather conditions, floral infection may lead to blossom blight, or may result in invisible latent infection (Tarbath *et al.*, 2014). When infected flower parts touch developing fruit, brown, quickly spreading lesions can develop on the fruit surface and the fruit can decay at a very early stage of development. On necrotic tissue, the pathogen sporulates profusely (Fig. 14.1D) under wet conditions, forming numerous conidia that are generally the most important dispersal propagules of *B. cinerea* (Holz *et al.*, 2004). Conidia can also infect developing fruit, leading to invisible (latent) or visible (quiescent) infections (Adaskaveg *et al.*, 2000; Tarbath *et al.*, 2014). Quiescent infections of sweet cherry fruit may appear as small necrotic flecks or as reddish halos that surround tannish spots or lesions on the fruit surface. Most quiescent infections eventually become aggressive rots in the field and, if undetected, in stored and marketed produce, while some of the quiescent infections may remain non-viable over time (Adaskaveg *et al.*, 2000). The manner in which the pathogen's aggression is triggered and fruit rot development is enabled is still poorly understood (Jarvis, 1994). Many fruit-decaying fungi, as *B. cinerea*, are primary wound pathogens. Thus, postharvest decay develops mainly from injuries that occur before and, most importantly, during or after harvest (Förster *et al.*, 2007). Once fungal spores are deposited into these wounds, rapid fruit decay occurs, even at a temperature of 0°C during transport, storage and distribution (Wermund and Lazar, 2003). In addition, a high positive correlation was found in sweet cherry between the severity of cuticular fractures on the fruit surface and fruit decay incidence (Børve *et al.*, 2000).

*B. cinerea* is present wherever its hosts are grown, ranging from tropical and subtropical to cold areas. Due to a rapid rate of conidial germination, mycelial growth and conidia formation, as well as the ability to establish infection under a wide range of microclimate conditions, suppression of the pathogen is a very demanding challenge worldwide (Elad *et al.*, 2004). Traditionally, the detection and identification of *B. cinerea* in symptomatic tissue has been based on isolation and identification of the pathogen using the morphological characteristics of fungal cultures grown on artificial nutrient media. There are two conventional methods for the detection of quiescent and latent infections of *B. cinerea*: (i) overnight freezing incubation (Michailides *et al.*, 2000); and (ii) paraquat-dipped green fruit incubation techniques (Northover and Cerkauskas, 1994). A specific molecular marker for *B. cinerea*, based on a 757 bp nucleotide sequence, was characterized and a species-specific primer pair, $C_{729}$+/−, has been designed (Rigotti *et al.*, 2002).

Several factors predispose host tissue to infection by *B. cinerea*. They include damage made by insects, birds and humans, as well as abiotic influences (e.g. weather conditions, plant nutrition, chemical and cultural practices). Controlling insects, modifying microclimate (e.g. by orchard row orientation, special pruning, prevention of cuticular fracturing and fruit cracking) and reducing substrate availability for *B. cinerea* significantly affect pathogen survival and epidemic development (Børve *et al.*, 2000; Elad *et al.*, 2004). Covering sweet cherry trees with rain shields prior to harvest significantly reduced fruit decay caused by various fungi including *B. cinerea* (Børve and Stensvand, 2003). Studies of *B. cinerea*–cultivar interactions showed that breeding programmes should be focused on obtaining cultivars with morphological, anatomical and inherent host resistance mechanisms (Elad *et al.*, 2004). The possibility of biological control of the pathogen is still under investigation (Tanovic *et al.*, 2012). Ippolito and Nigro (2000) produced a list of biocontrol agents that significantly decreased postharvest fruit decay when applied prior to harvest.

Besides non-chemical control measures, several fungicides with preventative and curative effects are available on the market. However, in cherry production in temperate regions, there is usually no need for fungicide applications specifically targeting *B. cinerea*. Regularly applied fungicides for blossom blight and brown rot control are sufficiently effective against grey mould, due to the close biological similarity between the causal agents.

### 14.2.3 Anthracnose

Anthracnose or bitter rot disease on cherry fruit can be recognized by circular brown and sunken lesions, which may have orange sporulation. The incidence of anthracnose in sour cherry has been reported at economically important levels from many countries in Europe, but only once from the USA, in West Virginia (Peet and Taylor, 1948). On sweet cherry, the incidence is lower, but has been reported from many European countries. The causal agent of anthracnose was believed to be *Colletotrichum gloeosporioides*, but recently it was documented that *Colletotrichum acutatum* J.H. Simmonds was the most commonly found species in most northern European countries. Herbarium isolates from cherry collected in the period 1948–1991 in Norway and Denmark were also identified as *C. acutatum* (Sundelin *et al.*, 2015). *C. acutatum* isolates have been grouped and recently divided into 31 species belonging to the *C. acutatum* complex (Damm *et al.*, 2012). The disease on cherry follows a typical 'fruit hemibiotrophic' lifestyle (Peres *et al.*, 2005), although symptoms occur on fruit only. Affected fruit of sour cherry can dry out early in the season, but fruit of sour and sweet cherry typically decay close to harvest (Fig. 14.2A) or postharvest.

Inoculum of *C. acutatum* seems to be present in the orchards throughout the year. It is known to overwinter on cherry buds (Fig. 14.2B), and more on flower buds than on vegetative buds on sweet (Børve and Stensvand, 2006a) and sour cherry. In sour cherry, fruit stalks commonly remain on the trees due to the mechanical harvest of fruit, and these stalks were reported as an inoculum source (Magyar and Oros, 2012), as well as blighted flowers and mummified fruit. Later in the season, the pathogen can occur asymptomatically on leaves of both sweet (Børve *et al.*, 2010) and sour cherry. The incidence of sweet cherry leaves with naturally occurring asymptomatic populations and severity on the leaves increased during the growing season. There was a higher incidence of the pathogen on fruit spur leaves than on leaves on vegetative shoots. In inoculation experiments, *C. acutatum* managed to establish asymptomatic populations on cherry trees throughout the whole season (Børve and Stensvand, 2013). After incubation, a low incidence of naturally occurring *C. acutatum* infection was found on green fruit. Non-abscised aborted fruit incubated at the same times had a high incidence (Børve and Stensvand, 2004). The anamorph stage of the pathogen fully dominates in orchards and is spread by rain splash or by tools. Thus, inoculum dispersal is mainly local. In addition to the cherry trees itself, several other plant species can be hosts and potential sources of inoculum, including weeds, ornamentals, wild trees and other fruit crops.

Only the anamorph *C. acutatum* has been observed in cherry, never the teleomorph *Glomerella acutata*. However, in culture, perithecia were observed from a cherry isolate (Stensvand *et al.*, 2008). The conidia of *C. acutatum* are produced in and released by water splash from acervuli, black structures on the fruit. As well as observation of visible symptoms, both in the field and after incubation, the paraquat test (Cook, 1993) and freezing method (Børve *et al.*, 2010) can be used to identify asymptomatic populations of the pathogen. Molecular identification is well established to identify both *C. acutatum* and species within the *C. acutatum* complex (Damm *et al.*, 2012).

Not all commonly used synthetic fungicides control *Colletotrichum* spp. effectively and anthracnose can be important in

**Fig. 14.2.** Anthracnose disease on sweet cherry fruit (A) and on incubated buds of sweet cherry (B).

frequently sprayed sour cherry orchards. In sour cherry, spraying with preventative synthetic chemicals from petal fall and during the following 6 weeks (Olszak and Piotrowski, 1985) or spraying with a curative compound 1–2 weeks prior to harvest were reported as effective. Spraying twice at the green fruit stage was reported to be effective against anthracnose disease development in sweet cherry, but more spraying was needed to obtain sufficient disease reduction in sour cherry (Børve and Stensvand, 2006b).

Although the risk of inoculum spread is highest only within a small distance, inoculum removal, if possible, might be effective. In sour cherry, it might be difficult to remove fruit stalks or mummified fruit because of the tree size. In sweet cherry, the fruit are normally hand-picked, and due to the high value of the crop, all the fruit are picked. If mummified fruit occur during winter, they could be removed during pruning. During the growing season, non-abscised aborted fruit can contain inoculum (Børve and Stensvand, 2004) and their removal might reduce the inoculum for ripening fruit. In organic cherry growing, no specific treatment besides cultural control strategies exists. No specific control methods exist at postharvest.

### 14.2.4 *Mucor* rot

*Mucor* rot can be caused by different species within the genus *Mucor* but primarily by *Mucor piriformis* A. Fisch. It occurs in northern growing areas of both Europe and the USA. Preharvest *Mucor* rot (Fig. 14.3A) occurs occasionally, for example in Norway (Børve and Stensvand, 2003) and Germany (Palm and Kruse, 2008), and is reduced by the use of covers (Børve *et al.*, 2007). More common is postharvest *Mucor* rot (Fig. 14.4A). It is primarily associated with wounds on stone fruits. The pathogen is found in the soil, and insects can disperse *Mucor* spp. spores from fruit on the orchard floor or weeds to fruit (Michailides and Spotts, 1990). Control can be difficult due to the limited number of effective fungicides. Pyraclostrobin and boscalide (Signum®) can be effective to control the disease (Hauke *et al.*, 2004) and can be applied as a last spray prior to harvest. Preventative control measures include avoiding wounds on the fruit (mechanical wounds, or insect, wasp or bird damage) and sanitation with chlorine (Spotts and Peters, 1980). Storage at 1°C reduces the amount of infection compared with 4°C (Kupferman and Sanderson, 2001).

### 14.2.5 *Sclerotinia* rot

*Sclerotinia* rot in general has little importance, but when the blossom period is cool and wet, the disease can cause severe losses. The pathogen, *Sclerotinia sclerotiorum* (Libert) de Bary, survives as sclerotia on infected host tissues on the soil. The sclerotia produce apothecia, from which ascospores are

**Fig. 14.3.** Sweet cherry fruit with *Mucor* rot (A) and *Sclerotinia* rot (B).

**Fig. 14.4.** Fungal decay of sweet cherry postharvest by *Mucor* rot (A), *Rhizopus* rot (B), blue mould caused by *Penicillium expansum* (C) and *Cladosporium* rot (D).

released and spread by air currents. They can colonize dead or senescing flower parts of the cherry trees, such as petals and sepals. Light brown necrotic lesions develop on infected flowers. Often, necrosis spreads from an infected flower to the surrounding blossoms and they stick together in a cluster. Subsequently, the pathogen moves from infected flowers to adjacent young fruit. Sometimes, under favourable weather conditions, a white mycelium may develop on affected fruit and flower parts (Fig. 14.3B). There are only two reports of the occurrence of this pathogen on sweet cherries, from Oregon, USA (Serdani and Spotts, 2007) and Chile (Ferrada *et al.*, 2014).

The traditional method for identification of the pathogen is based on the morphological characteristics of fungal cultures (Serdani and Spotts, 2007; Ferrada *et al.*, 2014). Molecular methods are now also available.

Control of green fruit rot disease of stone fruits is difficult due to the extensive mycelial growth, persistence in soil and wide host range of the fungus. However, fungicides applied at the full bloom stage to control brown rot caused by *Monilinia* spp. are considered effective against green fruit rot disease caused by *S. sclerotiorum* (Strand, 1999).

### 14.2.6 Other fruit rots

The relative importance of other types of decay, including postharvest diseases, varies according to the field conditions during winter and the blossoming period, and the ripening stage of the fruit, as well as postharvest practices (e.g. time and temperature of storage, chemicals used, bruising). For example, the formation of double fruit or spurs provides sites of choice for infections particularly by *Alternaria*, *Aspergillus* and *Rhizopus* spp. (Ogawa *et al.*, 1995). Since most of these fungi are wound pathogens, control measures rely on a combination of hygiene, careful handling and fungicide usage. Indeed, the packinghouse environment is a major source of inoculum (Youssef *et al.*, 2013). Spores formed on discarded fruit and juice-contaminated packaging are readily transferred to healthy fruit via air currents, insects, washing water, etc. Moreover, since in many producing countries no fungicides are registered for postharvest application on sweet cherries, alternative control means are needed (Feliziani *et al.*, 2013).

*Rhizopus* rot, caused mainly by *Rhizopus stolonifer* (Ehrenb.) Vuill. may inflict heavy losses to sweet cherries in the absence of proper refrigeration. Due to its fast growth, *Rhizopus* is considered one of the most devastating pathogens during storage, and symptoms may be seen in 3–6 days (Bautista-Baños, 2014). *Rhizopus* survives on dead material and the asexual spores (sporangiospores) are disseminated in the air and are responsible for disease initiation. *Rhizopus* generally infects the fruit via injuries caused by hail, cracking and harvesting/handling. Infected areas are brownish, water-soaked and covered with a profuse white mycelium, which gives rise to globular sporangia with numerous sporangiospores, the latter causing a toning to grey–black (Shipper, 1984). The fruit becomes very soft because of the production of pectinolytic enzymes and, at an advanced disease stage, releases juices having a sour odour (Fig.14.4B). During storage, rotted fruit can cause infection of sound fruit by contact, causing extensive nesting and the formation of a thick mycelium that completely covers the fruit. *R. stolonifer* tolerates high temperatures, having an optimum at 25°C, and thus a warm, moist environment favours infection (Pierson, 1966). Control relies mainly on storage at approximately 0°C and at $CO_2$ ≥20%.

*Alternaria* rot is quite frequent on sweet cherries, where it can cause more than 15% loss of stored fruit (Ippolito *et al.*, 2005). Generally, it develops on ripe and over-ripe fruit, and on fruit harvested in the late season. The genus is identified by the morphology of the spores, which are typically septate and catenulate (Simmons, 2007). The main species associated with the disease is *Alternaria alternata* (Fr.) Keissl. The pathogen survives on dead material in the orchard, and its spores are disseminated in the air. Infection is usually associated with injuries occurring during harvesting and handling; however, in cherries, fungal penetration can also take place through cracking or splitting of the skin while the fruit is still on the tree (Barkai-Golan, 2001). Lesions are firm, slightly sunken, blackish-brown and, in the presence of high humidity, are covered by a dense mat of olive-green or dark conidia. Symptoms develop in the field, especially in rainy seasons, and may enlarge after harvest. However, *Alternaria* rot may also appear during storage, particularly on fruit stored for a long time. Since the most common fungicides are ineffective, its control relies mainly on avoiding mechanical injuries and cracks by watering calibration. Storage at 0°C and/or in modified atmospheres at 10% $CO_2$ was shown to reduce disease incidence (Serradilla *et al.*, 2013).

*Penicillium expansum* has a worldwide distribution and attacks numerous fruit and vegetables (Sanzani and Ippolito, 2013). It causes a rot commonly known as blue mould, especially on long-storage or over-ripe fruit. The disease is a classic soft rot, whose typical external symptoms are slightly sunken lesions, which spread rapidly, so that complete rotting of fruit can take 5–7 days at field temperatures. The lesions have a very sharp margin between healthy and diseased tissues, which can easily be separated. Rotted tissues are watery in texture and light brown in colour. Conidia tufts are initially white, becoming blue–green as the spores mature, thus giving rise to the common name of the disease (Fig. 14.4C). Decayed fruit have an earthy, musty odour. *P. expansum* is the main producer of the mycotoxin patulin, which, although associated mainly with pome fruit, is strongly produced even on sweet cherries (Sanzani *et al.*, 2013). Fruit infection is mainly through natural openings or wounds, caused by fruit splitting, insects and birds, or by poor handling. *P. expansum* spores are readily dispersed, and initial contamination of fruit occurs during the growing season from spores originating from saprophytic cultures in soil or orchard debris.

Quiescent infections are a major means of transport between field and the packinghouse. Secondary spread of infection can result from poor handling practices during harvesting, sorting and marketing. In particular, the use of washing water contaminated with spores of *P. expansum* can lead to market losses during storage (Baraldi *et al.*, 2003). Chilling (between −1 and 0°C) retards but does not stop growth. Pyrimethanil and thiabendazole have traditionally been used for blue mould control, but their intense use over the years has led to resistant *Penicillium* populations (Baraldi *et al.*, 2003).

*Cladosporium* rot affects cherries in all producing countries. The causal agent, *Cladosporium herbarum*, survives on dead plant material in the soil and produces an abundance of conidia, which are disseminated in the air, and constitute the most common component of the air microbiome (Barkai-Golan, 2001). *C. herbarum* is a weak pathogen, infecting fruit damaged by rain or rough handling. Fruit shaken from the tree and collected from the ground exhibit a particularly high incidence of infection. Lesions are limited in area, but extend deeply towards the stone. They are covered with a white mould that is then covered by a velvety mat of dark green spores (Fig. 14.4D). Due to its low minimal growth temperature of –4°C, *Cladosporium* can grow and infect fruit even in cold storage, although to a lower extent (Snowdon, 1990). Control is based mainly on the use of copper compounds.

Rots caused by *Aspergillus* spp. are mostly of minor importance, but occasionally can cause postharvest losses, sometimes because of mycotoxin production. The most common species is *Aspergillus niger*, which develops on fruit stored at high temperatures; masses of black spores form on the surface of infected fruit, with the potential for ochratoxin contamination (Ogawa *et al.*, 1995). *Aspergillus flavus* and *Aspergillus parasiticus* seldom occur on fresh fruit; however, sweet cherries appear to be a favourable substrate for production of the aflatoxins B1, B2, G1 and G2 (Llewellyn *et al.*, 1982). Their incidence may increase on improperly dried fruit during storage or if insect infestation occurs, especially in the presence of spurs or doubles. The fungus survives on plant debris in the soil at 25–30°C, and airborne spores infect fruit through wounds, insect punctures, splits, stem-end fractures, etc. (Barkai-Golan, 2001). Infections occur mainly in mature fruit. Since the pathogen does not grow at temperatures below 5°C, cold storage is effective in suppressing disease development.

## 14.3 Foliar Diseases

### 14.3.1 Cherry leaf spot

Cherry leaf spot (CLS) caused by *Blumeriella jaapii* (Rehm) v. Arx (anamorph stage: *Phloeosporella padi* (Lib.) v. Arx) is, next to brown rot, the most important disease of both sweet and sour cherries in temperate zones (Ogawa *et al.*, 1995). CLS is indigenous in the USA and Canada, but was first reported on cherries in Europe around 1940 (Blumer, 1958). It further spread to all cherry-growing areas in Europe (Jakobsen and Jørgensen, 1986) and throughout the cooler northern and north-eastern North America (Jones and Ehret, 1993). It seriously affects the foliage of sour and sweet cherries. Nearly all known sour and sweet cherry cultivars are susceptible to varying degree (Wharton *et al.*, 2003). In addition, CLS can affect various wild and cultivated hosts of the genus *Prunus*, including ornamentals.

Leaf spot is primarily a disease of the foliage, and consequently can often adversely affect the vigour and health of the trees. Characteristic symptoms include numerous, tiny, reddish-purple spots on the upper leaf surface that become necrotic and tend to coalesce (Fig. 14.5). The centre often drops out of the brown, circular lesions, leaving a shot-hole appearance. When affected, leaves turn yellow soon after symptoms develop, but the area around the infected spots frequently remains green, giving the leaf a mottled appearance. On sweet cherry leaves, the spots are often larger and nearly circular in shape. Within the spots, acervuli are formed on their underside, producing conidia, which appear during wet periods as light pink to

**Fig. 14.5.** Symptoms of cherry leaf spot disease on leaves of sour cherry trees.

white spore masses. These secondary spores are rain splashed to neighbouring foliage where they germinate to enable consecutive infections. Affected leaves often fall off early in the season, which may result in severe defoliation. Infection of fruit and fruit pedicels is rare, but fruit on trees defoliated by leaf spot before harvest fail to mature normally, are low in soluble solids and are less firm than fruit on healthy trees (Keitt *et al.*, 1937). The stems of infected fruit become girdled, causing fruit drop. Trees defoliated in mid-summer, particularly young non-bearing trees, are also less cold hardy, showing increased mortality during severe winters (Proffer *et al.*, 2006).

The fungus overwinters in infected leaves as stromata (mycelial aggregates). Apothecia bearing ascospores and winter acervuli forming conidia are produced on the stromata and are released under favourable conditions at about the time of bud burst the following spring. The ascospores stick to the leaf surface, germinate in a film of water and penetrate the leaf through the stomata on the lower side of the leaves (Jakobsen and Jørgensen, 1986). Primary infections continue until the supply of ascospores is exhausted, usually by early summer, while secondary infections are initiated by conidia produced from late spring to early autumn (Garcia and Jones, 1993). Ascospores may be discharged during and shortly after rainfall (Dimova *et al.*, 2014) and are dispersed by rain splash and wind. Thus, management of CLS disease is most effective with weather-based timing of fungicide applications.

CLS is relatively easy to identify on host tissue, but the extremely slow-growing fungus is difficult to isolate and identify in culture. Molecular tools are available to overcome this problem (Proffer *et al.*, 2006).

*B. jaapii* survives the winter in leaf litter on the orchard floor; therefore, removal of fallen infected leaves in autumn is recommended for reducing disease incidence in the following year. Urea application on infected leaves post-leaf fall reduces the production of ascospores and winter conidia the following spring. However, it must be carried out when the fungus is in its active saprotrophic growth phase, about 4 weeks after leaf fall (Green *et al.*, 2006).

In integrated sour cherry orchards, CLS management typically involves four to eight fungicide applications per year (Jones and Ehret, 1993; McManus *et al.*, 2007). Fungicide control programmes are initiated from the petal fall stage, when the leaves are the most susceptible, and continued on a 7–10-day schedule until late summer. Rotation of fungicides with different modes of action is recommended to prevent the selection of resistant strains in the field. DMI fungicides have exhibited excellent activity against this leaf spot fungus (Ogawa *et al.*, 1995), but resistance in *B. jaapii* populations is reported (Jones and Ehret, 1993; Proffer *et al.*, 2006). McManus *et al.* (2007) showed that spray programmes in which chlorothalonil was applied during the bloom to shuck-split stage and followed by copper-based fungicides in early cover sprays were highly effective in controlling CLS. The results of field trials conducted by Proffer *et al.* (2012) indicated the efficacy of fluopyram against CLS on sour cherry. The chemical from the succinate dehydrogenase inhibitor (SDHI) class of respiration inhibitor fungicides, which acts similarly to quinone outside inhibitor ($Q_oI$) fungicides, is efficient in CLS control.

CLS forecasting was developed by Eisensmith and Jones (1981). In Europe, automatic computer-based CLS forecasts are available (Pedersen *et al.*, 2012).

For organic cherry production, removal of fallen infected leaves in autumn is recommended for reducing leaf spot incidence the following year. Only a few approved fungicides are available for CLS control, such as lime sulfur and copper compounds. Another alternative is to plant less susceptible cultivars or cultivars tolerant to leaf spot, but their number in sour cherry is very limited.

### 14.3.2 Shot-hole disease

*Wilsonomyces carpophilus* (Lév.) Adaskaveg, Ogawa & Butler (Adaskaveg *et al.*, 1990) is the causal agent of a disease known as shot-hole, which is present in all stone-fruit-growing areas of the world. It can affect leaves, twigs, branches, buds, blossoms and fruit. Symptoms are non-specific because they are not due to the pathogen itself, but to a stress reaction of the host. The disease is a minor problem on sour cherry compared with sweet cherry.

Leaf spotting is the principal symptom, especially in wet years. Lesions are initially small circular areas, red–violet, surrounded by a narrow, light green to yellow halo, which expand to spots 3–7 mm in diameter. Later, the halo becomes reddish-brown. In warm, dry environments, leaf lesions abscise poducing 'shot-holes'. Affected leaves remain on the twigs until autumn, except for highly susceptible cultivars, in case of heavy attack, and if petioles are infected. Since dehiscence eliminates the infected tissues, sporodochia are generally not formed on leaves. These structures can be observed in the centre of lesions, which develop on twigs; they first appear as purplish spots 2–3 mm in diameter, and then enlarge becoming brown and dead in the centre and covered by small tufts of spores (sporodochia).

Lesions on twigs may remain small but can expand into elongated necrotic cankers, which are often covered with gummy exudate. The spots are initially similar to those on the leaves, but then, always keeping a central darker area, they expand and stretch to encircle considerable sections of the twig, especially if they merge. The part of the twig above the lesion can dry out. Diseased buds show dark brown to black scales and are often covered with a gummy exudate, which give them a shiny appearance. Bud infection may lead to blossom blight or canker development at the base of the flower peduncle.

The symptoms on the fruit vary depending on the stage of development. Fruit lesions are initially small purplish areas, which, near ripening, expand to spots 10 mm in diameter, involving half or more of the fruit surface area. Lesions on young fruit may become rough and corky, and with growth, the fruit tends to isolate or remove the damaged parts. The fungus rarely sporulates on fruit lesions, but on ripe fruit, saprophytic microorganisms can colonize affected fruit leading to rot.

*W. carpophilus* survives winter as a mycelium in twig cankers, leaf scars, and blighted and apparently healthy buds, and is dispersed to infection courts by splashing of water droplets. In temperate climates, the pathogen can even overwinter as conidia protected by gum exudates. Inside blighted buds, conidia may be produced for at least 18 months after bud infection (Ogawa and English, 1991). Conidia are mostly dispersed by water splash. A water film is needed for germination as well as mechanical perforation of the cuticle. At least 24 h of continuous moisture is necessary for infection; the incubation period is 3–8 days, but may be longer (15 days) depending on temperature and tissue; it is longer in twigs than in leaves and flowers. During the spring, rainfall provides the conditions for primary infections on leaves and blossoms; indeed, the disease is most harmful in very cool and wet springs, although it can occur and cause damage at any time during prolonged wet conditions due to rain or sprinkler irrigation (Evans *et al.*, 2008). Severe and widespread epidemics on sweet cherry have been recorded cyclically in conjunction with some of the above-mentioned weather conditions (Grove, 2002). No growth is observed in culture below 5°C or above 30°C, and the optimum temperature range is 15–21°C (Adaskaveg *et al.*, 1990). In the germination process, not all cells germinate at one time, providing a mechanism for extended viability of spores. Spores can germinate immediately after formation, and germination has been observed over a 1 h period (Ogawa and English, 1991).

Germinating conidia produce a germ tube, which directly penetrates the host tissue. The optimum temperature for germination is 18–21°C, but the conidia germinate over a relatively wide range of temperatures. Infections typically take place with a temperature ranging from 5 to 26°C, with an optimum around 15°C. The activity of the fungus during the summer depends on climatic conditions, being inactive when this season is dry and hot; however, as autumn rains begin, its activity is renewed. In areas with regular summer precipitation, infections may occur during a great portion of the growing season. The pathogen can remain for several years in a row, virtually harmless, leading to a fluctuation of the disease that generally is resumed in conjunction with mild winters followed by very humid and rainy springs. In addition, the incidence of the disease is closely related to the susceptibility of the host plants. Defoliation, especially if it happens every year, weakens the tree considerably and reduces yield, vigour and winter hardiness. The activity of the pathogen is strongly increased in years successive to severe winter frosts that have weakened the woody organs, reducing the plant's vitality.

Preventative measures include the use of less susceptible cultivars and pruning during the dormant season to remove infected tissues (buds and cankered twigs). The pruned infected tissue must be burned to reduce the inoculum pressure as the pathogen is well established in perennial infection. A moderate nitrogen supply may reduce twig susceptibility during the growing season and at the end of the season. Wetting of branches, twigs, leaves and fruit during irrigation should be avoided. Since protection of leaves and/or fruit is of major importance in sweet cherry, treatments at bud swell and at the end of the blossoming period are recommended. Depending on the climatic and weather conditions and presence of susceptible cultivars, one or two treatments immediately after fruit set are suggested. An application during leaf fall is also recommended to reduce the inoculum pressure and protect the buds during the dormant season. In the past, Bordeaux mixture and fixed copper material were the standard fungicide to control this disease. These products are now recommended only for autumn applications after leaf fall. Organic fungicides allowed in particular countries should be used when the trees are in leaf.

### 14.3.3 Silver leaf

*Chondrostereum purpureum* (Pers.) Pouzar is the causal agent of silver leaf. The disease is not common on sweet cherry although present in some cultivars such as 'Giorgia'. Its polyphagy allows attacks on about 175 species in 26 families, including other stone fruits and pome fruits. Silver leaf is recorded in most temperate zones.

*C. purpureum* generally causes a progressive silver/lead discoloration of the leaves on affected branches, which gives the common name of the disease. The metallic sheen of the leaves is due to separation of the upper epidermis from the palisade layer, and consequent infiltration of a layer of air, which interferes with normal reflection of light. However, on sweet cherry, the typical symptoms of silver leaf, easy observable on peach, are rarely found. Instead, leaves on infected shoots show yellow/chlorotic discoloration, and some of them may then become upward rolled with necrosis and blisters. In severely affected plants, the foliage is stunted, leaves may turn necrotic before normal leaf fall, defoliation starts from the basis of the twigs (acropetal pattern) and twigs desiccate. Discoloration of the heartwood is one of the most characteristic symptoms. Sections of silver-leaved shoots show browning of the woody tissue, but most pronounced wood discoloration and necrosis is found in large branches, the trunk and roots, which in transversal sections may develop in sectorial areas or over the whole xylem. Due to the presence of the causal agent in the xylem, symptoms are irregularly distributed throughout the tree, with some sectors apparently healthy and others diseased. Attacks in orchards tend to be randomly distributed, but sometimes infected plants are grouped or contiguous. Silver-leaf-like symptoms can be caused by high temperatures and water imbalances during summer and by mite attack [(*Aculus fockeui* (Nalepa & Trouessart)]. The real silver leaf appears at the beginning of the growth season, whereas silver-leaf-like symptoms are displayed in mid-summer. Trees, or parts of trees, may die within months or, after showing symptoms in spring, may recover temporarily or permanently. After the death of branches or tree, leathery basidiocarps occur on the surfaces. *C. purpureum* forms well-differentiated carpophores, at first white–orange and flat on the host bark, and then becoming like small ears with a free margin up to several centimetres across, greyish on the outside (upper surface) and red–purple to pale brown with a smooth hymenium in the inside. The red–purple colour gives the name (*purpureum*) to the species.

At the end of summer, on dead or dying trunks close to the soil or in shaded conditions, carpophores of the basidiomycota are abundantly produced (Fig. 14.6A). They can dry out, but recover and sporulate once the humidity rises again, easily spreading the disease in orchards in the autumn and winter months. Wind-borne basidiospores are deposited on fresh, exposed woody tissue, especially on large pruning wounds. The monokaryotic spores germinate, penetrate the wood, and there the hyphae from different spores hybridize and form dikaryotic (parasitic) mycelia colonizing the larger branches, roots and trunk. Infection may occur also through contact of diseased and healthy roots. The plant tissues may hinder development of the pathogen through the formation of gum. Once the pathogen has settled in the xylem, the mycelium may begin to spread outwards, reaching the cortex, differentiating the fruiting bodies and then completing the biological cycle. The fungus itself does not spread to the shoots or leaves. Toxins formed by the fungus in the colonized xylem are carried to the leaves to cause symptoms. The need for high humidity for carpophore formation, sporulation and infection generally limits serious disease to regions with a humid climate and mild winters; however, local microclimatic conditions may favour *C. purpureum* in any region. The time evolution of the disease is unpredictable, and it can take both a chronic course, in which the symptoms occur with varying degrees of intensity for many years, or an acute one. In the latter case, which frequently affects young plants, death can occur very quickly, even without the presence on the leaves of the typical symptoms of silver discoloration. Young vigorous trees are generally most affected. In nurseries, any affected tree has to be destroyed.

Control is very difficult because the pathogen has a wide host range of perennial plants, the inoculum is produced over a long period and it is virtually impossible to protect all wounded surfaces. As with any pathogen affecting the xylem, preventative control is the best way to manage the disease, since no curative measures are effective. Particularly important is a periodic inspection of the orchard and other woody plants nearby (especially poplars), in order to search for fruiting bodies, which must be burned or at least sterilized with chemicals. Pruning should be done only after spring, preferably

**Fig. 14.6.** Fruiting bodies of *Chondrostereum purpureum* (silver leaf) (A), and cherry leaf curl (B).

in hot and dry conditions when basidiospores are less abundant and wounds heal more quickly; also healthy plants should always be pruned before infected ones and cutting tools should be sterilized between moving from plant to plant. Good protection is obtained by treating all pruning wounds immediately with fungicidal paints. In organic agriculture, formulations based on *Trichoderma* spp. should be a good alternative to treat wounds instead of chemicals (Dubos and Ricard, 1974); this can be done at pruning with special secateurs carrying the antagonist. In the case of replanting in a soil that previously hosted diseased plants, it is necessary to disinfect the area with a soil fumigant formulation, such as metam sodium. Resistance has been reported for crops such as plums and apples (Grosclaude, 1971), but not for sweet cherry. Further precautions suggested are the use of clean nursery stock and metal poles instead of wooden poles in the orchards, since the latter can be colonized by the pathogen.

### 14.3.4 Cherry leaf curl

*Taphrina wiesneri* (Rathay) Mix is the causal agent of leaf curl and witches' broom, a disease that occurs everywhere sweet cherry is grown (Booth, 1981), but damage is rare. The most visible manifestation of the disease is the appearance of witches' broom caused by the contemporary growth of numerous twigs on the largest branches, with the latter more or less twisted and turned in all directions. Usually, witches' broom appears in apical parts of the plant and can be very large, sometimes more than 3 m in diameter. The phenomenon is the result of a transformation of latent buds to ready ones, developing in the same year of their formation. *T. wiesneri* probably disturbs the hormonal balances of the host by producing hormones through gene duplication and horizontal gene transfer (Masuya *et al.*, 2015). The branches with witches' broom symptoms have short internodes and are much more evident in winter, after the leaves have fallen, or in spring during flowering. Among the flowers regularly distributed on sound twigs, the witches' broom branches stand out as large parts of the canopy that are completely green and on which only leaves are produced (non-fruit-bearing branches).

The fungus also causes leaf cast with only a slight thickening of the blade. The leaves then become fleshy and greatly deformed due to a kind of roughness of the leaf tissue (Fig. 14.6B). Even the petiole is larger than normal. Later, the affected leaves turn yellow and take on a reddish hue, while on the undersurface, a white layer made by a palisade of asci develops in early spring, favoured by a prolonged period of high humidity. Occasionally, the palisade of asci also develops on the upper surface. Enzymes produced by the fungus are responsible for the symptoms described above. Diseased leaves may also occur on normal, non-broomed branches.

Isolations on nutrient agar from ascospores or blastospores produce yeast-like colonies. The optimum temperature for ascospore germination is 20–25°C, although they germinate and grow between 10 and 30°C (Booth, 1981). Ascospore dispersal by wind or water splash is from early to late spring, depending on the climate. *T. wiesneri* overwinters as mycelium in the infected organs. That is why the infection is repeated from year to year on the same plants, and even on the same branches.

Since the pathogen overwinters on bud scales and in infected wood, the best means for controlling *T. wiesneri* is pruning and burning of infected twigs and branches. The infected branches should be pruned at least 30 cm below the last visible symptom. Under-canopy irrigation should be applied instead of wetting the leaves, as cherry leaf curl can spread by water splash. Good results are also obtained with products based on copper (e.g. Bordeaux mixture), ziram or thiram, applied in autumn and just before bud burst.

### 14.3.5 Powdery mildew

Powdery mildew of cherry trees is a fungal disease of sporadic importance worldwide. The causal agent of disease is *Podosphaera clandestina* (Wallr.:Fr.) Lév. All sweet cherry cultivars are susceptible but to varying degrees

(Olmstead *et al.*, 2000). Leaves, shoots and fruit are susceptible to disease, and generally sour cherry is more affected than sweet cherry. The first foliar infections are observed approximately 4–6 weeks after bud break. Primary mildew symptoms are encountered first on leaves of sucker shoots at the tree base, on leaves originating from and positioned close to the main scaffold branches, or on leaves positioned near tree crotches (Grove and Boal, 1991). The initial symptoms, frequently observed several days after bloom, are light, irregular circular lesions on both sides of the leaves. As the disease develops, the incidence and severity of foliar symptoms increase, and abundantly sporulating colonies of the pathogen appear. During damp years, the mildew colonies are more prevalent; they may remain discrete or coalesce, giving the leaves a whitish, mealy appearance. Severely infected leaves curl upwards or blister and pucker, and eventually fall prematurely. Foliar infections throughout the summer result in the infection of nearly all terminal shoots. Severely infected shoots are stunted and may appear distorted and blighted.

Yield losses are most severe when fruit infections occur, particularly in orchards where foliar mildew is prevalent during the last 4–6 weeks of fruit development (Grove, 1991). Continuous rainfalls during fruit development may result in high incidences of fruit infection. Powdery mildew is mostly observed during summers with hot, dry weather. The most susceptible to infection are young fruit. Susceptibility decreases when soluble solids exceed 12–13%, since conidia germinate less well as fruit sugar concentration increases up to 15%. On ripe fruit, hyphae of the fungus appear in circular, slightly depressed areas on the fruit surface, which may be small and restricted or cover the entire fruit. Severe petiole infections may interfere with mechanical harvesting.

As the disease develops, numerous brown, spherical cleistothecia are formed. The fungus survives winter as cleistothecia (bearing ascospores) formed on senescent cherry leaves on the orchard floor or trapped in tree crotches, in partially decomposed leaf litter trapped in tree crotches and in bark fissures (Grove and Boal, 1991). From these structures, ascospores are released, starting from about 1 month before bud burst and continuing until after the bloom period (Grove and Boal, 1991). Ascospore release requires free water and can occur during wetness durations as short as 1 h (Grove, 1991). The optimal temperature for ascospore release is 15°C. Cleistothecia appear to be the principal source of primary inoculum for epidemics of sweet cherry powdery mildew (Grove and Boal, 1991). The sporulating primary mildew colonies are a potential source of secondary inoculum for the subsequent foliar and fruit infections. Secondary foliar infections progress throughout the fruit development period and after harvest. The optimum temperature for germination of conidia on leaves is 20°C; however, primary infections can be initiated at temperatures of at least 10°C with simultaneous leaf wetting, while for secondary infections, high humidity and temperatures of 21–26°C are optimal.

Powdery mildew disease is relatively easy to identify, but species identification is based on molecular methods (Santiago-Santiago *et al.*, 2014).

Fruit losses can be reduced by decreasing the amount of inoculum available for secondary infection (i.e. by adequate control of foliar mildew). The disease is controlled mainly by applications of DMI fungicides, applied at 10–14-day intervals. However, during years of high disease pressure, this programme fails to result in adequate control of foliar mildew, especially when the initial application is too late to delay the onset of epidemics (Grove, 1991). Repeated use of similar fungicides has led to the development of site-specific incidences of *P. clandestina* resistance to members of this class of fungicides. Currently, strobilurins are the most effective fungicides for powdery mildew suppression. However, both chemicals and timing are responsible for overall success, due to the fact that it is mainly secondary cycles of infection that drive powdery mildew epidemics; hence, summer cover sprays targeting mildew control are critical. Planting mildew-resistant cultivars and implementing pruning techniques that encourage open canopies are the main practices used by growers to reduce cherry powdery mildew infestation levels.

## 14.4 Diseases on Trees

### 14.4.1 *Phytophthora* crown and root rot

This disease is caused by various species of *Phytophthora*: *P. cambivora*, *P. megasperma*, *P. drechsleri*, *P. cryptogea*, *P. cinnamoni*, *P. citricola*, *P. cactorum*, *P. citrophthora*, *P. syringae* and other unidentified species (Wilcox and Mircetich, 1985a; Ogawa and English, 1991) with a wide host range. A weakened tree may become like a 'medium' for growth, and various species of *Phytophthora* can cause rots. Prolonged flooding of soil is the main predisposing factor (Wilcox, and Mircetich, 1985a). In well-managed orchards, the disease is rarely found; for example, in a survey of southern Italy, *Phytophthora* spp. was isolated only in 3% of the samples.

Symptoms on the canopy of infected plants are generally unspecific, such as poor growth, discoloration, yellowing and fall of leaves, dieback, wilt, and collapse of shoots and scaffolds. Young trees may also collapse and die shortly after resuming growth in the spring, often following an excessively wet autumn/winter. Well-established trees may die within weeks or months, but also after several growing seasons from the first symptoms. At the base of the affected plants (the collar and crown), more specific symptoms can be observed, such as decaying bark turning brown (Fig. 14.7A), depressed and necrotic. The discoloration may extend a little into the sapwood, but the pathogen is not capable of colonizing the xylem. Cankers can show cracks of the bark with or without gum exudation. If the outer bark is scraped, a distinct border between infected and sound tissues is clearly visible. Sectors of the canopy above the trunk cankers are those showing first symptoms. Cankers can girdle the tree, resulting in eventual death; if the canker develops around more than 50% of the trunk circumference, the plant should be replaced. Cankers start growing from subterranean crown tissue, extending vertically up the trunk. As a result of the attack on the root system, feeder roots are few and decayed, with dark brown to black discoloration of the cortex, while the stele initially remains white. Bark tissue of large roots shows decay similar to that described for collar tissue, but without gum exudation, and eventually dissolves in the soil. Plants weakened by environmental factors are more prone to root rot.

Molecular identification protocols have been established (Schena *et al.*, 2008). Isolation of the pathogen may be more difficult in summer, when symptoms are more apparent, than in spring and autumn.

*Phytophthora* spp. are common inhabitants of soil, but can also enter orchards in soil particles on roots or working tools, or by wind or in water. Sporangia are produced by mycelia between two events of soil saturation, when the soil has a good balance of

Fig. 14.7. *Phytophthora* crown rot on a sweet cherry tree (A), sporulation of *Phomopsis* sp. (B) and *Cytospora sp.* (C) on dead stone fruit twigs, and *Armillaria* root rot (D).

oxygen; zoospores are released by sporangia when the soil becomes saturated. In general, temperatures in spring and autumn are conducive to the attack. The pathogen can survive in plant tissues, but in adverse environmental conditions, depending on the species of *Phytophthora*, it can survive as oospores and chlamydospores in soil or host tissue for several years (Ogawa *et al.*, 1995).

Preventative measures are based on water management and the use of tolerant rootstocks. First, it is necessary to avoid poorly drained or unlevelled soil sites. Since zoospores are released during each irrigation, minimizing the frequency of soil saturation (irrigation after demand) helps in preventing infection (Wilcox and Mircetich, 1985b). Trees should be irrigated according to the evapotranspiration demand. The crown and trunk should never be wetted by irrigation, and planting on raised beds reduces the risk of disease (Ogawa *et al.*, 1995). More loss has been observed on Mahaleb compared with Mazzard rootstock (Wilcox and Mircetich, 1985a). Scions are more susceptible than rootstocks, and trees should be planted with a proper distance from the scion to the soil. Chemical control is difficult. Fumigation with Vapam® can be a solution in the nursery but not in the field for planting and replanting purposes. After planting, the systemic fungicides metalaxyl and fosetyl-aluminium have in some cases provided a certain level of protection (Wilcox and Mircetich, 1985a).

### 14.4.2 Constriction canker

Constriction canker is a fungal disease caused by *Phomopsis amygdali* (Del.) Tuset & Portilla that also affects peach, almond and plum. Leaves and twigs are usually the only plant parts affected, but fruit decay can also be observed. Generally, twig infection (Fig. 14.7B) is more common and more severe than leaf infection. In severe cases, pathogen development may lead to loss of larger tree limbs or even tree death (Lalancette *et al.*, 2003).

On the leaves of the host plants, the pathogen induces large, brown, circular or irregular spots (lesions). The centres of the lesions become sparsely dotted with black pycnidia. During hot weather, the pathogen is restricted within the leaf spot, but it grows into the veins as the leaves senesce in the autumn. On twigs and shoots, the symptoms are reddish-brown cankers centred on infected buds or nodes of 1-year-old shoots. Lesions first become visible early in the spring. As they enlarge, they cause girdling, wilting and desiccation of the shoots resulting in shoot blight (Fig. 14.7B). The number of newly blighted shoots may continue to increase in the summer (Ogawa *et al.*, 1995). Constrictions, caused by the mycotoxin fusicoccin, are formed at the base of the infected shoots (Rhouma *et al.*, 2008). The symptoms are sometimes confused with blossom blight caused by *Monilinia* spp. However, lesions of constriction canker are sunken and centred on a bud or node, rather than on a blossom. In addition, they usually have a zonate pattern that is visible on the surface and a small amount of gum exudate from infected tissues. In contrast, gumming is absent in *Monilinia* spp. infected tissues (Ogawa *et al.*, 1995).

The pathogen enters the shoots through fresh leaf scars in the autumn or through bud scale scars, stipule scars, fruit scars and blossoms, or even directly through young shoots in the spring (Uddin *et al.*, 1997; Lalancette and Polk, 2000; Lalancette and Robison, 2002; Rhouma *et al.*, 2008). Among the infection sites, leaf scars are considered the most important (Lalancette and Robison, 2001). The duration of the latent period, the time from infection establishment to the first canker symptoms appearance, is approximately 1 month (Lalancette *et al.*, 2003). Sporulation then begins with pycnidium formation on the cankers. When the pycnidia mature, they exude conidia during humid, wet weather (Lalancette and Robison, 2001). The research of Lalancette *et al.* (2003) showed that conidia production occurs over a broad temperature range and after relatively short periods of high relative humidity. These conditions occur frequently during autumn and spring, coinciding with the periods of host susceptibility. The fungus produces alpha and beta conidia in solitary, globose, dark brown to black pycnidia when cultured on potato dextrose agar.

*Phomopsis* spp. are generally recognized as secondary or opportunistic pathogens that affect plants that are stressed, dying or infected by other pathogens. However, some reports have indicated that *Phomopsis* spp. could be dangerous pathogens on a wide range of plants, and that *P. amygdali* is one of the most aggressive species of the genus (Bienapfl and Balci, 2013).

Fungicides, such as benzimidazoles, chlorothalonil, captan and captafol, help to prevent constriction canker if applied just prior to infection (Ogawa *et al.*, 1995). Some studies have demonstrated that chlorothalonil was the most efficient fungicide followed by captan, azoxystrobin and mycobutanil (Lalancette and Robison, 2001, 2002). Control measures were recommended for the autumn leaf abscission period, as it is the critical period when peak infection is highest (Lalancette and Robison, 2002).

However, chemical control itself may not reduce disease severity to acceptable levels (Schnabel and Lalancette, 2003). Removal of infected plants, pruning out infected branches, use of drip irrigation, and making environmental conditions less favourable for sporulation and infection may aid in managing this disease (Lalancette and Robison, 2001, 2002; Lalancette *et al.*, 2003). The best option for constriction canker management is careful pruning and destruction of cankered twigs, resulting in removal of the inoculum source from the orchard. Pruning prior to leaf fall, as a part of the summer pruning routine, can reduce the incidence of constriction canker by 40% (Schnabel and Lalancette, 2003).

### 14.4.3 *Leucostoma* canker

*Leucostoma* canker is a dieback disease caused by two closely related fungi, *Leucostoma cinctum* Höhn. (anamorph *Leucocytospora cincta* (Sacc.) Höhn.) and *Leucostoma persoonii* Hohn. (anamorph *Leucocytospora leucostoma* (Pers.) Höhn.). The pathogens are found primarily on peach, but they can cause cankers and twig dieback on plum, prune, sweet and sour cherry, apricot, other wild *Prunus* spp., apple and pear (Biggs, 1989; Biggs and Grove, 2005). They are frequently found in orchards as anamorphs (Biggs and Grove, 2005; Romanazzi *et al.*, 2012). Symptoms include dieback of twigs and branches, bark cankers, gummosis and eventually tree decline. Infections of small twigs appear as sunken, discoloured areas near winter-killed buds or leaf scars. Nodal infections are observed 2–4 weeks after bud break. Infected twig tissues become darker with time and may ooze amber-coloured gum (Biggs and Grove, 2005). Cankers that form on the main trunk, branch crotches, scaffold limbs and older branches are conspicuous symptoms that begin with exudation of a copious quantity of amber-coloured gum. Gum production is a natural plant response to any abiotic or biotic stress, but that caused by *Leucostoma* infection is excessive to the point of being injurious. As cankers age, the gum becomes dark brown to black, and the infected bark dries out and cracks open, resulting in exposure of blackened tissue beneath elliptical cankers along the length of the stem. Extension of branch or twig infection may result in foliar symptoms: leaves that turn yellow, droop, and may wilt and die. On dead twigs and branches, black pycnidia erupting through the bark are observed. From mature pycnidia, conidia are released in a polysaccharide matrix named cirrus (Fig. 14.7C) (Biggs, 1989; Biggs and Grove, 2005).

The pathogen infects plants only through mechanical or freeze wounds or dead tissue. The most common infection sites are pruning cuts, leaf scars, shade-weakened twigs inside canopies, insect injuries, brown rot cankers, and winter-injured buds, twigs and bark. Rodent injuries and wounds resulting from cultivation, picking ladders, wire mouse guards and broken limbs may also be pathways for invasion of the pathogen (Biggs, 1989). Winter injury is considered the main contributing factor (Kable *et al.*, 1967). Small twigs killed by *Leucostoma* spp. are pathways by which the pathogen can reach older limbs and initiate cankers, which may then result in the death of large portions of the tree (Biggs and Grove, 2005).

Conidia are considered to be the primary source of inoculum for most new infections, while the role of ascospores in the disease cycle has not been determined (Grove and Biggs, 2006). Dead wood is the major source of conidia and ascospores in the orchard. Most conidia are disseminated by splashing rain, boring insects or birds, but they could also be on pruning tools. *Leucostoma* spp. sporulates all year long but produces the most spores under cool, damp conditions of late autumn and early spring (from November to March) and, depending on rainfall, also in June (Schulz and Schmidle, 1983; Regner *et al.*, 1990).

Both pathogens are widespread, but generally, *L. cinctum* occurs in cooler areas, whereas *L. persoonii* is usually found in warmer climates. *Leucostoma* canker is an important disease in cooler climates and is of minor importance in warmer regions. In Europe, it is important in sweet cherry as part of the 'apoplexy' disease complex (Biggs, 1989). It is frequently misdiagnosed as bacterial canker, since many of the symptoms are indistinguishable and pycnidia may not readily form in infected bark tissue. An important distinguishing characteristic is that *Leucostoma* cankers are usually active for one or more years, whereas bacterial cankers are rarely active the following season (Regner *et al.*, 1990). For identification of the pathogen to the species level, the following criteria can be applied: (i) colour of the mycelium (*L. cinctum*: white turning to buff or olive–buff; *L. persoonii*: white turning to brown or darker brown); (ii) size and properties of the pycnidia (*L. cinctum*: large, 1–3 mm in diameter, white, felty, rarely if ever exuding cirri; *L. persoonii*: small, 1 mm or less in diameter, with beaks, usually dark, exuding cirri when mature); (iii) presence or absence of growth at 33°C (*L. cinctum*: optimum at 18–20°C with a maximum at 30°C; *L. persoonii*: optimum at 25–30°C, maximum at 32°C) (Romanazzi *et al.*, 2012).

Management is based on preventative measures designed to decrease winter injury and insect damage, promote optimum plant health, and facilitate fast healing of wounds and other injuries (Biggs and Grove, 2005). Before planting, site selection for good air and surface water drainage and distance from heavily diseased areas is of high importance. Trees should be inspected carefully after growth begins, and any dead branches should be removed immediately (Biggs, 1989).

Once established, *Leucostoma* canker is very difficult to manage. Consequently, effective disease control strategies are based on avoiding the factors that predispose trees to the disease such as winter injury and damage by insects or other pathogens. Removal of diseased, woody tree material from orchards and maintenance of tree vigour are important (Barakat and Johnson, 1997). Cankers should be removed from the tree and burned, buried or moved out of the orchard. Cankers on trunks and large limbs can be removed surgically in mid-summer when the trees heal most rapidly. Surgery should be performed in dry weather that is expected to persist for at least 3 days. During surgery, all diseased bark around the canker and about 4–5 cm of healthy tissue from the sides and ends should be removed. The practice of covering pruning cuts in spring with a thiram/latex mixture provides some degree of protection against fungal infection, but sites of surgery heal best if left uncovered. There are no fungicides available (Biggs and Grove, 2005).

### 14.4.4 *Verticillium* wilt

*Verticillium dahliae* Kleb. is the causal agent of *Verticillium* wilt of cherry. Commonly used rootstocks, such as Mahaleb, are susceptible. The pathogen has an extremely wide host range and the disease is spread in all sweet and sour cherry-growing areas, but it is not very common in sweet cherry orchards. In an Italian survey, *V. dahliae* was found in 5% of plants with symptoms of decline. In orchards planted in soils that had previously hosted crops that are particularly susceptible, such as Solanaceae, the disease incidence can be high.

Initial symptoms are a sudden wilting of leaves on one or more young shoots in individual branches, early in the summer. Leaves can become yellow or directly show upward rolling, turn dull, pale brown in colour, followed by wilting and necrosis.

The phenomenon may be so rapid that the normal process of abscission, leading to leaf drop, is lacking and blades wither, remaining attached to the twigs. In a chronic course, affected branches show yellow leaves, which eventually fall, and plants may produce sprouts from the foot or the lower portions of the trunk, although in the end they die. A cross-section of trunk and branches, and rarely of twigs, may show areas of the xylem with brown to black discoloration, which are generally less distinct on sweet cherries compared with other stone fruits such as almond and peach. The xylem vessels appear occluded by gummy material, which prevents the upward movement of the sap.

*Verticillium* wilt is considered to be a single-cycle disease, since inoculum is generally not produced in the same year in which the plant has been infected. Indeed, after infection it can remain latent inside the xylem for more than 1 year. Microsclerotia of the pathogen are able to withstand adverse soil environmental conditions, surviving for more than 10 years. Considering its wide range of hosts, including weeds, its inoculum may remain stable for a long time in the soil, depending on soil management and the cultivated species. However, flooding and high temperatures reduce viability. The fungus can be disseminated in many ways, but in sound soil, it arrives through the use of infected plant propagation material. Within the orchard, irrigation water, windblown soil, cultural practices and cutting tools may spread the inoculum among plants. The vascular disease appears more severe in wet soil. Disease development occurs at 21–27°C with an optimum at 24°C. During summer, in southern Italy, the isolation of the pathogen from infected xylem appeared erratic, probably due to its devitalization by the high temperature. When plants become weakened, for example due to nutrient deficiency, asphyxia or cold damage, they are more prone to the disease. Germinating conidia of the pathogen can gain entrance through wounds created by insects or cultural practice, or may penetrate intact roots or rootlets, the latter through the elongation zone. The fungus advances inter- or intracellularly through the epidermis, cortex and endodermis, and reaches the xylem tissue without causing obvious root-rot damage (Mace *et al.*, 1981). In the xylem, conidia and fragments of the mycelium are easily transported from the site of infection to the twigs through the transpiration stream. *V. dahliae* produces hyaline mycelium, which on potato dextrose agar appears at first white and flocculose, and then gradually darkens due to the formation of black, thick-walled microsclerotia. Conidiophores are abundant, more or less erect, hyaline, and verticillately branched, with three to four phialides arising at each node. Conidia, produced singly by phialides, are ellipsoidal to irregularly subcylindrical, hyaline, and one-celled but occasionally one-septate, measuring 2.5–8 × 1.4–3.2 μm. Dark resting mycelium is produced only in association with microsclerotia. Microsclerotia are dark brown to black, torulose or botryoidal, consisting of swollen almost globular cells that are variable in shape and size, being 15–50 μm (occasionally up to 100 μm) in diameter (Hawksworth and Talboys, 1970).

Preventative control relies on avoidance of planting the orchard on soils previously cultivated by crops susceptible to *V. dahliae*. No resistant rootstocks are available for sour and sweet cherry. Soil fumigation with Vapam or with analogous products may help in destroying inoculum of *V. dahliae*, but the treatment is very expensive and may provide only a temporary restoration.

### 14.4.5 *Armillaria* root rot

*Armillaria mellea* (Vahl) Kummer, *sensu lato*, is the causal agent of *Armillaria* root rot, which occurs worldwide on various species including fruit trees, vines, shrubs and forest trees, but also on herbaceous plants such as potatoes and strawberries. '*Sensu lato*' means several biological species belonging to four species of *Armillaria*: *A. mellea sensu stricto*, *A. ostoyae*, *A. bulbosa* and *A. tabescens* (Ogawa *et al.*, 1995). Except for *A. tabescens*, the species are characterized by the presence of a skin-like ring attached to the upper part of the stipe. *Armillaria* root rot is considered one of the most important soil-borne diseases of sweet cherry trees, especially in orchards on newly cleared forest land, near rivers or on replanted infested sites. In southern Italy, it is widespread in old sweet cherry orchards together with *Rosellinia necatrix* Berl. ex Prill. (causing white root rot) with which it shares the same non-specific symptoms on the canopy. The disease is also known as mushroom root rot, shoe-string root rot and oak root fungus disease, while the pathogen is commonly named honey fungus.

Old/weakened trees are generally more susceptible to the disease. The symptoms on the canopy can be considered non-specific and consist of poor shoot growth, undersized leaves, premature yellowing and dropping of leaves, and dieback of twigs and branches; in some cases, the foliage remains green until mid-summer, when, in the presence of high temperatures and drought, the whole tree may collapse with leaves attached. The pathogen spreads from the infected site through subterranean rhizomorphs or direct root contact, giving a typical circular pattern of decaying/dead trees from a focus of infection. Root rot is more common on light, well-drained, sandy or loamy soils and on the banks of rivers subject to flooding. More specific symptoms appear at the crown level of the affected trees, where white mycelial mats – flat, fan-shaped, felt-like plaques between the bark and the wood – appear. In decayed areas, the bark is easily detached from the wood, and the mycelium releases a strong smell of fresh mushroom. On the surface of root bark, reddish-dark brown to black rhizomorphs or 'shoestrings' may be observed (Fig. 14.7D). Rhizomorphs are also found in the soil around infected trees, representing one of the ways of spread from tree to tree; rotten roots remaining in the soil can initiate new infection cycles. Clusters of basidiocarps are found at the base of declining/dead trees in autumn. It seems that the basidiospores play no part in the infection cycle as takes place in the soil.

*Armillaria* root rot is difficult to control since the pathogen is located deep in the soil as rhizomorphs and the mycelium is protected by the dead wood. Digging a trench around infected trees was the traditional solution to prevent the spread of the pathogen, but is not done in intensive plantations. Common rootstocks used for sweet cherry are more or less all susceptible and their behaviour is influenced by the *Armillaria* sp. and rootstock/scion combination. Fumigants available on the market are not effective against the pathogen.

Prevention of disease is based on avoiding risky sites for plantations and on good soil drainage. If present, further spread can be avoided by artificial depletion of infected trees, complete removal of roots and crop rotation with graminaceous plants for 3–5 years.

## References


Adaskaveg, J.E., Ogawa, J.M. and Butler, E.E. (1990) Morphology and ontogeny of conidia in *Wilsonomyces carpophilus*, gen. nov., and comb. nov., causal pathogen of shot hole disease of *Prunus* species. *Mycotaxon* 37, 275–290.

Adaskaveg, J.E., Förster, H. and Thompson, D.F. (2000) Identification and etiology of visible quiescent infections of *Monilinia fructicola* and *Botrytis cinerea* in sweet cherry fruit. *Plant Disease* 84, 328–333.

Barakat, R.M. and Johnson, D.A. (1997) Expansion of cankers caused by *Leucostoma cincta* on sweet cherry trees. *Plant Disease* 81, 1391–1394.

Baraldi, E., Mari, M., Chierici, E., Pondrelli, M., Bertolini, P. and Pratella, G.C. (2003) Studies on thiabendazole resistance of *Penicillium expansum* of pears: pathogenic fitness and genetic characterization. *Plant Pathology* 52, 362–370.

Barkai-Golan, R. (2001) *Postharvest Diseases of Fruits and Vegetables: Development and Control*. Elsevier, Amsterdam.

Bautista-Baños, S. (2014) *Postharvest Decay: Control Strategies*. Academic Press, Elsevier, London.

Bienapfl, J.C. and Balci, Y. (2013) Phomopsis blight: a new disease of *Pieris japonica* caused by *Phomopsis amygdali* in the United States. *Plant Disease* 97, 1403–1407.

Biggs, A.R. (1989) Integrated approach to controlling *Leucostoma* canker of peach in Ontario. *Plant Disease* 73, 869–874.

Biggs, A.R. and Grove, G.G. (2005) *Leucostoma* canker of stone fruits. *The Plant Health Instructor*, DOI: 10.1094/PHI 2005-1220-01.

Biggs, A.R. and Northover, J. (1988) Influence of temperature and wetness duration on infection of peach and sweet cherry fruits by *Monilinia fructicola*. *Phytopathology* 78, 1352–1353.

Blumer, S. (1958) Beitrage zur Kenntnis von Cylindrosporium padi. *Phytopathologische Zeitschrifte* 33, 263–290.

Booth, C. (1981) *Taphrina wiesnieri*. Sheet no. 712. Commonwealth Mycological Institute, Surrey, UK.

Børve, J. and Stensvand, A. (2003) Use of a plastic rain shield reduces fruit decay and need for fungicides in sweet cherry. *Plant Disease* 87, 523–528.

Børve, J. and Stensvand, A. (2004) Non-abscised aborted sweet cherry fruits are vulnerable to fruit decaying fungi and may be sources of infection for healthy fruits. *Acta Agriculturae Scandinavica* 54, 31–37.

Børve, J. and Stensvand, A. (2006a) *Colletotrichum acutatum* overwinters on sweet cherry buds. *Plant Disease* 90, 1452–1456.

Børve, J. and Stensvand, A. (2006b) Timing of fungicide applications against anthracnose in sweet and sour cherry production in Norway. *Crop Protection* 25, 781–787.

Børve, J. and Stensvand, A. (2013) *Colletotrichum acutatum* can establish on sweet and sour cherry trees throughout the growing season. *European Journal of Horticultural Science* 78, 258–266.

Børve, J., Sekse, L. and Stensvand, A. (2000) Cuticular fractures promote postharvest fruit rot in sweet cherries. *Plant Disease* 84, 1180–1184.

Børve, J., Meland, M. and Stensvand, A. (2007) The effect of combining rain protective covering and fungicide sprays against fruit decay in sweet cherry. *Crop Protection* 26, 1226–1233.

Børve, J., Djønne, R. and Stensvand, A. (2010) *Colletotrichum acutatum* occurs asymptomatically on sweet cherry leaves. *European Journal of Plant Pathology* 127, 325–332.

Byrde, R.J.W. and Willetts, H.J. (1977) *The Brown Rot Fungi of Fruit: Their Biology and Control*. Pergamon Press, Oxford.

Cook, R.T.A. (1993) Strawberry black spot caused by *Colletotrichum acutatum*. In: Ebbels, D. (ed.) *Plant Health and the European Single Market*. BCPC Monographs, Vol. 54, British Crop Protection Enterprises, Brighton, UK, pp. 301–304.

Damm, U., Cannon, P.F., Woudenberg, J.H.C. and Crous, P.W. (2012) The *Colletotrichum acutatum* species complex. *Studies in Mycology* 73, 37–113.

Dimova, M., Titjnov, M., Arnaudov, V. and Gandev, S. (2014) Harmful effect of cherry leaf spot (*Blumeriella jaapii*) on sour cherry and influence on fruit yield. *Agroznanje* 15, 393–400.

Dubos, B. and Ricard, J.L. (1974) Curative treatments of peach trees against silver leaf disease (*Stereum purpureum*) with *Trichoderma viride* preparation. *Plant Disease Reporter* 58, 147–150.

Eisensmith, S.P. and Jones, A.L. (1981) Infection model for timing fungicide applications to control cherry leaf spot. *Plant Disease* 65, 955–958.

Elad, Y., Williamson, B., Tudzynski, P. and Delen, N. (2004) *Botrytis: Biology, Pathology and Control*. Kluwer Academic Publishers, Dordrecht, The Netherlands.

Evans, K., Frank, E., Gunnell, J.D. and Shao, M. (2008) Coryneum or shot hole blight. Utah Pests Fact Sheet. Available at: http://extension.usu.edu/files/publications/factsheet/coryneum-blight08.pdf (accessed at 2 December 2015).

Feliziani, E., Santini, M., Landi, L. and Romanazzi, G. (2013) Pre-and postharvest treatment with alternatives to synthetic fungicides to control postharvest decay of sweet cherry. *Postharvest Biology and Technology* 78, 133–138.

Ferrada, E.E., Diaz, G.A., Zoffoli, J.P. and Latorre, B.A. (2014) First report of blossom blight caused by *Sclerotinia sclerotiorum* on Japanese plum, nectarine and sweet cherry orchards in Chile. *Plant Disease* 98, 695.

Förster, H., Driever, G.F., Thompson, D.C. and Adaskaveg, J.E. (2007) Postharvest decay management for stone fruit crops in California using the "reduced-risk" fungicides fludioxonil and fenhexamid. *Plant Disease* 91, 209–215.

Fourie, P.H. and Holz, G. (2003) Germination of dry, airborne conidia of *Monilinia laxa* and disease expression on plum fruit. *Australasian Plant Pathology* 32, 19–25.

Garcia, S.M. and Jones, A.L. (1993) Influence of temperature on apothecial development and ascospores discharge by *Blumeriella jaapii*. *Plant Disease* 77, 776–779.

Green, H., Bengtsson, M., Duval, X., Pedersen, H.L., Hockenhull, J. and Larsen, J. (2006) Influence of urea on the cherry leaf spot pathogen, *Blumeriella jaapii*, and on microorganisms in decomposing cherry leaves. *Soil Biology and Biochemistry* 38, 2731–2742.

Grosclaude, C. (1971) Silver leaf disease of fruit trees. VIII. Contribution to the study of varietal resistance in plum. *Annals of Phytopathology* 3, 283–298.
Grove, G.G. (1991) Powdery mildew of sweet cherry: influence of temperature and wetness duration on release and germination of ascospores of *Podosphaera clandestina*. *Phytopathology* 81, 1271–1275.
Grove, G.G. (2002) Influence of temperature and wetness period on infection of cherry and peach foliage by *Wilsonomyces carpophilus*. *Canadian Journal of Plant Pathology* 24, 40–45.
Grove, G.G. and Biggs, A.R. (2006) Production and dispersal of conidia of *Leucostoma cinctum* in peach and cherry orchards under irrigation in eastern Washington. *Plant Disease* 90, 587–591.
Grove, G.G. and Boal, R.J. (1991) Overwinter survival of *Podosphaera clandestina* in Eastern Washington. *Phytopathology* 81, 385–391.
Hauke, K., Creemers, P., Brugmans, W. and van Laer, S. (2004) Signum, a new fungicide with interesting properties in resistance management of fungal diseases in strawberries. *Communications in Agricultural and Applied Biological Sciences* 69, 743–755.
Hawksworth, D.L. and Talboys, P.L. (1970) Verticillium dahliae. CMI Description of Pathogenic Fungi and Bacteria. Commonwealth Mycological Institute, Ferry Lane, Kew, Surrey, UK, 256, pp. 1–2.
Hily, J.M., Singer, S.D., Villani, S.M. and Cox, K.D. (2011) Characterization of the *cytochrome b* (*cyt b*) gene from *Monilinia* species causing brown rot of stone and pome fruit and its significance in the development of QoI resistance. *Pest Management Science* 67, 385–396.
Holtz, B.A., Michailides, T.J. and Hong, C.X. (1998) Development of apothecia from stone fruit infected and stromatized by *Monilinia fructicola* in California. *Plant Disease* 82, 1375–1380.
Holz, G., Coertze, S. and Williamson, B. (2004) The ecology of Botrytis on plant surfaces. In: Elad, Y., Williamson, B., Tudzynski, P. and Delen, N. (eds) *Botrytis: Biology Pathology and Control.*, Kluwer Academic Publishers, Dordrecht, The Netherlands, pp. 9–27.
Hrustić, J., Delibašić, G., Stanković, I., Grahovac, M., Krstić, B., Bulajić, A. and Tanović, B. (2015) *Monilinia* species causing brown rot of stone fruit in Serbia. *Plant Disease* 99, 709–717.
Hu, M.J., Cox, K.D., Schnabel, G. and Luo, C.X. (2011) *Monilinia* species causing brown rot of peach in China. *PLoS One* 6, e24990.
Ippolito, A. and Nigro, F. (2000) Impact of preharvest application of biological control agents on postharvest diseases of fresh fruits and vegetables. *Crop Protection* 19, 715–723.
Ippolito, A., Schena, L., Pentimone, I. and Nigro, F. (2005) Control of postharvest rots of sweet cherries by pre-and postharvest applications of *Aureobasidium pullulans* in combination with calcium chloride or sodium bicarbonate. *Postharvest Biology and Technology* 36, 245–252.
Jakobsen, H. and Jørgensen, K. (1986) Cherry leaf spot. Danish investigations of perennation state and primary infection in sour cherries. *Tidsskrift för Planteavl* 90, 161–175.
Jarvis, W.R. (1994) Latent infection in the pre- and postharvest environment. *HortScience* 29, 747–749.
Jenkins, P.T. (1968) The longevity of conidia of *Sclerotinia fructicola* (Wint.) Rehm under field conditions. *Australian Journal of Agriculture and Research* 21, 937–945.
Jones, A. and Ehret, G. (1993) Control or cherry leaf spot and powdery mildew on sour cherry with alternate-side applications of fenarimol, mycobutanil, and tebuconazole. *Plant Disease* 77, 703–706.
Kable, P.F., Fliegel, P. and Parker, K.G. (1967) Cytospora canker on sweet cherry in New York State: association with winter injury and pathogenicity to other species. *Plant Disease Report* 51, 155–157.
Keitt, G.W. (1948) Blossom and spur blight of sour cherry. *Phytopathology* 38, 857–882.
Keitt, G.W., Blodgett, E.C., Wilson, E.E. and Magie, R.O. (1937) *The Epidemiology and Control of Cherry Leaf Spot*. Research Bulletin 132, Agricultural Experiment Station of the University of Wisconsin, Wisconsin.
Kupferman, E. and Sanderson, P. (2001) Temperature management and modified atmosphere packing to preserve sweet cherry quality. Postharvest information network. Washington State University, Pullman, Washington. Available at: http://postharvest.tfrec.wsu.edu/emk2001b.pdf (accessed at 8 January 2016).
Lalancette, N. and Polk, D.F. (2000) Estimating yield and economic loss from constriction canker of peach. *Plant Disease* 84, 941–946.
Lalancette, N. and Robison, D.M. (2001) Seasonal availability of inoculum for constriction canker of peach in New Jersey. *Phytopathology* 91, 1109–1115.
Lalancette, N. and Robison, D.M. (2002) Effect of fungicides, application timing, and canker removal on incidence and severity of constriction canker of peach. *Plant Disease* 86, 721–728.
Lalancette, N., Foster, K.A. and Robison, D.M. (2003) Quantitative models for describing temperature and moisture effects on sporulation of *Phomopsis amygdali* on peach. *Phytopathology* 93, 1165–1172.
Llewellyn, G.C., Eadie, T. and Dashek, W.V. (1982) Susceptibility of strawberries, blackberries, and cherries to Aspergillus mold growth and aflatoxin production. *Journal of the Association of Official Analytical Chemists* 65, 659–664.

Mace, M.E., Bell, A.A. and Beckman, C.H. (1981) *Fungal Wilt Diseases of Plants*. Academic Press, New York.

Magyar, D. and Oros, G. (2012) Application of the principal component analysis to disclose factors influencing on the composition of fungal consortia deteriorating remained fruit stalks on sour cherry trees. In: Sanguansat, P. (ed.) *Principal Component Analysis – Multidisciplinary Applications*. InTech, Rijeka, Croatia, pp. 89–110.

Mari, M., Gregori, R. and Donati, I. (2004) Postharvest control of *Monilinia laxa* and *Rhizopus stolonifer* in stone fruit by peracetic acid. *Postharvest Biology and Technology* 33, 319–925.

Masuya, H., Kikuchi, T. and Sahashi, N. (2015) New insights to develop studies on Witch's broom caused by *Taphrina wiesneri*. *Journal of the Japanese Forest Society* 97, 153–157.

McManus, P., Proffer, T., Berardi, R., Gruber, B., Nugent, J., Ehret, G., Ma, Z. and Sundin, G. (2007) Integration of copper-based and reduced risk fungicides for control of *Blumeriella jaapii* on sour cherry. *Plant Disease* 91, 294–300.

Michailides, T. and Spotts, R. (1990) Postharvest diseases of pome and stone fruits caused by *Mucor piriformis* in the Pacific Northwest and California. *Plant Disease* 74, 537–543.

Michailides, T.J., Morgan, D.P. and Felts, D. (2000) Detection and significance of symptomless latent infection of *Monilinia fructicola* in California stone fruits [abstract]. *Phytopathology* 90, S53.

Northover, J. and Biggs, A.R. (1990) Susceptibility of immature and mature sweet and sour cherries to *Monilinia fructicola*. *Plant Disease* 74, 280–284.

Northover, J. and Biggs, A.R. (1995) Effect of conidial concentration of *Monilinia fructicola* on brown rot development in detached cherries. *Canadian Journal of Plant Pathology* 17, 205–214.

Northover, J. and Cerkauskas, R.F. (1994) Detection and significance of symptomless latent infections of *Monilinia fructicola* in plums. *Canadian Journal of Plant Pathology* 16, 30–36.

Ogawa, J.M. and English, H. (1991) *Diseases of Temperate Zone Tree Fruit and Nut Crops*, Vol. 3345. UCANR Publications, California.

Ogawa, J.M., Zehr, E.I., Bird, G.W., Ritchie, D.F., Uriu, K. and Uyemoto, J.K. (1995) *Compendium of Stone Fruit Diseases*. APS Press, St Paul, Minnesota.

Olmstead, J.W., Lang, G.A. and Grove, G.G. (2000) A leaf disk assay for screening sweet cherry genotypes for susceptibility to powdery mildew. *HortScience* 35, 274–277.

Olszak, M. and Piotrowski, S. (1985) Studies of the control of tart cherry anthracnose caused by *Colletotrichium gloeosporioides* (Penz.) Sacc. (*Glomerella cingulata* (Stonem.) et Schrenk). *Fruit Science Reports* 12, 155–161.

Palm, G. and Kruse, P. (2008) Einfluss der Überdachung von Süsskirschen auf das Aufplatzen der Fruchte und die Fruchfäulnis. *Mitteilungen des Obstbauversuchsringes des Alten Landes* 63, 154–157.

Papavasileiou, A., Testempasis, S., Michailides, T.J. and Karaoglanidis, G.S. (2015) Frequency of brown rot fungi on blossoms and fruit in stone fruit orchards in Greece. *Plant Pathology* 64, 416–424.

Pedersen, H.L., Jensen, B., Munk, L., Bengtsson, M.V. and Trapman, M. (2012) Reduction in the use of fungicides in apple and sour cherry production by preventative methods and warning systems. Pesticides Research No. 139. Environmental Protection Agency, Danish Ministry of the Environment, Denmark.

Peet, C.E. and Taylor, C.F. (1948) Apple anthracnose on sour cherry in West Virginia. *Phytopathology* 38, 20–21.

Peres, N.A., Timmer, L.W., Adaskaveg, J.E. and Correll, J.C. (2005) Lifestyles of *Colletotrichum acutatum*. *Plant Disease* 89, 784–796.

Pierson, C.F. (1966) Effect of temperature on growth of *Rhizopus stolonifer* on peaches and on agar. *Phytopathology* 56, 276–278.

Poniatowska, A., Michalecka, M. and Puławska, J. (2016) Genetic diversity and pathogenicity of *Monilinia polystroma* – the new pathogen of cherries. *Plant Pathology* 65, 723–733.

Proffer, T.J., Berardi, R., Ma, Z., Nugent, J.E., Ehret, G.R., McManus, P.S., Jones, A.L. and Sundin, G.W. (2006) Occurrence, distribution, and polymerase chain reaction-based detection of resistance to sterol demethylation inhibitor fungicides in populations of *Blumeriella jaapii* in Michigan. *Phytopathology* 96, 709–717.

Proffer, T.J., Lizotte, E., Rothwell, N.L. and Sundin, G.W. (2012) Evaluation of dodine, fluopyram and penthiopyrad for the management of leaf spot and powdery mildew of tart cherry, and fungicide sensitivity screening of Michigan populations of *Blumeriella jaapii*. *Pest Management Science* 69, 747–754.

Regner, K.M., Johnson, D.A. and Gross, D.C. (1990) Etiology of canker and dieback of sweet cherry trees in Washington State. *Plant Disease* 74, 430–433.

Rhouma, A., Ali Triki, M., Ouerteni, K. and Mezghanni, M. (2008) Chemical and biological control of *Phomopsis amygdali* the causal agent of constriction canker of almond in Tunisia. *Tunisian Journal of Plant Protection* 3, 69–78.

Rigotti, S., Gindro, K., Richter, H. and Viret, O. (2002) Characterization of molecular markers for specific and sensitive detection of *Botrytis cinerea* Pers.: Fr. in strawberry (*Fragaria* × *ananassa* Duch.) using PCR. *FEMS Microbiological Letters* 209, 169–174.

Ritchie, D.F. (2000) Brown rot of stone fruits. *The Plant Health Instructor*, DOI: 10.1094/PHI-I-2000-1025-01.

Romanazzi, G., Mancini, V. and Murola, S. (2012) First report of *Leucostoma cinctum* on sweet cherry and European plum in Italy. *Phytopathologia Mediterranea* 51, 365–368.

Santiago-Santiago, V., Tovar-Pedraza, J.M., Ayala-Escobar, V. and Alanis-Martinez, E.I. (2014) Occurrence of powdery mildew caused by *Podosphaera clandestina* on black cherry in Mexico. *Journal of Plant Pathology* 96, 431–439.

Sanzani, S.M. and Ippolito, A. (2013) Mycotoxin contamination on harvested commodities and innovative strategies for their detection and control. *Acta Horticulturae* 1053, 123–132.

Sanzani, S.M., Montemurro, C., Di Rienzo, V., Solfrizzo, M. and Ippolito, A. (2013) Genetic structure and natural variation associated with host of origin in *Penicillium expansum* strains causing blue mould. *International Journal of Food Microbiology* 165, 111–120.

Schena, L., Duncan, J.M. and Cooke, D.E.L. (2008) Development and application of a PCR-based 'molecular tool box' for the identification of *Phytophthora* species damaging forests and natural ecosystems. *Plant Pathology* 57, 64–75.

Schnabel, G. and Lalancette, N. (2003) Constriction canker management in peach. Southeastern Regional Peach Newsletter 3, University of Georgia, Athens, Georgia. Available at: https://njaes.rutgers.edu/peach/pestmanagement/constrictioncanker.pdf (accessed at 4 February 2016).

Schulz, U. and Schmidle, A. (1983) Zur Epidemiologie der Valsa-Krankheit. *Angewandte Botanik* 57, 99–107.

Serdani, M. and Spotts, R.A. (2007) First report of blossom blight and green fruit rot of sweet cherry caused by *Sclerotinia sclerotiorum* in Oregon. *Plant Disease* 91, 1058.

Serradilla, M.J., del Carmen Villalobos, M., Hernández, A., Martín, A., Lozano, M. and de Guía Córdoba, M. (2013) Study of microbiological quality of controlled atmosphere packaged 'Ambrunés' sweet cherries and subsequent shelf-life. *International Journal of Food Microbiology* 166, 85–92.

Shipper, M.A.A. (1984) A revision of the genus *Rhizopus*. I. The *Rhizopus stolonifer*-group and *Rhizopus oryzae*. *Studies in Mycology* 25, 1–19.

Simmons, E.G. (2007) *Alternaria: An Identification Manual*. CBS Fungal Biodiversity Centre, Utrecht, The Netherlands.

Snowdon, A.L. (1990) *Post-harvest Diseases and Disorders of Fruits and Vegetables*. Manson Publishing, London.

Spotts, R.A. and Peters, B.B. (1980) Chlorine and chlorine dioxide for control of d'Anjou pear decay. *Plant Disease* 64, 1095–1097.

Stensvand, A., Aamot, H.U., Strømeng, G.M., Talgø, V., Elameen, A., Børve, J. and Klemsdal, S.S. (2008) *Colletotrichum acutatum* from Norway frequently develops perithecia in culture. *Journal of Plant Pathology* 90 (Suppl.), 93.

Strand, L.L. (1999) *Integrated Pest Management for Stone Fruits*. University of California, Division of Agriculture and Natural Resources, California.

Sundelin, T., Strømeng, G.M., Gjærum, H.B., Amby, D.B., Ørstad, K., Jensen, B., Lund, O.S. and Stensvand, A. (2015) A revision of the history of the *Colletotrichum acutatum* species complex in the Nordic countries based on herbarium specimens. *FEMS Microbiology Letters* 362, fnv130.

Tamm, L., Minder, C.E. and Flückiger, W. (1995) Phenological analysis of brown rot blossom blight of sweet cherry caused by *Monilinia laxa*. *Phytopathology* 85, 401–408.

Tanovic, B., Hrustic, J., Grahovac, M., Mihajlovic, M., Delibasic, G., Kostic, M. and Indic, D. (2012) Effectiveness of fungicides and an essential-oil-based product in the control of grey mould disease in raspberry. *Bulgarian Journal of Agricultural Science* 18, 689–695.

Tarbath, M.P., Measham, P.F., Glen, M. and Barry, K.M. (2014) Host factors related to fruit rot of sweet cherry (*Prunus avium* L.) caused by *Botrytis cinerea*. *Australasian Plant Pathology* 43, 513–522.

Uddin, W., Stevenson, K.L. and Pardo-Schultheiss, R.A. (1997) Pathogenicity of a species of *Phomopsis* causing shoot blight on peach in Georgia and evaluation of possible infection courts. *Plant Disease* 81, 983–989.

Villarino, M., Melgarejo, P., Usall, J., Segarra, J. and De Cal, A. (2010) Primary inoculum sources of *Monilinia* spp. in Spanish peach orchards and their relative importance in brown rot. *Plant Disease* 94, 1048–1054.

Villarino, M., Larena, I., Martinez, F., Melgarejo, P. and De Cal, A. (2012) Analysis of genetic diversity in *Monilinia fructicola* from the Ebro Valley in Spain using ISSR and RAPD markers. *European Journal of Plant Pathology* 132, 511–524.

Wermund, U. and Lazar, E.L. (2003) Control of grey mould caused by the postharvest pathogen *Botrytis cinerea* on English sweet cherries "Lapins" and "Colney" by controlled atmosphere (CA) storage. *Acta Horticulturae* 599, 745–748.

Wharton, P.S., Iezzoni, A. and Jones, A.L. (2003) Screening cherry germ plasm for resistance to leaf spot. *Plant Disease* 87, 471–477.

Wilcox, W.F. and Mircetich, S.M. (1985a) Pathogenicity and relative virulence of seven *Phytophthora* spp. on Mahaleb and Mazzard cherry. *Phytopathology* 75, 221–226.

Wilcox, W.F. and Mircetich, S.M. (1985b) Effects of flooding duration on the development of *Phytophthora* root and crown rots of cherry. *Phytopathology* 75, 1451–1455.

Wittig, H.P.P., Johnson, K.B. and Pscheidt, J.W. (1997) Effect of epiphytic fungi on brown rot blossom blight and latent infections in sweet cherry. *Plant Disease* 81, 383–387.

Xu, X.M., Bertone, C. and Berrie, A. (2007) Effects of wounding, fruit age and wetness duration on the development of cherry brown rot in the UK. *Plant Pathology* 56, 114–119.

Youssef, K., Ligorio, A.M., Sanzani, S.M., Nigro, F. and Ippolito, A. (2013) Investigations on *Penicillium* spp. population dynamic in citrus packinghouse. *Journal of Plant Pathology* 93 (Suppl. 4), 61–62.

# 15 Bacterial Diseases

Joanna Puławska,[1]* Michael Gétaz,[2] Monika Kałużna,[1] Nemanja Kuzmanović,[3] Aleksa Obradović,[3] Joël F. Pothier,[2] Michela Ruinelli,[2] Donato Boscia,[4] Maria Saponari,[4] Anita Végh[5] and László Palkovics[5]

[1]*Research Institute of Horticulture, Skierniewice, Poland;* [2]*Zurich University of Applied Sciences, Institute of Natural Resource Sciences, Wädenswil, Switzerland;* [3]*University of Belgrade – Faculty of Agriculture, Belgrade, Serbia;* [4]*CNR – Institute for Sustainable Plant Protection, Bari, Italy;* [5]*Szent István University, Faculty of Horticultural Science, Department of Plant Pathology, Budapest, Hungary*

## 15.1 Introduction

Bacterial diseases are very often the major limitation in the production of cherries. Yield losses reaching 50% and tree death resulting from established pathogens are common under certain climatic conditions. The two major bacterial diseases that are spread in all cherry cultivation regions are crown gall and bacterial canker. The occurrence of other bacterial diseases on cherries is less common, but the observed symptoms and the severity suggest that they could also cause serious problems in cherry production in the future.

Bacterial diseases can be very difficult to control because of the limited number of efficient chemical products. Moreover, no sources of resistance to bacterial diseases in cherry cultivars have been identified. This is why prevention is the primary control strategy for bacterial pathogens. Prevention consists of quarantine measures, early pathogen detection systems, eradication procedures, and sanitation and hygiene in nursery propagation and orchard maintenance.

## 15.2 Crown Gall

### 15.2.1 Disease description

Crown gall is a widespread bacterial disease causing significant economic losses in the production of many plants throughout the world. It occurs on over 600 plant species, including agricultural crops, primarily fruit species (stone and pome fruits, nut trees) and grapevine, as well as some perennial ornamentals (de Cleene and de Ley, 1976). Although it occurs worldwide, crown gall is especially prevalent and destructive in temperate climate regions.

Symptoms on cherry trees correspond to the typical symptoms observed in most susceptible plant species. Since the underground plant parts are usually affected, the first changes mostly remain unnoticed. However, roundish or irregular-shaped galls are usually formed at or near the soil line, at the graft or bud union, or on roots and lower stems (Fig. 15.1). The galls can vary in size from several millimetres to more than 15 cm in diameter, and the most destructive are

* joanna.pulawska@inhort.pl

Fig. 15.1. Crown gall on *Prunus avium*.

those occurring on the main root or root collar. Young galls are small, round, light coloured and soft. During the season, tumours enlarge and become hard and dark brown, with a rough and cracked surface. Older galls or their parts may become spongy, decay and crumble.

Large galls may partly girdle the main roots or lower stem, inhibiting plant physiological functions such as the transport of water and nutrients. Affected plants may show stunted growth and poor yield. These plants are more sensitive to abiotic stresses, particularly frost injury. As the disease progresses, plants lose vigour and eventually die.

Crown gall is an important disease, particularly in fruit tree nurseries, where it is responsible for extensive economic losses since symptomatic plants are unmarketable and need to be eliminated (Puławska, 2010). In cherry orchards, losses from the disease are mostly sporadic. The disease seldom kills the plant, but it is often destructive in the first years after planting, when the disease can stunt the growth of young cherry plants (Hołubowicz *et al.*, 1988). The studies of Sobiczewski *et al.* (1991) showed that, in water-deficiency conditions, 1-year-old shoots of infected Mazzard cherries were 50% shorter and the crown diameter was 25% smaller when compared with healthy plants. However, in some cases, no significant differences between diseased and healthy cherry trees were found, especially in older trees (Garrett, 1987).

### 15.2.2 Pathogen

Crown gall is caused by bacteria belonging to the genera *Agrobacterium*, *Allorhizobium* and *Rhizobium* of the family *Rhizobiaceae*. They are aerobic, motile, non-spore forming, flagellated, Gram-negative, rod-shaped bacteria.

The taxonomy of bacteria causing crown gall is debatable and still not fully resolved. Initially, their taxonomy was based on pathogenic features and all bacteria causing crown gall were classified as *Agrobacterium tumefaciens*. However, the pathogenicity of tumorigenic strains is determined mainly by the presence of a conjugative tumour-inducing (Ti) plasmid in their genome. Since Ti plasmid acquisition or loss may lead to a change in taxonomic status, classification based on pathogenic properties was not considered stable. More recently, Young *et al.* (2001) proposed inclusion of all *Agrobacterium* spp. into the genus *Rhizobium*; however, new studies relying on genotypic characterization have led to a revised phylogeny of the family *Rhizobiaceae* (Mousavi *et al.*, 2015). So far, tumorigenic bacteria have been found within eight validly published species: *Agrobacterium arsenijevicii*, *Agrobacterium nepotum*, *Agrobacterium larrymoorei*, *Agrobacterium radiobacter*, *Agrobacterium rubi*, *Agrobacterium skierniewicense*, *Allorhizobium vitis* and *Rhizobium rhizogenes*. Additionally, tumorigenic strains were also identified among some genomic species within the *A. tumefaciens* species complex (formerly called *Agrobacterium* biovar 1) that have not yet formally been named (Costechareyre *et al.*, 2010).

Cherry trees are infected mostly by pathogenic bacteria of the *R. rhizogenes* and

*A. tumefaciens* species complex, which generally have a wide host range (Dhanvantari, 1978; Süle, 1978; López *et al.*, 1988; Sobiczewski, 1996; Puławska and Kałużna, 2012).

Crown gall causal agents are generally soil-inhabiting bacteria able to survive a long period of time in soil. Wounds made by cultivating practices or by abiotic or biotic factors represent an entry point for the pathogen. If the soil is contaminated with the pathogen, wounds caused by the routine practice of root pruning before planting are particularly important for initiation of the infection. The pathogen is disseminated in the field usually by cultivation tools and equipment or by movement in soil water.

Contaminated nursery material is another source of inoculum. In this way, the pathogen can be spread over long distances throughout exchange and trade of infected symptomatic or asymptomatic planting material. Inoculum may be present in young inconspicuous tumours that remain unnoticed. In addition, systemic infections in which the pathogen may be latently present within asymptomatic plants have been reported for some hosts, including cherry (Cubero *et al.*, 2006).

Infection of plants by tumorigenic bacteria is a highly complex multistage process of natural genetic transformation. It is the only known natural example of transkingdom DNA transfer. The infection process begins when bacteria causing crown gall are attracted by phenolic compounds (e.g. acetosyringone) released by wounded cells to which they attach. The same inducers activate expression of the virulence (*vir*) genes on the Ti plasmid. The *vir* genes control the transfer of the Ti plasmid fragment, called transfer DNA (T-DNA), and its integration into the plant genome (Zhu *et al.*, 2000). Oncogenes present in T-DNA encode the plant hormones auxin and cytokinin, whose overproduction leads to abnormal proliferation of plant cells and tumour formation. Once the plant cell is transformed, tumours continue to develop, even in the absence of the pathogen. In the later course of the disease, galls may detach and disintegrate, releasing the bacteria into the surrounding soil.

The T-DNA also includes genes responsible for the synthesis of a specific class of small-sized molecules, called opines, which are divided into several chemical classes (Dessaux *et al.*, 1998). In general, they serve as selective nutrient sources for tumorigenic bacteria and promote conjugal transfer of their Ti plasmids (Kerr *et al.*, 1977).

From a practical point of view, it is important to determine whether the soil in fields designated for nursery plantations is free from tumorigenic agrobacteria. It is also essential to detect and identify the causal agent if disease occurs in established orchards. Bacteria causing crown gall of fruit species may routinely be isolated on general media such as yeast mannitol agar, or more efficiently on selective and differential media such as MG+Te (Mougel *et al.*, 2001) and 1A+2E (Puławska and Sobiczewski, 2005). A crucial point in pathogen identification is to determine its tumorigenicity. A pathogenicity assay is the only reliable method for this purpose, although it is time consuming and laborious. However, application of molecular methods, primarily the polymerase chain reaction (PCR), has enabled rapid detection of pathogenicity-associated genes that are located mainly on the Ti plasmid. To this end, numerous PCR primers and protocols have been developed (Palacio-Bielsa *et al.*, 2009); nevertheless, the wide genetic diversity of Ti plasmids hampers reliable detection of the pathogen. Non-pathogenic agrobacteria lacking the Ti plasmid are very often prevalent in diseased plants, especially in older galls, which also makes diagnosis difficult.

Although the role of phenotypic tests in identification of this group of bacteria has recently been called into question (Ormeño-Orrillo and Martínez-Romero, 2013), some discriminative tests are still useful for initial screening of isolated bacteria, such as reduction of tellurite, presence of urease and esculinase hydrolysis enzymes, or production of 3-ketolactose from α-lactose in differentiation of strains belonging to the *A. tumefaciens* species complex. Nowadays, molecular-based techniques provide rapid identification of crown gall bacteria. For instance, multiplex-PCR based on the 23S rRNA gene enables identification and differentiation of four agrobacterial taxa (Puławska *et al.*, 2006). Moreover, sequence analysis of the housekeeping *recA* gene has allowed clear delineation of this group of bacteria (Costechareyre *et al.*, 2010).

The genome architecture of the agrobacteria and rhizobia is diverse and may include single chromosomes, multiple chromosomes and plasmids (Slater *et al.*, 2009). Variability in the number of plasmids and their size among agrobacteria and rhizobia may be very high. Ti plasmids are highly diverse genetic elements that show a mosaic structure composed of conserved and highly variable regions. Generally, little is known about the other plasmids occurring in agrobacteria and rhizobia, although some of them have been characterized (Otten *et al.*, 2008).

### 15.2.3 Control

Since the plants infected with crown gall bacteria are genetically transformed, traditional chemical treatments are ineffective for disease control. Once established in an orchard, the pathogen can be very difficult to eliminate. Therefore, special attention must be paid to preventative control measures. Visual inspection for tumours on nursery plants before planting in the field is a useful measure for preventing introduction of the pathogen. However, inconspicuous symptoms, presence of the pathogen on the surface of asymptomatic plants and latent infections may compromise the efficiency of this strategy.

In order to prevent outbreaks and spread of crown gall, it is essential to establish phytosanitary measures that regulate the production of pathogen-free planting material in nurseries. However, in many countries, crown gall bacteria are not considered quarantine pathogens, although they are commonly regarded as harmful, widespread pathogens that can reduce the value of propagation material (quality pathogens). Therefore, they are not subject to adequate phytosanitary control, especially in international trade.

Taking into account the long-term persistence of tumorigenic bacteria in soil, establishment of new plantations in fields with a history of crown gall should be strongly avoided. It is also recommended to test the soil for the presence of crown gall bacteria prior to planting. Moreover, it is important to use appropriate crop rotation. Generally, planting of monocots is recommended to reduce populations of tumorigenic bacteria in the soil. It is also suggested that alkaline soils (high pH) should be avoided, or that they should be acidified with physiologically acidic fertilizers. Heavy and poorly drained soils in frost-prone areas, which promote injuries of plants, are not advisable as planting sites. Since the pathogen can infect plants through wounds made by root-knot nematodes (Rubio-Cabetas *et al.*, 2001), planting in soil heavily infested with these pests should also be avoided.

Soil fumigants (e.g. methyl bromide, metam sodium/potassium, dazomet) were used to combat soil-borne plant pathogens for years. However, these compounds have recently been avoided due to increasing concerns about their negative effects on human health and the environment. In addition, their efficacy as a preplant crown gall management strategy in an open field is questionable. Due to their high toxicity, non-selectivity and environmental threat, use of these compounds is restricted in many countries. Solarization (solar heating) of soil has been reported as effective for reduction of crown gall bacteria (Raio *et al.*, 1997). Furthermore, in the case of cherry plants, solarization also suppressed gall development in glasshouse and field trials.

Planting of crown gall-resistant plants, particularly rootstocks, would be an efficient measure in preventing this disease. For this reason, various *Prunus* spp. rootstocks have been tested for susceptibility to crown gall (Pierronnet and Salesses, 1996; Bliss *et al.*, 1999). However, further efforts should be made in testing promising *Prunus* genotypes with a larger number of tumorigenic strains, considering the latest insights into genetic diversity and taxonomy of crown gall bacteria.

Reduction of wounding during cultural practices is necessary to prevent pathogen ingress and colonization of the host plant.

Heat treatment of *Prunus* spp. seedlings has been proven to be a good measure for healing of root-pruning wounds and reduction of crown gall incidence (Moore and Allen, 1986). In this respect, the incidence of naturally occurring crown gall on Mazzard cherry seedlings was reduced from 66% (unheated) to 6% in a commercial nursery. Disinfection of pruning tools is recommended to avoid the spread of bacteria in nurseries and field. If crown gall symptoms are observed in an orchard, infected plants should immediately be removed in order to prevent spread of the inoculum.

Biological control of crown gall using non-pathogenic *R. rhizogenes* (formerly classified as *A. radiobacter* as a non-pathogenic strain) strain K84 was one of the first examples of biological control of plant diseases and the first commercial biocontrol agent. Strain K84 produces the antibiotic agrocin 84, which is a major component of biocontrol, and at least two other antibacterial substances: agrocin 434 and the antibiotic-like substance 84 (ALS84). Dipping the root system into a suspension of strain K84 prior to planting efficiently prevented crown gall incidence on roses and various stone fruits, including cherry (Moore, 1976). Another advantage of strain K84 is its ability to effectively persist in the plant rhizosphere.

However, certain limitations in the application of strain K84 were reported. Primarily, only agrocinopine-catabolic agrobacteria such as strains harbouring a nopaline-type Ti plasmid are sensitive to agrocin 84. The conjugal transfer of pAgK84 from strain K84 into virulent agrobacteria was another limitation in use of this biocontrol agent, since this plasmid encodes the production of agrocin 84 and provides agrocin resistance. This limitation was overcome by the construction of a mutant strain, K1026, which was unable to transfer the plasmid. Since strain K1026 is considered a genetically modified organism, its use is restricted in many countries (e.g. all EU countries). However, these biopesticides are presently the most effective compounds to control crown gall, and several commercial products based on strain K84 or K1026 (e.g. Galltrol-A®, Norbac 84C, Diegall, NOGALL™) were available in some countries.

Control of crown gall can be also achieved by genetic engineering and the development of disease-resistant plants (Otten *et al.*, 2008). This approach was used for inhibition of bacteria or for blocking T-DNA transfer and integration, silencing T-DNA oncogenes by RNA interference (RNAi), and by inhibiting the synthesis of host proteins interacting with bacterial virulence proteins. However, transgenic cherry plants resistant to crown gall are not available on the market.

## 15.3 Bacterial Canker

### 15.3.1 Disease description

Bacterial canker occurs in stone fruit-growing areas all over the world (Agrios, 2005) and can be found on all stone fruit trees and in the last few years also on apples and pears. However, the greatest damage is caused in orchards and nurseries of cherries and apricots.

The disease symptoms can be observed on all aboveground organs of the trees, thus causing heavy yield reduction (up to 75%) and sometimes leading to death of entire trees, especially in nurseries and young orchards. The occurrence of the symptoms is connected to the two phases of the disease: winter, occurring on woody tissues, and summer, present on newly developed organs in vegetative period. Characteristic symptoms of the disease on the main trunk and branches of cherry trees include a sunken, dark brown dieback and cankers, often accompanied by gummy leaks (Fig. 15.2). The bacteria overwinter in and on infected buds, tissue around the traces of fallen leaves and at the margin of necrosis and cankers, thereby constituting the source of primary infection. In spring time, in favourable conditions (i.e. cool, humid weather), the bacteria multiply and are spread by wind, rain and insects on new developing organs. On the leaves, brown necrotic spots with a more or less regular shape, often surrounded by a clearly visible yellow 'halo', can be observed (Fig. 15.3A). With the development of the disease, the

**Fig. 15.2.** Disease symptoms of bacterial canker on the main trunk of a cherry tree.

necrotic tissue crumbles, and the leaves look like they have been shot through. In the case of flower infection, which occurs especially on susceptible varieties of cherries after the spring frosts, initially wilting and browning of developing flowers, followed by blackening and dieback can be seen (Fig. 15.3B). The dying flowers are the source of secondary infections – bacteria can spread to shoots and to branches, which can result in necrosis, dieback and canker formation. Sometimes strong subsidence and subcortical tissue necrosis are observed. On fruitlets and fruit, sunken black necrotic spots, sometimes covering a significant part of the fruit, can be found. In the case of major infection, such fruit are not good for consumption and lose their commercial value (Fig. 15.3C).

In disease development, predisposing factors include nematodes, low soil pH and freezing temperatures (Melakeberhan *et al.*, 1993). Moreover, epiphytic bacteria, which survive on the surface of trees, mainly on the leaves, often without establishing a parasitic relationship with the plant, during

**Fig. 15.3.** Disease symptoms of bacterial canker on cherry leaves, (A) flowers (B) and fruit (C).

the growing season and fall of foliage, are an important source of infection (infection by traces of the leaves) (Renick *et al.*, 2008).

### 15.3.2 Pathogen description

The causal agents of bacterial canker belong to the *Pseudomonas syringae* species complex, which comprises polyphagous bacteria that are able to infect more than 180 species of plants, both annual and perennial, including fruit trees, ornamental plants and vegetables. The *P. syringae* species complex comprises plant pathogens divided into more than 50 pathovars, defined based on their pathogenic ability (Young, 2010), and belonging to nine genomospecies, determined by DNA–DNA hybridization (Gardan *et al.*, 1999).

The bacteria causing bacterial canker on cherry trees are spread over three different genomospecies (gs): gs 1, *P. syringae* pv. *syringae* (*Pss*); gs 2, *P. syringae* pv. *morsprunorum* race 1 (*Pmp1*); and gs 3, *P. syringae* pv. *morsprunorum* race 2 (*Pmp2*) and *P. syringae* pv. *avii* (*Psa*) (Wormald, 1932; Freigoun and Crosse, 1975; Ménard *et al.*, 2003). Recently, a new species of bacteria isolated from bacterial canker symptoms on cherries (mainly sour cherry) and named *Pseudomonas cerasi* sp. nov. (non Griffin, 1911) was also described (Kałużna *et al.*, 2016b). In addition, another member of the *P. syringae* species complex, *P. syringae* pv. *cerasicola* (gs 2), has been found in association with bacterial gall disease on ornamental cherry trees in Japan (Kamiunten *et al.*, 2000). However, a limited number of reports about its presence is available.

Symptoms of bacterial canker can be similar to those caused by other pathogens or factors. Necrosis on the main trunk and branches can also be caused by *Leucostoma* (*Valsa*) and *Monilinia* spp., and gummosis on woody tissue can also result from the physiological response to damage caused by abiotic factors; necrotic spots on leaves are similar to those caused by *Prunus necrotic ring spot virus*, as well as the fungal pathogens *Clasterosporium carpophilum* and *Blumeriella jaapii*. For this reason, it is crucial to use a rapid and specific method of diagnosis that will allow detection and identification of the causal agent of bacterial canker.

The diagnostics of cherry-pathogenic *P. syringae* is commonly based on bacterial isolation on microbiological media followed by morphological, biochemical and physiological characterization, including pathogenicity tests (Vicente *et al.*, 2004; Kałużna and Sobiczewski, 2009). If grown on King's B medium, all current known *P. syringae* pathovars causing bacterial canker on cherry, with the exception of *Psa*, produce a fluorescent pigment visible under UV light (King *et al.*, 1954). The results obtained from the traditional LOPAT (Lelliott *et al.*, 1966), GATTa and L-lactate utilization tests (Lattore and Jones, 1979; Lelliott and Stead, 1987) enable the identification of *P. syringae* spp. and their discrimination into pathovars and races.

Some features of the bacteria that affect the pathogenicity and virulence of the strains and which can be helpful in strain identification include, for example, the presence of phytotoxins or ice nucleation activity (INA) in *in vitro* conditions. Syringomycin production, a feature shared among many *Pss* strains can be tested by checking the ability to inhibit growth of the yeast *Rhodotorula pilimanae* or the fungi *Geotrichum candidum* (Kałużna *et al.*, 2012) or *Aspergillus niger* (Hu *et al.*, 1998) on potato dextrose agar or peptone–glucose–NaCl agar plates. In a similar way, but using the Gram-positive bacterium *Bacillus megaterium* as an indicator strain, the production of syringopeptin can also be assessed (Grgurina *et al.*, 1996; Lavermicocca *et al.*, 1997). The ability of some strains to catalyse ice formation from supercooled water (INA) (Hirano *et al.*, 1978) is a feature of strains classified as *Pss*. So-called INA$^+$ strains represent a higher risk for cherry fruit tree crops when (even weak) frost occurs, especially in spring time, as the size of the damage caused is much larger (Sobiczewski and Jones, 1992).

Serological detection methods such as a slide agglutination test (Lyons and Taylor, 1990) or indirect enzyme-linked immunosorbent assay (ELISA) (Nemeth *et al.*, 1987) are rarely used for the identification of bacteria causing bacterial canker because of frequent cross-reactions and an unclear response in distinguishing isolates of *P. syringae* (Zamze *et al.*, 1986; Vicente *et al.*, 2004).

One of the first DNA-based molecular tools developed for taxonomic purposes for *P. syringae* is repetitive PCR (rep-PCR). The use of PCR primers corresponding to repetitive (ERIC, BOX and REP) and insertion (IS*50*) sequences has been shown to generate amplification patterns that can be used as a genetic fingerprint for inter-pathovar discrimination (Louws *et al.*, 1994; Weingart and Völksch, 1997). So far, rep-PCRs have been successfully applied to distinguish *Pss*, *Pmp1* and *Pmp2* isolated from diseased wild cherry trees (Vicente and Roberts, 2007), as well as on *P. syringae* strains isolated from different stone fruits, comprising sweet and sour cherry (Kałużna *et al.*, 2010a). Another fingerprinting PCR-based technique used for phytopathogenic *P. syringae* strains is the PCR melting profile (Kałużna *et al.*, 2010b). The resulting electrophoretic pattern can be used for determination of strain genetic variability, but also for classification of isolates to pathovars and races.

For over a dozen years, a tool called multi-locus sequence analysis (MLSA) has been used to redefine the phylogenetic relationships within the *P. syringae* species complex based on the similarity of concatenated sequences of selected housekeeping genes. This analysis revealed the presence of up to 13 phylogroups (PGs) (Sarkar *et al.*, 2006; Parkinson *et al.*, 2011; Berge *et al.*, 2014), largely corresponding to the genomospecies defined by DNA–DNA hybridization (Gardan *et al.*, 1999). For strains of *P. syringae* isolated from cherry, a set of four loci corresponding to the *rpoD*, *gyrB*, *gltA* (also known as *cts*) and *gapA* genes (Kałużna *et al.*, 2010a), as well as the *rpoB* gene (Ait Tayeb *et al.*, 2005), have commonly been used in MLSA. Thanks to the elucidation of genetic mechanisms involved in phytotoxin production and other pathogenicity-related factors, more specific PCR primers have been developed for the detection of genes involved in toxin synthesis: syringomycin (*syrB* and *syrD* genes) characteristic for strains of *Pss* (Sorensen *et al.*, 1998; Bultreys and Gheysen, 1999), coronatine (*cfl* gene) present in strains of *Pmp1* (Bereswill *et al.*, 1994) and yersiniabactin (*irp* gene) characteristic for *Pmp2* (Bultreys *et al.*, 2006). However, it should be noted that the determination of the presence of genes needed for toxin production is not reliable for identification in itself. In fact, strains of *Pmp1* and *Pss* that do not have the ability to produce coronatine or syringomycin, respectively, are quite common (Renick *et al.*, 2008; Kałużna *et al.*, 2010a).

More recently, a rapid and highly specific conventional and real-time PCR assay was developed for the identification, detection and discrimination of the cherry bacterial canker causal agents *Pmp1* and *Pmp2* (Kałużna *et al.*, 2016a). Additionally, primers for specific detection of individual phylogroups of the *P. syringae* species complex to which the bacterial canker causal agent belongs were also published (Borschinger *et al.*, 2016). In contrast to MLSA and rep-PCR, the use of such specific PCRs usually does not require comparison with reference strains, thus speeding up the process for identification and detection.

However, despite the economic significance of the disease, there is a severe lack of genomic information concerning bacterial cherry pathogens and generally of related stone fruit pathogens. This constitutes a real obstacle to accurate and rapid identification, but also in terms of elucidating the mechanisms behind the host variation, epidemiology and pathogenicity of these pathogens. Currently, a set of 12 strains belonging to each pathovar shown to be responsible for cherry bacterial canker were selected for *de novo* whole-genome sequencing using next-generation sequencing technology PacBio (Ruinelli *et al.*, 2015). Phylogenetic analysis based on the concatenated core gene set ($n = 2519$) of the sequenced strains and six additional reference strains confirmed that the different pathovars causing bacterial canker on cherry belong to three different PGs: PG1 (gs 3) comprising *Pmp2* and *Psa*; PG2 (gs 1) comprising *Pss*; and PG3 (gs 2) comprising *Pmp1* (Fig. 15.4).

### 15.3.3 Control

The most effective strategy for protection of cherry trees against bacterial canker is

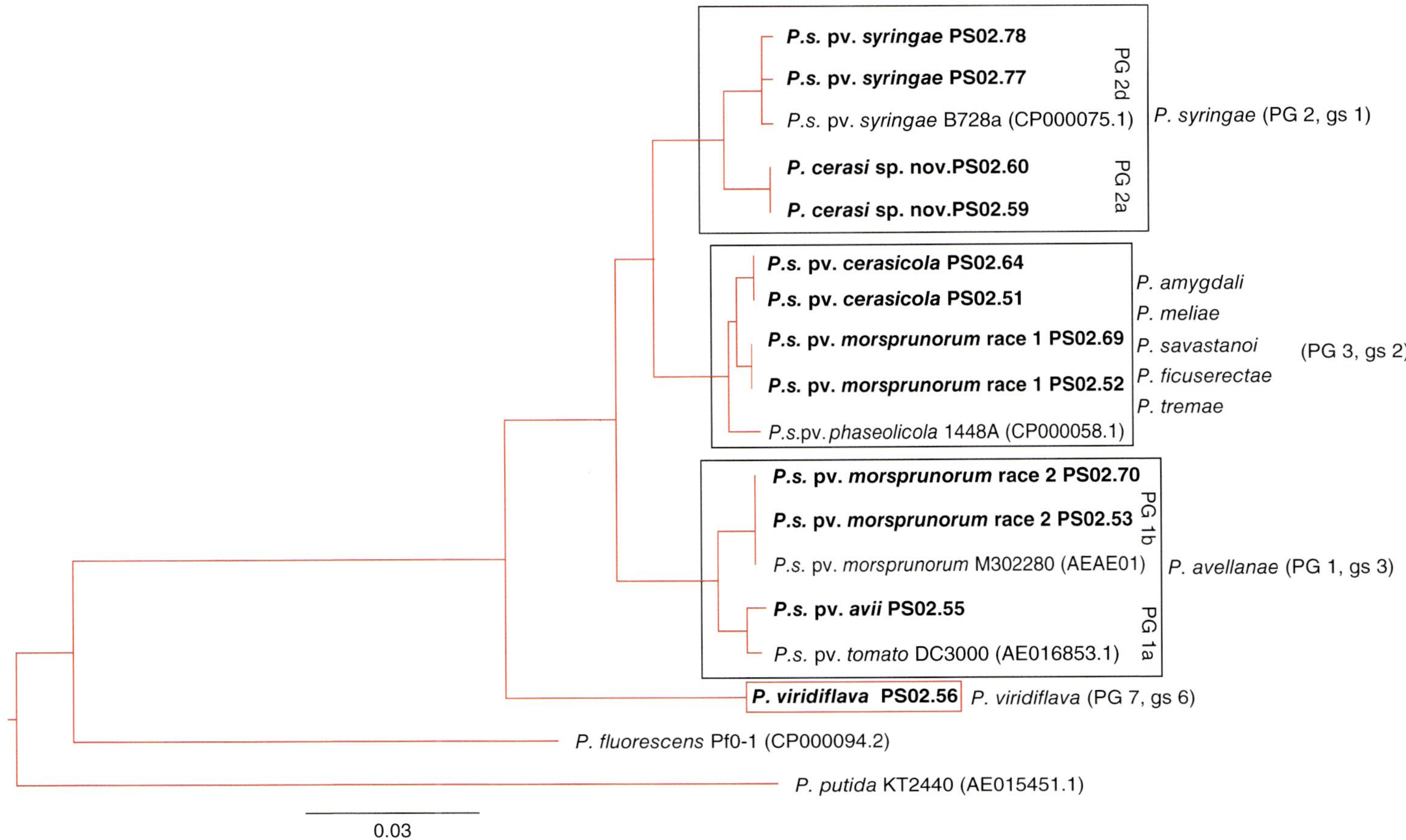


**Fig. 15.4.** Core genome phylogenetic analysis of strains isolated from diseased cherry that were selected for *de novo* whole-genome sequencing. The set of core genes ($n$ = 2159) as well as the phylogenetic relationships among the analysed strains were computed using EDGAR 1.3. Genome-sequenced strains isolated from cherry are indicated in bold. *Pseudomonas viridiflava* was isolated from cherry, but is not pathogenic. The GenBank accession numbers of the reference genomes are shown in parentheses. Phylogroups (PG) and genomospecies (gs) are shown. *P.s.*, *Pseudomonas syringae*. Bar, 0.03 nucleotide substitutions per site.

prevention. When new plantations are established, besides the selection of varieties, it is important to use healthy nursery material. This material should be absolutely free from the disease. In orchards, an important treatment is cutting and removal of infected shoots with excess of apparently healthy parts (done during dry weather) and securing the wound after cutting. In the case of severe infestation, the entire tree must be removed. The presence of nematodes as well as soil pH should be controlled, since both predispose cherry to bacterial canker.

For biological control, the use of Double Nickel 55 (*Bacillus amyloliquefaciens* D747) and Serenade ASO (*Bacillus subtilis* QST 713) were proposed. However, current results indicate that their effectiveness is not stable or is at best only moderate (Spotts and Wallis, 2008; Pscheidt and Ocamb, 2015).

For chemical protection against bacterial canker, reducing the incidence of bacteria on the surface of plants is done using copper-based compounds. These have good bacteriostatic and bactericidal activity, but only as surfactants. Three different active substances are mainly included in commercial products: copper oxide, copper oxychloride and copper hydroxide. In a programme for the protection of sweet and sour cherries, three main periods of copper spraying/treatments are recommended: (i) the leafless period to reduce pathogen populations on both the surface of cankers and emerging from dormant buds: (ii) blooming period; and (iii) the period of leaf fall to reduce leaf scar infection. In the case of susceptible varieties and in wet, warm weather favouring the spread of the pathogen, additional treatments immediately after flowering should be used.

Nanoparticles that contain silver and copper (Mondal and Mani, 2012; Abdel-Megeed, 2013) are increasingly appearing in various control applications. However, in the case of cherry bacterial canker, these tests have only been done in the laboratory (Ozaktan *et al.*, 2013).

## 15.4 Bacterial Leaf Spot (*Xanthomonas arboricola* pv. *pruni*)

### 15.4.1 Disease description

Symptoms caused by *Xanthomonas arboricola* pv. *pruni* were first described in Michigan, USA, in 1903 on plum (Smith, 1903). Nowadays, the disease is reported on all continents where stone fruits are grown, attacking a wide range of commercial, ornamental and forest *Prunus* spp. but more particularly stone fruit crops including cherry (Jami *et al.*, 2005; EPPO, 2006).

This bacterium can theoretically cause symptoms on the full range of species belonging to the genus *Prunus*. However, depending on the species, some differences in sensitivity can be observed. Although cherries – *Prunus avium* (sweet cherry) and *Prunus cerasus* (sour cherry) – belong to the group of minor hosts (Garcin *et al.*, 2005), this pathogen is generating an important economic loss for growers. As symptoms do not appear instantaneously and the pathogen can persist in buds, bacteria can be dispersed to other countries and areas by plant propagation before symptoms are noticed (Stefani, 2010).

This bacterium was also recently found in commercial production of ornamental cherry laurel (*Prunus laurocerasus*) in Europe (Marchi *et al.*, 2011; Bergsma-Vlami *et al.*, 2012; Tjou-Tam-Sin *et al.*, 2012). This semi-wild host is sometimes grown for human fruit consumption, as, for example, in the Black Sea region where the fruits are consumed to protect against diabetes (Ercisli, 2013).

Symptoms of bacterial leaf spot can be observed on different parts of plants: leaves, fruit, twigs and branches. *X. arboricola* pv. *pruni* causes variable symptoms and severities among *Prunus* spp. On cherries, documentation and reliable information are rare for observed symptoms, but it is reported that early fruit infection produces a distortion from the fruit and that bacteria could be found from the stone to the epidermis (EPPO, 2006). Leaf symptoms are similar to those on peach, but rarely as important. The first appearance of infection is located on the lower surface of the leaf where pale green to yellow circular or irregular areas appear. The spots become visible on the upper surface before they start to darken to a purple, brown or black spot. On *P. laurocerasus*, leaves show chlorotic spots and most have a necrotic brown centre with a distinct margin; the spot readily abscises, resulting in a 'shot-hole' appearance (Tjou-Tam-Sin *et al.*, 2012).

### 15.4.2 Pathogen

The species *X. arboricola* comprises phytopathogenic bacteria split into seven pathovars (Vauterin *et al.*, 1995; Janse *et al.*, 2001) involved in emerging diseases worldwide on a wide range of perennial plants such as poinsettia, poplar, hazelnut, *Prunus* spp. and *Juglans* spp. (Lamichhane, 2014). *X. arboricola* pv. *pruni* is listed as a quarantine organism in the EU phytosanitary legislation and in the European and Mediterranean Plant Protection Organization lists (EPPO A2 list). The phylogenetic relationship between pathovars has been determined using the housekeeping gene *rpoD* (Hajri *et al.*, 2012) or MLSA of housekeeping genes (Boudon *et al.*, 2005; Fischer-Le Saux *et al.*, 2015), highlighting a homogeneous group constituted from pv. *pruni* whereas all other pathovars seemed more heterogeneous (Vauterin *et al.*, 1995; Young *et al.*, 2008; Parkinson *et al.*, 2009; Hajri *et al.*, 2012).

The complete genome of a genotype-representative strain from Europe (Italy, CFBP 5530 isolated from symptomatic peach leaves) was sequenced (Pothier *et al.*, 2011b). The genome determination revealed the presence of a 41 kb ubiquitous plasmid (Pothier *et al.*, 2011c), confirmed the presence of a large type III effector repertoire (Hajri *et al.*, 2012) and proved to be useful for subtyping *X. arboricola* isolates (Bergsma-Vlami *et al.*, 2012; Cesbron *et al.*, 2014). The availability of several additional draft genomes of isolates from different hosts (Garita-Cambronero *et al.*, 2014; T. Fujikawa, 2014, unpublished data), some being potentially non-pathogenic (Garita-Cambronero *et al.*, 2016), but also of other *X. arboricola* isolates (Ibarra Caballero *et al.*, 2013; Ignatov *et al.*, 2015; Pereira *et al.*, 2015; Y.-H. Noh, 2014, unpublished data; Cesbron *et al.*, 2015; J. Harrison, 2015, unpublished data) will help the comprehension of pathogenesis processes associated with bacterial canker and spot disease of *Prunus*.

Detection of *X. arboricola* pv. *pruni* can be performed by visual inspection of leaves and fruit where necrotic lesion and canker development can be observed, but the bacterial origin of the symptoms will need to be identified.

Several detection and identification methods have been developed for *X. arboricola* pv. *pruni* in recent years (reviewed by Palacio-Bielsa *et al.*, 2012), improving the ability to detect the pathogen from fruit and trees. For example, conventional PCR (Park *et al.*, 2010), Bio-PCR (Ballard *et al.*, 2011), duplex-PCR (Pothier *et al.*, 2011a), multiplex-PCR (Pothier *et al.*, 2011c) and quantitative PCR (Palacio-Bielsa *et al.*, 2011) based on molecular sequences using specific targets for *X. arboricola* pv. *pruni* are common techniques used for high-sensitivity detection and identification.

A more recent method is a loop-mediated isothermal amplification (LAMP) (Bühlmann *et al.*, 2013), which has a higher specificity due to the use of three pairs of primers and allows testing directly in the field. This method aims to enhance the ability of on-site detection of the pathogen and therefore enhance diagnostic rapidity to prevent further spread of the bacteria. Airports and phytosanitary services at borders could then easily use this detection tool to test the import and export of plant material.

### 15.4.3 Control

Bacterial spot is very difficult to control on highly susceptible varieties, and under optimal environmental conditions for infection and disease, control can be difficult on moderately susceptible varieties. Control and management measures must be applied preventatively to successfully reduce losses from this disease (Ritchie, 1995). Strategies to control the spread of the disease and to limit its damage include maximum use of tolerant varieties of *Prunus*, prophylactic measures such as planting of healthy trees, and the use of chemical agents, and can be combined in disease-forecasting models (Socquet-Juglard, 2012). Good general hygiene measures prevent introduction and spread of the bacteria. The use of disinfected tools for pruning and grafting, as well as certified healthy rootstocks and buds, is highly recommended. Regular chemical sprays can help in reducing the amount of fruit and leaf infection. Copper applications could be the most efficient method to limit the spread of the bacterium; however, copper efficacy is only partial, and it is especially phytotoxic for stone fruit species, provoking leaf necrosis and defoliation (Ritchie, 1995). Different copper compounds including sulfates, oxychlorides and hydroxides with variable chemical formulations of copper are commonly used (Stefani, 2010). Due to the high correlation between climatic factors (especially heavy rains and high temperatures) and disease development and spread, developing reliable forecasting models for assessing the correct timing of spray applications could help stone fruit producers to control spread of this disease.

## 15.5 Other Diseases

### 15.5.1 Cherry leaf scorching associated with *Xylella fastidiosa*

*Xylella fastidiosa* (Wells *et al.*, 1987) is a Gram-negative, xylem-limited, slow-growing bacterium, transmitted by a number of xylem-feeding insect vectors. This plant pathogen has a very broad host range (EFSA, 2016) including monocotyledonous and dicotyledonous species, herbaceous and woody plants, cultivated crops and weeds, native flora, and riparian and landscape shade trees. Several of the most significant diseases caused by *X. fastidiosa* include Pierce's disease of grapevine, citrus variegated chlorosis, almond leaf scald, coffee leaf scorch and phony peach disease (Janse and Obradović, 2010; Purcell, 2013). More recently, infections caused by *X. fastidiosa* have been found in olive trees showing a new and severe disorder denoted olive quick decline, which affects this crop in Italy (Martelli *et al.*, 2016), Brazil (Coletta-Filho *et al.*, 2016) and Argentina (Haelterman *et al.*, 2015).

Based on DNA–DNA hybridization and MLSA typing (Schaad *et al.*, 2004; Scally *et al.*, 2005) there are four accepted subspecies (*X. fastidiosa* subsp. *fastidiosa*, *X. fastidiosa* subsp. *multiplex*, *X. fastidiosa* subsp. *sandyi* and *X. fastidiosa* subsp. *pauca*) and a putative subspecies (*X. fastidiosa* subsp. *morus*). This taxonomic subdivision is well supported by genetic and biological differences (i.e. host preference and largely non-overlapping host range). Thus, genotypic assignment to subspecies is helpful in allowing a preliminary inference into the general biology of a given isolate.

Although for a long time *X. fastidiosa* has been considered a plant pathogen mainly restricted to the Americas, the recent outbreaks in Italy (Martelli *et al.*, 2016) and France widen its geographical distribution and host range. Because a large number of European plant species have never been exposed to the bacterium, it is not known whether they would be hosts and, if so, whether they would be symptomatic or not. Results of surveys carried out in the areas of recent outbreaks have disclosed the presence of several unreported hosts of *X. fastidiosa* (European Commission, 2017).

Several *Prunus* spp. are listed as hosts of *X. fastidiosa* strains belonging to different subspecies. The major *X. fastidiosa*-induced diseases of stone fruits are almond leaf scorching, phony peach disease and plum leaf scald. The most characteristic symptoms of almond leaf scorching are leaf scorching followed by decreased productivity and a generalized decline, whereas phony peach disease elicits stunting of the shoots, shortening of the internodes and increased lateral branching, sometimes accompanied by early blooming. Leaves of plants affected by plum leaf scald look burnt or brown along the edges, and severely affected trees may decline and die.

*Prunus dulcis*, *P. avium* and *P. cerasifera* have been *X. fastidiosa* hosts in the EU outbreaks. Specifically, *X. fastidiosa* subsp. *pauca* strain CoDiRO was identified in symptomatic trees of *P. dulcis* and *P. avium* in Apulia (southern Italy), while the agent of *P. cerasifera* infections in France is *X. fastidiosa* subsp. *multiplex*. As for cherry, both sweet and sour cherry have been known hosts of *X. fastidiosa* subsp. *fastidiosa* in the Americas since the early 2000s (Hernandez-Martinez *et al.*, 2007; Nunney *et al.*, 2013). However, the records are few and the information on the symptoms associated with infections is limited, suggesting that cherry may not be a relevant host of this pathogen in the American continent.

More recently, sweet cherry was reported as a new host of *X. fastidiosa* subsp. *pauca* in Apulia (Saponari *et al.*, 2014). Infected plants show scanty vegetation and bud failure in the spring, and typical leaf scorching of mature leaves in summer (Fig. 15.5). Based on the observations so far carried out on these infected trees, the symptoms do not affect the entire canopy but are limited to a few branches. Diagnostic tests on these trees showed that the bacterium could be detected in the leaves only when collected at the onset of symptoms. Spring sampling when the trees are still symptomless failed to detect the infections. Conversely, on the same trees *X. fastidiosa* was readily detected in mature shoots collected throughout the whole year, even when the trees were in dormancy. MLST analysis disclosed that the sequence type of the strain infecting the Apulian cherry trees was ST53, the same as that infecting olives and several other hosts found in the same infected area (Loconsole *et al.*, 2016).

Preliminary vector transmission experiments have confirmed that *Philaenus spumarius* is so far the only ascertained vector of *X. fastidiosa* in Europe. It is able to acquire the bacterium from infected cherry, and infective *P. spumarius* can transmit the bacterium to healthy cherry plants. Although cherry is not a major crop in the area in Italy where *Xylella* epidemics have occurred, the economic relevance of this crop in the bordering provinces, in other Italian regions and in south-eastern and south-western Europe make it of great concern. *X. fastidiosa* could be a serious threat for the cherry industry, especially in the warmer Mediterranean Basin, where the environmental conditions are favourable to bacterial outbreaks (Bosso *et al.*, 2016).

There is no cure for *Xylella* infections, nor is there any known method to reduce bacterial inoculum in the field. Thus, controlling *X. fastidiosa*-induced diseases relies on prevention (i.e. implementation of quarantine measures, roughing infected plants in the new outbreak), use of resistant cultivars, management practices, and chemical and biological control of the vectors. However,

**Fig. 15.5.** Leaf scorching on mature leaves and initial symptoms (apical necrosis) on young leaves of a sweet cherry tree infected by *Xylella fastidiosa*.

considerable research efforts are under way for the development of novel strategies to control *X. fastidiosa* (Retchless *et al.*, 2014) aiming either at the pathogen (Chatterjee *et al.*, 2008) or at the vector such as RNAi to impact insect development (Rosa *et al.*, 2012), or at both partners and their interactions (Killiny *et al.*, 2012). Among the most promising approaches, the diffusible signal factor (DSF) signalling system of *X. fastidiosa* has been used as the basis for developing a confusion strategy, whereby the presence of DSF in the environment (i.e. the xylem stream) at all times should limit bacterial movement and plant colonization. If constitutively expressed in transgenic plants, DSF molecules should limit the growth of *X. fastidiosa* colonies and bacterial movement within plants, which, in turn, would affect the expression of disease symptoms.

A different approach relies on the inhibitory effect of *N*-acetylcysteine (Muranaka *et al.*, 2013), which, by breaking the disulfide bridges of the bacterial biofilm, exerts a sort of mucolytic action that facilitates water uptake with a symptom remission effect. *N*-Acetylcysteine adsorbed in organic fertilizer is an economically promising alternative worth more extensive studies aimed at controlling other *X. fastidiosa* strains.

### 15.5.2 Fire blight

*Erwinia amylovora* (Burrill) causes fire blight, a disease of 200 plant species within the family Rosaceae. This disease is considered the most serious bacterial disease of apples and pears all over the world (van der Zwet and Keil, 1979). In recent years, isolation of the pathogen from the dried spots of plum and apricot in the USA (Mohan, 2007), and natural infection of European plum in Germany (Vanneste *et al.*, 2002a) and apricot in Czech Republic (Korba and Sillerova, 2010) have been reported.

During 2011–2014, severe symptoms, similar to those of fire blight on apples and pears, were observed on young shoots of plum (*Prunus domestica* L. 'd'Agen'; Végh *et al.*, 2012), cherry plum (*Prunus cerasifera* 'Nigra'; Végh and Palkovics, 2014) and apricot (*Prunus armeniaca* 10/13 hybrid; Végh and Palkovics, 2013, 2014) trees in Hungary. The naturally infected shoots showed typical symptoms of fire blight, including terminal shoots with brown to black necrotic lesions. Similar symptoms develop on stone fruits during infection by *Monilinia laxa*. On the basis of the symptoms, colony type, biochemical tests, pathogenicity and 16S rRNA gene sequence homology, the pathogen was identified as *E. amylovora*. Although the occurrence of *E. amylovora* on stone fruits is not common, the observed symptoms and severity suggest that this disease could also cause a serious problem in cherry production in the future.

Diagnostic protocols for *E. amylovora* can differ between the methods based in laboratories. For reliable and rapid identification of the pathogen, immunofluorescence methods (Paulin, 1981) and PCR, real-time PCR and LAMP protocols using primers complementary to chromosomal or plasmid DNA have been developed (Kałużna *et al.*, 2013). As is the case for most bacterial diseases, control includes cultural practices such as sanitation of trees, obtained by prompt pruning to remove the parts of trees with symptoms. Pruning tools should be disinfected regularly during the pruning process and removed tree parts burnt. The number of chemicals of value for fire blight control is very limited; they belong to four categories: copper-containing compounds, antibiotics, growth regulators and elicitors. However, antibiotic use is restricted in several countries, and resistance has been reported. In biological plant protection, the most frequently used organisms are antagonistic bacteria, fungi and bacteriophages such as *Pantoea agglomerans*, *Pseudomonas fluorescens* and *Bacillus subtilis* (Johnson and Stockwell, 2000; Mercier and Lindow, 2001; Vanneste *et al.*, 2002b; Böszörményi *et al.*, 2009). Some yeasts (*Aureobasidium pullulans*, *Candida sake* and *Metschnikowia pulcherrima*) have proven effective against the pathogen *in vitro* and in field conditions (Seibold *et al.*, 2006). Bacteriophages can also play a role in the biological control of fire blight (Gill *et al.*, 2003; Müller *et al.*, 2010; Dömötör *et al.*, 2012).

Several studies on fire blight susceptibility of species, seedlings, cultivars and rootstocks have been carried out to identify resistant cultivars or sources of fire blight resistance; these sources of resistance are being used by breeding programmes in several countries for apple, pear and ornamentals.

After the identification of fire blight in *Prunus* spp. in Hungary, artificial infections were performed on cherry and sour cherry blossoms and fruit (sweet cherries 'Carmen', 'Linda 156' and 'Katalin 261'; sour cherries 'Kántorjánosi 3', 'Újfehértói Fürtös', 'Érdi Bőtermő', 'Debreceni Bőtermő' and 'Érdi Jubileum'). All inoculated blossoms of the cherry cultivars became infected and showed disease symptoms. The less susceptible cultivars were 'Érdi Bőtermő' and 'Érdi Jubileum' (A. Végh, unpublished data). In another study, sour cherry 'Northstar' and sweet cherry 'Lambert' and 'Lapins' showed higher susceptibility than other tested cultivars. The sour cherry 'Montmorency' did not develop any symptoms. The less susceptible sweet cherries cultivars were 'Bing' and 'Royal Ann' (Mohan and Bijman, 1999; Mohan *et al*., 2002).

## References


Abdel-Megeed, A. (2013) Controlling of *Pseudomonas syringae* by nanoparticles produced by *Streptomyces bikiniensis*. *Journal of Pure and Applied Microbiology* 7, 1121–1129.

Agrios, G.N. (2005) Plant diseases caused by prokaryotes: bacteria and Mollicutes. In: Agrios, G.N. (ed.) *Plant Pathology*, 5th edn. Elsevier Academic Press, Boca Raton, Florida, pp. 616–703.

Ait Tayeb, L.A., Ageron, E., Grimont, F. and Grimont, P.A.D. (2005) Molecular phylogeny of the genus *Pseudomonas* based on *rpoB* sequences and application for the identification of isolates. *Research in Microbiology* 156, 763–773.

Ballard, E.L., Dietzgen, R.G., Sly, L.I., Gouk, C., Horlock, C. and Fegan, M. (2011) Development of a Bio-PCR protocol for the detection of *Xanthomonas arboricola* pv. *pruni*. *Plant Disease* 95, 1109–1115.

Bereswill, S., Bugert, P., Volksch, B., Ullrich, M., Bender, C.L. and Geider, K. (1994) Identification and relatedness of coronatine-producing *Pseudomonas syringae* pathovars by PCR analysis and sequence determination of the amplification products. *Applied and Environmental Microbiology* 60, 2924–2930.

Berge, O., Monteil, C.L., Bartoli, C., Chandeysson, C., Guilbaud, C., Sands, D.C. and Morris, C.E. (2014) A user's guide to a data base of the diversity of *Pseudomonas syringae* and its application to classifying strains in this phylogenetic complex. *PLoS One* 9, e105547.

Bergsma-Vlami, M., Martin, W., Koenraadt, H., Teunissen, H., Pothier, J.F., Duffy, B. and van Doorn, J. (2012) Molecular typing of Dutch isolates of *Xanthomonas arboricola* pv. *pruni* isolated from ornamental cherry laurel. *Journal of Plant Pathology* 94 (Suppl. 1), 29–35.

Bliss, F.A., Schuerman, P.L., Almehdi, A.A., Dandekar, A.M. and Bellaloui, N. (1999) Crown gall resistance in accessions of 20 *Prunus* species. *HortScience* 34, 326–330.

Borschinger, B., Bartoli, C., Chandeysson, C., Guilbaud, C., Parisi, L., Bourgeay, J.F., Buisson, E. and Morris, C.E. (2016) A set of PCRs for rapid identification and characterization of *Pseudomonas syringae* phylogroups. *Journal of Applied Microbiology* 120, 714–723.

Bosso, L., Di Febbraro, M., Cristinzio, G., Zoina, A. and Russo, D. (2016) Shedding light on the effects of climate change on the potential distribution of *Xylella fastidiosa* in the Mediterranean basin. *Biological Invasions* 18, 1759–1768.

Böszörményi, E., Érsek, T., Fodor, A., Fodor, A.M., Földes, L.S., Hevesi, M., Hogan, J.S., Katona, Z., Klein, M.G., Kormány, A., Pekár, S., Szentirmai, A., Sztaricskai, F. and Taylor, R.A.J. (2009) Isolation and activity of *Xenorhabdus* antimicrobial compounds against the plant pathogens *Erwinia amylovora* and *Phytophthora nicotianae*. *Journal of Applied Microbiology* 107, 746–759.

Boudon, S., Manceau, C. and Notteghem, J.-L. (2005) Structure and origin of *Xanthomonas arboricola* pv. *pruni* populations causing bacterial spot of stone fruit trees in Western Europe. *Phytopathology* 95, 1081–1088.

Bühlmann, A., Pothier, J.F., Tomlinson, J.A., Frey, J.E., Boonham, N., Smits, T.H.M. and Duffy, B. (2013) Genomics-informed design of loop-mediated isothermal amplification for detection of phytopathogenic *Xanthomonas arboricola* pv. *pruni* at the intraspecific level. *Plant Pathology* 62, 475–484.

Bultreys, A. and Gheysen, I. (1999) Biological and molecular detection of toxic lipodepsipeptide-producing *Pseudomonas syringae* strains and PCR identification in plants. *Applied and Environmental Microbiology* 65, 1904–1909.

Bultreys, A., Gheysen, I. and de Hoffmann, E. (2006) Yersiniabactin production by *Pseudomonas syringae* and *Escherichia coli*, and description of a second yersiniabactin locus evolutionary group. *Applied and Environmental Microbiology* 72, 3814–3825.

Cesbron, S., Pothier, J., Gironde, S., Jacques, M.-A. and Manceau, C. (2014) Development of multilocus variable-number tandem repeat analysis (MLVA) for *Xanthomonas arboricola* pathovars. *Journal of Microbiological Methods* 100, 84–90.

Cesbron, S., Briand, M., Essakhi, S., Gironde, S., Boureau, T., Manceau, C., Fischer-Le Saux, M. and Jacques, M.-A. (2015) Comparative genomics of pathogenic and nonpathogenic strains of *Xanthomonas arboricola* unveil molecular and evolutionary events linked to pathoadaptation. *Frontiers in Plant Science* 6, 1126.

Chatterjee, S., Almeida, R.P.P. and Lindow, S. (2008) Living in two worlds: the plant and insect lifestyles of *Xylella fastidiosa*. *Annual Review of Phytopathology* 46, 243–271.

Coletta-Filho, H.D., Francisco, C.S., Lopes, J.R.S., de Oliveira, A.F. and da Silva, L.F.O. (2016) First report of olive leaf scorch in Brazil, associated with *Xylella fastidiosa* subsp. *pauca*. *Phytopathologia Mediterranea* 55, 130–135.

Costechareyre, D., Rhouma, A., Lavire, C., Portier, P., Chapulliot, D., Bertolla, F., Boubaker, A., Dessaux, Y. and Nesme, X. (2010) Rapid and efficient identification of *Agrobacterium* species by *recA* allele analysis: *Agrobacterium* recA diversity. *Microbial Ecology* 60, 862–872.

Cubero, J., Lastra, B., Salcedo, C.I., Piquer, J. and Lopez, M.M. (2006) Systemic movement of *Agrobacterium tumefaciens* in several plant species. *Journal of Applied Microbiology* 101, 412–421.

de Cleene, M. and de Ley, J. (1976) The host range of crown gall. *Botanical Review* 42, 389–466.

Dessaux, Y., Petit, A., Farrand, S.K. and Murphy, P.J. (1998) Opines and opine-like molecules involved in plant–*Rhizobiaceae* interactions. In: Spaink, H.P., Kondorosi, A. and Hooykaas, P.J.J. (eds) *The Rhizobiaceae, Molecular Biology of Model Plant-associated Bacteria*. Kluwer Academic Publishers, Dordrecht, The Netherlands, pp. 173–197.

Dhanvantari, B.N. (1978) Characterization of *Agrobacterium* isolates from stone fruits in Ontario. *Canadian Journal of Botany* 56, 2309–2311.

Dömötör, D., Becságh, P., Rákhely, G., Schneider, G. and Kovács, T. (2012) Complete genomic sequence of *Erwinia amylovora* phage PhiEaH2. *Journal of Virology* 86, 10899.

EFSA (2016) Update of a database of host plants of *Xylella fastidiosa*: 20 November 2015. *EFSA Journal* 14, 4378.

EPPO (2006) EPPO standards PM 7/64. Diagnostics *Xanthomonas arboricola* pv. *pruni*. *EPPO Bulletin* 36, 129–133.

Ercisli, S. (2013) Cherry genetic resources of Turkey. In: Proceedings of the COST Action FA1104: Sustainable production of high-quality cherries for the European market. COST Meeting, Pitesti, Romania, October 2013. Available at: https://www.bordeaux.inra.fr/cherry/docs/dossiers/Activities/Meetings/15-17%2010%202013_3rd%20MC%20and%20all%20WG%20Meeting_Pitesti/Presentations/05-Ercisli_Sezai.pdf (accessed 2 May 2016).

European Commission (2017) Commission database of host plants found to be susceptible to *Xylella fastidiosa* in the Union territory. Update 8. Available at: https://ec.europa.eu/food/sites/food/files/plant/docs/ph_biosec_legis_emergency_db-host-plants_update08.pdf (accessed 23 January 2017).

Fischer-Le Saux, M., Bonneau, S., Essakhi, S., Manceau, C. and Jacques, M.-A. (2015) Aggressive emerging pathovars of *Xanthomonas arboricola* represent widespread epidemic clones that are distinct from poorly pathogenic strains, as revealed by multilocus sequence typing. *Applied and Environmental Microbiology* 81, 4651–4668.

Freigoun, S.O. and Crosse, J.E. (1975) Host relations and distribution of a physiological and pathological variant of *Pseudomonas morsprunorum*. *Annals of Applied Biology* 81, 317–330.

Garcin, A., Rouzet, J. and Notteghem, J.-L. (2005) *Xanthomonas des Arbres Fruitiers à Noyau*. Centre Technique Interprofessionnel des Fruits et Légumes (CITFL), Paris, France.

Gardan, L., Shafik, H., Belouin, S., Broch, R., Grimont, F. and Grimont, P.A.D. (1999) DNA relatedness among the pathovars of *Pseudomonas syringae* and description of *Pseudomonas tremae* sp. nov. and *Pseudomonas cannabina* sp. nov. (ex Sutic and Dowson 1959). *International Journal of Systematic Bacteriology* 49, 469–478.

Garita-Cambronero, J., Sena-Vélez, M., Palacio-Bielsa, A. and Cubero, J. (2014) Draft genome sequence of *Xanthomonas arboricola* pv. *pruni* strain Xap33, causal agent of bacterial spot disease on almond. *Genome Announcements* 2, e00440-14.

Garita-Cambronero, J., Palacio-Bielsa, A., López, M.M. and Cubero, J. (2016) Draft genome sequence for virulent and avirulent strains of *Xanthomonas arboricola* isolated from *Prunus* spp. in Spain. *Standards in Genomic Sciences* 11, 1–10.

Garrett, C.M.E. (1987) The effect of crown gall on growth of cherry trees. *Plant Pathology* 36, 339–345.

Gill, J.J., Svircev, A.M., Smith, R. and Castle, A.J. (2003) Bacteriophages of *Erwinia amylovora*. *Applied and Environmental Microbiology* 69, 2133–2138.

Grgurina, I., Gross, D.C., Iacobellis, N.S., Lavermicocca, P., Takemoto, J.Y. and Benincasa, M. (1996) Phytotoxin production by *Pseudomonas syringae* pv. *syringae*: syringopeptin production by *syr* mutants defective in biosynthesis or secretion of syringomycin. *FEMS Microbiology Letters* 138, 35–39.

Haelterman, R.M., Tolocka, P.A., Roca, M.E., Guzman, F.A., Fernandez, F.D. and Otero, M.L. (2015) First presumptive diagnosis of *Xylella fastidiosa* causing olive scorch in Argentina. *Journal of Plant Pathology* 97, 393.

Hajri, A., Pothier, J.F., Fischer-Le Saux, M., Bonneau, S., Poussier, S., Boureau, T., Duffy, B. and Manceau, C. (2012) Type three effector gene distribution and sequence analysis provide new insights into the pathogenicity of plant-pathogenic *Xanthomonas arboricola*. *Applied and Environmental Microbiology* 78, 371–384.

Hernandez-Martinez, R., de la Cerda, K.A., Costa, H.S., Cockney, D.A. and Wong, F.P. (2007) Phylogenetic relationships of *Xylella fastidiosa* strains isolated from landscape ornamentals in Southern California. *Phytopathology* 97, 857–864.

Hirano, S.S., Maher, E.A., Kelman, A. and Upper, C.A. (1978) Ice nucleation activity of fluorescent plant pathogenic Pseudomonas. In: Station de Pathologie Végétale et Phytobactériologie INRA (ed.) *Proceedings of the 4th International Conference of Plant Pathogenic Bacteria*, 27 August–2 September, Angers, France, pp. 717–724.

Hołubowicz, T., Ugolik, M. and Weber, Z. (1988) Effect of crown gall on the growth and yield of cherry trees. In: Lipecki, J. (ed.) *International Symposium on Intensification of Production of Cherries and Plum*, Lublin, Poland, pp. 97–103.

Hu, F., Young, J. and Fletcher, M. (1998) Preliminary description of biocidal (syringomycin) activity in fluorescent plant pathogenic *Pseudomonas* species. *Journal of Applied Microbiology* 85, 365–371.

Ibarra Caballero, J., Zerillo, M.M., Snelling, J., Boucher, C. and Tisserat, N. (2013) Genome sequence of *Xanthomonas arboricola* pv. *corylina*, isolated from Turkish filbert in Colorado. *Genome Announcements* 1, e00246-13.

Ignatov, A.N., Kyrova, E.I., Vinogradova, S.V., Kamionskaya, A.M., Schaad, N.W. and Luster, D.G. (2015) Draft genome sequence of *Xanthomonas arboricola* strain 3004, a causal agent of bacterial disease on barley. *Genome Announcements* 3, e01572-14.

Jami, F., Kazempour, M.N., Elahinia, S.A. and Khodakaramian, G. (2005) First report of *Xanthomonas arboricola* pv. *pruni* on stone fruit trees from Iran. *Journal of Phytopathology* 153, 371–372.

Janse, J.D. and Obradović, A. (2010) *Xylella fastidiosa* – its biology, diagnosis, control and risks. *Journal of Plant Pathology* 92 (Suppl. 1), 35–48.

Janse, J.D., Rossi, M.P., Gorkink, R.F.J., Derks, J.H.J., Swings, J., Janssens, D. and Scortichini, M. (2001) Bacterial leaf blight of strawberry (*Fragaria (×) ananassa*) caused by a pathovar of *Xanthomonas arboricola*, not similar to *Xanthomonas fragariae* Kennedy & King. Description of the causal organism as *Xanthomonas arboricola* pv. *fragariae* (pv. nov., comb. nov.). *Plant Pathology* 50, 653–665.

Johnson, K.B. and Stockwell, V.O. (2000) Biological control of fire blight. In: Vanneste, J.L. (ed.) *Fire Blight: the Disease and its Causative Agent,* Erwinia amylovora. CABI Publishing, Wallingford, UK, pp. 235–251.

Kałużna, M. and Sobiczewski, P. (2009) Virulence of *Pseudomonas syringae* pathovars and races originating from stone fruit trees. *Phytopathologia* 54, 71–79.

Kałużna, M., Ferrante, P., Sobiczewski, P. and Scortichini, M. (2010a) Characterization and genetic diversity of *Pseudomonas syringae* from stone fruits and hazelnut using repetitive-PCR and MLST. *Journal of Plant Pathology* 92, 781–787.

Kałużna, M., Puławska, J. and Sobiczewski, P. (2010b) The use of PCR melting profile for typing of *Pseudomonas syringae* isolates from stone fruit trees. *European Journal of Plant Pathology* 126, 437–443.

Kałużna, M., Janse, J.D. and Young, J.M. (2012) Detection and identification methods and new tests as used and developed in the framework of COST 873 for bacteria pathogenic to stone fruits and nuts *Pseudomonas syringae* pathovars. *Journal of Plant Pathology* 94 (Suppl. 1), 117–126.

Kałużna, M., Puławska, J. and Mikiciński, A. (2013) Evaluation of methods for *Erwinia amylovora* detection. *Journal of Horticultural Research* 21, 65–71.

Kałużna, M., Albuquerque, P., Tavares, F., Sobiczewski, P. and Puławska, J. (2016a) Rapid and specific detection of *Pseudomonas syringae* pv. *morsprunorum* race 1 and 2, the causal agent of bacterial canker using classical and real-time PCRs. *Applied Microbiology and Biotechnology* 100, 3693–3711.

Kałużna, M., Willems, A., Pothier, J.F., Ruinelli, M., Sobiczewski, P. and Puławska, J. (2016b) *Pseudomonas cerasi* sp. nov. (non Griffin, 1911) isolated from diseased tissue of cherry. *Systematic and Applied Microbiology* 39, 370–377.

Kamiunten, H., Nakao, T. and Oshida, S. (2000) *Pseudomonas syringae* pv. *cerasicola*, pv. nov., the causal agent of bacterial gall of cherry tree. *Journal of General Plant Pathology* 66, 219–224.

Kerr, A., Manigault, P. and Tempe, J. (1977) Transfer of virulence *in vivo* and *in vitro* in *Agrobacterium*. *Nature* 265, 560–561.

Killiny, N., Rashed, A. and Almeida, R.P.P. (2012) Disrupting the transmission of a vector-borne plant pathogen. *Applied and Environmental Microbiology* 78, 638–643.

King, E.O., Raney, M.K. and Ward, D.E. (1954) Two simple media for the demonstration of pyocianin and fluorescin. *Journal of Laboratory and Clinical Medicine* 44, 301–307.

Korba, J. and Sillerova, S. (2010) First occurrence of fire blight infection on apricot (*Prunus armeniaca*) in Czech Republic. In*: 12th International Workshop on Fire Blight*, 16–20 August, Warsaw, Poland, p. 107.

Lamichhane, J.-R. (2014) *Xanthomonas arboricola* diseases of stone fruit, almond and walnut trees: progress toward understanding and management. *Plant Disease* 98, 1600–1610.

Lattore, B.A. and Jones, A.L. (1979) *Pseudomonas morsprunorum*, the cause of bacterial canker of sour cherry in Michigan and its epiphytic association with *P. syringae*. *Phytopathology* 69, 335–339.

Lavermicocca, P., Sante Iacobellis, N., Simmaco, M. and Graniti, A. (1997) Biological properties and spectrum of activity of *Pseudomonas syringae* pv. *syringae* toxins. *Physiological and Molecular Plant Pathology* 50, 129–140.

Lelliott, R.A. and Stead, D.E. (1987) Methods for the diagnosis of bacterial diseases of plants. In: Preece, T.F. (ed.) *Methods in Plant Pathology*, Vol. 2. Blackwell Scientific Publications, Oxford, pp. 37–131.

Lelliott, R.A., Billing, E. and Hayward, A.C. (1966) A determinative scheme for the fluorescent plant pathogenic pseudomonads. *Journal of Applied Bacteriology* 29, 470–489.

Loconsole, G., Saponari, M., Boscia, D., D'Attoma, G., Morelli, M., Martelli, G.P. and Almeida, R.P.P. (2016) Intercepted isolates of *Xylella fastidiosa* in Europe reveal novel genetic diversity. *European Journal of Plant Pathology* 146, 85–94.

López, M.M., Gorris, M.T. and Montojo, A.M. (1988) Opine utilization by Spanish isolates of *Agrobacterium tumefaciens*. *Plant Pathology* 37, 565–572.

Louws, F.J., Fulbright, D.W., Stephens, C.T. and de Bruijn, F.J. (1994) Specific genomic fingerprints of phytopathogenic *Xanthomonas* and *Pseudomonas* pathovars and strains generated with repetitive sequences and PCR. *Applied and Environmental Microbiology* 60, 2286–2295.

Lyons, N.F. and Taylor, J.D. (1990) Serological detection and identification of bacteria from plants by the conjugated *Staphylococcus aureus* slide agglutination test. *Plant Pathology* 39, 584–590.

Marchi, G., Cinelli, T. and Surico, G. (2011) Bacterial leaf spot caused by the quarantine pathogen *Xanthomonas arboricola* pv. *pruni* on cherry laurel in central Italy. *Plant Disease* 95, 74.

Martelli, G.P., Boscia, D., Porcelli, F. and Saponari, M. (2016) The olive quick decline syndrome in south-east Italy: a threatening phytosanitary emergency. *European Journal of Plant Pathology* 144, 235–243.

Melakeberhan, H., Jones, A.L., Sobiczewski, P. and Bird, G.W. (1993) Factors associated with the decline of sweet cherry trees in Michigan: nematodes, bacterial canker, nutrition, soil pH, and winter injury. *Plant Disease* 77, 266–271.

Ménard, M., Sutra, L., Luisetti, J., Prunier, J.P. and Gardan, L. (2003) *Pseudomonas syringae* pv. *avii* (pv. nov.), the causal agent of bacterial canker of wild cherries (*Prunus avium*) in France. *European Journal of Plant Pathology* 109, 565–576.

Mercier, J. and Lindow, S.E. (2001) Field performance of antagonistic bacteria identified in a novel laboratory assay for biological control of fire blight of pear. *Biological Control* 22, 66–71.

Mohan, S.K. (2007) Natural incidence of shoot blight in Pluot® caused by *Erwinia amylovora*. In: *11th International Workshop on Fire Blight*, 12–17 August, Portland, Oregon, p. 64.

Mohan, S.K. and Bijman, V.P. (1999) Susceptibility of *Prunus* species to *Erwinia amylovora*. *Acta Horticulturae* 489, 145–148.

Mohan, S.K., Bijman, V.P. and Fallahi, E. (2002) Field evaluation of *Prunus* species for susceptibility to *Erwinia amylovora* by artificial inoculation. *Acta Horticulturae* 590, 377–380.

Mondal, K.K. and Mani, C. (2012) Investigation of the antibacterial properties of nanocopper against *Xanthomonas axonopodis* pv. *punicae*, the incident of pomegranate bacterial blight. *Annals of Microbiology* 62, 889–893.

Moore, L.W. (1976) Prevention of crown gall on *Prunus* roots by bacterial antagonists. *Phytopathology* 67, 139–144.

Moore, L.W. and Allen, J. (1986) Controlled heating of root-pruned dormant *Prunus* spp. seedlings before transplanting to prevent crown gall. *Plant Disease* 70, 532–536.

Mougel, C., Cournoyer, B. and Nesme, X. (2001) Novel tellurite-amended media and specific chromosomal and Ti plasmid probes for direct analysis of soil populations of *Agrobacterium* biovars 1 and 2. *Applied and Environmental Microbiology* 67, 65–74.

Mousavi, S.A., Willems, A., Nesme, X., de Lajudie, P. and Lindström, K. (2015) Revised phylogeny of *Rhizobiaceae*: proposal of the delineation of *Pararhizobium* gen. nov., and 13 new species combinations. *Systematic and Applied Microbiology* 38, 84–90.

Müller, I., Jelkmann, W., Geider, K. and Lurz, R. (2010) Properties of *Erwinia amylovora* phages from North America and Germany and their possible use to control fire blight. *Acta Horticulturae* 896, 417–419.

Muranaka, L.S., Giorgiano, T.E., Takita, M.A., Forim, M.R., Silva, L.F.C., Coletta-Filho, H.D., Machado, M.A. and de Souza, A.A. (2013) *N*-Acetylcysteine in agriculture, a novel use for an old molecule: focus on controlling the plant–pathogen *Xylella fastidiosa*. *PLoS One* 8, e72937.

Nemeth, J., Emody, L. and Pacsa, S. (1987) The use of ELISA for the detection of *Pseudomonas syringae* pv. *phaseolicola* and *Xanthomonas campestris* pv. *phaseoli* in bean seed. In: Civerolo, E.L., Collmer, A., Davis, R.E. and Gillaspie, A.G. (eds) *Plant Pathogenic Bacteria*. Springer, The Netherlands, pp. 876–876.

Nunney, L., Vickerman, D.B., Bromley, R.E., Russell, S.A., Hartman, J.R., Morano, L.D. and Stouthamer, R. (2013) Recent evolutionary radiation and host plant specialization in the *Xylella fastidiosa* subspecies native to the United States. *Applied and Environmental Microbiology* 79, 2189–2200.

Ormeño-Orrillo, E. and Martínez-Romero, E. (2013) Phenotypic tests in *Rhizobium* species description: an opinion and (a sympatric speciation) hypothesis. *Systematic and Applied Microbiology* 36, 145–147.

Otten, L., Burr, T. and Szegedi, E. (2008) *Agrobacterium*: a disease-causing bacterium. In: Tzfira, T. and Citovsky, V. (eds) Agrobacterium*: From Biology to Biotechnology*. Springer, New York, pp. 1–46.

Ozaktan, H., Ertimurtaş, D. and Eğerci, K. (2013) The diagnosis and copper sensitivity of *Pseudomonas syringae* pv. *syringae* isolates from sweet cherry in Turkey. In: Proceedings of the COST Action FA1104: Sustainable production of high-quality cherries for the European market. COST Meeting, Pitesti, Romania, October 2013. Available at: https://www.bordeaux.inra.fr/cherry/page6-1-8.html (accessed 2 March 2016).

Palacio-Bielsa, A., Cambra, M.A. and López, M.M. (2009) PCR detection and identification of plant-pathogenic bacteria: updated review of protocols (1989–2007). *Journal of Plant Pathology* 91, 249–297.

Palacio-Bielsa, A., Cubero, J., Cambra, M.A., Collados, R., Berruete, I.M. and Lopez, M.M. (2011) Development of an efficient real-time quantitative PCR protocol for detection of *Xanthomonas arboricola* pv. *pruni* in *Prunus* species. *Applied and Environmental Microbiology* 77, 89–97.

Palacio-Bielsa, A., Pothier, J.F., Roselló, M., Duffy, B. and López, M.M. (2012) Detection and identification methods and new tests as developed and used in the framework of COST873 for bacteria pathogenic to stone fruits and nuts – *Xanthomonas arboricola* pv. *pruni*. *Journal of Plant Pathology* 94, 135–146.

Park, S., Lee, Y., Koh, Y., Hur, J.-S. and Jung, J. (2010) Detection of *Xanthomonas arboricola* pv. *pruni* by PCR using primers based on DNA sequences related to the *hrp* genes. *Journal of Microbiology* 48, 554–558.

Parkinson, N., Cowie, C., Heeney, J. and Stead, D. (2009) Phylogenetic structure of *Xanthomonas* determined by comparison of *gyrB* sequences. *International Journal of Systematic and Evolutionary Microbiology* 59, 264–274.

Parkinson, N., Bryant, R., Bew, J. and Elphinstone, J. (2011) Rapid phylogenetic identification of members of the *Pseudomonas syringae* species complex using the *rpoD* locus. *Plant Pathology* 60, 338–344.

Paulin, J.P. (1981) Overwintering of *Erwinia amylovora*: sources of inoculum in spring. *Acta Horticulturae* 117, 49–54.

Pereira, U.P., Gouran, H., Nascimento, R., Adaskaveg, J.E., Goulart, L.R. and Dandekar, A.M. (2015) Complete genome sequence of *Xanthomonas arboricola* pv. *jugliandis* 417, a copper-resistant strain isolated from *Juglans regia* L. *Genome Announcements* 3, e01126-15.

Pierronnet, A. and Salesses, G. (1996) Behaviour of *Prunus* cultivars and hybrids towards *Agrobacterium tumefaciens* estimated from hardwood cuttings. *Agronomie* 16, 247–256.

Pothier, J.F., Pagani, M.C., Pelludat, C., Ritchie, D.F. and Duffy, B. (2011a) A duplex-PCR method for species- and pathovar-level identification and detection of the quarantine plant pathogen *Xanthomonas arboricola* pv. *pruni*. *Journal of Microbiological Methods* 86, 16–24.

Pothier, J.F., Smits, T.H.M., Blom, J., Vorhölter, F., Goesmann, A., Pühler, A. and Duffy, B. (2011b) Complete genome sequence of the stone fruit pathogen *Xanthomonas arboricola* pv. *pruni*. *Phytopathology* 101, S144–S145.

Pothier, J.F., Vorhölter, F.-J., Blom, J., Goesmann, A., Pühler, A., Smits, T.H.M. and Duffy, B. (2011c) The ubiquitous plasmid pXap41 in the invasive phytopathogen *Xanthomonas arboricola* pv. *pruni*: complete sequence and comparative genomic analysis. *FEMS Microbiology Letters* 323, 52–60.

Pscheidt, J.W. and Ocamb, C.M. (2015) *Pacific Northwest Plant Disease Management Handbook* [online]. Oregon State University, Corvallis, Oregon. Available at: http://pnwhandbooks.org/plantdisease (accessed 29 January 2016).

Puławska, J. (2010) Crown gall of stone fruits and nuts, economic significance and diversity of its causal agents: tumorigenic *Agrobacterium* spp. *Journal of Plant Pathology* 92 (Suppl. 1), 87–98.

Puławska, J. and Kałużna, M. (2012) Phylogenetic relationship and genetic diversity of *Agrobacterium* spp. isolated in Poland based on *gyrB* gene sequence analysis and RAPD. *European Journal of Plant Pathology* 133, 379–390.

Puławska, J. and Sobiczewski, P. (2005) Development of a semi-nested PCR based method for sensitive detection of tumorigenic *Agrobacterium* in soil. *Journal of Applied Microbiology* 98, 710–721.

Puławska, J., Willems, A. and Sobiczewski, P. (2006) Rapid and specific identification of four *Agrobacterium* species and biovars using multiplex PCR. *Systematic and Applied Microbiology* 29, 470–479.

Purcell, A.H. (2013) Paradigms: examples from the bacterium *Xylella fastidiosa*. *Annual Review of Phytopathology* 51, 339–356.

Raio, A., Zoina, A. and Moore, L.W. (1997) The effect of solar heating of soil on natural and inoculated agrobacteria. *Plant Pathology* 46, 320–328.

Renick, L.J., Cogal, A.G. and Sundin, G.W. (2008) Phenotypic and genetic analysis of epiphytic *Pseudomonas syringae* populations from sweet cherry in Michigan. *Plant Disease* 92, 372–378.

Retchless, A.C., Labroussaa, F., Shapiro, L., Stenger, D.C., Lindow, S.E. and Almeida, R.P.P. (2014) Genomic insights into *Xylella fastidiosa* interactions with plant and insect hosts. In: Gross, D.C., Lichens-Park, A. and Kole, C. (eds) *Genomics of Plant-Associated Bacteria*. Springer, Berlin/Heidelberg, Germany, pp. 177–202.

Ritchie, D.F. (1995) Bacterial spot. In: Ogawa, J.M., Zehr, E.I., Bird, G.W., Ritchie, D.F., Uriu, K. and Uyemoto, J.K. (eds) *Compendium of Stone Fruit Diseases*. APS Press, St Paul, Minnesota, pp. 50–52.

Rosa, C., Kamita, S.G. and Falk, B.W. (2012) RNA interference is induced in the glassy winged sharpshooter *Homalodisca vitripennis* by actin dsRNA. *Pest Management Science* 68, 995–1002.

Rubio-Cabetas, M.-J., Minot, J.-C., Voisin, R. and Esmenjaud, D. (2001) Interaction of root-knot nematodes (RKN) and the bacterium *Agrobacterium tumefaciens* in roots of *Prunus cerasifera*: evidence of the protective effect of the Ma RKN resistance genes against expression of crown gall symptoms. *European Journal of Plant Pathology* 107, 433–441.

Ruinelli, M., Blom, J., Kałużna, M., Goesmann, A., Puławska, J., Duffy, B. and Pothier, J.F. (2015) Genomic investigation of cherry pathogenic *Pseudomonas syringae* pathovars. In: Joint Meeting with the 2nd International Workshop on Bacterial Diseases of Stone Fruits and Nuts, 21–24 April, Izmir, Turkey. Available at: https://www.bordeaux.inra.fr/cherry/docs/dossiers/Activities/Meetings/2015%2004%2021-24%20Meeting_Izmir/Presentations/Michela_Izmir2015.pdf (accessed 2 May 2016).

Saponari, M., Boscia, D., Loconsole, G., Palmisano, F., Savino, V.N., Potere, O. and Martelli, G.P. (2014) New hosts of *Xylella fastidiosa* strain CoDiRO in Apulia. *Journal of Plant Pathology* 96, 611.

Sarkar, S.F., Gordon, J.S., Martin, G.B. and Guttman, D.S. (2006) Comparative genomics of host-specific virulence in *Pseudomonas syringae*. *Genetics* 174, 1041–1056.

Scally, M., Schuenzel, E.L., Stouthamer, R. and Nunney, L. (2005) Multilocus sequence type system for the plant pathogen *Xylella fastidiosa* and relative contributions of recombination and point mutation to clonal diversity. *Applied and Environmental Microbiology* 71, 8491–8499.

Schaad, N.W., Postnikova, E., Lacy, G., Fatmi, M.B. and Chang, C.J. (2004) *Xylella fastidiosa* subspecies: *X. fastidiosa* subsp. [correction] *fastidiosa* [correction] subsp. nov, *X. fastidiosa* subsp. multiplex subsp. nov., and *X. fastidiosa* subsp. *pauca* subsp. nov. *Systematic Applied Microbiology* 27, 290–300 [Erratum: *Systematic Applied Microbiology* 27, 763].

Seibold, A., Giesen, N. and Jelkmann, W. (2006) Antagonistic activities of different yeast spp. against *Erwinia amylovora*. In: Zeller, W. and Ullrich, C. (eds) *Proceedings of the 1st International Symposium on Biological Control of Bacteria Plant Diseases*, No. 408. Arno Brynda, Berlin, Germany, pp. 254–257.

Slater, S.C., Goldman, B.S., Goodner, B., Setubal, J.C., Farrand, S.K., Nester, E.W., Burr, T.J., Banta, L., Dickerman, A.W., Paulsen, I., Otten, L., Suen, G., Welch, R., Almeida, N.F., Arnold, F., Burton, O.T., Du, Z., Ewing, A., Godsy, E., Heisel, S., Houmiel, K.L., Jhaveri, J., Lu, J., Miller, N.M., Norton, S., Chen, Q., Phoolcharoen, W., Ohlin, V., Ondrusek, D., Pride, N., Stricklin, S.L., Sun, J., Wheeler, C., Wilson, L., Zhu, H. and Wood, D.W. (2009) Genome sequences of three *Agrobacterium* biovars help elucidate the evolution of multichromosome genomes in bacteria. *Journal of Bacteriology* 191, 2501–2511.

Smith, E.F. (1903) Observations on a hither to unreported bacterial disease, the cause of which enters the plant through ordinary stomata. *Science* 17, 456–457.

Sobiczewski, P. (1996) Etiology of crown gall on fruit trees in Poland. *Journal of Fruit and Ornamental Plant Research* 4, 147–161.

Sobiczewski, P. and Jones, A.L. (1992) Effect of exposure to freezing temperatures on necrosis in sweet cherry shoots inoculated with *Pseudomonas syringae* pv. *syringae* or *P.s. morsprunorum*. *Plant Disease* 76, 447–451.

Sobiczewski, P., Karczewski, J. and Berczynski, S. (1991) Biological control of crown gall *Agrobacterium tumefaciens* in Poland. *Fruit Science Report* 18, 125–132.

Socquet-Juglard, D. (2012) Genetic determinism of *Xanthomonas arboricola* pv. *pruni* resistance and agronomic traits in apricot, and applications in marker-assisted selection. PhD dissertation, ETH Zurich, Zurich, Switzerland.

Sorensen, K.N., Kim, K.-H. and Takemoto, J.Y. (1998) PCR detection of cyclic lipodepsinonapeptide-producing *Pseudomonas syringae* pv. *syringae* and similarity of strains. *Applied and Environmental Microbiology* 64, 226–230.

Spotts, R.A. and Wallis, K. (2008) Evaluation of bactericides for control of bacterial canker of sweet cherry using detached leaves. Oregon State University, Hood River, Oregon. Available at: jhbiotech.com/docs/Study-Fosphite-Cherry.pdf (accessed 2 March 2016).

Stefani, E. (2010) Economic significance and control of bacterial spot/canker of stone fruits caused by *Xanthomonas arboricola* pv. *pruni*. *Journal of Plant Pathology* 92, 99–103.

Süle, S. (1978) Biotypes of *Agrobacterium tumefaciens* in Hungary. *Journal of Applied Bacteriology* 44, 207–213.

Tjou-Tam-Sin, N.N.A., van de Bilt, J.L.J., Bergsma-Vlami, M., Koenraadt, H., Westerhof, J., van Doorn, J., Pham, K.T.K. and Martin, W.S. (2012) First Report of *Xanthomonas arboricola* pv. *pruni* in ornamental *Prunus laurocerasus* in the Netherlands. *Plant Disease* 96, 759–759.

van der Zwet, T. and Keil, H.M. (1979) Fire blight – a bacterial disease of Rosaceous plants. In: *Agriculture Handbook 510*. US Department of Agriculture, Washington, DC, p. 200.

Vanneste, J.L., Lex, S., Vermeulen, M. and Berger, F. (2002a) Isolation of *Erwinia amylovora* from blighted plums (*Prunus domestica*) and potato roses (*Rosa rugosa*). *Acta Horticulturae* 590, 89–94.

Vanneste, J.L., Cornish, D.A., Yu, J. and Voyle, M.D. (2002b) A new biological control agent for control of fire blight which can be sprayed or distributed using honey bees. *Acta Horticulturae* 590, 231–235.

Vauterin, L., Hoste, B., Kersters, K. and Swings, J. (1995) Reclassification of *Xanthomonas*. *International Journal of Systematic Bacteriology* 45, 472–489.

Végh, A. and Palkovics, L. (2013) First occurrence of fire blight on apricot (*Prunus armeniaca*) in Hungary. *Notulae Botanicae Horti Agrobotanici Cluj Napoca* 41, 440–443.

Végh, A. and Palkovics, L. (2014) Kajszibarackon és cseresznyeszilván is megjelent a tűzelhalás betegség hazánkban. *Növényvédelem* 50, 319–324.

Végh, A., Némethy, Z., Hajagos, L. and Palkovics, L. (2012) First report of *Erwinia amylovora* causing fire blight on plum (*Prunus domestica*) in Hungary. *Plant Disease* 96, 759.

Vicente, J. and Roberts, S. (2007) Discrimination of *Pseudomonas syringae* isolates from sweet and wild cherry using rep-PCR. *European Journal of Plant Pathology* 117, 383–392.

Vicente, J.G., Roberts, S.J., Russell, K. and Alves, J.P. (2004) Identification and discrimination of *Pseudomonas syringae* isolates from wild cherry in England. *European Journal of Plant Pathology* 110, 337–351.

Weingart, H. and Völksch, B. (1997) Genetic fingerprinting of *Pseudomonas syringae* pathovars using ERIC-, REP-, and IS50-PCR. *Journal of Phytopathology* 145, 339–345.

Wells, J.M., Raju, B.C., Hung, H.Y., Weisburg, W.J., Mandelco, P.L. and Brenner, D.J. (1987) *Xylella fastidiosa* gen. nov., sp. nov.: Gram-negative, xylem-limited, fastidious plant bacteria related to *Xanthomonas* spp. *International Journal of Systematic Bacteriology* 37, 136–145.

Wormald, H. (1932) Bacterial diseases of stone fruit trees in Britain. IV. The organism causing bacterial canker of plum trees. *Transactions of the British Mycological Society* 17, 157–169.

Young, J.M. (2010) Taxonomy of *Pseudomonas syringae*. *Journal of Plant Pathology* 92, 5–14.

Young, J.M., Kuykendall, L.D., Martinez-Romero, E., Kerr, A. and Sawada, H. (2001) A revision of *Rhizobium* Frank 1889, with an amended description of the genus, and the inclusion of all species of *Agrobacterium* Conn 1942 and *Allorhizobium undicola* de Lajudie *et al.* 1998 as new combinations: *Rhizobium radiobacter*, *R. rhizogenes*, *R. rubi*, *R. undicola* and *R. vitis*. *International Journal of Systematic and Evolutionary Microbiology* 51, 89–103.

Young, J.M., Park, D.-C., Shearman, H.M. and Fargier, E. (2008) A multilocus sequence analysis of the genus *Xanthomonas*. *Systematic and Applied Microbiology* 31, 366–377.

Zamze, S.E., Smith, A.R.W. and Hignett, R.C. (1986) Composition of lipopolysaccharides from *Pseudomonas syringae* pv. *syringae* and a serological comparison with lipopolysaccharide from *P. syringae* pv. *morsprunorum*. *Journal of General Microbiology* 132, 3393–3401.

Zhu, J., Oger, P.M., Schrammeijer, B., Hooykaas, P.J., Farrand, S.K. and Winans, S.C. (2000) The bases of crown gall tumorigenesis. *Journal of Bacteriology* 182, 3885–3895.

# 16 Viruses, Viroids, Phytoplasmas and Genetic Disorders of Cherry

**Delano James,[1]* Mirosława Cieślińska,[2] Vicente Pallás,[3] Ricardo Flores,[3] Thierry Candresse[4] and Wilhelm Jelkmann[5]**

[1]*Sidney Laboratory – Centre for Plant Health, Canadian Food Inspection Agency, North Saanich, British Columbia, Canada;* [2]*Research Institute of Horticulture, Skierniewice, Poland;* [3]*Instituto de Biología Molecular y Celular de Plantas (UPV-CSIC), Universidad Politécnica de Valencia, Valencia, Spain;* [4]*Equipe de Virologie, UMR 1332 Biologie du Fruit et Pathologie, INRA et Université de Bordeaux, Villenave d'Ornon Cedex, France;* [5]*Julius Kuhn Institute, Institute for Plant Protection in Fruit Crops and Viticulture, Dossenheim, Germany*

## 16.1 Introduction

Cherries are infected by a range of viruses, viroids and phytoplasmas. Some of these cause severe diseases, having a significant impact on commercial cherry production. Viruses such as prune dwarf virus (PDV) and Prunus necrotic ringspot virus (PNRSV) can cause crop losses ranging from 18 to 30% (Cembali *et al.*, 2003). Severe damage by little cherry disease had devastating effects in British Columbia, Canada, over several decades, resulting in the initiation of major eradication programmes (Jelkmann and Eastwell, 2011). Cherry leaf roll virus (CLRV) infections can result in sour cherry crop losses as high as 91–98% (Büttner *et al.*, 2011). Phytoplasmas from a wide range of groups and subgroups infect cherry. Some of these, like the X-disease group and the aster yellows group, are associated with severe diseases that lead to tree decline and even death of the infected trees. Infections by some of these agents appear to cause no obvious symptoms and therefore no harm to the plants, as with viruses such as cherry virus A (CVA), some isolates of apple chlorotic leaf spot virus (ACLSV) or the viroids that infect cherry. It is possible, however, that some of these agents may have subtle effects that may affect plant performance and yield. Also, as a component of mixed infections, these mainly latent agents may elicit unexpected symptoms or enhance symptom expression. ACLSV is generally considered to infect cherry with no obvious symptoms; however in some cherry hosts, some isolates of ACLSV do cause symptoms that may have a significant impact on fruit quality (Fig. 16.1). Since disease development and expression are complex phenomena that are influenced by climatic conditions, pathogen isolate, pathogen complexes present in the plant, the cultivar of the host plant and the age of the host, it is advisable to screen for and avoid plant germplasm known to be infected with any of these pathogens. Some viruses and virus-like agents infecting cherry trees may

* delano.james@inspection.gc.ca

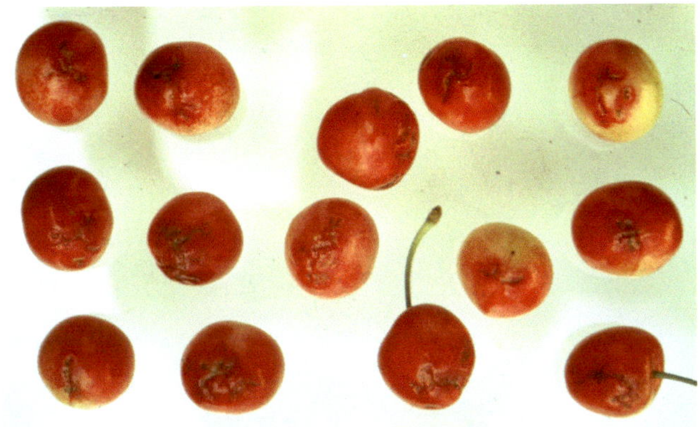

**Fig. 16.1.** Fruit necrosis symptoms in sweet cherry caused by a severe isolate of apple chlorotic leaf spot virus (ACLSV).

be transmitted by insects, pollen/seeds or nematodes, and in some cases there is evidence of transmission even if the mode is unknown. Phytoplasmas are transmitted by phloem-feeding insects such as leafhoppers and psyllids. This potential for transmission adds further challenges to the management and control of the agents and related diseases.

## 16.2 Viruses Spread by Cherry Pollen and/or Seeds

A significant number of plant viruses have evolved mechanisms that exploit the plants' own reproductive processes and can be transmitted by pollen and/or seed (Mink, 1993; Card *et al.*, 2007). Seed and pollen transmission are common mechanisms of spread for approximately 20% of plant viruses, which ensures their spread from generation to generation. Pollen-transmitted viruses are generally also transmitted by seed, but the reverse is not necessarily true. Virus-infected pollen can be a source of infection of other plants through the fertilized flower (horizontal transmission), or through the seed, which leads to infected seedlings (vertical transmission). Recently, it has been shown that vertical transmission selects for both reduced virulence and increased resistance in the host (Pagan *et al.*, 2014). Both means of transmission can sometimes be facilitated by thrips or other flower-visiting arthropods (Mink, 1993). Thus, the introduction of a poorly seed-transmitted virus into a new area can be catastrophic if an efficient biological facilitator is available.

### 16.2.1 Cherry leaf roll virus

Cherry leaf roll virus (CLRV) not only infects cherry but is common in many wild and cultivated woody and herbaceous plant species and causes diseases of economic importance. Foliar symptoms of CLRV on cherries include yellowing and chlorosis of leaves, leaf rolling and bunching of shoots. Plant death has been observed in England (reviewed by Büttner *et al.*, 2011). The disease caused by CLRV in sweet cherry (*Prunus avium*) was first described in Europe (Cropley, 1961) and later in the USA (Eastwell and Howell, 2010). Co-infection with PDV can cause the development of enations on the undersides of leaves, whereas co-infection with PNRSV can cause a rapid decline in both sour (*Prunus cerasus*) and sweet cherries. Unlike most other nepoviruses, CLRV is not transmitted by nematodes, but by seeds and pollen, as well as by grafting. CLRV is dispersed by human movement of infected seeds or plants.

CLRV belongs to the genus *Nepovirus*, family *Secoviridae* (Sanfaçon *et al.*, 2012). Members of this family have a bipartite genome of two single-stranded, positive-sense RNA molecules that are encapsidated separately in 28 nm isometric particles. The complete nucleotide sequence and genome organization of several CLRV isolates have been determined, including that of a cherry isolate (Eastwell *et al.*, 2012).

CLRV can be detected by biological assays, serological methods and molecular techniques (Büttner *et al.*, 2011). A restriction fragment length polymorphism (RFLP) assay was developed to differentiate CLRV isolates according to phylogenetic clades by examining restriction patterns from partial 3′-untranslated region genomic fragments (~420 bp) (Buchhop *et al.*, 2009). Although the distribution of the virus can be irregular, it can be detected in all parts of an infected plant, including flowers, young and old leaves, fruit, dormant wood and roots. For diagnostic purposes, optimal sampling is done in spring or early summer, but can vary from season to season, depending on the weather conditions (Rodoni *et al.*, 2011). It has been reported that CLRV is easily detectable throughout the year in the tissue of male inflorescences, leaf buds, leaves, single seeds and cortical tissues of young twigs (Werner *et al.*, 1997). The virus can also be detected in roots and meristems and within tubules in pollen, ovules and mature seeds (Jones, 1985).

As for most pollen-transmitted viruses, two main procedures to control CLRV are recommended: certification programmes that provide nurseries with virus-free material, and good orchard management practices with the aim of preventing or containing tree-to-tree spread. Good orchard management may include regular visual monitoring and removal of symptomatic trees. Vertical transmission of CLRV via seeds can be reduced by only using seeds collected from certified virus-free trees. EPPO (2006a) provides very good guidelines on the propagation of cultivars including seedling rootstocks.

### 16.2.2 Epirus cherry virus

There is no evidence to date that Epirus cherry virus (EpCV) is transmitted by cherry pollen or seed, but seed transmission was observed with EpCV-infected broad bean (*Vicia faba*) and *Nicotiana benthamiana* (Avgelis and Barba, 2011). It does therefore have the potential for seed transmission in other host species, hence its inclusion here. EpCV was first isolated from a sweet cherry from the Epirus region of Greece, showing symptoms that included very severe rasp leaf, stunting and deformed leaves of reduced size (Avgelis *et al.*, 1988; Avgelis and Barba, 2011). To date, the only other natural infection of the virus has been found in clover (*Trifolium resupinatum*) growing in the same orchard where the infected cherry tree was found (Avgelis and Barba, 2011).

EpCV is a species in the genus *Ourmiavirus*, with members that have unenveloped bacilliform virions (Rastgou *et al.*, 2012). The virus is easily mechanically transmitted via sap to a range of herbaceous plant species, and indexing on herbaceous indicators is recommended (Avgelis and Barba, 2011). Polyclonal antibodies have been developed against EpCV and have been used successfully for detecting the virus in gel double-diffusion tests and by immunosorbent electron microscopy (Avgelis *et al.*, 1988).

Since the virus was found naturally infecting clover and is easily mechanically transmitted to a range of herbaceous hosts, weed control might be an effective strategy for limiting any potential spread of this virus. The use of certified, pathogen-tested germplasm for propagation is recommended (Avgelis and Barba, 2011).

### 16.2.3 Prune dwarf virus

Prune dwarf virus (PDV) occurs mainly in sweet and sour cherry, where it causes chlorotic rings or mottle on young expanding leaves, and some shot holes. PDV is associated particularly with sour cherry yellows disease (Jones and Sutton, 1996). Fruit of affected cherry trees can be larger and firmer than fruit on healthy trees and can show chlorotic rings (Fig 16.2). Some PDV strains can form enations on the lower side of younger leaves (Çağlayan *et al.*, 2011). PDV is found worldwide in cherry. Remarkably, 38% of the cherry trees tested at the National Clonal Germplasm Repository in California were found to be positive for PDV (Osman *et al.*, 2012). PDV has been shown to be seed

Fig. 16.2. Chlorotic rings on sour cherry caused by prune dwarf virus (PDV).

transmitted, but more importantly, it is also pollen transmitted, which is influenced by all the factors that affect pollination (Çağlayan *et al.*, 2011). Transmission from cherry pollen to cucumber has been shown to be facilitated by thrips (Greber *et al.*, 1992). Moreover, seed transmission is very high in sweet cherry.

PDV belongs to the genus *Ilarvirus*, family *Bromoviridae*, and has the same tripartite genome organization and morphology as PNRSV, but has no serological relationships with this virus (see below and Pallás *et al.*, 2012). Similar to other ilarviruses, PDV is moderately immunogenic. Reagents and kits are commercially available for its detection. PDV isolates show wide serological variability. Although phylogenetic analysis of PDV coat protein gene sequences from Turkish isolates indicated the existence of cherry-specific phylogroups with four groups related to the host (Ulubas-Serçe *et al.*, 2009), the most recent phylogenetic analysis supported only two groups (Pallás *et al.*, 2012). Molecular hybridization, using single probes or polyprobes, and several different reverse transcriptase polymerase chain reactions (RT-PCRs) have been described for PDV detection (see James *et al.*, 2006, and references therein). A RT-PCR-based protocol for the multiplex detection of four viruses from sweet cherry, in which PDV and PNRSV were included, has recently been described (Zong *et al.*, 2014). *In situ* RT-PCR assays have demonstrated that, in addition to the leaves, PDV RNAs can be detected in flower buds, particularly in the generative and vegetative cells of pollen grains (Silva *et al.*, 2003).

Nurseries producing basic propagation material need to be separated from commercial orchards by an appropriate distance to prevent or limit contaminating pollen flow. Thermotherapy (24–32 days at 38°C) and/or apical meristem culture have been used to eliminate PDV (Diekmann and Putter, 1996). Compulsory sanitary controls during certification schemes should avoid any possible risk of reinfection while producing healthy propagated material.

### 16.2.4 Prunus necrotic ringspot virus

Prunus necrotic ringspot virus (PNRSV) naturally infects all *Prunus* spp., including both cultivated and wild species, but also hops and rose. PNRSV may cause a chlorotic to yellow line pattern mosaic and shot holes in leaves, a rugose mosaic in cherries, bud death in trees, and reduced bud take after grafting and delay of cherry fruit maturity (Hammond, 2011). In general, symptoms of PNRSV appear in the first year postinfection (acute or shock stage), and then the trees commonly become symptomless, although some strains cause recurrent symptoms annually. Infected sweet cherry trees produce smaller leaves with diffused chlorotic rings and/or spots, and necrotic lesions or spots, which give a tattered appearance to the blade upon detachment. The impact of PNSRV on sweet cherry trees has been estimated to be a 19% yield reduction (Albertini *et al.*, 1993). PNRSV is pollen and seed transmitted in cherry. The seed transmission rate of PNRSV in cherry is particularly high (up to 88.8%) (Kryczyński *et al.*, 1992). Fulton (1964) showed that PNRSV was seed transmitted in *Prunus pensylvanica* at a rate of between 60 and 70% after storage at 2°C for 4 years, and that this percentage dropped to below 5% after 6 years. Although the molecular mechanisms of pollen and seed transmission of PNRSV have not been addressed in cherry trees, studies in apricot have revealed that the virus invades pollen grains early, and infects the megaspore and generative cell of the bicellular pollen grains in nectarine (Aparicio *et al.*, 1999) and apricot (Amari *et al.*, 2007) trees. It was also shown that PNRSV drastically reduces pollen germination and delays pollen-tube growth. PNRSV infects all seed parts, including the embryo, at early developmental stages (Amari *et al.*, 2009). Cross-pollination with infected pollen to healthy plants revealed that PNRSV is not only able to infect seeds but also fruit. As indicated above for PDV, pollen transmission of PNRSV presents a serious risk for virus spread in an orchard, from infected trees to healthy trees via infected pollen.

PNRSV belongs to the genus *Ilarvirus*, family *Bromoviridae*, with members that are characterized by a tripartite genome and quasi-isometric particles (Pallás *et al.*, 2012). PNRSV infection requires the presence of a few molecules of the coat protein in the inoculum, a phenomenon known as genome activation (Pallás *et al.*, 2013). Sequence analysis of the RNA3 of seven PNRSV isolates from sweet cherry trees collected in the USA revealed single nucleotide and amino acid changes in movement protein and/or coat protein sequences, which correlated with serological relationships and pathotypes (Hammond and Crosslin, 1998). Phylogenetic analysis suggested clustering of the PNRSV variants into three groups named after the representative PNRSV isolates PV32, PV96 and PE5 (Pallás *et al.*, 2012, 2013). The majority of isolates from phylogroups PV96-II and PV32-I tend to exhibit latent/mild or chlorotic/necrotic symptoms, respectively.

Serology (Cambra *et al.*, 2011), molecular hybridization (Pallás *et al.*, 2011), RT-PCR (Hadidi *et al.*, 2011) and more recently DNA microarrays (Barba and Hadidi, 2011) have been used to detect PNRSV. Recent approaches have been described that allow the simultaneous detection of multiple plant viruses (multiplexing), which include PNRSV (reviewed by James *et al.*, 2006; Pallás *et al.*, 2009). RT-PCR-based methods have proven useful to differentiate closely related, but biologically distinct, cherry isolates of PNRSV (Hammond *et al.*, 1999).

The use of virus-tested certified plants in replant orchards or new orchards is the most important method for controlling PNRSV. Nurseries that produce basic propagation material need to be separated from commercial orchards by an adequate distance to prevent or limit contaminating pollen flow. Thermotherapy (24–32 days at 38°C) and/or apical meristem culture have been used to eliminate PNRSV (Manganaris *et al.*, 2003). Interestingly, cherry rootstocks have recently been engineered through RNA interference (RNAi) silencing to confer resistance to PNRSV (Song *et al.*, 2013). Inoculation of non-transgenic scions with PNRSV has revealed resistance of the scions grafted on the transgenic rootstocks (Zhao and Song, 2014).

## 16.3 Viruses Spread by Airborne Vectors

There are a number of viruses belonging to different taxonomic groups that are spread by airborne vectors such as aphids, thrips, mites and mealybugs. These viruses present a particular challenge for their control because if any infected plants are present in an orchard or if unidentified reservoirs of the virus exist in the field, these will act as potential sources for spread of the virus, if the appropriate vectors are present. Unidentified reservoirs may include symptomlessly infected cherry plants with low levels of infection that may still act as a virus source, cultivars or species that are latent hosts for the virus and never show any symptoms, and unknown wild woody hosts or weed species in the vicinity of cultivated cherry. The use of tested and virus-free germplasm for propagation is important. In some cases, control of the vector is also essential for disease control (see below).

### 16.3.1 Cherry mottle leaf virus

Cherry mottle leaf disease is caused by cherry mottle leaf virus (ChMLV). The symptoms may include irregular chlorotic mottling (Fig. 16.3), leaf distortion of terminal leaves, puckering, tattering, shot holes and a reduction of leaf size. Infected trees may appear stunted and display a rosette pattern due to

Fig. 16.3. Chlorotic mottling on 'Bing' cherry caused by cherry mottle leaf virus (ChMLV).

inhibition of terminal growth and shortened internodes. In some cultivars of cherry, both fruit quality and yield may be affected (Németh, 1986). Fruit from infected trees may be small and flavourless, with a delay of fruit ripening. Warm temperatures tend to suppress symptom development. Disease symptoms may differ depending on cultivar and region, but cherry mottle leaf disease is one of the most severe diseases of cherry in some regions (Cheney and Parish, 1976; Németh, 1986). ChMLV has a wide natural host range among *Prunus* spp. (Cheney and Parish, 1976). It is transmitted from woody plants to woody plants by budding and grafting, and by the bud/scale mite *Eriophyes inaequalis* (Oldfield, 1970). The virus is mechanically sap transmissible (James and Mukerji, 1993). The virus has been transmitted from infected herbaceous host *Chenopodium amaranticolor* plants to 'F 12/1' cherry rootstock by approach grafting (Li *et al.*, 1996). Li *et al.* (1996) report also the use of bark patches from infected 'F 12/1' to inoculate 'Bing' cherry trees.

ChMLV is a member of the genus *Trichovirus*, family *Betaflexiviridae* (Adams *et al.*, 2012). ChMLV is closely related to peach mosaic virus (PcMV) another member of the genus *Trichovirus*. Both are serologically related, with antibodies developed against either virus being able to detect the other virus (James, 2011a).

ChMLV may be detected using bioassays (herbaceous and woody indicator plants), by serological techniques such as dot-blot immunobinding assays, enzyme-linked immunosorbent assays (ELISAs) and Western blotting, and by nucleic acid-based techniques such as dot-blot hybridization and RT-PCR (James, 2011a). 'Bing' cherry is recommended as a woody indicator host for the detection of ChMLV (Németh, 1986). Symptoms on 'Bing' include irregular chlorotic mottling and leaf distortion, especially of the terminal leaves. The herbaceous hosts *C. amaranticolor* and *Chenopodium quinoa* are suitable indicator plants (James and Mukerji, 1993).

Control of the virus requires the use of virus-tested budwood, scion and rootstocks that are free of ChMLV. As the virus is mite transmitted and has a wide host range that includes *Prunus* spp. such as apricot and peach, care must be taken when cherry orchards are close to or surrounded by other *Prunus* orchards. ChMLV-positive plants in the field should be removed immediately, as the virus is transmitted efficiently by the mite vector, which lives and feeds on bitter cherry trees (*Prunus emarginata*) (Cheney and Parish, 1976), so removal of wild bitter cherry trees is advised to reduce mite populations. ChMLV elimination may be achieved by exposing the plants to 38°C for 40 days, excising 5 mm shoot tips and grafting these tips on to virus-free rootstocks (James, 2011a).

### 16.3.2 Little cherry virus 1 and little cherry virus 2

Little cherry was first reported in the early 1930s as an outbreak in different regions of British Columbia, Canada. It is a complex and serious viral disease occuring in both sweet and sour cherry. Symptoms on infected trees consist of small, angular and pointed fruit that do not fully ripen and are imperfectly coloured. Fruit have reduced sweetness and are unsuitable for commerce. In late summer and autumn, leaves show a characteristic red coloration or bronzing (Fig. 16.4). The intensity of symptoms is

**Fig. 16.4.** Little cherry disease: healthy 'Sam' (left) versus infected 'Sam' (right) showing red coloration of the leaves.

variety dependent, and infected trees are less vigorous. The disease has had a major economic impact on fruit production in Canada after its initial identification, and is distributed in cherry-production areas around the world (Jelkmann and Eastwell, 2011). There are two viruses related to the disease, little cherry virus 1 and 2 (LChV-1 and -2). Both have been identified using molecular methods (Rott and Jelkmann, 2005). While the primary causal agent of the disease is LChV-2, LChV-1 incites less severe symptoms. The viruses associated with little cherry disease seem to be limited to species within the genus *Prunus*. They occur in single and mixed infections. Sweet cherry is most affected by the disease, but symptoms are also visible on some sour cherry cultivars. Several ornamental *Prunus* spp. are often latently infected (Jelkmann and Eastwell, 2011). Symptomless infections with LChV-1 were found in plum, peach and almond (Matic *et al.*, 2007). Shirofugen stunting (Candresse *et al.*, 2013) and Kwanzan stunting (Matic *et al.*, 2009) have been associated with LChV-1 infection. LChV-1 and -2 were detected by RT-PCR in dodder (*Cuscuta europaea*) (Jelkmann and Eastwell, 2011). Successful artificial transmission of both viruses to *Nicotiana occidentalis* '37B' was achieved using dodder. LChV-2 is transmitted by at least two different species of mealybugs, the apple mealybug, *Phenacoccus aceris*, and the grape mealybug, *Pseudococcus maritimus* (Mekuria *et al.*, 2013). There is no known vector associated with LChV-1. Both viruses are readily bud and graft transmissible, which is probably the predominant mechanism for spread where vectors are absent or at low population density.

LChV-1 and LChV-2 are two distinct viruses in the family *Closteroviridae*. When molecular data for the mealybug-transmissible LChV-2 became available, it was assigned to the genus *Ampelovirus* (Rott and Jelkmann, 2005). Complete nucleotide sequences are available for both viruses (Jelkmann and Eastwell, 2011). LChV-1 has recently been assigned to the newly established genus *Velarivirus* (Martelli *et al.*, 2012). LChV-1 particles are long flexuous rods, approximately 1786–1820 nm in length. Estimates for LChV-2 particles showed dimensions of 11.2 × 1667 nm (Eastwell and Bernardy, 2001).

The classical method for the detection of little cherry disease in plant health and certification programmes is indexing with the woody indicators 'Sam' or 'Canindex I' (EPPO, 2001). While isolates of LChV-2 consistently cause pronounced leaf symptoms, LChV-1 infections display either weaker or no symptoms. Antisera have been produced from recombinant proteins for both viruses (Jelkmann and Eastwell, 2011), but serology has only been used for local surveys of LChV-2 in British Columbia, Canada. The most widely used detection method for both viruses is RT-PCR, with new primers/assays developed recently (Candresse *et al.*, 2013; Mekuria *et al.*, 2014; Katsiani *et al.*, 2015).

Control of little cherry disease is primarily dependent on the production and trade of healthy plant material in certification programmes. Valuable source plants of cherry varieties intended for fruit production should be tested rigorously for LChV-1 and -2. If source plants are infected, virus elimination can be accomplished by thermotherapy followed by meristem tip grafting. The technique may be combined with *in vitro* methods and/or chemical treatment (Panattoni *et al.*, 2013). Suitable recommendations for the production of propagation material are provided in certification schemes based on filiation for production of certified material (EPPO, 2001). If mealybug vectors are present, chemical control measures should be applied to avoid reinfection of nursery or field plants. A mealybug management programme might be necessary in orchards, depending on vector populations and disease pressure.

### 16.3.3 Plum pox virus

Plum pox disease or sharka is caused by plum pox virus (PPV), considered the most serious and important virus infecting stone fruit crops or *Prunus* spp. (Németh, 1986). There are nine strains of PPV (James *et al.*, 2013), but only two infect cherry naturally: strain Cherry (C; Nemchinov *et al.*, 1998) and strain Cherry Russian (CR; Chirkov *et al.*, 2013; Glasa *et al.*, 2013). Strain C isolates of PPV naturally infect both sweet and sour cherry cultivars. Symptoms on sweet cherry may include diffuse branch necrosis, leaf deformity, chlorotic and necrotic ring spots or notches on fruit, and fruit drop (Nemchinov *et al.*, 1998). Symptoms on sour cherry may include typical chlorotic ringspot symptoms on leaves, and depressions, necrosis and rings on fruit, with the rings on the fruit disappearing gradually during ripening (Nemchinov *et al.*, 1998). Strain CR isolates naturally infect sour cherry cultivars (Chirkov *et al.*, 2013; Glasa *et al.*, 2013) with symptoms that include leaf discoloration along veins, light green bands, leaf distortions, and pale green spots and rings typical of PPV infections. PPV is spread when infected material is used for budding or grafting. The virus is transmitted in a non-persistent manner by various species of aphids (Németh, 1986; Barba *et al.*, 2011).

PPV is a filamentous virus with a single-stranded RNA genome consisting of approximately 9.8–10 $\times 10^3$ nt, with a virus-encoded protein (VPg) covalently linked to the 5′ terminus and a poly(A) tail at the 3′ terminus of the genome (Barba *et al.*, 2011). The PPV genome contains a major open reading frame (ORF) encoding a polyprotein of approximately 355 kDa (Garcia *et al.*, 1994). A second ORF (named PIPO, for Pretty Interesting *Potyviridae* ORF), now known to be common among members of the genus *Potyvirus*, encodes a smaller protein that is expressed as a fusion protein (Chung *et al.*, 2008). The virus is a member of the genus *Potyvirus*, family *Potyviridae*.

PPV is a regulated and/or quarantine pest in all countries where *Prunus* spp. are cultivated, often requiring routine screening for the virus in nursery-propagated or imported germplasm. The virus may be detected using biological assays, serological tests or nucleic acid-based tests (Barba *et al.*, 2011). The International Standards for Phytosanitary Measures (ISPM) Diagnostic Protocol Annex #2: *Plum pox virus* describes a range of validated techniques recommended for PPV detection and identification (IPPC, 2012). The various PPV strains differ in their geographical distribution, host range, transmissibility and pathogenicity. Strain identification is therefore valuable in developing effective strategies for management and control of the virus.

Control is best achieved by using PPV-free plants. Both the rootstock and scion components of the tree should be tested and be PPV free, as the virus is graft transmitted. Once established in an orchard, the virus is difficult to eradicate, unless detected early, due to non-persistent aphid transmission. Depending on the age of the orchard, the susceptibility of the cultivars, the number of plants infected and the management objective, it may be necessary to remove individual infected trees or entire blocks of trees. In some situations, the planting of sentinel trees (susceptible varieties) have been advocated for early detection of infection if the virus is known to be present in the area. For further details on control, see Barba *et al.* (2011).

## 16.4 Viruses Spread by Soil/Soil-borne Vectors

Viruses spread by soil or soil-borne vectors present a special challenge for their control. Soil-borne virus vectors such as nematodes may retain the infectious virus or remain viruliferous for several years, even in the absence of any host plants (Bitterlin and Gonsalves, 1987). This means therefore that tree removal alone will not control infection in an orchard where the virus and the appropriate vectors are present; also care should be taken in the movement of trees from nematode-infested and diseased orchards (Welsh, 1976). Due to the wide host range of nematodes and/or nematode-borne viruses, particular attention should be paid to crop rotation and to the plant species present in infected orchards (Németh, 1986). Where and if still possible, soil fumigation may help in some cases (Welsh, 1976).

### 16.4.1 Cherry rasp leaf virus

Cherry rasp leaf virus (CRLV) is the causal agent of cherry rasp leaf disease (also known as leaf enation, ruffled leaf and cockscomb), which was first described in Colorado, USA, in 1935; the association of a virus with the disease was made in 1942 (Bodine and Newton, 1942). Symptoms on susceptible cherry cultivars such as 'Bing' include typical rasp-like leafy outgrowths or raised protuberances (Fig. 16.5) between the lateral veins and along the midrib, on the dorsal surface of the leaf (Németh, 1986). Spurs and branches on the lower part of the tree may die, making the tree appear open and bare, with consequent reductions in crop yield. Since the virus is nematode transmitted, initial infection may occur in the roots with symptoms limited to the lower sections of the plant at the early stages of infection. CRLV has a wide host range that includes sweet cherry (*P. avium*), sour cherry (*P. cerasus*), cherry rootstock (*Prunus mahaleb*) and also peach (*Prunus persica*) and apple (*Malus domestica*). CRLV is transmitted by grafting or budding and by the nematode *Xiphinema americanum* (Németh, 1986). Other *Xiphinema* spp. may transmit CRLV including *Xiphinema californicum* and *Xiphinema rivesi* (Brown *et al.*, 1994).

CRLV is the type member of the genus *Cheravirus*, family *Secoviridae* (Sanfaçon *et al.*, 2012). The virus genome is bipartite, consisting of RNA1 and RNA2 that are 6992–7034 nt and 3274–3315 nt, respectively (James, 2011b).

CRLV can be detected using woody or herbaceous indicator plants. 'Bing' (*P. avium*) is a good indicator plant for CRLV (Németh,

Fig. 16.5. Rasp-like outgrowths on leaves of 'Bing' cherry infected with cherry rasp leaf virus (CRLV).

1986). The symptoms induced on 'Bing' are very typical for CRLV and include the distinctive rasp-like enations (outgrowths or raised protuberances between the lateral veins and along the midribs) on the dorsal surface of the leaves. As suitable indicators for CRLV detection, Japanese flowering cherry 'Shirofugen' (*Prunus serrulata*) (Parish, 1977) and a number of herbaceous plant species (Stace-Smith and Hansen, 1976) have been recommended, but herbaceous indexing is of limited sensitivity. Serological assays used to detect the virus include ELISAs, gel diffusion and Western blotting, (James, 2011b). RT-PCRs have been developed for the detection of CRLV targeting both RNA1 and RNA2 of the virus genome (James, 2011b). RT-PCR targeting RNA1 may be more reliable for broad-spectrum detection of CRLV, as RNA2 appears to be more variable among CRLV isolates.

The use of virus-free germplasm is essential for controlling the spread of CRLV. Soil fumigation is required to eliminate viruliferous nematodes (Németh, 1986). The wide host range of the virus, which includes commercially cultivated cherry, apple and potato, means that care should be taken in crop rotation (to avoid inadvertent replication and maintenance of the virus) and in the proximity of cultivated crops to potential reservoirs. Wild herbaceous species such as *Taraxacum officinale* (dandelion), *Balsamorhiza* spp. (balsam root) and *Plantago* spp. (plantain) may serve as symptomless hosts for the virus, and removal of these species from orchards and surrounding areas is advised. Thermotherapy is an effective treatment for the elimination of CRLV from infected apple (James, 2011b). Treatment of apple plants at 37–38°C for 75 days and then excising 5 mm shoots tips from treated plants and cleft grafting the excised shoot tips on to apple seedlings produced virus-free plants.

### 16.4.2 Tomato ringspot virus

Tomato ringspot virus (ToRSV) has a very wide host range that includes pome fruits, stone fruits, small fruits, vegetables and ornamental species (Németh, 1986; Sanfaçon and Fuchs, 2011). The virus infects a wide range of *Prunus* spp. including cherries, peaches and plums, and is associated with cherry Eola rasp leaf and *Prunus* stem pitting disease of cherry (Németh, 1986; Gonsalves, 1995). Symptoms associated with Eola rasp leaf disease include infected trees gradually becoming bare, starting from the lower branches, and die back of some spurs, shoots and branches. Chlorotic spots may appear on the lower surface of the leaves during the shock phase of early infection. On the lower surface of the leaves, enations may also be observed that are similar to but smaller than those associated with cherry rasp leaf disease (Németh, 1986). In the case of the *Prunus* stem pitting disease of sweet and sour cherry, the symptoms are similar to those observed in peach and may include leaves that appear droopy, turn yellow or red prematurely, and absice prematurely (Gonsalves, 1995). 'Bing' cherry may show severe pitting, while in some other varieties such as 'Royal Ann', no symptoms may be observed (Gonsalves, 1995). ToRSV can be transmitted by grafting, budding, the dagger nematode complex *X. americanum sensu lato* and by mechanical sap inoculation (Németh, 1986; Sanfaçon and Fuchs, 2011).

The virions are isometric and the virus has a bipartite single-stranded RNA genome with a genomic organization typical for members of the genus *Nepovirus*, family *Secoviridae* (Sanfaçon *et al.*, 2012).

There are several woody indicator plants that can be used in screening for ToRSV including: *P. persica* 'Lovell', 'Elberta' or 'GF305'; *Prunus tomentosa* IR 473/1 or IR 474/1, and a wide range of herbaceous indicator plants (Németh, 1986; Sanfaçon and Fuchs, 2011). A number of antibodies against ToRSV have been developed; however, none will detect all isolates of the virus due to its variability (Sanfaçon and Fuchs, 2011). Double-antibody sandwich ELISA was used effectively for the detection of ToRSV in crude plant extracts (Powell *et al.*, 1991). RT-PCR, when suitable primers are used, is another reliable assay for the detection of ToRSV (Griesbach, 1995; Rowhani *et al.*, 1998). The use of immunocapture RT-PCR simplifies sample preparation and is suitable for large-scale surveys (Rowhani *et al.*, 1998).

As with all nematode-transmitted viruses, the use of certified virus-free germplasm for cultivation and propagation is recommended to prevent movement and avoid introducing the virus in areas where it is absent and/or where the nematode vectors might be present. Soil fumigation can be used to eliminate viruliferous nematodes (Németh, 1986). The use of virus-resistant and nematode-resistant rootstocks is also recommended (Sanfaçon and Fuchs, 2011). 'Mariana 2624' (*Prunus cerasifera* × *Prunus munsoniana*) appears to be highly resistant to ToRSV infection (Hoy and Mircetich, 1984; Kommineni *et al.*, 1998).

## 16.5 Viruses with No Known Vector

This section describes several economically important viruses with no known vectors identified, but sometimes with disease distribution patterns that may suggest the existence of a vector, as in the case of cherry twisted leaf disease (see below). Identification and removal of plants in the orchard infected with any of these viruses is often an effective method for control, increasing the chances for eradication in these situations. The use of virus-free germplasm is always a preferred option, as symptoms may not become evident for some time, and planting of infected material is the main pathway for introduction or spread.

### 16.5.1 American plum line pattern virus

American plum line pattern virus (APLPV) is an ilarvirus associated with plum line pattern disease (Németh, 1986; EPPO, 2006b; Myrta *et al.*, 2011a). Cherry is also a host for the virus, with symptoms on sweet cherry (*P. avium*) that include oak-leaf patterns and yellow or white lines on the leaves (Myrta *et al.*, 2011a). In oriental flowering cherry (*P. serrulata*), discoloured areas may appear in shades of white, yellow or pink, sometimes with large rings but often with oak leaf-type patterns (Németh, 1986; Myrta *et al.*, 2011a). In addition to the species described above, other natural hosts for APLPV include several species of *Prunus* (Németh, 1986; Myrta *et al.*, 2011a). Several herbaceous species have also been identified as natural hosts, such as *Chenopodium* spp., *Crotalaria* spp., *Cucumis* spp. and *Cucurbita* spp. (Myrta *et al.*, 2011a). The virus can be sap transmitted, and is also spread by budding and grafting (Németh, 1986). There is no evidence of vector transmission.

APLPV is a spherical virus belonging to the genus *Ilarvirus*, family *Bromoviridae* (Bujarski *et al.*, 2012). The virus has a tripartite genome consisting of RNA1 (encoding the methyltransferase and helicase protein), RNA2 (encoding the polymerase) and RNA3, which encodes the movement protein and coat protein, with the coat protein translated from a subgenomic fragment derived from RNA3 (Bujarski *et al.*, 2012).

Woody indicator hosts for APLPV include 'GF305' peach seedlings maintained at 20°C for 3 months, or 'Shirofugen' flowering cherry, which may require monitoring for up to 2 years (EPPO, 2006b). The virus can be mechanically sap transmitted to a wide range of herbaceous indicator hosts (EPPO, 2006b). ELISAs can be used for the detection of APLPV (Al Rwahnih *et al.*, 2004a), and more sensitive molecular assays including dot-blot hybridization and RT-PCR have also been developed (Myrta *et al.*, 2011a). March to May are the best months for the detection of the virus in Japanese plum leaves and the virus may be detected in dormant wood from December to February. The virus can be detected in leaves, flowers, fruit and cortical tissue, with leaves being better for detection than flowers in spring, while mature fruit are better than leaves in summer (Al Rwahnih *et al.*, 2004a).

For control, it is recommended that planting material free of the virus is used (Myrta *et al.*, 2011a). Myrta *et al.* (2011a) indicated that it is doubtful whether roguing of infected trees should be recommended; however, since APLPV is an ilarvirus and some ilarvirus species are mechanically transmitted by thrips feeding on pollen grains that contain the virus (Bujarski *et al.*, 2012), it may indeed be prudent to remove any diseased trees observed in an orchard. There is no evidence that APLPV is present in or spread by pollen.

### 16.5.2 Cherry green ring mottle virus

Cherry green ring mottle virus (CGRMV) is the causal agent of cherry green ring mottle disease, which affects mainly sour cherry (*P. cerasus*) cultivars such as 'Montmorency' and English 'Morello' and some sour and sweet cherry hybrids (*P. cerasus* × *P. avium*) (Parker *et al.*, 1976; Németh, 1986). Infected 'Montmorency' shows characteristic symptoms that may include yellow and green mottling of mature leaves that may drop soon after symptoms appear, irregular necrotic spots, and asymmetric distortion of the leaf blade resulting from a developmental disorder of the leaf tissue along the midrib and major lateral veins and constricting chlorosis (Parker *et al.*, 1976; Németh, 1986). The affected fruit tends to be misshapen, bitter, off-flavour and not marketable (Jelkmann *et al.*, 2011). Some strains of CGRMV may cause subepidermal net-like patterns of necrosis in fruit (Parker *et al.*, 1976). The natural host range of CGRMV includes: sour cherry (*P. cerasus*), sweet cherry (*P. avium*), sour and sweet cherry hybrids (*P. cerasus* × *P. avium*), Mahaleb cherry (*P. mahaleb*), oriental flowering cherry (*P. serrulata*), peach (*P. persica*) and apricot (*Prunus armeniaca*) (Parker *et al.*, 1976; Németh, 1986). CGRMV causes roughening of the bark of the oriental flowering cherry 'Kwanzan' (Chamberlain *et al.*, 1971) and has been reported to be associated with cherry rough bark disease (Parker *et al.*, 1976). Sweet cherry is a symptomless host of CGRMV, although natural infections are common among cultivars such as 'Black Republican', 'Bing', 'Deacon', 'Lambert' and 'Napoleon' (Parker *et al.*, 1976). The only known means of spread of CGRMV is by grafting and budding (Parker *et al.*, 1976; Németh, 1986).

CGRMV is a filamentous virus with a positive-sense, single-stranded RNA genome of 8372 nt, excluding a poly(A) tail at the 3′ terminus (Jelkmann *et al.*, 2011). The virus was originally classified as an unassigned species in the family *Betaflexiviridae* (Zhang *et al.*, 1998; Adams *et al.*, 2012). There was a proposal to create a new genus *Robigovirus* in the family *Betaflexiviridae* that would include CGRMV (Villamor et al., 2015), and this has now been adopted (ICTV, 2015).

Flowering cherry (*P. serrulata*) 'Kwanzan' and 'Shirofugen' are recommended as woody indicator plants for the detection of CGRMV (Németh, 1986; Jelkmann *et al.*, 2011). Symptoms develop usually 2–3 months after grafting, if the virus is present. There may be different strains of CGRMV, or different conditions of CGRMV infection that may on 'Kwanzan' cause cherry symptoms that are mild (leaves almost normal but distorted and curled), moderate (leaves smaller than normal, distorted and curled; shortened internodes with stunting) or severe (very distorted leaves that may fall prematurely leaving partially bare stems; pronounced stunting with very short internodes; dieback of some shoots; dark brown and severely roughened and cracked barks) (Chamberlain *et al.*, 1971). CGRMV has been detected by ELISA and Western blotting (Zhang *et al.*, 1998). Several RT-PCR assays have also been developed for detection of the virus (Jelkmann *et al.*, 2011).

Control of CGRMV is achieved using virus-free rootstock and scion germplasm. Infected trees should be removed from the field as soon as possible (EPPO, 2001), as there is the possibility of transmission by root grafting (Parker *et al.*, 1976). Since the virus is not seed transmitted, seedling rootstocks may be safe to use (Németh, 1986). Thermotherapy that includes treatment at 38°C for 6 weeks may be effective for the elimination of CGRMV (Ramsdell, 1995).

### 16.5.3 Cherry necrotic rusty mottle virus and cherry rusty mottle virus

Cherry necrotic rusty mottle (CNRM) and cherry rusty mottle (CRM) were first reported as graft-transmissible diseases in North America. Symptoms appear on leaves and are generally highly variable, depending on climatic conditions, virus isolates and cherry varieties. For CNRM, sensitive sweet cherry varieties show irregularly distributed brown angular necrotic spots on leaves, which turn to shot-holes later in the season. CRM has been described as either an 'American' or a 'European' type. Generally observed symptoms are light green or yellow mottling of basal leaves that become bright yellow, brown or red during the season. Early leaf drop occurs with severe forms of the disease, which can also result in tree decline and dieback. Other syndromes such as Lambert mottle, bark blister and Frogmore virus canker have been described (Rott and Jelkmann, 2011). Sweet cherry (*P. avium*) is the main natural host of the viruses associated with the diseases. The viruses can readily be transmitted by grafting and budding, which is believed to be the major means of spread for propagation material produced and traded outside of certification programmes.

Complete nucleotide sequences have been obtained for cherry necrotic rusty mottle virus (CNRMV) (Rott and Jelkmann, 2001a) and cherry rusty mottle associated virus (CRMaV) (Rott *et al.*, 2004). The two viruses share approximately 70% identity and are closely related to cherry green ring mottle virus (CGRMV). These viruses have been assigned to the genus *Robigovirus* in the family *Betaflexiviridae* (ICTV, 2015). The sequence of another virus occurring in sweet cherry and associated with cherry twisted leaf disease was published recently and designated cherry twisted-leaf associated virus (CTLaV; see below) (James *et al.*, 2014). With an expanding number of partial and full genomic information available for these viruses, the new genus *Robigovirus* was proposed based on analysis of the replicase and coat protein genes as well as the whole genome (Villamor *et al.*, 2015).

RT-PCR-based detection tests for CNRMV and CRMV have been developed (Rott and Jelkmann, 2001b,c). Primers have been developed in attempts towards broad-spectrum detection of closely related viruses including CNRMV (Foissac *et al.*, 2005), and specific primers have been used to detect strains of CNRMV (Li and Mock, 2005). Primers ERMUP/LO detected CRMV in six of seven European isolates, but could also detect several CNRMV and CGRMV isolates. In the same study, primers NEG1U/L detected CRMV isolates (Rott and Jelkmann, 2001b). The most common indicator plants are *P. avium* 'Sam' for detection of CNRMV and 'Bing' and Mazzard 'F 12/1' for detection of CRMV. These indicators were recommended as the preferred detection method in certification systems (EPPO, 2001).

As with other viruses in deciduous fruit crops for which no natural vectors are known, use of virus-free propagation material and prompt removal of infected trees are the most effective means of limiting spread (EPPO, 2001). If valuable germplasm is infected, virus elimination from cherry varieties and rootstocks can be accomplished by thermotherapy followed by meristem tip grafting. The technique may be combined with *in vitro* methods and/or chemical treatment (Panattoni *et al.*, 2013).

### 16.5.4 Cherry twisted leaf associated virus

There is strong evidence that cherry twisted leaf associated virus (CTLaV) may be the causal agent of cherry twisted leaf disease (ChTL) (James, 2011c; James *et al.*, 2014). The disease was first described in British Columbia, Canada, in 1943 (Lott, 1943), and association with a virus was assumed based on the symptoms observed and the graft/bud transmission of the causal agent (Lott, 1943). Symptoms of ChTL on 'Bing' cherry include abrupt bending of the midrib of the leaf, which results in typical twisting of the leaf. Distortion with downwards and side curling of the leaves may be observed, and parts of the midrib and lateral veins may become necrotic (Hansen and Cheney, 1976). Trees may appear stunted, and shortened internodes may cause the spurs to appear bunchy. The fruit may appear deformed, and pedicel necrosis may be observed in some ChTL-affected trees. A number of cherry cultivars are symptomless hosts and some *Prunus* spp. are immune (Hansen and Cheney, 1976). Natural hosts for CTLaV are sweet cherry (*P. avium*) and western chokecherry (*Prunus virginiana* var. *demissa*) (Hansen and Cheney, 1976; Németh, 1986); however, some isolates of the virus may be associated with apricot ring pox disease (Hansen and Cheney, 1976), and hence apricot (*P. armeniaca*) may be a natural host for CTLaV. The virus is transmitted from woody plant to woody plant by budding and grafting (Hansen and Cheney, 1976; Németh, 1986), while Keane and May (1963) showed that the virus may be transmitted by root grafting with an incubation period of 1–2 years before symptoms appear. No vector has been identified for CTLaV, but its pattern of distribution suggests a vectored mode of transmission.

CTLaV is a filamentous virus with a single-stranded RNA genome consisting of 8431 nt, excluding a poly(A) tail at the 3′ terminus of the genome (James *et al.*, 2014). CTLaV has been assigned to the genus *Robigovirus* in the family *Betaflexiviridae* (ICTV, 2015), together with CGRMV, CNRMV and CNMV.

The standard method for the detection of CTLaV is by indexing using 'Bing' cherry (Németh, 1986) in which the virus induces the typical and distinctive symptoms of twisted leaves. James *et al.* (2014) described an RT-PCR assay that was used for the detection of some CTLaV isolates. The oligonucleotide primers CTLV3-F1/CTLV-3R target a 559 bp region extending from the C terminus of the triple gene block protein 3 to the N terminus of the coat protein. The primers detected the CTLaV isolates but also amplified the corresponding region of the closely related CGRMV. Isolates of CGRMV may be distinguished by sequencing the amplicon and comparing the sequence to a known CTLaV isolate, which gives approximately 85–99% similarity for CTLaV isolates, but only 66–67% similarity for CGRMV.

Control of CTLaV is best achieved by using virus-free scion and rootstock germplasm. Since several cherry cultivars are symptomless hosts for the virus, screening of all cherry cultivars should be carried out either by woody indexing using 'Bing' cherry or by RT-PCR. Removal of wild chokecherry trees near orchards will help eliminate potential reservoirs of the virus.

### 16.5.5 Carnation Italian ringspot virus/Petunia asteroid mosaic virus/tomato bushy stunt virus

Carnation Italian ringspot virus (CIRV), Petunia asteroid mosaic virus (PAMV) and tomato bushy stunt virus (TBSV) have been associated with cherry destructive (detrimental) canker disease (CDC) (Jelkmann, 2011). Later detailed studies seem to indicate that isolates of PAMV may have been misidentified as TBSV, and that PAMV may more probably be the cause of the disease. Lesemann *et al.* (1989) described the detection of CIRV in a sweet cherry tree that showed symptoms similar to CDC except that shoot necrosis was limited to the shoot tips, and the severe stunting observed with PAMV infection was not associated with CIRV infection. Both PAMV and CIRV may cause CDC-like symptoms, and may very well be synergistic if mixed infections occur. Symptoms of CDC include sharp twisting of leaf blades resulting from the necrosis of the midrib and main veins, one-sided shoot necrosis that causes bending of the shoot, shoot dieback with replacement by lateral buds resulting in a zig-zag growth of shoots, brittle shoots and branches, and bark cankers with a strong flow of gum (Németh, 1986). Flower pedicles tend to be short and twisted because of necrosis, and extensive loss of flowers may result in poor fruit setting. Fruit symptoms vary with cultivars, but the most common symptoms include malformed fruit with sunken circular pits with necrotic flesh under the pits (Németh, 1986). The stones from affected fruit are often spotted and malformed, with seeds that are aborted. The disease can be transmitted from woody host to woody host by budding and grafting (Németh, 1986). CIRV, PAMV and TBSV are members of the genus *Tombusvirus* (the type species is *Tomato bushy stunt virus*), family *Tombusviridae* (Rochon *et al.*, 2012). Virions of members of the genus are icosahedral in shape and possess positive-sense, single-stranded RNA genomes of approximately 4.7–4.8 kb.

Sweet cherry (*P. avium*) 'Lambert', 'Sam' and 'Van' are recommended as woody indicator plants for the detection of CDC (Németh, 1986; EPPO, 2001). A 2-year observation period is required and typical symptoms as described above are observed if the agent(s) is present. A number of herbaceous indicator plants have been identified (EPPO, 2001;

Jelkmann, 2011). Various serological assays have been developed for the detection of CIRV, PAMV and TBSV including agar gel double-diffusion tests, ELISAs and immunosorbent electron microscopy (Lesemann *et al.*, 1989).

For control, it is recommended that healthy or virus-free propagation material (scion and rootstock) is used (Németh, 1986; EPPO, 2001). Trees in the orchard may be monitored regularly and symptomatic trees removed to eliminate the risk of root grafting (Jelkmann, 2011).

## 16.6 Viruses that Infect Cherry with No Obvious Related Symptoms

Several viruses, mostly from the family *Betaflexiviridae*, infect sweet or sour cherry without apparently causing any significant symptoms. The clearest case is cherry virus A (CVA), a virus found frequently in cherry. It has not been associated with any obvious symptoms in cherry. There is no known vector of CVA and, given the absence of symptoms, its impact is likely to be negligible. A second but more complex situation concerns apple chlorotic leaf spot virus (ACLSV), a virus that infects a wide range of hosts in the sub-families Maloideae and Prunoideae, including cherries. ACLSV has a very broad genetic diversity, with marked differences in pathogenicity among isolates. While infection is frequently symptomless, severe symptoms, in particular on fruit, have been associated with some ACLSV isolates (Fig. 16.1). In the absence of obvious symptoms, ACLSV has been considered to have the potential to enhance the symptoms caused by other viral infections or the impact of abiotic stress. Like CVA, ACLSV has no known vector, but is widely distributed in cherry trees worldwide. Given the severity of the symptoms caused by some isolates, it is generally included in certification schemes. The third situation concerns viruses less prevalent in cherry and/or of more recent discovery and for which the ability to cause symptoms in cherry remains to be determined.

### 16.6.1 Apple chlorotic leaf spot virus

Apple chlorotic leaf spot virus (ACLSV) was first described in the USA from apple trees, after transmission on the *Malus platycarpa* indicator (Mink and Shay, 1959), and the name of the virus finds its origin in the symptoms observed in apple indicators. Subsequently, it was gradually realized that ACLSV is very widespread in both pome (e.g. apple, pear, quince) and stone (e.g. peach, apricot, almond, plum, Japanese plum, cherry) fruits, as well as in some ornamental rosaceous species (Myrta *et al.*, 2011b; Katsiani *et al.*, 2014). ACLSV is distributed worldwide and is probably present wherever susceptible fruit tree species are grown (Myrta *et al.*, 2011b). There is no known vector of ACLSV and no clear evidence suggesting that it might be transmitted by vectors in the field (Myrta *et al.*, 2011b). Infection by ACLSV is often symptomless, but some isolates may cause a range of severe symptoms and sometimes symptoms reminiscent of PPV (sharka) symptoms (Myrta *et al.*, 2011b). In cherry, latent infection is generally observed, but some isolates may cause severe leaf deformations and chlorotic arabesques along the veins, fruit necrosis (Fig. 16.1), bark splitting and decline (Németh, 1986; Myrta *et al.*, 2011b). The possibility that ACLSV infection may act synergistically and increase the severity of the symptoms caused by other co-infecting viruses or may even render the trees more susceptible to abiotic stress has also been considered (Desvignes, 1999).

ACLSV has flexuous filamentous particles that led to its initial assignment as a member of the genus *Closterovirus*. Analysis of its genome sequence and organization (German *et al.*, 1990) resulted in its classification as the type member of the new genus *Trichovirus*, family *Betaflexiviridae* (Adams *et al.*, 2012). The virus has a single-stranded, positive-sense RNA genome of approximately 7.5 kb that contains three overlapping ORFs (German *et al.*, 1990; Myrta *et al.*, 2011b). Complete or partial sequences of a wide range of ACLSV isolates have shown that it is a virus with very large sequence variability and significant biological variability (Al Rwahnih *et al.*, 2004b; Myrta *et al.*, 2011b). Genomic sequences have been obtained from isolates responsible for plum pseudopox (Jelkmann, 1996) and for strong deforming mosaics in peach and cherry fruits (German *et al.*, 1997).

A wide range of techniques is available for the detection of ACLSV (Myrta *et al.*, 2011b). These include biological indexing by grafting on woody indicators, serological detection using polyclonal or monoclonal antibodies and commercial ELISA kits, and molecular techniques such as molecular hybridization and RT-PCR. A number of primer pairs for ACLSV detection have been published; however, given the high variability of the virus, it is likely that some of these primer pairs may lack polyvalence and may fail to detect some ACLSV isolates. The primers described by Candresse *et al.* (1995) and Menzel *et al.* (2002) have proven to have a relatively broad spectrum of detectability. ACLSV can be efficiently detected by polyvalent degenerate oligonucleotides (PDO) RT-PCR (Foissac *et al.*, 2005). Direct comparison of various techniques has illustrated the sensitivity of RT-PCR assays compared with other approaches (Spiegel *et al.*, 2006), which is of paramount importance for a virus with fluctuating and sometimes low titres in infected *Prunus* spp. (Myrta *et al.*, 2011b). Multiplexed detection of ACLSV together with other common fruit tree-infecting viruses and viroids is also possible using RT-PCR or molecular hybridization with polyprobes (Menzel *et al.*, 2003; Herranz *et al.*, 2005).

Given the absence of vectored transmission, the easiest and most efficient strategy for control of ACLSV is the production and use of certified 'virus-free' or 'virus-tested' planting materials (Myrta *et al.*, 2011b). The wide availability of efficient detection methods makes this approach relatively direct, and ACLSV detection is frequently integrated within certification schemes. ACLSV can be eliminated from Prunoideae or Maloideae propagation materials using a range of techniques including thermotherapy, chemotherapy or meristem tip grafting, alone or in combination (Myrta *et al.*, 2011b).

### 16.6.2 Other viruses

There are viruses detected in cherry that appear to have no significant effect on the health status of sweet and sour cherry trees. Included among them are the following.

Cherry virus A (CVA) was discovered inadvertently in Germany from a sweet cherry tree also infected with LChV-1 (Jelkmann, 1995). Since this initial characterization, it has been reported in cherry material in countries in Europe, North America and Asia, and is likely to be present wherever sweet and sour cherries are grown (Marais *et al.*, 2011). In addition to cherry, it has been found infecting, although at a much lower prevalence, several other *Prunus* spp. including apricot and peach (Marais *et al.*, 2011). CVA has never been convincingly associated with any type of symptoms in cherry (Marais *et al.*, 2011, 2012). It has not been reported to contribute to synergistic effects in cases of mixed infection. CVA can be detected by serology and molecular hybridization (Jelkmann, 1995; Marais *et al.*, 2011, 2012). Primers developed recently, with a much broader detection spectrum, may be more reliable (Marais *et al.*, 2011, 2012; Zong *et al.*, 2015). CVA can also be detected efficiently using the broad-spectrum PDO RT-PCR technique developed by Foissac *et al.* (2005).

Another such virus is cucumber mosaic virus (CMV), which is one of the viruses with the broadest host range, encompassing more than 1000 species of monocots and dicots (Scholthof *et al.*, 2011). It has been reported occasionally from sweet cherry (Tan *et al.*, 2010) and from flowering cherry (Kishi *et al.*, 1973). In flowering cherry ('Someiyoshino') it did not cause specific symptoms (Kishi *et al.*, 1973). In sweet cherry, although some chlorotic mottling and deformation of leaves were observed, CMV was only detected in a portion (33%) of the symptomatic samples, suggesting that it was not associated with the symptoms (Tan *et al.*, 2010).

Plum bark necrosis stem pitting associated virus (PBNSPaV) was described initially in California on Japanese plum (*Prunus salicina* 'Black Beaut') with symptoms of bark necrosis and stem pitting (Uyemoto and

Teviotdale, 1996; Boscia *et al.*, 2011). Further investigations showed it to have a relatively wide host range among *Prunus* spp., including cherry, peach, plum, almond and apricot (Boscia *et al.*, 2011). In cherry, PBNSPaV has been associated with symptoms of stem pitting, thickened corky bark, trunk deformations and bark cracking and chlorosis or chlorotic spotting of leaves (Boscia *et al.*, 2011). Artificial inoculation of a Japanese plum isolate resulted in the development of symptoms in several *Prunus* hosts including 'Colt' cherry but similarly inoculated 'Bing', Mazzard and 'Shirofugen' flowering cherry remained symptomless (Marini *et al.*, 2002). The associations of PBNSPaV with stem pitting and other bark disorders in cherry remain tentative and further investigations are needed. For detection, RT-PCR primers have been developed that should cover the full spectrum of PBNSPaV variability (Marais *et al.*, 2014). Serological reagents are also available, but probably afford only limited sensitivity (Boscia *et al.*, 2011).

Prunus virus T (PrVT) was recently identified in a sweet cherry tree ('Tardiva di Roccamonfina') from Italy using a metagenomics approach (Marais *et al.*, 2015). In further screening of a wide collection of *Prunus* spp., it was detected in a plum tree (*Prunus domestica*) and in a myrobalan plum tree (*P. cerasifera*) from Azerbaijan. There is currently no information as to its potential for spread. Due to the mixed infection status of the few trees in which it has been detected so far, it has not been possible to associate it with specific symptoms in cherry (Marais *et al.*, 2015). An efficient RT-PCR assay is available for the detection of PrVT (Marais *et al.*, 2015).

There are publications describing sweet cherry (*P. avium*) and sour cherry (*P. cerasus*) as hosts for tobacco mosaic virus (TMV); however, it is doubtful if this is of any economic significance (Gilmer, 1976; Németh, 1986). No symptoms were associated with TMV infections of any *Prunus* spp. TMV is easily mechanically sap transmitted, and there are a number of recommended herbaceous indicator plants (Németh, 1986). The virus can also be detected using serological techniques such as ELISA (van Regenmortel and Burckard, 1980), and by RT-PCR (Kumar *et al.*, 2011).

## 16.7 Viroids Infecting Cherry

Three viroids have been reported infecting cherry: peach latent mosaic viroid (PLMVd) (family *Avsunviroidae)*, hop stunt viroid (HSVd) and apple scar skin viroid (ASSVd) (family *Pospiviroidae)*. None has been conclusively associated with any cherry disease, and therefore their economic importance for cherry appears to be limited. However, apart from cherry acting as a reservoir for these viroids, the possibility that they could be involved in synergistic interactions in co-infections with some viruses cannot be dismissed.

Viroid diseases affect many important crops. Typical symptoms elicited by viroids in crops that do show symptoms include: leaf chlorosis and epinasty, internode shortening, bark cracking, flower and fruit skin discoloration or deformation, enlarged stones and tuber malformation. Most viroids are transmitted mechanically, and some through seed or pollen.

Viroids are small (250–400 nt), single-stranded, circular RNAs that, despite lacking protein-coding ability, can infect certain plants and frequently cause specific diseases (Diener, 2003). About 30 different viroids have been molecularly characterized, with their biological properties (e.g. host range and pathogenicity) studied to varying extents. They have been classified into two families: (i) *Pospiviroidae*, grouping those with a central conserved region and nuclear replication through a rolling-circle mechanism with only RNA intermediates; and (ii) *Avsunviroidae*, encompassing four viroids without a central conserved region but with hammerhead ribozymes (i.e. catalytic RNA motifs) that replicate in plastids through a slightly different rolling-circle mechanism. Replication is catalysed by host-encoded enzymes, which are supplemented in the second family by ribozymes 'encoded' in both polarity strands. The resulting progeny moves from cell to cell through plasmodesmata and reaches distal parts via the vasculature, for which they must recruit additional host proteins (Flores *et al.*, 2005).

The main prophylactic measure for viroids infecting vegetatively propagated hosts, such as cherry, is the use of viroid-free propagation material. Regular disinfection of pruning tools is recommended to prevent mechanical transmission. Thermotherapy or *in vitro* micrografting may remove viroids from part of the plants, with reduction of the tip size usually increasing efficacy.

### 16.7.1 Peach latent mosaic viroid

Peach latent mosaic viroid (PLMVd) was originally described and characterized in peach (Hernández and Flores, 1992) in which most isolates are symptomless, although some induce leaf mosaics/blotches and an extreme chlorosis (peach calico), fruit discolorations and deformations, delays in foliation, flowering and ripening, and reduction of the tree lifespan (Flores *et al.*, 2006).

In sweet cherry, PLMVd was first detected by dot blotting and Northern blot hybridization in one-half and two-thirds of the samples from Romania and Italy, respectively (Hadidi *et al.*, 1997). Sequencing of an RT-PCR full-length product from one cherry isolate revealed a size of 337 nt displaying a 91–92% sequence similarity with PLMVd isolates from peach (Hadidi *et al.*, 1997). Subsequently, PLMVd was detected in Italy when sweet cherry budwood was grafted on seedlings of the peach indicator ‘GF305’ (used as a bioamplification host). The plants remained symptomless, but molecular hybridization tests showed the presence of PLMVd in some of the inoculated ‘GF305’ plants (Crescenzi *et al.*, 2002). In a number of large-scale surveys, PLMVd was detected in peach cultivars, but never in any of the sweet or sour cherry trees tested (Michelutti *et al.*, 2005; Mandic *et al.*, 2008; Lin *et al.*, 2011). It would seem therefore that PLMVd infection of cherry is rare.

### 16.7.2 Hop stunt viroid

Hop stunt viroid (HSVd) was first described as the causal agent of hop stunt disease (Sasaki and Shikata, 1977). It has a very broad range of natural hosts that include several stone and pome fruits. HSVd infecting cherry was first reported in Turkey in the course of a survey in which, out of 127 trees examined by RT-PCR, 21 reacted positively (16 from sweet cherry and five from sour cherry) (Gazel *et al.*, 2008). HSVd, in mixed infections with ASSVd, was later detected by tissue-printing hybridization and RT-PCR in sweet cherry in Greece, with sequencing revealing a size of 297–298 nt (Kaponi *et al.*, 2012). HSVd was not detected in sweet or sour cherry in large-scale surveys carried out in Serbia and Canada (Michelutti *et al.*, 2005; Mandic *et al.*, 2008).

### 16.7.3 Apple scar skin viroid

Apple scar skin viroid (ASSVd) was isolated initially and characterized in apple (Hashimoto and Koganezawa, 1987), with some susceptible cultivars displaying scarred skin and cracking, and others dapple symptoms (dapple apple) (Hadidi and Barba, 2011). ASSVd was first detected in sweet cherry in Greece by tissue-printing hybridization and RT-PCR. Cloning and sequencing revealed sequences of 327–340 nt sharing 96–99% similarity with Indian apple isolates of ASSVd. The viroid was graft transmitted to cherry rootstocks, and transmission was confirmed by RT-PCR (Kaponi *et al.*, 2013). ASSVd has also been detected in Himalayan wild cherry (*Prunus cerasoides*) by RT-PCR, with sequencing of the amplicons showing 92% sequence similarity with sweet and wild cherry isolates from Greece, and 98% with an Indian isolate from apple (Walia *et al.*, 2012).

## 16.8 Phytoplasmas Infecting Cherry

Phytoplasmas are wall-less, non-helical, Gram-positive prokaryotic parasites of plants and insects constituting one of the groups within the class *Mollicutes*. ‘Phytoplasma’, formerly known as mycoplasma-like organism, has been designated a new taxon named ‘*Candidatus* Phytoplasma’ (IRPCM, 2004). Thirty-three ‘*Candidatus* Phytoplasma’ spp. have been formally described (Bertaccini *et al.*, 2014). They cause considerable loss of yield of many commercially important crops such as rice, potato, maize, cassava, legume, sesame, soybean, grapevine, and pome and stone fruits (Bertaccini *et al.*, 2014). Symptoms associated with phytoplasma infection include virescence/phyllody (development of green leaf-like structures instead of flowers) and distortion resulting in flower sterility, leaf discoloration and malformation, proliferation of axillary buds leading to the development of witches' broom, abnormal internode elongation and stunting (Bertaccini, 2007). The pathogens survive and multiply in the phloem tissue of the infected host plant and are spread through vegetative propagation of infected plant material and by grafting. They are also transmitted from plant to plant by phloem-feeding insects, mainly leafhoppers and less frequently psyllids, belonging to the families *Cicadellidae*, *Cixiidae*, *Psyllidae*, *Delphacidae* and *Derbidae* (Weintraub and Beanland, 2006).

### 16.8.1 European stone fruit yellows phytoplasma

The diseases apricot chlorotic leaf roll, plum leptonecrosis, peach yellowing, and declining of plum, peach and almond were found to have a common aetiology. Thus, a single name was assigned: European stone fruit yellows (ESFY) (Lorenz *et al.*, 1994). Further studies showed that the respective diseases are associated with '*Candidatus* Phytoplasma prunorum' ('*Ca.* P. prunorum') infection (Seemüller and Schneider, 2004). Infection with '*Ca.* P. prunorum' does not normally lead to devastation of sweet and sour cherry orchards, as trees are often latently infected or show only mild symptoms (Giunchedi *et al.*, 1982; Kison and Seemüller, 2001). In the Czech Republic, '*Ca.* P. prunorum' was identified in a sweet cherry tree showing symptoms of stunting, leaf rolling and yellowing, and in a symptomatic sour cherry tree showing small leaves, sparse foliage, oversized fruit, reduced vigour and dieback (Navràtil *et al.*, 2001). '*Ca.* P. prunorum' was also detected in sour cherry trees surveyed in East Bohemia, Czech Republic, showing leaf roll and yellowing symptoms, sparse foliage, small leaves and fruit (Ludvíková *et al.*, 2011). Stunting, chlorotic leaf roll, short internodes, wilting and dieback of branches were observed on several cherry trees infected with '*Ca.* P. prunorum' in Poland (Fig. 16.6) (Cieślińska and Morgaś, 2011). Some of the positively tested cherries became symptomless over time. '*Ca.* P. prunorum' is transmitted by *Cacopsylla pruni* (Carraro *et al.*, 1998). As '*Ca.* P. prunorum' causes economically devastating fruit tree diseases, it is included in EPPO's A2 list of pests, recommended for regulation as quarantine pests (http://www.eppo.int/).

Routinely, two-step (nested) PCR protocols are used for amplification of phytoplasmal DNA fragments, as concentration of these agents in infected plants is usually very low. The first round of PCR may be conducted with universal primers such as P1/P7 (Deng and Hiruki, 1991; Schneider *et al.*, 1995), followed by PCR with the universal nested primers R16F2n/R16R2 (Lee *et al.*, 1993; Gundersen and Lee, 1996). RFLP analysis of the amplicons is required for the accurate identification of different species and strains of phytoplasmas (Lee *et al.*, 1995; Seemüller *et al.*, 1998). Primers derived from ribosomal RNA gene sequences of '*Ca.* P. prunorum' can also amplify DNA from other phytoplasmas classified to the apple proliferation group (Kison *et al.*, 1997). ESFY phytoplasmas can be distinguished from other phytoplasmas, including '*Candidatus* Phytoplasma mali' and '*Candidatus* Phytoplasma pyri', by RFLP analysis of the amplicons using suitable restriction enzymes (Marcone *et al.*, 1996).

'*Ca.* P. prunorum'-specific primers have been designed using sequences derived from the 16S rRNA gene and 16S/23S intergenic spacer region (Yvon *et al.*, 2009), or the putative nitroreductase gene and an intergenic region (Jarausch *et al.*, 1998). Real-time PCR has been used for specific detection and quantification of '*Ca.* P. prunorum' in naturally infected stone fruit trees and insects (Jarausch *et al.*, 2010).

It is important to use healthy plant material and to avoid planting stone fruit orchards in areas where '*Ca.* P. prunorum' and its potential vector occur. It was shown that some wild *Prunus* spp., hackberry, ash and dog rose surrounding apricot chlorotic leaf roll-affected orchards in southern France were reservoirs of ESFY phytoplasma

Fig. 16.6. Cherry infected with the phytoplasma '*Candidatus* Phytoplasma prunorum' and showing symptoms of stunting and chlorotic yellowing (right) versus phytoplasma-free cherry (left).

(Jarausch *et al.*, 2001). A study by Poggi Pollini *et al.* (2007) where diverse pesticides were applied to control the vector *C. pruni* failed to limit the spread of ESFY in the Trentino region of Italy. Tissue culture can be used to eliminate phytoplasmas from plants. *In vitro* thermotherapy and meristem tip culture techniques have been applied successfully for the elimination of '*Ca.* P. prunorum' from apricot shoots (Bertaccini *et al.*, 2014).

### 16.8.2 X-disease group phytoplasmas (16SrIII)

X-disease is a serious problem in cherry and peach orchards in North America (Rawlins and Horne, 1931). Phytoplasmas classified to the X-disease group (16SrIII), with the suggested name '*Candidatus* Phytoplasma pruni' (IRPCM Phytoplasma/Spiroplasma Working Team – Phytoplasma Taxonomy Group, 2004), were identified in the USA in cherries and peach, and in chokecherry, which is a potential reservoir (Lee *et al.*, 1992; Kirkpatrick *et al.*, 1995). Associated symptoms include sparse canopies, small leaves and small pointed fruit (Uyemoto and Luhn, 2006). The X-disease phytoplasma was reported in a number of *Prunus* spp. including Japanese plums, almond, bitter cherry and chokecherry, and also a number of weed species (Uyemoto and Kirkpatrick, 2011). The X-disease-related phytoplasma is transmitted by various leafhopper species (Uyemoto and Kirkpatrick, 2011).

The primers R16(III)F2/R1, which are based on the 16S rRNA gene sequence, were designed for specific amplification of the X-disease phytoplasmas (Lee *et al.*, 1994). PCR/RFLP and sequence analyses of ribosomal protein operons allow identification and discrimination of the X-disease phytoplasmas from peach and chokecherry (Gundersen *et al.*, 1996). Southern blot hybridization can be used to identify phytoplasmas belonging to the X-disease group (Lee *et al.*, 1992).

Removal of infected plants and phytoplasma reservoir hosts is the most effective method. Eradication of cherry trees infected with X-disease in California protected the orchards from further spreading of the phytoplasma (Uyemoto *et al.*, 1998). Significant reduction of X-disease symptoms in sweet cherry orchards was achieved by cutting down the symptomatic trees and spraying with diazinon to prevent the infected leafhoppers (Uyemoto *et al.*, 1998). Treating cherry and peach trees with tetracycline reduced X-disease in some orchards (Lee *et al.*, 1987).

### 16.8.3 Aster yellows group phytoplasmas (16SrI)

'*Candidatus* Phytoplasma asteris' (Lee *et al.*, 2004a), classified in subgroup 16SrI-B, has been reported in stone fruit trees in many countries. It was detected in cherries, peach, apricot, European plum, blackthorn and almond in the Czech Republic (Navràtil *et al.*, 2001; Fialová *et al.*, 2004), in apricot, plum, nectarine and Japanese plum in Italy (Lee *et al.*, 1998a) and in apricot in Spain (Schneider *et al.*, 1993). Navràtil *et al.* (2001) reported that sweet cherry showing stunting, leaf rolling and yellowing, and sour cherry trees with small leaves, reduced vigour and dieback, were infected principally by '*Ca.* P. asteris'. Some sour cherry trees in Poland showing dieback, shoot proliferation and small leaves (Fig. 16.7) tested positive for '*Ca.* P. asteris' (Cieślińska and Smolarek, 2015). The leafhoppers species *Macrosteles*, *Euscelis*, *Scaphytopius* and *Aphrodes* are the major vectors of '*Ca.* P. asteris' (Lee *et al.*, 2004a).

The specific primers R16(I)F1/R1 (Lee *et al.*, 1994) are used routinely in nested PCR, following P1/P7 amplification. Several other genes have been analysed including *tuf*

**Fig. 16.7.** Sour cherry tree infected with '*Candidatus* Phytoplasma asteris' and showing symptoms of dieback, shoot proliferation and small leaves.

(Schneider *et al.*, 1993) and *rp* (Lee *et al.*, 2004a) to identify and characterize aster yellows phytoplasmas infecting *Prunus* spp.

The incidence of aster yellows disease can be reduced significantly by using healthy plant material, eradication of weeds from the field, roadways and fences, and by chemical control of the leafhopper vectors in the crop and on weeds. Application of tetracycline may be appropriate for the treatment of particularly valuable trees, but is not registered in some countries for this purpose (CAB International, 2017).

### 16.8.4 Elm yellows group phytoplasmas (16SrV)

Cherry lethal yellows (CLY) disease was reported on Chinese cherry in Sichuan Province, China (Zhu and Hou, 1989). The diseased cherry trees developed diffuse yellowing of the leaves, defoliated prematurely, produced little or no fruit, and died within 3–4 years. An electron microscopy study suggested that CLY may be caused by a phytoplasma (Zhu and Shu, 1992). Further analysis of the sequences of the 16S rRNA and ribosomal protein genes showed that the phytoplasma belonged to the elm yellows phytoplasma group and represented a new subgroup, 16SrV-B (Lee *et al.*, 1998b). This subgroup has been named collectively as '*Candidatus* Phytoplasmas ziziphi' and classified to the 16SrV-B subgroup (Jung *et al.*, 2003).

Primers R16(V)F1/R1 (Lee *et al.*, 1994) can be used for specific amplification of a 16S rRNA gene fragment of the elm yellows group of phytoplasmas that includes '*Candidatus* Phytoplasma ulmi' (16SrV-A), '*Ca.* P. ziziphi' (16SrV-B), '*Candidatus* Phytoplasma rubi' (16SrV-E), '*Candidatus* Phytoplasma balanitae' (16SrV-F) and phytoplasmas associated with flavescence dorée disease (16SrV-C and -D subgroups). The *rp* and *secY* operons were used for molecular characterization of '*Ca.* P. ziziphi' from sweet cherry with CLY symptoms (Lee *et al.*, 2004b) and from peach with peach yellow symptoms (Thakur *et al.*, 1998). No vector was identified for CLY, but very quick spreading of the disease was observed in orchards in Sichuan Province, China (Zhu *et al.*, 2011).

Control is best achieved by eradicating infected plants and by using phytoplasma-free plant material. Phytosanitary certification is required for the international movement of propagation material, to avoid inadvertent spread of phytoplasmas.

### 16.8.5 Other phytoplasmas reported to infect cherry

Sweet cherry trees growing in Italy displayed symptoms that included curled, small, chlorotic leaves that turned red prematurely and also symptoms of dieback. The symptoms observed were associated with phytoplasmas from several groups including the stolbur group (16SrXII-A) (Paltrinieri *et al.*, 2001). Phytoplasmas associated with stolbur- and bois noir-related diseases are classified as a novel taxon, '*Candidatus* Phytoplasma solani' (Quaglino *et al.*, 2013).

Wilting, floral and phloem necrosis, and dying of cherry were observed in southwestern Slovenia (Mehle *et al.*, 2007). A phytoplasma was detected using 4',6-diamidino-2-phenylindole (DAPI) staining and electron microscopy, and subsequently identified as apple proliferation phytoplasma ('*Ca.* P. mali', 16SrX-A) based on PCR/RFLP sequence analyses of a 16S rRNA gene fragment. In the Czech Republic, '*Ca.* P. mali' was detected in sweet cherry showing stunting, leaf rolling and yellowing (Navràtil *et al.*, 2001).

Pear decline phytoplasma ('*Ca.* P. pyri', 16SrX-C) was identified in declining sweet cherry trees in north-central Italy, based on PCR/RFLP analysis of the 16S rRNA gene fragment (Paltrinieri *et al.*, 2001). Using the same approach, it was shown that the phytoplasma associated with chlorotic leaf roll and wilting of a sweet cherry tree in Poland belonged to the 16SrX-C subgroup (Cieślińska and Morgaś, 2011).

Two different phytoplasmas were identified in sweet cherry trees growing in the central regions of Iran (Zirak *et al.*, 2010). '*Ca.* P. asteris' was detected in some trees showing leaf rolling and witches' broom symptoms. Sequence analysis of the 16S rRNA gene and 16S–23S intergenic spacer region indicated that a tree with rosetting symptoms was infected with a phytoplasma related to the peanut witches' broom group phytoplasmas (16SrII).

A phytoplasma strain related to ash yellows group (16SrVII, '*Candidatus* Phytoplasma fraxini') was reported in China in sweet cherry imported from Israel and showing fasciation symptoms (Li *et al.*, 1997). Restriction enzyme analysis with *Hha*I and *Taq*I, used for digestion of the 16S rRNA gene fragment, allowed differentiation of '*Ca.* P. fraxini' from phytoplasmas classified to the other 16Sr groups (Lee *et al.*, 1998b).

## 16.9 Virus-like Diseases of Cherry with Unknown Aetiology

Several diseases of cherry are described that are known to be transmissible by budding or grafting, but the aetiology is unknown, as the causal agent(s) of the disease has never been isolated. Some of these diseases can have a significant impact on the host. Without the isolation and characterization of the causal agent, a bioassay using woody indicators is the only tool that can be used for their diagnosis.

### 16.9.1 Cherry freckle fruit

Cherry feckle fruit symptoms are observed on the fruit and include fruit that are smaller than usual with necrotic brown flecks, spotting and fruit distortion (Németh, 1986). Late ripening of the fruit is observed and the leaves of affected trees are without symptoms. Natural hosts for the disease include sweet cherry (*P. avium*) 'Bing' and 'Lambert'. Sour cherry 'Montmorency' (*P. cerasus*) is recommended as a woody indicator host for freckle fruit disease (Németh, 1986).

### 16.9.2 Cherry rusty spot

This disease was first described in cherries in New Zealand (Wood, 1972). Symptoms include rust-coloured spots on leaves formed in spring and purple zones surrounding the spots that break away leaving holes. Distinctive rusty spots on Mazzard 'F 12/1' cherry rootstock indicator and the absence of any symptoms on 'Lambert' cherry and oriental flowering cherry (*P. serrulata*) differentiates this disease from diseases associated with PNRSV, CNRMV and the European rusty mottle viruses (Wood, 1972). Sweet cherry (*P. avium*) is the only known natural host of the disease. Cherry rusty spot disease can be detected using the Mazzard 'F 12/1' rootstock as a woody indicator host (Wood, 1972; Németh, 1986).

### 16.9.3 Cherry short stem

Cherry short stem disease was first observed in sweet cherry (*P. avium*) in Montana, USA, in 1958 (Afanasiev, 1963). Symptoms appear on leaves, fruit and fruit stems with typical symptoms including shortened fruit stems accompanied by twisted leaves (Németh, 1986). The only known natural host for the disease is sweet cherry, but the agent was successfully transmitted to a range of *Prunus* spp. including apricot and peach (Németh, 1986). The agent appears to spread naturally in the field, but the vector is not known (Parish and Cheney, 1976). The recommended woody indicator is 'Bing' cherry (Németh, 1986).

### 16.9.4 Cherry stem pitting

Cherry (*Prunus*) stem pitting disease has been associated with ToRSV as described above. It is possible, however, that there are as yet uncharacterized agents associated with the disease (Mircetich *et al.*, 1978). Stem pitting-affected trees have abnormally thickened and spongy bark, poor terminal growth and colour, and leaves that are often cupped or rolled (Mircetich *et al.*, 1978). The disease symptomatology is influenced by the scion/rootstock combination (Mircetich *et al.*, 1978).

#### 16.9.5 Spur cherry

Spur cherry symptoms vary with the host cultivar (Németh, 1986). Severe symptoms include downward curling of the leaves or epinasty, and bark necrosis on shoots. Affected plants tend to be stunted because of shortened internodes, with shoots that are thicker, rougher and more brittle at their terminal ends. Natural hosts for the cherry spur agent are sweet cherry cultivars. 'Bing' cherry has been recommended as an indicator, and typically displays shortening of the internodes (Németh, 1986).

## 16.10 Disorders of Cherry Attributed to Genetic Abnormalities

There are a number of disorders of cherry where no bacterium, fungus, virus or virus-like agent has been associated with the symptoms observed. Attempts to transmit the disorders observed from woody symptomatic trees to healthy trees, by budding and/or grafting, were unsuccessful, thus eliminating the possible involvement of viruses, viroids or phytoplasmas. The disorders observed are considered to be associated with some genetic abnormality and are observed mostly in young trees perhaps with rootstocks or scions derived from affected trees (Németh, 1986).

#### 16.10.1 Cherry crinkle leaf

Cherry crinkle leaf (CCL) is considered an important disorder of sweet cherry and has been reported in the USA, and also in British Columbia, Canada (Németh, 1986; Southwick and Uyemoto, 1999). Leaves that develop earlier in the season are more severely affected than leaves that develop later in the season. A tree may display severe symptoms one year but appear completely healthy the next year, with boron deficiency perhaps influencing the course of the disorder (Southwick and Uyemoto, 1999).

#### 16.10.2 Cherry deep suture

Cherry deep suture is widespread in the cherry-growing regions of the western USA and Canada (Southwick and Uyemoto, 1999). Nursery trees often display reduced growth and if transplanted do not develop normally (Németh, 1986). The fruit on affected trees tends to be small and may be abnormal in shape, and ripening is delayed.

#### 16.10.3 Cherry variegated leaf

Cherry variegated leaf disorder may affect certain sweet and sour cherry cultivars and the associated symptoms are characteristic (Németh, 1986). Significant growth reduction is observed in trees with severe symptoms, and the fruit are smaller and ripen later.

#### 16.10.4 Sour cherry rosette

Trees affected by sour cherry rosette possess narrow and deformed leaves with margins that are deeply and heavily serrated (Németh, 1986). Shortened internodes result in the leaves forming rosettes.

#### 16.10.5 Sour cherry leaf constriction

Abnormalities include chlorotic bands along one or more secondary veins, lack of tissue development near the leaf margin and deep indentations on both sides of the leaf blade (Németh, 1986).

## References


Adams, M.J., Candresse, T., Hammond, J., Kreuze, J.F., Martelli, G.P., Namba, S., Pearson, M.N., Ryu, K.H., Saldarelli, P. and Yoshikawa, N. (2012) Family *Betaflexiviridae*. In: King, A.M.Q., Adams, M.J., Carstens, E.B. and Lefkowitz, E.J. (eds) *Virus Taxonomy – Ninth Report of the International Committee on Taxonomy of Viruses*. Elsevier Academic Press, London, pp. 920–941.

Afanasiev, M.M. (1963) 'Short stem' conditions in Lambert cherry. *Phytopathology* 53, 1137.

Albertini, A., Giunchedi, L., Dradi, D. and Benini, A. (1993) Effetto di infezioni virali su alcune varieta di ciliegio. *Rivista di Frutticultura* 2, 61–64.

Al Rwahnih, M., Myrta, A., Herranz, M.C. and Pallás, V. (2004a) Monitoring *American plum line pattern virus* in plum by ELISA and dot-blot hybridisation throughout the year. *Journal of Plant Pathology* 86, 167–169.

Al Rwahnih, M., Turturo, C., Minafra, A., Saldarelli, P., Myrta, A., Pallás, V. and Savino, V. (2004b) Molecular variability of *Apple chlorotic leaf spot virus* in different hosts and geographical regions. *Journal of Plant Pathology* 86, 117–122.

Amari, K., Burgos, L., Pallás, V. and Sánchez-Pina, M.A. (2007) Prunus necrotic ringspot virus early invasion and its effects on apricot pollen grains performance. *Phytopathology* 97, 892–899.

Amari, K., Burgos, L., Pallás, V. and Sánchez-Pina, M.A. (2009) Vertical transmission of *Prunus* necrotic ringspot virus: Hitch-hiking from gametes to seedling. *Journal of General Virology* 90, 1767–1774.

Aparicio, F., Sánchez-Pina, M.A., Sánchez-Navarro, J.A. and Pallás, V. (1999) Location of *Prunus necrotic ringspot ilarvirus* within pollen grains of infected nectarine trees: evidence from RT-PCR, dot-blot and *in situ* hybridisation. *European Journal of Plant Pathology* 105, 623–627.

Avgelis, A. and Barba, M. (2011) *Epirus cherry virus*. In: Hadidi, A., Barba, M., Candresse, T. and Jelkmann, W. (eds) *Virus and Virus-like Diseases of Pome and Stone Fruits*. APS Press, St Paul, Minnesota, pp. 151–152.

Avgelis, A., Rumbos, J. and Barba, M. (1988) Epirus cherry virus, a new virus isolated from cherry in Greece. *Acta Horticulturae* 235, 245–246.

Barba, M. and Hadidi, A. (2011) DNA microarrays and other future trends in detection and typing of viruses, viroids and phytoplasmas. In: Hadidi, A., Barba, M., Candresse, T. and Jelkmann, W. (eds) *Virus and Virus-like Diseases of Pome and Stone Fruits*. APS Press, St Paul, Minnesota, pp. 363–374.

Barba, M., Hadidi, A., Candresse, T. and Cambra, M. (2011) *Plum pox virus*. In: Hadidi, A., Barba, M., Candresse, T. and Jelkmann, W. (eds) *Virus and Virus-like Diseases of Pome and Stone Fruits*. APS Press, St Paul, Minnesota, pp. 185–197.

Bertaccini, A. (2007) Phytoplasmas: diversity, taxonomy, and epidemiology. *Frontiers in Bioscience* 12, 673–689.

Bertaccini, A., Duduk, B., Paltrinieri, S. and Contaldo, N. (2014) Phytoplasmas and phytoplasma diseases: a severe threat to agriculture. *American Journal of Plant Science* 5, 1763–1788.

Bitterlin, M.W. and Gonsalves, D. (1987) Spatial distribution of *Xiphinema rivesi* and persistence of tomato ringspot virus and its vector in soil. *Plant Disease* 71, 408–411.

Bodine, E.W. and Newton, J.H. (1942) The rasp leaf of cherry. *Phytopathology* 32, 179–181.

Boscia, D., Myrta, A. and Uyemoto, J.K. (2011) *Plum bark necrosis stem pitting-associated virus*. In: Hadidi, A., Barba, M., Candresse, T. and Jelkmann, W. (eds) *Virus and Virus-like Diseases of Pome and Stone Fruits*. APS Press, St Paul, Minnesota, pp. 177–183.

Brown, D.J.F., Halbrendt, J.M., Jones, A.T., Vrain, T.C. and Robbins, R.T. (1994) Transmission of three North American nepoviruses by populations of four distinct species of the *Xiphinema americanum* group. *Phytopathology* 84, 646–649.

Buchhop, J., von Bargen, S. and Büttner, C. (2009) Differentiation of *Cherry leaf roll virus* isolates from various host plants by immunocapture-reverse transcription-polymerase chain reaction-restriction fragment length polymorphism according to phylogenetic relations. *Journal of Virological Methods* 157, 147–154.

Bujarski, J., Figlerowicz, M., Gallitelli, D., Roossinck, M.J. and Scott, S.W. (2012) Family *Bromoviridae*. In: King, A.M.Q., Adams, M.J., Carstens, E.B. and Lefkowitz, E.J. (eds) *Virus Taxonomy – Ninth Report of the International Committee on Taxonomy of Viruses*. Elsevier Academic Press, London, pp. 965–976.

Büttner, C., von Bargen, S., Bandte, M. and Myrta, A. (2011) *Cherry leaf roll virus*. In: Hadidi, A., Barba, M., Candresse, T. and Jelkmann, W. (eds) *Virus and Virus-like Diseases of Pome and Stone Fruits*. APS Press, St Paul, Minnesota, pp. 119–125.

CAB International (2017) *Candidatus* Phytoplasma asteris (yellow disease phytoplasmas). Invasive Species Compendium, CAB International, Wallingford, UK. Available at: http://www.cabi.org/isc/datasheet/7642 (accessed 23 January 2017).

Çağlayan, K., Ulubas-Serce, C., Gazel, M. and Varveri, C. (2011) *Prune dwarf virus*. In: Hadidi, A., Barba, M., Candresse, T. and Jelkmann, W. (eds) *Virus and Virus-like Diseases of Pome and Stone Fruits*. APS Press, St Paul, Minnesota, pp. 199–206.

Cambra, M., Boscia, D., Gil, M., Bertolini, E. and Olmos, A. (2011) Immunology and immunological assays applied to the detection, diagnosis and control of fruit tree viruses. In: Hadidi, A., Barba, M., Candresse, T. and Jelkmann, W. (eds) *Virus and Virus-like Diseases of Pome and Stone Fruits*. APS Press, St Paul, Minnesota, pp. 303–314.

Candresse, T., Lanneau, M., Revers, F., Grasseau, N., Macquaire, G., German, S., Malinowski, T. and Dunez, J. (1995) An immunocapture PCR assay adapted to the detection and the analysis of the molecular variability of apple chlorotic leaf spot virus. *Acta Horticulture* 386, 136–147.

Candresse, T., Marais, A., Faure, C. and Gentit, P. (2013) Association of *Little cherry virus 1* (LChV1) with the shirofugen stunt disease and characterization of the genome of a divergent LChV1 isolate. *Phytopathology* 103, 293–298.

Card, S.D., Pearson, M.N. and Clover, G.R.G. (2007) Plant pathogens transmitted by pollen. *Australasian Plant Pathology* 36, 455–461.

Carraro, L., Osler, R., Loi, N., Emancora, P. and Refatti, E. (1998) Transmission of European stone fruit yellows phytoplasma by *Cacopsylla pruni*. *Journal of Plant Pathology* 80, 233–239.

Cembali, T., Folwell, R.J., Wandschneider, P., Eastwell, K.C. and Howell, W.E. (2003) Economic implications of a virus prevention program in deciduous tree fruits in the US. *Crop Protection* 22, 1149–1156.

Chamberlain, E.E., Atkinson, J.D., Wood, G.A. and Hunter, J.A. (1971) Occurrence of cherry green ring mottle virus in New Zealand. *New Zealand Journal of Agricultural Research* 14, 499–508.

Cheney, P.W. and Parish, C.L. (1976) Cherry mottle leaf. In: Gilmer, R.M., Moore, J.D., Nyland, G., Welsh, M.F. and Pine, T.S. (eds) *Virus Diseases and Noninfectious Disorders of Stone Fruits in North America*. Handbook Number 437. US Department of Agriculture, Washington, DC, pp. 216–218.

Chirkov, S., Ivanov, P. and Sheveleva, A. (2013) Detection and partial molecular characterization of atypical plum pox virus isolates from naturally infected sour cherry. *Archives of Virology* 158, 1383–1387.

Chung, B.Y.-W., Miller, W.A., Atkins, J.F. and Firth, A.E. (2008) An overlapping essential gene in the *Potyviridae*. *Proceedings of the National Academy of Sciences USA* 105, 5897–5902.

Cieślińska, M. and Morgaś, H. (2011) Detection and identification of '*Candidatus* Phytoplasma prunorum', '*Candidatus* Phytoplasma mali' and '*Candidatus* Phytoplasma pyri' in stone fruit trees in Poland. *Journal of Phytopathology* 159, 217–222.

Cieślińska, M. and Smolarek, T. (2015) Molecular diversity of phytoplasmas infecting cherry trees in Poland. *Phytopathogenic Mollicutes* 5, 31–32.

Crescenzi, A., Piazzolla, P. and Hadidi, A. (2002) First report of peach latent mosaic viroid in sweet cherry in Italy. *Journal of Plant Pathology* 84, 168.

Cropley, R. (1961) Cherry leaf-roll virus. *Annals of Applied Biology* 49, 524–529.

Deng, S. and Hiruki, C. (1991) Genetic relatedness between two nonculturable mycoplasmalike organisms revealed by nucleic acid hybridization and polymerase chain reaction. *Phytopathology* 81, 1475–1479.

Desvignes, J.C. (1999) *Virus Diseases of Fruit Trees*. Centre Techniques Interprofessionnel des Fruits et Légumes (CTIFL), Paris.

Diekmann, M. and Putter, C.A.J. (1996) *Technical Guidelines for the Safe Movement of Germplasm*. Stone Fruits No. 16. Food and Agriculture Organization of the United Nations/International Plant Genetic Resources Institute, Rome.

Diener, T.O. (2003) Discovering viroids: a personal perspective. *Nature Reviews Microbiology* 1, 75–80.

Eastwell, K.C. and Bernardy, M.G. (2001) Partial characterization of a closterovirus associated with apple mealybug-transmitted little cherry disease in North America. *Phytopathology* 91, 268–273.

Eastwell, K.C. and Howell, W.E. (2010) Characterization of *Cherry leafroll virus* in sweet cherry in Washington State. *Plant Disease* 94, 1067.

Eastwell, K.C., Mekuria, T.A. and Howell, W.E. (2012) Complete nucleotide sequences and genome organization of a cherry isolate of cherry leaf roll virus. *Archives of Virology* 157, 761–764

EPPO (2001) Certification scheme for cherry. *EPPO Bulletin* 31, 447–461.

EPPO (2006a) Pathogen tested olive trees and rootstocks. *EPPO Bulletin* 36, 77–83.

EPPO (2006b) American plum line pattern virus. *EPPO Bulletin* 36, 157–160.

Fialová, R., Navràtil, M., Válová, P., Lauterer, P., Kocourek, F. and Poncarová-Voráčková, Z. (2004) Epidemiology of European stone fruit yellows phytoplasma in the Czech Republic. *Acta Horticulturae* 657, 483–487.

Flores, R., Hernández, C., Martínez de Alba, A.E., Daròs, J.A. and di Serio, F. (2005) Viroids and viroid–host interactions. *Annual Review of Phytopathology* 43, 117–139.

Flores, R., Delgado, S., Rodio, M.E., Ambrós, S., Hernández, C. and di Serio, F. (2006) *Peach latent mosaic viroid*: not so latent. *Molecular Plant Pathology* 7, 209–221.

Foissac, X., Svanella-Dumas, L., Gentit, P., Dulucq, M.J., Marais, A. and Candresse, T. (2005) Polyvalent degenerate oligonucleotides reverse transcription-polymerase reaction: a polyvalent detection and characterization tool for trichoviruses, capilloviruses, and foveaviruses. *Phytopathology* 95, 617–625.

Fulton, R.W. (1964) Transmission of plant viruses by grafting, dodder, seed and mechanical inoculation. In: Corbett, M.K. and Sisler, H.D. (eds) *Plant Virology*. University of Florida Press, Gainesville, Florida, pp. 39–67.

Garcia, J.A., Riechmann, J.L., Lain, S., Martin, M.T., Guo, H., Simon, L., Fernandez, A., Dominguez, E. and Cervera, M.T. (1994) Molecular characterization of plum pox potyvirus. *EPPO Bulletin* 24, 543–553.

Gazel, M., Ulubas, C. and Caglayan, K. (2008) Detection of hop stunt viroid in sweet and sour cherry trees in Turkey by RT-PCR. *Acta Horticulturae* 795, 955–958.

German, S., Candresse, T., Lanneau, M., Huet, J.C., Pernollet, J.C. and Dunez, J. (1990) Nucleotide sequence and genomic organization of apple chlorotic leaf spot closterovirus. *Virology* 179, 104–112.

German, S., Delbos, R.P., Candresse, T., Lanneau, M. and Dunez, J. (1997) Complete nucleotide sequence of the genome of a severe cherry isolate of apple chlorotic leaf spot trichovirus (ACLSV). *Archives of Virology* 142, 833–841.

Gilmer, R.M. (1976) Tobacco mosaic virus infection in *Prunus*. In: Gilmer, R.M., Moore, J.D., Nyland, G., Welsh, M.F. and Pine, T.S. (eds) *Virus Diseases and Noninfectious Disorders of Stone Fruits in North America*. Handbook Number 437. US Department of Agriculture, Washington, DC, p. 256.

Giunchedi, L., Poggi Pollini, C. and Credi, R. (1982) Susceptibility of stone fruit trees to the Japanese plum tree decline causal agent. *Acta Horticulturae* 130, 285–290.

Glasa, M., Prikhodko, Y., Predajna, L., Nagyova, A., Shneyder, Y., Zhivaeva, T., Subr, Z., Cambra, M. and Candresse, T. (2013) Characterization of sour cherry isolates of *Plum pox virus* from the Volga Basin in Russia reveals a new cherry strain of the virus. *Phytopathology* 103, 972–979.

Gonsalves, D. (1995) Tomato ringspot virus. In: Ogawa, J.M., Zehr, E.I., Bird, G.W., Ritchie, D.F., Eriu, K. and Uyemoto, J.K. (eds) *Compendium of Stone Fruit Diseases*. APS Press, St Paul, Minnesota, pp. 70–71.

Greber, R.S., Teakle, D.S. and Mink, G.I. (1992) Thrips-facilitated transmission of prune dwarf and Prunus necrotic ringspot viruses from cherry pollen to cucumber. *Plant Disease* 76, 1039–1041.

Griesbach, J.A. (1995) Detection of tomato ringspot virus by polymerase chain reaction. *Plant Disease* 79, 1054–1056.

Gundersen, D.E. and Lee, I.M. (1996) Ultrasensitive detection of phytoplasmas by nested-PCR assays using two universal primers. *Phytopathologia Mediterranea* 35, 144–151.

Gundersen, D.E., Lee, I.M., Schaff, D.A., Harrison, N.A., Chang, C.J., Davis, R.E. and Kingsbury, D.T. (1996) Genomic diversity and differentiation among phytoplasma strains in 16S rRNA groups I (Aster yellows and related phytoplasmas) and III (X-disease and related phytoplasmas). *International Journal of Systematic Bacteriology* 46, 64–75.

Hadidi, A. and Barba, M. (2011) *Apple scar skin viroid*. In: Hadidi, A., Barba, M., Candresse, T. and Jelkmann, W. (eds) *Virus and Virus-like Diseases of Pome and Stone Fruits*. APS Press, St Paul, Minnesota, pp. 57–62.

Hadidi, A., Giunchedi, L., Shamloul, A.M., Poggi-Pollini, C. and Amer, M.A. (1997) Occurrence of peach latent mosaic viroid in stone fruits and its transmission with contaminated blades. *Plant Disease* 81, 154–158.

Hadidi, A., Olmos, A., Pasquini, G., Barba, M., Martin, R.R. and Shamloul, A.M. (2011) Polymerase chain reaction for detection of systemic plant pathogens. In: Hadidi, A., Barba, M., Candresse, T. and Jelkmann, W. (eds) *Virus and Virus-like Diseases of Pome and Stone Fruits*. APS Press, St Paul, Minnesota, pp. 341–359.

Hammond, R.W. (2011) *Prunus necrotic ringspot virus*. In: Hadidi, A., Barba, M., Candresse, T. and Jelkmann, W. (eds) *Virus and Virus-like Diseases of Pome and Stone Fruits*. APS Press, St Paul, Minnesota, pp. 207–213.

Hammond, R.W. and Crosslin, J.M. (1998) Virulence and molecular polymorphism of Prunus necrotic ringspot virus isolates. *Journal of General Virology* 79, 1815–1823.

Hammond, R.W., Crosslin, J.M., Pasini, R., Howell, H.E. and Mink, G.I. (1999) Differentiation of closely related but biologically distinct cherry isolates of Prunus necrotic ringspot virus by polymerase chain reaction. *Journal of Virological Methods* 80, 203–212.

Hansen, A.J. and Cheney, P.W. (1976) Cherry twisted leaf. In: Gilmer, R.M., Moore, J.D., Nyland, G., Welsh, M.F. and Pine, T.S. (eds) *Virus Diseases and Noninfectious Disorders of Stone Fruits in North America*. Handbook Number 437. US Department of Agriculture, Washington, DC, pp. 222–225.

Hashimoto, J. and Koganezawa, H. (1987) Nucleotide sequence and secondary structure of apple scar skin viroid. *Nucleic Acids Research* 15, 7045–7051.

Hernández, C. and Flores, R. (1992) Plus and minus RNAs of peach latent mosaic viroid self-cleave *in vitro* via hammerhead structures. *Proceedings of the National Academy of Sciences USA* 89, 3711–3715.

Herranz, M.C., Sánchez-Navarro, J.A., Aparicio, F. and Pallás, V. (2005) Simultaneous detection of six stone fruit viruses by non-isotopic molecular hybridization using a unique riboprobe or 'polyprobe'. *Journal of Virological Methods* 124, 49–55.

Hoy, J.W. and Mircetich, S.M. (1984) Prune brownline disease: susceptibility of prune rootstocks and tomato ringspot virus detection. *Phytopathology* 74, 272–276.

ICTV (2015) Virus Taxonomy: 2015 Release. International Committee on Taxonomy of Viruses. Available at: http://www.ictvonline.org/virustaxonomy.asp (accessed 24 January 2017).

IPPC (2012) International Standards for Phytosanitary Measures (ISPM) 27, Annex 2, DP2: *Plum pox virus*. Adopted by the 7th Commission on Phytosanitary Measures, March 2012. International Plant Protection Convention, Rome. Available at: https://www.ippc.int/en/ (accessed 23 January 2017).

IRPCM Phytoplasma/Spiroplasma Working Team – Phytoplasma Taxonomy Group (2004) '*Candidatus* Phytoplasma', a taxon for the wall-less, non-helical prokaryotes that colonize plant phloem and insects. *International Journal of Systematic and Evolutionary Microbiology* 54, 1243–1255.

James, D. (2011a) *Cherry mottle leaf virus*. In: Hadidi, A., Barba, M., Candresse, T. and Jelkmann, W. (eds) *Virus and Virus-like Diseases of Pome and Stone Fruits*. APS Press, St Paul, Minnesota, pp. 127–131.

James, D. (2011b) *Cherry rasp leaf virus*. In: Hadidi, A., Barba, M., Candresse, T. and Jelkmann, W. (eds) *Virus and Virus-like Diseases of Pome and Stone Fruits*. APS Press, St Paul, Minnesota, pp. 137–141.

James, D. (2011c) *Cherry twisted leaf virus*. In: Hadidi, A., Barba, M., Candresse, T. and Jelkmann, W. (eds) *Virus and Virus-like Diseases of Pome and Stone Fruits*. APS Press, St Paul, Minnesota, pp. 143–146.

James, D. and Mukerji, S. (1993) Mechanical transmission, identification, and characterization of a virus associated with mottle leaf in cherry. *Plant Disease* 77, 271–275.

James, D., Varga, A., Pallas, V. and Candresse, T. (2006) Strategies for simultaneous detection of multiple plant viruses. *Canadian Journal of Plant Pathology* 28, 16–29.

James, D., Varga, A. and Sanderson, D. (2013) Genetic diversity of *Plum pox virus*: strains disease and related challenges for control. *Canadian Journal of Plant Pathology* 35, 431–441.

James, D., Varga, A. and Lye, D. (2014) Analysis of the complete genome of a virus associated with twisted leaf disease of cherry reveals evidence of a close relationship to unassigned viruses in the family *Betaflexiviridae*. *Archives of Virology* 159, 2463–2468.

Jarausch, W., Lansac, M., Saillard, C., Broquaire, J.M. and Dosba, F. (1998) PCR assay for specific detection of European stone fruit yellows phytoplasmas and its use for epidemiological studies in France. *European Journal of Plant Pathology* 104, 17–27.

Jarausch, W., Jarausch-Wertheim, B., Danet, J.L., Broquaire, J.M., Dosba, F., Saillard, C. and Garnier, M. (2001) Detection and identification of European stone fruit yellows and other phytoplasmas in wild plants in the surroundings of apricot chlorotic leaf roll-affected orchards in southern France. *European Journal of Plant Pathology* 107, 209–217.

Jarausch, W., Fuchs, A. and Jarausch, B. (2010) Establishment of a quantitative real-time PCR assay for the specific quantification of '*Ca*. Phytoplasma. prunorum' in plants and insects. *Julius-Kühn-Archiv* 427, 392–394.

Jelkmann, W. (1995) *Cherry virus A*: cDNA cloning of dsRNA, nucleotide sequence analysis and serology reveal a new plant capillovirus in sweet cherry. *Journal of General Virology* 76, 2015–2024.

Jelkmann, W. (1996) The nucleotide sequence of a strain of *Apple chlorotic leaf spot virus* (ACLSV) responsible for plum pseudopox and its relation to an apple and plum bark split strain. *Phytopathology* 86, 101–101.

Jelkmann, W. (2011) Cherry detrimental canker. In: Hadidi, A., Barba, M., Candresse, T. and Jelkmann, W. (eds) *Virus and Virus-like Diseases of Pome and Stone Fruits*. APS Press, St Paul, Minnesota, pp. 111–114.

Jelkmann, W. and Eastwell, K.C. (2011) *Little cherry virus*-1 and -2. In: Hadidi, A., Barba, M., Candresse, T. and Jelkmann, W. (eds) *Virus and Virus-like Diseases of Pome and Stone Fruits*. APS Press, St Paul, Minnesota, pp. 153–160.

Jelkmann, W., Rott, M. and Uyemoto, J.K. (2011) *Cherry green ring mottle virus*. In: Hadidi, A., Barba, M., Candresse, T. and Jelkmann, W. (eds) *Virus and Virus-like Diseases of Pome and Stone Fruits*. APS Press, St Paul, Minnesota, pp. 115–117.

Jones, A.T. (1985) Cherry leaf roll virus. In: *CMI/AAB Descriptions of Plant Viruses*, No. 306. Commonwealth Mycological Institute, Kew, UK.

Jones, A.L. and Sutton, T.B. (1996) Sour cherry yellows. In: *Diseases of Tree Fruits in the East (NCR045)*. Michigan State University Press, East Lansing, Michigan, p. 94.

Jung, H.Y., Sawayanagi, T., Kakizawa, S., Nishigawa, H., Wei, W., Oshima, K., Miyata, S., Ugaki, M., Hibi, T. and Namba, S. (2003) '*Candidatus* Phytoplasma ziziphi', a novel phytoplasma taxon associated with jujube witches'-broom disease. *International Journal of Systematic and Evolutionary Microbiology* 53, 1037–1041.

Kaponi, M.S., Luigi, M. and Kyriakopoulou, P.E. (2012) Mixed infections of pome and stone fruit viroids in cultivated and wild trees in Greece. *New Disease Reports* 26, 8.

Kaponi, M.S., Sano, T. and Kyriakopoulou, P.E. (2013) Natural infection of sweet cherry trees with apple scar skin viroid. *Journal of Plant Pathology* 95, 429–433.

Katsiani, A.T., Maliogka, V.I., Candresse, T. and Katis, N.I. (2014) Host range studies, genetic diversity and evolutionary relationships of ACLSV isolates from ornamental, wild and cultivated rosaceous species. *Plant Pathology* 63, 63–71.

Katsiani, A.T., Maliogka, V.I., Amoutzias, G.D., Efthimiou, K.E. and Katis, N.I. (2015) Insights into the genetic diversity and evolution of *Little cherry virus 1*. *Plant Pathology* 64, 817–824.

Keane, F.W.L. and May, M.F. (1963) Natural root grafting in cherry and spread of cherry twisted leaf virus. *Canadian Plant Disease Survey* 43, 54–60.

Kirkpatrick, B.C., Uyemoto, J.K. and Purcell, A.H. (1995) X disease. In: Ogawa, J.M., Zehr, E.I., Bird, G.W., Ritchie, D.F., Eriu, K. and Uyemoto, J.K. (eds) *Compendium of Stone Fruit Diseases*. APS Press, St Paul, Minnesota, pp. 57–58.

Kishi, K., Abiko, K. and Takanashi, K. (1973) Studies on the virus diseases of stone fruit. VII. *Cucumber mosaic virus* isolated from *Prunus* trees. *Annals of the Phytopathological Society of Japan* 39, 297–304.

Kison, H. and Seemüller, E. (2001) Differences in strain virulence of the European stone fruit yellows phytoplasma and susceptibility of stone fruit trees on various rootstocks to this pathogen. *Journal of Phytopathology* 149, 533–541.

Kison, H., Kirkpatrick, B.C. and Seemüller, E. (1997) Genetic comparison of peach yellow leaf roll agent with European stone fruit yellows phytoplasma of the apple proliferation group. *Plant Pathology* 46, 538–544.

Kommineni, K.V., Gillet, J.M. and Ramsdell, D.C. (1998) A study of tomato ringspot virus and prune brownline resistance in twenty-five rootstock–scion combinations. *HortTechnology* 8, 349–353.

Kryczyński, S., Szyndel, M.S., Stawiszynska, A. and Piskorek, W. (1992) The rate and the way of Prunus necrotic ringspot virus spread in sour cherry orchard and in the rootstock production. *Acta Horticulturae* 309, 105–110.

Kumar, S., Udaya Shankar, A.C., Nayaka, S.C., Lund, O.S. and Prakash, H.S. (2011) Detection of *Tobacco mosaic virus* and *Tomato mosaic virus* in pepper and tomato by multiplex RT-PCR. *Letters in Applied Microbiology* 53, 359–363.

Lee, R.F., Nyland, G. and Lowe, S.K. (1987) Chemotherapy of cherry buckskin and peach yellow leafroll diseases: an evaluation of two tetracycline formulations and methods of application. *Plant Disease* 71, 119–121.

Lee, I.-M., Gundersen, D.E., Davis, R.E. and Chiykowski, L.N. (1992) Identification and analysis of a genomic strain cluster of mycoplasmalike organisms associated with Canadian peach (eastern) X-disease, western X-disease, and clover yellow edge. *Journal of Bacteriology* 174, 6694–6698.

Lee, I.-M., Hammond, R.W., Davis, R.E. and Gundersen, D.E. (1993) Universal amplification and analysis of pathogen 16S rDNA for classification and identification of mycoplasmalike organisms. *Phytopathology* 83, 834–842.

Lee, I.-M., Gundersen, D.E., Hammond, R.W. and Davis, RE. (1994) Use of mycoplasmalike organism (MLO) group-specific oligonucleotide primers for nested-PCR assays to detect mixed-MLO infections in a single host plant. *Phytopathology* 84, 559–566.

Lee, I.-M., Zhu, S., Gundersen, D.E., Zhang, C. and Hadidi, A. (1995) Detection and identification of a new phytoplasma associated with cherry lethal yellows in China. *Phytopathology* 85, 1179.

Lee, I.-M., Gundersen-Rindal, D.E. and Bertaccini, A. (1998a) Phytoplasma: ecology and genomic diversity. *Phytopathology* 88, 1359–1366.

Lee, I.-M., Gundersen-Rindal, D.E., Davis, R.E. and Bartoszyk, I.M. (1998b) Revised classification scheme of phytoplasmas based on RFLP analyses of 16S rRNA and ribosomal protein gene sequences. *International Journal of Systematic Bacteriology* 48, 1153–1169.

Lee, I.-M., Gundersen-Rindal, D.E., Davis, R.E., Bottner, K.D., Marcone, C. and Seemüller, E. (2004a) '*Candidatus* Phytoplasma asteris', a novel phytoplasma taxon associated with aster yellows and related diseases. *International Journal of Systematic and Evolutionary Microbiology* 54, 1037–1048.

Lee, I.-M., Martini, M., Marcone, C. and Zhu, S.F. (2004b) Classification of phytoplasma strains in the elm yellows group (16SrV) and proposal of '*Candidatus* Phytoplasma ulmi' for the phytoplasma associated with elm yellows. *International Journal of Systematic and Evolutionary Microbiology* 54, 337–347.

Lesemann, D.E., Kunze, L., Krischke, G. and Koenig, R. (1989) Natural occurrence of carnation Italian ringspot virus in a cherry tree. *Journal of Phytopathology* 124, 171–174.

Li, R. and Mock, R. (2005) An improved reverse transcription-polymerase chain reaction (RT-PCR) assay for the detection of two cherry flexiviruses in *Prunus* spp. *Journal of Virological Methods* 129, 162–169.

Li, H.H., Qiu, B.S., Shi, C.L., Jin, K.X., Zhou, Q. and Huang, X.J. (1997) PCR amplification of 16S rDNA of phytoplasma associated with cherry fasciated disease and RFLP analysis. *Forest Research* 10, 478–481.

Li, T.S.C., Eastwell, K.C. and Hansen, A.J. (1996) Transmission of cherry viruses by approach grafting from herbaceous to woody hosts. *Canadian Journal of Plant Pathology* 18, 429–432.

Lin, L., Li, R., Mock, R. and Kinard, G. (2011) Development of a polyprobe to detect six viroids of pome and stone fruit trees. *Journal of Virological Methods* 171, 91–97.

Lorenz, K.H., Dosba, F., Poggi Pollini, C., Llacer, G. and Seemüller, E. (1994) Phytoplasma diseases of *Prunus* species in Europe are caused by genetically similar organisms. *Zeitschrift für Pflanzenkrankheiten und Pflanzenschutz* 101, 567–575.

Lott, T.B. (1943) Transmissible twisted leaf of sweet cherry. *Science of Agriculture* 23, 439–441.

Ludvíková, H., Fránová, J. and Suchá, J. (2011) Phytoplasmas in apricot, peach and sour cherry orchards in East Bohemia, Czech Republic. *Bulletin of Insectology* 64 (Suppl.), 67–68.

Mandic, B., Al Rwahnih, M., Myrta, A., Gómez, G. and Pallás, V. (2008) Incidence and genetic diversity of peach latent mosaic viroid and hop stunt viroid in stone fruits in Serbia. *European Journal of Plant Pathology* 120, 167–176.

Manganaris, G.A., Economou, A.S., Boubourakas, I.N. and Katis, N.I. (2003) Elimination of PPV and PNRSV through thermotherapy and meristem-tip culture. *Plant Cell Reports* 22, 195–200.

Marais, A., Candresse, T. and Jelkmann, W. (2011) *Cherry virus A*. In: Hadidi, A., Barba, M., Candresse, T. and Jelkmann, W. (eds) *Virus and Virus-like Diseases of Pome and Stone Fruits*. APS Press, St Paul, Minnesota, pp. 147–150.

Marais, A., Svanella-Dumas, L., Barone, M., Gentit, P., Faure, C., Charlot, G., Ragozzino, A. and Candresse, T. (2012) Development of a polyvalent RT-PCR detection assay covering the genetic diversity of *Cherry capillovirus A*. *Plant Pathology* 61, 195–204.

Marais, A., Faure, C., Couture, C., Bergey, B., Gentit, P. and Candresse, T. (2014) Characterization by deep sequencing of divergent *Plum bark necrosis stem pitting associated virus* (PBNSPaV) isolates and development of a broad-spectrum PBNSPaV-specific detection assay. *Phytopathology* 104, 660–666.

Marais, A., Faure, C., Mustafayev, E., Barone, M., Alioto, D. and Candresse, T. (2015) Characterization by deep sequencing of Prunus virus T, a novel *Tepovirus* infecting *Prunus* species. *Phytopathology* 105, 135–140.

Marcone, C., Ragozzino, A. and Seemüller, E. (1996) European stone fruit yellows phytoplasma as the cause of peach vein enlargement and other yellows and decline diseases of stone fruits in Southern Italy. *Journal of Phytopathology* 144, 559–564.

Marini, D.B., Zhang, Y.P., Rowhani, A. and Uyemoto, J.K. (2002) Etiology and host range of a *Closterovirus* associated with plum bark necrosis-stem pitting disease. *Plant Disease* 86, 415–417.

Martelli, G.P., Agranovsky, A.A., Bar-Joseph, M., Boscia, D., Candresse, T., Coutts, R.H.A., Dolja, V.V., Hu, J.S., Jelkmann, W., Karasev, A.V., Martin, R.R., Minafra, A., Namba, S. and Vetten, H.J. (2012) Family *Closteroviridae*. In: King, A.M.Q., Adams, M.J., Carstens, E.B. and Lefkowitz, E.J. (eds) *Virus Taxonomy – Ninth Report on the International Committee on Taxonomy of Viruses*. Elsevier Academic Press, London, pp. 987–1001.

Matic, S., Myrta, A. and Minafra, A. (2007) First report of *Little cherry virus 1* in cherry, plum, almond and peach in Italy. *Journal of Plant Pathology* 89, S75–S75.

Matic, S., Minafra, A., Boscia, D., da Cunha, A.T.P. and Martelli, G.P. (2009) Production of antibodies to Little cherry virus 1 coat protein by DNA prime and protein boost immunization. *Journal of Virological Methods* 155, 72–76.

Mehle, N., Brzin, J., Boben, J., Hren, M., Frank, J., Petrovič, N., Gruden, K., Dreo, T., Žežlina, I., Seljak, G. and Ravnikar, M. (2007) First report of '*Candidatus* Phytoplasma mali' in *Prunus avium*, *P. armeniaca* and *P. domestica*. *Plant Pathology* 56, 721.

Mekuria, T.A., Smith, T.J., Beers, E., Watson, G.W. and Eastwell, K.C. (2013) First report of transmission of *Little cherry virus 2* to sweet cherry by *Pseudococcus maritimus* (Ehrhorn) (Hemiptera: *Pseudococcidae*). *Plant Disease* 97, 851–851.

Mekuria, T.A., Zhang, S.L. and Eastwell, K.C. (2014) Rapid and sensitive detection of *Little cherry virus 2* using isothermal reverse transcription-recombinase polymerase amplification. *Journal of Virological Methods* 205, 24–30.

Menzel, W., Jelkmann, W. and Maiss, E. (2002) Detection of four apple viruses by multiplex RT-PCR assays with coamplification of plant mRNA as internal control. *Journal of Virological Methods* 99, 81–92.

Menzel, W., Zahn, V. and Maiss, E. (2003) Multiplex RT-PCR-ELISA compared with bioassay for the detection of four apple viruses. *Journal of Virological Methods* 110, 153–157.

Michelutti, R., Myrta, A. and Pallás, V. (2005) A preliminary account on the sanitary status of stone fruits at the clonal genebank in Harrow, Canada. *Phytopathologia Mediterranea* 44, 71–74.

Mink, G.I. (1993) Pollen- and seed-transmitted viruses and viroids. *Annual Review of Phytopathology* 31, 375–402.

Mink, G.I. and Shay, J.R. (1959) Preliminary evaluation of some Russian apple varieties as indicators for apple viruses. *Plant Disease* 254, 13–17.

Mircetich, S.M., Moller, W.J. and Nyland, G. (1978) Stem pitting disease of cherries and other stone fruits. *California Agriculture* pp. 19–20.

Myrta, A., Herranz, M.C., Choueiri, E. and Pallas, V. (2011a) *American plum line pattern virus*. In: Hadidi, A., Barba, M., Candresse, T. and Jelkmann, W. (eds) *Virus and Virus-like Diseases of Pome and Stone Fruits*. APS Press, St Paul, Minnesota, pp. 85–90.

Myrta, A., Matic, S., Malinowski, T., Pasquini, G. and Candresse, T. (2011b) *Apple chlorotic leaf spot virus* in stone fruits. In: Hadidi, A., Barba, M., Candresse, T. and Jelkmann, W. (eds) *Virus and Virus-like Diseases of Pome and Stone Fruits*. APS Press, St Paul, Minnesota, pp. 85–90.

Navràtil, M., Válová, P., Fialová, R., Petrová, K., Fránová, J., Nebesářová, J., Poncarová-Voráčková, Z. and Karešová, R. (2001) Survey for stone fruit phytoplasmas in Czech Republic. *Acta Horticulturae* 550, 377–382.

Nemchinov, L., Crescenzi, A., Hadidi, A., Piazzolla, P. and Verderevskaya, T. (1998) Present status of the new cherry subgroup of plum pox virus (PPV-C). In: Hadidi, A., Khetarpal, R.K. and Koganezawa, H. (eds) *Plant Virus Disease Control*. APS press, St Paul, Minnesota, pp. 629–638.

Németh, M. (1986) *Virus, Mycoplasma and Rickettsia Diseases of Fruit Trees*. Martinus Nijhoff Publishers, Dordrecht, The Netherlands, and Akademini Kiado, Hungary.

Oldfield, G.N. (1970) Mite transmission of plant viruses. *Annual Review of Entomology* 15, 343–380.

Osman, F., Al Rwahnih, M., Golino, D., Pitman, T., Cordero, F., Preece, J.E. and Rowhani, A. (2012) Evaluation of the phytosanitary status of the *Prunus* species in the national clonal germplasm repository in California: survey of viruses and viroids. *Journal of Plant Pathology* 94, 249–253.

Pagan, I., Montes, N., Milgroom, M.G. and Garcia-Arenal, F. (2014) Vertical transmission selects for reduced virulence in a plant virus and for increased resistance in the host. *PLoS One* 10, e1004293.

Pallás, V., Sanchez-Navarro, J., Varga, A., Aparicio, F. and James, D. (2009) Multiplex polymerase chain reaction (PCR) and real-time multiplex PCR for the simultaneous detection of plant viruses. In: Burns, R. (ed.) *Methods in Molecular Biology: Plant Pathology*. Humana Press, Totowa, New Jersey, pp. 193–208.

Pallás, V., Faggioli, F., Aparico, F. and Sanchez-Navarro, J.A. (2011) Molecular hybridization techniques for detecting and studying fruit tree viruses and viroids. In: Hadidi, A., Barba, M., Candresse, T. and Jelkmann, W. (eds) *Virus and Virus-like Diseases of Pome and Stone Fruits*. APS Press, St Paul, Minnesota, pp. 335–342.

Pallás, V., Aparicio, F., Herranz, M.C., Amari, K., Sanchez-Pina, M.A., Myrta, A. and Sanchez-Navarro, J.A. (2012) Ilarviruses of *Prunus* spp.: a continued concern for fruit trees. *Phytopathology* 102, 1108–1120.

Pallás, V., Aparicio, F., Herranz, M.C., Sanchez-Navarro, J.A. and Scott, S.W. (2013) The molecular biology of ilarviruses. *Advances in Virus Research* 87, 139–181.

Paltrinieri, S., Martini, M., Stefani, E., Pondrelli, M., Fideghelli, C. and Bertaccini, A. (2001) Phytoplasma infection in peach and cherry in Italy. *Acta Horticulturae* 550, 365–370.

Panattoni, A., Luvisi, A. and Triolo, E. (2013) Elimination of viruses in plants: 20 years of progress. *Spanish Journal of Agricultural Research* 11, 173–188.

Parish, C.L. (1977) A relationship between flat apple disease and cherry rasp leaf disease. *Phytopathology* 67, 982–984.

Parish, C.L. and Cheney, P.W. (1976) Short stem. In: Gilmer, R.M., Moore, J.D., Nyland, G., Welsh, M.F. and Pine, T.S. (eds) *Virus Diseases and Noninfectious Disorders of Stone Fruits in North America*. Handbook Number 437. US Department of Agriculture, Washington, DC, pp. 250–251.

Parker, K.G., Fridlund, P.R. and Gilmer, R.M. (1976) Green ring mottle. In: Gilmer, R.M., Moore, J.D., Nyland, G., Welsh, M.F. and Pine, T.S. (eds) *Virus Diseases and Noninfectious Disorders of Stone Fruits in North America*. Handbook Number 437. US Department of Agriculture, Washington, DC, pp. 193–199.

Poggi Pollini, C., Bianchi, L., Forno, F., Franchini, S., Giunchedi, S., Gobber, M., Mattedi, L., Miorelli, P., Pignatta, D., Profaizer, D., Ratti, C. and Reggiani, N. (2007) Investigation on European stone fruit yellows in experimental apricot orchards in province of Trentino (Italy). *Bulletin of Insectology* 60, 323–324.

Powell, C.A., Hadidi, A. and Halbrendt, J.M. (1991) Detection and distribution of tomato ringspot virus in infected nectarine trees using ELISA and transcribed RNA probes. *HortScience* 26, 1290–1292.

Quaglino, F., Zhao, Y., Casati, P., Bulgari, D., Bianco, P.A., Wei, W. and Davis, R.E. (2013) '*Candidatus* Phytoplasma solani', a novel taxon associated with stolbur- and bois noir-related diseases of plants. *International Journal of Systematic and Evolutionary Microbiology* 63, 2879–2894.

Ramsdell, D.C. (1995) Sour cherry green ring mottle virus. In: Ogawa, J.M., Zehr, E.I., Bird, G.W., Ritchie, D.F., Eriu, K. and Uyemoto, J.K. (eds) *Compendium of Stone Fruit Diseases*. APS Press, St Paul, Minnesota, pp. 76–77.

Rastgou, M., Turina, M. and Milne, R.G. (2012) Genus *Ourmiavirus*. In: King, A.M.Q., Adams, M.J., Carstens, E.B. and Lefkowitz, E.J. (eds) *Virus Taxonomy – Ninth Report of the International Committee on Taxonomy of Viruses*. Elsevier Academic Press, London, pp. 1177–1180.

Rawlins, T.E. and Horne, W.T. (1931) Buckskin, a destructive graft-infectious disease of the cherry. *Phytopathology* 21, 331–335.

Rochon, D., Lommel, S., Martelli, G.P., Rubino, L. and Russo, M. (2012) Family *Tombusviridae*. In: King, A.M.Q., Adams, M.J., Carstens, E.B. and Lefkowitz, E.J. (eds) *Virus Taxonomy – Ninth Report of the International Committee on Taxonomy of Viruses*. Elsevier Academic Press, London, pp. 1111–1138.

Rodoni, B., Mackie, J. and Constable, F. (2011) National diagnostic protocol for cherry leaf roll virus, CLRV (cherry and walnut strains). Department of Agriculture, Australian Government. Available at: http://plantbiosecuritydiagnostics.net.au/wordpress/wp-content/uploads/2015/03/NDP-10-Cherry-Leaf-Roll-Virus-V1.2.pdf (accessed 23January 2017).

Rott, M.E. and Jelkmann, W. (2001a) Complete nucleotide sequence of cherry necrotic rusty mottle virus. *Archives of Virology* 146, 395–401.

Rott, M.E. and Jelkmann, W. (2001b) Detection and partial characterization of a second closterovirus associated with little cherry disease, *Little cherry virus-2*. *Phytopathology* 91, 261–267.

Rott, M.E. and Jelkmann, W. (2001c) Development of PCR primer pairs for the characterization and detection of several related filamentous viruses of cherry. *Acta Horticulturae* 550, 199–205.

Rott, M.E. and Jelkmann, W. (2005) Little cherry virus-2: sequence and genomic organization of an unusual member of the *Closteroviridae*. *Archives of Virology* 150, 107–123.

Rott, M.E. and Jelkmann, W. (2011) *Cherry necrotic rusty mottle* and *Cherry rusty mottle viruses*. In: Hadidi, A., Barba, M., Candresse, T. and Jelkmann, W. (eds) *Virus and Virus-like Diseases of Pome and Stone Fruits*. APS Press, St Paul, Minnesota, pp. 133–136.

Rott, M.E., Johnson, V.L. and Belton, M.P. (2004) Characterization of a new foveavirus associated with cherry rusty mottle disease. *Phytopathology* 94, S89–S89.

Rowhani, A., Biardi, L., Routh, G., Daubert, S.D. and Golino, D.A. (1998) Development of a sensitive colorimetric-PCR assay for detection of viruses in woody plants. *Plant Disease* 82, 880–884.

Sanfaçon, H. and Fuchs, M. (2011) *Tomato ringspot virus*. In: Hadidi, A., Barba, M., Candresse, T. and Jelkmann, W. (eds) *Virus and Virus-like Diseases of Pome and Stone Fruits*. APS Press, St Paul, Minnesota, pp. 41–48.

Sanfaçon, H., Iwanami, T., Karasev, A.V., van der Vlugt, R., Wellink, J., Wetzel, T. and Yoshikawa, N. (2012) *Secoviridae*. In: King, A.M.Q., Adams, M.J., Carstens, E.B. and Lefkowitz, E.J. (eds) *Virus Taxonomy – Ninth Report of the International Committee on Taxonomy of Viruses*. Elsevier Academic Press, London, pp. 881–899.

Sasaki, M. and Shikata, E. (1977) On some properties of hop stunt disease agent, a viroid. *Proceedings of the Japanese Academy Series B* 53, 109–112.

Schneider, B., Ahrens, U., Kirkpatrick, B.C. and Seemüller, E. (1993) Classification of plant-pathogenic mycoplasmalike organisms using restriction-site analysis of PCR-amplified 16S rDNA. *Journal of General Microbiology* 139, 519–527.

Schneider, B., Seemüller, E., Smart, C.D. and Kirkpatrick, B.C. (1995) Phylogenetic classification of plant pathogenic mycoplasma-like organisms or phytoplasmas. In: Razin, S. and Tully, J.G. (eds) *Molecular and Diagnostic Procedures in Mycoplasmology*, Vol. 1. Academic Press, San Diego, California, pp. 369–380.

Scholthof, K.B.G., Adkins, S., Czosnek, H., Palukaitis, P., Jacquot, E., Hohn, T., Hohn, B., Saunders, K., Candresse, T., Ahlquist, P., Hemenway, C. and Foster, G.D. (2011) Top 10 plant viruses in molecular plant pathology. *Molecular Plant Pathology* 12, 938–954.

Seemüller, E. and Schneider, B. (2004) Taxonomic description of '*Candidatus* Phytoplasma mali', '*Candidatus* Phytoplasma pyri' and '*Candidatus* Phytoplasma prunorum', the causal agents of apple proliferation, pear decline and European stone fruit yellows, respectively. *International Journal of Systematic and Evolutionary Microbiology* 54, 1217–1226.

Seemüller, E., Marcone, C., Laurer, U., Ragozzino, A. and Göschl, M. (1998) Current status of molecular classification of the phytoplasmas. *Journal of Plant Pathology* 80, 3–26.

Silva, C., Tereso, S., Nolasco, G. and Oliveria, M.M. (2003) Cellular location of Prune dwarf virus in almond sections by *in situ* reverse transcription-polymerase chain reaction. *Phytopathology* 93, 278–285.

Song, G.-Q., Sink, K.C., Walworth, A.E., Cook, M.A., Allison, R.F. and Lang, G.A. (2013) Engineering cherry rootstocks with resistance to Prunus necrotic ring spot virus through RNAi-mediated silencing. *Plant Biotechnology Journal* 11, 702–708.

Southwick, S.M. and Uyemoto, J. (1999) *Cherry Crinkle-Leaf and Deep Suture Disorders*. Publication 8007. Communications Service, University of California, Oakland, California.

Spiegel, S., Thompson, D., Varga, A., Rosner, A. and James, D. (2006) Evaluation of reverse transcription – polymerase chain reaction assays for detecting *Apple chlorotic leaf spot virus* in certification and quarantine programs. *Canadian Journal of Plant Pathology* 28, 280–288.

Stace-Smith, R. and Hansen, A.J. (1976) Cherry rasp leaf virus. In: *CMI/AAB Descriptions of Plant Viruses*, No. 156. Wm. Culross and Sons, Perthshire, UK.

Tan, H.D., Li, S.Y., Du, X.F. and Seno, M. (2010) First report of *Cucumber mosaic virus* in sweet cherry in the People's Republic of China. *Plant Disease* 94, 378–1378.

Thakur, P.D., Handa, A., Chowfla, S.C. and Krczal, G. (1998) Outbreak of a phytoplasma disease of peach in the Northwestern Himalayas of India. *Acta Horticulturae* 472, 737–742.

Ulubas-Serçe, Ç., Ertunç, F. and Öztürk, A. (2009) Identification and genomic variability of Prune dwarf virus variants infecting stone fruit in Turkey. *Journal of Phytopathology* 157, 298–305.

Uyemoto, J.K. and Kirkpatrick, B.B. (2011) *X-disease phytoplasma*. In: Hadidi, A., Barba, M., Candresse, T. and Jelkmann, W. (eds) *Virus and Virus-like Diseases of Pome and Stone Fruits*. APS Press, St Paul, Minnesota, pp. 243–245.

Uyemoto, J.K. and Luhn, C.F. (2006) In-season variations in transmission of cherry X-phytoplasma and implication in certification programs. *Journal of Plant Pathology* 88, 317–320.

Uyemoto, J.K. and Teviotdale, B.L. (1996) Graft-transmission of the causal agent of a bark necrosis-stem pitting disease of Black Beaut plum (*Prunus salicina*). *Phytopathology* 86, 111–112.

Uyemoto, J.K., Bethell, R.E., Kirkpatrick, B.C., Munkvold, G.P., Marois, J.J. and Brown, K.W. (1998) Eradication as a control measure for X-disease in California cherry orchards. *Acta Horticulturae* 472, 715–721.

van Regenmortel, M.H.V. and Burckard, J. (1980) Detection of a wide spectrum of tobacco mosaic virus strains by indirect enzyme-linked immunosorbent assay (ELISA). *Virology* 106, 327–334.

Villamor, D.E., Susaimuthu, J. and Eastwell, K.C. (2015) Genomic analyses of cherry rusty mottle group and cherry twisted leaf-associated viruses reveal a possible new genus within the family *Betaflexiviridae*. *Phytopathology* 105, 399–408.

Walia, Y., Dhir, S., Bhadoria, S., Hallan, V. and Zaidi, A.A. (2012) Molecular characterization of apple scar skin viroid from Himalayan wild cherry. *Forest Pathology* 42, 84–87.

Weintraub, P.G. and Beanland, L. (2006) Insect vectors of phytoplasmas. *Annual Review of Entomology* 51, 91–111.

Welsh, M.F. (1976) Control of stone fruit virus diseases. In: Gilmer, R.M., Moore, J.D., Nyland, G., Welsh, M.F. and Pine, T.S. (eds) *Virus Diseases and Noninfections Disorders of Stone Fruits in North America*. Agriculture Handbook No. 437, Agricultural Research Service, US Department of Agriculture, Washington, DC, pp. 10–15.

Werner, R., Mühlbach, H.P. and Büttner, C. (1997) Detection of cherry leaf roll nepovirus (CLRV) in birch, beech and petunia by immunocapture RT-PCR using a conserved primer pair. *European Journal of Forest Pathology* 27, 309–318.

Wood, G.A. (1972) Cherry rusty spot virus. *New Zealand Journal of Agricultural Research* 15, 155–159.

Yvon, M., Thébaud, G., Alary, R. and Labonne, G. (2009) Specific detection and quantification of the phytopathogenic agent '*Candidatus* Phytoplasma prunorum'. *Molecular and Cellular Probes* 23, 227–234.

Zhang, Y., Kirkpatrick, B.C., Smart, C.D. and Uyemoto, J.K. (1998) cDNA cloning and molecular characterization of cherry green mottle virus. *Journal of General Virology* 79, 2275–2281.

Zhao, D. and Song, G.-Q. (2014) Rootstock-to-scion transfer of transgene-derived small interfering RNAs and their effect on virus resistance in nontransgenic sweet cherry. *Plant Biotechnology Journal* 12, 1319–1328.

Zhu, S.F. and Hou, L. (1989) An outbreak of a lethal yellows disease on Chinese cherry. *Plant Quarantine* 3, 148.

Zhu, S.F. and Shu, X. (1992) A mycoplasma-like organism on cherry with lethal yellows. *Acta Phytopathologica Sinica* 22, 25–28.

Zhu, S.F., Bertaccini, A., Lee, I.-M., Paltrinieri, S. and Hadidi, A. (2011) Cherry lethal yellows and decline phytoplasmas. In: Hadidi, A., Barba, M., Candresse, T. and Jelkmann, W. (eds) *Virus and Virus-like Diseases of Pome and Stone Fruits*. APS Press, St Paul, Minnesota, pp. 255–257.

Zirak, L., Bahar, M. and Ahoonmanesh, A. (2010) Characterization of phytoplasmas related to '*Candidatus* Phytoplasma asteris' and peanut WB group associated with sweet cherry diseases in Iran. *Journal of Phytopathology* 158, 63–65.

Zong, X., Wang, W., Wei, H., Wang, J., Chen, X., Xu, L., Zhu, D., Tan, Y. and Liu, Q. (2014) A multiplex RT-PCR assay for simultaneous detection of four viruses from sweet cherry. *Scientia Horticulturae* 180, 118–122.

Zong, X., Wang, W., Wei, H., Wang, J., Yan, X., Hammond, R.W. and Liu, Q. (2015) Incidence of sweet cherry viruses in Shandong Province, China and a case study on multiple infection with five viruses. *Journal of Plant Pathology* 97, 61–68.

# 17 Fruit Chemistry, Nutritional Benefits and Social Aspects of Cherries

Manuel Joaquín Serradilla,[1] Milica Fotirić Akšić,[2] George A. Manganaris,[3] Sezai Ercisli,[4] David González-Gómez[5] and Daniel Valero[6]

[1]Scientific and Technological Research Centre of Extremadura (CICYTEX), Junta de Extremadura, Badajoz, Spain; [2]Faculty of Agriculture, University of Belgrade, Belgrade, Zemun, Serbia; [3]Cyprus University of Technology, Lemesos, Cyprus; [4]Agricultural Faculty, Ataturk University, Erzurum, Turkey; [5]Teacher Training College, University of Extremadura, Cáceres, Spain; [6]EPSO, University Miguel Hernández, Alicante, Spain

## 17.1 Introduction

Cherry nutritional composition, phytochemical content and antioxidant capacity should be considered on a cultivar/genotype basis since many new cultivars that enter into the market through the breeding programmes (Sansavini and Lugli, 2008) have apparent differences for both qualitative and phytochemical antioxidants contents (Ballistreri *et al.*, 2013; Goulas *et al.*, 2015). The breeding programmes have led to the release of numerous cultivars, where the main attributes considered were bearing habits, ripening period, fruit size and yield, increased fertility, reduced susceptibility to environmental damage and diseases, extension of seasonality, especially for early-ripening genotypes, and resistance to cracking. However, to the best of our knowledge, phytochemical status and nutritional properties are not being evaluated through the breeding programmes.

This chapter focuses on the physicochemical characteristics (soluble solids, pH, titratable acidity, volatile compounds) and nutritional (e.g. carbohydrates, proteins, lipids, sugars, organic acids, minerals, vitamins) and non-nutritional (other constituents with biological properties beyond nutrition) composition of sweet and sour cherry fruits. Non-nutrient food constituents include phytochemicals, phytonutrients, plant secondary metabolites, and bioactive and health-promoting compounds.

## 17.2 Fruit Chemistry

The chemical characteristics of sweet (*Prunus avium* L.) and sour cherries (*Prunus cerasus* L.) have been widely reviewed, not only because they largely affect the sensory quality of the fruit, but also because they have a strong influence on consumer acceptance (Crisosto *et al.*, 2003). Furthermore, physicochemical studies are also relevant for producers for proper design of the harvesting and postharvest technology for sweet cherry production in the world (Hayaloglu and Demir, 2015). The large and diverse reported values of pomological characteristics of cherries denote how these

[1] manuel.serradilla@juntaex.es

properties are highly influenced not only by the cultivar, but also by other environmental variables, such as climatological conditions and geographical origin (Faniadis *et al.*, 2010; Tomás-Barberán *et al.*, 2013).

Both sweet and sour cherries present a very low caloric content: 63.0 kcal (263.34 kJ) per 100 g for sweet cherry and 50.0 kcal (209 kJ) per 100 g for sour cherry (USDA ARS, 2016). They are also considered an excellent source of numerous nutrients and phytochemicals (McCune *et al.*, 2011), which is one of the major reasons for their increasing popularity in the human diet. In addition, many epidemiological studies have established that their regular consumption is associated with health benefits and the well-being of individuals (Ferretti *et al.*, 2010; McCune *et al.*, 2011).

### 17.2.1 Total soluble solids

The value of total soluble solids (TSS) in sour and sweet cherries may reach up to 24.5 g per 100 g fresh weight (FW) (Table 17.1). This parameter is an important factor in determining the consumer's acceptability (Crisosto *et al.*, 2003; Valero and Serrano, 2010). For sweet cherry, reported values range from as low as 12.3 g per 100 g FW in 'Van' (González-Gómez *et al.*, 2010) to 24.5 g per 100 g FW in 'Salmo' (Girard and Kopp, 1998). Such differences in TSS can be attributed to microclimatic conditions, rootstock selection and planting system, as well as differences in the physiological stage adopted as a harvesting criterion (Goulas *et al.*, 2015). Finally, it has also been reported that TSS must be above the threshold of 14.0–16.0 g per 100 g FW as acceptable for marketing cherries (Crisosto *et al.*, 2003).

In the case of sour cherry, the average values found in commercial cultivars are around 15.0 g per 100 g FW, while only a few cultivars are above the threshold of 17.0 g per 100 g FW (Grafe and Schuster, 2014). Interestingly, autochthonous sour cherry genotypes in Portugal were found to have TSS values in the range of 17.4–22.8 g per 100 g FW (Rodrigues *et al.*, 2008). Hungarian sour cherry cultivars also showed appreciably higher levels of TSS (up to 23.1 g per 100 g FW in cultivar 'Pipacs 1') (Papp *et al.*, 2010).

### 17.2.2 Titratable acidity

Titratable acidity (TA) is one of the most important attributes in cherry, since it is also

**Table 17.1.** Standard physicochemical attributes of sweet and sour cherries.

| Species | TSS (mg per 100 g FW) | References | TA (g malic acid per 100 g FW) | References | Maturation index (TSS/TA) | References |
|---|---|---|---|---|---|---|
| Sweet cherry (*Prunus avium* L) | 12.3–24.5 | Girard and Kopp (1998); González-Gómez *et al.* (2010) | 0.7–1.2 | Serradilla *et al.* (2016) | 8.6–24.4 | Usenik *et al.* (2010); Serradilla *et al.* (2012) |
| Sour cherry (*Prunus cerasus* L.) | 15–23.1 | Papp *et al.* (2010); Grafe and Schuster (2014) | 1.3–3.1 | Rodrigues *et al.* (2008); Papp *et al.* (2010); Rakonjac *et al.* (2010); Damar and Ekşi (2012); Grafe and Schuster (2014) | 5.8–15.8 | Wojdyło *et al.* (2014) |

FW, fresh weight; TSS, total soluble solids; TA, titratable acidity.

directly related to the acceptability by consumers, and it is a highly cultivar-dependent parameter. Sweet cherries are considered as mildly acidic fruits with pH values between 3.7 and 4.2, while sour cherries range from pH 3.1 to 3.6 (Serradilla *et al.*, 2016). Regarding TA, important differences have been observed between sweet and sour cherry and among cultivars. For sweet cherries, TA ranges from 0.7 to 1.2 g malic acid per 100 g FW (Table 17.1), with cultivars such as 'Lapins' showing low TA values, while 'Sweetheart' contains higher values. In the case of sour cherry, several studies reported a TA range of 1.4–2.9 g malic acid per 100 g FW (Damar and Ekşi, 2012). Grafe and Schuster (2014) found TA from 1.3 to 3.1 g malic acid per 100 g FW ('Spinell' and 'Topas', respectively). A similar range was determined in Portuguese (Rodrigues *et al.*, 2008), Hungarian (Papp *et al.*, 2010) and Serbian (Rakonjac *et al.*, 2010) germplasm collections.

### 17.2.3 Maturation index

The maturity index (TSS/TA ratio) is one of the major analytical measures of fruit quality, and it is widely accepted that it directly affects the perception of sweetness and flavour, and thus consumer acceptance of the cherry fruit (Crisosto *et al.*, 2003). In this sense, Guyer *et al.* (1993) observed that, as the TSS/TA ratio of cherry fruits increases, so does the consumer perception of sweetness. Compared with sweet cherry, sour cherry is characterized by a higher acidity level, resulting in lower TSS/TA ratios. For sweet cherry, reported values range between 19.0 in Turkish cultivars (Hayaloglu and Demir, 2015) to 29.0 in some Canadian sweet cherries (Girard and Kopp, 1998), while values around 40.0 have been monitored in certain cultivars (Usenik *et al.*, 2010; Serradilla *et al.*, 2012). For sour cherry, evaluation of the physicochemical composition of 33 sour cherries revealed a range in TSS/TA ratio from 5.8 to 15.3 (Wojdyło *et al.*, 2014), while in Hungarian cultivars it varied from 9.6 to 15.8 (Table 17.1). Cultivars with higher TSS/TA ratios (≥11.0), contributing to a balanced flavour, have been considered an optimal choice for fresh consumption (Papp *et al.*, 2010). The new German cultivars 'Achat' (12.2) and 'Spinell' (14.8) (Schuster *et al.*, 2014) and the new Serbian cultivar 'Lenka' (17.7) (Fotirić Akšić *et al.*, 2015), resulting from different breeding programmes in Germany and Serbia, were particularly selected for fresh consumption due to their sweet sour cherry taste.

### 17.2.4 Volatile compounds

In terms of sensory quality, aroma and flavour are becoming key factors that determine the choice to purchase a fruit, although the compounds that contribute to the flavour of fresh fruit comprise only 0.001–0.01% of the fruit's fresh weight (Zhang *et al.*, 2007; Valero and Serrano, 2010). It is well known that the aroma of the fruit is the result of a complex mixture of esters, alcohols, aldehydes, ketones and terpenoid compounds (Li *et al.*, 2008; Valero and Serrano, 2010). In the case of cherries, their aroma has been studied extensively and comprises free and glycosidically volatile compounds (Girard and Kopp, 1998; Serradilla *et al.*, 2012; Wen *et al.*, 2014). Among the free volatile compounds, more than 100 have been identified such as hexanal, (E)-2 hexenal and benzaldehyde, which are among the predominant volatile flavour constituents in both sweet and sour cherries (Schmid and Grosch, 1986; Poll *et al.*, 2003; Serradilla *et al.*, 2016). In this sense, it has been reported that volatile compounds such as decanal, nonanal and (Z)-3-hexenal were identified as important odorants in 'Lapins', 'Rainier' and 'Stella' (Girard and Kopp, 1998). On the other hand, the aromatic carbonyl benzaldehyde has been determined at the highest level in sour cherry (Levaj *et al.*, 2010). Alcohols were the second largest class, including compounds such as benzyl alcohol, 1-hexanol and (E)-2-hexen-1-ol for sweet cherries. In contrast, according to Levaj *et al.* (2010), the alcohols identified in sour cherries, aside from 1-hexanol, were 1-butanol and 2-phenylethanol. Other compounds such as acids have also been identified, mainly linear and branched acids, esters, monoterpenes (C10), sesquiterpenes (C15) and diterpenes (C20), in both sweet and sour cherries (Levaj *et al.*, 2010; Serradilla *et al.*, 2016). Aside from free volatile compounds, Wen *et al.* (2014) also reported that glycosidically bound aromatic compounds, integrated mainly by alcohols and terpenes, contribute markedly to the aroma of cherries.

## 17.3 Nutritional Composition

### 17.3.1 Water

Water is considered the predominant component of cherries, followed by carbohydrates, proteins and lipids (Serradilla *et al.*, 2016). The water content of sweet and sour cherry genotypes, as fleshy fruits, is around 80–83% (Serradilla *et al.*, 2016) and 81–88% (Filimon *et al.*, 2011), respectively. In general, the water content of sweet cherries is lower than that obtained from other stone fruits such as peaches with 88%, plums with 87% or apricots with 86% (USDA ARS, 2016).

### 17.3.2 Carbohydrates, proteins and lipids

Carbohydrates are the most abundant macronutrients found in cherries (Pacifico *et al.*, 2014; Bastos *et al.*, 2015). Although differences can be observed among cultivars, in general terms fruits exhibit moderate amounts of carbohydrates between 12.2 and 17.0 g per 100 g edible portion for sweet cherry, while sour cherry fruit has an average value of 12.2 g per 100 g edible portion (USDA ARS, 2016). In addition, within the genus *Prunus*, cherry fruit is a moderate source of dietary fibre, accounting for 1.3–2.1 g per 100 g edible portion (McCune *et al.*, 2011).

For sweet cherries, the protein content is between 0.8 and 1.4 g per 100 g edible portion (Serradilla *et al.*, 2016). However, for sour cherries, the protein content is below 1.0 g per 100 g edible portion (Ferretti *et al.*, 2010).

In general, the fat content of sweet and sour cherries is low and below 1.0 g per 100 g edible portion, particularly saturated fat as cherries are a cholesterol-free fruit (Ferretti *et al.*, 2010; McCune *et al.*, 2011; Pacifico *et al.*, 2014).

### 17.3.3 Sugars

Among these compounds, simple sugars (glucose, fructose and sorbitol) are the most relevant (Usenik *et al.*, 2008, 2010; Serradilla *et al.*, 2011; Ballistreri *et al.*, 2013; Pacifico *et al.*, 2014), although trace amounts of sucrose were also identified in sweet cherries, ranging from 0.1 to 1.2 mg per 100 g FW (Esti *et al.*, 2002; Usenik *et al.*, 2008; Ballistreri *et al.*, 2013). The major sugar in sweet and sour cherries is glucose, whose range varies from 6.0 to 10.0 g per 100 g FW, depending on the genotype and environmental conditions (Papp *et al.*, 2010; Ballistreri *et al.*, 2013). The second most abundant sugar is fructose. Its content ranges from 5.0 to 7.6 g per 100 g FW for sweet cherry and from 3.5 to 4.9 g per 100 g FW for sour cherry (Papp *et al.*, 2010; Ballistreri *et al.*, 2013). In fact, in both cases, these authors reported that the genotypes with a higher glucose content had also a higher fructose level. Aside from glucose and fructose, the content of sorbitol for sweet cherries ranged between 0.9 and 26.7 mg per 100 g FW, showing quantities similar to other fruits such as apples, pears, peaches and prunes (Usenik *et al.*, 2008; Ballistreri *et al.*, 2013).

### 17.3.4 Organic acids

The type of organic acid is an important factor in determining fruit acidity (Valero and Serrano, 2010), with malic acid being the principal organic acid in cherries, with values of 360.0–1277.0 mg per 100 g FW, accounting for more than 98% of the total organic acid content. It is also possible to find, as minor constituents, citric, succinic, shikimic, fumaric and oxalic acids (Usenik et al., 2008; Ballistreri et al., 2013; Serradilla et al., 2016). Additionally, Ballistreri et al. (2013) found a high correlation between the total content of organic acids and TA levels in sweet cherry, reflecting the influence of the different content of organic acids on TA.

### 17.3.5 Minerals

Sweet cherry is considered to be a good source of dietary potassium with approximately 260.0 mg potassium per 100 g edible portion (McCune et al., 2011). For sour cherry, potassium is also the main mineral with 200.0 mg per 100 g edible portion. Cherries also contain other minerals in low concentrations such as calcium, phosphorus, magnesium and sodium (USDA ARS, 2016). In sweet cherries, calcium concentration ranged between 13.0 and 20.0 mg per 100 g edible portion, while phosphorus levels varied between 15.0 and 18.0 mg per 100 g edible portion, magnesium between 8.0 and 13.0 mg per 100 g edible portion, and sodium between 1.0 and 8.0 mg per 100 g edible portion. Sour cherries showed a content of calcium that ranged between 9.0 and 14.0 mg per 100 g edible portion, magnesium between 7.0 and 10.0 mg per 100 g edible portion, and phosphorus between 9.0 and 20.0 mg per 100 g edible portion (Mitić et al., 2012; USDA ARS, 2016).

### 17.3.6 Vitamins

Cherries are an excellent source of vitamins, especially vitamin C (7.0–50.0 mg per 100 g edible portion), followed by vitamin E (0.1 mg per 100 g edible portion) and vitamin K (2.0 μg per 100 g edible portion) (McCune et al., 2011). In addition, sour cherries are characterized by a higher content of vitamin A (64.0 mg of retinol activity equivalent (RAE) per 100 g edible portion), whereas the vitamin A content in sweet cherries is around 3.0 mg RAE per 100 g edible portion (Serradilla et al., 2016).

## 17.4 Phytochemical Composition and Antioxidant Activity

### 17.4.1 Carotenoids

Carotenoids are the most widely distributed group of pigments naturally accumulating in large quantities, and are known for their structural diversity and various functions, including the brilliant red, orange and yellow colours of edible fruits (Valero and Serrano, 2010). Within the major phytochemicals found in sweet cherries are carotenoids (β-carotene, lutein, α-carotene, β-cryptoxanthin, zeaxanthin and phytoene). Sweet cherries contain important amounts of carotenoids, mainly β-carotene (38.0 μg per 100 g FW) and lutein/zeaxanthin (85.0 μg per 100 g FW) (Tomás-Barberán et al., 2013). Sour cherries contain some carotenoids, in particular β-carotene (770.0 μg per 100 g FW), and to a lower extent lutein and zeaxanthin (85.0 μg per 100 g FW) (Ferretti et al., 2010).

### 17.4.2 Phenolic compounds

Phenolic compounds, as well as their health-promoting properties, also play a key role in cherry quality attributes since they contribute to colour, taste, aroma and flavour (Tomás-Barberán and Espín, 2001). Despite its significant economic impact as a temperate fruit crop, few comprehensive studies have dealt with physicochemical and phytochemical aspects of different cultivars (Ballistreri et al., 2013). Most of these studies have focused on analyses solely at harvest time, although alterations in the phytochemical content during the postharvest period have also been reported (Valero et al., 2011).

Cherry polyphenols include phenolic acids (hydroxycinnamic and hydroxybenzoic acids) and flavonoids (anthocyanins, flavonols and flavan-3-ols) (Fig. 17.1). Those secondary metabolites are involved in antioxidative defence of plants against biotic and abiotic stresses such as high and low temperatures, drought, alkalinity, salinity,

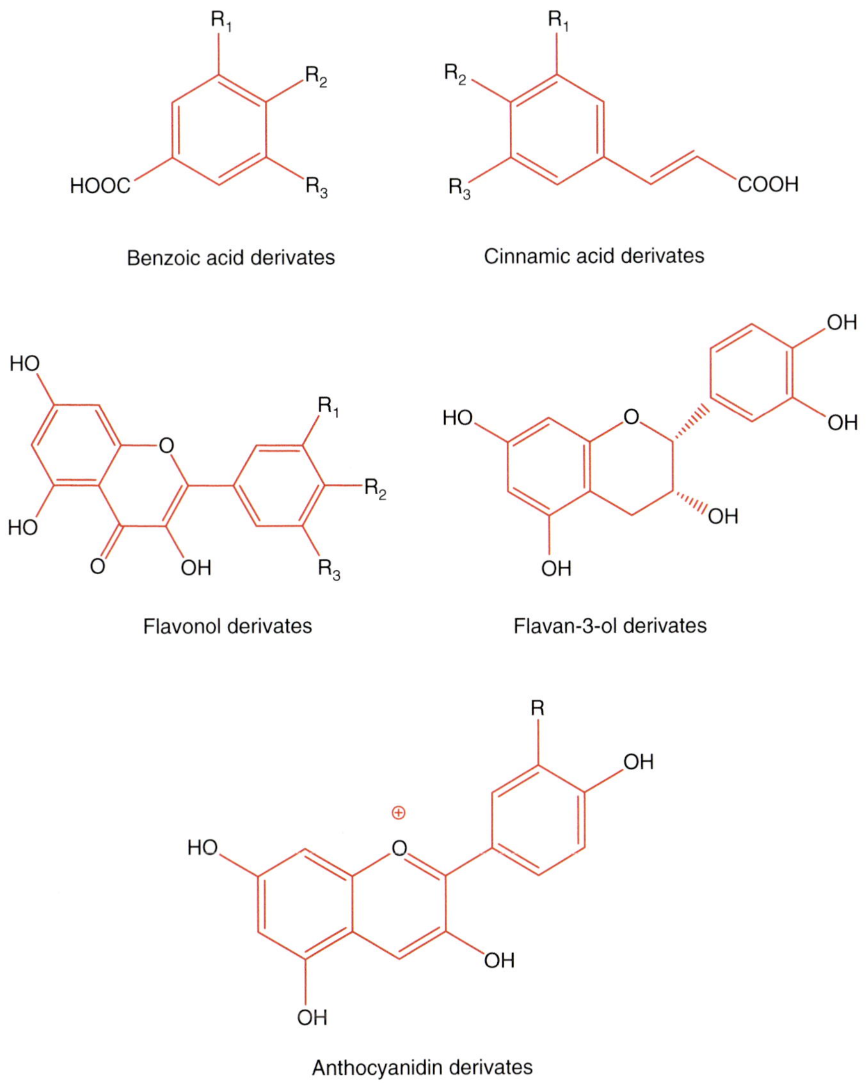


**Fig. 17.1.** The main phytochemical compounds present in sweet and sour cherries. (Drawn using ChemDraw Ultra v.12.0, CambridgeSoft©.)

UV stress and pathogen attack (Viljevac *et al.*, 2012). The highest levels of total polyphenolic compounds are found in the skin of cherry fruits, followed by the flesh and pit (Chaovanalikit and Wrolstad, 2004). Current epidemiological studies strongly support a contribution of polyphenols in the prevention of cardiovascular diseases, cancers, diabetes, insomnia, obesity and osteoporosis, as well as neurodegenerative diseases (Kang *et al.*, 2003; Kim *et al.*, 2005; Pigeon *et al.*, 2010).

A wide range of concentrations of total phenolic content (TPC) has been reported both in sweet and sour cherries (Ballistreri *et al.*, 2013; Tomás-Barberán *et al.*, 2013; Serradilla *et al.*, 2016). The most important results for both species are represented in Table 17.2, which show that sour cherry exhibits higher TPC than sweet cherry.

**Table 17.2.** Total phenolic content of sweet and sour cherries. Data are expressed on a fresh weight (FW) or dry weight basis (DW).

| Species | Total phenolic content (mg per 100 g FW) | Total phenolic content (mg per 100 g DW) | References |
|---|---|---|---|
| Sweet cherry (*Prunus avium* L.) | 44.3–192.0 | 440.0–1309.0 | Usenik *et al.* (2008); Serra *et al.* (2011); Ballistreri *et al.* (2013); Tomás-Barberán *et al.* (2013) |
| Sour cherry (*Prunus cerasus* L.) | 74.0–754.0 | 1539.0–2983.0 | Kim *et al.* (2005); Bonerz *et al.* (2007); Dragović-Uzelac *et al.* (2007); Kirakosyan *et al.* (2009); Khoo *et al.* (2011); Wojdyło *et al.* (2014); Alrgei *et al.* (2016) |

### *Phenolic acids*

Phenolic acids or phenolcarboxylic acids are types of aromatic secondary plant metabolites, widely spread throughout the plant kingdom. They contribute to food quality and organoleptic properties, and they belong to two subgroups: the hydroxybenzoic and the hydroxycinnamic acids.

Small amounts of hydroxybenzoic acids have been found in sweet cherries (Mattila *et al.*, 2006). With respect to sour cherry, Díaz-García *et al.* (2013) found gallic acid, 3,4-dihydroxybenzoic acid and vanillic acid, in accordance with the results of Chaovanalikit and Wrolstad (2004).

In contrast to hydroxybenzoic acid content, sweet cherries are rich in derivatives of hydroxycinnamic acids, which are the dominant polyphenols in sweet cherry fruit (Tomás-Barberán *et al.*, 2013; Martínez-Esplá *et al.*, 2014). The major hydroxycinnamic acids in sweet cherry are neochlorogenic and *p*-coumaroylquinic acid, followed by chlorogenic acid (Serradilla *et al.*, 2016). According to Mozetiĉ *et al.* (2006), the ratio of neochlorogenic acid to *p*-coumaroylquinic acid is characteristic of each sweet cherry cultivar. Additionally, regarding neochlorogenic acid content, sweet cherry cultivars can be classified into three groups, the first group ranging between 40.0 and 128.0 mg per 100 g (e.g. 'Bing'), the second group between 20.0 and 40.0 mg per 100 g FW (e.g. '0900 Ziraat') and the third group ranging from 4.0 to 20.0 mg per 100 g FW (e.g. 'Sweetheart') (Ballistreri *et al.*, 2013). On the other hand, *p*-coumaroylquinic acid content ranges from 0.8 to 131.0 mg per 100 g FW (Table 17.3) (Serradilla *et al.*, 2016). Currently, this acid is increasingly receiving attention for its health-promoting potential due to its ability to inhibit low-density lipoprotein (Tomás-Barberán *et al.*, 2013). Finally, Ballistreri *et al.* (2013) reported that chlorogenic acid concentrations in 24 sweet cherry cultivars were between 0.2 and 8.7 mg per 100 g FW.

Regarding sour cherry cultivars, the hydroxycinnamic acids found in cultivars 'Érdi Bőtermő' and 'Aode' grown in Chinese agroecological conditions were neochlorogenic acid, 4-coumaroylquinic acid, caffeoylquinic acid, chlorogenic acid and 3′,5′- dicaffeoylquinic acid (Cao *et al.*, 2015), where neochlorogenic and chlorogenic acid were dominant. An earlier study (Kim *et al.*, 2005) showed that the amount of chlorogenic acid in sour cherries was between 0.6 and 5.8 mg per 100 g FW, while neochlorogenic acid ranged from 6.7 to 27.8 mg per 100 g FW.

Similarly, Wojdyło *et al.* (2014) determined that in almost all 33 sour cherry cultivars studied, neochlorogenic acid (~47%) was the major hydroxycinnamic acid derivative, followed by chlorogenic acid (~30%) and *p*-coumaroylquinic acid (~19%). In 'Oblačinska' sour cherry clones, chlorogenic acid, the most widespread natural plant dietary antioxidant, varied from 0.8 to 3.7 mg per 100 g FW (Alrgei *et al.*, 2016). Levaj *et al.* (2010) determined the derivatives of caffeic, *p*-coumaric and chlorogenic acid in both 'Maraska' and 'Oblačinska' sour cherry cultivars (Fig. 17.2).

**Table 17.3.** Standard phytochemical attributes of sweet and sour cherries.

| | Anthocyanins | | | Hydroxycinnamic acids | | | Flavonols | Flavan-3-ols | |
|---|---|---|---|---|---|---|---|---|---|
| Species | CY 3-*O*-GLU | CY 3-*O*-RUT | References | NCHL | PCQ | References | QUER | EPIC | References |
| Sweet cherry (*Prunus avium* L) | 0.1–35.0[a] | 2.0–243.0[a] | Gao and Mazza (1995); Usenik *et al.* (2008); Ballistreri *et al.* (2013) | 4.0–128.0[a] | 0.8–131.0[a] | Ballistreri *et al.* (2013); Serradilla *et al.* (2016) | 2.0–6.0[a] | 0.4–14.8[a] | Usenik *et al.* (2008); González-Gómez *et al.* (2010) |
| Sour cherry (*Prunus cerasus* L) | 0.9–1.3[a] | 9.5–17.1[a] | Jakobek *et al.* (2007) | 9.4–12.6[a] | | Mitić *et al.* (2012) | 0.03–0.8[a] | 18.0–283.0[b] | Wojdyło *et al.* (2014) |
| | 2.0–9.9[c] | 35.4–85.5[c] | Mitić *et al.* (2012) | 212.0–998.0[c] | 191.0–999.0[c] | Bonerz *et al.* (2007) | | | |
| | 10.1[c] | 93.0[c] | Damar and Ekşi (2012) | | | | | | |

CY 3-*O*-GLU, cyanidin 3-*C*-glucoside; CY 3-*O*-RUT, cyanidin 3-*O*-rutinoside; NCHL, neochlorogenic acid; PCQ, *p*-coumaroylquinic acid; QUER, quercetin; EPIC, epicatechin.

[a]mg per 100 g FW.

[b]mg per 100 g DW.

[c]mg $L^{-1}$ of juice.

**Fig. 17.2.** Sour cherries (cultivar 'Oblačinska').

### *Flavonoids*

The flavanoids or bioflavonoids are a class of plant secondary metabolites. They include anthoxanthins (flavones and flavonols), flavanones, flavanonols, flavans and anthocyanidins. They play a role in protection against UV radiation, as well as being natural pigments, enzyme inhibitors, and precursors of toxic substances, flavour components and antioxidants, and they also provide resistance to pathogens (Piccolella *et al.*, 2008). Their functionality in human health has been proved in numerous studies suggesting protective effects against cardiovascular diseases, cancers and other age-related diseases (Yao *et al.*, 2004). The main flavonoids present in cherries are provided below.

ANTHOCYANINS. The common anthocyanidins, which are responsible for the attractive colour of cherries, are cyanidin, pelargonidin, peonidin, delphinidin, petunidin and malvidin (Valero and Serrano, 2010). For their quantitative and qualitative analysis, the main methodology used has been the technique of high-performance liquid chromatography (HPLC) coupled to a diode array detector (DAD) or a single quadrupole mass spectrometer equipped with an atmospheric pressure electrospray ionization source (API-ES-MS) (González-Gómez *et al.*, 2010; Serra *et al.*, 2011). Anthocyanins such as cyanidin 3-*O*-rutinoside, cyanidin 3-*O*-glucoside, peonidin 3-*O*-rutinoside and pelargonidin 3-*O*-rutinoside have been reported in sweet cherries (Gonçalves *et al.*, 2004; González-Gómez *et al.*, 2010). However, Tomás-Barberán *et al.* (2013) and Serradilla *et al.* (2016) reported that cyanidin 3-*O*-rutinoside and cyanidin 3-*O*-glucoside are the predominant anthocyanins in sweet cherries.

For sweet cherries, total anthocyanin content ranges from a few milligrams per 100 g FW in light-coloured (score of 3 on the CTIFL colour chart) to about 300 mg per 100 g FW in dark cherries (score of 5) (Gao and Mazza, 1995; Wang *et al.*, 1997; Valero and Serrano, 2010). In general, light-coloured and dark-coloured red sweet cherry cultivars contain cyanidin 3-*O*-rutinoside (2.0–243.0 mg per 100 g FW) and cyanidin 3-*O*-glucoside (0.1–35.0 mg per 100 g FW) (Table 17.3), as the primary and secondary anthocyanin, respectively (Gao and Mazza, 1995; Usenik *et al.*, 2008; Ballistreri *et al.*, 2013).

The total anthocyanin content of sour cherries was reported to be between 27.8 and 80.4 mg per 100 g FW (Blando *et al.*, 2004). However, total anthocyanin content and the anthocyanin fractions differ according to the sour cherry cultivar (Wang *et al.*, 1997; Kim *et al.*, 2005; Simunic *et al.*, 2005). Several 'Oblačinska' clones, studied by Alrgei *et al.* (2016), showed substantial levels of total anthocyanin content (over 100.0 mg cyanidin 3-*O*-glucoside per 100 g FW). Several types of anthocyanin compounds were also determined in sour cherries by HPLC-DAD/API-ES-MS, but the most prevalent were those that are derivatives of cyanidin (cyanidin 3-*O*-glucosylrutinoside, cyanidin 3-*O*-sophoroside, cyanidin 3-*O*-rutinoside, cyanidin 3-*O*-glucoside, cyanidin 3-*O*-xylosylrutinoside and cyanidin 3-*O*-arabinosylrutinoside) (Blando *et al.*, 2004; Chaovanalikit and Wrolstad, 2004; Kim *et al.*, 2005; Bonerz *et al.*, 2007; Cao *et al.*, 2015). According to Kirakosyan *et al.* (2009), total cyanidins in 'Montmorency' cherries are about 93% of total anthocyanins, while in Balaton™ (syn. 'Újfehértói Fürtös') they are about 94%. Finally, Mulabagal *et al.* (2009) reported that the cultivars 'Montmorency' and 'Batalon' are characterized by exhibiting cyanidin 3-*O*-glucosylrutinoside and cyanidin 3-*O*-rutinoside at a ratio of 3/1.

Peonidin 3-*O*-rutinoside, peonidin- 3-*O*-glucoside and pelargonidin 3-*O*-glucoside were also found in sour cherry fruit but in much lower concentration (Kirakosyan *et al.*, 2009), while the Hungarian sour cherry cultivars included in the study of Ficzek *et al.* (2011) showed very low concentrations of delphinidin. Jakobek *et al.* (2009) quantified the content of cyanidin 3-*O*-rutinoside (56.9 mg per 100 g FW), cyanidin 3-*O*-glucosylrutinoside (940.1 mg per 100 g FW), cyanidin 3-*O*-sophoroside (18.6 mg per 100 g FW) and cyanidin 3-*O*-glucoside (7.0 mg per 100 g FW).

Recently, the sweet cherry fruit nutraceutical profile has been monitored using an array of instrumental techniques, including spectrophotometric assays, HPLC and nuclear magnetic resonance (NMR) (Goulas *et al.*, 2015). In particular, NMR spectroscopy allows a rapid screening of specific primary and secondary metabolites of sweet cherries; Goulas *et al.* (2015) showed that the resonance of H-4 can be used to discriminate anthocyanins in fruit extracts as it appears at 8.2–8.6 p.p.m., a non-overcrowded region of the spectrum. The resonance of H-4 is dependent on the substitution of the anthocyanin skeleton and the discrimination of anthocyanins in a complex mixture is feasible. In a subsequent step, cyanidin 3-*O*-rutinoside was used to study the effect of pH on the chemical shift of H-4. The data indicated that the chemical shift of the diagnostic peak (H-4) is strongly influenced by pH, highlighting the need for pH adjustment of the sample. Finally, a pH value of 3.0 was selected to obtain $^{1}$H-NMR spectra, since a sharp peak of H-4 was recorded and it is also the nearest pH to the actual pH of sweet cherry fruit at harvest (Goulas *et al.*, 2015).

The fact that cherries contain significant levels of anthocyanins has attracted much attention. One of the best-known properties of anthocyanins in general is their strong antioxidant activity in metabolic reactions, due to their ability to scavenge oxygen free radicals and other reactive species. Likewise, Wang *et al.* (1999) reported that sour cherry anthocyanins have an anti-inflammatory effect in cases of rheumatoid arthritis. Seeram *et al.* (2001) found that anthocyanins originating from sour cherries have an inhibitory effect on COX-1 and COX-2 enzymes, which trigger inflammation, offering some protection against colon cancer (Kang *et al.*, 2003) and against type II diabetes by increasing insulin excretion (Jayaprakasam *et al.*, 2005). Studies have shown that numerous factors such as harvest season, variety, stage of harvesting, climatic conditions and growing season can affect the composition and concentration of individual as well as total anthocyanins (Sass-Kiss *et al.*, 2005).

FLAVONOLS. Flavonols are very important bioactive compounds, crucial for human health (Knekt *et al.*, 2000). A total of six flavonols have been quantified in sweet cherry fruit, with quercetin being the predominant one, fluctuating from 2.0 to 6.0 mg per 100 g FW (Table 17.3) (Usenik *et al.*, 2008; Bastos *et al.*, 2015; Serradilla *et al.*, 2016). This compound has been reported to have a great ability to act as a free-radical scavenger and therefore is associated with the prevention of degenerative diseases caused by oxidative stress, such as cardiovascular disease and cancer (Tomás-Barberán *et al.*, 2013).

In sour cherry, Kirakosyan *et al.* (2009) claimed quercetin, kaempferol and isorhamnetin rutinoside to be the main flavonol compounds. Levaj *et al.* (2010) showed that both quercetin and kaempferol were present in ‘Maraska’ (5.4 and 3.0 mg per 100 g FW, respectively) and ‘Oblačinska’ sour cherry (3.8 and 1.3 mg per 100 g FW, respectively). The same flavonols in sour cherry were determined by Jakobek *et al.* (2007), Piccolella *et al.* (2008), Ferretti *et al.* (2010) and Liu *et al.* (2011). In addition, Alrgei *et al.* (2016) determined the quantities of myricetin, pinobanksin and galangin in specific ‘Oblačinska’ sour cherry clones.

FLAVAN-3-OLS. In cherries, (+)-catechin and (–)-epicatechin are the main flavan-3-ols identified (Serra *et al.*, 2011). Cherry fruit stores flavan-3-ols in much lower amounts than the rest of the polyphenols. In general, for sweet cherry, (–)-epicatechin contents are higher than (+)-catechin, ranging from 0.4 mg per 100 g FW (‘Lapins’) to 15.0 mg per 100 g FW ('Larian') (Usenik *et al.*, 2008; González-Gómez *et al.*, 2010). With respect to (+)-catechin, concentrations range from 2.9 mg per 100 g FW ('0900 Ziraat') to 9.0 mg per 100 g FW ('Noir de Guben') (Kelebek and Selli, 2011). The levels of these two compounds have been reported to be greatly influenced by agronomic and environmental conditions, as well as by genotype (Serradilla *et al.*, 2016).

For sour cherry, Usenik *et al.* (2010) reported the existence of procyanidin B2 and procyanidin dimer. As stated by Wojdyło *et al.* (2014) in a study with 33 sour cherry cultivars, procyanidin B1, procyanidin dimer, procyanidin trimer and procyanidin tetramer were found, and together ranged from 403.6 to 1215.7 mg per 100 g dry weight (DW). Besides procyanidins, Levaj *et al.* (2010) also found (+)-catechin, (–)-epicatechin and (+)-gallocatechin as monomers in both 'Maraska' and 'Oblačinska' sour cherry, which is in agreement with similar identification by Tsanova-Savova *et al.* (2005) and Chaovanalikit and Wrolstad (2004). As reported by Wojdyło *et al.* (2014), concentrations of (–)-epicatechin ranged from 18.0 to 283.0 mg per 100 g DW and of (+)-catechin from 4.0 to 116.0 mg per 100 g DW in 33 sour cherry cultivars. The highest monomer levels were found in 'Dradem', 'Meteor Korai' and 'Winer' fruits, while the lowest were found in 'Wanda', 'Lucyna' and 'Wifor' fruits. In contrast, catechin (1.4–1.6 mg per 100 g FW) is the only flavan-3-ol found in 'Érdi Bőttermő' and 'Aode' (Cao *et al.*, 2015) grown in China. In one study, no flavon-3-ols were reported in sour cherry cultivars (Kim *et al.*, 2005).

### 17.4.3 Indolamines

The indolamine melatonin (MLT; *N*-acetyl-5-methoxytryptamine) is an endogenous hormone found to be present in all vertebrates (Reiter, 1993). MLT is synthesized from tryptophan via 5-hydroxytryptophan, serotonin and *N*-acetylserotonin in the vertebrate pineal gland (Fig. 17.3). MLT has been shown to possess a great number of health benefits (Reiter *et al.*, 1997). The most well-known function of MLT in mammals is regulation of the sleep–wake cycle (Baker and Driver, 2007). Its other functions in humans range from sexual maturation to depression and antioxidative defence (Macchi and Bruce, 2004). As well as these properties, MLT has been reported to be a potent free-radical scavenger and a broad-spectrum antioxidant (Hardeland *et al.*, 2006). In addition, MLT detoxifies a variety of free radicals and reactive oxygen intermediates, including the hydroxyl radical, peroxynitrite anion, singlet oxygen and nitric oxide.

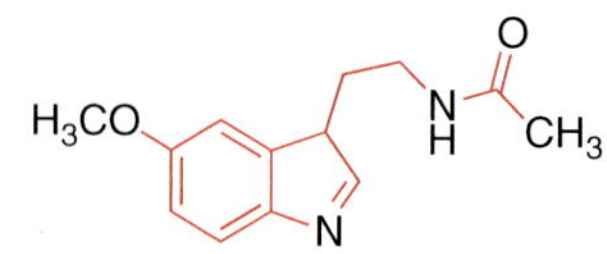


**Fig. 17.3.** The chemical structure of melatonin (*N*-acetyl-5-methoxytryptamine).

The presence of MLT is not restricted to the animal kingdom. This indolamine is also found in a wide variety of plants and fruits (Feng *et al.*, 2014). The MLT biosynthetic route starts with the primary metabolite shikimate; this metabolite serves as the precursor of tryptophan, which through different metabolic pathways concludes in the synthesis of melatonin (Kurkin, 2003). MLT consumed in plant products is absorbed, enters the circulation and has physiological effects via receptor- or non-receptor-mediated processes. A number of reports are available describing the positive health effects of MLT intake form cherry derivatives (Garrido *et al.*, 2009, 2012, 2013; Zhao *et al.*, 2013).

In recent years, there has been particular interest in determining and quantifying the presence of MLT in different cherry species and cultivars, since the abundance of MLT in cherries is strongly correlated with the species and fruit cultivar, and some research also indicates that MLT abundance is related to fruit maturity (Burkhardt *et al.*, 2001; González-Gómez *et al.*, 2009; Kirakosyan *et al.*, 2009). The data shown in Table 17.4 highlight the significantly higher amounts of MLT found in sour cherries.

**Table 17.4.** Melatonin concentrations reported for different sweet and sour cherry cultivars. Results are expressed as ng $g^{-1}$ dry weight (DW).

| Species | Cultivar | Amount reported (ng $g^{-1}$ DW) | Reference |
|---|---|---|---|
| Sweet cherry (*Prunus avium* L.) | 'Burlat' | 0.2 | González-Gómez *et al.* (2009) |
| | 'Sweetheart' | 0.1 | González-Gómez *et al.* (2009) |
| | 'Pico Negro' | 0.12 | González-Gómez *et al.* (2009) |
| | 'Navalinda' | 0.03 | González-Gómez *et al.* (2009) |
| | 'Van' | 0.01 | González-Gómez *et al.* (2009) |
| | 'Ambrunés' | 0.1 | González-Gómez *et al.* (2009) |
| | 'Pico Colorado' | 0.1 | González-Gómez *et al.* (2009) |
| Sour cherry (*Prunus cerasus* L.) | 'Montmorency' | 5.6–19.6 | Burkhardt *et al.* (2001) |
| | 'Montmorency' | 12.3 | Kirakosyan *et al.* (2009) |
| | Balaton™[a] | 1.1–2.2 | Burkhardt *et al.* (2001) |
| | Balaton™[a] | 2.9 | Kirakosyan *et al.* (2009) |

[a]Syn. 'Újfehértói Fürtös'.

### 17.4.4 Antioxidant activity

Antioxidant activity has been widely investigated using different methodological approaches. In cherry, antioxidant potential has been associated with ascorbic acid, phenols and anthocyanins (Chaovanalikit and Wrolstad, 2004; Serrano *et al.*, 2005, 2009). In addition, cherries are characterized by the total antioxidant activity (TAA) in both hydrophilic and lipophilic fractions by measuring the scavenging capacity of 2,2′-azino-bis(3-ethylbenzothiazoline-6-sulfonic acid) radicals ($ABTS^{\bullet +}$) (Tomás-Barberán *et al.*, 2013). For sweet cherries, it has been reported that dark-coloured cultivars, such as 'Sonata', exhibited higher concentrations in both fractions compared with light-coloured cultivars, such as 'Brooks'. In addition, for all cultivars, hydrophilic TAA is higher than lipophilic TAA, showing that polyphenols or hydrophilic compounds are the major contributors to antioxidant activity (Tomás-Barberán *et al.*, 2013). According to Ballistreri *et al.* (2013), the TAA of 24 sweet cherry cultivars ranged from 646.0 to 3166.0 μmol Trolox equivalents (TE) per 100 g FW.

The total antioxidant capacity of 34 sour cherries was determined as between 900.0 and 6300.0 μmol TE per 100 g FW using an ABTS assay, where 'Fanal' and 'Heimanns' exhibited strong antioxidant capacity (Khoo *et al.*, 2011). As determined by Wojdyło *et al.* (2014), the antioxidant activity of 33 sour cherry cultivars, evaluated using an oxygen radical absorbance capacity (ORAC) assay, was 8130.0–38,110.0 μmol TE 100 g DW. Among Hungarian sour cherry cultivars, 'Pipacs 1' presented an outstanding antioxidant capacity (21,850.0 μmol ascorbic acid $l^{-1}$), given the fact that it is an amarelle-type sour cherry with yellowish fruit flesh, and hence with an appreciably low anthocyanin content (Papp *et al.*, 2010). Tsuda *et al.* (1994) demonstrated that cyanidin 3-*O*-glucoside shows very strong antioxidant activity. According to Heinonen *et al.* (1998), using liposomes as model membranes, anthocyanins (especially malvidin with strong antioxidant activity and cyanidin, delphinidin and pelargonidin, which have pro-oxidant activity) and hydroxycinnamates isolated from sweet cherries are more active compared with those from other berries (e.g. blackberries, red raspberries, blueberries or strawberries).

## 17.5 Preharvest Factors Affecting Quality and Nutritional Compounds

Sweet cherry fruit is of prime importance worldwide with high commercial acceptability. However, sweet cherry is highly perishable after harvest; therefore, advanced fast precooling followed by cold storage is a necessary postharvest tool to maintain fruit quality until consumption (Manganaris *et al.*, 2007). As mentioned earlier, the main factors determining the consumer's acceptability are TSS, acidity and colour (Crisosto *et al.*, 2003). Producers use a number of parameters to establish the optimum time for harvesting, the most reliable being skin colour (Romano *et al.*, 2006). Red colour development in sweet cherry is used as an indicator of quality and ripening, and is due to the accumulation and profile of anthocyanins (Díaz-Mula *et al.*, 2009). In addition, as mentioned earlier, in recent studies, an inverse association between fruit and vegetable intake and chronic diseases, such as different types of cancer and cardiovascular diseases, has been demonstrated in numerous epidemiological studies in which phytochemicals have been indicated to be responsible for this observed protective effect (Schreiner and Huyskens-Keil, 2006). Among these compounds, special interest has been focused on anthocyanins and other polyphenolics, carotenoids and vitamins C and E.

Consumer choice and preference for sweet cherries is influenced mainly by factors such as convenience, culture, price, appearance, taste and, in recent years, also their nutrient value and content of bioactive compounds. Accordingly, there are different preharvest factors that influence the content of bioactive compounds at the time of harvest, the most important being cultivar, temperature and light intensity, ripening stage at harvest and some preharvest treatments such as salicylate derivatives, oxalic acid and abscisic acid.

### 17.5.1 Influence of cultivar

Differences in phenolic contents were found among cultivars, with concentrations ranging between 98.0 and 200.0 mg per 100 g FW (Díaz-Mula *et al.*, 2008). 'Brooks' cherry has the lowest anthocyanin content (40.0 mg per 100 g), while 'Cristalina' shows the highest (225.0 mg per 100 g FW). The main phenolic compounds in sweet cherry fruits are anthocyanins, which also differed in concentration depending on cultivar. Those cultivars with the lowest anthocyanins ('Brooks', 'Somerset', 'Prime Giant' and 'Sweetheart') are considered as light-coloured cultivars (score of 3 on the CTIFL colour chart), while those with the highest anthocyanin content ('Cristalina and 'Sonata') are classified as dark-coloured (score of 5), showing a direct relationship between colour parameters and anthocyanin concentration (Díaz-Mula *et al.*, 2008). The most abundant phenolic acids in sweet cherry are derivatives of hydroxycinnamic acid such as caffeic acid and *p*-coumaric acid. The most common colourless phenolics in sweet cherries are neochlorogenic acid (3′-caffeilquinic acid) and *p*-coumaroylquinic acid (Mozetiĉ *et al.*, 2002; Chaovanalikit and Wrolstad, 2004). The hydroxycinnamates are increasingly receiving attention for their potential health-promoting effects through their potent antioxidant action, their ability to inhibit low-density lipoprotein oxidation, and their chemopreventative properties (e.g. inhibitory effects on tumour promotion and the ability to block the formation of mutagenic compounds such as nitrosamines), as demonstrated by *in vitro* studies (Boots *et al.*, 2008; McCune *et al.*, 2011). The ability of phenolics to act as free-radical scavengers suggests that they could play a beneficial role in reducing reactive oxygen species (i.e. hydrogen peroxide, superoxide anion) associated with chronic diseases such as cardiovascular disease and cancer (Wilms *et al.*, 2005). Sweet cherry cultivars have a considerable influence on the antioxidant capacity in both hydrophilic and lipophilic extracts as measured by the scavenging capacity of $ABTS^{\bullet+}$ radicals. According to Díaz-Mula *et al.* (2008), hydrophilic TAA is usually higher than lipophilic TAA for all studied cultivars (~80% of TTA in 'Cristalina' and ~50% in 'Prime Giant'), showing that the major contributors to antioxidant activity are hydrophilic compounds, such as polyphenols and anthocyanins. Antioxidant vitamins, such as tocopherols, and carotenoids are lipophilic compounds that might contribute to lipophilic TAA.

According to the National Cancer Institute (2004), sweet cherry contains important amounts of carotenoids. Although carotenoids are another important bioactive constituent in fruits (Valero and Serrano, 2010), almost no evidence exists on their occurrence in sweet cherry. Valero *et al.* (2011) quantified carotenoids in two sweet cherry cultivars ('Prime Giant' and 'Cristalina'), and found different concentrations in both cultivars, with 'Prime Giant' having significantly higher total carotenoids (1.1 mg per 100 g FW) than 'Cristalina' (0.6 mg per 100 g FW). Leong and Oey (2012) reported individual contents in sweet cherry with concentration of 2.0 mg per 100 g DW for β-carotene, β-cryptoxanthin and α-carotene, and 1.0 mg per 100 g DW for lycopene and lutein.

In sweet cherry, differences in vitamin C concentration at time of harvest have been reported. Thus, cultivar '4-70' had 28.2 mg per 100 g FW (Serrano *et al.*, 2005), while 'Souvenir', 'Samba' and 'Prime Giant' showed ascorbic acid values of 3.98, 2.30 and 5.95 mg per 100 g FW, respectively (Schmitz-Eiberger and Blanke, 2012). In sour cherry, according to Wojdyło *et al.* (2014), the content of ascorbic acid within 33 sour cherry cultivars differed greatly, ranging from 5.5 mg per 100 g FW ('Kelleris 14') to 22.1 mg per 100 g FW ('Morina'), although its content in most of the analysed cultivars was lower than 10.0 mg per 100 g FW.

### 17.5.2 Temperature and light intensity

Light intensity increases the levels of ascorbic acid, and different growing temperatures (day/night) also affect the TPC. High-temperature growth conditions (25/30°C) significantly enhance the anthocyanin and TPC (Wang, 2006). Recently, there has been increasing interest in growing cherries under plastic greenhouses, especially in cold areas. This cultivation system can influence canopy and soil temperature, and the quantity and quality of transmitted, reflected or absorbed light (Ferretti *et al.*, 2010). The highest levels of nutrients and bioactive components were found in the year characterized by the highest temperature and greatest solar radiation exposure (McCune *et al.*, 2011).

### 17.5.3 Ripening stage

Fruit ripening is a highly coordinated, genetically programmed process occurring at the later stages of fruit development and involving a series of physiological, biochemical and sensory changes leading to an edible ripe fruit with desirable quality parameters (Valero and Serrano, 2010).

In sweet cherry, the ripening process is characterized by colour changes, from green to red, which can be followed by the evolution of $L^*$, $a^*$ and $b^*$ parameters, and are due to the accumulation and profile of anthocyanins. In fact, red colour development in sweet cherry is used as indicator of quality and ripening of fresh cherry (Esti *et al.*, 2002; Serrano *et al.*, 2005; Mozetiĉ *et al.*, 2006). Harvesting is usually performed based on the attainment of acceptable fruit size, fruit firmness, colour and concentration of soluble solids. However, there is little available information about the changes in the content of health-promoting compounds during sweet cherry development and ripening on the tree. Serrano *et al.* (2005) reported changes in the concentrations and activities of antioxidants of sweet cherry at 14 different stages of ripeness, with total anthocyanins increasing exponentially from stage 8 to the maximum value at stage 14 (63.3 mg cyanidin equivalent activity per 100 g FW). TAA decreased from stage 1 to stage 8, and increased again from stage 8 to stage 14, coinciding with the TPC and the accumulation of anthocyanins. TAA reached its maximum activity at stage 14, with average ascorbic acid equivalent activity of 50.0 mg per 100 g FW. Thus, harvesting sweet cherries at stage 12 of ripening when the fruit reaches maximum size would support the development of the highest organoleptic, nutritional and functional quality attributes.

Gonçalves *et al.* (2004) investigated total phenolics in four cherry cultivars at two ripening stages and found the lowest total phenolics in cultivar 'Van' at the partially ripe stage (69.0 mg per 100 g FW), compared with the highest in cultivar 'Saco' at the fully ripened stage (264.0 mg per 100 g FW). Similarly, as maturity progressed in Turkish sweet cherry (unknown cultivar), the total phenolics also increased (Mahmood *et al.*, 2013). In red-coloured fruits, total phenols generally increase during the ripening stage due to the maximal accumulation of anthocyanins and flavonols.

### 17.5.4 Preharvest treatments

Signalling molecules, such as salicylic acid (SA) and methyl jasmonate, are endogenous plant growth substances that may play a key role in plant growth and development, and in responses to environmental stresses. The effects of SA or acetylsalicylic acid (ASA) treatments (at 0.5, 1.0 and 2.0 mM concentrations) during on-tree cherry growth and ripening were studied in ‘Sweetheart’ and ‘Sweet Late’ cultivars, and showed that treated cherries had higher concentrations of total phenolics and total anthocyanins, as well as higher antioxidant activity, in both the hydrophilic and lipophilic fractions (Giménez *et al.*, 2014). On average, treated fruit had 10–15% more phenolics, 15–20% more anthocyanins and 40–60% more antioxidant activity. The authors postulated that preharvest treatments with SA or ASA could be promising tools to improve sweet cherry quality and the health-beneficial effects for consumers.

In sour cherry (‘Cigány’), trees were sprayed with 250.0 mg $L^{-1}$ ethephon 1 week before the anticipated commercial harvest. Fruit from ethephon-sprayed trees had significantly lower soluble solids concentration (SSC), anthocyanin content, antioxidant activity and firmness than those from non-sprayed controls. The ethephon spray did not affect TPC, although its content tended to be higher in fruit from non-treated controls. TA, pH and SSC/TA ratio were not affected by the ethephon spray (Khorshidi and Davarynejad, 2010).

Abscisic acid (ABA) is a plant growth regulator, and plays a variety of important roles throughout the life cycle of a plant. These roles include seed development and dormancy, the plant response to environmental stresses and fruit ripening. ABA concentration is very low in unripe fruit, but increases as the fruit ripens, so it is believed that ABA plays an important role in regulating the rate of fruit ripening. ABA application 36 days after full blossom increased the total sugar content of fruit and stimulated the accumulation of anthocyanin in sweet cherry. In contrast, ABA and ethephon applications decreased the malic acid content, whereas applications 30 days after full blossom failed to reduce the malic acid levels (Kondo and Inoue, 1977). These results suggest that ABA may be closely related to the maturation of cherry fruit, and that the effects of ABA and ethephon on maturation may vary with the time of application. It was found that ABA content increased rapidly at the straw-coloured stage and reached its highest level 4 days before commercial harvest time. During the straw-coloured stage, the application of exogenous ABA induced its accumulation, anthocyanin biosynthesis and an increase in the maturity index (TSS/TA), thereby promoting fruit ripening (Luo *et al.*, 2014).

Oxalic acid (OA), as a final metabolic product in plants, has many physiological functions, the main one being related to the induction of systemic resistance against diseases caused by fungi, bacteria and viruses by increasing defence-related enzyme activities and secondary metabolites such as phenolics. Trees of ‘Sweetheart’ and ‘Sweet Late’ sweet cherry cultivars treated with OA at 0.5, 1.0 and 2.0 mM increased fruit size at harvest, manifested by higher fruit volume and weight. Other quality parameters, such as colour and firmness, were also increased by OA treatments, which were accompanied by increases in total anthocyanins, total phenolics and antioxidant activity (Martínez-Esplá *et al.*, 2014). At the time of harvest, treated cherries had 15–20% more phenolics and 25–30% more anthocyanins, while increases of 70–80% were obtained for TAA.

## 17.6 Postharvest Factors Affecting Quality and Nutritional Compounds

The sweet cherry horticultural production chain involves a number of steps: production, harvesting, precooling, cooling, selection, grading, packaging, transport, distribution and consumption. Extension of the postharvest life of sweet cherry depends on three factors: (i) a reduction in dehydration and weight loss: (ii) slowing down the physiological processes of maturation and senescence; and (iii) avoiding the onset and rate of microbial growth. To control these three factors, the main tools are refrigeration and controlling the relative humidity. The optimum temperature for harvest and handling of cherries is between 10 and 20°C (outside this temperature range, more pitting is observed), while the optimum storage temperature is 0°C, with a relative humidity of 90–95% (Romano *et al.*, 2006). Thus, storage at low temperatures is the main postharvest treatment to reduce sweet cherry metabolism, maintain quality and prolong the storability in those perishable fruit and vegetables considered to be non-chilling sensitive, such as sweet cherry fruit. Some evidence exists on the changes in bioactive compounds and antioxidant activity during cold storage, although no general trends have been found. Thus, loss of health-beneficial compounds (phenolics and ascorbic acid) has been found in table grapes, broccoli, pomegranate and apple (Serrano *et al.*, 2011), in which the loss of phenolics was highly dependent on cultivar. However, increases in phytochemicals were reported for sweet cherry during cold storage, although different behaviour has been reported depending on storage temperature. Gonçalves *et al.* (2004) studied the phenolic compounds hydroxycinnamates, anthocyanins, flavonols and flavan-3-ols of 'Burlat', 'Saco', 'Summit' and 'Van' sweet cherry cultivars harvested at two different ripening stages and stored under different cold conditions. Phenolic acid content generally decreased with storage at 1–2°C and increased with storage at 15 ± 5°C. Anthocyanin levels increased at both storage temperatures, while flavonol and flavan-3-ol contents remained quite constant.

The maturity stage at harvest also determines the antioxidant potential after cold storage of sweet cherries. In a study on 11 cherry cultivars harvested at three ripening stages (S1, S2 and S3), significant increases in anthocyanin content were found during cold storage and subsequent shelf life at 20°C, the accumulation of anthocyanins during storage being attributed to normal sweet cherry ripening (Serrano *et al.*, 2009). HPLC-DAD chromatograms revealed that in all cultivars the main anthocyanins were cyanidin 3-*O*-rutinoside, followed by cyanidin 3-*O*-glucoside and pelargonidin 3-*O*-rutinoside, which increased with ripening from S1 to S3. With respect to total phenolics, an increase in total phenolic compounds as maturity advanced was observed (from S1 to S3) for all cultivars. As mentioned above, neochlorogenic acid was the predominant hydroxycinnamic acid followed by *p*-coumaroylquinic acid, and both increased significantly from S1 to S3 and during storage.

In recent years, particular attention has been paid to the use of natural safe compounds as postharvest treatments to improve the content of bioactive compounds during storage of sweet cherries. Thus, 'Cristalina' and 'Prime Giant' cherries harvested at the commercial ripening stage and treated with SA, ASA or OA at 1 mM before storage under cold temperature showed beneficial effects on maintenance of organoleptic quality by a delay of the postharvest ripening process, manifested by lower acidity, colour changes and firmness losses. This delay was also manifested by a delay in the accumulation of total phenolics, anthocyanins and antioxidant activity (Valero *et al.*, 2011).

Another postharvest treatment with beneficial effects on reducing postharvest ripening of sweet cherry has been the use of edible coatings. In this sense, 'Sweetheart' cherry coated with sodium alginate at several concentrations (1, 3 or 5%, w/v) delayed the evolution of parameters related to postharvest ripening, such as colour, softening and loss of acidity, and reduced respiration rate. In addition, the edible coatings showed a positive effect on maintaining higher concentration of total phenolics and TAA, which decreased in control fruit associated with the over-ripening and senescence processes (Díaz-Mula *et al.*, 2012). Since the ingestion of fruit and vegetables with higher amounts of phenolics has antioxidant activity '*in vivo*' by increasing the plasma antioxidants (Fernández-Panchón *et al.*, 2008), the use of alginate as an edible coating led to fruits with higher proportion of functional properties than control ones. However, no data exist on the bioavailability and bioconversion of phenolic compounds after the intake of sweet cherry fruit, and thus more research is needed on this issue.

## 17.7 Medicinal, Traditional (Folk) and Other Usage

As described earlier, sweet cherry fruit contains fibre, vitamin C, carotenoids and anthocyanins, each of which may help play a role in cancer prevention. Medicine can be prepared from the stalks of sweet cherry drupes, which are astringent, antitussive and diuretic (Baytop, 1984). The hard, reddish-brown wood (cherry wood) is valued as a hardwood for woodturning, and for making cabinets and musical instruments (Baytop, 1984). In Turkey, sarma, a famous dish traditionally prepared from grape leaves, can also be made out of sweet cherry leaves. Sweet cherry leaves are rolled around a filling usually based on ground meat. It is found in the cuisines of the former Ottoman Empire from the Middle East to the Balkans and central Europe.

The fruit and stem of the sour cherry are also used to produce medicine and food. Sour cherry is used for osteoarthritis, muscle pain, gout, to increase urine production, and to help digestion (McCune *et al.*, 2011). Sour cherries are eaten as a food or flavouring. Sour cherry fruit contains ingredients that reduce inflammation, protect from oxidative stress in neuronal cells (Wang *et al.*, 1999) and enhance muscle recovery (Connolly *et al.*, 2006). They also contain MLT, which helps to regulate sleep patterns (Pigeon *et al.*, 2010). With regard to sour cherry anthocyanins, *in vitro* studies have demonstrated that they are able to reduce the proliferation of human colon cancer cells in culture (Kang *et al.*, 2003).

## 17.8 Conclusions

Sweet and sour cherries are popular temperate fruits due mainly to their excellent organoleptic characteristics, especially sweet cherries. In addition, they are important sources of nutrient and bioactive food components, mainly sour cherries, and are potentially beneficial to health, and for this reason should be included as an essential part of the human diet.

## References


Alrgei, H.O., Dabić, D., Natić, M., Rakonjac, V., Milojković-Opsenica, D., Tešić, Ž. and Fotirić Akšić, M. (2016) Chemical profile of major taste- and health-related compounds of (Oblačinska) sour cherry. *Journal of the Science of Food and Agriculture* 96, 1241–1251.

Baker, F.C. and Driver, H.S. (2007) Circadian rhythms, sleep, and the menstrual cycle. *Sleep Medicine* 8, 613–622.

Ballistreri, G., Continella, A., Gentile, A., Amenta, M., Fabroni, S. and Rapisarda, P. (2013) Fruit quality and bioactive compounds relevant to human health of sweet cherry (*Prunus avium* L.) cultivars grown in Italy. *Food Chemistry* 140, 630–638.

Bastos, C., Barros, L., Dueñas, M., Calhelha, R.C., Queiroz, M.J.R.C., Santos-Buelga, C. and Ferreira, I.C.F.R. (2015) Chemical characterisation and bioactive properties of *Prunus avium* L.: the widely studied fruits and the unexplored stems. *Food Chemistry* 173, 1045–1053.

Baytop, T. (1984) *Therapy with Medicinal Plants in Turkey (Past and Present)*. Nobel Press, Istanbul, Turkey.

Blando, F., Gerardi, C. and Nicoletti, I. (2004) Sour cherry (*Prunus cerasus* L.) anthocyanins as ingredients for functional foods. *Journal of Biomedicine and Biotechnology* 5, 253–258.

Bonerz, D., Wurth, K., Dietrich, H. and Will, F. (2007) Analytical characterization and the impact of ageing on anthocyanin composition and degradation in juices from five sour cherry cultivars. *European Food Research and Technology* 224, 355–364.

Boots, A.W., Haenen, G.R. and Bast, A. (2008) Health effects of quercetin: from antioxidant to neutraceutical. *European Journal of Pharmacology* 585, 325–337.

Burkhardt, S., Dun Xian, T., Manchester, L.C., Hardeland, R. and Reiter, R.J. (2001) Detection and quantification of the antioxidant melatonin in Montmorency and Balaton tart cherries (*Prunus cerasus*). *Journal of Agricultural and Food Chemistry* 49, 4898–4902.

Cao, J., Jiang, Q., Lin, J., Li, X., Sun, C. and Chen, K. (2015) Physicochemical characterisation of four cherry species (*Prunus* spp.) grown in China. *Food Chemistry* 173, 855–863.

Chaovanalikit, A. and Wrolstad, R.E. (2004) Total anthocyanins and total phenolics of fresh and processed cherries and their antioxidant properties. *Journal of Food Science* 69, 67–72.

Connolly, D.A.J., McHugh, M.P. and Padilla-Zakour, O.I. (2006) Efficacy of a tart cherry juice blend in preventing the symptoms of muscle damage. *British Journal of Sports Medicine* 40, 679–683.

Crisosto, C.H., Crisosto, G.M. and Metheney, P. (2003) Consumer acceptance of 'Brooks' and 'Bing' cherries is mainly dependent on fruit SSC and visual skin color. *Postharvest Biology and Technology* 28, 159–167.

Damar, I. and Ekşi, A. (2012) Antioxidant capacity and anthocyanin profile of sour cherry (*Prunus cerasus* L.) juice. *Food Chemistry* 135, 2910–2914.

Díaz-García, M.C., Obón, J.M., Castellar, M.R., Collado, J. and Alacid, M. (2013) Quantification by UHPLC of total individual polyphenols in fruit juices. *Food Chemistry* 138, 938–949.

Díaz-Mula, H.M., Castillo, S., Martínez-Romero, D., Valero, D., Zapata, P.J., Guillén, F. and Serrano, M. (2008) Sensory, nutritive and functional properties of sweet cherry as affected by cultivar and ripening stage. *Food Science and Technology International* 15, 535–543.

Díaz-Mula, H.M., Valero, D., Zapata, P.J., Guillén, F., Castillo, S., Martínez-Romero, D. and Serrano, M. (2009) The functional properties of sweet cherry as a new criterion in a breeding program. *Acta Horticulturae* 839, 275–280.

Díaz-Mula, H.M., Serrano, M. and Valero, D. (2012) Alginate coatings preserve fruit quality and bioactive compounds during storage of sweet cherry fruit. *Food and Bioprocess Technology* 5, 2990–2997.

Dragović-Uzelac, V., Levaj, B., Bursać, D., Pedišić, S., Radojčić, I. and Biško, A. (2007) Total phenolics and antioxidant capacity assays of selected fruits. *Agriculturae Conspectus Scientificus* 72, 279–284.

Esti, M., Cinquanta, L., Sinesio, F., Moneta, E. and Di Matteo, M. (2002) Physicochemical and sensory fruit characteristics of two sweet cherry cultivars after cool storage. *Food Chemistry* 76, 399–405.

Faniadis, D., Drogoudi, P.D. and Vasilakakis, M. (2010) Effects of cultivar, orchard elevation, and storage on fruit quality characters of sweet cherry (*Prunus avium* L.). *Scientia Horticulturae* 125, 301–304.

Feng, X., Wang, M., Zhao, Y., Han, P. and Dai, Y. (2014) Melatonin from different fruit sources, functional roles, and analytical methods. *Trends in Food Science and Technology* 37, 21–31.

Fernández-Panchón, M.S., Villano, D., Troncoso, A.M. and García-Parrilla, M.C. (2008) Antioxidant activity of phenolic compounds: from *in vitro* results to *in vivo* evidence. *Critical Reviews in Food Science and Nutrition* 48, 649–671.

Ferretti, G., Bacchetti, T., Belleggia, A. and Neri, D. (2010) Cherry antioxidants: from farm to table. *Review Molecules* 15, 6993–7005.

Ficzek, G., Végvári, G., Sándor, G., Stéger-Máté, M., Kállay, E., Szügyi, S. and Tóth, M. (2011) HPLC evaluation of anthocyanin components in the fruits of Hungarian sour cherry cultivars during ripening. *Journal of Food, Agriculture and Environment* 9, 30–35.

Filimon, R.V., Beceanu, D., Niculaua, M. and Arion, C. (2011) Study on the anthocyanin content of some sour cherry varieties grown in Iaşi area, Romania. *Cercetări Agronomice in Moldova* 1, 81–91.

Fotirić Akšić, M., Nikolić, T., Zec, G., Cerović, R., Nikolić, M., Milivojević, J. and Radivojević, D. (2015) 'LENKA', a new sour cherry cultivar from Serbia. In: *Proceedings from the Third Balkan Symposium on Fruit Growing*, 16–18 September, Belgrade, Serbia, S1-P14, p. 27.

Gao, L. and Mazza, G. (1995) Characterization, quantification, and distribution of anthocyanins and colorless phenolic in sweet cherries. *Journal of Agricultural and Food Chemistry* 43, 343–346.

Garrido, M., Espino, J., González-Gómez, D., Lozano, M., Cubero, J., Toribio-Delgado, A.F., Maynar-Mariño, J.I., Terrón, M.P., Muñoz, J.L., Pariente, J.A., Barriga, C., Paredes, S.D., and Rodríguez, A.B. (2009) A nutraceutical product based on Jerte Valley cherries improves sleep and augments the antioxidant status in humans. *e-SPEN* 4, 321–323.

Garrido, M., Espino, J., González-Gómez, D., Lozano, M., Barriga, C., Paredes, S.D. and Rodríguez, A.B. (2012) The consumption of a Jerte valley cherry product in humans enhances mood, and increases 5-hydroxyindoleacetic acid but reduces cortisol levels in urine. *Experimental Gerontology* 47, 573–580.

Garrido, M., González-Gómez, D., Lozano, M., Barriga, C., Paredes, S.D., Rodríguez, A.B. (2013) Characterization and trials of a Jerte valley cherry product as a natural antioxidant-enriched supplement. *Italian Journal of Food Science* 25, 90–97.

Giménez, M.J., Valverde, J.M., Valero, D., Guillén, F., Martínez-Romero, D., Serrano, M. and Castillo, S. (2014) Quality and antioxidant properties on sweet cherries as affected by preharvest salicylic and acetylsalicylic acids treatments. *Food Chemistry* 160, 226–232.

Girard, B. and Kopp, T.G. (1998) Physicochemical characteristics of selected sweet cherry cultivars. *Journal of Agricultural and Food Chemistry* 46, 471–476.

Gonçalves, B., Landbo, A.K., Knudse, D., Silva, A.P., Moutinho-Pereira, J., Rosa, E. and Meyer, A.S. (2004) Effect of ripeness and postharvest storage on the phenolic profiles of cherries (*Prunus avium* L.). *Journal of Agricultural and Food Chemistry* 52, 523–530.

González-Gómez, D., Lozano, M., Fernández-León, M.F., Ayuso, M.C., Bernalte, M.J. and Rodríguez, A.B. (2009) Detection and quantification of melatonin and serotonin in eight sweet cherry cultivars (*Prunus avium* L.). *European Food Research and Technology* 229, 223–229.

González-Gómez, D., Lozano, M., Fernández-León, M.F., Bernalte, M.J., Ayuso, M.C. and Rodríguez, A.B. (2010) Sweet cherry phytochemicals: identification and characterization by HPLC-DAD/ESI-MS in six sweet-cherry cultivars grown in Valle del Jerte (Spain). *Journal of Food Composition Analysis* 23, 533–539.

Goulas, V., Minas, I.S., Kourdoulas, P.M., Lazaridou, A., Molassiotis, A.N., Gerothanasis, I. and Manganaris, G.A. (2015) $^1$H NMR metabolic fingerprinting to probe temporal postharvest changes on qualitative attributes and phytochemical profile of sweet cherry fruit. *Frontiers in Plant Science* 6, 959.

Grafe, C. and Schuster, M. (2014) Physicochemical characterization of fruit quality traits in a German sour cherry collection. *Scientia Horticulturae* 180, 24–31.

Guyer, D.E., Sinha, N.K., Chang, T.S. and Cash, J.N. (1993) Physico-chemical and sensory characteristics of selected Michigan sweet cherry (*Prunus avium* L.) cultivars. *Journal of Food Quality* 16, 355–370.

Hardeland, R., Pandi-Perumal, S.R. and Cardinali, D.P. (2006) Melatonin. *International Journal of Biochemistry and Cell Biology* 38, 313–316.

Hayaloglu, A.A. and Demir, N. (2015) Physicochemical characteristics, antioxidant activity, organic acid and sugar contents of 12 sweet cherry (*Prunus avium* L.) cultivars grown in Turkey. *Journal of Food Science* 80, 564–570.

Heinonen, M., Meyer, A.S. and Frankel, E.N. (1998) Antioxidant activity of berry phenolics on human low-density lipoprotein and liposome oxidation. *Journal of Agricultural and Food Chemistry* 46, 4107–4111.

Jakobek, L., Šeruga, M., Medvidović-Kosanović., M. and Novak., I. (2007) Anthocyanin content and antioxidant activity of various red fruit juices. *Deutsche Lebensmittel Rundschau* 103, 58–64.

Jakobek, L., Šeruga, M., Šeruga, B., Novak, I. and Medvidović-Kosanović, M. (2009) Phenolic compound composition and antioxidant activity of fruits of *Rubus* and *Prunus* species from Croatia. *International Journal of Food Science and Technology* 44, 860–868.

Jayaprakasam, B., Vareed, S.K., Olson, L.K. and Nair, M.G. (2005) Insulin secretion by bioactive anthocyanins and anthocyanidins present in fruits. *Journal of Agricultural and Food Chemistry* 53, 28–31.

Kang, S.Y., Seeram, N.P., Nair, M.G. and Bourquin, L.D. (2003) Tart cherry anthocyanins inhibit tumor development in Apc$^{Min}$ mice and reduce proliferation of human colon cancer cells. *Cancer Letters* 194, 13–19.

Kelebek, H. and Selli, S. (2011) Evaluation of chemical constituents and antioxidant activity of sweet cherry (*Prunus avium* L.) cultivars. *International Journal of Food Science and Technology* 46, 2530–2537.

Khoo, G.M., Clausen, M.R., Pedersen, B.H. and Larsen, E. (2011) Bioactivity and total phenolic content of 34 sour cherry cultivars. *Journal of Food Composition and Analysis* 24, 772–776.

Khorshidi, S. and Davarynejad, G. (2010) Influence of preharvest ethephon spray on fruit quality and chemical attributes of 'Cigány' sour cherry cultivar. *Journal of Biodiversity and Environmental Sciences* 4, 133–141.

Kim, D.O., Heo, H.J., Kim, Y.J., Yang, H.S. and Lee, C.Y. (2005) Sweet and sour cherry phenolics and their protective effects on neuronal cells. *Journal of Agricultural and Food Chemistry* 53, 9921–9927.

Kirakosyan, A., Seymour, E.M., Urcuyo Llanes, D.E., Kaufman, P.B. and Bolling, S.F. (2009) Chemical profile and antioxidant capacities of tart cherry products. *Food Chemistry* 115, 20–25.

Knekt, P., Isotupa, S., Rissanen, H., Heliovaara, M., Jarvinen, R., Hakkinen, S., Aroma, A. and Reunanen, A. (2000) Quercetin intake and the incidence of cerebrovascular disease. *European Journal of Clinical Nutrition* 54, 415–417.

Kondo, S. and Inoue, K. (1977) Abscisic acid (ABA) and 1-aminocyclopropane-1-carboxylic acid (ACC) content during growth of 'Satohnishiki' cherry fruit, and effect of ABA and ethephon application on fruit quality. *Journal of Horticultural Science and Biotechnology* 72, 221–227.

Kurkin, V.A. (2003) Phenylpropanoids from medicinal plants: distribution, classification, structural analysis, and biological activity. *Chemistry of Natural Compounds* 39, 123–153.

Leong, S.Y. and Oey, I. (2012) Effects of processing on anthocyanins, carotenoids and vitamin C in summer fruits and vegetables. *Food Chemistry* 133, 1577–1587.

Levaj, B., Dragović-Uzelac, V., Delonga, K., Kovacevic Ganic, K., Banovic, M. and Bursac Kovacevic, D. (2010) Polyphenols and volatiles in sour cherries, berries and jams. *Food Technology and Biotechnology* 48, 538–547.

Li, X.L., Kang, L., Hu, J.J., Li, X.F. and Shen, X. (2008) Aroma volatile compound analysis of SPME headspace and extract samples from crabapple (*Malus* sp.) fruit using GC–MS. *Agricultural Science in China* 7, 1451–1457.

Liu, Y., Liu, X., Zhong, F., Tian, R., Zhang, K., Zhang, X. and Li, T. (2011) Comparative study of phenolic compounds and antioxidant activity in different species of cherries. *Journal of Food Science* 76, 633–638.

Luo, H., Dai, S.J., Ren, J., Zhang, C.X., Ding, Y., Li, Z., Sun, Y., Ji, K., Wang, Y.P., Li, Q., Chen, P., Duan, C., Wang, Y. and Leng, P. (2014) The role of ABA in the maturation and postharvest life of a nonclimacteric sweet cherry fruit. *Journal of Plant Growth Regulation* 33, 373–383.

Macchi, M.M. and Bruce, J.N. (2004) Human pineal physiology and functional significance of melatonin. *Frontiers in Neuroendocrinology* 25, 177–195.

Mahmood, T., Anwar, T.F., Bhatti, I.A. and Iqbal, T. (2013) Effect of maturity on proximate composition, phenolics and antioxidant attributes of cherry fruit. *Pakistan Journal of Botany* 45, 909–914.

Manganaris, G.A., Ilias, I.F., Vasilakakis, M. and Mignani, I. (2007) The effect of hydrocooling on ripening related quality attributes and cell wall physicochemical properties of sweet cherry fruit (*Prunus avium* L.). *International Journal of Refrigeration* 30, 1386–1392.

Martínez-Esplá, A., Zapata, P.J., Valero, D., García-Viguera, C., Castillo, S. and Serrano, M. (2014) Preharvest application of oxalic acid increased fruit size, bioactive compounds, and antioxidant capacity in sweet cherry cultivars (*Prunus avium* L.). *Journal of Agricultural and Food Chemistry* 62, 3432–3437.

Mattila, P., Hellström, J. and Törrönen, R. (2006) Phenolic acids in berries, fruits, and beverages. *Journal of Agricultural and Food Chemistry* 54, 7193–7199.

McCune, L.M., Kubota, C., Stendell-Hollins, N.R. and Thomson, C.A. (2011) Cherries and health: a review. *Critical Reviews in Food Science and Nutrition* 51, 1–12.

Mitić, M.N., Obradović, M.V., Kostić, D.A., Micić, R.J. and Pecev, E.T. (2012) Polyphenol content and antioxidant activity of sour cherries from Serbia. *Chemical Industry and Chemical Engineering Quarterly* 18, 53–62.

Mozetiĉ, B., Trebŝe, P. and Hribar, J. (2002) Determination and quantitation of anthocyanins and hydroxycinnamic acids in different cultivars of sweet cherries (*Prunus avium* L.) from Nova Gorica Region (Slovenia). *Food Technology and Biotechnology* 40, 207–212.

Mozetiĉ, B., Simĉiĉ, M. and Trebŝe, P. (2006) Anthocyanins and hidroxycinnamic acids of Lambert Compact cherries (*Prunus avium* L.) after cold storage and 1-methylcyclopropene treatments. *Food Chemistry* 97, 302–309.

Mulabagal, V., Lang, G.A., Dewitt, D.L., Dalavoy, S.S. and Nair, M.G. (2009) Anthocyanin content, lipid peroxidation and cyclooxygenase enzyme inhibitory activities of sweet and sour cherries. *Journal of Agricultural and Food Chemistry* 57, 1239–1246.

National Cancer Institute (2004) *Glycemic Index Database Based on CS-FII 96 Data*. DHQ Nutrient Database, Applied Research Program, National Cancer Institute, Bethesda, Maryland.

Pacifico, S., Di Maro, A., Petriccione, M., Galasso, S., Piccolella, S., Di Giuseppe, A.M.A., Scortichini, M. and Monaco, P. (2014) Chemical composition, nutritional value and antioxidant properties of autochthonous *Prunus avium* cultivars from Campania Region. *Food Research International* 64, 188–199.

Papp, N., Szilvássy, B., Abrankó, L., Szabó, T., Pfeiffer, P., Szabó, Z., Nyéki, J., Ercisli, S., Stefanovits-Bányai, E. and Hegedüs, A. (2010) Main quality attributes and antioxidants in Hungarian sour cherries: identification of genotypes with enhanced functional properties. *International Journal of Food Science and Technology* 45, 395–402.

Piccolella, S., Fiorentino, A., Pacifico, S., D'Abrosca, B., Uzzo, P. and Monaco, P. (2008) Antioxidant properties of sour cherries (*Prunus cerasus* L): role of colorless phytochemicals from the methanolic extract of ripe fruits. *Journal of Agricultural and Food Chemistry* 56, 1928–1935.

Pigeon, W.R., Carr, M., Gorman, C. and Perlis, M.L. (2010) Effects of a tart cherry juice beverage on the sleep of older adults with insomnia: a pilot study. *Journal of Medicinal Food* 13, 1–5.

Poll, L., Petersen, M.B. and Nielsen, G.S. (2003) Influence of harvest year and harvest time on soluble solids, titratable acid, anthocyanin content and aroma components in sour cherry (*Prunus cerasus* L. cv. 'Stevnsbaer'). *European Food Research and Technology* 216, 212–216.

Rakonjac, V., Fotirić Akšić, M., Nikolić, D., Milatović, D. and Čolić, S. (2010) Morphological characterization of 'Oblačinska' sour cherry by multivariate analysis. *Scientia Horticulturae* 125, 679–684.

Reiter, R.J. (1993) The melatonin rhythm: both a clock and a calendar. *Experientia* 49, 654–664.

Reiter, R.J., Carneiro, R.C. and Oh, C.S. (1997) Melatonin in relation to cellular antioxidative defense mechanisms. *Hormone and Metabolic Research* 29, 363–372.

Rodrigues, L.C., Morales, M.R., Fernandes, A.J.B. and Ortiz, J.M. (2008) Morphological characterization of sweet and sour cherry cultivars in a germplasm bank at Portugal. *Genetic Resources and Crop Evolution* 55, 593–601.

Romano, G.S., Cittadini, E.D., Pugh, B. and Schouten, R. (2006) Sweet cherry quality in the horticultural production chain. *Stewart Postharvest Review* 6, 2.

Sansavini, S. and Lugli, S. (2008) Sweet cherry breeding programs in Europe and Asia. *Acta Horticulturae* 795, 41–57.

Sass-Kiss, A., Kiss, J., Milotay, P., Kerek, M.M. and Toth-Markus, M. (2005) Differences in anthocyanin and carotenoid content of fruits and vegetables. *Food Research International* 38, 1023–1029.

Schmid, W. and Grosch, W. (1986) Quantitative-analysis of the volatile flavor compounds having high aroma values from sour (*Prunus cerasus* L.) and sweet (*Prunus avium* L.) cherry juices and jams. *Zeitschrift für Lebensmittel-Untersuchung und -Forschung* 183, 39–44.

Schmitz-Eiberger, M.A. and Blanke, M.M. (2012) Bioactive components in forced sweet cherry fruit (*Prunus avium* L.), antioxidative capacity and allergenic potential as dependent on cultivation cover. *LWT – Food Science and Technology* 46, 388–392.

Schreiner, M. and Huyskens-Keil, S. (2006) Phytochemicals in fruit and vegetables: health promotion and postharvest elicitors. *Critical Reviews in Plant Sciences* 25, 267–278.

Schuster, M., Grafe, C., Wolfram, B. and Schmidt, H. (2014) Cultivars resulting from cherry breeding in Germany. *Erwerbs-Obstbau* 56, 678–772.

Seeram, N.P., Momin, R.A., Nair, M.G. and Bourquin, L.D. (2001) Cyclooxygenase inhibitory and antioxidant cyanidin glycosides in cherries and berries. *Phytomedicine* 8, 362–369.

Serra, A.T., Duarte, R.O., Bronze, M.R. and Duarte, C.M.M. (2011) Identification of bioactive response in traditional cherries from Portugal. *Food Chemistry* 125, 318–325.

Serradilla, M.J., Lozano, M., Bernalte, M.J., Ayuso, M.C., López-Corrales, M. and González-Gómez, D. (2011) Physicochemical and bioactive properties evolution during ripening of 'Ambrunés' sweet cherry cultivar. *LWT – Food Science and Technology* 44, 199–205.

Serradilla, M.J., Martín, A., Ruiz-Moyano, S., Hernández, A., López-Corrales, M. and Córdoba, M.G. (2012) Physicochemical and sensorial characterisation of four sweet cherry cultivars grown in Jerte Valley (Spain). *Food Chemistry* 133, 1551–1559.

Serradilla, M.J., Hernández, A., López-Corrales, M., Ruiz-Moyano, S., Córdoba, M.G. and Martín, A. (2016) Composition of the cherry (*Prunus avium* L. and *Prunus cerasus* L.; Rosaceae). In: Simmonds, M.S.J. and Preedy, V.R. (eds) *Nutritional Composition of Fruit Cultivars*. Academic Press, London, pp. 127–147.

Serrano, M., Guillén, F., Martínez-Romero, D., Castillo, S. and Valero, D. (2005) Chemical constituents and antioxidant activity of sweet cherry at different ripening stages. *Journal of Agricultural and Food Chemistry* 5, 2741–2745.

Serrano, M., Díaz-Mula, H., Zapata, P.J., Castillo, S., Guillén, F., Martínez-Romero, D., Valverde, J.M. and Valero, D. (2009) Maturity stage at harvest determines the fruit quality and antioxidant potential after storage of sweet cherry cultivars. *Journal of Agricultural and Food Chemistry* 57, 3240–3246.

Serrano, M., Díaz-Mula, H.M. and Valero, D. (2011) Antioxidant compounds in fruits and vegetables and changes during postharvest storage and processing. *Stewart Postharvest Review* 1, 1.

Simunic, V., Kovac, S., Gaso-Sokac, D., Pfannhauser, W. and Murkovic, M. (2005) Determination of anthocyanins in four Croatian cultivars of sour cherries (*Prunus cerasus*). *European Food Research and Technology* 220, 575–578.

Tomás-Barberán, F.A. and Espín, J.C. (2001) Phenolic compounds and related enzymes as determinants of quality in fruits and vegetables. *Journal of the Science of Food and Agriculture* 81, 853–876.

Tomás-Barberán, F.A., Ruiz, D., Valero, D., Rivera, D., Obón, C., Sánchez-Roca, C. and Gil, M. (2013) Health benefits from pomegranates and stone fruit, including plums, peaches, apricots and cherries. In: Skinner, M. and Hunter, D. (eds) *Bioactives in Fruit: Health Benefits and Functional Foods*. Wiley, Hoboken, New Jersey, pp. 125–167.

Tsanova-Savova, S., Ribarova, F. and Gerova, M. (2005) (+)-Catechin and (-)-epicatechin in Bulgarian fruits. *Journal of Food Composition and Analysis* 18, 691–698.

Tsuda, T., Watanabe, M., Ohshima, K., Norinobu, S., Choi, S.W., Kawakishi, S. and Osawa, T. (1994) Antioxidative activity of the anthocyanin pigments cyanidin 3-*O*-β-d-glucoside and cyanidin. *Journal of Agricultural and Food Chemistry* 42, 2407–2410.

USDA ARS (2016) National Nutrient Database for Standard Reference, Release 26. Nutrient Data Laboratory, US Department of Agriculture, Agricultural Research Service. Available at: https://ndb.nal.usda.gov/ (accessed 2 January 2016).

Usenik, V., Fabčič, J. and Štampar, F. (2008) Sugars, organic acids, phenolic composition and antioxidant activity of sweet cherry (*Prunus avium* L.). *Food Chemistry* 107, 185–192.

Usenik, V., Fajt, N., Mikulic-Petkovsek, M., Slatnar, A., Štampar, F. and Veberic, R. (2010) Sweet cherry pomological and biochemical characteristics influenced by rootstock. *Journal of Agricultural and Food Chemistry* 58, 4928–4933.

Valero, D. and Serrano, M. (2010) *Postharvest Biology and Technology for Preserving Fruit Quality*. CRC Press, Boca Raton, Florida.

Valero, D., Díaz-Mula, H.M., Zapata, P.J., Castillo, S., Guillén, F., Martínez-Romero, D. and Serrano, M. (2011) Postharvest treatments with salicylic acid, acetylsalicylic acid or oxalic acid delayed ripening and enhanced bioactive compounds and antioxidant capacity in sweet cherry. *Journal of Agricultural and Food Chemistry* 59, 5483–5489.

Viljevac, M., Dugalic, K., Jurkovic, V., Mihaljevic, I., Tomaš, V., Puškar, B., Lepeduš, H., Sudar, R. and Jurković, Z. (2012) Relation between polyphenols content and skin colour in sour cherry fruits. *Journal of Agricultural Science* 57, 57–67.

Wang, H., Nair, M.G., Iezzoni, A.F., Strasburg, G.M., Booren, A.M. and Gray, J.I. (1997) Quantification and characterization of anthocyanins in Balaton tart cherries. *Journal of Agricultural and Food Chemistry* 45, 2556–2560.

Wang, H., Nair, M.G., Strasburg, G.M., Booren, A.M. and Gray, J.I. (1999) Antioxidant polyphenols from tart cherries (*Prunus cerasus*). *Journal of Agricultural and Food Chemistry* 47, 840–844.

Wang, S.Y. (2006) Pre-harvest conditions and antioxidant capacity in fruits. *Acta Horticulturae* 712, 1–10.

Wen, Y.-Q., He, F., Zhu, B.-Q., Lan, Y.-B., Pan, Q.-H., Li, C.-Y., Reeves, M.J. and Wang, J. (2014) Free and glycosidically bound aroma compounds in cherry (*Prunus avium* L.). *Food Chemistry* 152, 29–36.

Wilms, L.C., Hollman, P.C., Boots, A.W. and Kleinjans, J.C. (2005) Protection by quercetin and quercetin-rich fruit juice against induction of oxidative DNA damage and formation of BPDE-DNA adducts in human lymphocytes. *Mutation Research* 582, 155–162.

Wojdyło, A., Nowicka, P., Laskowski, P. and Oszmiański, J. (2014) Evaluation of sour cherry (*Prunus cerasus* L.) fruits for their polyphenol content, antioxidant properties, and nutritional components. *Journal of Agricultural and Food Chemistry* 62, 12332–12345.

Yao, L.H., Jiang, Y.M., Shi, J., Tomás-Barberán, F.A., Datta, N., Singanusong, R. and Chen, S.S. (2004) Flavonoids in food and their health benefits. *Plant Foods for Human Nutrition* 59, 113–122.

Zhang, X., Jiang, Y.M., Peng, F.T., He, N.B., Li, Y.J. and Zhao, D.C. (2007) Changes of aroma components in Hongdeng sweet cherry during fruit development. *Agricultural Science in China* 6, 1376–1382.

Zhao, Y., Tan, D.X., Lei, Q., Chen, H., Wang, L., Li, Q., Gao, Y. and Kong, J. (2013) Melatonin and its potential biological functions in the fruits of sweet cherry. *Journal of Pineal Research* 55, 79–88.

# 18 Fruit Harvest Methods and Technologies

M.D. Whiting[1]* and R.L. Perry[2]
[1]Irrigated Agriculture Research and Extension Center, Washington State University, Prosser, Washington, USA; [2]Michigan State University, East Lansing, Michigan, USA

## 18.1 Introduction

Harvest is a time of heightened anxiety for cherry growers, as their livelihood depends upon successful collection and delivery of a highly perishable fruit within a matter of days. Sweet and sour cherries are susceptible to various types of damage during harvest and handling. In addition, sweet cherry harvest is an extraordinarily labour-intensive operation, requiring large crews to collect fruit at optimum maturity and during a short harvest window. Long-standing practices for harvest of fresh market sweet cherries (Looney *et al.*, 1996) and mechanical harvest of both sweet and sour cherries for processing (Brown and Kollár, 1996) are still common in most cherry production regions. However, given the strong fresh market demand for sweet cherries, significantly increased production in many countries over the past 20 years, increasing costs and decreasing availability of harvest labour, and concerns for labour safety, there is renewed interest in alternative technologies to improve harvest efficiency and labour safety. Similarly, the need for alternative mechanical harvest technologies that can facilitate higher harvest efficiencies, earlier yields and improved fruit quality for processed cherries is increasingly important for economic sustainability. This chapter will review the current harvest processes, including the potential for mechanical harvest of fresh market-quality sweet cherries.

## 18.2 Harvest Maturity

The ability to properly judge fruit maturity is important to optimize shelf life and consumer acceptance of the fruit. Harvest decisions must consider that fruit quality will not improve beyond harvest. Fruit harvested too immature tend to have low soluble solids and poor flavour. In contrast, fruit harvested too late tend to be soft and have a short shelf life. The challenge of securing sufficient harvest labour further complicates harvest timing. Growers may begin harvest prior to the fruit reaching optimum harvest maturity in order to secure a labour force. Conversely, if the labour force is insufficient, fruit harvest may extend beyond their ideal maturity. The optimum harvest window (i.e. period of time in which fruit remain at optimum harvest maturity and storability) is perceived to be quite narrow for sweet cherry, perhaps

* mdwhiting@wsu.edu

a matter of days, particularly during periods of high temperature.

Commercial maturity generally is determined by the degree of red coloration of the fruit exocarp. A dark red/mahogany exocarp colour is thought to be ideal (Kappel *et al.*, 1996), and bright red or light mahogany fruit do not appeal to consumers (Crisosto *et al.*, 2003). The influence of harvest time on the sensory properties of the fruit is significant. Chauvin *et al.* (2009) harvested 'Sumtare' (Sweetheart™) fruit at three different harvest times: early (3 days before commercial harvest), at commercial timing and late (5 days after commercial harvest), and evaluated consumer acceptance of the fruit. Overall fruit acceptance was highest for those harvested at commercial maturity, with no difference between fruit harvested early or late. It is likely that fruit harvested early were downgraded for colour (too red) and being less sweet, while the late-harvested fruit were rated lower due to being softer. In contrast, 'Lapins' exhibited a 10-day window for harvest with no fruit quality loss or significant differences in fruit appearance (Drake and Elfving, 2002), although the authors did not assess consumers' responses to the fruit.

There can be tremendous variability in maturity among fruit in a tree, particularly where winters are mild or climatic conditions during flowering are unfavourable for a uniform bloom and pollination period. This further complicates harvest timing decisions, especially for dark sweet cultivars that typically are strip-picked (i.e. all fruit harvested in a single pass). Harvest decisions are a compromise between harvesting fruit with excellent flavour and consumer appeal (i.e. later harvest with high soluble solids) and harvesting fruit with high firmness. Crop load, fruit-to-leaf ratio, fruit position, light exposure and use of gibberellic acid (Patten and Proebsting, 1986; Whiting and Lang, 2004; Einhorn *et al.*, 2013) all affect fruit quality.

## 18.3 Hand Harvest for Fresh Market Sweet Cherries

Harvest costs constitute the single greatest expense for sweet cherry growers, accounting for 50–60% of total costs of production (Seavert *et al.*, 2008). Currently, all fresh market sweet cherries are harvested manually. This process is one of the most labour-demanding operations among all temperate tree fruits, due to the high fruit number per tree, relatively small fruit size and, in older orchards, the large and complex tree canopies. In traditional tree architectures, fruit are accessed with tall (e.g. 4–5 m) ladders that are heavy, difficult to position, and require significant skill for efficient and safe use. The procedures for sweet cherry harvest comprise two basic steps: (i) pickers remove fruit manually, placing them into some form of portable receptacle such as a picking bucket or tote; and (ii) pickers transfer or deposit harvested fruit from the portable receptacle into a larger receptacle, such as a bin, receive credit for their harvest and return to picking.

Fruit are removed with the pedicel intact by twisting it at the junction between the spur and the cluster. Low pedicel–fruit retention force can be a problem since it is challenging to harvest fruit with the pedicel intact, especially when fruit clusters are substantial. Experienced pickers will use both hands independently, removing fruit in singles or clusters for placement in their picking bucket, depending on the cultivar. There are no studies on the biomechanics of the harvest motion for sweet cherries, although the biomechanics of manual harvest of apples is being studied to improve the development of robotic picking mechanisms (M. Karkee, Prosser, Washington, USA, 2014, personal communication). Care must be taken during harvest to prevent removal of the entire spur, which would reduce the future fruiting potential of the tree. In addition, in most production regions, it is unacceptable to 'milk' the cherries off the tree (i.e. remove the fruit at the pedicel–fruit abscission zone, which may lead to tearing depending on cultivar and stage of ripeness), thereby picking stem-free fruit. This is largely a wholesale market-driven decision, since harvest efficiency of experienced pickers can be improved by up to 30% by permitting hand harvest of stem-free cherries (M. Whiting, unpublished data). In most countries, fruit picked in clusters remain as clusters until they are separated to singles by the cluster cutter at the packing facility (see Chapter 19, this volume). In some operations, generally those with smaller orchards and less total production, pickers are instructed to split clusters of fruit apart into singles. This process reduces harvest efficiency.

In the USA, management of the orchard picking crew usually includes a 'checker', an employee whose responsibilities include ensuring that pickers deliver a full bucket of fruit, that pickers empty their fruit into the bin gently and in the best location (to minimize handling of fruit within the bin), and that each picker is credited for their work. This credit involves marking each picker's daily harvest ticket for each bucket of fruit. Pickers are usually paid piece rate (i.e. per bucket), receiving a prenegotiated price for a full bucket of fruit. However, when harvesting delicate blush cultivars, such as 'Rainier' or 'Early Robin', pickers are paid an hourly wage so that there is no incentive to harvest the fruit so quickly that bruising might occur.

The collection of fruit from pickers' buckets has changed in recent years with the goals of: (i) minimizing fruit handling; and (ii) filling large bins quickly so that they may be removed from the orchard and cooled as soon as possible. The most widely adopted process in the USA involves teams of pickers (usually 10–15) who empty their picking buckets into a larger tote or bin in the orchard. Pickers then resume harvesting with the same bucket. The most commonly utilized bins are made from non-porous plastic (114 cm long × 122 cm wide × 42 cm high) and are considered to be full with the contents from 17–18 buckets (~180 kg). An optimized picking crew can fill a bin in about 15 min. A second bin can be stacked on the full bin and filled in turn. In the case of blush cultivars, pickers harvest fruit into rectangular totes that are unharnessed, and the entire tote is placed in the bin (i.e. fruit are not emptied); pickers resume harvest with a new tote. Full bins or totes are collected by either a tractor with rear or front forks or, more commonly, a mechanical bin trailer capable of collecting up to eight bins when double-stacked. Full bins are transported to a loading area where fruit are typically kept cool and loaded into a refrigerated transport.

There are biological, technological and sociological factors that affect the rate of sweet cherry harvest. However, there have been few studies on sweet cherry harvest efficiency, despite its importance in production budgets. This is attributable, in part, to the difficulty in collecting reliable, accurate data during harvest. A recent study showed that hand harvest rates varied significantly by cherry training system (Ampatzidis and Whiting, 2013). The highest mean (± standard error) harvest rates ($0.94 \pm 0.02$ kg min$^{-1}$ and $0.78 \pm 0.03$ kg min$^{-1}$, respectively) were recorded in orchards trained to the Upright Fruiting Offshoots (UFO) system, facilitated by the planar, simplified tree architecture and the proportion of fruit that were accessible without ladders. The third highest picking rate was recorded in the Kym Green Bush (KGB) system ($0.73 \pm 0.04$ kg min$^{-1}$), a fully pedestrian orchard. The harvest rate of slower pickers was improved to a greater extent (+132%) than for skilled pickers (+83%) when comparing these systems with traditional tree architectures (i.e. multiple-leader open centre). Furthermore, harvest efficiency varied significantly among pickers, and picking rates of individual pickers varied by more than 100%, probably due to variability in fruit density within trees, tree size and fruit accessibility (Fig. 18.1). For trees with average crop loads, there was no discernible diurnal trend in picking efficiency within a day, and no relationship between fruit yield per tree and individual harvest rate expressed as kilograms picked per hour (Y. Ampatzidis and M.D. Whiting, unpublished data) (Fig. 18.2). This suggests that the time required to harvest an orchard will be directly proportional to the crop size and will depend primarily upon the efficiency of the picking crew and the tree training system.

Experienced orchard managers estimate how many pickers are required to harvest fruit at optimum maturity based on crop load and weather forecasts. Since fruit are highly perishable, inefficiency in the harvest and handling process can have detrimental effects on sweet cherry quality and storability. The window for harvesting cherries at optimum maturity varies by cultivar and environmental conditions, but is generally considered to be a matter of days. To optimize harvest efficiency, the number of

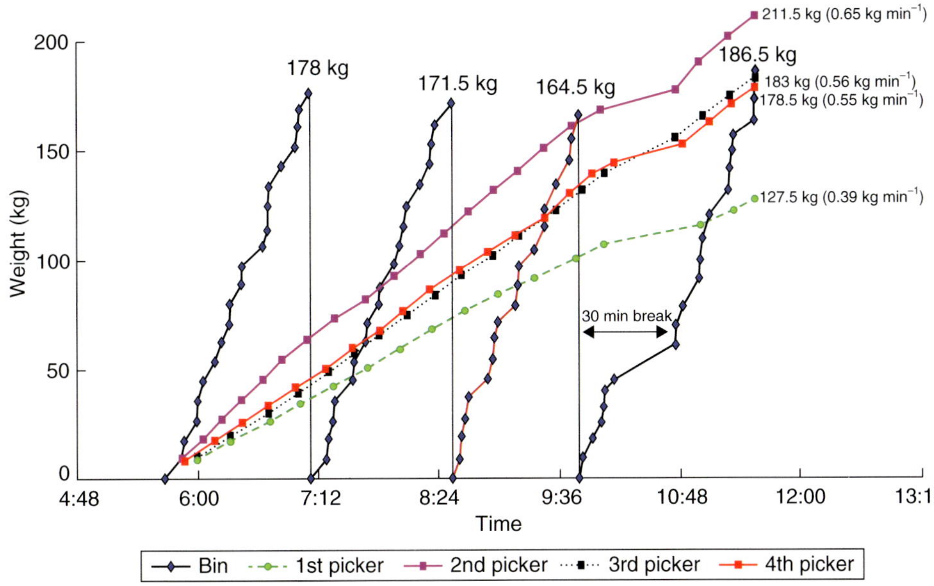


**Fig. 18.1.** Sweet cherry harvest efficiency of four individual pickers, and their combined (bin) harvest efficiency for mature 'Bing' on Mazzard sweet cherry trees trained to a three- to four-leader open-centre architecture in Washington State, USA.

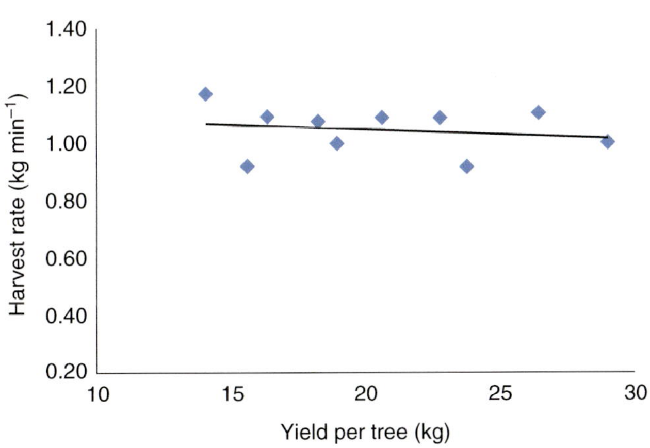


**Fig. 18.2.** Relationship between picker harvest rate (kg $min^{-1}$) and crop load (kg per tree) for mature 'Chelan' sweet cherry trees trained to a steep-leader architecture. Each data point is a single tree (Y. Ampatzidis and M.D. Whiting, unpublished data).

workers and equipment necessary for harvesting, handling and transport, as well as the execution of field operations, should be planned. Ampatzidis *et al.* (2012) developed modified machine repair models to improve sweet cherry harvest efficiency that may be used by orchard managers to determine the ideal size for picking crews (based on harvest rate) to minimize queues and maximize the time spent picking.

## 18.4 Innovations in Mechanical Harvest for Fresh Market Sweet Cherries

The extraordinary labour requirements, increasing labour costs and decreasing labour availability for sweet cherry harvest is driving interest in alternative, efficient harvest technologies, particularly in the Pacific Northwest USA. In addition, a study of ergonomics during manual apple harvest revealed that overhead picking motions and carrying of heavy totes and ladders over long distances are sources of physical strain to pickers, often causing injury (Fulmer *et al.*, 2002). Hofmann *et al.* (2006) documented a strong and compelling need to develop interventions to reduce the number of ladder-related injuries in orchards. An assessment of workers' compensation claims over 5 years in tree fruit-growing regions of Washington State, USA, found that nearly half of all claims were ladder related, and that claims related to ladders were the most expensive in terms of medical aid and time loss. Development of a mechanized harvest system for fresh market sweet cherries will require a transformation of operations across harvest equipment, cultivars, tree training systems, packing and marketing.

Mechanized harvest systems have been adopted for most tree nuts and several tree fruits, including citrus, olive, and sour and sweet cherry, but only for processed fruits, not fresh market fruits. These commonly involve trunk-shaker harvesters used with large tree canopies, which result in unacceptable levels of fruit damage for fresh markets (Halderson, 1966). Recent research at Washington State University has investigated the potential use of fully mechanical systems and mechanical-assist technologies to harvest fresh market sweet cherries. There are significant horticultural obstacles to the commercial adoption of mechanized harvest technologies, including canopy architecture and pedicel–fruit retention force (i.e. fruit abscission).

### 18.4.1 Engineering considerations for mechanical harvest

There are several important engineering components for effective mechanical harvest of fresh market sweet cherries. At the simplest level, they may be considered as related to fruit removal and fruit capture. Fruit removal issues include the means of force transmittal (actuation) and the positioning of the actuator mechanism. Fruit collection aspects include the catching surface and transfer system to the bin filler. Tanagaki *et al.* (2008) described the development and testing of a robotic system for selective harvest of sweet cherry using a three-dimensional vision system for spectral reflectance and an end effector that clasped the fruit at the pedicel. They used potted trees aligned to a planar system comprised of single upright trunks and a rail on which the robot harvester could travel. However, they concluded that visibility of fruit and fruit removal remained significant challenges. Subsequent research has focused on mass harvest technologies for entire trees or portions of trees, rather than a robotic approach to selectively harvest individual fruit.

#### *Prototype full-tree harvest technology*

The experimental harvester developed at the US Department of Agriculture (USDA) and evaluated in orchards in Washington State has shown promise for harvesting stem-free sweet cherries suitable for fresh markets and comparable to commercial hand harvest (Peterson *et al.*, 2003; Peterson, 2005). This harvester has two mirror-image self-propelled units that operate opposite each other on each side of the tree row (Fig. 18.3). The harvesters have overlapping padded catch frames that form a seal at the tree trunk and row middle for collecting fruit as they drop from the angled trunk to the angled catch frame. The force applied from actuation on one side of the row has the potential to remove fruit from the opposite side of the Y-trellis, so it is important for both units to operate in concert along the row. The spring-loaded catching pans permit the units to travel directly down the row, improving harvest speed over the previous design (Peterson and Wolford, 2001) that required stopping at each tree and extending a catching pan to seal around the trunks. Machine harvest rates with the previous system were up to 80 trees $h^{-1}$; with the spring-loaded catching pans, harvest rates nearly doubled (158 trees $h^{-1}$), equivalent in test orchards to 1590 kg $h^{-1}$ (Peterson *et al.*, 2003). However, harvest efficiency varies significantly with crop load, and more recent tests have nearly doubled the rate again

**Fig. 18.3.** The US Department of Agriculture (USDA) prototype harvester for fresh market sweet cherries: paired left and right side-of-row harvesters in a Y-trellis orchard in Washington State, USA. (From Peterson *et al.*, 2003.)

(M. Whiting, unpublished data). At approximately 1 kg $h^{-1}$ for hand harvest (see above), picking efficiency has the potential to be increased more than 26-fold with a fully mechanical approach.

In mass harvest of sweet cherries, the mechanics for applying energy to trunks or fruiting limbs include hydraulically operated shakers (e.g. Norton *et al.*, 1962), a rapid displacement actuator (RDA) that impacts limbs with a 2.5 cm thick × 7.5 cm diameter rubber disk (Peterson *et al.*, 2003), and a padded continuous vibration mechanism (Larbi *et al.*, 2015). The trunk and limb shakers induced significant bark injury at the shaking point, which can foster disease infection at the site of injury, as well as fruit damage (Norton *et al.*, 1962; Halderson, 1966). The RDA effectively removed fruit with minimal tree damage. 'Bing' fruit removal rates always exceeded 90% (Peterson *et al.*, 2003). The vibratory actuation method achieved 83–85% fruit removal rates (Larbi *et al.*, 2015). Fruit removal under vibration varied with the shaking frequency and duration, with multiple shaking intervals at 14–18 Hz for 2–5 s removing up to 81% of fruit (Zhou *et al.*, 2013). Fruit removal is dependent not only on the pedicel–fruit removal force (PFRF), but also fruit position, fruit proximity to actuation point, limb angle and the calibre of the wood (Smith and Whiting, 2010; Chen *et al.*, 2012; Du *et al.*, 2012; Zhao *et al.*, 2013). Up to 97% of 'Skeena' fruit were removed when limbs were vibrated at a position both near the base and at the terminus of the branch (Zhou *et al.*, 2014).

Fruit separated from the tree by the impact of the actuator are caught on a padded catching conveyor set at an angle parallel to that of the planar canopy to minimize drop height (Fig. 18.4). Harvested fruit are conveyed to the top of the catch frame where they are transferred to a collecting conveyor that runs to the bin filler apparatus. The catching frame material, angle of orientation and distance from the canopy all affect fruit damage during mechanical harvest (Zhou *et al.*, 2016a). Maximum impact force of fruit reaching the catching surface increased linearly with distance from the catching frame, and only at a drop height greater than 1 m did significant damage result when fruit landed on the padded, angled catching frames. Peterson *et al.* (2003) reported damage was comparable for mechanically and hand-harvested fruit. Using the same catching frame but with vibratory actuation, Larbi *et al.* (2015) found harvest-induced fruit damage rates of 8–12% for the cultivars 'Skeena',

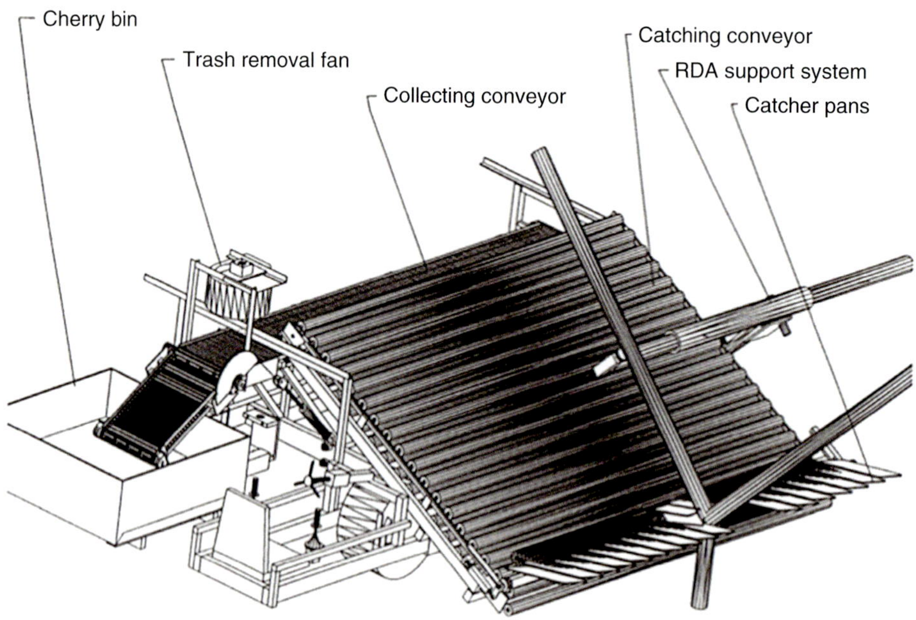


**Fig. 18.4.** The USDA prototype harvester for fresh market sweet cherries. Schematic of the left side-of-row harvester with angled catch conveyor and harvest force actuator (RDA). (From Peterson *et al.*, 2003.)

'Selah' and 'Sweetheart', with the variation related to canopy architectures.

Fruit capture rates impact overall harvest efficiency and production economics since fruit that are removed but not captured represent irrecoverable yield reductions. An economic sensitivity analysis illustrated the importance of yield reductions on the net present value of a mechanical harvester (Seavert and Whiting, 2011): decreasing fruit capture by 15% (i.e. from 100 to 85%) revealed only a slight reduction in the net present value. It was concluded that a loss of this extent should not deter growers from adopting such a mechanical harvest system, due to the overwhelming potential economic benefit for mechanical harvest to reduce harvest costs – machine harvest costs were estimated to be approximately one-tenth of hand harvest costs (Seavert and Whiting, 2011). The USDA harvester described above captured ~88% of the fruit removed (Peterson *et al.*, 2003). A multi-row harvest trial with the USDA harvester and RDA system had an 8% yield reduction due to fruit removed but not recovered (D. Peterson and M. Whiting, unpublished data). Fruit fell to the ground primarily from beyond the front or rear of the catching conveyor, and the authors proposed extending the catching frame to eliminate these losses. Modifications to vibrational actuation have reduced the losses from fruit missing the catching frames, probably related to less energy being used to remove fruit (Larbi *et al.*, 2015).

### Prototype partial-tree harvest technology

An alternative to a fully mechanical harvest system for fresh market sweet cherries is to utilize a mechanically assisted shake-and-catch harvest system (Zhou *et al.*, 2014). This requires teams of workers, one operating a handheld portable shaker and another one or two carrying lightweight catching frames. Mechanical force transfer is accomplished with a battery-powered, modified reciprocating saw in which the blade has been replaced with a V-shaped aluminium 'hook' for attaching to a range of limb sizes. This utilizes a vibratory actuation similar to that of the full machine harvest prototype

(Larbi *et al.*, 2015). In field evaluations, harvest rates of 'Skeena' trained to a Y-trellised architecture were approximately fourfold greater than hand harvest rates in the same orchard (Ampatzidis *et al.*, 2012), despite requiring a three-person team. Preliminary tests revealed the importance of pruning to open 'windows' in the canopy for improved accessibility of the catching frame, removing weak and pendent wood, and regulating the energy (i.e. frequency and stroke) of the shaker to restrict force transmission to the portion of the canopy directly above the catching frame. Short, higher-intensity impacts at higher vibration frequencies (e.g. 18 Hz) were more damaging than multiple, low-intensity impacts at 14 Hz (Zhou *et al.*, 2016b). At 18 Hz, fruit impact during harvest was about 25% of that recorded at 14 Hz, yet fruit damage was significantly lower at 14 Hz. In addition, fruit-to-fruit impacts were approximately fourfold more prevalent than fruit-to-limb impacts during harvest (Zhou *et al.*, 2016b). As with the full-tree harvester, fruit removal rate was related to the proximity of fruit to the actuation site (Zhou *et al.*, 2016a); multiple actuation locations resulted in more than 90% fruit removal. The versatility of a shake-and-catch harvest system is particularly of interest for some existing orchards of moderately sized trees that might require only modest corrective pruning.

### 18.4.2 Horticultural considerations for mechanical harvest

#### *Canopy architecture*

Despite more than a decade of research into fully mechanized harvest of fresh market sweet cherries, no commercial system has yet been adopted. This is due, in part, to a lack of compatible tree architectures. Mechanical harvest of fresh market sweet cherries requires a Y- or V-trellised compact fruiting wall architecture to facilitate the unimpeded drop from the canopy to the catching surface (Fig. 18.5). The planes of the angled fruiting wall should be relatively flat (e.g. 55° from horizontal) compared with orchards designed for manual harvest, which typically are 70–80° from horizontal. This minimizes fruit-to-limb contact and bruise damage

**Fig. 18.5.** Y-trellised orchard at Washington State University designed for mechanical harvest with 'Skeena' and 'Selah' sweet cherry on 'GiSelA 12'. Row spacing is 4.5 m and trees are spaced 1 m apart in the row.

during harvest. At this flat orientation, however, new shoots tend to fill the middle of the Y, and intra-canopy shading can become excessive. Such shoots must be removed postharvest or in the dormant season.

There are examples of Y- and V-trellised orchards in most production regions around the world, although they tend to be too upright for mechanical harvest and usually are established at a row spacing (less than ~4.5 m) that does not accommodate the current prototype harvester (Fig. 18.3). Evaluations of prototype mechanical harvest in some Y-trellised orchards demonstrated harvest rates up to 158 trees $h^{-1}$ when the limbs were readily accessible to the harvester actuator for transfer of harvester force and visible to the operator so that the actuator could be positioned quickly (Peterson *et al.*, 2003). In contrast, the same mechanical harvest system harvested only 45 trees $h^{-1}$ in orchards trained to the Spanish bush or central leader architectures. Low harvest rates were due to the relative inaccessibility of the branches that required actuation and poor visibility from the operator's fixed position. In these ineffective architectures, up to 25% of the fruit were not harvested because they were not accessible to the actuator (Peterson *et al.*, 2003). In addition, poor placement of the actuator led to significant bark tearing during harvest (D. Peterson and M. Whiting, unpublished data). Adoption of machine vision systems to control actuation could potentially improve harvest rate and reduce operator error and tree damage. Amatya *et al.* (2016) identified potential mechanical harvest actuation points in the canopy using vision systems with nearly 90% accuracy.

Key canopy architecture considerations for mechanical harvest include: (i) a compact, single-plane, angled fruiting wall at 55–60° from horizontal: (ii) high visibility of actuation points for the operator; and (iii) absence of weak, particularly pendent limbs, for most efficient transfer of the harvest force from the point of actuator contact with the canopy to the points of fruit attachment. The upright, unbranched limbs of the UFO canopy architecture are ideal for mechanized harvest when trained to a Y-configuration, yielding better fruit removal rates (up to 98%) and harvest efficiency than the more traditional Tatura-type Y-trellised systems that bear fruit predominantly on horizontal limbs (M. Whiting, unpublished data). Irrespective of architecture, fruiting wood needs to be sufficiently robust for energy transfer; small-calibre wood and limbs with a high degree of lateral branching are undesirable. Ultimately, the energy applied and transferred along a limb to the fruiting sites must be sufficient to overcome the PFRF to achieve fruit removal.

### Pedicel–fruit retention force

Sweet cherry cultivars exhibit significant genetic variability in PFRF, from less than 300 g to over 1 kg (Zhao *et al.*, 2013), and cultivars with high PFRF are not suitable for mechanized harvest. At high PFRF, excessive energy is required to remove the fruit, often resulting in damage to the mesocarp or, in some cases, removal of entire spurs rather than individual fruit. It appears that the ideal PFRF for effective mechanical harvest and minimal fruit damage is less than 400 g, irrespective of actuation method (Peterson *et al.*, 2003; M. Whiting, unpublished data), although this varies somewhat with fruit position, branch calibre and proximity to the actuation site. The use of ethephon (2-chloroethylphosphonic acid) to reduce PFRF is a standard practice to facilitate mechanized harvest of sweet and sour cherries for processing. Sweet cherry cultivars vary significantly in their response to ethephon rates and timings (Bukovac *et al.*, 1971; Wirch *et al.*, 2009) and some cultivars, such as 'Chelan', may exhibit no reduction in PFRF with ethephon treatment (Smith and Whiting, 2010).

The use of ethephon can have negative effects on fruit quality, the most common being a reduction in firmness (Bukovac *et al.*, 1971; Smith and Whiting, 2010), as well as gummosis, terminal meristem senescence and defoliation (Bukovac *et al.*, 1969; Bukovac, 1979). These responses can be dependent on cultivar, application rate and timing, and environment (Li *et al.*, 1994; Smith and Whiting, 2010; Zhao *et al.*, 2013).

Smith and Whiting (2010) proposed three classes of genotypes based on their abscission characteristics: (i) not inducible and therefore unsuitable for mechanical harvest due to a high natural PFRF and poor response to ethephon (e.g. 'Chelan', 'Cowiche'); (ii) inducible and therefore suitable for mechanical harvest when treated with ethephon (e.g. 'Bing', 'Lapins'); and (iii) auto-abscising, exhibiting a natural decline in PFRF to levels suitable for mechanical harvest (e.g. 'Skeena', 'Selah'). Considering the potential negative effects of ethephon on sweet cherry fruit quality, the greatest potential for mechanical harvest of fresh market sweet cherries lies in high-quality cultivars that exhibit the auto-abscission trait. From assessments of many cultivars and $F_1$ seedlings, Zhao *et al.* (2013) concluded that PFRF is a quantitative trait, and there were low correlations between PFRF and key fruit quality attributes such as size and firmness, suggesting that it is possible to combine low PFRF and excellent fruit quality traits through strategic hybridization.

### Marketability of mechanically harvested fruit

Since successful mechanical harvest of sweet cherries is tied to PFRF, a very high percentage of fruit are harvested without the pedicel. Of the two abscission zones of sweet cherry fruit, the retention force at the pedicel–fruit junction generally is less than at the pedicel–peduncle junction, particularly at harvest (Bukovac, 1971). Therefore, the marketability of 'stem-free' sweet cherries presents a potential roadblock to the further development and commercial adoption of mechanical harvest technologies. Market research has shown that consumers rate the presence/absence of the fruit stem as the least important factor for purchase decisions of sweet cherries; price and shelf life are the most significant factors (Koutsimanis *et al.*, 2012). Yet, wholesale buyers in the marketing chain tend to be resistant to stem-free fruit, since a green stem is an indicator of likely fruit shelf life.

The ability to mechanically harvest fruit quickly (compared with the difficulties of obtaining and managing a large labour force for a short, specific period of time) provides the potential to harvest a greater proportion of fruit at a fully ripe stage of development, which can increase fruit size and yield, as well as sugar content and peak flavour. There are also potential storage and marketing advantages of stem-free fruit, including reduced pitting, ease of electronic sorting and packing, and expanded opportunities for packaging such as heat-sealed, bio-based packages (Drake *et al.*, 1989; Koutsimanis *et al.*, 2015). Postharvest evaluations of 'Bing' and 'Skeena' fruit, with or without pedicels, demonstrated that weight loss during storage and marketing occurs predominantly via the fruit exocarp and not the pedicel (Smith and Whiting, 2011). Marketing evaluations with a major US retailer were positive, with consumers willing to pay the same price for stem-free sweet cherries when marketed in 2 kg clamshells. In a recent trial, 1 kg plastic bags of stem-free 'Skeena' fruit harvested with a prototype shake-and-catch system (see below) sold as well, or better, than fruit harvested with stems over a 4-week marketing period (M. Whiting, unpublished data). The 'Picota'-brand sweet cherries marketed from Spain have gained regional acceptance as high-quality, stem-free fruit. 'Ambrunés' is the most widely grown cultivar marketed under the 'Picota' designation, and it is hand-harvested naturally stem-free, having the advantages of requiring less harvest labour and improved postharvest quality (Serradilla *et al.*, 2011).

## 18.5 Innovations in Mechanical Harvest for Processed Sour Cherries

While horticultural and genetic innovations have resulted in the conversion of fresh market sweet cherry orchards to high-density production systems, providing increased early and cumulative yields, fruit quality and profitability, the sour cherry industry has only just begun to explore the potential for higher density systems. This has been due, for the most part, to the lower value of sour cherries for processing markets, the significant investment in trunk-shaker harvesters for existing orchards and the long duration to begin returns on investment (ROI) due to the need for trees to attain a size that is adequate for trunk shaking. Consequently, typical sour cherry orchards harvested mechanically are planted on seedling Mahaleb or Mazzard rootstocks at 4.5–6 × 5.5–6 m spacing to accommodate the trunk-shaker harvester and catch frame. This creates large spherical canopies (5 m or more high) with no branches below 1.5 m in height to provide the clearance for accommodating the clamping mechanism of the shaker. Harvest does not begin until year 6 or 7, with gibberellic acid treatments applied during flower bud induction to minimize bloom and fruiting to reduce competition with canopy and trunk growth. Trunk damage from the shaker attachment also tends to shorten the productive life of the orchard. The standard low-density sour cherry production harvester systems in Europe and the USA have been described elsewhere (Brown and Kollár, 1996; Calleson, 1997).

An experimental continuous harvest system for sour cherries was described by Brown and Kollár (1996) based on modified trunk-shaker technology developed by D. Peterson at the USDA Agricultural Research Service in Kearneysville, West Virginia (Peterson, 1984). While this system was never commercialized, in the late 1990s, Wawrzyńczak *et al.* (1998) proposed a new type of continuous mechanical harvest for sour cherries. Commercial sour cherry production in Eastern Europe typically utilized genetically compact Morello-type cultivars such as 'Schattenmorelle', planted at 4 × 5 m for hand- or limb-shaker machine harvest. Researchers in Poland began exploring the planting of trees at higher densities (1670 trees ha$^{-1}$) and using over-the-row (OTR) berry harvesters with rotary-tine-tower technology with the goal of reducing labour costs (Mika *et al.*, 2011). During the past 10 years, research on this type of mechanical harvest has spread to the US sour cherry industries in Michigan (R. Perry, East Lansing, Michigan, USA, 2008, personal communication) and Utah (B. Black, Logan, Utah, 2010, personal communication), as well as a unique sour cherry growing area in Canada (B. Bors, Saskatoon, Saskatchewan, 2009, personal communication).

### 18.5.1 Engineering considerations for OTR mechanical harvest

Scientists and engineers in Poland initiated the adoption of rotary-tine-tower harvesting technology (typically used for berry harvesting) to machines suitable for OTR sour cherry harvest (Wawrzyńczak *et al.*, 1998). These are self-propelled machines that straddle the plant row, subjecting each plant to two vibrating, freely rotating spindles of many tines (10–20 mm in diameter and 45–60 cm long) to transmit a moderate harvest force relatively uniformly throughout the canopy for fruit removal as the machine passes overhead (Fig. 18.6). Multiple rows of fibreglass, plastic or steel tines are affixed to the two intersecting vertical spindles like the spokes on a bicycle wheel (Fig. 18.7). As the 'tunnel' of the harvester moves over the plant canopy, it funnels the branches between the two towers of vibrating tines to separate the fruit from the shoots. The tines can vibrate in a horizontal, vertical or orbital motion (depending on harvester model), actuated by counterweights. The harvester operator is able to control vibration amplitude, frequency and attitude (stroke strength), and speed of travel.

Fruit separated from the branches fall a relatively short distance to slightly inclined, spring-loaded plastic 'fish scales' that seal against each other and around the tree trunk as the harvester moves down the row (Fig. 18.7). These divert the fruit to conveyors on each side that carry the fruit to the lugs, bins or tanks (Fig. 18.8) for collection and transport out of the orchard to the processing plant. Harvester tunnels designed for berry crops typically are limited to canopies 1.2–1.4 m wide by 2.4 m high, although as interest in OTR harvest of sour cherries has increased, machines are beginning to be modified for somewhat larger canopies. In addition to self-propelled OTR harvesters, smaller

Fig. 18.6. A self-propelled dual-spindle rotary-tine over-the-row (OTR) harvester for sour cherries in Michigan, USA (Littau Harvesters, Oregon, USA).

Fig. 18.7. The dual-spindle rotary-tine (Oxbo 9000) harvest force transmission mechanism, fish-scale catch plates and dual conveyors for sour cherry fruit removal to lugs, bins or tanks.

OTR or half-row harvesters designed to be pulled by a tractor have also become available. Commercial examples of OTR harvesters that have been tested in the USA include those manufactured by Oxbo International Company (http://www.oxbo-corp.com) and Littau Harvesters (http://www.littauharvester.com) (R. Perry, East

**Fig. 18.8.** The conveyor system for collecting mechanically harvested sour cherries in lugs, bins or tanks (Littau Harvesters, Oregon, USA).

Lansing, Michigan, USA, 2008, personal communication). A commercial self-propelled OTR harvester in Poland is manufactured by Weremczuk Agromachines (http://www.aroniaharvest.com), which also makes a smaller tractor-pulled half-row harvester that collects fruit from half of the tree on each pass down the row. This small harvester has been used in Canada for the genetically compact sour cherries described further below (Bors, 2009).

Since OTR harvesters have the ability to harvest fruit continuously down the tree hedgerow, their efficiency is much higher than that for trunk-shaker mechanical harvesters. An experienced trunk-shaker operator can harvest a 1.0 ha orchard in 3.7 h, depending on crop size (in a typical orchard of 278 trees ha$^{-1}$). In comparison, an experienced rotary-tine OTR operator can harvest a 1.0 ha orchard (at 1300–1700 trees ha$^{-1}$) in about 1 h when young (at 1.5–2.4 km h$^{-1}$ maximum speed) and about 2 h when mature (at 1.0–1.5 km h$^{-1}$ maximum speed), depending on row spacing and crop load (R. Perry and B. Black, unpublished data). Thus, not only can younger trees be harvested for earlier ROI, but orchards can be harvested more quickly (depending on crop load and row spacing) and with less damage to the permanent tree structure (i.e. the trunk). Furthermore, harvesters can be outfitted with lights for harvest during cooler temperatures at night, if desired. The average yield of mature 'Montmorency' trees in Michigan is about 9 t ha$^{-1}$ (McManus, 2012) and in Utah about 17 t ha$^{-1}$ in good years (B. Black, Logan, Utah, 2010, personal communication). As early as year 3 or 4, high-density 'Montmorency' orchard yields can reach 9 t ha$^{-1}$ with summer hedging or 12–14 t ha$^{-1}$ without hedging (R. Perry, unpublished data); however, optimized yield levels at maturity and OTR orchard longevity have yet to be determined.

The lower maximum distance for fruit to fall to the catch frame also reduces the potential for fruit damage compared with trunk-shaker harvest. The prototype harvester designed at the Institute of Horticulture (Skierniewice, Poland) required four persons to operate and removed 83–95% of the crop at 1.3–2.6 t h$^{-1}$ when operated at a speed of 0.8 km h$^{-1}$ (Mika *et al.*, 2011).

The fruit quality was suitable for processing, but was less than that of hand-harvested fruit; however, the harvest rate for a picking crew of four persons was only 0.05 t $ha^{-1}$. Fruit was readily removed when PFRF was below 3.0 N. In the USA, 'Montmorency' harvest with trunk shakers usually requires the application of ethephon 10–14 days in advance of harvest to concentrate fruit ripening and accelerate reduction of the PFRF to between 150 and 300 g (1.5–3.0 N) for fruit shaking and removal, since the harvest force must be transmitted from the trunk through the scaffolds to the branches and fruiting shoots. However, OTR harvesters have successfully removed 'Montmorency' fruit at PFRF values of up to 600 g (6.0 N) (R. Perry, unpublished data) and with removal of up to 98% of the crop (Pullano, 2013). Thus, the potential exists for eliminating the use of ethephon, which can also hasten fruit softening, if synchronization of fruit colouring is not required. Alternatively, with adjustment of the vibrating force, multiple pass harvests could be made to selectively remove the ripest fruit at each pass. Fruit quality has been equivalent to that of traditional trunk-shaker harvesters, and by 2016, Michigan's sour cherry industry had established 35 ha of high-density 'Montmorency' sour cherry for OTR harvest.

### 18.5.2 Tree considerations for OTR mechanical harvest

Plant spacing for sour cherry orchards designed for OTR harvest has been recommended at 4 × 1.5–2.0 m (Mika *et al.*, 2011). Even with typically vigorous rootstocks, trees established at close spacing result in reduced vigour due to increased competition for water and nutrients. Robinson (2007) demonstrated this inverse relationship between tree vigour and plant density per hectare for apple. The goal for high-density sour cherry orchard design is to create relatively narrow (2.0–2.5 m) hedgerows of small, bush-like trees with alleys just large enough to accommodate tractor equipment and one-half of the OTR harvester that straddles the narrow hedgerow. Since many sour cherry genotypes are relatively precocious, this ability to harvest young trees provides a much earlier ROI than trunk-shaker orchards. Additional means under study for reducing sour cherry tree vigour, to maintain bush-like trees small enough for OTR harvest, include: root pruning, summer hedging at 40–45 days after bloom, planting leaders at an oblique angle such as the UFO training system, the use of growth-inhibiting plant growth regulators such as prohexadione calcium (e.g. Apogee, Regalis), and dwarfing rootstocks.

Hedgerow canopy development and maintenance is a current area of critical research focused not only on maintaining the bush-like tree size suitable for OTR harvest, but also the optimization of renewal of fruit-bearing structure. Sour cherry genotypes vary in bearing habit, with some forming more fruiting spurs while others bear predominantly on shoot growth from the previous season. In the former, good canopy light distribution must be promoted to maintain fruitful spurs throughout the canopy. In the latter, fruiting on 1-year-old shoots results in the potential for extensive portions of canopy to subsequently develop blind nodes, for which optimized renewal strategies have yet to be reported. Canopy management studies in Poland with Morello cultivars, such as 'Debreceni Bőtermő', 'Nefris', 'English Morello' and 'Sokówka Serocka' on *Prunus mahaleb* rootstock, have focused on leader and branch renewal ('spindle') protocols comparable to those used in high-density apples (Mika *et al.*, 2011). Trees were trained to a 2.5–3.0 m high central leader, with removal of any strong shoots that might compete with the leader. This resulted in a narrow canopy of thin, pliable lateral branches. For annual renewal pruning, three or four 3-year-old or older branches were cut back to stubs on the leader for regrowth and renewal, and the crowns were thinned moderately. Fruiting was primarily (74–99%) on 1-year-old shoots, depending on cultivar.

Canopy management studies for OTR harvesters in the USA generally are focused on the dominant industry cultivar, 'Montmorency', and are being conducted at Michigan State University and Utah State University (Pullano, 2013; Lehnert, 2015). The Michigan studies have examined the potential to also partially mechanize pruning and fruitwood renewal, including summer hedging, winter hedging and root pruning, as well as the use of dwarfing rootstocks and oblique leader canopy architectures for tree size control. Thus far, yields in years 4 and 5 have surpassed typical yields in mature traditional sour cherry orchards (R. Perry, unpublished data). 'Montmorency' on seedling *P. mahaleb* rootstocks trained to a multiple-leader bush or an oblique cordon-like trunk with multiple upright, highly branched leaders had lower vigour and maintained smaller canopies than trees trained to a central leader (N. Rothwell and G.A. Lang, unpublished data). In other tree fruit studies, root pruning in the spring (full bloom) significantly reduced canopy vigour of apple, sweet and sour cherry by 15–30% (Brunner, 1986; Ferree, 1992; Toldam-Andersen *et al.*, 2007). Six years of root pruning in apple consistently reduced shoot length, trunk circumference and fruit size, except in years with satisfactory soil moisture (Ferree, 1992). In hedgerow 'Montmorency' sour cherry trees on sandy soils in Michigan, root pruning reduced canopy and fruit size by 20 and 10%, respectively (R. Perry, unpublished data).

In all OTR sour cherry canopy training systems, strategies for branch recycling are critical for maintaining productivity as well as appropriate tree size. 'Montmorency' trees on *P. mahaleb* seedling rootstock reached their size limits for OTR harvest, 4.5 m in height and 3.2 m in width, in the sixth growing season (R. Perry, East Lansing, Michigan, USA, 2008, personal communication). Early summer hedging at 40–45 days postbloom and root pruning (at prebloom) show promise for maintaining a compact canopy with minimal yield suppression. Winter hedging reduced subsequent yield significantly. Hedging at 45 days postbloom has been shown to improve yield and canopy light penetration for 'Montmorency' (Flore and Layne, 1990). The Utah studies utilize standard and dwarfing rootstocks and spindle-type tree canopies at various plant spacings. Trees were developed with one to four permanent leaders positioned parallel to the tractor alley, centred in the middle of the row to accommodate the OTR harvester, to form single spindle trees, in-row V-shaped (two-leader) trees, or in-row candelabra (three- or four-leader) canopies. Unlike spindle training for apples, renewal pruning cuts in 'Montmorency' must be at least 7–10 cm in length to induce adequate regrowth (B. Black, unpublished data).

In Michigan, own-rooted 'Montmorency' trees lacked the moderate precocity of trees grafted on seedling *P. mahaleb* rootstocks, while trees on 'GiSelA 6' were somewhat more precocious and trees on 'GiSelA 5' and 'GiSelA 3' were significantly more precocious (N. Rothwell, and G.A. Lang, unpublished data); comparable rootstock precocity has also been documented in Utah (B. Black, unpublished data). 'Montmorency' on experimental hybrid rootstocks from the Michigan State University (MSU) cherry rootstock selection programme were even more precocious (R. Perry and A. Iezzoni, unpublished data). Trees on the MSU rootstocks 'Cass' and 'Lake' were not only more productive, but also 60% smaller than trees on seedling *P. mahaleb* (Fig. 18.9).

Sour cherry production for OTR harvest in Canada has focused on own-rooted fruiting genotypes from the University of Saskatchewan breeding programme that are naturally compact hybrids of *Prunus cerasus* and *Prunus fruticosa* (Bors, 2005, 2009). These are cold hardy, naturally bushy plants suitable for growing on the Canadian prairies and harvested with smaller OTR harvesters than those under study in the USA, which can also be used to harvest raspberries, blueberries, haskaps and saskatoons. A trial with several of these genotypes, including 'Carmine Jewel' and 'Crimson Passion', was established in Michigan for comparison with the industry standard 'Montmorency' as well as two other compact selections (R. Perry, East Lansing, Michigan, USA, 2008, personal communication).

Fig. 18.9. 'Montmorency' sour cherry hedgerows on *Prunus mahaleb* seedling rootstock (left) and the dwarfing precocious Michigan State University hybrid rootstock 'Cass' (right) after six growing seasons in Michigan, USA.

## References


Amatya, S., Karkee, M., Gongal, A., Zhang, Q. and Whiting, M.D. (2016) Detection of cherry tree branches with full foliage in planar architecture for automated sweet cherry harvesting. *Biosystems Engineering* 146, 3–15.

Ampatzidis, Y. and Whiting, M.D. (2013) Training system affects sweet cherry harvest efficiency. *HortScience* 48, 547–555.

Ampatzidis, Y., Zhang, Q. and Whiting, M.D. (2012) Comparing the efficiency of future harvest technologies for sweet cherry. *Acta Horticulturae* 965, 195–198.

Bors, B. (2009) Mechanical harvesting of Haskap, Saskatoons, and dwarf sour cherries using the Joanna Harvester in 2009. Available at: http://www.fruit.usask.ca/articles/mechanicalharvest2009.pdf (accessed 28 July 2016).

Bors, R.H. (2005) Dwarf sour cherry breeding at the University of Saskatchewan. *Acta Horticulturae* 667, 135–140.

Brown, G.K. and Kollár, G. (1996) Harvesting and handling sour and sweet cherries for processing. In: Webster, A.D. and Looney, N.E. (eds) *Cherries: Crop Physiology, Production and Uses*. CAB International, Wallingford, UK, pp. 443–469.

Brunner, T. (1986) Growth regulation based on self-regulating system of fruit trees. *Acta Horticulturae* 179, 291–292.

Bukovac, M.J. (1971) The nature and chemical promotion of abscission in maturing cherry fruit. *HortScience* 6, 385–388.

Bukovac, M.J. (1979) Machine-harvest of sweet cherries: effect of ethephon on fruit removal and quality of the processed fruit. *Journal of the American Society for Horticultural Science* 104, 289–294.

Bukovac, M.J., Zucconi, F., Larsen, R.P. and Kesner, C.D. (1969) Chemical promotion of fruit abscission in cherries and plums with special reference to 2-chloroethyl phosphonic acid. *Journal of the American Society for Horticultural Science* 94, 226–230.

Bukovac, M.J., Zucconi, F., Wittenbach, V.A., Flore, J.A. and Inoue, H. (1971) Effects of 2-chloroethyl phosphonic acid on development and abscission of maturing sweet cherry (*Prunus avium* L.). *Journal of the American Society for Horticultural Science* 96, 777–781.

Calleson, O. (1997) Orchard systems for sour cherry. *Acta Horticulturae* 451, 653–660.

Chauvin, M.A., Whiting, M.D. and Ross, C.F. (2009) The influence of harvest time on sensory properties and consumer acceptance of sweet cherries. *HortTechnology* 19, 748–754.

Chen, D., Du, X., Zhang, Q., Whiting, M., Scharf, P. and Wang, S. (2012) Performance evaluation of mechanical cherry harvesters for fresh market grade fruit. *Applied Engineering in Agriculture* 28, 483–489.

Crisosto, C.H., Crisosto, G.M. and Metheney, P. (2003) Consumer acceptance of 'Brooks' and 'Bing' cherries is mainly dependent on fruit SSC and visual skin color. *Postharvest Biology and Technology* 28, 159–167.

Drake, S.R. and Elfving, D.C. (2002) Indicators of maturity and storage quality of 'Lapins' sweet cherry. *HortScience* 12, 687–690.

Drake, S.R., Williams, M.W. and Fountain, J.B. (1989) Stemless sweet cherry (*Prunus avium* L.) – fruit quality and consumer purchase. *Journal of Food Quality* 11, 411–416.

Du, X., Chen, D., Zhang, Q., Scharf, P.A. and Whiting, M.D. (2012) Dynamic responses of sweet cherry trees under vibratory excitations. *Biosystems Engineering* 111, 305–314.

Einhorn, T.C., Wang, Y. and Turner, J. (2013) Sweet cherry fruit firmness and postharvest quality of late-maturing cultivars are improved with low-rate, single applications of gibberellic acid. *HortScience* 48, 1010–1017.

Ferree, D. (1992) Time of pruning influences vegetative growth, fruit size, biennial bearing, and yield of 'Jonathan' apple. *Journal of the American Society for Horticultural Science* 117, 198–202.

Flore, J. and Layne, D. (1990) The influence of tree shape and spacing on light interception and yield in sour cherry (*Prunus cerasus* cv. Montmorency). *Acta Horticulturae* 285, 91–96.

Fulmer, S., Punnett, L., Slingerland, D.T. and Earle-Richardson, G. (2002) Ergonomic exposures in apple harvesting: preliminary observations. *American Journal of Industrial Medicine* (Suppl. 2), 42, 3–9.

Halderson, J.L. (1966) Fundamental factors in mechanical cherry harvesting. *Transactions of the American Society of Agricultural Engineers* 9, 681–684.

Hofmann, J., Snyder, K. and Keifer, M. (2006) A descriptive study of workers' compensation claims in Washington State orchards. *Occupational Medicine* 56, 251–257.

Kappel, F., Fisher-Fleming, B. and Hogue, E. (1996) Fruit characteristics and sensory attributes of an ideal sweet cherry. *HortScience* 31, 443–446.

Koutsimanis, G., Getter, K.L., Behe, B., Harte, J. and Almenar, E. (2012) Influences of packaging attributes on consumer purchase decisions for fresh produce. *Appetite* 59, 270–280.

Koutsimanis, G., Harte, J. and Almenar, E. (2015) Freshness maintenance of cherries ready for consumption using convenient, microperforated, bio-based packaging. *Journal of the Science of Food Agriculture* 95, 972–982.

Larbi, P.A., Karkee, M., Amatya, S., Zhang, Q. and Whiting, M.D. (2015) Modification and field evaluation of an experimental mechanical sweet cherry harvester. *Applied Engineering in Agriculture* 31, 387–397.

Lehnert, R. (2015) Cherries harvested as berries. *Good Fruit Grower* 66, 34–35.

Li, S., Andrews, P.K. and Patterson, M.E. (1994) Effects of ethephon on the respiration and ethylene evolution of sweet cherry (*Prunus avium* L.) fruit at different development stages. *Postharvest Biology and Technology* 4, 235–243.

Looney, N.E., Webster, A.D. and Kupferman, E.M. (1996) Harvest and handling sweet cherries for the fresh market. In: Webster, A.D. and Looney, N.E. (eds) *Cherries: Crop Physiology, Production and Uses*. CAB International, Wallingford, UK, pp. 411–441.

McManus, J. (2012) Grower decision support tool for conversion to high-efficiency tart cherry orchard system. MS thesis, Michigan State University, East Lansing, Michigan.

Mika, A., Wawrzyńczak, P., Buler, Z., Krawiec, A., Białkowski, P., Michalska, B., Plaskota, M. and Gotowicki, B. (2011) Results of experiments with densely-planted sour cherry trees for harvesting with a continuously moving combine harvester. *Journal of Fruit and Ornamental Plant Research* 19, 31–40.

Norton, R., Claypool, L.L., Leonhard, S.J., Adrian, P.A., Fridley, R.B. and Charles, F.M. (1962) Mechanical harvesting of sweet cherries: 1961 tests show promise and problems. *California Agriculture* 16, 8–10.

Patten, K.D. and Proebsting, E.L. (1986) Effect of different artificial shading durations and natural light intensities on fruit quality of 'Bing' sweet cherries. *Journal of the American Society for Horticultural Science* 111, 360–363.

Peterson, D.L. (1984) Mechanical harvester for high density orchards. In: *Fruit, Nut, and Vegetable Harvesting Mechanization*, Special Publication 5-84. American Society of Agricultural Engineers, St Joseph, Michigan, pp. 46–51.

Peterson, D.L. (2005) Harvest mechanization progress and prospects for fresh market quality deciduous tree fruits. *HortTechnology* 15, 72–75.

Peterson, D.L. and Wolford, S.D. (2001) Mechanical harvester for fresh market quality stemless sweet cherries. *Transactions of the American Society of Agricultural Engineers* 44, 481–485.

Peterson, D.L., Whiting, M.D. and Wolford, S.D. (2003) Fresh-market quality tree fruit harvester. Part 1: sweet cherry. *Applied Engineering in Agriculture* 19, 539–543.

Pullano, G. (2013) Researchers eye high-density tart cherry harvest. *Fruit Growers News* 52, 11–12.

Robinson, T. (2007) Effects of tree density and tree shape on apple orchard performance. *Acta Horticulturae* 732, 405–414.

Seavert, C. and Whiting, M.D. (2011) Comparing the economics of mechanical vs. hand harvest of sweet cherry. *Acta Horticulturae* 903, 725–730.

Seavert, C., Freeborn, J. and Long, L. (2008) *Orchard Economics: Establishing and Producing High-Density Sweet Cherries in Wasco County*. Oregon State University Extension Service Publication EM 8802-E, Corvallis, Oregon.

Serradilla, M.J., Lozano, M., Bernalte, M.J., Ayuso, M.C., López-Corrales, M. and González-Gómez, D. (2011) Physicochemical and bioactive properties evolution during ripening of 'Ambrunés' sweet cherry cultivar. *LWT – Food Science and Technology* 44, 199–205.

Smith, E. and Whiting, M. (2010) Effect of ethephon on sweet cherry pedicel-fruit retention force and quality is cultivar dependent. *Plant Growth Regulation* 60, 213–223.

Smith, E.D. and Whiting, M.D. (2011) The pedicel's role in postharvest weight loss of two sweet cherry cultivars. *Acta Horticulturae* 903, 935–939.

Tanagaki, K., Fujiura, T., Akase, A. and Imagawa, J. (2008) Cherry-harvesting robot. *Computers and Electronics in Agriculture* 63, 65–72.

Toldam-Andersen, T.B., Jensen, N.L. and Dencker, I. (2007) Effects of root pruning in sour cherry (*Prunus cerasus*) 'Stevnsbaer'. *Acta Horticulturae* 732, 439–442.

Wawrzyńczak, P., Cianciara, Z. and Krzewiński, J. (1998) A new concept of mechanical harvest of sour cherries. *Journal of Fruit and Ornamental Plant Research* 6, 123–128.

Whiting, M.D. and Lang, G.A. (2004) 'Bing' sweet cherry on the dwarfing rootstock Gisela 5: crop load effects on fruit quality, vegetative growth, and carbon assimilation. *Journal of the American Society for Horticultural Science* 129, 407–415.

Wirch, J., Kappel, F. and Scheewe, P. (2009) The effect of cultivars, rootstocks, fruit maturity and gibberellic acid on pedicel retention of sweet cherries (*Prunus avium* L.). *Journal of the American Pomological Society* 63, 108–114.

Zhao, Y., Athanson, B., Whiting, M. and Oraguzie, N. (2013) Pedicel-fruit retention force in sweet cherry (*Prunus avium* L.) varies with genotype and year. *Scientia Horticulturae* 150, 135–141.

Zhou, J., He, L., Zhang, Q., Du, X., Chen, D. and Karkee, M. (2013) Evaluation of the influence of shaking frequency and duration in mechanical harvest of sweet cherry. *Applied Engineering in Agriculture* 29, 607–612.

Zhou, J., He, L., Zhang, Q. and Karkee, M. (2014) Effect of excitation position of a handheld shaker on fruit removal efficiency and damage in mechanical harvesting of sweet cherry. *Biosystems Engineering* 125, 36–44.

Zhou, J., He, L., Karkee, M. and Zhang, Q. (2016a) Effect of catching surface and tilt angle on bruise damage of sweet cherry due to mechanical impact. *Computers and Electronics in Agriculture* 121, 282–289.

Zhou, J., He, L., Karkee, M. and Zhang, Q. (2016b) Analysis of shaking-induced cherry fruit motion and damage. *Biosystems Engineering* 144, 105–114.

# 19 Postharvest Biology and Handling for Fresh Markets

**Juan Pablo Zoffoli,[1]* Peter Toivonen[2] and Yan Wang[3]**
*[1]Pontificia Universidad Católica de Chile, Santiago, Chile; [2]Summerland Research and Development Centre, Agriculture and Agri-Food Canada, Summerland, British Columbia, Canada; [3]Mid-Columbia Agricultural Research and Extension Center, Oregon State University, Oregon, USA*

## 19.1 Introduction

Sweet cherry is an edible drupe that can be classified according to the physical or pomological characteristics of the fruit. The Bigarreau and Duroni groups (Italy) include cultivars with firm flesh, while the Guigne (France), Gean (England) and Tenerine groups (Italy) include soft and tender flesh. Only Bigarreau cherries are firm enough for commercial use, being better able to withstand the rigours of harvest, postharvest handling and long-distance transport. Fruit have either dark- or light-coloured flesh. Dark cherries are red to reddish-purple or mahogany in colour, whereas light cherries (so-called white) are yellow, usually with a pink to red partial blush on the yellow skin. Fruit vary in shape from round to oval to heart-shaped, and their pedicels vary in length from 2 to 8 cm (Fogle *et al.*, 1973).

Sweet cherry production in the northern hemisphere is harvested from late April to early September. In the USA, the season starts in the hot San Joaquin Valley of southern California and gradually extends to the cooler states of Oregon and Washington. Late cultivars, grown under moderate weather conditions in Canada and at high elevations in Oregon and Washington, are mainly responsible for late-season production (July to August). US production is currently about 2:1 for domestic versus export markets, but export production is expanding as better prices are available in some more lucrative overseas markets such as in China, Taiwan and Hong Kong. Postharvest storage extends from a week (domestic market) up to about 30 days (long-distance export) including a holding period, shipping by sea and distribution to distant markets (e.g. China).

Production in the southern hemisphere extends from November to February and is dominated by Chile, although it accounts for only 4% of the overall world production. The primary market for Chilean exports is China, with a postharvest storage period of 45 days at 0°C by sea freight. New Zealand (Central Otago) and Australia (Tasmania) appear in the China market in December and they are the main suppliers for the later part of the season, until February (mainly by air freight).

This chapter will review the physiology of sweet cherry fruit growth and maturation, and the critical factors involved in the postharvest handling and deterioration of the fruit.

* zoffolij@uc.cl

## 19.2 Physiology of Fruit Growth and Maturation

The sweet cherry fruit has a hard endocarp (pit), an exocarp (skin) with stomata and an edible mesocarp (flesh) formed from the ovary wall. The epidermis is a single layer of cells covered by a thin cuticle, which is continuous except where interrupted by stomata (Tukey and Young, 1939; Bukovac *et al.*, 1999; Knoche *et al.*, 2000, 2001). Sweet cherry hypodermal cells are described as small compared with the mesocarp cells, and have thicker cell walls. Cells of the mesocarp have tangential elongation near the epidermis and radial elongation near the pit (Glenn and Poovaiah, 1989). The cuticle layer has been characterized as a protective barrier against pathogens, fractures of this layer being the main route for pathogen entry (Børve *et al.*, 2000).

Fruit development is usually described as having three stages (see also Chapter 2, this volume). Stage I, which starts after flowering, pollination and fertilization, is characterized by an increase in fruit size accompanied by vigorous cell division in the mesocarp. During this period, the epidermal cells also increase rapidly in number and later show increases in size and wall thickening. Epidermal cell number is 50% less at this stage than at stage II, and the average cell size increases by up to 2.4-fold (Knoche *et al.*, 2004). In stage II, fruit volume growth slows, the endocarp lignifies into the hard pit and the final development of the embryo occurs. In stage III, the fruit again increases rapidly in size as the mesocarp cells enlarge and ripening occurs (Yamaguchi *et al.*, 2004). There is also a marked decrease in the mass per unit of area of the cuticle membrane in the cheek region of cherries (Knoche *et al.*, 2001, 2004). At the same time, the cells of the hypodermal layer increase their size in the tangential direction. During this time, there is also a decrease in the thickness of the cell walls, and intercellular spaces become less abundant. The length of the three stages depends on cultivar. In 'Bing', stage I extends from 1 to 41 days after full bloom (DAFB), stage II from 41 to 52 DAFB and stage III from 53 to 87 DAFB (Zhao *et al.*, 2013). In general, the later the ripening dates of a cultivar, the longer the duration of stage II development (Azarenko *et al.*, 2008; Zhao *et al.*, 2013).

Sweet cherry is classified as a non-climacteric fruit because there is no marked peak of ethylene production during maturation (Li *et al.*, 1994). However, some studies have shown an early ethylene peak when the fruit turns from green to whitish (Eccher and Noè, 1998; Zhao *et al.*, 2013) and a later ethylene peak during maturation (Remón *et al.*, 2006), although neither peak correlates with respiratory activity. Neither fruit respiration nor loss of firmness is altered in the presence of external ethylene (Li *et al.*, 1994). Also, fruit skin colour is not affected when treated with the ethylene perception inhibitor 1-methylcyclopropene (Gong *et al.*, 2002; Mozetič *et al.*, 2006). These data indicate that the regulation of these changes occurs independently of ethylene.

The direct precursor of ethylene biosynthesis, 1-aminocyclopropane-1-carboxylate (ACC), has been reported in sweet cherry fruit (Kondo and Inoue, 1997), but the ACC oxidase gene transcript, responsible for ethylene synthesis, was not found (Ren *et al.*, 2011). Ethephon treatment stimulates ethylene biosynthesis, but ethylene production increases only transiently after exposure to ethylene for 6 h, and then decreases to below detectable levels (Gong *et al.*, 2002; Ren *et al.*, 2011). The role of abscisic acid (ABA) in ethylene biosynthesis has been investigated in sweet cherry, and it has been shown that the expression of PacNCED1 (a cDNA encoding 9-*cis*-epoxycarotenoid dioxygenase, NCED – a key enzyme in ABA biosynthesis) increases at the beginning of maturation and peaks about 4 days before harvest. This is consistent with ABA accumulation during ripening. Application of exogenous ABA increases the ABA content, induces the expression of PacNCED1, and promotes ripening via enhancement of colour formation and sugar accumulation (Ren *et al.*, 2011).

Sweet cherry fruit have moderate respiratory activity compared with other fruit species (Kader, 1992). Fruit respiration rate decreases continuously through all three phases of maturation (Sekse, 1988; Li *et al.*,

1994) as fruit weight, volume, soluble solids content (SSC) and titratable acidity (TA) all increase (Zhao *et al.*, 2013). Respiration rate can vary from 30 to 60 mg $CO_2$ $kg^{-1}$ $h^{-1}$ at 20°C (Wills *et al.*, 1983; Crisosto *et al.*, 1993) and the rate depends on cultivar and stage of maturation (Wang and Long, 2014).

Fruit firmness increases during stage I, reaching a maximum value at the end of stage I, but then decreases during stage II and eventually reaches a minimum value during stage III. Fruit firmness can vary from 25 N $mm^{-1}$ (20–40 days after anthesis) to 5 N $mm^{-1}$ at harvest (Muskovics *et al.*, 2006). SSC, TA and dry matter content are reported to increase rapidly from stage II to stage III. SSC increases from 8–12% to 17.5–20%, while TA rises from 0.43 to 0.77 g malic acid per 100 ml of juice (Remón *et al.*, 2006), depending on cultivar and agronomic practices (Muskovics *et al.*, 2006).

A change in exocarp colour from green to red is visible evidence of the beginning of fruit maturation. Chlorophyll degrades and anthocyanins accumulate, resulting in the yellow background colour and mahogany as blush. Because of the good correlations between sugar accumulation, decreases in firmness and increases in mass and red colour development, ripening can be followed by chromometric measurement of skin colour using the $L$ (luminosity), $a$ and $b$ parameters and the calculations of $C$ (chroma), $h°$ (hue) and $a/b$. The red skin colour evolves from bright red to dark (Fig. 19.1). For example, the evolution of 'Brooks' skin colour starts from full light red to become full dark red, described by decreases in $h°$ from 26.15° to 11.80°, in $L$ from 41.35 to 29.11 and in $C$ from 42.30 to 23.77 (Crisosto *et al.*, 2003). Therefore, the $L$ value and $h°$ diminish as maturation takes place, $C$ displays an increase at the beginning and then a decrease with advancing maturation, and $a/b$ values increase linearly during this period (Mozetič *et al.*, 2006; Muskovics *et al.*, 2006; Remón *et al.*, 2006; Díaz-Mula *et al.*, 2009).

The main anthocyanins of sweet cherry are 3-rutinoside and 3-glucoside of cyanidin. The content of anthocyanins for more intensely coloured cultivars ranges between 0.28 and 2.97 mg $g^{-1}$, and from 0.02 to 0.41 mg $g^{-1}$ in those with less coloration (Gao and Mazza, 1995; Serrano *et al.*, 2009).

## 19.3 Postharvest Characteristics of Sweet Cherry Fruit

Sweet cherry is recognized as a high-value crop and is grown extensively in cool-temperate regions worldwide because of high demand, prompted by the fruit's attractive appearance and distinctively pleasing flavour. Sweet cherry is harvested ripe with no reserve of starch in the flesh, and therefore, unlike some other fruit (e.g. apples, kiwifruit), the content of soluble carbohydrates does not increase during storage. However, when small increases in SSC are measured during postharvest, generally they are attributable to dehydration.

Since sweet cherry fruit are harvested ripe, postharvest deterioration is primarily associated with the respiration rate and mechanical properties of the tissue in the specific environment around the fruit. Sweet cherry fruit have high respiratory activity compared with apple, but lower compared with strawberry fruit at 20°C (Fig. 19.2). Skin darkening and loss of flavour and acidity are promoted by respiration, whereas pedicel shrivelling and browning and loss of weight are related mainly to the surrounding physical environment of the fruit.

Sweet cherries grow over a relatively short period from spring to mid-summer (60–70 days) when unstable weather conditions (e.g. humidity, warm temperatures) may promote skin cracking. Preharvest factors such as canopy management (Lang *et al.*, 2004), crop load (Zoffoli *et al.*, 2008), fertilization (Crisosto *et al.*, 1995) and water supply (Sekse, 1995) affect sweet cherry postharvest performance in different ways. Fungal infection and tissue softening become critical during maturation. High temperatures (>30°C) during harvest usually result in softer fruit compared with cool temperatures (Sekse *et al.*, 2009), thus reducing the harvest window and increasing the overlap between harvest periods of different cultivars, as well as shortening postharvest storage life at 0°C and subsequent shelf life.

**Fig. 19.1.** Evolution of skin and flesh colour of sweet cherry fruit from bright red to dark red. Chrome (C) and hue (H) values are indicated.

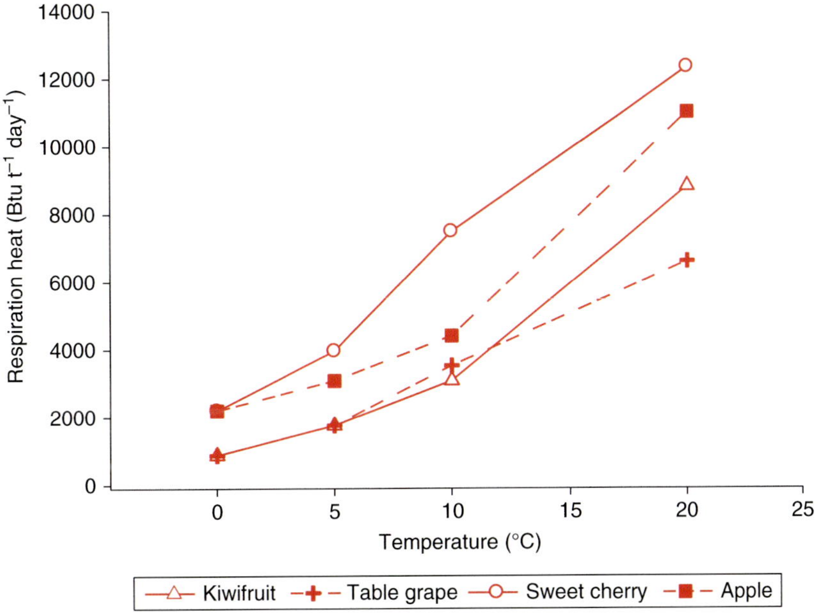


**Fig. 19.2.** Respiration heat (Btu t$^{-1}$ day$^{-1}$) of different fruit species (kiwifruit, table grape, sweet cherry and apple) determined at 0, 5, 10 and 20°C. (Adapted from Postharvest Technology, UC-Davis Research and Information Center: Produce Fact Sheets, http://postharvest.ucdavis.edu/Commodity_Resources/Fact_Sheets/).

Postharvest handling of sweet cherry must be focused on reducing the time in the field after harvest, avoiding exposure to sun and high temperatures and maintaining high relative humidity until arrival at the packing house, ensuring rapid fruit cooling, reducing the time to sorting and packing, and increasing the humidity around the fruit by use of suitable packaging in bags, clamshell containers or boxes. Shortening the period to packing into a modified-atmosphere bag enhances the fruit's postharvest storage performance. A relatively stable low temperature close to 0°C is the most important for maintaining fruit quality during storage or shipping. All other techniques, such as modified-atmosphere packaging (MAP), are considered to be supplementary to good temperature management.

### 19.3.1 Fruit quality traits and market requirements

Appearance is the primary factor that drives sweet cherry purchases by consumers, with skin colour and green stems together with price being the key decision criteria for buyers. Skin colour preferences vary among consumers; dark red to mahogany colour is the key to marketing 'Bing' in American markets, regardless of consumer ethnicity (Crisosto *et al.*, 2003). Similar results were found in Norway and the UK (Sekse and Lyngstad, 1996; Wermund and Fearne, 2000). A glossy red mahogany appearance is required for Chinese consumers during their New Year celebration. Large, bicolour cultivars such as 'Rainier' are preferred in the off-season in Japan (Ito and Clever, 2012). Size is a key marketing criterion, with higher prices per kilogram being associated with larger fruit (29–30 mm diameter and larger) (Kappel *et al.*, 1996).

Among the various flavour traits, sweetness is considered by consumers to be the most important, with a lack of 'cherry' flavour and sourness being major causes of consumer complaint (Turner *et al.*, 2008). Sweetness is due to glucose and fructose levels, while sourness is due to excessively high levels of malic acid (Serrano *et al.*, 2005; Usenik *et al.*, 2008). Relationships between analytical assessments of cherry quality and sensory evaluation have been studied. A moderate correlation ($r = 0.78$) was found between perceived sweetness and the SSC/TA ratio, and there were high correlations between sourness and TA ($r = 0.82$), and between sourness and the SSC/TA ratio (Cliff *et al.*, 1995). The importance of balance between sweetness and sourness was demonstrated in 'Brooks' and 'Bing' such that reduced consumer acceptance was found when TA was >0.6% and when a minimum of 16% SS was not reached (Crisosto *et al.*, 2003). To assure consumer acceptance, minimum quality standards have been proposed or adopted for some cultivars based on SSC and TA (Drake and Fellman, 1987; Guyer *et al.*, 1993; Dever *et al.*, 1996). In sweet cherry cultivars, the sum of sugars (glucose, fructose, sucrose and sorbitol) ranges from 125 to 265 g kg$^{-1}$ fresh weight (FW) and the sum of organic acids (malic, citric, shikimic and fumaric) ranges from 3.67 to 8.66 g kg$^{-1}$ FW (Usenik *et al.*, 2008).

Fruit firmness is another important quality attribute of sweet cherries that is valued by consumers along with crispness. Cherry fruit firmness can be measured by various devices (Mitcham *et al.*, 1998; Garcia-Ramos *et al.*, 2005), but most common are the FirmTech 2 automated force gauge (Bioworks, Wamego, Kansas, USA) and analogue or digital handheld penetrometers (e.g. Durofel DFT 100 or Agrosta 100, Agrosta SARL, Serqueux, France). Firmness is reported in Newtons (N), but usually is measured as g mm$^{-1}$ (FirmTech 2) in North and South America or as Durofel index values in Europe. These are converted as 0.01(FirmTech value in g mm$^{-1}$) = N or 9.8{exp[(Durofel value − 59.32)/14.89]} = N (Polenta *et al.*, 2005), respectively. Mitcham *et al.* (1998) and Clayton *et al.* (1998) found that the automated FirmTech machine provided the greatest precision and consistency among several tested, including the Durofel and other penetrometers that were twice as variable or more.

Firmness not only influences eating quality, but also affects storage performance. Firmness correlates with susceptibility to mechanical damage and infection by microorganisms. While crispness and firmness are distinct measures, they are highly correlated. An acceptable firmness range from 2.52 to 4.75 N has been identified, lying between 'slightly too soft' and 'slightly too firm' in a 'just-about-right' rating by untrained panelists (Hampson *et al.*, 2014). Firmness has been associated with a number of underlying factors contributing to texture, including cell wall strength, cell-to-cell adhesion, cell wall- and pectin-related enzymes (Choi *et al.*, 2002), cell turgor, tissue anatomy and environmental conditions during maturation (see Chapter 11, this volume). Firm fruit with good mechanical properties are required to cope with the high speed flow during operation of most packing lines; otherwise, pitting will appear later at the market.

Secondary metabolic compounds such as anthocyanins and other polyphenols, carotenoids, and vitamins C and E, all of which have nutraceutical properties valuable for mitigating the risk of chronic diseases such as various cancers and cardiovascular disorders, have been shown to be present and active in sweet cherry cultivars (Serrano *et al.*, 2005; Vursavus *et al.*, 2006; Vangdal *et al.*, 2007; Díaz-Mula *et al.*, 2009; Mulabagal *et al.*, 2009; McCune *et al.*, 2011). In cell-culture studies, constituents of sweet cherry have been shown to inhibit cyclooxygenase, which is responsible for inflammatory responses (Seeram *et al.*, 2001). In sweet cherry cultivars, total phenolic content ranges from 44.3 to 87.9 mg gallic acid equivalents per 100 g FW and antioxidant activity ranges from 8.0 to 17.2 mg ascorbic acid equivalents per 100 g FW (Usenik *et al.*, 2008). The dominant polyphenols in sweet cherry are caffeolytartaric acid and 3-*p*-coumaroylquinic acid (Gonçalves *et al.*, 2004).

### 19.3.2 Cultivar traits relative to postharvest performance

Genetic enhancement of sweet cherry traits has focused mainly on fruit size, firmness and flavour, but little on characteristics that directly affect the postharvest quality of the fruit such as low respiration rate or high tolerance to mechanical damage and atmospheric stress (low $O_2$ and/or high $CO_2$). For example, the production and storage of high-quality fruit, obtained at a late harvest window stage, requires genetic material with a high efficiency for accumulating carbohydrates in combination with fruit having a low respiration rate. Since sugar levels are relatively stable during postharvest storage but TA tends to decline, it might be advantageous for genotypes bred for long-term shipping and storage to accumulate higher-than-normal levels of both sugars and acids.

Postharvest characteristics of cultivars are difficult to generalize because of their high dependence on orchard management practices and the environmental conditions in which the cultivar is grown. However, fruit characteristics relevant to postharvest performance will be discussed for some key cultivars as examples (Drake and Fellman, 1987; Cliff *et al.*, 1995; Dever *et al.*, 1996; Drake and Elfving, 2002; Kappel *et al.*, 2002; Crisosto *et al.*, 2003; Toivonen *et al.*, 2004; Kappel and Toivonen, 2005; Harb *et al.*, 2006; Kappel *et al.*, 2006; Remón *et al.*, 2006; Agulheiro-Santos *et al.*, 2014).

'Early Burlat' is a very old cultivar, attractive because of its early harvest time, but the fruit is small, soft and very susceptible to mechanical damage. It is difficult to store longer than 15 days at 0°C. The soft texture renders this cultivar incompatible with water flume handling systems that are present in all mechanized cherry packing lines.

'Summit' and 'Newstar' have large fruit with good flavour, but with a soft texture susceptible to pitting. 'Sumleta' (Sonata™) is considered a cultivar with large fruit having high acidity at harvest that is maintained during storage, with low SSC and soft texture. The fruit is susceptible to pitting and is not suitable for long-term storage (15 days maximum).

'Brooks' is a light red cherry, the fruit is large and firm with high SSC, but it is very susceptible to postharvest cracking due to absorption of condensed moisture in packaging, and therefore MAP is not recommended.

'Santina' has large fruit, with low acid and sugar contents, suitable for long-term storage (45 days at 0°C, with MAP), but it has a tendency to develop a pebbled texture (see Fig. 19.4) on the surface of overripe fruit after long-term storage.

'Bing' has been the standard cultivar for fresh market producers in the Pacific Northwest and California, USA. It is considered to be a long-term storage cultivar with high sugar content, firm texture and excellent flavour. Fruit are suitable for long storage periods for up to 45 days at 0°C in MAP.

'Rainier' is a bicolour fruit having a yellow flesh and background colour with red blush on the skin, and it is sweet, large and firm, but susceptible to rubbing injuries. The packing line needs to be adjusted to avoid friction, which leads to fruit surface discoloration.

It has good storage performance for 45 days at 0°C in MAP.

'Lapins' fruit have good size and flavour. It is the most planted self-fertile cultivar worldwide. Conditions at harvest are important because under high temperatures, the fruit softens rapidly on the tree and becomes susceptible to pitting. Overripe fruit tend to develop a pebbled texture during storage. Fruit can be stored in MAP for up to 45 days at 0°C.

'Sumtare' (Sweetheart™) is the most reliable cultivar for containerized ocean shipment up to 45 days at 0°C in MAP. The fruit is large, with good flavour and firmness. Its tendency to overcrop, however, can lead to pitting and decay susceptibility.

## 19.4 Postharvest Deterioration

### 19.4.1 Softening

As noted above, fruit firmness is a critical factor in determining and maintaining sweet cherry quality during handling and shipping. The middle lamella is a morphologically distinct layer, rich in pectin polysaccharides between the primary cell walls of adjoining cells. The middle lamellae and primary cell walls are subject to structural changes during ripening, which lead to cell separation and tissue softening (Bartley and Knee, 1982). During softening, an increase in the content of soluble pectin polysaccharides is observed (Bartley and Knee, 1982). The major difference between soft and crisp cherries is the degree of polymerization of pectin side chains, being higher in crisp fruit and lower in soft fruit (Bartley and Knee, 1982). The enzymes polygalacturonase, pectin methylesterase and β-galactosidase have been detected in sweet cherry fruit, but the activity of polygalacturonase is low at ripening. β-Galactosidase was detected at early ripening stages, and changes in pectin methylesterase have not been associated with changes in fruit traits during maturation (Barrett and Gonzalez, 1994). A full understanding of the mechanism of cherry fruit softening remains elusive.

The phenomenon of softening during storage is controversial. Excessive softening is described as a common problem during long-term storage in some cultivars, but varies depending on the season (Kappel *et al.*, 2002). However, increases in firmness are reported under modified-atmosphere cold storage of 'Sweetheart' (Meheriuk *et al.*, 1997), 'Bing' (Chen *et al.*, 1981), 'Lapins' and 'Skeena' (Wang *et al.*, 2015) and 'Rainier' (Drake and Fellman, 1987). The mechanisms that affect postharvest firmness are not fully understood. A relationship between reduced turgor and softening has been suggested (Glenn and Poovaiah, 1987). Fruit moisture loss, water allocation in the fruit, skin toughness and cell wall modifications all may also be associated with this phenomenon (Wang and Long, 2014).

### 19.4.2 Decay

Brown rot and grey mould are the major causes of pre- and postharvest decay in sweet cherries in many regions of the world (see Chapter 14, this volume). These are caused by *Monilinia* spp. and *Botrytis cinerea*, respectively. Visible and non-visible quiescent infections of both fungi have been described in sweet cherry fruit (Adaskaveg *et al.*, 2000). These observations implicate the importance of preharvest practices to reduce infection and the quantitative detection of quiescent infections in the prediction of both diseases during the postharvest period. Some other decays common in cherries include: *Penicillium* spp., *Mucor* spp., *Rhizopus stolonifer*, *Alternaria* spp. and *Cladosporium* spp.

Postharvest decays usually are due to preharvest infections and these are often associated with skin fracture. Infections are further caused and exacerbated during packing and storage by contaminated water and humid conditions. Børve (2014) reported that brown rot is increased from 13 to 28% and *Mucor* rot from 11 to 26% by free-water contact in packing. Moreover, fruit packaging such as use of polyliners creates a saturated environment. While it reduces postharvest spoilage by reducing water loss, it also favours condensation and thus promotes fungal decay.

An atmosphere high in $CO_2$ hinders germination of the spores and reduces both the risk and the progress of infection. Thus, lesion size due to *Monilinia* spp. was reduced when fruit was stored in an atmosphere rich in $CO_2$ (15–20%) and decay was completely prevented when stored in 30% $CO_2$ (Tian *et al.*, 2001). However, the effect did not persist during a period at room temperature after removal from the high-$CO_2$ condition (de Vries-Paterson *et al.*, 1991; Zoffoli and Rodríguez, 2014b).

The control of these pathogens in the postharvest continuum using synthetic fungicides such as fludioxonil can become necessary, but the limited product registrations restrict this approach. An alternative control using food preservatives such as sodium bicarbonate shows interesting possibilities as postharvest dip treatments (Karabulut *et al.*, 2001, 2005). Natural compounds with antimicrobial activity and plant defence-eliciting properties also have been studied, including chitosan, a natural polysaccharide derived from chitin. Chitosan has been demonstrated *in vitro* and in field trials to be effective in reducing storage decay in sweet cherries to an extent comparable to that achieved with the fungicide Fenhexamid (Feliziani *et al.*, 2013).

### 19.4.3 Dehydration

Dehydration is a physical process under vapour pressure deficit where vapour-phase water moves from a region of high water potential to a low water potential. Dehydration during storage is affected by several factors, including temperature, relative humidity and air movement.

Water loss from plant surfaces (including fruit and pedicels) is best described by Fick's Law. The water flux across a cuticular membrane is directly proportional to the surface area, to the conductance properties of the surface, and to the difference in water vapour concentration between the inside and outside. The conductance (or permeance) of a plant surface measures the ease with which water can pass through it. The conductance for the surface of sweet cherry fruit is $1.15 \times 10^{-4}$ m s$^{-1}$ (Knoche *et al.*, 2000) and for the pedicel surface is $8.7 \times 10^{-4}$ m s$^{-1}$ (Athoo *et al.*, 2015). Water fluxes increase with increasing temperature. Fruit skin permeance increases from minimum values at the cheek to moderate levels at the ventral suture, to maximum values at the stylar cavity (Knoche *et al.*, 2000). Although water loss is mainly via the cuticular membrane route, losses from stomata cannot be excluded. Epicuticular and cuticular waxes account for most of the diffusion resistance to water loss. Postharvest water loss is both from the fruit and from the pedicel, with pedicel loss being most evident and most likely to affect consumer acceptability of the product.

The main factors contributing to postharvest water loss are high temperature and low humidity. Thus, lowering fruit temperature quickly after harvest and maximizing the humidity around the product are essential for minimizing fruit dehydration. Schick and Toivonen (2002) evaluated reflective tarps for covering the top of the bin, to shade the fruit as soon as it is harvested and during transport to the packing facility. They demonstrated more uniform lower temperatures and higher humidity with tarped compared with uncovered fruit. This now commercial practice significantly reduces stem dehydration before fruit is delivered to the packing house, as well as reducing heat build-up in harvested fruit.

Increasing the humidity in packed fruit, as with the use of perforated liners or sealed plastic or MAP bags, is effective for reducing water loss (Sharkey and Peggie, 1984; Kappel *et al.*, 2002; Harb *et al.*, 2006; Khorshidi *et al.*, 2011; Agulheiro-Santos *et al.*, 2014). Water loss from unpackaged 'Napoleon' sweet cherries was up to 48 times higher than from packaged fruit under the same storage conditions (Esturk *et al.*, 2012).

### 19.4.4 Surface pitting

Sweet cherry fruit are damaged by both compression and impact forces during harvesting, hauling from field to packing facility, and online sorting and processing. Both types of damage lead to symptoms described as surface pitting (Fig. 19.3). Large depressions (bruising) of the tissue can be seen on the shoulders of fruit and can be induced by pickers crushing the fruit in their hand during harvest. Other small indentations appear as pits (4–8 mm) (Porritt *et al.*, 1971; Kappel *et al.*, 2006) due to postharvest handling impacts with other fruit (especially with pedicels of other fruit) or with hard surfaces. Damage is especially severe when impact forces are concentrated over a small area of the fruit. Studies of handling damage often highlight deficiencies in packing line design resulting from increased velocity at the moment of impact. The greatest damage is often caused by cluster cutters and shower-type hydrocoolers (Thompson *et al.*, 1997). The height of drop on to the sorting belt has also been identified as a critical point for mechanical damage (Grant and Thompson, 1997; Candan *et al.*, 2014).

Laboratory and field studies have shown that visible symptoms of surface pitting usually are not evident before shipping, but develop during the first 10 days or so of storage at 0°C (Porritt *et al.*, 1971). Pitting affects not only the cosmetic appearance of the fruit but also shortens shelf life and reduces market quality at point of sale. Increased respiration, premature decay and softening during storage are all associated with severity of pitting (Ogawa *et al.*, 1972; Mitchell *et al.*, 1980).

Pitting incidence varies from year to year, among cultivars and even among trees

**Fig. 19.3.** Surface pitting in sweet cherries produced by mechanical damage. Symptoms differ according to the origin of the damage during harvest and packing: shoulder depression due to incorrect picking (A), impact damage produced by rotating blades of the cluster cutter separator (B), impact damage produced in a hydraulic cluster separator (C) and pedicel (stem) puncture (D).

of the same cultivar (Porritt *et al.*, 1971; Facteau and Rowe, 1979). Cultivar differences in pitting susceptibility have been studied under local conditions (Toivonen *et al.*, 2004; Kappel and Toivonen, 2005; Kappel *et al.*, 2006). Fruit weight and SSC are inversely related to the incidence of surface pitting (Facteau and Rowe, 1979), so riper fruit are less severely affected. Also, firmer fruit generally show less pitting (Facteau and Rowe, 1979; Lidster *et al.*, 1980; Facteau, 1982; Toivonen *et al.*, 2004; Kappel and Toivonen, 2005).

Flesh temperature at the time of injury is important, with fruit expressing more severe pitting at 0°C than at higher temperatures (Lidster and Tung, 1980; Crisosto *et al.*, 1993; Candan *et al.*, 2014). The incidence of pitting doubles when fruit is handled during the packing process at 2°C instead of at the slightly warmer temperature of 5°C (Zoffoli and Rodríguez, 2014a). Overcropping also renders fruit more sensitive to mechanical damage, while fruit thinning reduces that sensitivity (Zoffoli *et al.*, 2008).

Gibberellic acid (GA) has been shown in many studies to make the tissue firmer at harvest (Facteau and Rowe, 1979; Facteau, 1982; Facteau *et al.*, 1985). However, there is little information available on the influence of preharvest GA application on postharvest fruit quality of sweet cherry. GA lengthened the storability of 'Bing' (Zhang and Whiting, 2011) and reduced the incidence of surface pitting of 'Lambert' (Facteau and Rowe, 1979) and 'Bing' (Clayton *et al.*, 2003) in severe pitting years. 'Sweetheart' cherries treated with GA at 10 or 30 p.p.m. were significantly firmer and had less stem browning at the end of cold storage than untreated fruit; the effects were rate dependent (Horvitz *et al.*, 2003). A single application of 25 p.p.m. GA at colour break (from straw-coloured to pink) or pit hardening reduces the incidence and severity of pitting and stem browning after 40 days at 0°C in late-ripening cultivars such as 'Skeena', 'Lapins' and 'Sweetheart' (Einhorn *et al.*, 2013). The reduction in pitting susceptibility by GA application is related to enhanced firmness of the treated cherry fruit.

### 19.4.5 Pebbling

Pebbling (alligator skin) is a physiological disorder associated with sweet cherry fruit stored for a long time (45 days) at 0°C, although it is not exclusive to postharvest handling since it is possible to detect even at harvest on the tree. Pebbling is expressed as a uniform skin roughness that can cover a large surface of the fruit. It is expressed on the surface without compromising the flesh quality (Fig. 19.4). Observations indicate that the firmness is not affected with the disorder and it is more frequent to find in fruit of certain cultivars, such as 'Santina', 'Lapins' and 'Sweetheart', that are harvested at a more advanced stage of maturity.

The damage appears with high variation among lots of the same cultivar, and therefore preharvest factors such as nutrition and irrigation could be involved, but more research needs to be done in order to clarify the origin of the disorder.

## 19.5 Postharvest Handling and Packaging

The postharvest handling of sweet cherry is focused on avoiding dehydration, reducing fruit metabolism, and preventing infection with microorganisms during fruit sorting and packing. Postharvest management, therefore, involves several technologies, starting from choosing the right time for harvest, rapidly reducing fruit temperature, increasing relative humidity with the use of packaging materials, and keeping the cool chain during transport and marketing to assure high fruit quality upon arrival to the consumer. A summary of the postharvest handling operations, with the identification of critical points to optimize fruit quality through 45 days of storage at 0°C, is provided in Table 19.1.

### 19.5.1 Harvest index

The time of harvest has a big impact on both fruit quality and the rate of fruit deterioration

Fig. 19.4. Pebbling (alligator skin), a physiological disorder in sweet cherry fruit after a long-term storage period of 45 days at 0°C in modified-atmosphere packaging.

during storage. Sweet cherry fruit are non-climacteric and therefore harvested ripe, near to the senescent stage of development, which limits the window for marketing. The skin colour change during fruit maturation is highly related to the increase in SSC and fruit metabolism. Therefore, fruit harvested at a dark skin colour shortens the period of storage but improves fruit flavour and acceptability to the consumer. In contrast, fruit acceptability is lower at a lighter colour, but storability is improved. The sweet cherry industry has a standard colour chart for harvest, which was developed by the Centre Technique Interprofessionel de Fruits et Legumes (CTIFL, Paris). The chart is a stepwise colour progression from light pink to red (stage 1) to very dark, almost black (stage 7).

Skin deterioration (decay and pebbling) and pedicel browning are more critical for cherries harvested at an overmature stage and held in long-term storage. Adjusting the time of harvest of a cultivar with its storage requirements becomes critical for providing high eating quality of the fruit after long-term storage or transport. An example of this is the evaluation done with 'Sweetheart' grown in British Columbia, Canada, and harvested at different stages of maturity, for which a superior eating quality after 6 weeks at 0°C was demonstrated when the fruit was harvested at late stage of maturity (Toivonen, 2015). However, late-harvested fruit of 'Lapins' and 'Sweetheart' declined in storage potential relative to fruit skin lustre, colour and stem browning under the growing conditions in the Pacific Northwest USA; therefore, to balance eating quality and storage potential, the harvest timings for 'Lapins' and 'Sweetheart' are optimized at CTIFL 5.5 and 4.5, respectively, under those growing conditions (Wang and Einhorn, 2017).

### 19.5.2 Packing line operation

Packing houses designed for handling sweet cherry fruit consist of a fruit receiving area, a hydrocooler zone, a packing line, post-packing forced-air cooling facilities and refrigerated holding rooms (Grant and Thompson, 1997). Once fruit is received at the packing house, it can be packed immediately or stored for packing later. When the fruit is not packed within 24 h of harvest, it should be hydrocooled to a pulp temperature of 0–2°C and held in a cold room until it can be packed. To begin the sorting and packing process, fruit is dumped manually or automatically into a water tank at the beginning of the line

**Table 19.1.** Main operations of the sweet cherry postharvest handling process to maximize the storage period at 0°C. Postharvest management of the critical points are indicated for each operation.

| Operation | Critical points | Postharvest management |
|---|---|---|
| Harvest | • Reference colour harvest index for individual cultivar<br>• Planning of the period of harvest operation<br>• Select the appropriated picking harvest materials<br>• Temperature and humidity<br>• Picker training | • Adjust the colour index for harvesting the cultivars specifically for the intended storage period<br>• Protect the harvested fruit from sun using reflective tarps on the top of the boxes<br>• Avoid pitting by using trained pickers and correct harvest methods<br>• Determine the number of pickers required and the frequency of harvest<br>• Separate cracked and non-marketable fruit |
| Transport | • Select the appropriated type of transport | • Reduce the transport time to <4 h from picking to the packing facility<br>• Protect the fruit from sun during transport<br>• Long-distance transport requires considering the installation of an in-field hydrocooler; a cold chamber and refrigerated truck will be required |
| Sorting and packing | • Quality-control procedures<br>• Cultivar characteristics<br>• Critical points for pitting<br>• Fruit pulp temperature<br>• Packaging materials<br>• Fruit quality in the commercial boxes | • Check the maturity and quality of the fruit<br>• Define suitability of cultivar for water flow operation<br>• Select fruit at different points of the packing line and define critical risk of pitting<br>• Determine the number of people for the sorting table; adjust it in relation to type of defect and the effectiveness of the electronic defect sorting machine<br>• Match the quality of the commercial boxes with the expectations of the market<br>• Avoid pulp temperature <2°C while the fruit is running through the packing line to reduce pitting<br>• Pulp temperature must be <6°C at the sealing point of MA bags |
| Hydrocooling | • Temperature and time of operation<br>• Water sanitation | • The fruit must be cooled down to 0–2°C by hydrocooling process if the fruit will be stored for >24 h<br>• Keep the pH to 7 and 80 p.p.m. of free chlorine for water sanitation or keep ORP >650 mV |
| Forced-air cooling | • Time and temperature and time of operation | • Packaged fruit pulp temperature must be forced-air cooled to 0–1°C, using forced-air cooling<br>• Do not mix perforated liner with MA bag<br>• Maximize cooling operation using a well-sealed tunnel, aligning the box perforations of the boxes with high air volume at high static pressure |
| Transport by sea container | • Layout of the pallets<br>• Air flow and temperature | • Assure –0.5°C air delivery temperature of the container, and a minimum air exchange with the internal and external air of the container for transporting palletized MA bags<br>• Provide a uniform temperature from bottom-air delivery, by covering the space not occupied by the palletized fruit |

MA, modified atmosphere; ORP, oxidation–reduction potential.

and is moved with a flow of water in a flume system throughout the line. Cluster cutting and manual or electronic sorting are the main steps for grading fruit size and colour before packing.

The most popular system to separate cherry fruit and their pedicel individually from multi-fruit clusters is with the use of cluster cutters that align the fruit using mechanical roller blades with height adjustment to cut the upper end of the pedicel. However, a hydraulic cluster separating system (Facheux Grading, St Martin d'Auxigny, France) is also commercially available and used in smaller-scale operations. Blade-type cluster cutters can be the source of significant damage to the fruit, although hydraulic cluster cutters can also lead to impact damage (Fig. 19.3).

Sweet cherry defect separation can be done manually at a high labour cost, or using an automated electronic system that relies on spectral reflectance, imaging processing and pattern recognition programmes that recognize defects such as bruising, cracking, different types of russets, deformations, fruit with soft patches and decayed tissue. Size grading is done mechanically using diverging or parallel metallic pipes. However, most of these systems have been replaced more recently with sophisticated optical/electronic sorting systems that are able to sort fruit by colour and size with good accuracy.

Boxes are filled with fruit automatically at the end of the conveyors with different combinations of size and colour, depending on the market's specifications. Fruit can be packed directly by tight fill in boxes or in individual bags and clamshells of different sizes to be distributed inside the 5 or 10 kg boxes, lined with perforated, non-perforated or modified-atmosphere bags. The perforated and non-perforated bags are used mainly for air transportation, and MAP is destined for sea freight transport by vessels or shipping containers. The layout of the operations is summarized in Fig. 19.5.

### 19.5.3 Modified-atmosphere packaging

MAP is being used extensively with sweet cherry to maximize postharvest performance at 0°C (Lurie and Aharoni, 1997; Remón *et al.*, 2000; Kupferman and Sanderson, 2005). In passive MAP systems, which are widely used, the final atmosphere is achieved by fruit consumption of $O_2$ and release of $CO_2$ during natural respiration. The rate of $CO_2$ and $O_2$ exchange by the film determines the steady-state concentrations of these gases inside the bag. The equilibrium is reached 3–5 days after the bag is sealed (Zoffoli and Rodríguez, 2014b).

Research conducted in controlled-atmosphere storage demonstrates high tolerance of cherry fruit to high concentrations of $CO_2$, up to 20% with 5% $O_2$, for 12 weeks (Mattheis *et al.*, 1997). A combination of 5% $O_2$ and 10% $CO_2$ inhibited the enzymatic activities of polyphenol oxidase and peroxidase and reduced malondialdehyde content (Tian *et al.*, 2004). Fruit firmness was improved and darkening of fruit colour was reduced with 0.5–2% $O_2$ (Chen *et al.*, 1981) or 20–25% $CO_2$ (Patterson, 1982). 'Sweetheart' storage life was extended up to 6 weeks at 1°C with 2% $CO_2$ and 5% $O_2$ (Remón *et al.*, 2003). Levels of $CO_2$ higher than 30% have been associated with brown skin discoloration of 'Bing' cherries (Kader, 1997) and 18% $CO_2$ and 2% $O_2$ with internal browning in 'Regina' fruit after 7 weeks at 1°C (Harb *et al.*, 2003).

Low $O_2$ is necessary for reducing respiration rate, and progressive reductions in $O_2$ concentrations less than 10% produce a logarithmic reduction in respiration rate (Wang and Long, 2014). Fermentation or development of 'off-flavour' is the critical issue associated with commercial storage temperatures for sweet cherry under MAP. The fermentation induction points for 'Bing' and 'Sweetheart' have been estimated for 1 and 3–4% $O_2$ at 0 and 20°C, respectively (Wang and Long, 2014). Ethanol and acetaldehyde accumulate to significant levels when a low concentration of $O_2$ (1.5%) is combined with a high concentration of $CO_2$ (11.5–12%) (Goliáš *et al.*, 2007). The accumulation of these compounds is responsible for loss in fruit acceptability after long-term storage. 'Regina' fruit stored at 6, 12 or 18% $CO_2$ and 2% $O_2$ achieved the highest acceptability values after 4 weeks at 1°C, but the high acceptability only persists for fruit stored under the lowest concentration of $CO_2$ after 7 weeks at 1°C (Harb *et al.*, 2003).

The commercial application of MAP entails the retention of relatively higher concentrations of $O_2$ to avoid fermentation as discussed above. With MAP, development of low $O_2$ levels leading to fermentation will develop under temperature fluctuations that

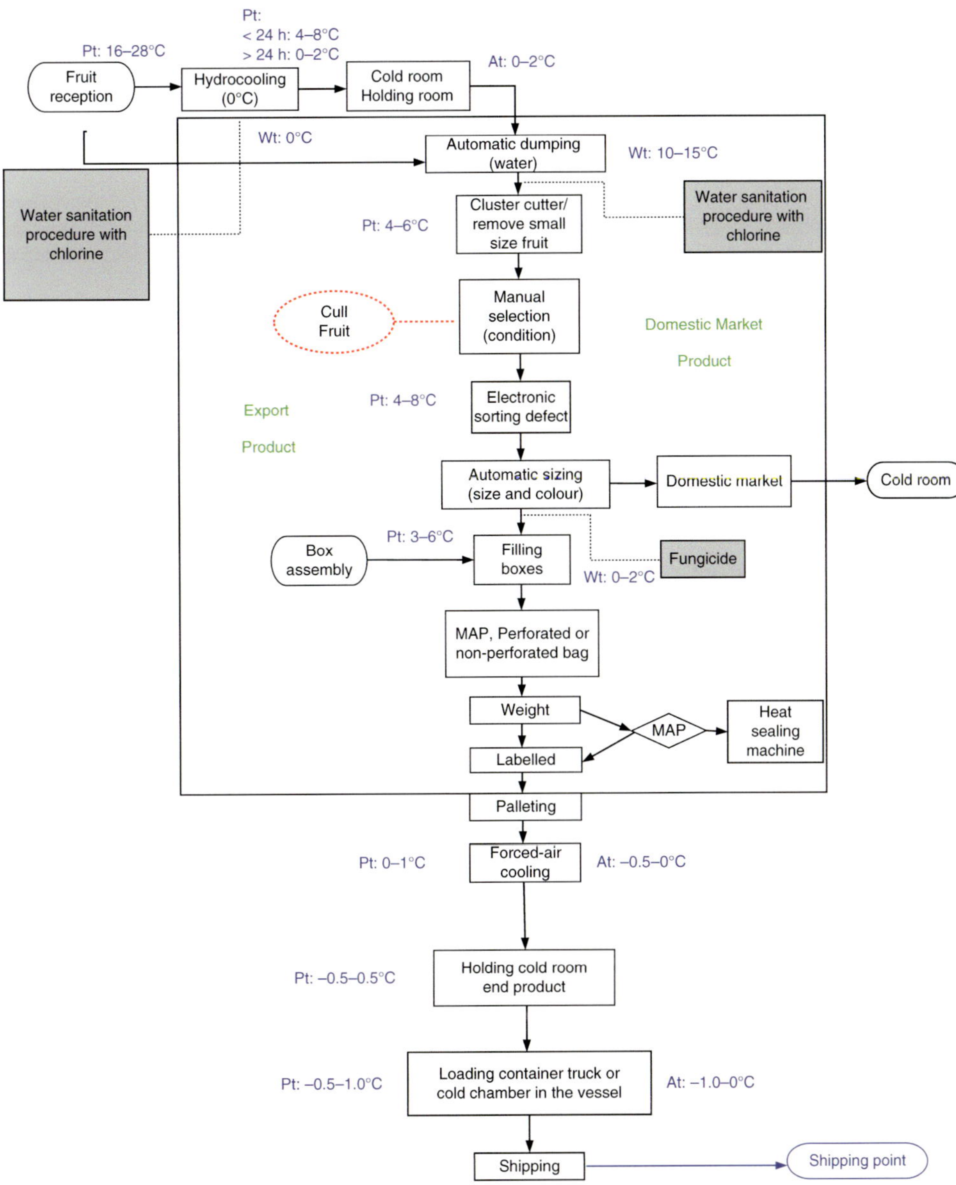


**Fig. 19.5.** Layout and operations of a sweet cherry processing line during postharvest handling for domestic market and export from fruit reception to shipping point. Quality-control (QC) points, fruit pulp temperature (Pt), water temperature (Wt) and air temperature (At) are indicated along the process.

often occur under commercial transport and distribution of the fruit (Wang and Long, 2014; Wang *et al.*, 2015). This is because, with temperature fluctuations, respiration rates of the cherries can increase, lowering $O_2$ levels below that for which the MAP film was designed. Normally, MAP liners used with cherries have a temperature specification provided by the manufacturer.

Among different MAPs, those that generate a steady-state concentration of 1.8–8.0% $O_2$ and 7.3–10.3% $CO_2$ reduce respiration and maintain fruit acidity and flavour. In contrast, an $O_2$ concentration at an equilibrium level higher than 10% maintains firmness but leads to loss of flavour, which is similar to storage in a macro-perforated bag (Wang and Long, 2014). Therefore, a safe concentration between 5 and 8% $O_2$ and 7–10% $CO_2$ should be targeted for commercial operations in order to maximize the benefits of MAP technology, while also reducing the risk of generating fermentation conditions. Cultivars may respond to MAP differently due to differences in respiration activity and fermentation induction point. For example, 'Skeena' fruit accumulated a higher content of ethanol from fermentation than 'Lapins' in the same MAP conditions (Wang *et al.*, 2015).

MAP not only modifies both internal $O_2$ and $CO_2$ concentrations within packages, but also increases the relative humidity surrounding the fruit. High relative humidity retards shrivelling, but small temperature fluctuations can cause condensation that promotes decay.

The fruit packed in a properly designed MAP bag maintains red skin colour similar to that at harvest and suppresses decay; however, this technology does not reduce surface pitting (Zoffoli and Rodríguez, 2014b; Wang *et al.*, 2015). Other positive effects such as the retention pedicel greenness and fruit turgidity are more associated with the high humidity environment around the fruit as opposed to the headspace gas composition generated by the MAP bag (Wang *et al.*, 2015).

### 19.5.4 Cooling operations

Rapid cooling of sweet cherries after harvest is the best way to maintain fruit quality and maximize postharvest life. The main objectives of cooling are to remove the field heat and reduce the respiration heat, both minimizing subsequent water loss from the fruit and retarding development of decay. The cool chain should be established as soon as possible after harvest and maintained at just above –0.5°C to avoid condensation and keep fruit quality deterioration to a minimum.

#### *Hydrocooling*

Hydrocooling is one of the most common fast cooling methods (Looney *et al.*, 1996) in which the exchange of heat from the fruit is achieved by direct contact with cold water (Fig. 19.6). The positive effect of hydrocooling results from the fact that cooling is faster, more uniform, and that the fruit are washed and can therefore be disinfected during the process. To optimize cooling efficiency, the hydrocooler is operated continuously, at maximum capacity, and must be located inside a cold room or covered with insulation to avoid heat transfer from surrounding air. Sweet cherry fruit is well adapted to hydrocooling when fruit is harvested under dry conditions. If fruit are harvested under rainy conditions, they become very susceptible to postharvest cracking and this requires that resident time in water and hydrocooling be minimized. Adding calcium at appropriate rates in hydrocooling water increases fruit tissue calcium content and reduces postharvest cracking (Wang and Long, 2015), as well as improving storage/shipping quality (Wang *et al.*, 2014).

#### *Forced-air cooling*

Forced-air cooling is the standard method (Looney *et al.*, 1996) for rapid removal of heat from packed fruit. The fast cooling procedure using a forced-air facility is mandatory in packed and palletized sweet cherries. This requires that forced-air cooling is applied to palletized boxes of sweet cherries since the fruit temperature is most commonly above 2°C after hydrocooling and packing (Toivonen, 2014). The cold air, in this type of cooling, infiltrates through the packaging by creating a slight pressure gradient to cause the air to flow through the vent of the container and favour close contact between the cold air and warm fruit. The pressure difference can vary between 0.6 and 6.4 cm $H_2O$. Fruit packed in a modified-atmosphere bag are the most difficult to cool, since the non-vented liner renders the strongest resistance to air movement through boxes on the pallet. Proper box designs that allow high surface contact

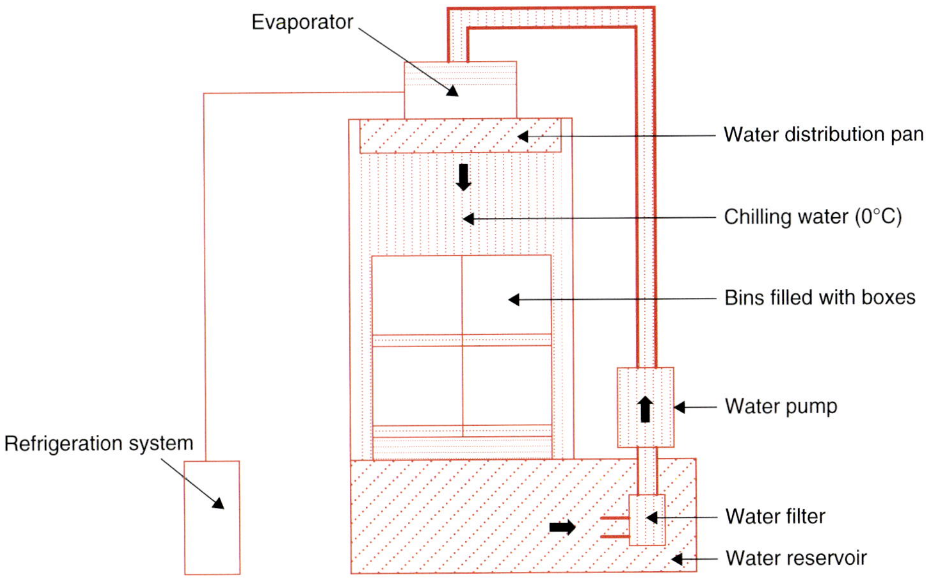


**Fig. 19.6.** End view schematic of a continuous flow-type hydrocooler for bins filled with sweet cherry boxes. (Adapted from Thompson and Chen, 1989.)

of the air with the bag are required to optimize cooling time. Pulp temperature between 0 and 1°C is necessary to restrict the heat produced by fruit respiration and avoid fermentation under MAP during storage or long-distance shipping. Stringent management of the lowest core temperature possible is necessary to ensure the best quality retention for sweet cherries to be shipped long distances in refrigerated containers.

### *Room cooling*

Room cooling is a common method used upon receipt of fruit at the packing house handling or for cold-packed fruit waiting for shipment (Looney *et al.*, 1996). It is far less efficient in reducing fruit temperature than forced-air cooling. If room cooling is to be used, it is recommended that boxes are stacked to facilitate airflow between them (Fig. 19.7). When comparing forced-air cooling to this method of stacking in a room cooling situation, the times to reach target temperatures are similar. Air temperature should be held below 0°C, preferably –0.5 to –1.0°C in order to cool flesh temperature near the pit to 0°C.

### *Packing house water sanitation*

The use of sanitizer in hydrocooling and packing line flume water is required to reduce microbial populations and prevent further potential fruit infections. Chlorine is commonly used; its antimicrobial activity depends on the concentration, the organic matter content of the water, the pH of the water, and time and temperature of exposure.

The main forms of chlorine are sodium hypochlorite (NaOCl), calcium hypochlorite ($Ca(OCl)_2$) and chlorine gas ($Cl_2$). In solution, all forms of chlorine produce hypochlorous acid, which disassociates into hypochlorite ion ($OCl^-$) with a $pK_a$ of 7.5 at 25°C. The hypochlorous acid and hypochlorite ion are the forms of free chlorine that can be monitored. Adjusting the free chlorine concentration between 80 and 100 p.p.m. assures a low microbiological count in the water. The non-dissociated form of hypochlorous acid has the highest antimicrobial activity and should be achieved by adjusting the pH between 6 and 7.5. Lowering the pH (<6) induces the release of $Cl_2$ gas, which is irritating to workers. A $Cl_2$ sensor is required when chlorine is used under closed systems with workers. Automatic monitoring and

**Fig. 19.7.** Stacking method of sweet cherry boxes for increasing room cooling efficiency.

adjustment of the pH and chlorine concentration is possible through commercial in-line oxidation–reduction potential (ORP) instruments (Suslow, 2004b). ORP values are proportional to antimicrobial activity and provide a rapid and single-value assessment of the disinfection potential in the water. Organic matter in water inactivates hypochlorous acid, reducing the amount of free chlorine and the activity against pathogens. Therefore, the evaluation of free chlorine and pH of the water must be considered as critical monitoring points in the hydrocooling and flume water.

Among the major forms of chlorine, $Ca(OCl)_2$ is most used in US cherry packing houses in recent years, since a build-up of $Na^+$ in water may damage cherry stems while an accumulation of $Ca^{2+}$ in water may benefit fruit quality (Wang *et al.*, 2014; Wang and Long, 2015). Chlorine dioxide is an alternative to chlorine sanitation. Its disinfection activity is not affected by pH or organic matter, and it is active at lower concentrations. Chlorine dioxide-generating systems allow electronically controlled dosing *in situ*; however, stabilized liquid formulations are available, making the application easier and cheaper (Zoffoli *et al.*, 2005).

Ozone is a naturally occurring gas and strong oxidizer. It has a short life in water and does not leave any residues. The effective concentration for preventing pathogen growth in water is 1.5 p.p.m. Ozone is inactivated by the organic matter. Its solubility increases at low water temperature and its activity is not influenced by pH (Suslow, 2004a). It also can be monitored and adjusted in-line using ORP instrumentation (Suslow, 2004b).

Peracetic acid is a strong disinfectant with a wide spectrum of antimicrobial activity (Kitis, 2004). Its activity is not influenced by organic matter or pH, and it leaves no residues in the water. Its disinfectant activity is based on the release of active hydrogen peroxide and acetic acid (Liberti and Notarnicola, 1999). No phytotoxicity was found for sweet cherries using the peracetic acid in the hydrocooling process (Kupferman, 2008), but every formulation needs to be tested for phytotoxicity because differences in the chemistry of each commercial product exist for release of hydrogen peroxide and acetic acid. The antifungal activity of peracetic acid was reported at 125 p.p.m. in naturally infected fruit when dipped in cold (5°C) water (Mari *et al.*, 2004).

The antifungal activity among different sanitizers is related to oxidative power, which is highest for ozone followed by peracetic acid, chlorine dioxide and sodium hypochlorite. However, local regulations and practical aspects of the use of each sanitizer must be considered.

Most synthetic chemicals are prohibited (such as the fungicide fludioxonil) or restricted in postharvest handling operation of organic fruit. Of particular interest are the sanitizers used in postharvest water. While chlorine, ozone and peracetic acid all can be used, chlorine can be used only in specific limits for organic fruit. For examples, the California Certified Organic Farmers (CCOF) permits a threshold of 10 p.p.m. residual chlorine measured downstream of the wash water (effluent water). There are no international harmonized standards for postharvest chemical use in organic commodities, although the USA and the European Union accept each other's organic standards as equivalent.

### 19.5.5 Quarantine treatments

Sweet cherries for export must be treated after harvest to satisfy quarantine restrictions imposed by some importing countries. Methyl bromide, a strong oxidant, is still approved (as of 2016) in some countries (e.g. Australia, Japan and the USA, among others) for postharvest and quarantine preshipment fumigation use. Although its uses for other tasks, such as soil fumigation, were phased out in developed countries by 2005 and in developing countries by 2015, Critical Use Exemptions (CUE) have been granted on a case-by-case basis by regulatory agencies such as the Environmental Protection Agency in the USA (where its complete phase-out is scheduled for 2018; EPA, 2016). The treatment schedules approved for sweet cherries are 64 g $m^{-3}$ methyl bromide per 2 h at 6–12°C, 48 g $m^{-3}$ methyl bromide per 2 h at 12–17°C, 40 g $m^{-3}$ methyl bromide per 2 h at 17–22°C or 32 g $m^{-3}$ methyl bromide per 2 h at >22°C. Fumigation treatment at 32 g $m^{-3}$ of methyl bromide per 2 h at 24°C was proved to be effective in killing larvae of *Cydia pomonella* (codling moth) in 'Bing', 'Lambert', 'Rainier' and 'Van' fruit (Anthon *et al.*, 1975; Gaunce *et al.*, 1981; Moffitt *et al.*, 1992). After fumigation and ventilation, residue of methyl bromide in fruit decreases quickly (Hansen *et al.*, 2000a).

The methyl bromide fumigation process increases pedicel browning (Anthon *et al.*, 1975; Hansen *et al.*, 2000b; Feng *et al.*, 2004). However, this damage seems to be related to the high temperatures necessary for the treatment efficacy rather than methyl bromide itself. Fruit firmness was not affected by fumigation (Hansen *et al.*, 2000b). Radiofrequency heating, gamma irradiation and controlled-atmosphere heat treatment have potential as alternative quarantine treatments to methyl bromide for sweet cherries (Neven and Drake, 2000; Monzón *et al.*, 2006). Cold treatment can be used for control of Mediterranean fruit fly in sweet cherry. The treatments include one of the following protocols: 10 days at 0°C, 11 days at 0.6°C, 12 days at 1.1°C, 14 days at 1.7°C or 17 days at 2.2°C.

### 19.5.6 Long-distance marine transport

Shipping sweet cherries by boat to distant export markets instead of by air freight is increasing worldwide (Toivonen, 2014; Wang *et al.*, 2015). While transport and distribution take 2–3 days by air freight, the time by ocean shipping to export markets may range from 30 to 40 days. With protracted transport, significant arrival issues can occur, including flavour loss and off-flavour development (Wang *et al.*, 2015). The high perishability of sweet cherries requires a stable low temperature to slow down fruit respiration and physiological activities during postharvest life. Temperature fluctuation is common in commercial storage and shipping, and is the main cause of producing fruit off-flavours as a result of anaerobic respiration in MAP. During ocean shipping, cherries in MAP are loaded in refrigerated marine containers or inside the cold chambers of the vessel.

The refrigerated marine containers are designed to maintain fruit temperature during transport via independent refrigerated units powered by 220 V or 440 V three-phase electricity, allowing direct plugging in to the electric power system on the vessel or in the port. Temperature readers are installed to record the supply and return air temperatures. Refrigerated containers are equipped with a bottom-air delivery system. Air from the refrigeration unit flows from the floor through the packages vertically and returns horizontally across the top of the load and back to the refrigeration unit to complete the air cycle. Air space and venting are mandatory to allow vertical airflow from the floor to the pallet load. Inadequate ventilation or poor alignment of container vents on a tightly stacked unitized pallet can deter air flow through fruit mass (LaRue and Johnson, 1989). Full coverage of the floor by the produce or a solid material is necessary.

## 19.6 Outlook and Challenges

Breeding/selecting new cultivars with superior fruit quality and storage/shipping potential is essential for extending postharvest quality and reducing postharvest losses of sweet cherries. Pitting continues to be the leading cause of product rejection and price modification by buyers and receivers in both domestic and international markets, and therefore causes significant economic losses for cherry growers. Producing cherries with high resistance to mechanical damage is desirable.

Globalization of sweet cherry marketing is expected to continue in the future. Currently, the majority of sweet cherries for long-distance international trading are shipped via air freight. The current debate on $CO_2$ emissions, food miles and shipping costs may shift their major transport from air to sea. The adoption of sea freight by the cherry industry has begun, but the arrival quality of fruit has not been consistent among cultivars, production lots and years.

New postharvest technologies need to be adapted for the transport phase, which comprises the majority of the postharvest period. Better knowledge and optimization of the vertical air movement and proper packaging design inside sea freight containers are required. The sweet cherry industries that utilize long shipping periods currently are highly dependent on postharvest fungicide applications. An integrated sanitation approach, combining pre- and postharvest practices with antifungal activity products generally recognized as safe (GRAS), needs to be addressed for a more sustainable future.

Pre- and postharvest factors (e.g. climatic, agronomic, cultivar, maturity, calcium nutrition, temperature, $O_2$, $CO_2$, hypobaric, logistic) affecting cherry fruit arrival quality warrant further research. The continued registration of methyl bromide for postharvest quarantine treatment is not guaranteed. It is relatively easy to develop quarantine treatments in the laboratory. However, efficient, safe and cost-effective alternative methods to methyl bromide at commercial application scales are needed. Evaluation of the impact of pre- and postharvest factors and quarantine conditions on fruit postharvest quality should include flavour quality in addition to fruit appearance, texture, and pitting.

## References


Adaskaveg, J.E., Förster, H. and Thompson, D.F. (2000) Identification and etiology of visible quiescent infections of *Monilinia fructicola* and *Botrytis cinerea* in sweet cherry fruit. *Plant Disease* 84, 328–333.

Agulheiro-Santos, A.C., Palma, V., Rato, A.E., Machado, G., Lozano, M. and González-Gómez, D. (2014) Quality of 'Sweetheart' cherry under different storage conditions. *Acta Horticulturae* 1020, 101–110.

Anthon, E.W., Moffitt, H.R., Couey, H.M. and Smith, L.O. (1975) Control of codling moth in harvested sweet cherries with methyl bromide and effects upon quality and taste of treated fruit. *Journal of Economic Entomology* 68, 524–526.

Athoo, T.O., Winkler, A. and Knoche, M. (2015) Pedicel transpiration in sweet cherry fruit: mechanisms, pathways, and factors. *Journal of the American Society for Horticultural Science* 140, 136–143.

Azarenko, A.N., Chozinski, A. and Brewer, L.J. (2008) Fruit growth curve analysis of seven sweet cherry cultivars. *Acta Horticulturae* 795, 561–566.

Barrett, D.M. and Gonzalez, C. (1994) Activity of softening enzymes during cherry maturation. *Journal of Food Science* 59, 574–577.

Bartley, I.M. and Knee, M. (1982) The chemistry of textural changes in fruit during storage. *Food Chemistry* 9, 47–58.

Børve, J. (2014) Fungal contamination of fruit in sweet cherry grading lines. *Acta Horticulturae* 1020, 127–130.

Børve, J., Sekse, L. and Stensvand, A. (2000) Cuticular fractures promote postharvest fruit rot in sweet cherries. *Plant Disease* 84, 1180–1184.

Bukovac, M.J., Knoche, M., Pastor, A. and Fader, R.G. (1999) The cuticular membrane: a critical factor in rain-induced cracking of sweet cherry fruit. *Journal of the American Society for Horticultural Science* 34, 549.

Candan, A.P., Raffo, M.D., Calvo, G. and Gomila, T. (2014) Study of the main points of impact during cherry handling and factors affecting pitting sensitivity. *Acta Horticulturae* 1020, 137–141.

Chen, P.M., Mellenthin, W.M., Kelly, S.B. and Facteau, T.J. (1981) Effects of low oxygen and temperature on quality retention of 'Bing' cherries during prolonged storage. *Journal of the American Society for Horticultural Science* 106, 533–535.

Choi, C., Wiersma, P.A., Toivonen, P. and Kappel, F. (2002) Fruit growth, firmness and cell wall hydrolytic enzyme activity during development of sweet cherry fruit treated with gibberellic acid (GA3). *Journal of Horticultural Science and Biotechnology* 77, 615–621.

Clayton, M., Biasi, B. and Mitcham, B. (1998) Devices for measuring firmness of cherries. *Perishables Handling Quarterly* 95, 2–4. Available at: http://ucce.ucdavis.edu/files/datastore/234-52.pdf (accessed 24 January 2017).

Clayton, M., Biasi, W.V., Agar, I.T., Southwick, S.M. and Mitcham, E.J. (2003) Postharvest quality of 'Bing' cherries following preharvest treatment with hydrogen cyanamide, calcium ammonium nitrate, or gibberellic acid. *HortScience* 38, 407–411.

Cliff, M.A., Dever, M.C., Hall, J.W. and Girard, B. (1995) Development and evaluation of multiple regression models for prediction of sweet cherry liking. *Food Research International* 28, 583–589.

Crisosto, C.H., Garner, D., Doyle, J. and Day, K.R. (1993) Relationship between fruit respiration, bruising susceptibility, and temperature in sweet cherries. *HortScience* 28, 132–135.

Crisosto, C.H., Mitchell, F.G. and Johnson, S. (1995) Factors in fresh market stone fruit quality. *Postharvest News and Information* 6, 17–21.

Crisosto, C.H., Crisosto, G.M. and Metheney, P. (2003) Consumer acceptance of 'Brooks' and 'Bing' cherries is mainly dependent on fruit SSC and visual skin color. *Postharvest Biology and Technology* 28, 159–167.

de Vries-Paterson, R.M., Jones, A.L. and Cameron, A.C. (1991) Fungistatic effects of carbon dioxide in a package environment on the decay of Michigan sweet cherries by *Monilinia fructicola*. *Plant Disease* 75, 943–946.

Dever, M.C., MacDonald, R.A., Cliff, M.A. and Lane, W.D. (1996) Sensory evaluation of sweet cherry cultivars. *HortScience* 31, 150–153.

Díaz-Mula, H.M., Castillo, S., Martínez-Romero, D., Valero, D., Zapata, P.J., Guillén, F. and Serrano, M. (2009) Sensory, nutritive and functional properties of sweet cherry as affected by cultivar and ripening stage. *Food Science and Technology International* 15, 535–543.

Drake, S.R. and Elfving, D.C. (2002) Indicators of maturity and storage quality of 'Lapins' sweet cherry. *HortTechnology* 12, 687–690.

Drake, S.R. and Fellman, J.K. (1987) Indicators of maturity and storage quality of 'Rainier' sweet cherry. *HortScience* 22, 283–285.

Eccher, T. and Noè, N. (1998) Respiration of cherries during ripening. *Acta Horticulturae* 464, 501.

Einhorn, T.C., Wang, Y. and Turner, J. (2013) Sweet cherry fruit firmness and postharvest quality of late-maturing cultivars are improved with low-rate, single applications of gibberellic acid. *HortScience* 48, 1010–1017.

EPA (2016) Methyl bromide. Environmental Protection Agency, Washington, DC. Available at: https://www.epa.gov/ods-phaseout/methyl-bromide (accessed 23 January 2017).

Esturk, O., Ayhan, Z. and Ustunel, M.A. (2012) Modified atmosphere packaging of 'Napoleon' cherry: effect of packaging material and storage time on physical, chemical, and sensory quality. *Food Bioprocess Technology* 5, 1295–1304.

Facteau, T.J. (1982) Relationship of soluble solids, alcohol-insoluble solids, fruit calcium, and pectin levels to firmness and surface pitting in 'Lambert' and 'Bing' sweet cherry fruit. *Journal of the American Society for Horticultural Science* 107, 151–154.

Facteau, T.J. and Rowe, K.E. (1979) Factors associated with surface pitting of sweet cherry. *Journal of the American Society for Horticultural Science* 104, 707–710.

Facteau, T.J., Rowe, K.E. and Chestnut, N.E. (1985) Firmness of sweet cherry fruit following multiple applications of gibberellic acid. *Journal of the American Society for Horticultural Science* 19, 127–129.

Feliziani, E., Santini, M., Landi, L. and Romanazzi, G. (2013) Pre-and postharvest treatment with alternatives to synthetic fungicides to control postharvest decay of sweet cherry. *Postharvest Biology and Technology* 78, 133–138.

Feng, X., Hansen, J.D., Biasi, B., Tang, J. and Mitcham, E.J. (2004) Use of hot water treatment to control codling moths in harvested California 'Bing' sweet cherries. *Postharvest Biology and Technology* 31, 41–49.

Fogle, H.W., Snyder, J.C., Baker. H, Cameron, H.R., Cochran, L.C., Schomer, H.A. and Yang, H.Y. (1973) *Sweet Cherries: Production, Marketing, and Processing*. Agriculture Handbook 442. US Department of Agriculture, Washington, DC.

Gao, L. and Mazza, G. (1995) Characterization, quantitation, and distribution of anthocyanins and colorless phenolics in sweet cherries. *Journal of Agricultural and Food Chemistry* 43, 343–346.

Garcia-Ramos, F.J., Valero, C., Homer, I., Ortiz-Cañavate, J. and Ruiz-Altisent, M. (2005) Non-destructive fruit firmness sensors: a review. *Spanish Journal of Agricultural Research* 3, 61–73.

Gaunce, A.P., Madsen, H.F. and McMullen, R.D. (1981) Fumigation with methyl bromide to kill larvae and eggs of the codling moth in Lambert cherries. *Journal of Economic Entomology* 74, 154–157.

Glenn, G.M. and Poovaiah, B.W. (1987) Role of calcium in delaying softening of apples and cherries. *Postharvest Pomology Newsletter* 5, 10–19.

Glenn, M.G. and Poovaiah, B.W. (1989) Cuticular properties and postharvest calcium applications influence cracking of sweet cherries. *Journal of the American Society for Horticultural Science* 114, 781–788.

Goliáš, J., Nemcová, A., Čanek, A. and Kolenčíková, D. (2007) Storage of sweet cherries in low oxygen and high carbon dioxide atmospheres. *Horticultural Science* 34, 26–34.

Gonçalves, B., Landbo, A.K., Knudse, D., Silva, A.P., Moutinho-Pereira, J., Rosa, E. and Meyer, A.S. (2004) Effect of ripeness and postharvest storage on the phenolic profiles of cherries (*Prunus avium* L.). *Journal of Agricultural and Food Chemistry* 52, 523–530.

Gong, Y., Fan, X. and Mattheis, J.P. (2002) Responses of 'Bing' and 'Rainier' sweet cherries to ethylene and 1-methylcyclopropene. *Journal of the American Society for Horticultural Science* 127, 831–835.

Grant, J. and Thompson, J. (1997) Packing-line modifications reduce pitting and bruising of sweet cherries. *California Agriculture* 51, 31–35.

Guyer, D.E., Sinha, N.K., Chang, T.S. and Cash, J.N. (1993) Physiochemical and sensory characteristics of selected Michigan sweet cherry (*Prunus avium* L.) cultivars. *Journal of Food Quality* 16, 355–370.

Hampson, C.R., Stanich, K., McKenzie, D.L., Herbert, L., Lu, R., Li, J. and Cliff, M.A. (2014) Determining the optimum firmness for sweet cherries using Just-About-Right sensory methodology. *Postharvest Biology and Technology* 91, 104–111.

Hansen, J.D., Sell, C.R., Moffit, H.R., Leesch, J.G. and Hartsell, P.L. (2000a) Residues in apples and sweet cherries after methyl bromide fumigation. *Pest Management Science* 56, 555–559.

Hansen, J.D., Drake, S.R., Moffitt, H.R., Albano, D.J. and Heidt, M.L. (2000b) Methyl bromide fumigation of five cultivars of sweet cherries as a quarantine treatment against codling moth. *HortTechnology* 10, 194–198.

Harb, J., Streif, J. and Saquet, A. (2003) Impact of controlled atmosphere storage conditions on storability and consumer acceptability of sweet cherries 'Regina'. *Journal of Horticultural Science and Biotechnology* 78, 574–579.

Harb, J., Saquet, A.A., Bisharat, R. and Streif, J. (2006) Quality and biochemical changes of sweet cherries cv. 'Regina' stored in modified atmosphere packaging. *Journal of Applied Botany and Food Quality* 80, 145–149.

Horvitz, S., Godoy, C., Lopez Camelo, A.F., Yommi, A. and Godoy, C. (2003) Application of gibberellic acid to 'Sweetheart' sweet cherries: effects on fruit quality at harvest and during cold storage. *Acta Horticulturae* 628, 311–316.

Ito, K. and Clever, J. (2012) *Japan: Stone Fruit Annual*. Global Agricultural Information Network, Report No. JA2020. USDA Foreign Agricultural Service, Washington, DC.

Kader, A.A. (1992) Modified atmospheres during transport and storage. In: *Postharvest Technology of Horticultural Crops*, 2nd edn. University of California, Davis, California, pp. 85–92.

Kader, A.A. (1997) A summary of CA requirements and recommendations for fruits other than apples and pears. In: Kader, A.A. (ed.) *Proceedings of the 7th International Controlled Atmosphere Research Conference*, Vol. 3. University of California Davis, California, pp. 1–34.

Kappel, F. and Toivonen, P. (2005) Resistance of advanced sweet cherry selections and cultivars to fruit surface pitting. *Acta Horticulturae* 667, 515–522.

Kappel, F., Fisher-Fleming, B. and Hogue, E. (1996) Fruit characteristic and sensory attributes of an ideal sweet cherry. *HortScience* 31, 443–446.

Kappel, F., Toivonen, P., McKenzie, D.L. and Stan, S. (2002) Storage characteristics of new sweet cherry cultivars. *HortScience* 37, 139–143.

Kappel, F., Toivonen, P., Stan, S. and McKenzie, D.L. (2006) Resistance of sweet cherry cultivars to fruit surface pitting. *Canadian Journal of Plant Science* 86, 1197–1202.

Karabulut, O.A., Lurie, S. and Droby, S. (2001) Evaluation of the use of sodium bicarbonate, potassium sorbate and yeast antagonists for decreasing postharvest decay of sweet cherries. *Postharvest Biology and Technology* 23, 233–236.

Karabulut, O.A., Arslan, U., Ilhan, K. and Kuruoglu, G. (2005) Integrated control of postharvest disease of sweet cherry with yeast antagonists and sodium bicarbonate applications within a hydrocooler. *Postharvest Biology and Technology* 37, 135–141.

Khorshidi, S., Davarynejad, G., Tehranifar, A. and Fallahi, E. (2011) Effect of modified atmosphere packaging on chemical composition, antioxidant activity, anthocyanin, and total phenolic content of cherry fruits. *Horticulture, Environment and Biotechnology* 52, 471–481.

Kitis, M. (2004) Disinfection of wastewater with peracetic acid: a review. *Environment International* 30, 47–55.

Knoche, M., Peschel, S., Hinz, M. and Bukovac, M.J. (2000) Studies on water transport through the sweet cherry fruit surface: characterizing conductance of the cuticular membrane using pericarp segments. *Planta* 212, 127–135.

Knoche, M., Peschel, S., Hinz, M. and Bukovac, M.J. (2001) Studies on water transport through the sweet cherry fruit surface: II. Conductance of the cuticle in relation to fruit development. *Planta* 213, 927–936.

Knoche, M., Beyer, M., Peschel, S., Oparlakov, B. and Bukovac, M.J. (2004) Changes in strain and deposition of cuticle in developing sweet cherry fruit. *Physiologia Plantarum* 120, 667–677.

Kondo, S. and Inoue, K. (1997) Abscisic acid (ABA) and 1-aminocyclopropane-l-carboxylic acid (ACC) content during growth of 'Satonishiki' cherry fruit, and the effect of ABA and ethephon application on fruit quality. *Journal of Horticultural Science* 72, 221–227.

Kupferman, E. (2008) *Evaluation of Sweet Cherry Fruit and Stem Damage When Applying Peroxyacetic Acid or Sodium Hypochlorite After Harvest*. Washington State University – Tree Fruit Research And Extension Center, Wenatchee, Washington.

Kupferman, E. and Sanderson, P. (2005) Temperature management and modified atmosphere packing to preserve sweet cherry fruit quality. *Acta Horticulturae* 667, 523–528.

Lang, G.A., Olmstead, J.W. and Whiting, M.D. (2004) Sweet cherry fruit distribution and leaf populations: modeling canopy dynamics and management strategies. *Acta Horticulturae* 636, 591–599.

LaRue, J.H. and Johnson, R.S. (eds) (1989) *Peaches, Plums and Nectarines – Growing and Handling for Fresh Market*. Publication No. 3331, University of California Division of Agriculture and Natural Resources, Oakland, California.

Li, S., Andrews, P.K. and Patterson, M.E. (1994) Effects of ethephon on the respiration and ethylene evolution of sweet cherry (*Prunus avium* L.) fruit at different development stages. *Postharvest Biology and Technology* 4, 235–243.

Liberti, L. and Notarnicola, M. (1999) Advanced treatment and disinfection for municipal wastewater reuse in agriculture. *Water Science and Technology* 40, 235–245.

Lidster, P.D. and Tung, M.A. (1980) Effects of fruit temperature at time of impact damage and subsequent storage temperature and duration on the development of surface disorders in sweet cherries. *Canadian Journal of Plant Science* 60, 555–559.

Lidster, P.D., Muller, K. and Tung, M.A. (1980) Effects of maturity on fruit composition and susceptibility to surface damage in sweet cherries. *Canadian Journal of Plant Science* 60, 865–871.

Looney, N.E., Webster, A.D. and Kupferman, E.M. (1996) Harvest and handling sweet cherries for the fresh market. In: Webster, A.D. and Looney, N.E. (eds) *Cherries: Crop Physiology, Production and Uses*. CAB International, Wallingford, UK, pp. 443–470.

Lurie, S. and Aharoni, N. (1997) Modified atmosphere storage of cherries. In: Kader, A. (ed.) *Proceedings of the 7th International Controlled Atmosphere Research Conference*, Vol. 3. University of California Davis, California, pp. 149–152.

Mari, M., Gregori, R. and Donati, I. (2004) Postharvest control of *Monilinia laxa* and *Rhizopus stolonifer* in stone fruit by peracetic acid. *Postharvest Biology and Technology* 33, 319–325.

Mattheis, J.P., Buchanan, D.A. and Fellman, J.K. (1997) Volatile constituents of Bing sweet cherry fruit following controlled atmosphere storage. *Journal of Agricultural and Food Chemistry* 45, 212–216.

McCune, L.M., Kubota, Ch., Stendell-Hollis, N. and Thomson, C.A. (2011) Cherries and health: a review. *Critical Reviews in Food Science and Nutrition* 51, 1–12.

Meheriuk, M., Mckenzje, L.D., Girard, B., Moyls, A.L., Weintraub, S., Hocking, R. and Kopp, T. (1997) Storage of 'Sweetheart' cherries in sealed plastic film. *Journal of Food Quality* 20, 189–198.

Mitcham, E.J., Clayton, M. and Biasi, W.V. (1998) Comparison of devices for measuring cherry fruit firmness. *HortScience* 33, 723–727.

Mitchell, F.G., Mayer, G. and Kader, A.A. (1980) Injuries cause deterioration of sweet cherries. *California Agriculture* 34, 14–15.
Moffitt, H.R., Drake, S.R., Toba, H.H. and Hartsell, P.L. (1992) Comparative efficacy of methyl bromide against codling moth (Lepidoptera: Tortricidae) larvae in 'Bing' and 'Rainier' cherries and confirmation of efficacy of a quarantine treatment for 'Rainier' cherries. *Journal of Economic Entomology* 85, 1855–1858.
Monzón, M.E., Biasi, B., Simpson, T.L., Johnson, J., Feng, X., Slaughter, D.C. and Mitcham, E.J. (2006) Effect of radio frequency heating as a potential quarantine treatment on quality of 'Bing' sweet cherry fruit and motility of codling moth larvae. *Postharvest Biology and Technology* 40, 197–203.
Mozetič, B., Simčič, M. and Trebše, P. (2006) Anthocyanins and hydroxycinnamic acids of 'Lambert' compact cherries (*Prunus avium* L.) after cold storage and 1- methylcyclopropene treatment. *Food Chemistry* 97, 302–309.
Mulabagal, V., Lang, G.A., DeWitt, D.L., Dalavoy, S.S. and Nair, M.G. (2009) Anthocyanin content, lipid peroxidation and cyclooxygenase enzyme inhibitory activities of sweet and sour cherries. *Journal of Agricultural and Food Chemistry* 57, 1239–1246.
Muskovics, G., Felfoldi, J., Kovacs, E., Perlaki, R. and Kallay, T. (2006) Changes in physical properties during fruit ripening of Hungarian sweet cherry (*Prunus avium* L.) cultivars. *Postharvest Biology and Technology* 40, 56–63.
Neven, L.G. and Drake, L.R. (2000) Comparison of alternative postharvest quarantine treatments for sweet cherries. *Postharvest Biology and Technology* 20, 107–114.
Ogawa, J.M., Bose, E., Manji, B.T. and Schreader, W.R. (1972) Bruising of sweet cherries resulting in internal browning and increased susceptibility to fungi. *Phytopathology* 62, 579–580.
Patterson, M.E. (1982) CA storage of cherries. In: Richardson, D.G. and Meheriuk, M. (eds) *Proceedings of the 3rd National Controlled Atmosphere Research Conference*, Oregon State University Symposium Series No. 1. Timber Press, Beaverton, Oregon, pp. 149–154.
Polenta, G., Budde, C. and Murray, R. (2005) Effects of different pre-storage anoxic treatments on ethanol and acetaldehyde content in peaches. *Postharvest Biology and Technology* 38, 247–253.
Porritt, S.W., Lopatecki, L.E. and Meheriuk, M. (1971) Surface pitting – a storage disorder of sweet cherries. *Canadian Journal of Plant Science* 51, 409–414.
Remón, S., Ferrer, A., Marquina, P., Burgos, J. and Oria, R. (2000) Use of modified atmospheres to prolong the postharvest life of Burlat cherries at two different degrees of ripeness. *Journal of the Science of Food and Agriculture* 80, 1545–1552.
Remón, S., Marquina, P., Peiró, J.M. and Oria, R. (2003) Storage potential of Sweetheart cherry in controlled atmospheres. *Acta Horticulturae* 600, 763–769.
Remón, S., Ferrer, A., Venturini, M.E. and Oria, R. (2006) On the evolution of key physiological and physico-chemical parameters throughout the maturation of 'Burlat' cherry. *Journal of the Science of Food and Agriculture* 86, 657–665.
Ren, J., Chen, P., Dai, S.J., Li, P., Li, Q., Ji, K. and Leng, P. (2011) Role of abscisic acid and ethylene in sweet cherry fruit maturation: molecular aspects. *New Zealand Journal of Crop and Horticultural Science* 39, 161–174.
Schick, J.L. and Toivonen, P.M. (2002) Reflective tarps at harvest reduce stem browning and improve fruit quality of cherries during subsequent storage. *Postharvest Biology and Technology* 25, 117–121.
Seeram, N.P., Momin, R.A., Nair, M.G. and Bourquin, L.D. (2001) Cyclooxygenase inhibitory and antioxidant cyanidin glycosides in cherries and berries. *Phytomedicine* 8, 362–369.
Sekse, L. (1988) Respiration of plum (*Prunus domestica* L.) and sweet cherry (*P. avium* L.) fruits during growth and ripening. *Acta Agriculturae Scandinavica* 38, 317–320.
Sekse, L. (1995) Cuticular fracturing in fruits of sweet cherry (*Prunus avium* L.) resulting from changing soil water contents. *Journal of Horticultural Science* 70, 631–635.
Sekse, L. and Lyngstad, L. (1996) Strategies for maintaining high quality in sweet cherries during harvesting, handling and marketing. *Acta Horticulturae* 410, 351–356.
Sekse, L., Meland, M., Reinsnos, T. and Vestrheim, S. (2009) Cultivar and weather conditions determine pre-and postharvest fruit firmness in sweet cherries (*Prunus avium* L.). *European Journal of Horticultural Science* 74, 268–274.
Serrano, M., Guillén, F., Martínez-Romero, D., Castillo, S. and Valero, D. (2005) Chemical constituents and antioxidant activity of sweet cherry at different ripening stages. *Journal of Agricultural and Food Chemistry* 53, 2741–2745.
Serrano, M., Díaz-Mula, H.M., Zapata, P.J., Castillo, S., Guillén, F., Martínez-Romero, D., Valverde, J.M. and Valero, D. (2009) Maturity stage at harvest determines the fruit quality and antioxidant potential after storage of sweet cherry cultivars. *Journal of Agricultural and Food Chemistry* 57, 3240–3246.

Sharkey, P.J. and Peggie, I.D. (1984) Effects of high-humidity storage on quality, decay and storage life of cherry, lemon and peach fruits. *Scientia Horticulturae* 23, 181–190.

Suslow, T.V. (2004a) *Ozone Applications for Postharvest Disinfection of Edible Horticultural Crops*. Publication 8133. University of California, Division of Agriculture and Natural Resources, Davis, California.

Suslow, T.V. (2004b) *Oxidation-Reduction Potential (ORP) for Water Disinfection Monitoring, Control, and Documentation*. Publication 8149. University of California, Division of Agriculture and Natural Resources, Davis, California.

Thompson, J.F. and Chen, Y.L. (1989) Energy use in hydrocooling stone fruit. *Applied Engineering in Agriculture* 5, 568–572.

Thompson, J.F., Grant, J.A., Kupferman, E.M. and Knutson, J. (1997) Reducing cherry damage in postharvest operations. *HortTechnology*, 7, 134–138.

Tian, S., Fan, Q., Xu, Y., Wang, Y. and Jiang, A. (2001) Evaluation of the use of high $CO_2$ concentrations and cold storage to control of *Monilinia fructicola* on sweet cherries. *Postharvest Biology and Technology* 22, 53–60.

Tian, S., Jiang, A., Xu, Y. and Wang, Y. (2004) Responses of physiology and quality of sweet cherry fruit to different atmospheres in storage. *Food Chemistry* 87, 43–49.

Toivonen, P.M.A. (2014) Relationship of typical core temperatures after hydrocooling on retention of different quality components in sweet cherry. *HortTechnology* 24, 457–462.

Toivonen, P.M. (2015) Integrated analysis for improving export of sweet cherries and how a small industry can compete by focusing on premium quality. *Acta Horticulturae* 1079, 71–82.

Toivonen, P.M., Kappel, F., Stan, S., McKenzie, D.L. and Hocking, R. (2004) Firmness, respiration, and weight loss of 'Bing', 'Lapins' and 'Sweetheart' cherries in relation to fruit maturity and susceptibility to surface pitting. *HortScience* 39, 1066–1069.

Tukey, H.B. and Young, O. (1939) Histological study of the developing of the sour cherry. *Botanical Gazette* 4, 723–749.

Turner, J., Seavert, C., Colonna, A. and Long, L.E. (2008) Consumer sensory evaluation of sweet cherry cultivars in Oregon, USA. *Acta Horticulturae* 795, 781–786.

Usenik, V., Fabcic, J. and Stampar, F. (2008) Sugars, organic acids, phenolic composition and antioxidant activity of sweet cherry (*Prunus avium* L.). *Food Chemistry* 107, 185–192.

Vangdal, E., Sekse, L. and Slimestad, R. (2007) Phenolics and other compounds with antioxidative effect in stone fruit – preliminary results. *Acta Horticulturae* 734, 123–131.

Vursavus, K., Kebelek, H. and Selli, S. (2006) A study on some chemical and physico-mechanic properties of three sweet cherry varieties (*Prunus avium* L.) in Turkey. *Journal of Food Engineering* 74, 568–575.

Wang, Y. and Einhorn, T. (2017) Harvest maturity and crop load influence pitting susceptibility and postharvest quality deterioration of sweet cherry (*Prunus avium* L.). *Acta Horticulturae* (in press).

Wang, Y. and Long, L.E. (2014) Respiration and quality responses of sweet cherry to different atmospheres during cold storage and shipping. *Postharvest Biology and Technology* 92, 62–69.

Wang, Y. and Long, L.E. (2015) Physiological and biochemical changes related to postharvest splitting of sweet cherries affected by calcium application in hydro-cooling water. *Food Chemistry* 181, 241–247.

Wang, Y., Xie, X. and Long, L.E. (2014) The effect of postharvest calcium application in hydro-cooling water on tissue calcium content, biochemical changes, and quality attributes of sweet cherry fruit. *Food Chemistry* 160, 22–30.

Wang, Y., Bai, J. and Long, L.E. (2015) Quality and physiological responses of two late-season sweet cherry cultivars 'Lapins' and 'Skeena' to modified atmosphere packaging (MAP) during simulated long distance ocean shipping. *Postharvest Biology and Technology* 110, 1–8.

Wermund, U. and Fearne, A. (2000) Key challenges facing the cherry supply chain in the UK. *Acta Horticulturae* 536, 613–624.

Wills, R.B., Scriven, F.M. and Greenfield, H. (1983) Nutrient composition of stone fruit (*Prunus* spp.) cultivars: apricot, cherry, nectarine, peach and plum. *Journal of the Science of Food and Agriculture* 34, 1383–1389.

Yamaguchi, M., Sato, I., Takase, K., Watanabe, A. and Ishiguro, M. (2004) Differences and yearly variation in number and size of mesocarp cells in sweet cherry (*Prunus avium* L.) cultivars and related species. *Journal of the Japanese Society of Horticultural Science* 73, 12–18.

Zhang, C. and Whiting, M.D. (2011) Pre-harvest foliar application of Prohexadione-Ca and gibberellins modify canopy source–sink relations and improve quality and shelf-life of 'Bing' sweet cherry. *Plant Growth Regulation* 65, 145–156.

Zhao, Y., Collins, H.P., Knowles, N.R. and Oraguzie, N. (2013) Respiratory activity of 'Chelan', 'Bing' and 'Selah' sweet cherries in relation to fruit traits at green, white-pink, red and mahogany ripening stages. *Scientia Horticulturae* 161, 239–248.

Zoffoli, J.P., Latorre, B.A., Daire, N. and Viertel, S. (2005) Effectiveness of chlorine dioxide as influenced by concentration, pH, and exposure time on spore germination of *Botrytis cinerea, Penicillium expansum* and *Rhizopus stolonifer*. *Ciencia e Investigación Agraria* 32, 142–148.

Zoffoli, J.P., Muñoz, S., Valenzuela, L., Reyes, M. and Barros, F. (2008) Manipulation of 'Van' sweet cherry crop load influences fruit quality and susceptibility to impact bruising. *Acta Horticulturae* 795, 877–882.

Zoffoli, J.P. and Rodríguez, J. (2014a) Fruit temperature affects physical injury sensitivity of sweet cherry during postharvest handling. *Acta Horticulturae* 1020, 111–114.

Zoffoli, J.P. and Rodríguez, J. (2014b) Effect of active and passive modified atmosphere packaging of sweet cherry. *Acta Horticulturae* 1020, 115–119.

# 20 Processing for Industrial Uses

**Martin Jensen***
*Aarhus University, Aarslev, Denmark*

## 20.1 Introduction

The season for fresh cherry fruit is short in one cultivation area, and even when combining different growing regions of the world, it is not possible to supply fresh fruit throughout the year.

Storage of fresh cherry fruit is limited by a short shelf life of 2–4 weeks at 0°C with 90–95% relative humidity (Manganaris *et al.*, 2007; Valero, 2015). Processing of fruit to prolong the shelf life is an important way to offer a diverse array of cherry products year-round. The large diversity of products requires a similar diversity in processing technology. This review provides an overview of some key processing steps, and highlights selected recent research and developments in sweet and sour cherry processing targeting selected products, and focuses on how processing methods influence product quality.

## 20.2 Raw Fruit Quality

### 20.2.1 Characterization of raw fruit quality and cultivar variation

A detailed discussion of chemical and sensory quality of cherries is presented in Chapter 17 (this volume). Cultivar studies of fruit quality have demonstrated a large variation in primary and secondary metabolites as well as size, structure, firmness, colour, bioactive compounds, sensory attributes and consumer preferences in sweet and sour cherry. Cultivars should not only be evaluated by their raw fruit quality but also by their ability to form new compounds or resist degradation of important compounds during processing and storage of products. The large variation in quality often only refers to released cultivars or selected promising lines, whereas the much larger variation found in breeding germplasm, including local landraces, may provide resources that could also be exploited. The emphasis on breeding for yield, visual quality, sugar and acidity may in the future be supplemented by a search for more product-specific cultivars and more focus on aroma and sensory aspects.

### 20.2.2 Causes of variation in quality

Even if genetics determine the potential for fruit quality in a certain cultivar, it is well known that quality varies from year to year and from site to site, as the fruit quality

* Martin.Jensen@food.au.dk

reflects both the cultivar's potential and the environmental influence. Using fully mature fruit for processing is normally critical to obtain good-quality products. A 1–2-week delayed harvest may give a broader and deeper flavour. In contrast, fruit used for brining are sometimes harvested slightly before full maturity to ensure firmer fruit tissue. The industry needs to compensate for any variation to obtain a certain product quality. Concentrating, diluting, mixing different cultivars, adding concentrate or extra sugar, acid and aroma are part of the tool box to achieve this. Changes in cherry fruit quality with different maturity have been reported in numerous studies including Burkhardt *et al.* (2001), Gonçalves *et al.* (2004), Serrano *et al.* (2005, 2009) and Serradilla *et al.* (2012). Fungal infections in fruit are detrimental for taste and quality and a stringent sorting of fruit is imperative for a good quality. Cracking of fruit often leads to infections or oxidation of fruit compounds that will affect taste negatively. Postharvest storage may significantly affect quality, as even short storage of cherries at high temperatures may result in initiation of fermentation. Bonerz *et al.* (2007) stated that optimal harvest time and fast processing give low levels of spoiling indicators such as acetic acid, lactic acid and ethanol. Taken together, cultivars that continuously provide high-quality fruit are of prime interest to growers and the processing industry.

### 20.2.3 Preservation and loss of quality

Freshly harvested cherry fruit naturally contain bacteria, yeasts, fungi and enzymes that may rapidly reduce the quality of the raw fruit during processing. Methods to reduce the occurrence and activity of these factors are thus important to consider at all steps leading to the final product.

A brief blanching of whole sour or sweet cherry fruit will reduce the loss of quality. Blanching could involve heating the fruit at 85°C for 3–4 min in a water bath followed by rapid cooling. This reduced the activity of polyphenol oxidase and peroxidase enzymes by 95% in cherry fruit and thereby reduced the enzyme-mediated loss of colour, phenolic compounds and browning of the juice (Gao *et al.*, 2012). Surface contamination with microorganisms is also reduced by blanching. For cold-pressed and non-pasteurized products, blanching may have a positive effect on shelf life.

Pasteurization may be obtained by a number of thermal, physical and chemical methods. Thermal treatments will affect many aspects of juice quality; however, loss of anthocyanins and aroma compounds is of most concern as these compounds are temperature unstable (Patras *et al.*, 2010). For juice, traditionally low temperature, long time (LTLT) and high temperature, short time (HTST) treatments have been used to pasteurize products. The HTST treatment is generally found to result in significantly less loss of quality and typically involves applying 90–95°C for 25–30 s in flash pasteurizers (Rupasinghe and Yu, 2012). Szalóki-Dorkó *et al.* (2015) studied the degradation of monomeric anthocyanins in juice from two sour cherry cultivars at a constant temperature of 70, 80 or 90°C over 4 h and found a loss in anthocyanins of 19, 29 and 46%, respectively, in 'Kantorjanosi' and 18, 29 and 38%, respectively, in 'Érdi Bőtermő' juice. The shorter the treatment time and lower the temperature, the less the reduction in anthocyanins. The estimated half-life of anthocyanins in juice at 80°C varied from 5.2 to 7.9 h. Zoric *et al.* (2014) investigated pasteurization of freeze-dried 'Marasca' sour cherry paste at 9.7% moisture content and at 80–120°C for 5–50 min, and found that the half-life at 80°C varied between different anthocyanins, ranging from 32.10 min for cyanidin 3-glucosylrutinoside to 45.69 min in cyanidin 3-rutinoside and that anthocyanins generally were more sensitive to heat than phenolic acids.

Thermal conductivity of canned sour cherry pomace at different moisture contents was investigated by Greiby *et al.* (2014) in order to provide a detailed model of the heat transfer parameters during pasteurization. Similarly, Márquez *et al.* (2003) modelled heat transfer parameters during thermal pasteurization at 85°C of sweet and sour cherries stored in glasses with 25% sucrose syrup. This knowledge is important to optimize pasteurization methods of fruit, fruit mash and canned fruit in order to ensure the least possible damage to the product by heating and the lowest possible use of processing energy.

Several other pasteurization methods that avoid using high temperatures have been attempted. Hosseinzadeh Samani *et al.* (2015) combined a mild microwave heating to approximately 50°C with an ultrasound treatment in sour cherry juice inoculated with *Escherichia coli* and were able to eliminate the bacteria. Gamma radiation of sour cherry juice to reduce microbial load was studied by Arjeh *et al.* (2015). While a reduction in microbial contamination was seen after 3 kGy irradiation, they also found a significant loss of anthocyanins, ascorbic acid and other quality parameters after irradiation and also a decline during a 60-day storage period at 4°C.

High-pressure processing (HPP) treatment of juice and puree with no or little heating is of growing interest as a method for pasteurization. HPP treatment of sweet cherry juice at 400 MPa for 5 min or 550 MPa for 2 min at 10°C was compared with thermal pasteurization at 70°C for 30 s (Queirós *et al.*, 2015). All treatments reduced microbial load to non-detectable levels during a 4-week refrigerated storage period. The HPP samples showed a slightly higher anthocyanin concentration and also less loss of total phenolics during storage. Bayındırlı *et al.* (2006) studied HPP treatment of sour cherry juice inoculated with *Staphylococcus aureus*, *E. coli* and *Salmonella enteritis* and found that 350 MPa for 5 min at 40°C completely inactivated the pathogens. Polyphenol oxidase enzymes were more resistant to degradation than bacteria and needed higher temperatures or longer HPP exposure to be eliminated compared with the bacteria tested.

Recently, new electro-technologies such as pulsed electric fields (PEFs) have been introduced as a method for pasteurization of juice and inactivation of enzymes (Evrendilek *et al.*, 2012). Evrendilek *et al.* (2008) showed that sour cherry juice at pH 3.1 inoculated with *Penicillium expansum* and treated with a PEF at 30 kV cm$^{-1}$ field strength for 218 μs completely inhibited spore germination. Similarly, Evrendilek *et al.* (2009) found that PEF treatment of sour cherry juice at 20 kV cm$^{-1}$ for 123 μs inhibited spore germination of *Botrytis cinerea* inoculated into the juice. Altuntas *et al.* (2010) tested PEF treatment on seven bacteria and fungi inoculated into sour cherry juice and also investigated the effect on the quality of the juice. Although the survival of all bacteria and fungi was reduced significantly at increasing electric field strengths up to 30 kV cm$^{-1}$ and with longer treatment time up to 200 μs, these conditions were not enough to completely eliminate most of the pathogen species. Generally a reduction in survival of bacteria and fungi but not a complete elimination was found in most PEF studies (Evrendilek *et al.*, 2012). None of the PEF treatments affected quality parameters like Brix, pH, total acidity, *L–a–b* colour, ascorbic acid or anthocyanin concentration in the juice significantly.

Cold gas phase plasma treatments is a method that generates reactive oxygen species by exposing for example argon gas to a strong electric field to generate ionized gas that juice can be exposed to. Garofulić *et al.* (2015) used a short exposure of 3 min to an argon gas phase including a resultant heating to about 50°C of sour cherry juice and reported a higher content of anthocyanins and phenolic acids than by a traditional pasteurization at 80°C for 2 min. The higher level of anthocyanins was believed to be a result of dissociation of small-sized agglomerates or particles by the treatment.

While pasteurization by heating is still the most widely used method for cherry products, membrane filtration techniques such as ultrafiltration, microfiltration or nanofiltration for clearing or concentrating juices are also commercially used to obtain sterile products (Echavarría *et al.*, 2011; Rupasinghe and Yu, 2012). Filters with a pore size of less than 45 μm are necessary to remove bacteria, yeasts and fungi. Bagger-Jørgensen *et al.* (2002) showed the effect of juice temperature, flow rate and filter pore size in microfiltration of sour cherry juice on the transmembrane pressure, juice turbidity, protein, sugar and total phenolic compounds. While turbidity was reduced significantly, proteins, sugars and phenolic compounds were not affected. Fouling of membranes often reduces filtration capacity and a depectination by enzymes, use of proteases and or precentrifuging is recommended before filtration in sour cherry juice (Şahin and Bayindirli, 1993; Meyer *et al.*, 2001; Pinelo *et al.*, 2010). Membranes with widely different pore sizes are available and the choice of pore size depends on the microorganisms to be removed. The risk of removal of wanted compounds should be considered when applying filter techniques.

Although efficient pasteurization of cherry products has been accomplished, there can be a significant loss of product quality during storage due to oxidation, light reactions and other chemical reactions between compounds. Bonerz *et al.* (2007) investigated ageing up to 6 months at 20°C in the dark of pasteurized 100% sour cherry juices. Total anthocyanins declined up to 75% after 6 months, resulting in reduced colour saturation, whereas hue was less affected. Colourless polyphenols and antioxidant capacity did not decline. Only catechin concentration decreased slightly. They found half-lives of individual anthocyanins to vary between 28.6 and 54 days for peonidin 3-rutinoside and 72 to 94 days in cyanidin 3-(2G-glucosylrutinoside), depending on which cultivar the anthocyanins originated from.

In pasteurized sweet and sour cherry jam stored at 20°C for 3 months, Poiana *et al.* (2011) found a 22% reduction in vitamin C, 18% reduction in total phenolics and 21% reduction in monomeric anthocyanins, whereas ferric reducing antioxidant power (FRAP) antioxidant capacity did not decrease significantly. Rababah *et al.* (2011) similarly showed a decrease in total anthocyanin concentration of cherry jam stored at 25°C for 5 months, whereas total phenolics did not change significantly. In this study, 2,2-diphenylpicrylhydrazyl (DPPH) antioxidant capacity was reduced significantly over time.

### 20.2.4 Dealing with toxins: amygdalin and cyanide risk

Cherry seeds contain amygdalin, a cyanogenic glucoside that upon crushing enzymatically can be converted to hydrogen cyanide and benzaldehyde by the enzyme β-glucosidase. Crushing seeds during processing will release these compounds into the fruit mash. While benzaldehyde is an important and non-harmful aroma component of cherries, hydrogen cyanide is toxic and could potentially be harmful if present in high doses in a product. During heating in pasteurization, this compound is normally lost. Crushing a small percentage of stones to achieve an almond-like taste profile due to benzaldehydes is a normal procedure for some products. In non-heated processing methods where breakdown or evaporation of hydrogen cyanide is less likely to occur, and high enzymatic conversion can take place, more care is needed to avoid the presence of toxic cyanides. In Europe, the European Economic Community council regulation No. 1576/89 states that the maximum hydrocyanic acid content allowed in stone fruit spirits is 70 mg $l^{-1}$ (Balcerek and Szopa, 2012).

## 20.3 Preprocessing Operations

### 20.3.1 Cleaning, sorting and advanced grading of raw fruit by quality

Freshly harvested fruit of sweet and sour cherry are washed to remove impurities, and hydrocooled briefly in water at 0°C as soon as possible to maintain the highest possible quality, including better firmness before the sorting steps. Depending on regulations, a disinfectant or conditioner to improve firmness may be added to the water. Mechanical removal of pedicels is done using cylindrical rubber rollers that pull off and discard the pedicel. The fruit may then be aligned in tray rows or on running conveyor belts and sorted for fruit size by machine vision systems often based on laser technology and digital picture analysis. Some row systems have rotating devices for single fruit that ensure a full surface scan of the entire fruit. Additional sorting may include scanning fruit for mechanical damage, fruit cracks, fungal attack and abnormalities such as double fruit and pitted fruit. Laser, visual and near infrared (NIR)-based techniques may be used in these operations that run at high speed and can deal with large quantities of fruit in a relatively short time. NIR-based sorting technology focusing on fruit quality such as colour (Pappas *et al.*, 2011), soluble sugars and firmness may be applied to obtain high and uniform fruit quality (Carlini *et al.*, 2000; Lu, 2001). A number of companies today provide advanced technical machinery for integrated cleaning, sorting and grading of both sweet and sour cherries for both fresh products and the processing industry.

### 20.3.2 Stone removal

Sweet and sour cherries with their pits removed are used for canned and brined products, but also for dried and infused fruit. Metal pins are used to push out the stones from fruit that are held in a fixed position. Machines with high capacity that leave only small fruit scars and result in minimal losses of fruit tissue and juice are optimal. Short-term precooling before pit removal makes the fruit more firm and results in better pitted fruit quality. Efficiency in pit removal needs to be very high, since even rare cases of stone remnants and splinters in final products could be damaging to consumer health and product acceptability. Haff *et al.* (2013) developed an X-ray sorting machine to determine if there were stones or stone parts left in the fruit after pitting. Using air jet streams, such fruit could be eliminated from further processing.

## 20.4 Processing into Products

### 20.4.1 Matching cultivars and raw fruit quality to product type and high-value products

To optimize processing and product quality of different cherry products, it is required that the raw fruit quality is suited to the specific processing end product (Fig. 20.1). The hundreds of sweet and sour cherry cultivars available worldwide show large variation in technical and intrinsic quality, and the choice of cultivar is therefore the most

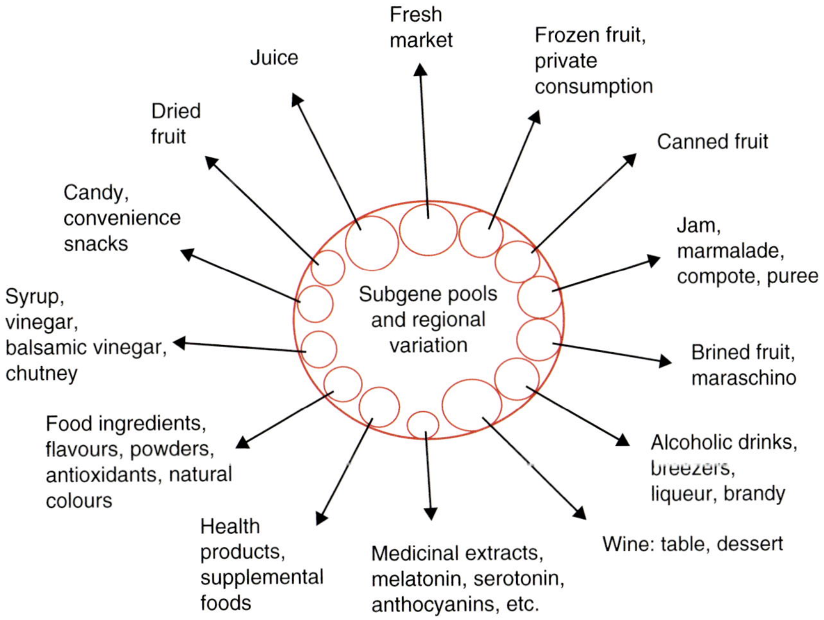


**Fig. 20.1.** Better matching of individual genetic pools of cherry with specific products may optimize processing and product quality. (Reproduced with permission from M. Jensen.)

important way to ensure successful processing and a high-quality product. Different consumer segments and regional preferences may, however, require different taste and quality, and thus no one quality can be assigned to be the best for all markets. A comprehensive and detailed review of the characteristics of raw cherry fruit that are necessary or most well adapted to processing into different high-quality products seems to be missing in the scientific literature. However, several single studies indicated which cultivars were most adapted to processing into different products and which characters were considered to be of high importance. Bors (2011), for example, tested five cultivars and seven selections of sour cherry for their characteristics and suitability for canned maraschino, individually quick frozen (IQF), dried cherries, sugar-infused cherries, wines and cherry jerky. Optimal match of cultivars to the different products was based on the technical quality of raw fruit and sensory quality of the final products. Preferred cultivar characteristics for each product were listed.

Will *et al.* (2005) and Bonerz *et al.* (2007) reported detailed analyses of juice in five sour cherry cultivars. Damar and Ekşi (2012) characterized the juice quality of 11 cultivars of native Turkish sour cherry, and Clausen *et al.* (2011) studied cold-pressed juice quality of seven sour cherry cultivars by nuclear magnetic resonance ($^1$H-NMR) and a sensory description of the aroma profile. Cultivars for juice should have a high juice yield, soft fruit and show uniform maturity, and preferentially should have high Brix, medium to high acidity, a balanced sugar/acidity ratio, high colour and high antioxidant capacity. The characteristic aroma for cherry should be strong, and cultivars with a pungent and bitter taste should be avoided. Many of these quality requirements are similar for other beverages such as wine, liqueur and brandy. Nikićević *et al.* (2011) compared five sour cherry cultivars in their chemical and sensorial characteristics adapted to cherry brandy and found that cultivars with a high ability to produce the aroma compounds benzaldehyde and linalool were preferred.

For jam products, fruit should preferably be more firm than for juice, have high sugar content, low to medium acidity, high colour and a high level of aroma compounds. Kim and Padilla-Zakour (2004), for example, compared jams from four sour cherry cultivars and concluded that the cultivars differed in their ability to retain high levels of anthocyanins and phenolics after processing. Berger (1991) argued that the large differences in the ability to produce aroma compounds in different cultivars should be exploited for improved jam taste. For dried fruit products, fruit flesh must be quite firm and stable during processing. Konopacka *et al.* (2014) studied the suitability of nine sour cherry cultivars on their performance in osmodehydrated–convective drying of fruit for use as snacks. Only cultivars that matured uniformly and had even fruit size gave consistently high dried fruit quality. Cultivars with a good sugar/acid balance and high content of anthocyanins and total phenolics were preferred. Juhnevica *et al.* (2011) tested five sour cherry cultivars for their ability to produce high-quality candied fruit and found that cultivars should have a mild sweet and sour taste with good structure. A high phenolic content and very high soluble solids were negatively correlated with consumer preference. Toivonen *et al.* (2006) compared nine sweet cherry cultivars for their adaptiveness to produce a novel fresh-cut cherry product where the stones were removed and other food products could be inserted in the cut-out cavity.

### 20.4.2 IQF fruit

Freezing of whole fruit with or without stones is an efficient way to preserve a high quality of fruit for a long time. Such fruit can be thawed and used for different industrially processed products, but may also be targeted to private consumers directly. On a worldwide basis, a large percentage of sweet and sour cherry fruit is frozen. For both private consumers and industry, it is more convenient if fruit are frozen individually. Such fruit, called IQF fruit, are typically frozen by blowing cold air up through a slowly running mesh conveyor belt in a fluid bed system that continuously moves the fruit, thereby preventing the fruit from clumping together during freezing (Barbosa-Cánovas *et al.*, 2005). Linear freezing tunnels or spiral belt freezers are common technologies. During the first part of the freezing, only single layers of fruit are frozen to quickly freeze the surface. Later, full freezing can be done in thicker layers. Frozen fruit are packed in airtight plastic bags and kept frozen at −18 to −23°C. Flushing fruit with gases from liquid $CO_2$ or liquid nitrogen may give faster freezing of IQF fruit. For other products, a plate freezer may be used, where pulp or paste is frozen between two metal surfaces cooled on the outside by a coolant. Some cherries are frozen with sugar added, making them more convenient for use in pastries and other foods. It is important to remember that frozen fruit kept at about −20°C will not maintain its quality over prolonged periods of time and therefore preferably should be used within 8–12 months of storage (Barbosa-Cánovas *et al.*, 2005).

### 20.4.3 Fruit juice, nectars and concentrates

Cherry juice is produced from both sweet and sour cherry and may be a 100% pure extract, or diluted from concentrate. Raw extracted juice can be dehydrated by different methods into a concentrate better suited for storage or transport. For production of nectars, sucrose syrup solutions are normally added to the juice or concentrate (Toydemir *et al.*, 2013b). Juice may be made from fresh fruit or from frozen stored fruit. The hot-pressing methods use enzymes and high temperatures including pasteurization, whereas cold pressing normally does not use enzymes or heat in an attempt to preserve the maximal quality of the raw fruit. If the fruit is not heat pasteurized, the cold-pressed juice needs to be pasteurized by other means to extend the shelf life longer than a week or two stored at 5°C. Cultivars with a high juice content and low firmness, high sugar and medium to high colour content, and medium to high total acidity are preferred for juice production. A good balance between sugar content and acidity and a positive sensory profile is considered to be very important for high sensory quality (Clausen *et al.*, 2011).

Production of cherry juice involves a number of steps. A standard process could be as follows (see McLellan and Padilla-Zakour, 2004, for more technical details). Freshly harvested fully mature fruit are washed, cleaned from impurities, and damaged and diseased fruit are carefully removed. A brief blanching at 85°C for 3 min in a water bath may reduce microbial surface contamination and also reduce the activity of polyphenol oxidase and peroxidase enzymes. This also makes the pericarp more easily degradable (Gao *et al.*, 2012). Blanching resulted in higher concentrations of anthocyanins, total phenols, ascorbic acid and soluble solids in the extracted sour cherry juice compared with non-blanched treatments.

Fruit of firm varieties may be sorted in colour and sugar content before stones are removed in pit-removal machines. For sour cherry and very soft sweet cherries, stones may be extracted in a hammer mill. A certain percentage of stones may be crushed and entered into the mash to ensure the benzaldehyde aroma in the juice. Pitted fruit are then macerated briefly. Pure pectinases or a mix of different enzymes and possibly proteases (Pinelo *et al.*, 2010) may be applied to the mash at 50°C for 1–2 h while stirring the mash. Following this, pressing of juice is performed by a number of potential pressing methods and pomace is separated. The pressing technology will not be described here. The raw juice may then be pasteurized to completely inactivate all enzymes. The juice is then filtered, typically in several steps of lower and lower pore size in order to remove remnants of pulp, cloudiness and minor dissolved particles. A sedimentation period traditionally was used in the past, but high efficiency in filtration steps may eliminate the need for this. Depending on whether the juice should be cloudy or clear, a clarification step could be done using a very small pore size membrane that also sterilizes the juice (Bagger-Jørgensen *et al.*, 2002; Pinelo *et al.*, 2010). Pasteurization of the final juice product may be done before filling in bottles or by thermal heating of bottles after filling.

The effect of processing methods on quality parameters in juice from different cultivars has been investigated mainly for sour cherry and to a lesser degree in sweet cherry. Schüller *et al.* (2015) studied the quality of juice from nine sweet cherry cultivars made by a hot vapour juice extraction and then stored at −20°C or room temperature for 30 days. Juice quality varied a lot among cultivars, and total phenolics ranged from 553 to 1757 mg $l^{-1}$ and anthocyanins between 85 and 1095 mg $l^{-1}$. The cultivars differed in their ability to retain their quality after storage. Will *et al.* (2005) compared juice from five sour cherry cultivars stored at 20°C for 6 months. Detailed analysis of the juice for more than 30 parameters showed significant differences between cultivars in many parameters. Brix varied from 13 to 18, total acidity from 15 to 23 g $l^{-1}$, total phenolics from 1707 to 5498 mg $l^{-1}$, anthocyanins from 262 to 410 mg $l^{-1}$ and Trolox equivalent antioxidant capacity (TEAC) activity from 16 to 44 mmol $l^{-1}$ Trolox. 'Stevnsbaer Birgitte' had the highest values in most of these parameters. After 6 months of storage at 20°C, about 50% of the anthocyanins were lost. In a parallel study, Bonerz *et al.* (2007) compared juice from the same cultivars and roughly confirmed the overall results obtained by Will *et al.* (2005). Anthocyanins exhibited a loss of up to 75% after 6 months storage at 20°C in the dark, whereas TEAC antioxidant capacity did not change. The half-life of the anthocyanins in the juice was, on average, between 50 and 60 days.

Damar and Ekşi (2012) characterized the juice quality of 11 cultivars of native Turkish sour cherry. Juice was prepared by a manual press, pasteurized at 85°C, cooled to 45°C and treated with 0.1 g Pectinex® Be Colour $kg^{-1}$ for 1 h. The juice was filtered in a multilayer filter cloth, bottled and pasteurized for 10 min at 85°C. Brix varied from 16 to 26°, total acidity from 16 to 26 g $l^{-1}$, total phenolics from 1510 to 2550 mg $l^{-1}$, total anthocyanin content from 350 to 633 mg $l^{-1}$ and TEAC antioxidant capacity from 20 to 38 mmol $l^{-1}$.

Clausen *et al.* (2011) studied the cold-pressed juice quality of seven sour cherry cultivars by $^{1}$H-NMR and detailed sensory description of the aroma profile. The seven cultivars divided into two sensory groups with 'Sumandinka' and 'Debrecini' both having a sweeter taste compared with 'Stevnsbaer' clones and 'Fanal', which were more sour, bitter and astringent. 'Stevnsbaer' clones had a high intensity taste of benzaldehyde and were very high in colour. An NMR model based on these two groups could explain 82% of the variation. The clustering pattern of the cultivars into groups was closely linked to the genetic relationships of the cultivars and clones tested.

Repajić *et al.* (2015) characterized quality parameters of 'Marasca' and 'Oblačinska' sour cherry during commercial processing, as fresh fruit, after juice pressing, after filtering juice and in final concentrate. The fruit were mashed and heated to 45–50°C, treated with 20–40 ml $t^{-1}$ of pectolytic enzymes for 1 h and pressed. The juices were pasteurized at 85°C for 2 min, cooled to 50°C and treated with 0.02–0.03 g $l^{-1}$ of pectolytic enzyme for 2 h, precipitated and vacuum- and plate-filtered. The juice was finally evaporated in a four-stage evaporator with aroma recovery to obtain concentrated juice with 65°Brix. The dry weight mass fraction concentration of total phenolics was significantly higher in the pressed juice (20.5 mg $g^{-1}$) compared with fresh fruit (12.4 mg $g^{-1}$) and decreased non-significantly in filtered juice (18.6 mg $g^{-1}$) and was significantly lower in concentrate (16.8 mg $g^{-1}$). Total monomeric anthoycanins were lowest in fresh fruit (2.42 mg $g^{-1}$), and significantly higher in pressed juice (4.03 mg $g^{-1}$), and even higher in filtered juice (4.64 mg $g^{-1}$), whereas the concentrate was between fresh and pressed juice (3.42 mg $g^{-1}$). 'Marasca' had a higher anthocyanin and total phenolics content than 'Oblačinska'.

Toydemir *et al.* (2013a) studied the quality aspects and polyphenolic compounds at 22 sampling points during an industrial-scale processing of 'Kutahya' sour cherry nectar. Detailed process, yield data and metabolomics data were provided for the processing. Out of 193 compounds revealed by the analysis, 38 compounds were identified in the samples and only seven were affected significantly by processing, and of these, five were phenolic compounds. Overall, 87% of the major anthocyanin compound cyanidin 3-(2G-glucosylrutinoside) was recovered in the nectar, whereas only 62% of procyanidins were recovered. Procyanidins were high in pomace, whereas anthocyanins were almost eliminated in pomace. A similar juice process as above was used by Toydemir *et al.* (2013b) to study the changes in phenolics, antioxidant capacity and *in vitro* gastrointestinal digestion to reveal serum availability of antioxidants. The concentration of total phenolics, total flavonoids and total anthocyanins initially increased significantly after pressing, but after filtration all were reduced to below the level in the fresh fruit. Antioxidant capacity measured by CUPRAC (cupric reducing antioxidant capacity) and ABTS (2,2'-azinobis(3-ethylbenzothiazoline-6-sulphonic acid)) methods were found to correlate well with total phenolics, whereas DPPH showed less correlation. Anthocyanins were responsible for 61% of the total antioxidant capacity in fruit and 57% in nectar, and the most abundant anthocyanin alone accounted for 50 and 40%, respectively. Phenolic acids and flavonols provided 10–15% of the antioxidant capacity. Anthocyanin availability in the simulated gastrointestinal digestion system was found to be much higher from the nectar than from fresh fruit. The higher stability of anthocyanins was considered to be caused by the high sucrose concentration.

The diversity of the compounds present in sour cherry juice is high. For example, Toydemir *et al.* (2013b) identified a new compound in sour cherry juice, ascorbic acid glucoside, for the first time. Similarly, Rentzsch *et al.* (2007) found several new yellow-coloured pigments, 5-carboxypyranoanthocyanins, in beverages of sour cherry that were formed during storage of sour cherry juice and beverages. The concentration was very low, and it only contributed slightly to the colour of sour cherry juice (0.6%) compared with monomeric (16.6%) and polymeric (82.8%) pigments.

Production of cherry concentrates can be done by several methods, including evaporating by heating, vacuum drying, freeze drying, cryo-concentration, membrane filtration, reverse osmosis and spray drying (Aider and de Halleux, 2008). Heating often reduces quality, since a gradient is needed that causes degradation of sensitive compounds near the heating surface. Membrane methods, especially reverse osmosis, are efficient and economical methods to reach concentrates of up to 30–40% dry-matter content. At higher dry-matter levels, the high-pressure costs, damage and fouling of membranes make these techniques less interesting (Aider and de Halleux, 2008).

Normohamadpor Omran *et al.* (2013) compared the thermal concentration of ultrafiltrated sour cherry juice using a three-stage column evaporator running at 70–90°C and at 45 KPa vacuum and a feed rate of 70 l $h^{-1}$ with cryo-concentration done at −10 or −20°C in three steps with thawing at room temperature. During thawing, juice with a higher sugar content will thaw first and run off. They found that depectinized and ultrafiltrated juice could be concentrated to a Brix value of 46° with no difference between −10 and −20°C. Clearing of juice by electroflotation before ultrafiltration allowed cryo-concentration to reach a Brix of 52°. Cryo-concentration is of interest since it maintains a very high quality of the concentrate compared with thermal methods where vitamins, aroma and other compounds may be degraded. Normohamadpor Omran *et al.* (2013), for example, showed that the antioxidant capacity of sour cherry juice after cryo-concentration was much higher than after thermal concentration. Aider and de Halleux (2008) also studied cryo-concentration at −10 or −20°C of cherry juice and found that they could go from 15 g dry matter per 100 g to 45 g after three cycles, with much less loss of aroma and vitamin C than after thermal concentrating.

### 20.4.4 Marmalade, jam, jelly, compote, fruit sauce and puree

There is a very large diversity and variation in marketed marmalades, jams, compotes and jellies of sweet and sour cherry from low-priced industrial products to high-priced gourmet products sold for several fold higher prices per kilogram. Furthermore, the market is segmented into subareas including organic products and low-sugar products for diabetics. Thus, one processing recipe cannot explain all products. However, some general comments and trends can be pointed out. The quality of the raw fruit will determine the quality and taste obtained in a product. A strong focus should therefore be on ensuring fully mature fruit of the highest sensory and chemical qualities possible, improved by further sorting. A medium to high firmness of fruit flesh, a high colour and sugar content, and medium to low total acidity may be preferred cultivar characteristics. However, the aroma and sensory experience is considered to be more important for high-priced products. In addition consumers are increasingly expecting that quality factors originate from the cultivar itself or from the effect of processing and not from added flavours. Low-processed natural foods are therefore gaining popularity.

An example of a procedure of jam production of sour cherry would include the following steps, as used by Kim and Padilla-Zakour (2004). Fully mature fruit of sour cherries are harvested, immediately sorted and washed before the pits are removed. Fruit is then coarsely ground in a food processor for 30 s. Jam formulation is based on 50% fruit, 48% sugar and 2% pectin mix (pectin, dextrose, fumaric acid). Fruit is cooked briefly to inactivate enzymes and partly disintegrate fruit tissue. Pectin mix is added and boiled for 2 min, and then the pH may be adjusted with a few drops of 50% citric acid with a pH target of 3–3.2. Sugar is added and the mixture is boiled to a concentration of 65–68°Brix. Jam is hot packed at 90°C in glass jars, immediately sealed with lids and the jars inverted for 5 min to sterilize the lids followed by air cooling.

Heat pasteurization of bulk jam traditionally has reduced anthocyanin content dramatically and also reduced phenols in jams. Poiana *et al.* (2011) compared the quality of frozen sweet and sour cherry fruit with jam cooked for 20 min at 80°C with 45°Brix and pasteurized for 10 min at 80°C. They found a reduction of more than 70% in vitamin C for sour cherry and 54% for sweet cherry, a 25–30% reduction in total phenolics in the cherry jams, up to over a 90% reduction in total anthocyanins and a 20–30% reduction in antioxidant capacity. During storage, the anthocyanins were reduced even further. A similar large decrease in total phenolic compounds and anthocyanins after bulk cooking and pasteurizing cherry jam was found by Rababah *et al.* (2011) and also by Kim and Padilla-Zakour (2004) in jams from four sour cherry cultivars. Sensory studies on loss of aroma during jam production are few, but large losses of some aroma compounds are found due to cooking, whereas other compounds such as benzaldehyde and linalool, both important characteristic cherry flavours, were produced at higher concentration during processing (Berger, 1991). Berger (1991) further emphasized the large cultivar variation in the ability to produce these compounds, which may be exploited. Alternative methods of achieving smooth uniform jam texture and good taste without a long time cooking and with low-temperature pasteurization need to be investigated. Thin-layer pasteurization of jams in a plate heater may allow short-time pasteurization to be used.

### 20.4.5 Canned and brined fruit

Canning of sweet and sour cherry fruit is used to provide long preservation of whole or parts of fruit that are easy to apply as ingredients in other foods or as part of pastries or ready-to-eat desserts. Fresh fruit are washed, sorted for damaged fruit, the pedicels removed, and in most cases, the pits removed. Fruit is filled into cans/glasses and then hot syrup from 16–45°Brix is added, filling the can (Kaack *et al.*, 1996). Some products use concentrated sour cherry juice with high Brix as filling to optimize flavour. For use in pastries, sour cherries are supplemented with a starch filling. After filling, the headspace air is exhausted and the can is sealed and thermally processed for pasteurization as fast as possible and immediately cooled to avoid heat damage to the product (Chaovanalikit and Wrolstad, 2004a). Thermal heat transfer and heating requirements for canned fruit have been modelled depending on can size and particle size of sweet and sour cherries, and allow for designing optimal processing and energy use (Márquez *et al.*, 2003). Even if airtight, the quality and colour of canned fruit will degrade during storage, with degradation increasing as the temperature increases. Storage at low temperature until shipping for selling is recommended.

Processing of brined cherry fruit, for example of maraschino cherries, is described in detail by Watters and Woodroof (1986) and by McLellan and Padilla-Zakour (2005), and largely follows the procedure described by Chaovanalikit and Wrolstad (2004a). Cleaned good-quality firm cherries with or without stems are placed in glass jars and covered with a brine solution with 2.2% sodium metabisulfite, 2% calcium chloride and 0.1% citric acid with a pH of 3.0 to bleach fruit into whitish or yellowish colour. Fruit are covered with brine and an air-permeable plastic wrap is used to close the jar. The cherries may be stored in the brine for several months up to a year. Recovered bleached cherry fruit are washed for 5 days in running cold water to reduce $SO_2$ levels to below 200 p.p.m. Following this, sucrose is added to 48% (for maraschino cherries) and 72–74% for candied cherries or glacé cherries (with stem) (Kaack *et al.*, 1996). Gardiner *et al.* (1993) described the process of producing glacé cherries. Artificial or natural colours and flavours are added to give the desired cocktail berry characteristic.

Chaovanalikit and Wrolstad (2004a,b) compared the quality of sweet cherries 'Bing', 'Royal Ann' and the sour cherry 'Montmorency' as fresh frozen (stored at −23 or −70°C) or after processing into canned fruit with light syrup or brined fruit including storage at 2 or 22°C for 12 months. Frozen fresh fruit stored well with only a slight loss of anthocyanins at −70°C (12%), whereas anthocyanins were dramatically reduced when stored at −23°C (88%). In particular, hydroxycinnamates and epicatechin were found to be reduced considerably at −23°C, whereas flavonol glycosides were almost unchanged. Canning overall did not decrease anthocyanins, but resulted in about a 50% transfer of anthocyanins into the syrup and accordingly less in the fruit. There was a significant loss in anthocyanins after storage for 5 months at 22°C (42%) and significantly less at 2°C (12%), while total phenolics remained almost constant. During brining and washing, almost all anthocyanins and phenolics were lost from fruit, and some leached into the brine or were degraded. The acidic condition during brining was found to convert some of the cyanidin 3-rutinoside into cyanidin 3-glucoside, thereby changing the anthocyanin profile (Chaovanalikit and Wrolstad, 2004b).

Ou *et al.* (2012) compared phytochemicals, antioxidant capacity and anti-inflammatory activity of frozen fruit, juice concentrate, and dried and canned 'Montmorency' sour cherries. Frozen fruit had the highest anthocyanin content, followed by cherry juice concentrate, dried cherries and canned fruit. Calculated on a per-serving basis, the relative amount of anthocyanin per serving of the products with the same order were rated as 3.9/1.7/1.2/1, demonstrating that minimal processing retains the highest concentrations and content per serving. The distribution of monomeric and different size groups of polymeric proanthocyanidins were reported for the four products; total phenolics (gallic acid equivalents $g^{-1}$) were 9.36 for juice concentrate, 7.45 for dried fruit, 4.18 for frozen fruit and 3.57 for canned fruit. Oxygen radical absorbance capacity (ORAC; μmol Trolox $g^{-1}$) of products showed values of 128 (concentrate), 68 (dried fruit), 20 (frozen fruit) and 17 (canned fruit).

### 20.4.6 Dried fruit products and processes

A number of dried or semi-dried cherry products are marketed, including whole fruit without stones, as raisins or infused fruit as snacks, smaller fruit parts as ingredients, candied fruit and gums, and dried fruit powders. The traditional drying method of fruit uses convective air drying where fruit placed in single layers on trays or shelves are exposed to a heated and sometimes dried air stream at a controlled air speed.

Gazor *et al.* (2014) and Gazor and Roustapour (2015) studied air drying of sour cherry fruit from 75% initial moisture content to 17% final moisture content at 50, 60 or 70°C in an airflow velocity of 1 m $s^{-1}$ and with or without different pretreatments of fruit. They proposed a model for the drying kinetics of sour cherries. Dipping fruit for 1 min in pure boiling water, 20% NaCl boiling water or a 2% ethyl oleate solution before drying significantly reduced drying time and energy expenditure. Taste acceptance was better after drying at 50°C than at 70°C (Gazor *et al.*, 2014). Doymaz (2007) and Doymaz and İsmail (2011) previously found similar overall results based on drying of ethyl oleate-pretreated sour cherries. Pirone *et al.* (2014) studied air drying at 70°C, 8% relative humidity and 4 m $s^{-1}$ air speed of sweet cherry 'Napolitana' fruit following six pretreatments: (i) water vapour blanch for 1.5 min at 100°C; (ii) blanch and dip in 10% citric acid for 5 min; (iii) blanch and dip in 10% citric acid and 2.5% calcium lactate for 5 min; (iv) freezing fruit at –18°C; (v) pitted fruit; and (vi) control fruit that was not pretreated. Drying kinetics was modelled and compared between treatments. After drying, fruit resembled raisins in texture and had water activity ($a_w$) of <0.6, thus being microbiologically stable. Blanching reduced drying time significantly and, combined with dipping, resulted in better fruit quality. Franceschinis *et al.* (2015) investigated the effect of blanching or sugar infusion prior to air drying or freeze drying of discs or dices of sweet cherry used for snacks or ingredients. Air drying caused a darker colour than freeze drying. Blanched freeze-dried discs retained high anthocyanin and phenolics, while blanched and air-dried discs had the highest antiradical activity. Sugar infusion caused a decrease in anthocyanin concentration.

Wojdyło *et al.* (2014) compared drying of sour cherry fruit by vacuum-microwave drying in the range of 120–480 W with freeze drying or air drying at 50, 60 and 70°C. Vacuum-microwave drying initially at 480 W but reduced to 120 W at low moisture content gave as good quality as air drying at 50°C. Vacuum drying is characterized by high drying rates, low drying temperatures and an oxygen-deficient environment, and it therefore maintains a high quality of dried products. In addition, the energy cost is low. Šumić *et al.* (2013) studied vacuum drying of frozen 'Meteor' sour cherry fruit at temperatures between 46 and 74°C and at 17–583 mbar pressure and found that drying at 54°C and 148 mbar was optimum to achieve a high content of vitamins, total phenolics and anthocyanins and a high antioxidant capacity. Vacuum freeze drying at temperatures from –30 to –55°C where water is removed from fruit by sublimation may preserve the raw quality even better than vacuum drying at positive temperatures (Ivancevic *et al.*, 2012). Costs are, however, typically larger and this means that dried products would need to be sold at higher prices if this technology is to be economical viable. Kirakosyan *et al.* (2009) compared a number of dried products, i.e. dried fruit or dried IQF powder, with fresh-frozen fruit and found that anthocyanins and phenolics were fairly well preserved in freeze-dried powders compared with fresh-frozen fruit, but that drying eliminated the melatonin.

Osmodehydration is an osmotic-driven process used to reduce the water content by about 50% in fruit, which are then dried further with other methods. This works best in frozen or pretreated fruit, since diffusion of water through the pericarp is facilitated. According to Yadav and Singh (2014), a solution with, for example, sucrose at 40°Brix, held at 40°C for 132 min is efficient to obtain this in many fruit species. They also provided a detailed review of the mechanisms, methods and results obtained for a number of fruit species. Klewicki *et al.* (2009) provided models for sorption isotherms for osmodehydrated and then freeze-dried or convectively dried sour cherry fruit of 'English Morello' and demonstrated that the equilibrium water content differed slightly in the two drying methods. This means that safe storage water activity ranged from 0.54 in freeze-dried fruit to 0.63 in convectively dried fruit. Water content in the monolayer is related to safe storage and was 17 g per 100 g dry matter in sour cherry. Nowicka *et al.* (2015a) studied the effect of osmodehydration of sour cherry fruit for 90 min at 40°C in 40°Brix sucrose or different juice concentrates as a first step in drying, followed by convective drying and finally microwave-vacuum drying to obtain fully dried fruit. Osmodehydration in juice concentrate in some cases resulted in an improved polyphenol content and antioxidant capacity of dried fruit, but also a reduction in the content of anthocyanins compared with no pretreatment. Sensory evaluation could be slightly improved for cherries treated in juice concentrate. Using a combined drying strategy reduced the drying period and produced high-quality fruit. Konopacka *et al.* (2008) also found a reduction in anthocyanins, but an improved taste sensation by using apple–sour cherry juice concentrates in osmodehydration of sour cherries. Nowicka *et al.* (2015b) compared a similar combined process on sour cherries frozen with or without their stones, and thawed with or without their stones. Optimal osmodehydration treatment was 180 min in 40°Brix apple juice, which reduced the water content by 50%, and was then followed by 90 min of convective air drying (50°C, 0.8 m $s^{-1}$) and finally vacuum-microwave drying (4–6 kPa, 360 W). Thawed fruit with stones removed dried faster, but freezing the fruit with the stone before drying was found to give the best overall fruit quality.

Konopacka *et al.* (2014) studied the suitability of nine sour cherry cultivars for their performance in osmodehydrated–convective drying of fruit for use as snacks. They found that most cultivars were too variable in maturity and raw fruit quality to give consistently high dried fruit quality. 'Nefris' was found to be the most adapted cultivar, with a good sugar/acid balance, a high content of anthocyanins and total phenolics, and uniform fruit size.

Infusion of fruit is a technique that allows removal of water to make fruit stable for storage, while at the same time adding other compounds into the fruit during a multi-sequential osmotic dehydration. For example, this process may use a 25% sucrose solution with other compounds added, and three or more sequential vacuum and released vacuum periods over several days (Jacob and Paliyath, 2012). During normal pressure, the sucrose and added compounds such as oils, vitamins, preservatives, health compounds and aromatic compounds are infused into the fruit. Infusion is most easily obtained in cut fruit since the skin will inhibit or delay movement of compounds into the fruit. Infused dried fruit may have a very attractive taste and provide a softer texture than air-dried fruit; therefore, this process is highly interesting for the production of fruit snacks or candy.

Fruit leather is a dried product that, according to Gardiner *et al.* (1993), is prepared by cooking cherries in water at a 2/1 (w/w) ratio for 4 min and then draining. Cooled cherries are then blended to a smooth paste and mixed with 10% (w/w) sucrose and 0.25% (w/w) citric acid. The paste is spread in a thin layer on a tray and dried at 70°C for 6 h. The final fruit leather is around 18% sugar and is soft. The organoleptic appearance did not change during storage for 6 months.

Cherry beef jerky is another name for fruit leather, and Bors (2011) studied three different recipes for making jerky on a range of sour cherry cultivars in Canada and found that jerky made from dried infused fruit gave better quality than that from wet infused or frozen fruit. All tested cultivars could be used, but 'Romeo' was rated best for infused dried fruit sensory quality.

Juhnevica *et al.* (2011) tested five sour cherry cultivars for their ability to produce high-quality candied fruit. The process involved removing stones from the fruit and soaking the fruit in syrup to obtain a 40% sucrose content in fruit mass. The fruit were then stored at 4°C for 48 h, drained of syrup and dried at 45–50°C in forced ventilation until a 40–43% moisture content was obtained. The cultivar 'Shokoladnica' was the preferred raw material for these products, resulting in a mild sweet and sour taste with good structure. A high phenolic content and high soluble solids were negatively correlated with consumer preference.

Finally, a method of drying cherry juice concentrate by spray drying into powders should be mentioned. High sugar and acid content makes this process difficult in many fruit species (Krishnaiah *et al.*, 2014). Karaca *et al.* (2016) studied the processing conditions and formulation effects on spray drying of sour cherry concentrate with 65% soluble solids. Concentrate was mixed with a carrier to a slurry of 40°Brix and kept at 43°C before injecting in the spray dryer. Drying air was dehumidified to below 5% relative humidity. The moisture content of the powder was between 1 and 2%. Powders should be stored in sealed packages to be stable. They found that a high yield of over 85% was obtained by drying at 150°C inlet temperature, 30% pump setting and 25% sour cherry content and using maltodextrin DE12 as carrier.

### 20.4.7 Fruit wine, liqueurs and brandy

A number of different types of alcoholic beverages are made from sweet and especially sour cherry.

Sweet cherry fruit is generally fairly high in sugar so is suitable for fermentation, but with an acidity that is too low to give a balanced taste. Sour cherry in general has high acidity, and some varieties have high sugar contents. In some cases, a sugar source needs to be added to assure a satisfactory fermentation, increased stability and the desired alcohol content in the finished wine. The high content of malic acid in sour cherry may need to be reduced somewhat by controlled malolactic fermentation. Use of preservatives such as $SO_2$ or sorbic acid, or sterile filtration, may be necessary to obtain a stable wine and avoid microbial spoilage. Traditionally, several beverages were made and are still made by adding alcohol to clarified, concentrated sour cherry juice or nectar to produce alcoholic cherry wine or beverages of varying percentages of alcohol, often sweet tasting and aiming for dessert drinks. Such products are not truly wines in the sense that no fermentation of fruit has been done.

True wines produced by fermented sour cherry fruit have recently gained more interest since consumers ask for better quality and more diversity in products, and premium prices may be obtained by such products. Traditionally, fermented wines were fairly sweet and high in alcohol (15–16%), aiming for the dessert wine market. However, there is currently a trend towards developing table wines with slightly lower alcoholic percentages. The vinification processes may follow different grape-wine traditions. The process of sour cherry winemaking could be as follows. Fruit of single cultivars or mixes of uniform sour cherry cultivars with high sugar (Brix of at least 20–22°), high acidity and colour, and with an optimal aroma and flavour profile, are harvested by machine 1–2 weeks later than normal harvest time for juice to obtain a maximum full flavour, aroma and the highest possible sugar content. Sorting of fruit to obtain the highest possible fruit quality is extremely important to avoid an off-taste in the wine. The fruit are washed in cold water to reduce any surface contamination and impurities. If the fruit cannot be processed at once, freezing storage of the fruit is possible. This will make pressing easier later following thawing, since the cell membranes are disrupted, but will also somewhat change the quality profile of the juice. Following maceration of fruit, cold pressing of juice may be done at 5–10°C or hot pressing of juice after enzymatic treatment at around 50°C. Heating will potentially give a slight change in aroma towards a 'canned aroma'. Stones may be removed during the juice extraction step. Many winemakers now use the pulp mash directly for fermentation for some days or a few weeks before pressing to obtain a more complex taste. Stones and skin are then removed by filtering after fermentation (Pantelić *et al.*, 2014). Fermentation is done by using appropriate pure yeast strains at recommended optimal concentrations. Temperature during fermentation should be kept relatively low at 12–15°C to achieve a slower fermentation and obtain more volatile esters for a better aroma profile. The duration of the fermentation depends on temperature and may last from days to several weeks. The pressed fermented must is filtered or clarified by sedimentation over some days. Stabilization of the wine may be obtained by using microfiltration techniques with a pore size of less than 45 μm or adding $SO_2$. Sour cherry wine normally should not be aged too long due to lower stability than most grape red wines. Sour cherry wine may be stored in steel tanks or in oak barrels for improved flavour before it is bottled.

Pantelić *et al.* (2014) compared a sour cherry wine produced from 'Oblačinska' with five grape red wines and found that total phenolics and total anthocyanins were in the same range, whereas the concentrations of gallic acid were about 20–30 times lower in sour cherry wine than in grape red wine and catechins about four to ten times lower. *p*-Coumaric acid and caffeic acid were much higher in sour cherry wine. Xiao *et al.* (2014) analysed nine commercial sweet cherry wines from China by advanced chemical fingerprinting and sensory analysis and used multivariate analysis to discriminate between the wines. They identified 75 volatiles, including 29 esters, 22 alcohols, eight acids, three ketones, five aldehydes and eight other compounds. Principal component analysis of the sensory attributes revealed three groups characterized by 'sweet, aromatic aroma', 'sweet, esters, green, bitter and fermented' and 'sour, alcoholic, fruity'. Correlations were found between the concentration of individual volatiles and sensory attributes. They concluded that the production process of the cherry wines had a large influence on the aroma profile, but that both fruit maturation and the ageing of wines contributed to the profile. Niu *et al.* (2011) also characterized five commercial cherry wines from China with gas chromatography–olfactometry and correlated volatile metabolites to sensory attributes. They found that the wines differed significantly as described by the terms, fruity, sour, woody, fermentation, cameral and floral, and also found a clear association between volatile metabolites concentration and sensory profile.

Sun *et al.* (2011) compared volatiles, soluble solids, acidity and phenolic compounds in sour cherry wine with the use of six different commercial yeast strains of *Saccharomyces cerevisiae*, and found that some strains resulted in higher amounts of volatile esters and acids, and other strains resulted in higher concentrations of alcohols, thus affecting the sensory quality. In addition, the concentration after fermentation of soluble sugars, acidity, total phenolics and total anthocyanins varied among wines produced by different yeast strains. Sun *et al.* (2014) studied the effect of using two non-*Saccharomyces* wine yeasts together with two strains of *S. cerevisiae* in a multi-starter fermentation on fermentation behaviour and aroma of sour cherry wines. The non-*Saccharomyces* yeasts gave much slower fermentation when used alone. When mixed yeasts were used, *S. cerevisiae* in some cases inhibited the proliferation of the non-*Saccharomyces*. The alcohol percentage obtained and fermentation period did not differ with the yeast mix used, whereas reducing sugars, total acidity and the volatile profile varied depending on the yeast mixes used. Supplementing *S. cerevisiae* with the two non-*Saccharomyces* species *Torulaspora delbrueckii* or *Metschnikowia pulcherrima* provided a richer aroma profile to the cherry wine.

Cherry liqueur and brandy are popular products with strong traditions in central, southern and south-eastern Europe. Nikićević *et al.* (2011) investigated how five different sour cherry cultivars influenced the chemical and sensorial characteristics of cherry brandy. Thirty-two volatile aroma compounds were identified. Ethyl octanoate and ethyl decanoate were the most abundant esters, and together with linalool and benzaldehyde were considered to be very important for the aroma of distilled cherry products. Linalool and benzaldehyde varied significantly between cultivars. The sensory rating evaluated 'Celery's 16' cultivar as the best brandy, showing the highest content of benzaldehyde and linalool.

### 20.4.8 Exploitation of side streams: extraction of ingredients

The side streams from processing of several cherry products include considerable amounts of pomace, stones including seeds and to a minor degree pedicel. In the past, these have not been exploited, but have mainly been discarded as waste. The current increasing focus on using all side streams from food production combined with knowledge of what resources are available in these side streams will probably result in much more extensive exploitation of these resources in the future. Pomace contains colours, phenolics with a high antioxidant capacity, health compounds, pectins, and other fibres and polysaccharides that may be exploited. Similarly, seeds and stones may contain oils and flavour in the seed and carbon products in the stone that may be used for products.

Pomace of sour cherry following juice pressing may reach 15–28% weight of raw fruit (Toydemir *et al.*, 2013a). Yılmaz *et al.* (2015) studied the extraction of anthocyanins from pomace of sour cherry after juice pressing. They found that a 51% ethanol, 75°C, 12 ml $g^{-1}$ solvent-to-solid ratio was optimal for the extraction of anthocyanins from vacuum-dried and ground pomace powder. Equilibrium concentrations were reached after 80–100 min. The extract also had a high content of polyphenols.

Kołodziejczyk *et al.* (2013) extracted anthocyanins and phenolics with water at 70°C followed by column extraction using ethanol, evaporation and freeze drying of extract, and found an 80% yield of anthocyanins, hydroxycinnamic acids and flavonols, while the flavanol yield was 30%. Garofulić *et al.* (2013) showed that microwave-assisted extraction of anthocyanins from frozen and thawed whole sour cherry 'Marasca' fruit was faster and gave higher yields than traditional extraction methods. Grigoras *et al.* (2012) used similar techniques on sweet cherries and further purified anthocyanins by preparative high performance liquid chromatography to obtain highly pure anthocyanins. Extraction of antioxidants from sour cherry as a potential ingredient has also been attempted (Piccolella *et al.*, 2008) as well as polysaccharides from sour cherry pomace (Kosmala *et al.*, 2009). Phenolics including anthocyanins are thought to be some of the bioactive ingredients that provide beneficial effects on muscle restitution, anti-inflammation and detoxification as seen in *in vitro* and clinical trials (Ferretti *et al.*, 2010).

Other bioactive compounds such as melatonin and serotonin involved in sleep disorders (Howatson *et al.*, 2012) occur in small and varying concentrations in fruit and may in the future be extracted and concentrated to achieve a stronger clinical effect (Burkhardt *et al.*, 2001; González-Gómez *et al.*, 2009; Garrido *et al.*, 2014). In general, interest in the production of health-related products based on cherries is anticipated to increase considerably in the future.

Cherry stones, consisting of seeds and endocarp, typically represent 4–8% of fruit fresh weight. A large quantity of stones arises following processing of canned, frozen or juiced fruit. Bak *et al.* (2010) isolated and characterized possible bioactive compounds in ground dry seed kernels of sour cherry after laboratory extraction by *n*-hexane using a Soxhlet apparatus and vacuum evaporation at 40°C and found a 32–36% oil content and 64–68% solid fraction. Mahmoud *et al.* (2014) evaluated the effect of such seed extracts on rheumatoid arthritis markers *in vitro* using human cells and found positive anti-inflammatory effects. Yılmaz and Gökmen (2013) studied sour cherry seeds and their content of oil, protein and fibres in detail and compared different extraction methods and the effect of roasting of seeds on the quality of products. They found oleic acid (46%) and linoleic acid (41%) to be the dominant compounds in the oil. Roasting seeds at 160°C for 30 min decreased the tocopherol content, but dramatically increased the total phenolic content in the oil. Seeds also contain a high content of amygdalin, which may be used to produce the naturally derived benzaldehyde. This aroma ingredient can then be used instead of synthetically produced compounds for special consumer segments. The endocarp may, for example, find use in the production of activated carbon products or to replace industrial microplastic products as environmentally friendly grinding/eroding fibre particles in products.

## References


Aider, M. and de Halleux, D. (2008) Production of concentrated cherry and apricot juices by cryoconcentration technology. *LWT – Food Science and Technology* 41, 1768–1775.

Altuntas, J., Evrendilek, G.A., Sangun, M.K. and Zhang, H.Q. (2010) Effects of pulsed electric field processing on the quality and microbial inactivation of sour cherry juice. *International Journal of Food Science and Technology* 45, 899–905.

Arjeh, E., Barzegar, M. and Sahari, M.A. (2015) Effects of gamma irradiation on physicochemical properties, antioxidant and microbial activities of sour cherry juice. *Radiation Physics and Chemistry* 114, 18–24.

Bagger-Jørgensen, R.I.C.O., Casani, S. and Meyer, A.S. (2002) Microfiltration of red berry juice with thread filters: effects of temperature, flow and filter pore size. *Journal of Food Process Engineering* 25, 109–124.

Bak, I., Lekli, I., Juhasz, B., Varga, E., Varga, B., Gesztelyi, R., Szendrei, L. and Tosaki, A. (2010) Isolation and analysis of bioactive constituents of sour cherry (*Prunus cerasus*) seed kernel: an emerging functional food. *Journal of Medicinal Food* 13, 905–910.

Balcerek, M. and Szopa, J. (2012) Ethanol biosynthesis and hydrocyanic acid liberation during fruit mashes fermentation. *Czech Journal Food Science* 30, 144–152.

Barbosa-Cánovas, G.V., Altunakar, B. and Mejía-Lorío, D.J. (2005) *Freezing of Fruits and Vegetables: An Agribusiness Alternative for Rural and Semi-rural Areas*, Vol. 158. Food and Agriculture Organization of the United Nations, Rome.

Bayındırlı, A., Alpas, H., Bozoğlu, F. and Hızal, M. (2006) Efficiency of high pressure treatment on inactivation of pathogenic microorganisms and enzymes in apple, orange, apricot and sour cherry juices. *Food Control* 17, 52–58.

Berger, R. (1991) Fruits I. In: Maarse, H. (ed.) *Volatile Compounds in Foods and Beverages*, Vol. 44. Marcel Dekker, New York, pp. 283–304.

Bonerz, D., Würth, K., Dietrich, H. and Will, F. (2007) Analytical characterization and the impact of ageing on anthocyanin composition and degradation in juices from five sour cherry cultivars. *European Food Research and Technology* 224, 355–364.

Bors, B. (2011) Variety variations in new and existing U of S sour cherries. ADF Research – Final report. University of Saskatchewan, Saskatoon, Saskatchewan, pp. 1–39. Available at: http://www.agriculture.gov.sk.ca/apps/adf/ADFAdminReport/20090405.pdf (accessed 20 January 2016).

Burkhardt, S., Tan, D.X., Manchester, L.C., Hardeland, R. and Reiter, R.J. (2001) Detection and quantification of the antioxidant melatonin in Montmorency and Balaton tart cherries (*Prunus cerasus*). *Journal of Agricultural and Food Chemistry* 49, 4898–4902.

Carlini, P., Massantini, R. and Mencarelli, F. (2000) Vis-NIR measurement of soluble solids in cherry and apricot by PLS regression and wavelength selection. *Journal of Agricultural and Food Chemistry* 48, 5236–5242.

Chaovanalikit, A. and Wrolstad, R.E. (2004a) Total anthocyanins and total phenolics of fresh and processed cherries and their antioxidant properties. *Journal of Food Science* 69, 67–72.

Chaovanalikit, A. and Wrolstad, R.E. (2004b) Anthocyanin and polyphenolic composition of fresh and processed cherries. *Journal of Food Science* 69, 73–83.

Clausen, M.R., Pedersen, B.H., Bertram, H.C. and Kidmose, U. (2011) Quality of sour cherry juice of different clones and cultivars (*Prunus cerasus* L.) determined by a combined sensory and NMR spectroscopic approach. *Journal of Agricultural and Food Chemistry* 59, 12124–12130.

Damar, İ. and Ekşi, A. (2012) Antioxidant capacity and anthocyanin profile of sour cherry (*Prunus cerasus* L.) juice. *Food Chemistry* 135, 2910–2914.

Doymaz, İ. (2007) Influence of pretreatment solution on the drying of sour cherry. *Journal of Food Engineering* 78, 591–596.

Doymaz, İ. and İsmail, O. (2011) Drying characteristics of sweet cherry. *Food and Bioproducts Processing* 89, 31–38.

Echavarría, A.P., Torras, C., Pagán, J. and Ibarz, A. (2011) Fruit juice processing and membrane technology application. *Food Engineering Reviews* 3, 136–158.

Evrendilek, G.A., Tok, F.M., Soylu, E.M. and Soylu, S. (2008) Inactivation of *Penicillum expansum* in sour cherry juice, peach and apricot nectars by pulsed electric fields. *Food Microbiology* 25, 662–667.

Evrendilek, G.A., Tok, F.M., Soylu, E.M. and Soylu, S. (2009) Effect of pulsed electric fields on germination tube elongation and spore germination of *Botrytis cinerea* inoculated into sour cherry juice, apricot and peach nectars. *Italian Journal of Food Science* 21, 171–182.

Evrendilek, G.A., Baysal, T., Icier, F., Yildiz, H., Demirdoven, A. and Bozkurt, H. (2012) Processing of fruits and fruit juices by novel electrotechnologies. *Food Engineering Reviews* 4, 68–87.

Ferretti, G., Bacchetti, T., Belleggia, A. and Neri, D. (2010) Cherry antioxidants: from farm to table. *Molecules* 15, 6993–7005.

Franceschinis, L., Sette, P., Schebor, C. and Salvatori, D. (2015) Color and bioactive compounds characteristics of dehydrated sweet cherry products. *Food and Bioprocess Technology* 1–14.

Gardiner, M.A., Beyer, R. and Melton, L.D. (1993) Sugar and anthocyanidin content of two processing-grade sweet cherry cultivars and cherry products. *New Zealand Journal of Crop and Horticultural Science* 21, 213–218.

Gao, J., Wang, B., Feng, X. and Zhu, Y. (2012) Effect of blanching on quality of sour cherry (*Prunus cerasus* L. cv. CAB) juice. *American Eurasian Journal of Agricultural and Environmental Sciences* 12, 123–127.

Garofulić, I.E., Dragović-Uzelac, V., Jambrak, A.R. and Jukić, M. (2013) The effect of microwave assisted extraction on the isolation of anthocyanins and phenolic acids from sour cherry Marasca (*Prunus cerasus* var. Marasca). *Journal of Food Engineering* 117, 437–442.

Garofulić, I.E., Jambrak, A.R., Milošević, S., Dragović-Uzelac, V., Zorić, Z. and Herceg, Z. (2015) The effect of gas phase plasma treatment on the anthocyanin and phenolic acid content of sour cherry Marasca (*Prunus cerasus* var. Marasca) juice. *LWT – Food Science and Technology* 62, 894–900.

Garrido, M., Rodríguez, A.B., Lozano, M., Hernández, M.T. and González-Gómez, D. (2014) Formulation and characterization of a new nutraceutical product based on sweet cherries (*Prunus avium* L.) grown in the Jerte Valley of Spain. *Acta Horticulture* 1020, 149–152.

Gazor, H.R. and Roustapour, O.R. (2015) Modeling of drying kinetic of pretreated sour cherry. *International Food Research Journal* 22, 476–481.

Gazor, H.R., Maadani, S. and Behmadi, H. (2014) Influence of air temperature and pretreatment solutions on drying time, energy consumption and organoleptic properties of sour cherry. *Agriculturae Conspectus Scientificus* 79, 119–124.

Gonçalves, B., Landbo, A.K., Knudsen, D., Silva, A.P., Moutinho-Pereira, J., Rosa, E. and Meyer, A.S. (2004) Effect of ripeness and postharvest storage on the phenolic profiles of cherries (*Prunus avium* L.). *Journal of Agricultural and Food Chemistry* 52, 523–530.

González-Gómez, D., Lozano, M., Fernández-León, M.F., Ayuso, M.C., Bernalte, M.J. and Rodríguez, A.B. (2009) Detection and quantification of melatonin and serotonin in eight sweet cherry cultivars (*Prunus avium* L.). *European Food Research and Technology* 229, 223–229.

Greiby, I., Mishra, D.K. and Dolan, K.D. (2014) Inverse method to sequentially estimate temperature-dependent thermal conductivity of cherry pomace during nonisothermal heating. *Journal of Food Engineering* 127, 16–23.

Grigoras, C.G., Destandau, E., Zubrzycki, S. and Elfakir, C. (2012) Sweet cherries anthocyanins: an environmental friendly extraction and purification method. *Separation and Purification Technology* 100, 51–58.

Haff, R.P., Pearson, T.C. and Jackson, E.S. (2013) One dimensional Linescan X-ray detection of pits in fresh cherries. *American Journal of Agricultural Science and Technology* 1, 18–26.

Hosseinzadeh Samani, B., Khoshtaghaza, M.H., Minaee, S. and Abbasi, S. (2015) Modeling the simultaneous effects of microwave and ultrasound treatments on sour cherry juice using response surface methodology. *Journal of Agricultural Science and Technology* 17, 837–846.

Howatson, G., Bell, P.G., Tallent, J., Middleton, B., McHugh, M.P. and Ellis, J. (2012) Effect of tart cherry juice (*Prunus cerasus*) on melatonin levels and enhanced sleep quality. *European Journal of Nutrition* 51, 909–916.

Ivancevic, S., Mitrovic, D., Brkic, M. and Cvijanovic, D. (2012) Specificities of fruit freeze drying and product prices. *Economics of Agriculture* 59, 461–471.

Jacob, J.K. and Paliyath, G. (2012) Infusion of fruits with nutraceuticals and health regulatory components for enhanced functionality. *Food Research International* 45, 93–102.

Juhnevica, K., Ruisa, S., Seglina, D. and Krasnova, I. (2011) Evaluation of sour cherry cultivars grown in Latvia for production of candied fruits. In: Straumite, E. (ed.) *Proceedings of the 6th Baltic Conference on Food Science and Technology: 'Innovations for Food Science and Production'*. FOODBALT, LLU, Jelgava, Latvia, pp. 19–22.

Kaack, K., Spayd, E.E. and Drake, S.R. (1996) Cherry Processing. In: Webster, A.D. and Looney, N.E. (eds) *Cherries. Crop Physiology, Production and Uses*. CAB International, Wallingford, UK, pp. 473–481.

Karaca, A.C., Guzel, O. and Ak, M.M. (2016) Effects of processing conditions and formulation on spray drying of sour cherry juice concentrate. *Journal of the Science of Food and Agriculture* 96, 449–455.

Kim, D.O. and Padilla-Zakour, O.I. (2004) Jam processing effect on phenolics and antioxidant capacity in anthocyanin-rich fruits: cherry, plum, and raspberry. *Journal of Food Science* 69, 395–400.

Kirakosyan, A., Seymour, E.M., Llanes, D.E.U., Kaufman, P.B. and Bolling, S.F. (2009) Chemical profile and antioxidant capacities of tart cherry products. *Food Chemistry* 115, 20–25.

Klewicki, R., Konopacka, D., Uczciwek, M., Irzyniec, Z., Piasecka, E. and Bonazzi, C. (2009) Sorption isotherms for osmo-convectively-dried and osmo-freeze-dried apple, sour cherry and blackcurrant. *Journal of Horticultural Science and Biotechnology* 1, 75.

Kołodziejczyk, K., Sójka, M., Abadias, M., Viñas, I., Guyot, S. and Baron, A. (2013) Polyphenol composition, antioxidant capacity, and antimicrobial activity of the extracts obtained from industrial sour cherry pomace. *Industrial Crops and Products* 51, 279–288.

Konopacka, D., Jesionkowska, K., Mieszczakowska, M. and Płocharski, W. (2008) The usefulness of natural concentrated fruit juices as osmotic agents for osmo-dehydrated dried fruit production. *Journal of Fruit and Ornamental Plant Research* 16, 275–284.

Konopacka, D., Markowski, J., Płocharski, W. and Rozpara, E. (2014) New or lesser known cultivar selection as a tool for sensory and nutritional value enhancement of osmo-convectively dried sour cherries. *LWT – Food Science and Technology* 55, 506–512.

Kosmala, M., Milala, J., Kołodziejczyk, K., Markowski, J., Mieszczakowska, M., Ginies, C. and Renard, C.M.G.C. (2009) Characterization of cell wall polysaccharides of cherry (*Prunus cerasus* var. Schattenmorelle) fruit and pomace. *Plant Foods for Human Nutrition* 64, 279–285.

Krishnaiah, D., Nithyanandam, R. and Sarbatly, R. (2014) A critical review on the spray drying of fruit extract: effect of additives on physicochemical properties. *Critical Reviews in Food Science and Nutrition* 54, 449–473.

Lu, R. (2001) Predicting firmness and sugar content of sweet cherries using near-infrared diffuse reflectance spectroscopy. *Transactions of the American Society of Agricultural Engineers* 44, 1265–1274.

Mahmoud, F., Haines, D., Al-Awadhi, R., Dashti, A.A., Al-Awadhi, A., Ibrahim, B., Al-Zayer, B., Juhasz, B. and Tosaki, A. (2014) Sour cherry (*Prunus cerasus*) seed extract increases heme oxygenase-1 expression and decreases proinflammatory signaling in peripheral blood human leukocytes from rheumatoid arthritis patients. *International Immunopharmacology* 20, 188–196.

Manganaris, G.A., Ilias, I.F., Vasilakakis, M. and Mignani, I. (2007) The effect of hydrocooling on ripening related quality attributes and cell wall physicochemical properties of sweet cherry fruit (*Prunus avium* L.). *International Journal of Refrigeration* 30, 1386–1392.

Márquez, C.A., Salvadori, V.O., Mascheroni, R.H. and De Michelis, A. (2003) Application of transfer functions to the thermal processing of sweet and sour cherries preserves: influence of particle and container sizes. *Food Science and Technology International* 9, 69–76.

McLellan, M.R. and Padilla-Zakour, O.I. (2004) Sweet cherry and sour cherry processing. In: Barret, D.M., Somogyi, L.P. and Ramaswamy, S. (eds) *Processing Fruits Science and Technology*, 2nd edn. CRC Press, Boca Raton, Florida, pp. 497–512.

Meyer, A.S., Köser, C. and Adler-Nissen, J. (2001) Efficiency of enzymatic and other alternative clarification and fining treatments on turbidity and haze in cherry juice. *Journal of Agricultural and Food Chemistry* 49, 3644–3650.

Nikićević, N., Veličković, M., Jadranin, M., Vučković, I., Novaković, M., Vujisić, L., Stankovic, M., Urosevic, I. and Tešević, V. (2011) The effects of the cherry variety on the chemical and sensorial characteristics of cherry brandy. *Journal of the Serbian Chemical Society* 76, 1219–1228.

Niu, Y., Zhang, X., Xiao, Z., Song, S., Eric, K., Jia, C., Yu, H. and Zhu, J. (2011) Characterization of odor-active compounds of various cherry wines by gas chromatography–mass spectrometry, gas chromatography–olfactometry and their correlation with sensory attributes. *Journal of Chromatography B* 879, 2287–2293.

Normohamadpor Omran, M., Pirouzifard, M.K., Aryaey, P. and Hasan Nejad, M. (2013) Cryo-concentration of sour cherry and orange juices with novel clarification method; Comparison of thermal concentration with freeze concentration in liquid foods. *Journal of Agricultural Science and Technology* 15, 941–950.

Nowicka, P., Wojdyło, A., Lech, K. and Figiel, A. (2015a) Chemical composition, antioxidant capacity, and sensory quality of dried sour cherry fruits pre-dehydrated in fruit concentrates. *Food and Bioprocess Technology* 8, 2076–2095.

Nowicka, P., Wojdyło, A., Lech, K. and Figiel, A. (2015b) Influence of osmodehydration pretreatment and combined drying method on the bioactive potential of sour cherry fruits. *Food and Bioprocess Technology* 8, 824–836.

Ou, B., Bosak, K.N., Brickner, P.R., Iezzoni, D.G. and Seymour, E.M. (2012) Processed tart cherry products – comparative phytochemical content, *in vitro* antioxidant capacity and *in vitro* anti-inflammatory activity. *Journal of Food Science* 77, 105–112.

Pantelić, M., Dabić, D., Matijašević, S., Davidovic, S., Dojcinovic, B., Milojkovic-Opsenica, D., Tesic, Z. and Natic, M. (2014) Chemical characterization of fruit wine made from Oblačinska sour cherry. *Scientific World Journal* 2014, 454797.

Pappas, C.S., Takidelli, C., Tsantili, E., Tarantilis, P.A. and Polissiou, M.G. (2011) Quantitative determination of anthocyanins in three sweet cherry varieties using diffuse reflectance infrared Fourier transform spectroscopy. *Journal of Food Composition and Analysis* 24, 17–21.

Patras, A., Brunton, N.P., O'Donnell, C. and Tiwari, B.K. (2010) Effect of thermal processing on anthocyanin stability in foods; mechanisms and kinetics of degradation. *Trends in Food Science and Technology* 21, 3–11.

Piccolella, S., Fiorentino, A., Pacifico, S., D'Abrosca, B., Uzzo, P. and Monaco, P. (2008) Antioxidant properties of sour cherries (*Prunus cerasus* L.): role of colorless phytochemicals from the methanolic extract of ripe fruits. *Journal of Agricultural and Food Chemistry* 56, 1928–1935.

Pinelo, M., Zeuner, B. and Meyer, A.S. (2010) Juice clarification by protease and pectinase treatments indicates new roles of pectin and protein in cherry juice turbidity. *Food and Bioproducts Processing* 88, 259–265.

Pirone, B.N., de Michelis, A. and Salvatori, D.M. (2014) Pretreatments effect in drying behaviour and colour of mature and immature 'Napolitana' sweet cherries. *Food and Bioprocess Technology* 7, 1640–1655.

Poiana, M.A., Moigradean, D., Dogaru, D., Mateescu, C., Raba, D. and Gergen, I. (2011) Processing and storage impact on the antioxidant properties and color quality of some low sugar fruit jams. *Romanian Biotechnological Letters* 16, 6504–6512.

Queirós, R.P., Rainho, D., Santos, M.D., Fidalgo, L.G., Delgadillo, I. and Saraiva, J.A. (2015) High pressure and thermal pasteurization effects on sweet cherry juice microbiological stability and physicochemical properties. *High Pressure Research* 35, 69–77.

Rababah, T.M., Al-Mahasneh, M.A., Kilani, I., Yang, W., Alhamad, M.N., Ereifej, K. and Al-u'datt, M. (2011) Effect of jam processing and storage on total phenolics, antioxidant activity, and anthocyanins of different fruits. *Journal of the Science of Food and Agriculture* 91, 1096–1102.

Rentzsch, M., Quast, P., Hillebrand, S., Mehnert, J. and Winterhalter, P. (2007) Isolation and identification of 5-carboxy-pyranoanthocyanins in beverages from cherry (*Prunus cerasus* L.). *Innovative Food Science and Emerging Technologies* 8, 333–338.

Repajić, M., Bursać Kovačević, D., Putnik, P., Dragović Uzelac, V., Kušt, J., Čošić, Z. and Levaj, B. (2015) Influence of cultivar and industrial processing on polyphenols in concentrated sour cherry (*Prunus cerasus* L.) juice. *Food Technology and Biotechnology* 53, 215–222.

Rupasinghe, H.V. and Yu, L.J. (2012) Emerging preservation methods for fruit juices and beverages. In: El-Samragy, Y. (ed.) *Food Additive*. INTECH Open Access Publisher, Rijeka, Croatia, pp. 65–82.

Şahin, S. and Bayindirli, L. (1993) The effect of depectinization and clarification on the filtration of sour cherry juice. *Journal of Food Engineering* 19, 237–245.

Schüller, E., Halbwirth, H., Mikulic-Petkovsek, M., Slatnar, A., Veberic, R., Forneck, A., Stick, K. and Spornberger, A. (2015) High concentrations of anthocyanins in genuine cherry-juice of old local Austrian *Prunus avium* varieties. *Food Chemistry* 173, 935–942.

Serradilla, M.J., Martín, A., Ruiz-Moyano, S., Hernández, A., López-Corrales, M. and de Guía Córdoba, M. (2012) Physicochemical and sensorial characterisation of four sweet cherry cultivars grown in Jerte Valley (Spain). *Food Chemistry* 133, 1551–1559.

Serrano, M., Guillén, F., Martínez-Romero, D., Castillo, S. and Valero, D. (2005) Chemical constituents and antioxidant activity of sweet cherry at different ripening stages. *Journal of Agricultural and Food Chemistry* 53, 2741–2745.

Serrano, M., Díaz-Mula, H.M., Zapata, P.J., Castillo, S., Guillén, F., Martínez-Romero, D., Valverde, J.M. and Valero, D. (2009) Maturity stage at harvest determines the fruit quality and antioxidant potential after storage of sweet cherry cultivars. *Journal of Agricultural and Food Chemistry* 57, 3240–3246.

Šumić, Z., Tepić, A., Vidović, S., Jokić, S. and Malbaša, R. (2013) Optimization of frozen sour cherries vacuum drying process. *Food Chemistry* 136, 55–63.

Sun, S.Y., Jiang, W.G. and Zhao, Y.P. (2011) Evaluation of different *Saccharomyces cerevisiae* strains on the profile of volatile compounds and polyphenols in cherry wines. *Food Chemistry* 127, 547–555.

Sun, S.Y., Gong, H.S., Jiang, X.M. and Zhao, Y.P. (2014) Selected non-*Saccharomyces* wine yeasts in controlled multistarter fermentations with *Saccharomyces cerevisiae* on alcoholic fermentation behaviour and wine aroma of cherry wines. *Food Microbiology* 44, 15–23.

Szalóki-Dorkó, L., Végvári, G., Ladányi, M., Ficzek, G. and Stéger-Máté, M. (2015) Degradation of anthocyanin content in sour cherry juice during heat treatment. *Food Technology and Biotechnology* 53, 354.

Toivonen, P.M.A., Kappel, F., Stan, S., McKenzie, D.L. and Hocking, R. (2006) Factors affecting the quality of a novel fresh-cut sweet cherry product. *LWT – Food Science and Technology* 39, 240–246.

Toydemir, G., Capanoglu, E., Roldan, M.V.G., de Vos, R.C., Boyacioglu, D., Hall, R.D. and Beekwilder, J. (2013a) Industrial processing effects on phenolic compounds in sour cherry (*Prunus cerasus* L.) fruit. *Food Research International* 53, 218–225.

Toydemir, G., Capanoglu, E., Kamiloglu, S., Boyacioglu, D., De Vos, R.C., Hall, R.D. and Beekwilder, J. (2013b) Changes in sour cherry (*Prunus cerasus* L.) antioxidants during nectar processing and *in vitro* gastrointestinal digestion. *Journal of Functional Foods* 5, 1402–1413.

Valero, D. (2015) Recent developments to maintain overall sweet cherry quality during postharvest storage. *Acta Horticulturae* 1079, 83–94.

Watters, G.G. and Woodroof, J.G. (1986) Brining cherries and other fruits. In: Woodroof, J.D. and Lu, B.S. (eds) *Commercial Fruit Processing*, 2nd edn. AVI Publishing Co., New York, pp. 407–424.

Will, F., Hilsendegen, P., Bonerz, D., Patz, C.D. and Dietrich, H. (2005) Analytical composition of fruit juices from different sour cherry cultivars. *Journal of Applied Botany and Food Quality* 79, 12–16.

Wojdyło, A., Figiel, A., Lech, K., Nowicka, P. and Oszmiański, J. (2014) Effect of convective and vacuum–microwave drying on the bioactive compounds, color, and antioxidant capacity of sour cherries. *Food and Bioprocess Technology* 7, 829–841.

Xiao, Z., Liu, S., Gu, Y., Xu, N., Shang, Y. and Zhu, J. (2014) Discrimination of cherry wines based on their sensory properties and aromatic fingerprinting using HS-SPME-GC-MS and multivariate analysis. *Journal of Food Science* 79, 284–294.

Yadav, A.K. and Singh, S.V. (2014) Osmotic dehydration of fruits and vegetables: a review. *Journal of Food Science and Technology* 51, 1654–1673.

Yılmaz, C. and Gökmen, V. (2013) Compositional characteristics of sour cherry kernel and its oil as influenced by different extraction and roasting conditions. *Industrial Crops and Products* 49, 130–135.

Yılmaz, F.M., Karaaslan, M. and Vardin, H. (2015) Optimization of extraction parameters on the isolation of phenolic compounds from sour cherry (*Prunus cerasus* L.) pomace. *Journal of Food Science and Technology* 52, 2851–2859.

Zoric, Z., Dragovic-Uzelac, V., Pedisic, S., Kurtanjek, Z. and Garofulic, I.E. (2014) Kinetics of the degradation of anthocyanins, phenolic acids and flavonols during heat treatments of freeze-dried sour cherry Marasca paste. *Food Technology and Biotechnology* 52, 101–108.

# Index

Note: Page numbers in **bold** type refer to **figures**
Page numbers in *italic* type refer to *tables*